# ESSDERC 90

20th European Solid State Device Research Conference, Nottingham, 10–13 September 1990

Organized by the Institute of Physics
Co-sponsored by the IEE, the IEEE (EDS) and the European Physical Society

Edited by W Eccleston and P J Rosser

Adam Hilger
Bristol, Philadelphia and New York

© 1990 by IOP Publishing Ltd and individual contributors

All rights reserved. No part of this publication may be reproduced, stored in a retrieval system or transmitted in any form or by any means, electronic, mechanical, photo-copying, recording or otherwise, without the prior permission of the publisher. Multiple copying is only permitted under the terms of the agreement between the Committee of Vice-Chancellors and Principals and the Copyright Licensing Agency.

*British Library Cataloguing in Publication Data*

European Solid State Device Research Conference *(20th: 1990: Nottingham, England)*
ESSDERC 90.
1. Semiconductor devices
I. Title.  II. Eccleston, W.  III. Rossner, P. J.
621.3815

ISBN 0-7503-0065-5

*Library of Congress Cataloguing-in-Publication Data are available*

# ANDERSONIAN LIBRARY

## 0 4. APR 91

### UNIVERSITY OF STRATHCLYDE

Published under the Adam Hilger imprint by IOP Publishing Ltd
Techno House, Redcliffe Way, Bristol BS1 6NX, England
335 East 45th Street, New York, NY 10017-3483, USA
US Editorial Office: 1411 Walnut Street, Philadelphia, PA 19102

Printed in Great Britain by Galliard (Printers) Ltd, Great Yarmouth, Norfolk.

D
621.3815'2
ESS

# Contents

**MONDAY MORNING SESSION**

**Session 1IP room B1: Invited Papers**

08:45  Opening Address

09:00  1IP1
Progress in high-frequency heterojunction field effect transistors§    619
L F Eastman
Cornell University, USA

10:00  1IP2
Novel applications of porous silicon§    613
J Keen
Royal Signals and Radar Establishment, Great Malvern, UK

10:45  COFFEE


**Session 1A room B1: Silicon on Insulator I**

11:15  1A1
Characterisation of interface electron state distributions at directly bonded silicon/silicon interfaces    1
S Bengtsson and O Engström
Chalmers University of Technology, Göteborg, Sweden

11:30  1A2
Assessment of SIMOX material by optical waveguide losses    5
N M Kassim†, H P Ho†, T M Benson† and D E Davies‡
† University of Nottingham, UK, ‡ EOARD, London, UK

11:45  1A3
High density 3D CMOS circuits with ELO SOI technology    9
G Roos, B Hoefflinger and R Zingg
Institute for Microelectronics, Stuttgart, W Germany

12:00  1A4
Rounded edge mesa for submicron SOI CMOS process    13
O Le Néel, M D Bruni‡, J Galvier‡ and M Haond‡
† MATRA-MHS, Nantes, France, ‡ CNET, Meylan, France

§ Paper submitted late

## Session 1B room C4: Heterojunction Transistor Structures

11:15    1B1
**High quality pseudomorphic InGaAs/GaAs HFET structures grown by MBE**    17
C Wölk, F Berlec and H Brugger
Daimler Benz AG, Ulm, W Germany

11:30    1B2
**Modelling the parasitic effects in GaAs/GaAlAs heterojunction bipolar transistors**    21
J Dangla, M Filoche, A Koncyzkowska and E Caquot
Centre National d'Etudes des Télécommunications, Laboratoire de Bagneux, France

11:45    1B3
**Design, simulation and measurement of high efficiency microwave power heterojunction bipolar transistors**    25
J G Metcalfe, R W Allen, A J Holden, A P Long and R Nicklin
Plessey Research Caswell Ltd, Towcester, UK

12:00    1B4
**Microwave HBT fabrication by using a self-aligned technology with a perpendicular side-wall Mesa**    29
X Chen and Y Wu
Nanjing Electronic Devices Institute, Nanjing, China

12:15    1B5
**High-frequency properties and application of invertible GaInAsP/InP double-heterostructure bipolar transistors**    33
A Paraskevopoulos, H G Bach, H Schroeter-Janßen, G Mekonnen, H J Hensel and N Grote
Heinrich-Hertz-Institut für Nachrichtentechnik, Berlin GmbH, W Germany

## Session 1C Room B13: Advanced Silicon Device Concepts

11:15    1C1
**Frequency performances of MOS compatible silicon permeable base transistors and comparison with simulation results**    37
M Mouis[†], P Letourneau[‡] and G Vincent[§]
[†] Institut d'Electronique Fondamentale, Orsay, France, [‡] CNET, Meylan, France, [§] Université Joseph Fourier, Grenoble, France

11:30    1C2
**P-Channel etched-groove Si permeable base transistors**    41
A Gruhle and P A Badoz
CNET, Meylan, France

| | | |
|---|---|---|
| 11:45 | 1C3 | |
| | **A permeable base transistor on Si(100) with implanted $CoSi_2$-gate** | 45 |
| | A Schüppen, S Mantl, L Vescan and H Lüth | |
| | Institut für Schicht- und Ionentechnik, Jülich, W Germany | |
| 12:00 | 1C4 | |
| | **A novel compact model description of reverse-biased diode characteristics including tunnelling** | 49 |
| | G A M Hurkx, H C de Graaff, W J Kloosterman and M P G Knuvers | |
| | Philips Research Laboratories, Eindhoven, The Netherlands | |
| 12:15 | LUNCH | |

## MONDAY AFTERNOON SESSION 2

**Session 2IP room B1: Invited Paper**

| | | |
|---|---|---|
| 13:30 | 2IP1 | |
| | **Compositional analysis of submicron silicon in one, two and three dimensions** | 53 |
| | C Hill | |
| | Plessey Research Caswell Ltd, Towcester, UK | |

**Session 2A room B1: Silicon Processing I**

| | | |
|---|---|---|
| 14:15 | 2A1 | |
| | **TITAN-RTA: a 2D integrated equipment and process model for simulation of rapid thermal processing** | 61 |
| | S K Jones† and A Gérodolle‡ | |
| | †Plessey Research Caswell Ltd, Towcester, UK, ‡CNET, Meylan, France | |
| 14:30 | 2A2 | |
| | **Reflectivity measurements in a rapid thermal processor: application to silicide formation and solid phase regrowth** | 65 |
| | J-M Dilhac, N Nolhier and C Ganibal | |
| | Laboratoire d'Automatique et d'Analyse des Systèmes du CNRS, Toulouse, France | |
| 14:45 | 2A3 | |
| | **Temperature non-uniformities in a silicon test square during rapid thermal processing** | 69 |
| | A G O'Neill†, D Boys‡, S K Jones‡ and C Hill‡ | |
| | †University of Newcastle-upon-Tyne, UK, ‡Plessey Research Caswell Ltd, Towcester, UK | |

viii  *Contents*

15:00   2A4
**Prevention of boron penetration from $p^+$ poly gate by RTN produced thin gate oxide**    73
T Morimoto, H S Momose, K Yamabe and H Iwai
ULSI Research Center, Toshiba Corporation, Kawasaki, Japan

15:15   2A5
**Formation of $(Ti_xW_{1-x})$ $Si_2/(Ti_xW_{1-x})N$ contacts by rapid thermal silicidation**    77
H Norström†, K Maex†, J Vanhellemont†, G Brijs†, W Vandervorst and U Smith‡
† IMEC, Leuven, Belgium, ‡ Royal Institute of Technology, Kista, Sweden

15:30   TEA

16:00   2A6
**Contact-related effects on junction behaviour**    81
M L Polignano and N Circelli
SGS—Thomson Microelectronics, Agrate MI, Italy

16:15   2A7
**A new improved electrical vernier to measure mask misalignment**    85
D Morrow, A J Walton, W R Gammie, M Fallon, J T M Stevenson and R J Holwill
University of Edinburgh, UK

16:30   2A8
**New aspects of intrinsic gettering for CCD imagers**    89
N A Sobolev, I Yu Shapiro, V I Sokolov and E D Vasilyeva
Ioffe Institute, Academy of Sciences of the USSR, Leningrad, USSR

16:45   2A9
**Direct observation of the mask edge effect in boron implantation**    93
L Gong, J Lorenz and H Ryssel
Fraunhofer Arbeitsgruppe für Integrierte Schaltungen, Erlangen, W Germany

17:00   2A10
***In Situ* cleaning of silicon wafers for selective epitaxial growth (SEG)**    97
M R Goulding, P Kightley, P D Augustus and C Hill
Plessey Research Caswell Ltd, Towcester, UK

17:15   2A11
**Properties of Metal–$SiO_2$–Si diodes on HF/ethanol cleaned silicon substrates***
K Kassmi, J L Prom, F Gessin and G Sarrabayrouse
Laboratoire d'Automatique et d'analyse des Systemes, Toulouse, France

\* Paper not available at time of going to press

17:30 2A12
**Complementary silicon JFETs using novel ultra-shallow gate junctions** 101
Ö Grelsson, A Söderbärg and U Magnusson
University of Uppsala, Sweden

**Session 2B room C4: Compound Device Structure and Technology**

14:15 2B1
**Submicron pseudomorphic $Al_{0.2}Ga_{0.8}As/In_{0.25}Ga_{0.75}$ As-HFET made by conventional optical lithography for microwave circuit applications above 100 GHZ** 105
H Meschede†, J Kraus†, R Bertenburg†, W Brockerhoff†, W Prost†, K Heime‡, H Nickel§, W Schlapp§ and R Lösch§
† Univrersität -GH- Duisburg, W Germany, ‡RWTH-Achen, Institut für Halbleitertechnik, Aachen, W Germany, §Deutsche Bundespost Telekom, Darmstadt, W Germany

14:30 2B2
**High electron density and mobility InGaAs/InAlAs modulation doped structures grown on InP** 109
F Gueissaz, R Houdré and M Ilegems
Institut de Micro-Optoélectronique, et Ecole Polytechnique Fédérale, Lausanne, Switzerland

14:45 2B3
**Surface Characteristics of plasma treated $WN_x$/GaAs Contacts from C-V Measurements** 113
P E Bagnoli†, A Paccagnella‡, A Callegari§ and F Fantini∥
† University of Pisa, Italy, ‡ University of Padova, Italy, §I.B.M., T J Watson Research Center, NY, USA, ∥ Scuola Superiore S. Anna, Pisa, Italy

15:00 2B4
**Gate technologies for AlInAs/InGaAs HEMTs** 117
T D Hunt, J Urquhart, J Thompson, R A Davies and R H Wallis
Plessey Research Caswell Ltd, Towcester, UK

15:30 TEA

16:00 2B5
**A parametric investigation of the reactive ion etching of InP in $CH_4/H_2$ plasmas using response surface methodology** 121
D J Thomas and S J Clements
STC Technology Ltd, Harlow, UK

16:15    2B6
**Effects of the passivation process on the electrical characteristics of GaInAs planar photodiodes**    125
F Ducroquet†, G Guillot†, J C Renaud‡ and A Nouailhat§
† Laboratoire de Physique de la Matière, Villeurbanne, France, ‡ CNET, Bagneux, France, § CNET, Meylan, France

16:30    2B7
**The application of selective electroless plating for microelectronics applications**    129
C NiDheasuna†, T Spalding† and A Mathewson‡
† University College, Cork, Ireland, ‡ National Microelectronics Research Centre, Cork, Ireland

16:45    2B8
**p-Type AuMn ohmic contact on GaAs: integration in a HBT processing technology**    133
C Dubon-Chevallier, A M Duchenois, A C Papadopoulo, L Bricard, F Héliot and P Launay
CNET, Bagneux, France

**Session 2C room B13: Bulk MOS devices and circuits**

14:15    2C1
**A fast method of parameter extraction for MOS transistors**    137
P R Karlsson and K O Jeppson
Chalmers University of Technology, Göteborg, Sweden

14:30    2C2
**A charge and capacitance model for modern MOSFETs**    141
T Smedes and F M Klaasen
Eindhoven University of Technology, The Netherlands

14:45    2C3
**Electron velocity overshoot in sub-micron silicon MOS transistors**    145
P J H Elias, Th G van de Roer and F M Klaassen
Eindhoven University of Technology, The Netherlands

15:00    2C4
**New short-channel effects on Nitrided Oxide gate MOSFETs**    149
H S Momose, T Morimoto, S Takagi, K Yamabe, S Onga and H Iwai
ULSI Research Center, Toshiba Corporation, Kawasaki, Japan

15:15    2C5
**Analytical model for circuit simulation with quarter micron MOSFETs: Subthreshold characteristics**    153
M Miura-Mattausch and H Jacobs
Siemens AG, Munich, W Germany

15:30    TEA

| | | |
|---|---|---|
| 16:00 | 2C6 | |
| | **Small-signal modelling of MOSFET for circuit design applications** | 157 |
| | V Altschul, E Finkman and D Lubzens | |
| | Technion—Israel Institute of Technology, Haifa, Israel | |
| 16:15 | 2C7 | |
| | **Numerical simulation of MOS devices with non-degenerate gate** | 161 |
| | P Habaš and S Selberherr | |
| | Institute for Microelectronics, Vienna, Austria | |
| 16:30 | 2C8 | |
| | **Series resistance effects on EPROM programming** | 165 |
| | R Bez, D Cantarelli, P Cappelletti, A Maurelli and L Ravazzi | |
| | SGS-Thomson Microelectronics, Agrate, MI, Italy | |
| 16:45 | 2C9 | |
| | **A flash EEPROM cell scaling including tunnel oxide limitations** | 169 |
| | K Yoshikawa†, S Mori†, E Sakagami†, N Ari‡, Y Kaneko† and Y Oshima† | |
| | † Semiconductor Device Engineering Laboratory, Toshiba Corporation, Kawasaki, Japan, ‡ Toshiba Microelectronics Corporation, Kawasaki, Japan | |
| 17:00 | 2C10 | |
| | **A new 0.5 $\mu m^2$ DRAM cell with internal charge gain investigated by 2D transient device simulation** | 173 |
| | R Richter†, K E Ehwald†, B Heinemann†, W E Matzke†, H Gajewski‡ and W Winkler† | |
| | † Institute for Physics of Semiconductors, Academy of Sciences of the GDR, Frankfurt, GDR, ‡ Karl-Weierstrass-Institute of Mathematics, Academy of Sciences of the GDR, Berlin, GDR | |
| 17:15 | 2C11 | |
| | **A novel flash erase EEPROM memory cell with asperities aided erase** | 177 |
| | A A M Amin | |
| | King Fahd University of Petroleum and Minerals, Dhahran, Saudi Arabia | |

## TUESDAY MORNING SESSION

**Session 3IP room B1: Invited Paper**

| | | |
|---|---|---|
| 09:00 | 3IP1 | |
| | **Circuit level models for VLSI components** | 181 |
| | F M Klaassen | |
| | Philips Research Laboratories, Eindhoven, The Netherlands | |

## Session 3A Room B1: Silicon Processing II

**10:00**   3A1
**Non-destructive 2D doping profiling by the numerical inversion of $CV$ measurement**   193
G J L Ouwerling and M Kleefstra
Delft University of Technology, The Netherlands

**10:15**   3A2
**Low Leakage current evaluations for process characterizations**   197
P Girard, B Pistoulet and P Nouet
Laboratoire d'Automatique et de Microélectronique de Montpellier, France

**10:30**   COFFEE

**11:00**   3A3
**Computer simulation of oxygen precipitation in CZ-silicon during rapid thermal anneals**   201
M Schrems†, P Pongratz‡, M Budil†, H W Pötzl†, J. Hage§, E Guerrero§ and D Huber§
† Institut für Allgemeine Elecktrotechnik und Elektronik, Vienna, Austria, ‡ Institut für Angewandte Physik, Vienna, Austria, § Wacker-Chemitronic GmbH, Burghausen, W Germany

**11:15**   3A4
**Simulation of arsenic and boron diffusion during rapid thermal annealing in silicon**   205
M Heinrich, M Budil and H W Pötzl
Institut für Allgemeine Elecktrotechnik und Elektronik, Vienna, Austria

**11:30**   3A5
**An Improved model of plasma etching including temperature dependence: comparison between simulation and experimental results**   209
A Gérodolle, J Pelletier and S Drouot
CNET, Meylan, France

**11:45**   3A6
**Simulation of a Polysilicon LPCVD Reactor**   213
Ch Hopfmann, J I Ulacia F, and Ch Werner
Siemens AG, Munich, W Germany

**12:00**   3A7
**Channeling of boron in silicon: experiments and simulation**   217
C Hobler†, H Pötzl†, R Schork‡, J Lorenz‡, S Gara§ and G Stingeder§
† Institut für Allgemeine Elecktrotechnik und Elektronik, Vienna, Austria, ‡ Fraunhofer-Arbeitsgruppe für Integrierte Schaltungen, Erlangen, W Germany, § Institut für Analysche Chemie, Technische Universität, Vienna, Austria

12:15   3A8
Towards the limit of ion implantation and rapid thermal annealing as a
technique for shallow junction formation                                221
J L Altrip†, A G R Evans†, J R Logan‡ and C Jeynes§
† University of Southampton, UK, ‡ Lucas Automotive Ltd, Soihull,
UK, § University of Surrey, Guildford, UK


**Session 3B Room C4: GaAs FET Structures**

10:00   3B1
**A two-dimensional approach to the noise simulation of GaAs MESFETs**   225
G Ghione
Politecnico di Milano, Italy

10:15   3B2
**3D Integration of GaAs MESFET and varactor diode for a VCO-MMIC**   229
M Joseph†, B Roth†, F Scheffer†, H Meschede†, A Beyer† and
K Heime‡
† Halbleiteriechnik/Halbleiter technologie, University of Duisburg,
W Germany, ‡ Institut für Halbleitertechnik, Aachen, W Germany

10:30   COFFEE

11:15   3B3
**High performance 0·5 μm GaAs MESFET for MMIC applications**   233
A Belache and S Gourrier
Laboratoires d'Electronique Philips, Limeil-Brevannes, France

11:30   3B4
**GaAs FET and HFET on InP substrate**   237
A Clei, R Azoulay, N Draida, S Biblemont and C Joly
CNET, Bagneux, France

11:45   3B5
**Analysis of the breakdown phenomena in GaAs MESFETs**   241
J Ashworth† and P Lindorfer‡
† Siemens AG, Munich, W Germany, ‡ TU Vienna, Austria


**Session 3C Room B13: Phenomena in Advanced MOS Structures**

10:00   3C1
**Dynamic hot-carrier stress on submicron n- and p-channel transistors**   245
C Bergonzoni, G Dalla Libera and A Nannini
SGS-Thomson Microelectronics, Agrate MI, Italy

xiv  Contents

10:15   3C2
Hot-carrier experiments on scaled NMOS transistors   249
R Woltjer, G M Paulzen, P H Woerlee, C A H Juffermans and H Lifka
Philips Research Laboratories, Eindhoven, The Netherlands

10:30   COFFEE

11:00   3C3
Noise characterisation of Silicon MOSFETs degraded by F–N injection   253
C Nguyen-Duc, G Ghibaudo and F Balestra
Laboratoire de Physique des Composants à Semiconducteurs, ENSERG, Grenoble, France

11:15   3C4
Hot-carrier-induced degradation in short-channel silicon-on-insulator MOSFETs   257
T Ouisse†, S Cristoloveanu‡, G Reimbold§ and G Borel†
† Thomson-TMS, Grenoble, France, ‡ Laboratoire de Physique des Composants à Semiconducteurs, Grenoble, France, § LETI-CENG, Grenoble, France

11:30   3C5
Study of the enhanced hot-electron injection in split-gate transistor structures   261
J Van Houdt, P Heremens, J S Witters, G. Groeseneken and H E Maes
IMEC, Leuven, Belgium

11:45   3C6
Electron conduction and charge trapping behaviour of $SiO_2$ prepared by plasma anodisation   265
J F Zhang†, P Watkinson†, S Taylor†, W Eccleston† and N D Young‡
† University of Liverpool, UK, ‡ Philips Research Laboratory Redhill, UK

12:00   3C7
A study of multiplication-induced breakdown in buried-channel p-MOSFETs   269
T Skotnicki, G Merckel and A Merrachi
CNET, Meylan, France

12:15  3C8
Electrical characterization of ferroelectric thin films for integration into VLSI  273
D M Swanston, D J Johnson, D T Amm, E Griswold and M Sayer
Queen's University, Kingston, Canada

12:30  LUNCH


## TUESDAY AFTERNOON SESSION

**Session 4IP room B1: Invited Paper**

13:30  4IP1
Recent trends in multilayer process simulation for submicron technologies  277
A Poncet, A Gérodolle and S Martin
CNET, Meyan, France


**Session 4A room B1: Hot Carriers in MOS Devices**

14:15  4A1
Comparison of hot-carrier degradation in n- and p-MOSFETs with various nitride-oxide gate films  287
H Iwai, H S Momose, T Morimoto, S Takagi and K Yamabe
ULSI Research Centre, Toshiba Corporation, Kawasaki, Japan

14:30  4A2
Duty cycles in digital logic applications: a realistic way of considering hot-carrier reliability  291
W Weber, M Brox, T Künemund, D Schmitt-Landsiedel, Q. Wang
Siemens AG, Munich, W Germany

14:45  4A3
Annealing of fixed oxide charge induced by hot-carrier stressing  295
M Brox and W Weber
Siemens AG, Munich, W Germany

15:00  4A4
Gate oxide integrity and hot-carrier degradation of $TaSi_2$ $P^+$ polycide gate MOSFETs  299
U Schwalke, W Hänsch, M Kerber, A Lill and F Neppl
Siemens AG, Munich, W Germany

## Contents

15:15 **4A5**
**The effects of gate and drain biases on the stability of low temperature poly-Si TFTs**   303
N D Young and A Gill
Philips Research Laboratories, Redhill, UK

15:30 **TEA**

16:00 **4A6**
**Optimisation of a 5 nm ONO-multilayer-dielectric for 64 Mbit DRAMs**   307
A Spitzer, H Reisinger and W Hönlein
Siemens AG, Munich, W Germany

16:15 **4A7**
**Optimised and reliable drain structure for 0.5 μm n-channel devices**   311
G Guegan, G Reimbold and M Lerme
D LETI CENG, Grenoble, France

16:30 **4A8**
**The use of a buried lightly doped drain (BLEDD) for improved NMOS hot-carrier reliability***
K J Barlow
Plessey Research Caswell Ltd, Northants, UK

16:45 **4A9**
**Simulation of SOI-like kink effects in a 'Horseshoe-Drain' MOSFET for 16M and 64M DRAM applications**   315
R Subrahmanyan, M Orlowski and H Kirsch
Motorola Inc., Austin, TX, USA

**Session 4B room C4: Compound Semiconductor Symposium**

14:15 **4BIP1**
**The future of epitaxy**   319
P Balk
Delft University of Technology, The Netherlands

14:45 **4BIP2**
**Directions for optoelectronic circuits§**   607
J R Hayes, J Gimlett, W-P Hong, G-K Chang, J B D Soole and R Bhat
Bellcore, Red Bank, NJ, USA

15:15 Discussion

---

\* Paper not available at time of going to press
§ Paper submitted late

15:30 TEA

16:00 4BIP3
**GaInAs-based transistors and circuits for millimetre-wave applications***
D Pons and P Briere
Thomson-CSF, Paris, France

16:30 4BIP4
**High packing density techniques for cost effective multifunction GaAs MMICs** 327
A A Lane and F A Myers
Plessey Research Caswell Ltd, UK

17:00 Discussion


**Session 4C room B13: Silicon Bipolar Processing**

14:15 4C1
**2D Computer simulation of emitter resistance in presence of interfacial oxide break-up in polysilicon emitter bipolar transistors** 333
J S Hamel†, D J Roulston†, P Ashburn‡, D Gold§ and C R Selvakumar†
† University of Waterloo, Ontario, Canada, ‡ University of Southampton, UK, § University of Oxford, UK

14:30 4C2
**Tunnelling in implanted emitter-base junctions in a low-power UHF process** 337
B Schlicht and L Strobel
Philips GmbH, Hamburg, W Germany

14:45 4C3
**Low frequency noise of npn/pnp polysilicon emitter bipolar transistors** 341
N Siabi-Shahrivar, W Redman-White, P Ashburn and I Post
University of Southampton, UK

15:00 4C4
**Small geometry effects in CMOS compatible self-aligned 'etched-polysilicon' emitter bipolar transistors** 345
G Giroult-Matlakowski†, A Marty†, N Degors‡, A Chantre‡ and A Nouailhat‡
† Motorola Semiconductor, Toulouse, France, ‡ CNET, Meylan, France

* Paper not available at time of going to press

## Contents

15:15    4C5
**The application of a selective implanted collector to an advanced bipolar process**    349
M C Wilson
Plessey Research Caswell Ltd, Towcester, UK

15:30    TEA

16:00    4C6
**SiC pn structures grown by container-free LPE (CF LPE) and semiconductor devices based on these structures**    353
V A Dmitriev, Ya V Morozenko, A M Strel'chuk, V E Chelnokov and A E Cherenkov
Ioffe Institute, Academy of Sciences of the USSR, Leningrad, USSR

16:15    4C7
**The application of limited reaction processing to the deposition of silicon carbide layers**    357
F H Ruddell, D W McNeill, B M Armstrong and H S Gamble
Queen's University, Belfast, UK

## WEDNESDAY MORNING SESSION

### Session 5IP room B1: Invited Paper

09:00    5IP1
**Degradation and wearout of thin dielectic layers during charge injection**    361
M M Heyns
IMEC, Leuven, Belgium

### Session 5A room B1: Silicon Bipolar and Power Devices

10:00    5A1
**Measurement and Simulation of degradation effects in high voltage DMOS devices**    369
P Dickinger, G Nanz and S Selberherr
Institute for Microelectronics, Technical University Vienna, Austria

10:15    5A2
**Technology and design of SIPOS films used as field plates for high voltage planar devices**    373
D Jaumet†, G Charitat‡, A Peyre-Lavigne† and P Rossel‡
† Motorola Semiconductors, Toulouse, France, ‡ Laboratoire d'Automatique et d'Analyse des Systèmes, Toulouse, France

10:30    COFFEE

| | | |
|---|---|---|
| 11:00 | **5A3**<br>**High-speed radiation-hardened ECL circuits on bonded SOI wafers**<br>K Ueno, M Kawano and Y Arimoto<br>Fujitsu Ltd, Kawasaki, Japan | 377 |
| 11:15 | **5A4**<br>**Epitaxial regrowth in double-diffused polysilicon emitters**<br>J D Williams†, P Ashburn†, N E Moisewitsch†, D P Gold‡,<br>J Whitehurst‡, G R Booker‡ and G R Wolstenholme†<br>† University of Southampton, UK, ‡ University of Oxford, UK | 381 |
| 11:30 | **5A5**<br>**Sidewall effects in submicron bipolar transistors**<br>S Decoutere, L Deferm, C Claeys and G Declerck<br>IMEC, Leuven, Belgium | 385 |
| 11:45 | **5A6**<br>**BASIC II: A super self-aligned technology for high-performance bipolar applications**<br>A Pruijmboom, A C L Jansen, H G R Maas, P H Kranen, R A van Es, R Dekker and J W A van der Velden<br>Philips Research Laboratories, Eindhoven, The Netherlands | 389 |
| 12:00 | **5A7**<br>**Silicon-based pseudo-heterojunction bipolar transistors**<br>Z A Shafi and P Ashburn<br>University of Southampton, UK | 393 |
| 12:15 | **5A8**<br>**A 12 V BICMOS technology for mixed analog–digital applications, with high performance vertical pnp**<br>C Mallardeau, P Keen, A Monroy, J C Marin, F Dell'Ova, P A Brunel, D Celi and M Roche<br>SGS-Thomson Microelectronics, Grenoble, France | 397 |

**Session 5B room C4: Optical Properties and Devices**

| | | |
|---|---|---|
| 10:00 | **5B1**<br>**MBE HEMT-compatible diode lasers**<br>J Ebner, J E Lary, G W Eliason and T K Plant<br>Oregon State University, USA | 401 |
| 10:15 | **5B2**<br>**Optical characterisation of InGaAs/InP and InGaAs/InGaAsP MQW structures for optoelectronic applications**<br>K Wolter, R Schwedler, F Reinhardt, R Kersting, X Q Zhou, D Grützmacher and H Kurz<br>Institute of Semiconductor Electronics, Aachen, W Germany | 405 |
| 10:30 | COFFEE | |

xx   *Contents*

11:00  5B3
**Strongly directional emission from AlGaAs/GaAs light emitting diodes** 409
A Köck†, E Gornik†, M Hauser‡ and W Beinstingl‡
† Walter Schottky Institut, TU Munich, W Germany, ‡ Institut für Experimentalphysik, Universität Innsbruck, Austria

11:15  5B4
**Detection of Near IR radiation by SiGe material** 413
B Sopko, J Pavlu, I Prochazka and I Macha
Technical University, Prague, Czechoslovakia

11:30  5B5
**In-line control with scanning photoluminescence of infrared photodiode arrays fabricated on lattice mismatched InGaAs/InP heterostructures***
S Krawczyk†, K Schohe†, C Klingelhofer†, B Vilotitch‡, C Lenoble‡, M Villard‡, X Hugon‡, D Regaud§, F Ducroquet§
† Laboratoire d'Electronique, Ecole Centrale de Lyon, France, ‡ Thomson CMS, St Egreve, France, § Laboratoire de Physique de la Matière, Lyon, France

11:45  5B6
**High-reliability semi-conductor laser amplifiers** 417
S J Fisher, C P Skrimshire, R N Shaw, J R Farr, P C Spurdens, H J Wickes and W J Devlin
British Telecom Research Laboratories, Ipswich, UK

12:00  5B7
**Band to band absorption coefficients in heavily doped Si and SiGe** 421
A Nathan†, S C Jain‡, D R Briglio†, J M McGregor† and D J Roulston†
† University of Waterloo, Ontario, Canada, ‡ Oxford University, UK

**Session 5C Room B13: Silicon on Insulator II**

10:00  5C1
**Design considerations for 0·5 micron ultra thin film sub-micron SOI transistors by two-dimensional simulation** 425
G A Armstrong†, W D French† and J R Davis‡
† Queen's University, Belfast, UK, ‡ British Telecom Research Laboratories, Ipswich, UK

10:15  5C2
**The influence of substrate bias fixed charge in the buried insulator on the gain of the parasitic bipolar inherent in silicon-on-insulator MOSFETs** 429
L J McDaid†, S Hall†, W Eccleston† and J C Alderman‡
† University of Liverpool, UK, ‡ Plessey Research Caswell Ltd, Towcester, UK

* Paper not available at time of going to press

Contents xxi

10:30 COFFEE

11:00 5C3
**Single-transistor latch induced degradation in thin-film SOI MOSFETs: implications for sub-micron SOI MOSFETs** 433
R J T Bunyan†, M J Uren†, L McDaid‡, S Hall‡, W Eccleston‡, N Thomas§ and J R Davis§
† Royal Signals and Radar Establishment, Great Malvern, UK,
‡ University of Liverpool, UK, § British Telecom Research Laboratories, Ipswich, UK

11:15 5C4
**Intrinsic gate capacitances of SOI MOSFETs: Measurement, modelling, floating substrate effects** 437
D Flandre†, F Van de Wiele†, P G A Jespers† and M Haond‡
† Université Catholique de Louvain, Belgium, ‡ CNET, Meylan, France

11:30 5C5
**Identification and characterisation of noise sources in SIMOX MOSFETs** 441
T Elewa†, B Boukriss†, H Haddara‡, A Chovet† and S Cristoloveanu†
† Laboratoire de Physique des Composant à Semiconducteurs, Grenoble, France, ‡ Ain-Shams University, Cairo, Egypt

11:45 5C6
**Improvement of output impedance in SOI MOSFETs** 445
M-H Gao, J-P Colinge, S-H Wu and C Claeys
IMEC, Leuven, Belgium

12:00 5C7
**Temperature behaviour of CMOS devices built on SIMOX substrates** 449
J Belz, G Burbach, H Vogt and W Zimmermann
Fraunhofer Institute of Microelectronic Circuits and Systems, Duisburg, W Germany

12:15 5C8
**Thin SIMOX SOI material for half-micron CMOS** 453
H Lifka and P H Woerlee
Philips Research Laboratories, Eindhoven, The Netherlands

12:30 LUNCH


WEDNESDAY AFTERNOON

Conference excursion and banquet

## THURSDAY MORNING SESSION

**Session 6IP room B1: Invited Papers**

09:00    6IP1
        **Microcontamination§**     625
        T Ohmi and T Shibata
        Tohoku University, Japan

10:00    6IP2
        **Nanometric structures on compound semiconductors***
        C D Wilkinson
        University of Glasgow, UK

10:45    COFFEE

**Session 6A Room B1: DRAM Technology**

11:15    6A1
        **A new stacked capacitor cell for 64 Mbit DRAMs**     457
        Cheon Soo Kim, Jin Ho Lee, Kyu Hong Lee, Dae Yong Kim, Jin Hyo Lee and Chung Duk Kim
        Electronics and Telecommunication Research Institute, Daejeon, Korea

11:30    6A2
        **Stacked capacitor cell technology for 16M DRAM using double self-aligned contacts**     461
        M Fukumoto, Y Naito, K Matsuyama, H Ogawa, K Matsuoka, T Hori, H Sakai, I Nakao, H Kotani, H Iwasaki and M Inoue
        Semiconductor Research Centre, Matsushita Electric Industry Co Ltd, Osaka, Japan

11:45    6A3
        **Buried stacked capacitor cells for 16M and 64M DRAMs**     465
        J Dietl, L DoThanh, K H Küsters, L Kusztelan, H M Mülhoff, W Müller and F X Stelz
        Siemens AG, Munich, W Germany

12:00    6A4
        **Coupling of different leakage paths between trench capacitors**     469
        W Bergner and R Kircher
        Siemens AG, Munich, W Germany

\* Paper not avilable at time of going to press
§ Paper submitted late

**Session 6B room C4: Heterojunction Bipolar Technology**

11:15 6B1
Temperature dependence of DC characteristics of Si/SiGe
heterojunction bipolar transistors 473
A S R Martin, M A Gell, A A Reeder, D J Godfrey, M E Jones,
C J Gibbings and C G Tuppen
British Telecom Research Laboratories, Ipswich, UK

11:30 6B2
Self aligned Si/SiGe heterojunction bipolar transistors grown by
molecular beam epitaxy on diffused $n^+$-buried layer structures 477
P Narozny, D Köhlhoff, H Kibbel and E Kasper
Daimler-Benz Research Centre, Ulm, W Germany

11:45 6B3
Fabrication and characteristics of a MBE-grown InAlAs/InGaAs
heterojunction bipolar transistor using an embedded collector 481
L M Su, H Künzel, R Gibis, G Mekonnen, W Schlaak and N Grote
Heinrich-Hertz-Institut für Nachrichtentechnik Berlin GmbH, W
Germany

12:00 6B4
Performance of AlGaAs/GaAs heterostructure bipolar transistors
grown by MOVPE 485
B Willén, D Haga and G Landgren
Swedish Institute of Microelectronics, Kista, Sweden

**Session 6C room B13: Physical Processes in Silicon Device Structures**

11:15 6C1
A Monte Carlo simulator including generation–recombination
processes 489
L Reggiani†, T Kuhn†, L Varani†, D Gasquet‡, J C Vaissiere‡ and
J P Nougier‡
† Universita' di Modena, Italy, ‡ Université des Sciences et Technique
du Languedoc, Montpellier, France

11:30 6C2
Treatment of thermomagnetic effects in semiconductor device
modelling 493
S Rudin and H Baltes
Institute of Quantum Electronics, ETH Zürich, Switzerland

11:45 6C3
On the modelling of mobility in silicon MOS transistors 497
A Emrani, G Ghibaudo and F Balestra
Laboratoire de Physique des Composants à Semiconducteurs,
ENSERG, Grenoble, France

12:00 6C4
**The hysteresis behaviour of silicon p-n diodes at liquid helium temperature** 501
B Dierickx, E Simoen, L Deferm and C Claeys
IMEC, Leuven, Belgium

12:15 6C5
**Influence of generation–recombination process on transient transport in p-type silicon at 77 K***
N Nemar, J C Vaissiere, J P Nougier
Université Montpellier II, Sciences et Techniques du Lanquedoc, Montpellier, France

12:30 LUNCH

### Session 7IP room B1: Invited Paper

13:30 7IP1
**Recent trends in InP-based optoelectronic components** 505
O Hildebrand
SEL Research Centre, Stuttgart, W Germany

### Session 7A room B1: Silicon Circuits

14:15 7A1
**A high-energy ion implanted BICMOS process with compatible EPROM structures** 515
R C M Wijburg, G J Hemink, J Middelhoek and H Wallinga
University of Twente, Enschede, The Netherlands

14:30 7A2
**A high performance VLSI structure–SOI/SDB complementary buried channel MOS (CBCMOS) IC** 519
Q-Y Tong, X-L Xu, H-Z Zhang
Southeast University, Nanjing, China

14:45 7A3
**A well concept for field-implant free isolation and width independent n-MOSFET threshold and reliability** 523
Ch Zeller, C Mazure, A Lill and M Kerber
Siemens AG, Munich, W Germany

15:00 7A4
**Simulation of EPROM writing** 527
C Fiegna[†], F Venturi[†], M Melanotte[‡], E Sangiorgi[§] and B Riccò[†]
[†] DEIS University of Bologna, Italy, [‡] SGS-Thomson Microelectronics, Agrate MI, Italy, [§] University of Udine, Italy

15:15 TEA

* Paper not available at time of going to press

| | | |
|---|---|---|
| 15:45 | 7A5<br>**Field isolation and active devices for 16 Mbit DRAMs**<br>H-M Mühlhoff, J Dietl, P Küpper and R Lemme<br>Siemens AG, Munich, W Germany | 531 |
| 16:00 | 7A6<br>**Dimensional characterisation of poly buffer LOCOS in comparison with suppressed LOCOS**<br>N A H Wils, P A van der Plas and A H Montree<br>Philips Research Laboratories, Eindhoven, The Netherlands | 535 |
| 16:15 | 7A7<br>**Modelling and simulation of silicon-on-sapphire MOSFETs for analogue circuit design**<br>R Howes†, W Redman-White†, K G Nichols†, S J Murray‡ and P J Mole§<br>† Southampton University, UK, ‡ Marconi Electronic Devices Ltd, Lincoln, UK, § GEC Hirst Research Centre, Wembley, UK | 539 |
| 16:30 | 7A8<br>**ELSIMA: ELDO short-channel IGFET model for analog applications**<br>T Pedron and G Merckel<br>CNET, Meylan, France | 543 |
| 16:45 | 7A9<br>**A fully modular 1 μm CMOS technology incorporating EEPROM, EPROM and interpoly capacitors**<br>P J Cacharelis, M J Hart, G R Wolstenholme, R D Carpenter, I F Johnson and M H Manley<br>National Semiconductor Corporation, Santa Clara, CA, USA | 547 |

**Session 7B room C4: Physics of Optical and Related Devices**

| | | |
|---|---|---|
| 14:15 | 7B1<br>**Optical switches and heterojunction bipolar transistors in InP for monolithic integration**<br>N Shaw, P J Topham, M J Wale<br>Plessey Research Caswell Ltd, Towcester, UK | 551 |
| 14:30 | 7B2<br>**Numerical modelling based comparison of the submicrometre III–V compounds MESFET's performance**<br>V Ryzhii and G. Khrenov<br>Institute of Physics and Technology, USSR Academy of Sciences, Moscow, USSR | 555 |
| 14:45 | 7B3<br>**Small signal analysis of resonant tunnel diodes in the bistable mode**<br>A Zarea, A Sellai, M S Raven, D P Steenson, J M Chamberlain, M Henini and O H Hughes<br>University of Nottingham, UK | 559 |

15:00 7B4
**Planar InP/InGaAs avalanche photodiodes fabricated without a guard ring using silicon implantation and two-stage atmospheric pressure MOVPE** 563
M D A MacBean, P M Rodgers, T G Lynch, M D Learmouth and R H Walling
British Telecom Research Laboratories, Ipswich, UK

15:15 TEA

15:45 7B5
**Estimation of noise figure for conventional and multilayered avalanche photodiodes using the lucky drift model** 567
J S Marsland†, R C Woods‡, C A Brownhill‡ and S Gould‡
† University of Liverpool, UK, ‡ University of Sheffield, UK

16:00 7B6
**10 Gbit/s on-chip photodetection with self-aligned silicon bipolar transistors** 571
J Popp† and H Philipsborn‡
† Siemens AG, Munich, W Germany, ‡ University of Regensburg, W Germany

16:15 7B7
**A Sensitive MNMOS structure for optical storage** 575
M S Shivaraman and O Engström
Chalmers University of Technology, Göteborg, Sweden

## Session 7C room B13: MOS-based Structures and Measurements

14:15 7C1
**A new charge pumping procedure to measure Interface trap energy distributions on MOSFETs** 579
G Van den bosch, G Groeseneken, P Heremans and H E Maes
IMEC, Leuven, Belgium

14:30 7C2
**Interface states extracted from gated diode and charge pumping measurements** 583
F Hofmann
Siemens AG, Munich, W Germany

14:45 7C3
**Extended static $CV$-procedure to investigate minority carrier generation in MOS capacitors** 587
M Kerber and U Schwalke
Siemens AG, Munich, W Germany

| | | |
|---|---|---|
| 15:00 | 7C4<br>**Local temperature distribution in Si-MOSFETs studied by micro-raman spectroscopy**<br>R Ostermeir†, K Brunner†, G Abstreiter† and W Weber‡<br>†Walter Schottky Institut, Technische Universität, Munich, W Germany, ‡Siemens AG, Munich, W Germany | 591 |
| 15:15 | TEA | |
| 15:45 | 7C5<br>**Defects in highly doped silicon investigated by combined current and capacitance DLTS**<br>G I Andersson and O Engström<br>Chalmers University of Technology, Göteborg, Sweden | 595 |
| 16:00 | 7C6<br>**A four million pixel CCD image sensor**<br>T H Lee, B C Burkey and R P Khosla<br>Eastman Kodak Company, Rochester, NY, USA | 599 |
| 16:15 | 7C7<br>**Unified model of the enhancement-mode MOS transistor**<br>W J Kordalski<br>Technical University of Gdańsk, Poland | 603 |
| 16:30 | END OF CONFERENCE | |
| | Author index | 633 |

# Preface

The European Solid State Device Conference has emerged in recent years as the premier European meeting covering all types of semiconductor devices. The 20th meeting continues this trend with invited and contributed papers on topics of current importance to both the compound semiconductor and silicon communities. From a total of almost 300 submitted papers and 540 authors, the three sub-committees selected 139 contributions on devices and technology. The conference plays a major role as a forum for the presentation of work completed in the main European programmes within ESPRIT as well as within national and other international initiatives.

A particular feature of the 20th conference is the inclusion within the four days of a special symposium of invited contributions on aspects of compound semiconductor growth and the properties of a wide range of high performance optical, microwave and other devices. In the compound semiconductor field, the importance of heterojunction structures in non-optical as well as optical devices is clear. These III–V and II–VI structures are finding increasingly strong competition from Si/Ge based structures and therefore papers on bipolar and field effect structures in the two materials systems were placed together to encourage the interchange of ideas across this divide.

The sessions on silicon devices reflect the increasing concerns of the device and circuit communities with the effects of hot electrons in sub-micron devices where the field strengths are extremely high. There is particular interest in thin dielectric layers, where their integrity is very dependent on the method of preparation.

Silicon on Insulator for MOS is a well-developed technology and the potential advantages of this technology for other devices is apparent from a number of contributions.

The conference continues the strong tradition, within Europe, of advanced silicon bipolar work.

It is with great pleasure that we take the opportunity to thank Jacki Butler, Maureen Clarke and the many staff of the Institute of Physics who have provided an excellent service to the Conference, and to Valerie Wilson of the University of Liverpool who copied and distributed the abstracts. Thanks are also due to the sub-committees and to Joe Mun of STC Technology Limited who organised the special symposium on compound devices.

W Eccleston<br>P Rosser

# TIMETABLE

## MONDAY 10 September 1990

| Time | | | |
|---|---|---|---|
| 08:45 | OPENING ADDRESS | | |
| 09:00 | 1IP1: | Invited Paper EASTMAN | |
| 10:00 | 1IP2: | Invited Paper KEEN | |
| 10:45 | COFFEE | | |
| 11:15 | Session 1A | Session 1B | Session 1C |
| 12:30 | LUNCH | | |
| 13:30 | 2IP1: | Invited Paper HILL | |
| 14:15 | Session 2A | Session 2B | Session 2C |
| 15:30 | TEA | | |
| 16:00 | Session 2A cont. | Session 2B cont. | Session 2C cont. |
| 17:45 | ALL SESSIONS FINISHED | | |

## TUESDAY, 11 September 1990

| Time | | | |
|---|---|---|---|
| 09:00 | 3IP1: | Invited Paper KLAASSAN | |
| 10:00 | Session 3A | Session 3B | Session 3C |
| 10:30 | COFFEE | | |
| 11:00 | Session 3A cont. | Session 3B cont. | Session 3C cont. |
| 12:30 | LUNCH | | |
| 13:30 | 4IP1: | Invited Paper PONCET | |
| 14:15 | Session 4A | Session 4B | Session 4C |
| 15:30 | TEA | | |
| 16:00 | Session 4A cont. | Session 4B cont. | Session 4C cont. |
| 17:15 | ALL SESSIONS FINISHED | | |

## COMPOUND SEMICONDUCTOR SYMPOSIUM

| Time | SESSION 4B | |
|---|---|---|
| 14:15 | 4BIP1: | Invited Paper BALK |
| 14:45 | 4BIP2: | Invited Paper HAYES et al |
| 15:30 | TEA | |
| 16:00 | 4BIP3: | Invited Paper PONS/BRIERE |
| 16:30 | 4BIP4: | Invited Paper LANE/MYERS |

## WEDNESDAY, 12 September 1990

| Time | | | |
|---|---|---|---|
| 09:00 | 5IP1: | Invited Paper HEYNS | |
| 10:00 | Session 5A | Session 5B | Session 5C |
| 10:30 | COFFEE | | |
| 11:00 | Session 5A cont. | Session 5B cont. | Session 5C cont. |
| 12:30 | LUNCH | | |
| 14:00 | AFTERNOON EXCURSION | | |

## THURSDAY, 13 September 1990

| Time | | | |
|---|---|---|---|
| 09:00 | 6IP1: | Invited Paper OHMI | |
| 10:00 | 6IP2: | Invited Paper WILKINSON | |
| 10:45 | COFFEE | | |
| 11:15 | Session 6A | Session 6B | Session 6C |
| 12:30 | LUNCH | | |
| 13:30 | 7IP1: | Invited Paper HILDEBRAND | |
| 14:15 | Session 7A | Session 7B | Session 7C |
| 15:15 | TEA | | |
| 15:45 | Session 7A cont. | Session 7B cont. | Session 7C cont. |
| 17:00 | END OF CONFERENCE | | |

*Paper presented at ESSDERC 90, Nottingham, September 1990*
*Session 1A1*

# Characterisation of interface electron state distributions at directly bonded silicon/silicon interfaces

Stefan Bengtsson and Olof Engström

Department of Solid State Electronics, Chalmers University of Technology,
S-412 96 Göteborg, Sweden

Abstract Measurement methods for characterizing the electrical properties of directly bonded Si/Si n/n-type or p/p-type interfaces are presented. The density of interface states in the bandgap of the semiconductor and the density of interface charges at the bonded interface are determined from measurements of current and capacitance vs applied voltage.

1. Introduction

Direct bonding of silicon or oxidized silicon wafers opens new possibilities for novel device geometries and structures. The wafer bonding is achieved through chemical treatment of the silicon wafer surfaces followed by wafer contacting and a temperature step above 700° C (Lasky 1986, Shimbo et al 1986, Bengtsson and Engström 1990). Power devices (Ohashi et al 1987) and sensors (Stemme and Stemme 1989) as well as silicon-on-insulator (SOI) materials (Lasky 1986, Maszara et al 1988, Haisma et al 1989) have been prepared. When bonded interfaces are incorporated in a device structure the presence of interface charges captured into electron states is of critical importance for the functioning of the device. For $Si/SiO_2$ and $SiO_2/SiO_2$ interfaces standard methods for MOS structures can be used to characterize interface state distributions and charges (Lasky 1986, Bengtsson and Engström 1989). The electrical properties of bonded Si/Si interfaces can be controlled by the chemical treatment used during the joining process (Bengtsson and Engström 1988, 1989). Characterization techniques are of importance for their optimization. In this paper we present methods of obtaining energy distributions of interface states and the total charges present at bonded Si/Si interfaces.

2. Theoretical description

Charge present at electron states of an interface between two n-type surfaces may give rise to an energy barrier as illustrated in Fig. 1 for two equally doped semiconductors of doping concentration N (Bengtsson and Engström 1989). When a voltage V is applied to the structure, as shown in the figure, the occupation level of the interface state distribution is expected to change. This changes the ratio between the two voltages $V_1$ and $V_2$ on each side of the interface. For the situation in Fig. 1, the decrease of $V_1$ is opposed by the build up of negative charge in the interface states. By considering the charge transport as a thermoionic emission across the interface energy barrier, the barrier height, the interface charge and its energy distribution in the silicon bandgap can be determined. A relation between the applied voltage V, the potential barrier $V_1$ and the interface charge $Q_s$ can be obtained as

$$V = \left( \frac{Q_s}{\sqrt{2\varepsilon\varepsilon_0 q N}} - \sqrt{V_1} \right)^2 - V_1. \qquad (1)$$

The total current density across the interface can be expressed as

$$i = A^* T^2 \left( \frac{N}{N_c} \right) \exp\left( -\frac{qV_1}{kT} \right) \left( 1 - \exp\left( -\frac{qV}{kT} \right) \right) \qquad (2)$$

© 1990 IOP Publishing Ltd

where A* is the effective Richardson constant and $N_c$ is the effective density of states in the conduction band of silicon. The potential barrier $V_0$ at zero bias is found from Eq. (1) with V= 0 and $V_1 = V_0$:

$$V_0 = \left(\frac{1}{8\varepsilon\varepsilon_0}\right)\left(\frac{Q_{s0}^2}{qN}\right). \qquad (3)$$

When $V_1$ and $Q_s$ are known as functions of the applied bias V, the derivative of the interface charge $Q_s$ with respect to the position of the Fermi level at the interface (Fig. 1), $dQ_s/dE$, is obtained as

$$\frac{1}{q}\frac{dQ_s}{dE}(\Delta E) = \frac{dQ_s}{dV_1}\frac{1}{q}. \qquad (4)$$

Here the argument $\Delta E = E_c - E_T$ denotes the energy distance from the conduction band edge to the trap energy level as shown in Fig. 1.

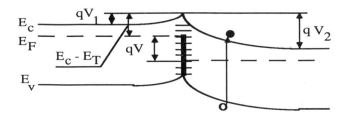

Fig. 1. Band diagram of an interface between two bonded n-type silicon wafers with a voltage V applied to the structure.

The capacitance of the structure can be simulated from the same basic model. The capacitance at zero bias $C_0$ of an interface having the density of interface charge $Q_{s0}$ is given by

$$C_0 = \frac{\varepsilon\varepsilon_0}{2x} = \frac{qN\varepsilon\varepsilon_0}{Q_{s0}} = \sqrt{\frac{qN\varepsilon\varepsilon_0}{8V_0}} \qquad (5)$$

where x is the width of the space charge regions on either side of the interface. When the structure is biased the total capacitance C is given by

$$\frac{1}{C} = \sqrt{\frac{2}{q\varepsilon\varepsilon_0 N}}\left(\sqrt{V_1} + \sqrt{V_2}\right) = \frac{Q_s}{q\varepsilon\varepsilon_0 N}. \qquad (6)$$

Because of the build up of the interface charge $Q_s$ with the applied bias, a change in the bias mainly influences $V_2$ (see Fig. 1), while $V_1$ remains close to the equilibrium value $V_0$. Under these conditions Eq. (6) can be linearized giving

$$\left(\frac{2C_0}{C} - 1\right)^2 = 1 + \frac{V}{V_0}. \qquad (7)$$

Thus, by plotting the left hand side of Eq. (7) against the applied voltage V a straight line which intersects the voltage axis at $V = -V_0$ is expected.

3. Sample preparation

Three inch n-type silicon wafers were dipped in a 2% HF : $H_2O$ solution followed by the standard RCA cleaning procedure and a treatment in warm 65% $HNO_3$ giving hydrophilic wafer surfaces. The wafers were rinsed in water, dried and then brought together at room temperature in a mechanical fixture to form 0.1/0.1 Ωcm and 50/50 Ωcm structures. The wafers were released from the fixture and loaded into an annealing furnace on a flat boat. Annealing was performed at 1000° C for 2 hours in oxygen. Phosphorus was deposited at 1025° C for 15 minutes as contact doping followed by deposition of aluminum. Test samples were cut from the wafers and the edges of the samples were etched to reduce surface leakage currents.

## 4. Measurement examples

Results for a 50/50 Ωcm sample (area 0.04 cm$^2$) and a 0.1/0.1 Ωcm sample (area 0.07 cm$^2$) are presented. From the current-voltage characteristic the dependence of the potential barrier $V_1$ and the interface charge $Q_s$ on the applied voltage V is obtained from Eqs. (1) and (2). The interface charge density $Q_s$ vs the applied voltage V is shown in Fig. 2 for the 50/50 Ωcm sample. The interface charge $Q_{s0}$ at zero bias is found to be about $7.3 \cdot 10^{-9}$ As/cm$^2$. From Eq. (3), the corresponding potential barrier $V_0$ is 0.50 V. For the 0.1/0.1 Ωcm sample $Q_{s0}$ was found to be $2.0 \cdot 10^{-7}$ As/cm$^2$, corresponding to a potential barrier $V_0$ of 0.43 V. Fig. 3 shows the quantity $(1/q) dQ_s/dE$ from Eq. (4).

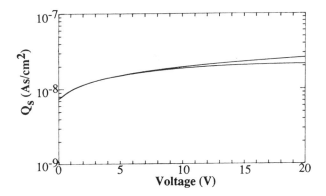

Fig. 2. The interface charge $Q_s$ vs the applied bias, for the 50/50 Ωcm sample, found from the i-V measurement (upper curve) and the C-V measurement (lower curve).

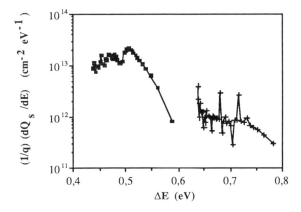

Fig. 3. The quantity $(1/q) dQ_s/dE$ for the 50/50 Ωcm sample (crosses) and for the 0.1/0.1 Ωcm sample (squares).

The interface charge $Q_s$ can be found also from the capacitance vs voltage measurement using Eq. (6). The voltage dependence of $Q_s$ for the 50/50$\Omega$cm sample is given in Fig. 2. The two values of $Q_s$ found from the i-V and C-V measurements are in good agreement. For the 0.1/0.1 $\Omega$cm sample $Q_{s0}$ was found to be $6.5 \cdot 10^{-8}$ As/cm$^2$. The difference in $Q_{s0}$ obtained by i-V and C-V techniques for the low-ohmic sample is probably due to errors in the capacitance measurement caused by the high current level in this sample. Fig. 4 shows the capacitance function on the left hand side of Eq. (7) vs the applied bias for the 50/50 $\Omega$cm sample. A potential barrier at zero bias of about 0.5 V is found from this analysis.

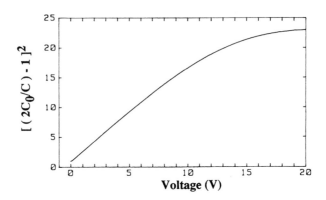

Fig. 4. $(2C_0/C - 1)^2$ (Eq. (7)) vs the applied voltage V for the 50/50 $\Omega$cm sample.

5. Discussion

Two measurement methods for characterizing directly bonded Si/Si structures in which the current is restricted by a barrier at the interface are proposed. By using samples of different doping levels different parts of the bandgap can be investigated. The derivative of the interface charge with respect to the energy position of the Fermi level in the bandgap is the density of interface states $D_{it}$ at the bonded interface if the total increase in $Q_s$ with the applied bias V is due to an increase in the number of electrons trapped at the interface. However, there are other possibilities of having charge trapping in the silicon volume close to the interface, such as electron states in the bandgap due to a small region of disordered structure or charge trapping at metal precipitates. The energy distributions of interface states obtained by the methods presented here differ considerably from the corresponding distributions obtained from Si/SiO$_2$ interfaces, bonded as well as thermally grown. The slope of the $dQ_s/dE$-curves (Fig. 3) is negative in the energy region near the silicon valence band edge where a corresponding graph for a Si/SiO$_2$ structure normally has a positive slope.

References

Bengtsson S and Engström O *1988 J. de Physique* **49** C4-63
Bengtsson S and Engström O *1989 J. Appl. Phys.* **66** 1231
Bengtsson S and Engström O *1990 J. Electrochem. Soc.* **137** 2297
Haisma J, Spierings G A C M, Bierman U K P and Pals J A *1989 Jpn. J. Appl. Phys.* **28** 1426
Lasky J B *1986 Appl. Phys. Lett.* **48** 78
Maszara W P, Goetz G, Caviglia A and McKitterick J B *1988 J. Appl. Phys.* **64** 4943
Ohashi H, Furukawa K, Atsuta M, Nakagawa A and Imamura K *1987 IEEE IEDM Tech. Dig.* 678
Shimbo M, Furukawa K, Fukuda K and Tanzawa K 1986 *J. Appl. Phys.* **60** 2987
Stemme E and Stemme G *1990 Sensors and Actuators* **A21** 336

*Paper presented at ESSDERC 90, Nottingham, September 1990*
*Session 1A2*

# Assessment of SIMOX material by optical waveguide losses

N Mohd Kassim, H P Ho, T M Benson,
Department of Electrical and Electronic Engineering
University of Nottingham

D E Davies
EOARD, London

Abstract

The propagation loss of single-mode optical waveguides in multiple implant SIMOX wafers has been used to assess the quality of the superficial silicon layer. Increased attenuation can be correlated to the creation of thermal and new donors. Some higher loss wafers also show interfacial roughness which gives rise to additional scattering losses.

In the III-V semiconductors high-quality lattice-matched multi-layers such as GaAs/GaAlAs or InGaAsP/InP can conveniently be used to form low-loss optical waveguides (McIlroy et al 1987, Kapon and Bhat 1987, Angenent et al 1989). For silicon, the layered material commonly encountered with it is $SiO_2$ so a $Si-SiO_2$ configuration is worth exploring for optical waveguide fabrication. A silicon equivalent to a III-V semiconductor heterojunction guide requires a buried-oxide confining layer. The SIMOX (separation by implanted oxygen) process, in which a buried oxide layer is created by ion-implantation of oxygen followed by thermal annealing, is one technology being developed for silicon-on-insulator MOS electronic components. The SIMOX process typically produces a buried oxide around 0.4 $\mu$m thick, with apparently sharp interfaces on top of which lies 0.1–0.2 $\mu$m of superficial silicon. This conveniently provides for a single mode slab waveguide at 1.3 $\mu$m with the thickness of the implanted oxide layer just sufficient to prevent leakage of confined modes into the silicon substrate. The quality of the silicon guiding layer is important and factors such as interface planarity, precipitates free carriers and defects could contribute to transmission loss. This paper shows that measurement of optical waveguide loss provides a sensitive means of material assessment.

As previously reported (Davies et al 1989) SIMOX wafers processed in a single implant-anneal stage show large optical losses of around 30dBcm$^{-1}$. It was noted at that time that substantially reduced losses could be observed in SIMOX material produced using a sequence of implant-anneal steps. This is in line with the substantially reduced losses could be observed in SIMOX material produced using a sequence of implant-anneal steps. This is is line with the substantially reduced dislocation densities reported by Hill et al (1988) and Van Ommen (1989) and the successful fabrication of bipolar devices by Platteter and Cheek (1988) in SIMOX material processed in this way. Here waveguiding in three sequentially processed SIMOX wafers is reported. Each wafer was implanted in three sequential stages with intermediate and final annealing as summarized in Table 1.

© 1990 IOP Publishing Ltd

|  | Sample A | Sample B | Sample C |
|---|---|---|---|
| Implantation Conditions<br>Beam Energy (keV)<br>Substrate Temperature (°C)<br>Total dose (cm$^{-2}$) | 160<br>600<br>$1.2 \times 10^{18}$ | 200<br>600<br>$1.5 \times 10^{18}$ | 200<br>640<br>$1.8 \times 10^{18}$ |
| Intermediate and final<br>anneal conditions | 6 hours<br>each at<br>1300°C | 2 hours<br>each at<br>1285°C | 6 hours<br>each at<br>1300°C |

Table 1 Implantation and Annealing Stages

Strip-loaded waveguides with the structure shown in Figure 1 were fabricated by reactive ion etching. The loading strips were formed parallel to the (001) and a series of widths in the range 3-5 μm was used. The additional surface oxide was deposited on samples A and C but an additional epitaxial layer was grown on sample B which allowed a thermal oxide to be grown whilst maintaining the silicon guiding layer thickness.

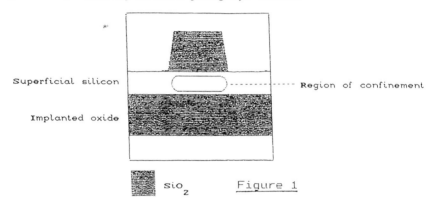

Figure 1

TE-like optical waveguiding at 1.3 μm was observed by end-fire coupling into the cleaved end of a waveguide via a X45 microscope objective. The overall thickness of the silicon wafer was reduced to about 100 μm to assist the cleaving process. The light transmitted was focused on to an IR camera and a Ge photodiode via another x45 objective. Because of the small guide thickness encountered coupling problems were anticipated and whilst excitation of both slab and pencil-beam guided modes was clear it was also inefficient. All the guides tested were single moded in both vertical and horizontal directions as expected.

The cleaved input and output facets form a Fabry-Perot cavity and the transitted power $I_T$ varies as (Regener and Sohler 1985)

$$I_T = \frac{\eta I_o T^2 e^{-\alpha L}}{(1-Re^{-\alpha L})^2 + 4Re^{-\alpha L} \sin^2\beta L}$$

where $I_o$ is the incident intensity, R the mode reflectivity, $\alpha$ the attenuation coefficient, T the transmittance, $\beta$ the propagaton constant of the guided mode, L the sample length and $\eta$ the coupling efficiency to the mode. $I_T$ is a maximum ($I_{TMAX}$) when $\sin^2 \beta L = 0$ and a minimum ($I_{TMIN}$) when $\sin^2 \beta L = 1$. It follows, Walker (1985), that

$$f(u) = \ln(\frac{1+U^{\frac{1}{2}}}{1-U^{\frac{1}{2}}}) = \alpha L - \ln R$$

where $U = I_{TMIN}/I_{TMAX}$.

The optical phase difference ($\beta L$) can be tuned by gently heating the sample to produce the periodic variation in transmission shown in Figure 2.

If R and L are known, measurement of U gives a non-destructive means of calculating $\alpha$. In our initial experiments f(u) was measured for various lengths L of each sample and $\alpha$ and R found from the slope and intercept of a plot of f(u) against L. TE attenuation values were 10.9dB/cm for sample A, 7.6 dBcm$^{-1}$ for sample B and 4.8 dBcm$^{-1}$ for sample C with a facet reflection coefficient of 0.26 in all cases.

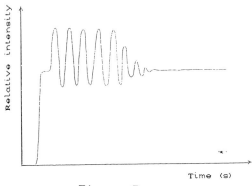

Figure 2

The losses were found to be independent of the rib width, indicating that the effect of scattering from strip edge roughness is negligible because of the strong vertical confinement. The Fabry-Perot technique has the advantages of being insensitive to $\eta$ and Io, always giving an upper bound to $\alpha$ (if for example cleaves are poor or higher order modes propagate) and becoming more accurate the smaller the waveguide loss. Loss values for samples A and B were confirmed by plotting intensity (with error bars representing Fabry-Perot variations) against sample length.

The electrical activity associated with the oxygen present in SIMOX material presents an additional potential loss mechanism. The so called thermal donors (active around 450°C) and new donors (active around 750°C) generate additional free carriers which Cristoloveanu et al (1985) showed can significantly modify the resistivity even after relatively short anneals. The presence of free-carriers should also manifest itself as an increase in optical waveguide loss, as can be gauged from data summarised by Soref and Bennett (1987). Waveguide specimens from Samples A and B were submitted to a series of 30 minute anneals in vacuum at successively increasing temperatures from 250°C to 800°C and then at 1200°C. After each anneal waveguide attenuation was measured and the results obtained are shown in Figure 3.

Sheet resistance measurements of Cristoloveanu et al (1985) made after similar short anneals, but in nitrogen, are superimposed on Figure 3. There is excellent colleration between the temperatures where loss and conductance increases occur. Figure 3 shows optical loss increases significantly larger than the 2dBcm$^{-1}$ anticipated using estimates of free carrier concentration made using the reported sheet resistance changes. Further studies of this are in hand. Losses presented in Figure 3 increase but never fall below the starting value. The loss values measured for sample A always exceed those for sample B. Cross-sectional TEM studies show that this can be attributed to scattering associated with observed interfacial roughness between the superficial silicon and buried oxide layers.

In a further experiment, sample C was subjected to extended anneals, in vacuum, at 400°C and 750°C. Figure 4 shows how optical losses increase with the time of the anneal.

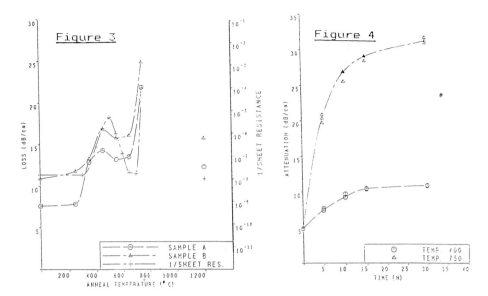

Figure 3

Figure 4

## Conclusion

Optical waveguides fabricated in SIMOX material formed by a multiple stage implant and anneal process show losses significantly below those for similar structures in material formed by a single implant and anneal. Observations of increased waveguide attenuation after long and short low-temperature anneals can be correlated with the presence of new and thermal donors. Interfacial roughness is also seen to contribute to increased attenuation.

These factors indicate that waveguide loss measurements provide an effective means of assessing the quality of SIMOX material.

## References

Angenent J H, Erman M, Auger J M, Gamonal R and Thigs P J A 1989 Elec Lett 25 629

Cristoloveanu S, Pumfrey J, Scheid E, Hemment PLF and Arrowsmith R P 1985 Elec Lett 21, 802

Davies D E, Burnham M, Benson T M, Mohd Kassim N and Seifouri M 1989, Proc SOS/SOI Technology Conference, 160

Hill D, Fraundorf P and Fraundorf G 1988 J Appl Phys 63, 4933

Kapon E and Bhat R 1987 Appl Phys Lett 50, 1628

McIlroy P W, Rodgers P M, Singh J S, Spurdens P C and Henning I D 1987 Elec Lett 23, 703

Platteter D G and Cheek T F 1988 IEEE Trans Nucl Sci 35, 1350

Regener R and Sohler W 1985 Appl Phys B36, 143

Soref R A and Bennett B R 1987 IEEE J Quantum Elec QE-23, 123

Walker R G 1985 Elec Lett. 21, 581

Van Ommen D H 1989 Nucl Inst and Methods in Phys Res B39, 194

## High density 3D, CMOS circuits with ELO SOI technology

### Gerhard Roos, Bernd Hoefflinger and René Zingg

Institute for Microelectronics Stuttgart IMS
Allmandring 30a
7000 Stuttgart 80, West Germany

**Abstract:** *A priori crystalline silicon–on–insulator–on–silicon structures were fabricated with 8 masks. With this three–dimensional CMOS technology, three high–quality transistor channels are stacked vertically. Dual-gate PMOS transistors on top of NMOS transistors deliver the same current for the same channel width. 3D CMOS test circuits have been built with a footprint of one third of their 2D bulk counterparts.*

### Introduction

The functional density of a chip can be increased by decreasing feature sizes more and more. However, the circuit density of logic functions depends more on the wiring, vias and contacts than on the transistor sizes. The large number of contacts, especially from metal to silicon also reduces yield and reliability [Che85]. A fundamentally different way of improving circuit density is by using a 3D approach, where transistors are stacked on top of each other [Hoe84]. Here, problems are usually related to the quality of the devices, to the lack of planarity and to the number of contacts and vias.

Vertical stacking of transistors is possible with a layered silicon–on–insulator–on–silicon (SOIS) structure. Here, the upper silicon film should preferably be a priori crystalline, vias should be self–aligned and free from electromigration, and the overall structure should be planarized for step coverage and density.

Reduced-temperature selective epitaxy and localized lateral overgrowth of silicon over oxide has been shown to meet all of the above criteria. Bulk NMOS with identical features as in standard 2D CMOS and thin film PMOS in the SOI with the same current as NMOS are reported in [Zin89a] and [Zin89c].

### Process

The actual process starts with an n+ silicon wafer, on which p–epitaxy is performed for bulk NMOS transistors. The n+ substrate is used as a ground plane, which reduces ground bounce and improves signal immunity [Gab88]. N+ sinkers (mask 1 in Figure 1) bring ground potential to certain NMOS transistors, offering ground ties everywhere and eliminating one half metal power plane as well as metal contacts with their electromigration problems. The 2D simulation results of the sinker are shown in Figure 2.

A standard NMOS is built with LOCOS, field implant and poly gate (masks 2,3). No trenches, wells and well ties are needed. Latch–up problems are eliminated. A plasma oxide is formed (mask 4) for the definition of the PMOS islands. After the growth of the second gate oxide, seed holes are opened with mask 5, from which silicon grows selec-

| # | 2D | 3D | # |
|---|---|---|---|
|   | --- | Sinker | 1 |
| 1 | Well | --- |   |
| 2 | LOCOS | LOCOS | 2 |
| 3 | Field | --- |   |
| 4 | Poly | Poly1 | 3 |
|   | --- | AAPMOS | 4 |
| 5 | SD n+ | --- |   |
|   | --- | Si-Via | 5 |
| 6 | SD p+ | --- |   |
|   | --- | Poly2 | 6 |
| 7 | Contat | Contact | 7 |
| 8 | Metal | Metal | 8 |

Table 1: Mask Count Table

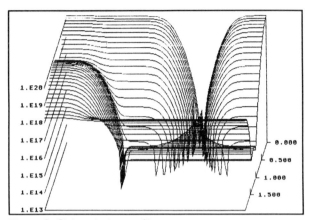

Figure 2: Sinker doping profile

tively up and laterally over the oxide and the buried poly gate. The overgrowth is planarized and thinned to 0.4 microns by chemo-mechanical polishing of silicon with the plasma oxide as spacer. Self-aligned silicon vias, n+-to-p+ contacts and crystalline islands for the thin-film PMOS transistors are thus obtained. After the growth of the third gate oxide, the top poly gate is formed (mask 6). It serves as a mask for the p+ sources and drains. Contact and metal layers (mask 7 and 8) complete the single-metal process. The planarity of the overall structure is actually better than that of a 2D bulk CMOS process. Including epitaxy and polishing, only industrial standard processes are used.

### Device Design and Results

2D and 3D test circuits have been built with identical design rules, which are a derivative of the MOSIS rules. Circuits with symmetric output characteristics are built with the same size of NMOS in the bulk and PMOS in the SOI. This is achieved by buffer inverters, which have PMOS transistors controlled by a dual gate [Zin89a].
The drain current of the PMOS transistor is shown in Figure 3b. It shows the current of the PMOS controlled by the bottom gate only and by the top gate only. If top and bottom gate both control the device, the measured current is about 70% higher than the sum from the two individual gates. This equals the NMOS bulk-transistor in Figure 3a and leads to symmetric output characteristics (Figure 4).

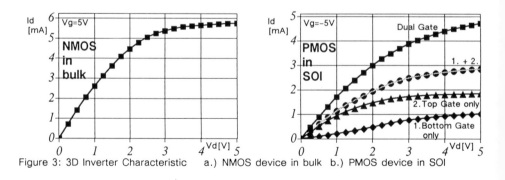

Figure 3: 3D Inverter Characteristic   a.) NMOS device in bulk   b.) PMOS device in SOI

Silicon on insulator I    11

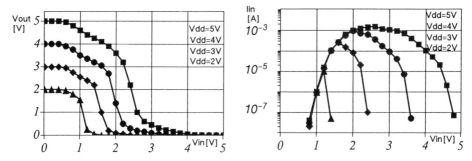

Figure 4: 3D CMOS Inverter    a.) Transfer Characteristics    b.) Current Characteristics

The area advantage of the 3D circuits, such as inverter, NAND range from 38% to 43% not yet reflecting the area gain due to the missing Vss metal busses. Figure 5 and 6 show the schematic, the layout and the cross-section of a NAND2 controlling 4 transistor channels with 3 poly gates.

The series connection of the 2 NMOS is done in the bulk diffusion, while the OR-function of the two PMOS is performed by one PMOS layer controlled by two gates. The top gate has the input A, the bottom gate the input B. This bottom gate also controls one NMOS transistor.

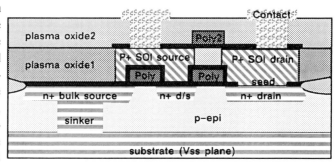

Figure 5: Cross-Section of 3D-NAND2

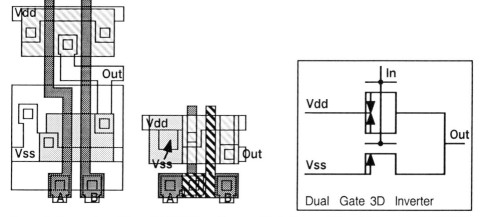

Figure 6: Comparison 2D vs 3D NAND2   a.)Layout 2D   b.)Layout 3D   c.)Schematic 3D NAND2

The micro-photograph in Figure 7 shows the physical realization of a master slave flip-flop. This design is not yet area optimized, but it demonstrates the usefulness of 3D stacked PMOS/NMOS transistors.

Figure 8 illustrates that the master slave flip-flop is built of inverters and selectors only.

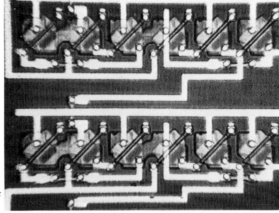

Figure 7: Micro-photograph of MS-FF

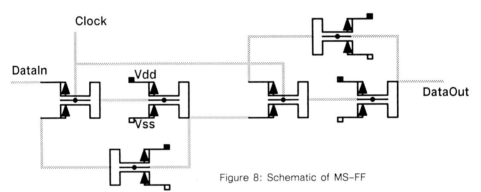

Figure 8: Schematic of MS-FF

## Conclusion

The examples show that 6 of the 8 mask layers in this 3D process are used for connection and wiring: sinker, poly1, seed, poly2, contact and metal1. The comparable 2D process uses only 3 out of 8. By using these connection possibilities and by applying circuit techniques which are especially suited for this 3D process, logic functions require between half and one third of the area of comparable 2D bulk circuits.

## Acknowledgement

The valuable discussions with V.Dudek and T. Schwederski are gratefully acknowledged.

## References

[Che85]  J.Chern, W.G.Oldham, N.Cheung, "Contact-Electromigration-Induced Leakage Failure in Aluminium-Silicon to Silicon Contacts", IEEE ED-32 (7), July 1985

[Gab88]  T.Gabara, "Reduced Ground Bounce and Improved Latch-Up Suppression Through Substrate Conduction", IEEE-SC-23 (5), pp1224-1232, October 1988

[Hoe84]  B.Hoefflinger, S.T.Liu and B.Vajdic, "A Three-Dimensional CMOS Design Methodology", IEEE SC-19(1), pp37-39, February 1984

[Zin89a]  R.P.Zingg, B.Hoefflinger and G.W.Neudeck, "High Quality Dual-Gate PMOS Devices in Local Overgrowth (LOG)", IEE Electronics Letters-25 (15), pp1009-1011, July 20th,1989

[Zin89c]  R.P.Zingg, B.Hoefflinger and G.W.Neudeck, "Stacked CMOS Inverter with Symmetric Device Performance", IEDM89, pp909-911, December 1989

# Rounded edge mesa for submicron SOI CMOS process

O. Le Néel*, M.D Bruni, J. Galvier and M. Haond

C.N.E.T., BP 98, 38243 Meylan Cedex, France
* MATRA-MHS, Route de Gachet, 44087 Nantes Cedex 03, France

**Abstract.** Different isolation features have been proposed for SOI: LOCOS, mesa, reoxidized mesa. Mesas allow a low width loss and a high integration density if an anisotropic etch is used. However, some isotropic step is necessary for the gate etch to avoid residues. We present here the Rounded Edge Mesa (**REM**) which allows an accurate control of the gate dimensions without residues. Characteristics of devices fabricated with 0.7 µm SOI CMOS process are presented.

SOI has received attention for its potential in submicron CMOS devices: no latch-up, total isolation, sharp subthreshold slope[1]. SOI material offers many posibilities by using suitable lateral isolation structure : LOCOS, MESA, Reoxidised-MESA.[2] Mesa structures allow a low width loss and a high integration density by using an anisotopical etch for the active zone defintion.
For the case of mesa isolation when an anisotropical etch is used for the gate definition, a specific overetch of the polysilicon is necessary in order to avoid residues wich might cause gate shorts : figure 1.

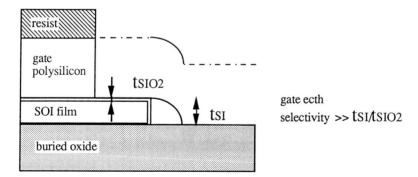

Figure 1: Gate etch process selectivity conditon to avoid shorts

The selectivity of the overetch of polysilicon on oxide must be higher than the $t_{Si}/t_{SiO2}$ ratio in order to suppress residues and preserve the SOI film integrity below the gate oxide. If this condition is not fullfilled, an aditional isotropical etch is necessary to eliminate the gate polysilicon residues,. This provides a lack of control of the gate dimensions. Another solution is to avoid the creation of residues by avoiding sharp active zone edge steps.We present in this paper the Rounded Edge Mesa (R.E.M) technique wich avoids gate shorts and allows an accurate control of the gate dimensions.

© 1990 IOP Publishing Ltd

The R.E.M process is schematically summarised in Figure 2. The active zone is defined by using a polysilicon/pad oxide mask with a thickness equal to the SOI film thickness(Fig.2.a). Boron is then implanted in the SOI film at the border of the active zone of the NMOS device in order to avoid the parasitic edge transistor. An additional 300nm polysilicon film is deposited atop (fig.2.b). Full trench anisotropic etch is then performed in a R.I.E process : when the SOI film is reached (fig 2.c), the structure presents polysilicon spacers around the polysilicon/pad oxide mask. The etch stops when the buried and pad oxides are detected (Fig.2.d). Since there is no selectivity between polysilicon and monosilicon etch, the polysilicon spacers are transfered in the SOI film. Notice that the masking layer is removed within the same step as the mesa definition. .

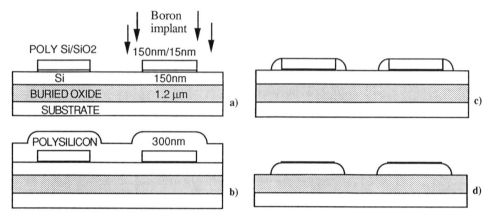

**Figure 2: Schematics of the R.E.M process**

Figure.3 presents a SEM photomicrograph of a flat view of the REM, after the gate definition step, wich shows the soft edge profile of the mesa.

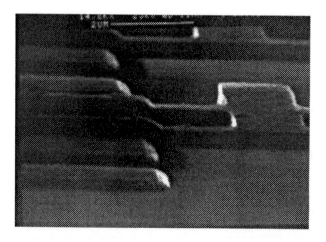

**Figure 3: Photomicrograph of a flat view of the R.E.M after gate definition**

The absence of residues was studied by measuring the resistance of 22 mm long 0.8 μm-spaced 0.8 μm-wide interdigitated polysilicon fingers running on mesas. Figure 4 presents 2 histograms of these measurements: a) vertical edge mesa, b) REM. The same gate etch process was used for both cases. The REM almost suppresses the occurence of shorts along the mesa: we detect 29% of shorts in case a) and less than 5% in case b). These last shorts are likely due to defects associated with the complete DLM CMOS process.

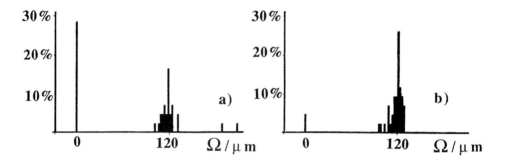

Figure 4: Histograms of the resistance of poly-Si fingers running on mesa edges : a) vertical b) R.E.M

We present in Figure 5 the width loss measured on NMOS and PMOS devices ($\Delta$Wn,$\Delta$Wp). It shows that with the REM technique, the width of the PMOS devices is oversized of about 200nm, corresponding roughly to the actual final mean size of the two spacers generated by the REM technique. For the case of the NMOS devices, the REM technique allows the doping of sidewalls without any width loss, as is the case in LOCOS or classical mesa processes.

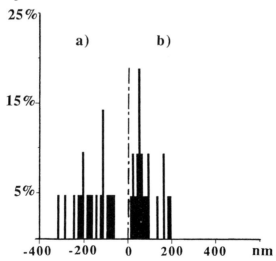

Figure 5: Histograms of the mask and electrical width difference for a) PMOS and b) NMOS devices.

In order to characterize this technique, we have run a 0.7 µm DLM CMOS process on ZMR SOI films thinned down to 150 nm. After the REM formation, the pad oxide is stripped and a 15nm gate oxide is grown. After a threshold voltage adjust implant a 380nm polysilicon film is deposited and implanted with phosphorus. LDDs are implanted both in N and PMOS devices, followed by a classical oxide spacers formation. The source and drain are then implanted before a 100/400nm intrinsic/doped PECVD oxide is deposited. It is reflowed at 1060°C /20s after contact opening. We have used sputtered W as first metal and Al-Si as a second metal. The intermetallic dielectric (PECVD oxide) is planarized.

Figure 6 presents $I_D$-$V_G$ characteristics of 0.55 µm effective channel lenght PMOS a) and NMOS b) transistors, carried out at Vd equal to ±.1V and ±3.5V, the transistor width is 20µm in both cases. The threshold voltage for these devices are -1V and .3V for PMOS and NMOS respectively. Notice the low subthreshold current which shows the good quality of the SOI substrate and of the CMOS process. The absence of subthreshold leakage or "bumps" in the subthreshold slope of the NMOS device confirms that this technique prevents the formation of sidewall parasitic channels.

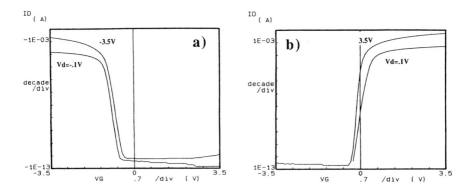

**Figure 6: Id vs Vg characteristics of 20/.55 transistors
a) PMOS b) NMOS**

In conclusion we, have demonstrated in this paper the capability of the R.E.M process as an inter-device isolation technique in a sub-micron SOI CMOS process. The R.E.M permits a good control of the gate dimensions without shorts and results in an oversize of the device active zones.

References:
1. M.Haond et al, Proceedings of the ESSDERC 89, p. 893, 1989.
2. J.P Colinge, IEEE Electron Device Letters, Vol.EDL.7, n°4, p. 244, 1986.

*Paper presented at ESSDERC 90, Nottingham, September 1990*
*Session 1B1*

# High quality pseudomorphic InGaAs/GaAs HFET structures grown by MBE

C. Wölk, F. Berlec and H. Brugger

Daimler Benz AG, Research Center, D-7900 Ulm, FRG

> Abstract: The growth of modulation-doped heterojunction-field effect transistor structures with a pseudomorphic InGaAs quantum well for high frequency device applications is reported. The In-concentrations are varied between 10% and 30%. The quantum well widths range from 8 nm to 16 nm. Depending on the sheet concentration one or two strong photoluminescence transitions with a high-energy cutoff are observed from the two-dimensional electron gas in the InGaAs channel. The transition energies shift to lower energies with higher In-concentration and/or larger quantum well width.

## 1. Introduction

Heterojunction field-effect transistors (HFET) with modulation-doped GaAs channels have demonstrated the high potential for low noise amplifiers. Improved device performance in the microwave and mm-wave frequency region is achieved by incorporation of a pseudomorphic InGaAs channel (Kao et al. 1989). This advance arises in part as a consequence of the higher two-dimensional (2D) electron gas (EG) density and the better confinement of carriers in the quantum well (QW) channel. We report about the growth of pseudomorphic HFET structures with different In-concentrations and QW widths and the characterization of the samples with low temperature (LT) photoluminescence (PL).

## 2. Epitaxial Structure and Growth

The HFET structures used for this study were grown by molecular-beam epitaxy (MBE) with a Varian Gen II system. Details about the sample structures are shown in Fig. 1. The epitaxial layer sequence consists of a short-period superlattice and an undoped GaAs buffer on a semi-insulating (100) GaAs substrate. Two different series (structure A and B) with In(y)Ga(1-y)As QW channels were grown to investigate the influence of the QW width ($L_z$) and the QW depth (In-content) on the optical, transport and device behaviour.

In structure A $L_z$ is varied between 8 nm and 16 nm at an In-concentration y = 18%. An additional energetic AlGaAs barrier towards the substrate side is incorporated for improved carrier confinement. In structure B the In-content is varied between y = 10% and 30% at a QW width $L_z$ = 12 nm. A thin GaAs cap layer is deposited on top of the InGaAs material. In all samples $L_z$ was kept below the critical thickness of the biaxially strained layer. Above the QW an appropriate spacer layer is incorporated followed by a continuously Si-doped AlGaAs region with different doping levels in

© 1990 IOP Publishing Ltd

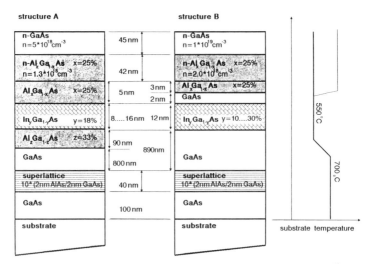

Fig. 1. Schematics of the layered HFET structures.

structure A and B. We used an Al-concentration of x = 25% to avoid degradation effects of device performance due to deep donor levels. The final layer consists of highly doped GaAs to protect AlGaAs from air exposure.

The first part of the growth is done with a substrate temperature $T_g$ = 700°C; just before the InGaAs QW $T_g$ was lowered to 550°C to prevent In desorption and three-dimensional growth. The subsequent layers of the samples, reported here, were also grown at that reduced $T_g$. Additional runs with higher $T_g$ for AlGaAs growth show no remarkable influence on the transport properties for room temperature (RT) device applications.

The MBE growth rates and compositions were determined from reflection high-energy electron diffraction (RHEED) intensity oscillations on a series of InGaAs epi-layers, single and multi quantum well (SQW, MQW) samples grown under identical conditions and characterized with LT PL.

3. Optical Characterization

The as-grown samples were characterized by LT PL with the use of a Fourier Transform Spectrometer (BIORAD FT-PL) and a liquid nitrogen cooled Ge-detector (Northcoast). The excitation source was an $Ar^+$-ion laser operating at 488 nm with power densities in the range between 1 and 10 W/cm$^2$. The samples were mounted on a coldfinger of a closed-cycle He cryostate. The sample temperature was about 20 K.

The quality of the epitaxial layers is directly reflected in the PL behaviour. Typical luminescence results from pseudomorphic HFET layers (structure A) with In(0.18)Ga(0.82)As QW's are shown in Fig. 2. Two groups of transition lines appear in the spectra. The dominant lines around 1.51 eV and 1.49 eV originate form the buffer layer and are due to carbon (C) related transitions and a bound-exciton (BE) peak. The half-width of the BE signal is about 1.3 meV (20 K!). The transitions are typical for high quality undoped GaAs material grown with MBE. A second group of transitions occurs in the range between 1.25 eV and 1.40 eV. We assign this signals to the radiative recombination of the 2D EG with photogenerated

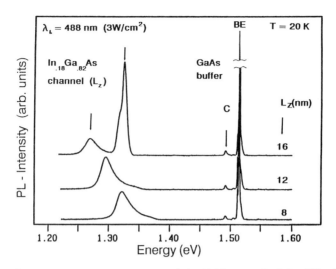

Fig. 2. PL spectra from HFETs with different InGaAs QW widths

holes in the well. The QW additionally confines the holes which greatly enhances the PL efficiency of the 2D EG in comparison with standard AlGaAs/GaAs HFET's (Kukushkin et al. 1988).

PL allows a direct investigation of the whole energy spectrum of the degenerate EG. The spectra of the samples with $L_z$ = 8 nm and 12 nm are dominated by one peak which broadens significantly on the high-energy side due to the band filling. We assign this peak to the lowest n1-electron and h1-heavy-hole subband transition. As we expect from the MBE growth parameters, the energetic position shifts to lower values with larger Lz due to a decrease of the subband energies. In the sample with $L_z$ = 16 nm a second even more intensive signal appears on the high energy side. There is some evidence that this peak is correlated with the n2-electron and n1-heavy-hole transition. A strong optical signal is expected due to the pronounced overlap of the n2-electron and h1-hole wavefunctions and the built-in band bending which violates the dipole selection rules.

All the spectra show a cutoff at the high-energy side. This is clearly seen on a logarithmic intensity scale, especially for samples with only one dominant peak (see Fig. 2). A continuous broadening of the PL-signals with higher 2D EG density is observed. From a detailed line shape analysis information about the density of carriers in the QW is obtained (Colvard et al. 1989). If we use the fermi-energy as the difference between the n1-h1 transition and the high-energy cutoff, which are marked by arrows in Fig. 2 for the 16 nm sample, we estimate a 2D EG density ns in the range between 1.5 and $1.6*10^{12}$ cm$^{-2}$ in reasonable agreement with ns = $1.5*10^{12}$ cm$^{-2}$ obtained from Shubnikov-de Haas (SdH) measurements.

In Fig. 3 typical PL spectra are shown from samples of structure B. All spectra show two strong transitions indicating high electron densities. The In-contents in the 12 nm wide QW's are varied between 10% and 27%. The energies shift to lower values with higher In-concentration. The energy gap of InGaAs shrinks with increasing In-cocentration and lateral biaxial strain in the pseudomorphic layers. The observed energetic difference of the n1-h1 transitions between samples with nominally y = 18% and 27% (10%)

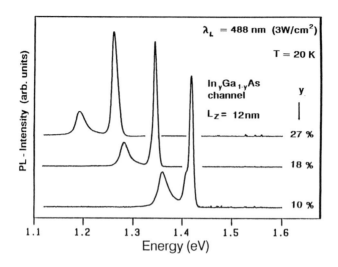

Fig. 3. PL spectra from pseudomorphic HFETs with different In-concentrations and constant QW width of 12 nm.

is about -92 meV (78 meV). This is in reasonable agreement with expected values for strained InGaAs bulk material after Andersson el al. (1988). To obtain reliable information about the absolute QW composition and width it is necessary to analyze the optical data with self-consistent subband calculations. Additionally SdH-measurements are performed on identical samples to get reliable $n_s$-values and to investigate the potential of PL to determine the 2D EG density in pseudomorphic HFET's (Brugger et al.).

## 4. Transport and Device Results

The densities and mobilities of the 2D EG were investigated with a stripping Hall technique at different temperatures and in addition with SdH not discussed in this paper in detail. The densities of the above mentioned samples range between 1.1 and $2*10^{12}$ cm$^{-2}$. There are significant improvements in $n_s$ compared to standard AlGaAs/GaAs HFET's (y = 0) without any loss in mobility for RT applications.

HFET transistor devices are fabricated from the above mentioned material. First results show excellent DC and HF performance. The sample with y = 18% and $L_z$ = 12 nm with gate lengths of 0.25 µm using a electron-beam lithography exhibits a RT drain current as high as 680 mA/mm, an extrinsic transconductance of 700 mS/mm, a peak current gain cutoff frequency ($f_T$) of 100 GHz and a power gain cutoff frequency ($f_{max}$) of 200 GHz. The noise figure (NF) is below 1 dB at 18 GHz (Narozny et al. 1990).

References:

Andersson T G el al. 1988, Phys. Rev. B37, 4032
Brugger H et al. (to be published)
Colvard C, Nouri N, Lee H and Ackley D 1989, Phys. Rev. B39, pp 8033-8036
Kao M Y et al. 1989, IEEE Electron Devices 10, 580
Kukushkin I V, v Klitzing K and Ploog K 1988, Phys. Rev. B37, pp 8509-8512
Narozny P et al. 1990 (unpublished results)

# Modelling the parasitic effects in GaAs/GaAlAs heterojunction bipolar transistors

J. Dangla, M. Filoche, A. Koncyzkowska, E. Caquot

Centre National d'Etudes des Télécommunications, Laboratoire de Bagneux
196, rue H. Ravera, 92220 BAGNEUX, FRANCE

Abstract : A modelling approach and an experimental characterization of various GaAs/GaAlAs HBT parasitic effects are presented. They include the emitter base offset voltage, resistance effects, recombination currents and the outdiffusion of the p-dopant.

## 1 Introduction

GaAs/GaAlAs heterojunction bipolar transistor (HBT) has been proved to be very efficient for gigabit logic, microwave and linear applications, because of very high cutoff frequencies, good reproducibility of threshold voltage, low 1/f noise and high power capabilities [1] [2] [3]. These properties are due to the good transport parameters of GaAs, to the vertical structure of the device and to the GaAlAs/GaAs heterojunction used to build the emitter-base junction. Very high base doping levels are then possible (higher than $10^{19}$ cm$^{-3}$), which leads to very low base resistance (less than 200 $\Omega$/square).

These very attractive properties are sometimes screened by parasitic effects due to a non-adequate design of the device or to some non-optimized technological steps. In order to take full advantage of the HBT, these parasitic effects must be firstly understood and quantified and then minimized. The aim of this communication is to present a modelling approach and an experimental characterization of various HBT parasitic effects.

## 2 Offset of emitter-collector voltage

Because of the use of an emitter-base heterojunction and of a base-collector homo-junction, which have not the same turn-on characteristics, an offset of the emitter-collector voltage has been observed. This effect is amplified by the fact that the two junctions have not exactly the same area. When double heterojunction bipolar transistors are used with the same GaAlAs emitter and collector layer characteristics, this effect can be neglected. In order to minimize this effect on single heterojunction bipolar transistors, numerical simulations have been performed (figure 1) for various kinds of heterojunction. The aim of these simulations is to design the emitter-base heterojunction so as to obtain the same turn-on characteristics than the collector-base homojunction. These simulations show that it is impossible to eliminate completely this effect. Nevertheless, if one introduces an aluminum graded layer in the emitter with a 15 nm

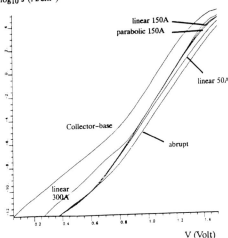

Fig 1 Forward I(V) characteristics for
- collector base junction
- emitter base heterojunction (abrupt, linear grading 50-150-300 nm, parabolic 150 nm)

thickness, this turn-on voltage can be reduced. It must be noted that the shape (linear or parabolic) of this graded layer does not change the result [4].

## 3 Resistance effects

Access resistances are of the highest importance for both DC and AC behaviour of HBTs. They are mainly due to ohmic contacts and to the emitter, base and collector layers properties. When compared with GaAs FETs, HBTs require a specific p-type contact for the base [5], which must be particularly optimized. For analog application, as Analog to Digital Converters, the reproducibility of the threshold voltage $V_{BE}$ of the comparator is very important. At normal operating point,

$$V_{BE} = n\frac{kT}{q}\text{Ln}\left(\frac{I_C}{I_S}\right) + \left(R_E + \frac{R_B}{\beta}\right)I_C$$

where $R_E$ and $R_C$ are respectively emitter and base series resistance, $\beta$ is DC common emitter current gain, while $I_S$ and n are respectively the saturation current and the non ideality factor of the heterojunction and $I_C$ the collector current. $\frac{kT}{q}$ is the thermal voltage.

From this equation two terms clearly appear.

• The first one is related to the emitter-base junction characteristics. For abrupt heterojunction, this term strongly depends on interface characteristics and it is difficult to obtain a good reproducibility on the wafer. For Al-graded emitter base heterojunction, the same equation can be used as in the case of a conventional homojunction bipolar transistor. As a consequence, it depends only on the bulk parameters. In this case, the reproducibility of its value throughout the wafer is very good (experimental results have shown that its standard deviation can be smaller than 1 mV).

• The second term, which can be more important, is directly related to the emitter and base resistances reproducibilities. Self-aligned processes, high base doping level allow to reduce the base resistance and consequently contribute to increase the reproducibility of the threshold voltage.

For gigabit ECL digital applications, the gain of the transfer characteristic of the elemental gate must be as high as possible, in order to use the lowest logical swing. From the threshold equation above, it is possible to compute a new analytical expression of this gain which takes into account series resistance.

$$\left(\frac{dv_{out}}{dv_{in}}\right)_{max} = \frac{v_l}{2\left(\frac{kT}{q} + \left(\frac{R_B}{\beta} + R_E\right)I_C\right)}$$

where $v_l$ is the logical swing value.

Without series resistance, this gain is proportional to the swing value. It is obvious that this value can be degraded by emitter and base resistances. For the same collector current, it is possible to reduce series resistance by increasing the length of emitter and base fingers. The base resistance can be reduced by use of two base fingers instead of only one, highly doped base [6] or self aligned processes. If the gain remains too low, the logical swing must then be increased (figure 2). (Note that, in this case, the reproducibility of the threshold voltage is less important.)

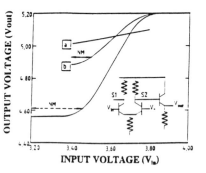

Fig 2 Transfer characteristics of an ECL gate for three case :
a / a non optimized HBT
b / an optimized one
c / the same HBT as b / but with a higher logic swing

## 4 Recombination effects

The current gain of an HBT is very sensitive to the recombination currents. These currents can be due to the emitter base interface, to the surface effects or to the bulk itself. For feedback purpose on technological process, it is very important to know the origin of these currents. An experimental method, based on a scale of transistors, has been developed, which allows to identify it. A jumbo HBT with large emitter area and relaxed design rules is considered as a reference ($HBT_{ref}$). This HBT is large enough in order to minimize the surface recombination current but not too large in order to avoid emitter crowding effect and thermal effect. The gain of two other HBTs are then carefully measured :

• Another jumbo HBT with an emitter stripe finger ($HBT_1$), which has an emitter perimeter three times higher than $HBT_{ref}$ in order to increase the surface recombination current.

• $HBT_2$ which uses the same design rules as $HBT_{ref}$ but its emitter-base area is three times larger.

If the maximum current gain is almost the same on each structure, that means that the bulk recombination in the base layer is the limiting effect. If surface recombination current is predominant, the maximum gain of $HBT_1$ is lower than for the other ones while, if space charge current is important, the largest HBT ($HBT_2$) has the lowest current gain.

## 5 Outdiffusion of base doping

During some high temperature steps of the fabrication process, the high concentration base doping species can diffuse into the GaAlAs emitter layer [7]. The current gain is then adversely affected. When low current gain is observed, it is quite impossible to know if its value is due to this effect or to recombinations. With the help of numerical simulation, we have developed a method based on electrical measurements to quantify this diffusion [8]. The value of the threshold voltage (which is very sensitive to this phenomena) is monitored and, when compared with simulation results, allows to characterize

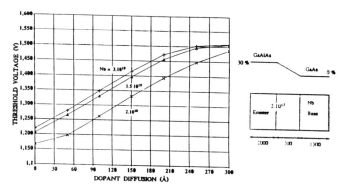

fig 3 Numerical simulation of the influence on the threshold voltage of the p-type dopant diffusion into the emitter for three different base doping level

this effect. Careful comparison with Polaron or SIMS measurements has been used to validate this method. It shows that this new method is very precise for graded heterojunction and is able to characterize outdiffusion lengths which are less than 50 Ångström (figure 3). For abrupt heterojunction, it is necessary to know precisely the profile of the aluminium because the threshold voltage is strongly dependant on this parameter.

## 6 Conclusion

GaAs/GaAlAs HBT parasitic effects are mainly due to process steps which can be widely improved and are representative of infancy defaults of the fabrication. The constant progress of the performances of this device during the last years shows that a careful optimization of epitaxial layers growth, metal deposition, wet or dry etch, surface passivation and transistor design has been permanently done. This communication has described how precise electric measurements on well-suited test devices, when correlated with adequate modelling results, contribute to these improvements, which will allow to take full advantage of the intrinsic performances of the device.

## 7 Acknowledgments

The authors wish to thank M. Laporte and P. Calmel for the test, C. Chevallier and P. Launay for the fabrication, F. Alexandre and R. Azoulay for the epitaxial growth, P. Desrousseaux for the design and M. Bon and H. Wang for fruitful discussions.

[1] Asbek P.M et Al. 1989 IEEE Proc. of Bipolar Circuits and technology Meeting, Minneapolis, pp 65-69

[2] Yamaushi Y. et al 1989, $11^{th}$ GaAs IC Symposium, San Diego -Californie

[3] Kim M.E. et al 1989 IEEE Trans. on Microwave Theory and Techniques, vol. 37, pp 1286-1303

[4] Dangla J. et al 1984 Proc. of ESSDERC, Lille (North-Holland), pp 366-370

[5] Dubon-Chevallier C. et al 1986, Journal of Applied Physics, Vol 59, pp 3783-3786

[6] lievin J.L. and al 1986 IEEE Electron Device letters, Vol. EDL-7, NO. 2, pp 129-131

[7] Ilegems M. 1977, Journal of Applied Physics, 48, pp 1278-1287

[8] Dangla J. et al 1990 to be published in Electronic Letters

*Paper presented at ESSDERC 90, Nottingham, September 1990*
*Session 1B3*

## Design, simulation and measurement of high efficiency microwave power heterojunction bipolar transistors

J G Metcalfe, R W Allen, A J Holden, A P Long and R Nicklin
Plessey Research Caswell Ltd., Towcester, Northants., NN12 8EQ, England.

Abstract. In this paper we describe the design, simulation and measured performance of high efficiency power HBTs. Devices with emitter lengths of 1.7mm operated under pulse bias (0.2µs, 1% duty cycle) can deliver output powers at 1 dB gain compression of up to 8W at 4GHz. The associated gain is 5dB and the peak power added efficiency around 40%. This represents the highest power reported for a power HBT.

### 1. Introduction

In recent years GaAs/AlGaAs hetero-junction bipolar transistors (HBTs) have demonstrated high power densities and high power added efficiencies at both C- and X-band frequencies [1,2]. To achieve the full potential of such devices it is necessary to consider the various aspects of device design, their interaction and the constraints set by material properties/device processing and the required operating conditions.

### 2. Design

The design of power HBTs must consider the choice of semiconductor layer parameters and device geometry from both microwave performance and thermal management perspectives. In the absence of thermal effects the output power will be limited by the base-collector breakdown voltage and an operating current density limited by the roll-off in gain at high current values ("Kirk effect"). High breakdown voltages (> 30V) can be achieved by a wide (1µm) low doped collector region, while the peak current density (~$2 \times 10^5 Acm^{-2}$) may be raised by increasing the doping in the collector region. A further consideration arises when the microwave performance is considered since $F_T$ and $F_{max}$ will be reduced for wide heavily doped collector regions. The device geometry, defined in terms of the number, length and spacing of the emitter fingers, will be determined by the operating requirements i.e. output power, frequency and CW/pulsed operation. The potential for high power density and power added efficiency will only be realised if both adequate heat sinking and thermal layout are included in the design.

### 3. Fabrication

GaAs/AlGaAs HBTs were fabricated from epitaxial layers grown on 2 inch n+ substrates by metal-organic chemical vapour deposition [1]. The layer structure includes a 0.1µm wide, $4 \times 10^{18} cm^{-3}$ doped base and a 1µm wide, $10^{16} cm^{-3}$ n-type doped collector. Base and isolation implants were used to maintain a near planar surface geometry. The use of an n+ substrate enables the collector contact to be formed on the back of the wafer which facilitates the emitter-base contact arrangement on the front face.

A multiple emitter-base arrangement is used for high current operation. Two geometries were considered. A cell geometry consists of groups of three closely spaced (13µm) emitter fingers each 2.5µm wide by 100µm long and with a range of spacings, from 20µm, between cells. An individual emitter approach has a number of widely spaced single emitters (48µm apart).The former design is intended for pulsed operation with short pulse lengths and low duty cycles. Heat sinking is achieved by thinning the substrate to 100µm and solder mounting the chips using an integral plated heat sink technology.

## 4. Simulation

To aid the design of these transistors a comprehensive suite of modelling software has been developed. 1D hetero-junction transport simulation is used to establish the basic device performance. A parameter extraction program, GUMPOON, can use a-priori semiconductor layer parameters to calculate the transport properties and the associated device parasitic capacitance and resistance components - these latter parameters can also be determined from measurements of test structures provided on the wafers. The output from this program is fed into the Gummel-Poon bipolar device model in the circuit simulation program SPICE. The model can be enhanced to include measured values of D.C. gain, junction ideality, base-collector avalanche breakdown [3] and high current effects at the base-collector junction. It can provide small signal S-parameters for comparison with measurements or for post processing to give values for microwave gain, Ft etc. SPICE can also be used to calculate the large signal performance of the device.

High current effects in HBTs have been analysed within a simple model which assumes saturated drift motion in the collector and simple carrier diffusion in the base. However, due to the large fields and field variations such a simple model will not be entirely valid as both velocity overshoot and mobility dominated transport are possible under certain conditions. We define three currents:

$$I_1 = q\,N_c\,v_{eff}\,A_c, \qquad I_2 = V_{cb}\Big/\Big(r_c + \frac{(W_c - x_0)}{qN_c\mu_n A_c}\Big), \qquad I_0 = q\,v_{eff}\,A_c\left[\frac{2\varepsilon\varepsilon_0 V_{cb}}{qW_c^2} + N_c\right]$$

where $V_2 = q\,N_c\,W_c^2/(2\varepsilon\varepsilon_0)$.

$N_c$, $W_c$, $A_c$, $v_{eff}$ and $\mu_n$ are the collector doping, total width, area, effective carrier velocity at the base-collector interface and mobility respectively. The zero bias depletion width and external collector resistance are $x_0$ and $r_c$. It can be shown that the current $I_0$ is the current at which the peak field has moved completely across the collector and the field has become zero at the collector-base junction. This can be seen as the onset point for base pushout, increase in base width and consequent degradation of gain and $f_t$. Beyond this point the zero field point moves into the collector giving the edge of the neutral base a new position at $x=x_1$ and a new base width of $W_b+x_1$. This can be calculated as:

$$x_1 = 0.0 \quad I_c < I_0, \qquad x_1 = W_c\left(1 - \sqrt{\left\{\frac{(I_0 - I_1)}{(I_c - I_1)}\right\}}\right) \qquad I_c > I_0$$

The effect of an increased base width is twofold. It increases the Gummel number $Q_B$, which in turn reduces the collector current, and also increases the diffusion contribution to the transit time at high currents. The effect on $Q_B$ is taken care of within the SPICE device model by including a parasitic collector current to reduce the effective $h_{fe}$ for $I_c > I_0$. The contribution to the device performance from the part of the output power generated in this region of the rf cycle is assumed to be small so the effect on the transit time is ignored. The standard parameters are not considered to be adequate and a separate model has been developed to add in to the SPICE circuit.

SPICE is run under the control of another program which allows the calculation of the 1dB gain compressed output power from the device at particular matching and bias conditions. The transistor re-biases itself as the power is increased and the model allows either constant current or constant voltage operation. The input and output match and the bias levels can be optimised to give the maximum output power from the device, with a minimum specified gain.

To achieve the high power densities for which power HBTs are capable (5W/mm of emitter) attention must be paid to the thermal layout of the device and the provision of adequate heat sinking. We have assessed the thermal performance of the device, under both pulsed and CW operation, using a computationally efficient, finite difference based simulation. This can be used to calculate, for a given power dissipation, the temperature at various critical points within the device structure. Figure 1 shows the calculated thermal response of the device at several points in the structure to a 1μs pulse at 10% duty cycle with 3.5W dissipation.

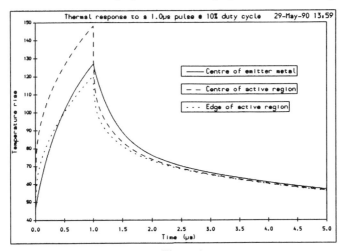

Figure 1

6. Microwave measurements

A comparison is made in figure 2 between the device microwave gain calculated from both measured and modelled s-parameters at a bias of $V_{cb}$=2V and $I_c$=200mA.

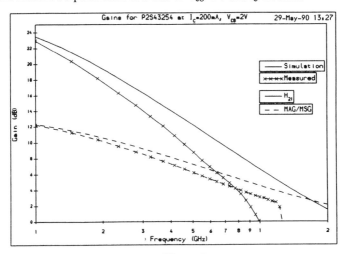

Figure 2

Figure 3 shows the measured and simulated power curve and power added efficiency for a device, consisting of 2 cells each having 3 emitter fingers, operated under pulse bias at 4GHz. The measured output power at 1dB gain compression was 34.7dBm (2.95W) the associated gain was 5.7dB and power added efficiency 40%.

Figure 4 shows the measured results for a device consisting of 17 individual emitter fingers operated under pulse bias.

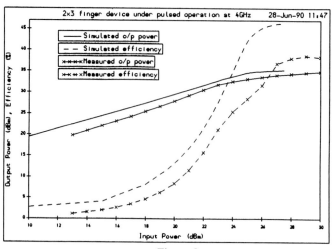

Figure 3

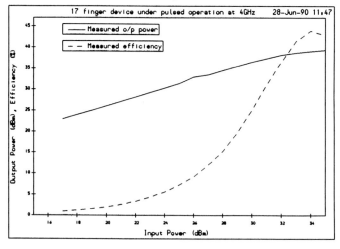

Figure 4

### 7. Acknowledgements

This work has been supported and sponsored by the Procurement Executive, Ministry of Defence (Royal Signals and Radar Establishment). We are grateful to B.A. Hollis for device processing.

### 8. References

1) A.P. Long, A.J. Holden, J.G.Metcalfe, R. Nicklin and R.C. Goodfellow. Proceedings Workshop 1, Microwave Heterostructure Devices and Applications, 19th European Microwave conference, 8th September 1989, p.44.

2) B. Bayraktaroglu, R.D. Hudgens, M. Khatibzadeh, H.Q. Tserng. IEEE MTT-S Digest, 1989, p. 1057.

3) J.G. Metcalfe, R.C. Hayes, A.J. Holden, A.P. Long, Journal de Physique, Colloque C4, supplement au n° 9, Tome 49, septembre 1988, pp. 579-582.

# Microwave HBT fabrication by using a self-aligned technology with a perpendicular side-wall

Xiaojian CHEN    Ying WU
(Nanjing Electronic Devices Institute, P.O. Box 1601, Nanjing, PRC)

Abstract. A HBT self-aligned technology with a perpendicular side-wall mesa has been adopted in microwave HBT fabrication by virtue of a high selectivity chemical wet etchant. Principal features and technological processes of the method are discussed. The gap of ~0.1μm between the emitter mesa edge and base contact metallization edge has been formed repeatedly by the method. The experimental results of the developed microwave HBT for test purpose is given.

## 1. Introduction

In the microwave HBT design and fabrication both vertical layer structure and traverse dimensions need to be controlled accurately. Therefore, besides the multi-layer heterostructure wafer grown by MBE or MOCVD, various self-aligned methods using dielectric side-wall and dry-process technology have been employed in HBT fabrication, leading to excellent experimental results. However, these methods have some problems in applications such as process complexity, the surface damage suffered from dry-etching and the parasite caused by dielectric film, which are continuously puzzled questions for device designers.

For the sake of elliminating the mentioned effects initiated by the dielectric side-wall mesa, a different method with uncomplex process has been used by authors, named HBT self-aligned technology with a perpendicular side-wall (PSW). It does not adopt any dielectric side-wall for controlling device traverse dimension, but the gap between the emitter mesa edge and the base contact metallization edge (one of the key dimensions effecting HBT high frequency performance), $l_{eb}$, can be made extremely narrow (~0.1μm), especially promising for mm-wave HBT developing.

## 2. PSW Method

Microwave performances of HBT are close related with maximum oscillating frequency:

$$f_{max} = (f_T/8\pi R_b C_{bc})^{\frac{1}{2}} \qquad (1)$$

Where, $R_b$ and $C_{bc}$ are base resistance and collector junction capacitance, respectively. Cut-off frequency $f_T$ is a reflection of carrier total transfer time, mainly effected by the vertical layer structure design of the device; however $R_b$ and $C_{bc}$ tightly depend on the dimensions of HBT traverse structure. It can be seen from Fig. 1 that $R_b$ is composed of three parts with no account of metallization contact resistance (chen X J and Wu Y 1990):

$$R_b = r_{bo} + r_b' + r_b'' \qquad (2)$$

and
$$r_{bo} = \frac{\rho_b W_e}{12 L_B l_e}$$

$$r_b' = \frac{\rho_b L_{eb}}{L_B' l_e} \qquad (3)$$

$$r_b'' = \frac{\rho_b L_B'}{6 l_c' l_e}$$

where, $W_e$, $l_e$ are the width and length of emitter strip, respectively; $L_B$ is $p^+$-GaAs layer thickness of intrinsic base (just right under the emitter junction) and $L_B'$ means that of extrinsic base; $l_c'$ represents the width of the base metallization layer covered the extrinsic base. The intrinsic base resistance $r_{bo}$ can be minimized by higher base doping level; on the contrary, $r_b'$ might be much larger than $r_{bo}$ due to considerable $l_{eb}$, and become a main contribution to $R_b$. Therefore, it is one of most important measures for improving $f_{max}$ in HBT fabrication to decrease $l_{eb}$. For this reason, most authors have used self-aligned HBT technology with a dielectric side-wall emitter mesa and the dry-etching technique. In this method the thickness of dielectric film formed side-wall is used to define the gap width $l_{eb}$, which

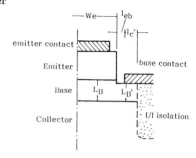

Fig. 1  Schematic of base resistance for Microwave HBT

shows advantages in its reliability and controllability, and disadvantages as follows: (1) the possible surface damage caused by dry-etching operation of both emitter mesa and dielectric film; (2) comparative complexity of the process in which various dry apparatuses (plsma CVD, RIE, RIBE and IM) for repeated growth and removal of dielectric film are needed; and (3) nonignorable dielectric parasite at microwave and mm-wave frequencies.

The authors has used a different self-aligned HBT technology in which minimum $l_{eb}$ was obtained by virtue of the shape rather than the dielectric thickness of mesa's side-wall. The shematic process flow of the method is shown as Fig. 2. First, etching heteroemitter mesa is performed by a selective chemical wet-etchant. Next, a base ohmic metallization evaporation is immediately completed before removing the resist mask of emitter mesa. After lifting-off the resist, an extreme narrow air-gap ($l_{eb}$) is obtained. After these, both base and collector electrodes are fabricated in conventional manner. The advantages of the method are obvious: (1) because of no use of dielectric in the process all drawbacks arising thereform are removed; (2) free from any possible surface damage caused by dry-etching because of a chemical wet etchant with high selectivity adopted for etching the emitter mesa.

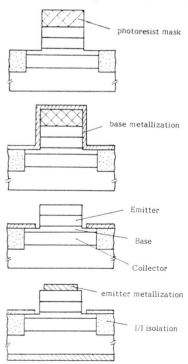

The shape of the emitter mesa's side-wall might be quite different, depending on the employed etching method and leading to considerable difference in $l_{eb}$. Generally the side-wall shape would have three types: slopping, perpendicular and concave. In our technology, the slopping side-wall would short-circuit the emitter junction by base ohmic metallization, but the concave side-wall would result in larger $l_{eb}$, and hence the perpendicular is most suitable (Fig. 3). It's clear that such a mesa with a strict perpendicular

Fig. 2  Schematic process flow of PSW method

side-wall only can be achieved along with a special crystallographic orientation by chemical wet etchant rather than the dry-etching method. Here the emitter strip mesa is aligned accurately with <001> direction, the side-wall of hetero-emitter-junction mesa etched on (001) GaAs wafer is naturally perpendicular to its upper surface, hence the formed $l_{eb}$ is narrowest with

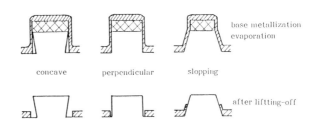

Fig. 3  Effect of the side-wall shapes of emitter mesa on $l_{eb}$ formation

the aid of lifting-off base contact metallization from the mesa. Experimental $l_{eb}$ is ~0.1μm (Fig. 4.), better than that obtained generally from the dielectric-sidewall self-aligned technology.

### 3. Experimental Result

By use of PSW self-aligned technology AlGaAs/GaAs microwave HBT is fabricated for test purpose. In order to simplify the process, the large-dimension device (emitter junction area 10x24μm$^2$, collector junction area 14x26μm$^2$) is designed and an n$^+$-GaAs substrate is used for easily making the collector electrode onto its back. The heterojunction layer structure of MOCVD wafer is shown as Fig. 5 in which the emitter junction with an abrupt band-gap heterojunction is adopted for increasing the

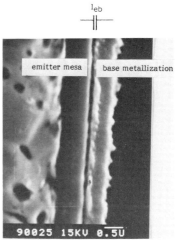

Fig. 4  Photo of the gap $l_{eb}$ resulted from PSW technology

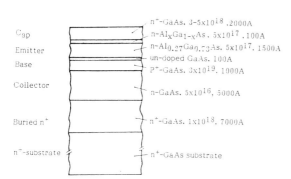

Fig. 5  Representative HBT epitaxial structure

quasi-ballistic velocity of the carrier transferring the base. The main difficulty in fabrications is the choice of emitter mesa's etchant. Generally, KI/I$_2$ redox system has been used as a selective etchant of AlGaAs layer (Tijburg et al 1976) but with an limited selectivity of ~10:1 (Asbeck P M et al 1989). However, the selectivity of ~100:1 (enough to stop etching at p$^+$-GaAs layer), comparable to that of high selective dry-etching tyechnique, has been repeatedly achieved under our experimental condition but free from possible surface damage. The ohmic contact metallization of both emitter and collector is AuGeNi/Au, but $C_r$/Au for the base. Although the contact resistivity of $C_r$/Au in $p^+$-GaAs ($\sim 10^{-6}\Omega\cdot$cm) is slightly large compared with that of AuGe or AuBe, but its slower diffusion in base is farorable in ultra-thin base mm-wave HBT fabrication.

The extreme narrow $l_{eb}$ can be seen from the fabricated HBT chip (Fig. 6). The Typical common-emitter d.c. characteristic is given in Fig. 7 with $h_{FE}$ of $\geqslant 40$.

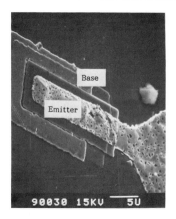

Fig. 6   Photo of typical developed HBT chip

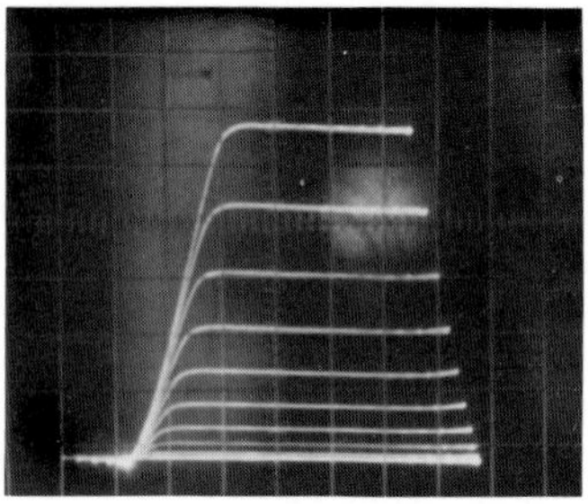

Fig. 7   Typical common-emitter d.c. characteristic of developed HBT

2mA/div, 0.5V/div, 50μA/step

Owing to the lack of suitable HBT package, the chip is mounted into a microstrip package with larger parasite. Microwave performance has been measured by HP8510 automatic network analyzer: $f_T > 10$GHz is obtained even at a low collector current density ($j_c$ $4\times 10^3$ A/cm$^2$), indicating the potential superiority of our PSW self-aligned technology in mm-wave HBT development.

4. Conclusion

A self-aligned technology with a perpendicular side-wall mesa (PSW) has been applied to microwave HBT fabrication. This method, in comparison with the common self-aligned method using dielectric-side-wall, shows the advantages of an uncomplicated process, non-destruction of etched surface and low parasite due to the elimination of dielectric-side-wall and the use of high selective chemical wet etching of hetero-emitter mesa. With PSW method, the gap between the emitter mesa edge and the base contact metallization edge could be reduced to $\sim 0.1$μm, promising for minimum base resistance and higher operation frequency that considerably benefits for mm-wave HBT development.

Acknowledgement

The authors would like to thank Prof. M.H. Jiang, Prof. Z.S. Shao and Dr. B.B. Huang (the Crystal Material Research Institute, Shan-Dong University) for their cordial support in MOCVD growth of HBT wafer.

References

Asbeck P M et al 1989 IEEE Trans. Elec. Dev. 36 pp2032-42
Chen X J and Wu Y 1990 6th Chinese Compound Semiconductors and Microwave/ Optoelectronic Devices Symp. Digest
Tijburg R P and Dongen T Van 1976 J Electrochem. Soc. 123 pp687-91

… Paper presented at ESSDERC 90, Nottingham, September 1990
Session 1B5

# High-frequency properties and application of invertible GaInAsP/InP double-heterostructure bipolar transistors

A. Paraskevopoulos, H.G. Bach, H. Schroeter-Janßen, G. Mekonnen, H.J. Hensel, N. Grote

*Heinrich-Hertz-Institut für Nachrichtentechnik Berlin GmbH, Einsteinufer 37, D-1000 Berlin 10, F.R.G.*

## Abstract

In this paper we report on invertible DHBTs fabricated on GaInAsP/InP. Localized Zn diffusion allowed for practically equal active transistor areas in the forward and inverse mode. Transit frequencies up to 6 GHz with collector currents over 100mA could be demonstrated on these devices. As an application, three transistors were monolithically integrated to form a laser driver circuit showing modulation rates up to 2.6 Gbit/s.

## Introduction

Double Heterostructure Bipolar Transistors (DHBTs) are recognized as an efficient structure for integrated circuits, especially in the GaInAsP/InP system which offers high-gain, high-speed and low surface recombination [Kroemer 1982, Nakano et al 1986, Shibata et al 1984, Kasahara et al 1986]. In this paper we report on the performance of invertible DHBTs, with emphasis on the bilateral operation, and their application to an integrated laser driver circuit.

## Device structure and fabrication

The DHBT structure investigated is schematically presented in fig. 1. The LPE-grown epitaxial double heterostructure consisted of a p-GaInAsP ($\lambda_g$ = 1.3 µm) base layer (p = 3 - 5 · $10^{17}$ $cm^{-3}$, 0.3 µm thick) sandwiched between the n-InP emitter and collector layers (n = 1 - 3 · $10^{17}$ $cm^{-3}$, 0.3 µm thick). A crucial technological measure to achieve bilateral operation with these DHBTs is the use of localized p diffusion, extending into the lower InP "collector" layer underneath the base contact. Thereby, a p-n homojunction is created, which suppresses electron injection into the extrinsic part of the base in the inverted structure. In this way, an almost equal active transistor area is attained for the "emitter up" (normal mode) and "emitter down" (inverse mode) configuration. Moreover, the diffusion region allows for lower base access resistance, as the base metal contact is deposited on the ternary cap layer, which is an important difference to the transistors reported by us previously [Bach et al 1987, Paraskevopoulos et al 1989]. Defining the diffusion time proved to be a complicated procedure, as Zn must diffuse over three heterointerfaces (InGaAs contact layer - InP, InP "emitter" - GaInAsP base, GaInAsP base - InP "collector"). Therefore, diffusion was first simulated by a precise

© 1990 IOP Publishing Ltd

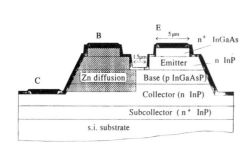
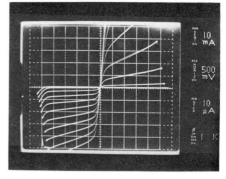

Fig. 1. Schematic cross section
of the DHBT structure

Fig. 2. Output I-V characteristics
normal and inverse mode

model [Weber 1990]; the experimental results demonstrated a quite satisfactory agreement with calculations.

After the p-diffusion, the "emitter" and base mesas were etched by RIE. In order to avoid leakage through the "emitter" layer, a 1.5 μm wide groove was etched separating the base and "emitter" contacts. The "collector" mesa was then etched down to the s.i. substrate isolating the devices from each other. By using a specially prepared photoresist, mesa edges with a small angle of inclination - for a mesa height over 2 μm - could be obtained independent from the crystaline orientation. Simultaneously deposited Ti-Au metallization was used for the p- and n-type contacts. The devices reported here had emitter areas of 5 • 60 μm$^2$.

## Results

Typical current gain values of theses devices were of the order of 1000 in the normal mode, whereas inverse gain values were found to be lower ($\approx 200$), a difference already observed previously [Bach et al 1987]. Collector currents well in excess of 100 mA could be easily reached for both modes of operation (fig. 2).

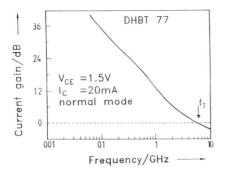

Fig. 3 Gain vs. frequency curve in normal mode

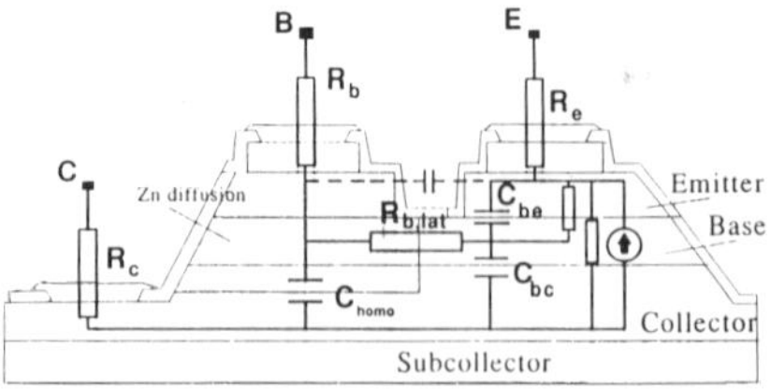

Fig. 4 Equivalent circuit of the DHBT structure

High frequency s-parameter measurements on these transistors showed transit frequencies of up to 6 GHz (fig. 3). Nevertheless, if compared to the characteristics obtained by us formerly on larger devices [ Paraskevopoulos et al 1989], these results lie somewhat lower than expected. In order to better understand this deviation, a simulation was performed using a detailed equivalent circuit, schematically presented in fig. 4. In combination with separate measurements of some of the elements in this model, a higher simulation precision could be achieved. It was thus shown that, although the capacitance values ($C_{homo}$, $C_{be}$, $C_{bc}$) were of the expected order, the resistance values were higher than projected, due to the low intrinsic base doping ($R_{b,lat}$) and also, presumably, to an only moderate contact resistivity ($R_c$). On the other hand, nearly symmetrical small-signal, high-frequency behaviour was observed for the normal and inverse mode (fig. 5), demonstrating that the active transistor area is effectively defined by the localized diffusion. A slight improvement was also produced by the "emitter" groove, which eliminates the respective homojunction capacitance, designed with dotted lines on fig. 4. A drastic decrease of the contact pads parasitic capacitance could also be observed due to their realization on the s.i. substrate.

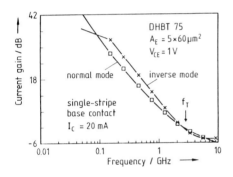

Fig. 5. High-frequency characteristics in normal and inverse mode

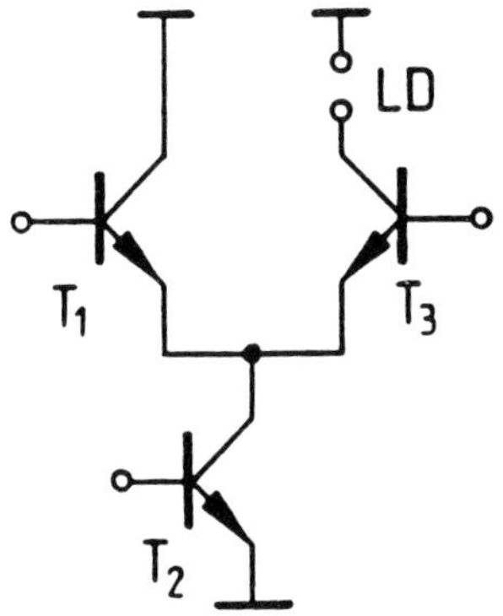

Fig. 6. Laser driver circuit

## Application (Integrated Laser Driver)

Using this device structure, three transistors were monolithically integrated to form a compact laser driver circuit (fig. 6), a concept presented last year [Paraskevopoulos et al 1989]. The differential amplifier scheme (transistors $T_1$ and $T_3$, with transistor $T_2$ as a common current source) was chosen, because it allows for high switching speed and good control of the driver output current. Operated in a differential input driving mode, this circuit is insensitive to supply voltage variation and shows good impendance matching properties in a 50 Ohm environment. Fabrication of this device is similar to that of the single transistors, with the $T_1$, $T_3$ transistors operating in inverse, and the $T_2$ in normal mode. As the high doped "subcollector" layer forms the common point of the three transistors, the integrated structure gains significantly in compactness, as seen in fig. 7.

To simulate a driven laser diode, a 10 Ohm resistor was chosen for the current modulation tests. In agreement with the measured figures on single DHBTs of the same wafers, the implemented bipolar driver circuits showed current modulation capability of up to 2.6 Gbit/s (fig. 8) with output currents of more than 50 mA and an AC transconductance of approximately 120 mS.

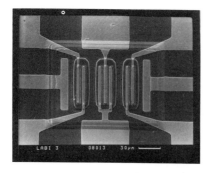

Fig. 7. SEM view of the integrated laser driver

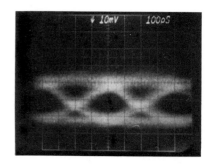

Fig. 8. Measured eye-pattern diagramme at 2.6 Gbit/s

## Summary

In conclusion, invertible DHBTs have been fabricated on GaInAsP/InP. An accurate control of localized Zn diffusion allowed for practically equal active transistor areas in the forward and inverse mode. Transit frequencies of up to 6 GHz, with collector currents over 100 mA could be demonstrated on these devices. Three transistors were monolithically integrated to form a laser driver circuit using a differential amplifier scheme with the "subcollector" layer as a common point. Modulation rates up to 2.6 Gbit/s were achieved with this integrated circuit, which more than doubles the performance of the devices reported last year. Further improvement of the high-frequency performance is expected using an MOVPE-grown, high-doped base configuration.

## References

Bach H G, Grote N and Fiedler F 1987 Proc. *17th ESSDERC* Bologna
 ed G. Soncini ( North Holland: Elsevier Sc. Publ.) pp. 883-886
Kasahara K, Suzuki A, Fujita S, Inomoto Y, Terakado T and Shikada M 1986
 Techn. Dig. *12th Europ. Conf. Opt. Comm. (ECOC)* **Vol. 1** p. 119
Kroemer 1982 *IEEE Proc.* **70** pp. 13-25
Nakano H, Yamashita S, Tanaka T, Hirao M, and Maeda M 1986
 *IEEE J.Lightw. Technol.* **LT-4** p 574
Paraskevopoulos A, Bach H G, Mekonnen G, Schroeter-Janßen H, Fiedler F and
 Grote N 1989 Proc *19th ESSDERC* W.Berlin ed A Heuberger et al
 ( Berlin: Springer Verlag) pp. 385-38
Shibata J, Natao I, Sassai Y, Kimura S, Hase N and Serizawa H 1984
 *Appl. Phys. Lett.* **45** p. 191
Weber R 1990 *Zn Diffusion over InGaAs(P)/InP heterojunctions* PhD Thesis
 TU Berlin

## Acknowledgements

This work was conducted under the ESPRIT program (Project 263).

## Frequency performances of MOS compatible silicon permeable base transistors and comparison with simulation results

M. Mouis[*], P. Letourneau, G. Vincent[o]

CNET / CNS, BP 98, F38243 Meylan Cedex, France
[*] Institut d'Electronique Fondamentale, CNRS URA 22, F91405 Orsay Cedex, France.
[o] Université Joseph Fourier, BP 53X, F38041 Grenoble Cedex, France

Abstract: Permeable Base Transistors (PBTs) with gate periodicity down to 0.6 µm have been fabricated using a MOS technology process. Both static and microwave measurements have been performed. The results obtained on the smallest structures are presented and compared with two-dimensional simulations.

We have previously presented the fabrication of etched-groove Silicon PBTs using a MOS technology process (Letourneau 89). Since this realization, device performance has been improved. The starting material is now a 0.3 Ωcm, 1 µm thick, n type Si epilayer on a 100 mm, (100) oriented, n+ degenerate wafer. To improve the top source contact, As is implanted on the front side. A 150 nm thermal oxide is then grown, followed by a 30 nm sputter deposited tungsten film.

The only one lithography pattern needed in our process is obtained by electron-beam exposure on a PMMA resist and is transferred by anisotropic ion etching into tungsten using SF6, into oxide with CHF3 and finally into silicon using SF6+O2. In this process, each layer is an excellent mask for the underlying one so that the walls of the silicon trenches are abrupt. The remaining $SiO_2$ is removed by HF dip and the result is a 0.3 µm deep grating connected to a large rectangular contact (Figure 1).

In order to obtain a self-alignment of the source and gate metal contacts and also to passivate the vertical walls, thin oxide spacers are realized. Due to the faster oxidation of the highly implanted top silicon, this oxide is obtained in two steps: a thin thermal oxidation is followed by a conformal plasma-enhanced deposition in order to minimize the thickness difference between the source and gate oxides. An anisotropic reactive ion etching leaves the source and gate silicon free of oxide, while a 40 nm oxide remains on the vertical walls. The final source contact and Schottky gate are simultaneously realized through the directional evaporation of a 40 nm platinum layer. An anneal follows to form 80 nm of PtSi on the source and gate while the vertical $SiO_2$ remains covered by an ultra thin platinum layer. This non-reacted platinum is selectively removed by aqua regia so that source and gate are electrically separated.

From capacitance-voltage measurements, the doping level in the active region was found to be around $2.10^{16}$ cm$^{-3}$. The static I-V characteristics were measured using tungsten test probes. However, with this mask geometry (without connection between the source lines) and with no additional lithography step, the probe can only contact a limited number of lines. Moreover, in order to be comparable with simulation results, the electrical characteristics were corrected for the source contact resistance deduced from the source to gate diode characteristic. Microwave measurements were also used to evaluate the frequency performance of the structure.

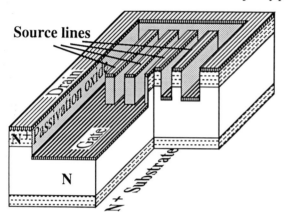

*Figure 1*
Perspective view of the PBT structure realized with one mask level.

In addition, two-dimensional simulations of the process and of the device have been performed using the CNET program Titan (Gerodolle 1989). The device simulation program solves consistently Poisson's equation and the continuity equation for the carrier densities. Currents are expressed in the drift-diffusion approximation, assuming stationary carrier transport. Since the main uncertainty concerns the actual distance between the metallic gate fingers in the measured devices, several values of this parameter, further noted d, were used in the simulation.

Figure 2 compares one measured transfer characteristic under low drain voltage with the simulation results when d is varied between 0.26 and 0.38 µm.

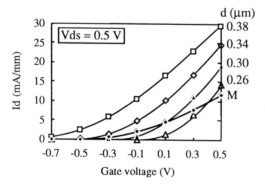

*Figure 2 :*
Transfer characteristics obtained by simulation (with different values of the inter-gate distance d) and by a static characterization of the device (M). The drain voltage is 0.5 V.

By comparing the threshold voltages, it is clear that the real value of d is only slightly larger than the 0.3 µm nominal value. It seems realistic to assume a 0.34 spacing between the gate fingers. However, the measured currents are significantly lower than the simulation results. Indeed the test probe probably does not contact all the source lines. A remarkable feature of the transfer characteristics is that they become almost parallel for large gate voltages. This is consistent with a MESFET type control mode of the channel : the current control is directly related to the modulation of the gate space-charge layer which is only dependent on the doping level. Therefore, the ratio between the measured and simulated values of the transconductance for $V_{gs} = 0.5$ V gives an evaluation of the proportion of contacted source lines. This evaluation is almost independent of the actual value of d and of its possible dispersion. It amounts to 45-50% of the total device area.

As can be seen in Figure 3, when a 45% reducing factor is applied to the simulation results, the I-V transfer characteristic of the device with d = 0.34 μm fits exactly the measured one on the whole range of gate voltage. The agreement is also excellent in the low drain voltage region of the drain characteristic (Figure 4). However, an increasing discrepancy is observed for drain voltages above the saturation voltage. This could be due to non-stationary effects which cannot be accounted for with this simulation method. In order to ascertain this explanation, a Monte-Carlo simulation would be necessary.

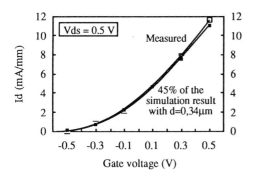

*Figure 3 :*
*Comparison of the measured transfer characteristic with the simulation result obtained with a 0.34μm inter-gate distance, assuming only 45% of the source lines are contacted during the measurement.*

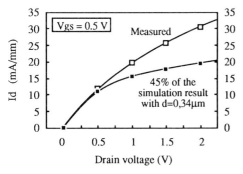

*Figure 4 :*
*Comparison of the measured and simulated drain characteristics under high gate voltage assuming only 45% of the source lines were contacted during the measurement.*

Table 1 compares the electrical parameters obtained from i) the static I-V measurements (corrected for the surface factor), ii) the microwave measurements, iii) the simulation results with d = 0.34 μm and iv) analytical evaluations, all for the same bias voltages ($V_{gs}= 0.5$ V; $V_{ds} = 2$ V). The analytical values of the capacitances were obtained with a plane capacitance model, using the nominal layer thicknesses and contact pad areas.

|  | Measured static characteristics | Microwave measurements | Simulation results | Analytical evaluations |
|---|---|---|---|---|
| Transconductance $g_m$ (mS/mm) | 83 | 90 | 53 |  |
| Drain conductance $g_d$ (mS/mm) | 18 |  | 9 |  |
| $C_{gs}$ (pF/mm) |  | < 1 | 0.33 | 0.32 |
| $C_{gd}$ (pF/mm or pF) |  | 3.3 pF (parasitic) | 0.12 pF/mm (intrinsic) | 3.3 pF (parasitic) |

*Table 1 :*
*Comparison of the electrical parameters of a 0.3-0.3 µm PBT obtained by different methods.*

Apart from the already mentioned too low results of the simulation for the drain conductance and transconductance for drain voltages above the saturation voltage, it is clear that the results are quite consistent. The experimental performances are obviously limited by the large gate to drain parasitic capacitance for this mask geometry. A simple equivalent circuit is drawn in Figure 5.

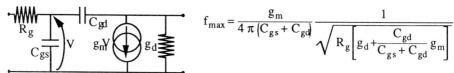

$$f_{max} = \frac{g_m}{4\pi(C_{gs}+C_{gd})} \frac{1}{\sqrt{R_g\left[g_d + \frac{C_{gd}}{C_{gs}+C_{gd}}g_m\right]}}$$

*Figure 5*
*Simplified equivalent circuit of the PBT and corresponding maximum oscillation frequency*

The gate resistance is calculated for a 80nm thick PtSi metallization (with a 40 µΩcm resistivity) and is corrected in order to account for its distributed nature. The constant parasitic capacitance $C_{gd}$ introduces a more severe limitation for the smallest devices due to its higher relative importance in comparison with the other gate length dependent parameters. Using the measured values of the electrical parameters, we calculate a $f_{max}$ value around 1.6 GHz for the 10x10 µm² devices and around 3.3 GHz for the 20x20 µm² devices. This compares well with the 2 to 3 GHz obtained from microwave measurements.

The parasitic gate capacitance can be easily eliminated by additional mask levels. PBTs with a new mask geometry are presently under fabrication. The residual gate pad area will be only 15 µm² compared to 220x70 µm² for the presented devices. This leads to negligible parasitic gate to drain capacitance. The calculated $f_{max}$ value is then over 100 GHz for the 10x10 µm² devices and over 55 GHz for the 20x20 µm² devices.

Gerodolle A. 1989 *TITAN V Two-Dimensional Process/Device Simulator*, Short Course, NASECODE VI, (Boole Press).
Letourneau P. 1989 *Si Permeable Base Transistor Realization Using a MOS Compatible Technology*, ESSDERC 89 (Boole Press) p 579.

# P-channel etched-groove Si permeable base transistors

A. Gruhle, P.A. Badoz
CNET, BP98, F-38243 Meylan Cedex, France

**Abstract:** For high-speed complementary logic using permeable base transistors (PBTs) p-channel devices are needed. For the first time the simulation and fabrication of this kind of transistors are reported. Two-dimensional computer modeling indicate that in general p-channel PBTs reach up to 75% of the transit frequency of their n-channel counterparts. First experimental devices with 0.3µm finger size exhibited a transconductance of 30mS/mm.

Permeable base transistors (PBTs) belong to the fastest known silicon devices. High frequency power PBTs have been reported /Rathman et al. 1988/ as well as ultrasmall low-power devices for digital integrated circuits /Gruhle et al. 1990/. For applications in the latter case the introduction of p-channel PBTs offers the possibility of a complementary logic similar to CMOS with ultra low power consumption. With the further development of epitaxial silicides as $CoSi_2$ /v.Känel et al. 1990/ and $ErSi_2$ for n- and p-type transistors, respectively, the 3-dimensional integration with stacked PBTs might be feasible in future.

We report for the first time the simulation and fabrication of p-channel PBTs of the etched-groove type with a minimum finger size of 0.3 µm. Different device geometries were modeled with two-dimensional numerical simulations. The results show that p-PBTs reach up to 75% of the transit frequency values of their n-channel counterparts. This surprising result is due to the fact that PBTs mainly operate in the region of velocity saturation, the limit of which is almost equal for electrons and holes. Fig.1 shows as an example the simulated results of an n- and p-type PBT of the same geometry (emitter-on-top, 0.2 µm finger width, $5·10^{16}$ $cm^{-3}$ emitter doping, $1·10^{16} cm^{-3}$ collector doping, groove depth 0.3 µm, collector layer thickness 0.4µm). While at low voltages the p-channel transistor lies well below the n-channel PBT it, considerably

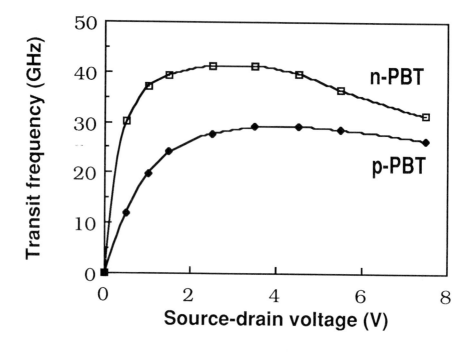

Fig.1: Simulated transit frequencies of p- and n-channel PBTs

approaches n-channel values for higher source-drain biases when current transport is governed by velocity saturation.

The first fabricated p-PBTs were built on p+ substrates with a 0.5 - 2 µm thick epilayer doped $3 \cdot 10^{15}$ cm$^{-3}$. The top layer received a shallow B implantation for the ohmic emitter contacts. PBT patterns were defined by lift-off of a PtCr mask using PMMA and e-beam exposure. The grooves were etched in a SF$_6$, O$_2$ RIE process leading to vertical sidewalls. A final CrPt evaporation formed the gates at the bottom of the grooves. Several layers of PIX-1400 polyimide were used to planarize the surface and to form low-capacitance contacts. Fig.2 shows a SEM micrograph of 0.2µm wide PBT fingers. In Fig.3 the output characteristics of a 0.3µm device can be seen. The overall finger length is 60µm. Due to the low collector layer doping the transistor exhibits triode-like characteristics and reaches only the modest transconductance of 30mS/mm. Devices with a thick collector layer (2µm) had even poorer performance, an indication of

Advanced silicon device concepts 43

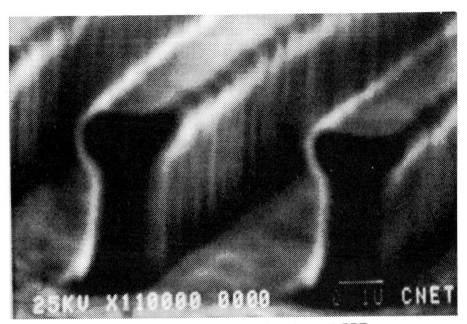

Fig.2: SEM micrograph of the fabricated p-channel PBT

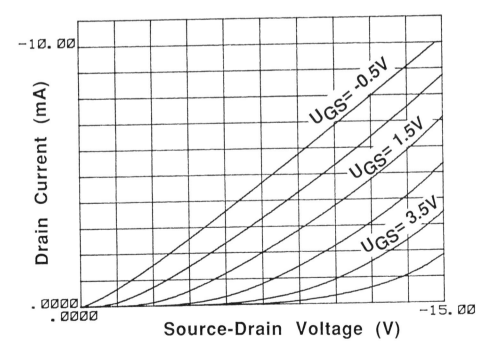

Fig.3: Output characteristics of a 60μm long p-channel PBT

the detrimental effect of the collector layer series resistance. PBTs with optimized layer thickness and doping according to Fig.1 are presently beeing fabricated. The well-known problem of sidewall depletion /Gruhle et al. 1990/ seems to be less pronounced with p-doped material, however a future sidewall passivation is recommended.

The authors accknowledge the contribution of F. Chevalier and R. Carré to this work.


References:
    Rathman D D, Niblack W K 1988, MTT-S Digest, 537
    Gruhle A, Beneking H 1990, IEEE Electr. Dev. Lett. EDL-11, 165
    v.Känel H et al. 1990, Thin Sol. Films 184, 295

# A permeable base transistor on Si(100) with implanted $CoSi_2$-gate

A Schüppen, S Mantl, L Vescan, and H Lüth

Institut für Schicht- und Ionentechnik (ISI)
Forschungszentrum Jülich, P.O.Box 1913, D-5170 Jülich, FRG

Abstract. A permeable base transistor (PBT) has been fabricated by local implantation of $^{59}Co$ into Si(100) with subsequent rapid thermal annealing and epitaxial growth of silicon by LPVPE. Transmission electron microscopy shows abrupt interfaces between the buried $CoSi_2$ and the adjacent silicon. Rutherford backscattering and channeling experiments with a minimum yield of 5.3% for the Co signal as well as a specific resistance of 13 µohmcm of the $CoSi_2$ layers demonstrate the good quality of the $Si/CoSi_2/Si$ heterostructure. $Si/CoSi_2$ Schottky diodes revealed ideality factors of 1.01, while PBTs with 1.5 µm gratings exhibited a maximum transconductance of 11mS/mm.

## 1. Introduction

Permeable base transistors (PBT) are certainly interesting high speed devices for digital logic circuits with regard to three dimensional integration, because of their very low power delay product (Bozler 1980) and their vertical geometry. Principally, there exist two different types of PBT: the etched-groove PBT (e.g. Rathman 1982, Gruhle 1988) and the PBT with embedded metal or silicide grating (e.g. Bozler 1980, Tung 1986, Ishibashi 1986). The main problem of the first one is to control the undesired space charge region at the side walls of the etched groove, while PBTs with buried gates don't have this disadvantage. PBTs with $CoSi_2$-gates have already been built, by evaporating Co on Si(111), thermal annealing, and subsequent MBE process (e.g. Ishibashi 1986, Nakamura 1989, v. Känel 1990).
In this paper we present the first PBT with buried $CoSi_2$-gate obtained by high dose $^{59}Co^+$-implantation on Si(100). Ion implantation and subsequent annealing to obtain a buried monocrystalline $CoSi_2$ layer is applicable to both (111) and (100) Si (White 1988). An important advantage of the Co implantation is, that the $CoSi_2$ formation is nearly independent of surface effects and preparation. This technique leads to a coherent, buried $CoSi_2$ layer under a smooth thin monocrystalline silicon top layer, which can be overgrown by MBE as well as by low pressure vapor phase epitaxy (LPVPE).

## 2. Device fabrication process

The important steps of the present device fabrication process are shown in Fig.1. For first investigations we used <100> oriented, n-type Si wafers with a resistance of 0.1 ohmcm. A conventional lithography step defined gratings on the thermal oxidized wafers with fingers and channel widths of 1.5 μm and 2 μm respectively. Reactive ion etching by $CHF_3$ plasma led to the $SiO_2$ mask for the ion implantation. The $Co^+$ implantation was performed with a 200keV medium current ion accelerator (EATON NV-3204) at a dose of $2 \cdot 10^{17}$ cm$^{-2}$. The substrates were tilted by 7° to avoid channeling and were heated to 350°C during implantation to recover beam induced damages (Kohlhof 1989). After implantation the $SiO_2$ mask was removed by wet chemical etching ($NH_4F/HF$) to exclude diffusion of Co from the $SiO_2$ into the silicon during the subsequent annealing process. Before rapid thermal annealing (RTA) a $SiO_2$ cap layer was evaporated to prevent the formation of holes in the silicon top layer (Kohlhof 1989). By RTA at 750°C for 30s and at 1150°C for 10s we achieved a coherent $CoSi_2$ layer buried under a thin monocrystalline silicon layer. After removing the $SiO_2$ silicon was grown by LPVPE (Vescan 1987) with a n-type doping concentration of $1 \cdot 10^{17}$ cm$^{-3}$ for the first 780nm and $1 \cdot 10^{19}$ cm$^{-3}$ for the upper layer to obtain an ohmic top contact. Then a second lithography step for the metallization of the top contact followed by using lift-off technique. The contacts were annealed at 300°C for 5 minutes. In order to perform the gate contact the silicon was mesa-etched by 44% solution of KOH, which does not attack the $CoSi_2$ layer. Finally the bottom contact was metallized.

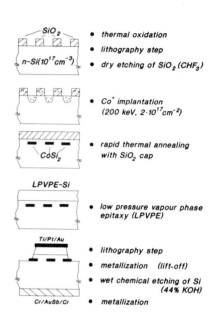

Fig. 1 Device fabrication steps for the implanted $CoSi_2$ PBT

## 3. Results

Implantation of $2 \cdot 10^{17}$ cm$^{-2}$ $Co^+$ into Si(100) with an energy of 200keV leads to a broad Co distribution with a peak concentration of 22%. During the RTA step the Co atoms rearrange to form a continuous 80 nm thick $CoSi_2$ layer buried under 60 nm Si. The Co distribution has sharpened to an almost ideal, rectangular profile in <100> direction as reavealed by Rutherford backscattering (RBS) (Fig.2). The ratio of RBS random and channeling spectra provides a minimum yield of 5.3% for the Co signals, (the peaks on the right hand in Fig. 2), showing the good single crystallinity of the buried layer.

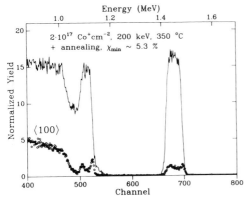

Fig. 2 Random (o) and aligned (-) RBS spectra after implantation and RTA of the Si/CoSi$_2$/Si heterostructure

Fig. 3 shows a transmission electron microscopy (TEM) picture of an implanted CoSi$_2$ structure after the RTA step and overgrowth of 480 nm Si by LPVPE. Some strain and misfit dislocations can be found around the buried CoSi$_2$, but the transition from Si to CoSi$_2$ is abrupt. At the edges of the implanted areas (111) facets are present as a result of the rhomic structure of the CoSi$_2$ in Si(100). These facets might be advantageous for high frequency performance of the PBT, because the gate length and the metallic finger surface are reduced without a proportional increase of the gate resistance. This photograph also demonstrates that the gaussian cobalt distribution shrinks to a coherent, crystalline CoSi$_2$ layer by the RTA process.

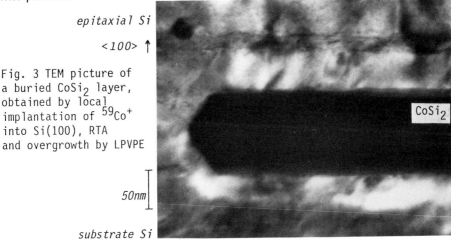

Fig. 3 TEM picture of a buried CoSi$_2$ layer, obtained by local implantation of $^{59}$Co$^+$ into Si(100), RTA and overgrowth by LPVPE

The measured electrical specific resistance of such a buried CoSi$_2$ layer has a value of only 13 µohmcm. I-V characteristics of the CoSi$_2$/Si Schottky diodes yielded ideality factors n of 1.01 to 1.3. The diodes have a barrier height $\emptyset_B$, obtained from saturation current, of $\emptyset_B$ = 0.63 ± 0.02 eV. These results also indicate the possibility of fabrication abrupt metal/semiconductor interfaces by Co$^+$ implantation into heated Si (100) substrates. It should be noted, that the I-V characteristics of the CoSi$_2$/Si-substrate diodes are of similar quality as those of the CoSi$_2$/Si-epilayer diodes. Fig.4 shows the output characteristics of a PBT with an effective gate width of 60 μm (three fingers, 20 μm each), which operates in both directions in common-source circuit.

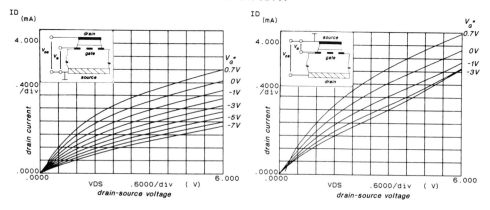

Fig. 4a Source at the backside       Fig. 4b Source at the top
Output characteristics of a Si/CoSi$_2$/Si PBT with implanted gate
gate width: 60 μm, channel width: 1.5 μm, parameter: gate voltage $V_G$

With the present experimental conditions, namely 1.5 μm spacings between the 1.5 μm broad metallic CoSi$_2$ fingers and a specific substrate resistance of 0.1 ohmcm, it is impossible to reach the pinch-off condition. The breakdown voltages of the device exceed 8 V, when the source contact is at the backside of the wafer (Fig. 4a) and the space charge region extends more deeply into the substrate. For source at the top (Fig. 4b) the breakdown voltage and the range of the gate voltages are lower. However, in this circuit the PBT reaches higher current levels, with a maximum transconductance of 11mS/mm. The asymmetry of the output characteristics upon inverse operation (Fig. 4a,b) may be due to differences in doping concentration and/or crystal quality of substrate and epitaxial layer. Nevertheless, the results demonstrate the feasibility of Co$^+$ implantation for fabricating a Si/CoSi$_2$/Si PBT in Si(100).

Acknowledgement
We would like to thank Mrs C. Dieker for the TEM photography, Mrs. S. Meesters, and Mr M. Gebauer for technical support.

Literature
Bozler C O and Alley G D 1980 *IEEE Trans. on Elec. Dev.* **ED-27** pp 1128-41
Gruhle A, Vescan L and Beneking H 1988 *Superlat. and Micstr.* **Vol.4** p 139
Ishibashi K and Furukawa S 1986 *IEEE Trans. on Elec. Dev.* **ED-33** pp 322-27
Känel v. H, Henz J, Ospelt M, Hugi J, Müller E, Onda N and Gruhle A 1990
    *Thin Solid Films* **Vol 84** pp 295-305
Kohlhof K, Mantl S, Stritzker B and Jäger W 1989 *Appl. Surf. Science* **38** pp 207-16
Nakamura N, Ohshima T, Nakagawa K and Miyao M 1989 *21st Conf. on Solid State Dev. and Mat.* Tokyo pp 85-88
Rathman D D, Economou N P, Silversmith D J, Mountain R W and Cabral S M
    1982 *IEDM Tech. Dig.* **IEDM 82** pp 650-53
Tung R T, Levi A F J and Gibson J M 1986 *Appl. Phys. Lett.* **48**(10) pp 635-7
Vescan L, Beneking H 1987 *J. Electrochem. Soc.* **87-2** pp 1478-9
White A E, Short K T, Dynes R C, Gibson J M and Hull R 1988 *Mat. Res. Soc. Sym. Proc.* **Vol 100** pp 3-15

Paper presented at ESSDERC 90, Nottingham, September 1990
Session 1C4

# A novel compact model description of reverse-biased diode characteristics including tunnelling

### G.A.M. Hurkx, H.C. de Graaff, W.J. Kloosterman and M.P.G. Knuvers
Philips Research Laboratories, 5600 JA Eindhoven, The Netherlands

### ABSTRACT

A new compact model description of reverse-biased diode characteristics is presented. This model includes tunnelling effects and avalanche breakdown. From comparison with both numerical simulations and measurements it is found that the model gives a good description of the I-V characteristics of reverse-biased diodes.

### 1 INTRODUCTION

At high dopant concentrations tunnelling effects strongly influence reverse-biased diode characteristics. Existing compact models such as that in SPICE give a constant very small reverse current up to the breakdown voltage. When tunnelling is important such a description is very inaccurate and in many cases inadequate. In this paper we present a new model for circuit simulation purposes which incorporates the following physical mechanisms: band-to-band tunnelling, trap-assisted tunnelling, Shockley-Read-Hall generation and avalanche breakdown.

### 2 MODEL DESCRIPTION

The steady-state hole continuity equation, including tunnelling and avalanche generation is given by:

$$\frac{dJ_p}{dx} = -qR_{SRH}(x) + \alpha_n(x)|J_n(x)| + \alpha_p(x)|J_p(x)| + (J_{bbt} + J_{tat})\delta(x). \quad (1)$$

In the above equation $R_{SRH}(x)$ is the Shockley-Read-Hall recombination rate, $J_{bbt}$ is the band-to-band tunnelling current density, $J_{tat}$ is the trap-assisted tunnelling current density [1,2] and $\alpha_n$ and $\alpha_p$ are the ionization coefficients for electrons and holes respectively. Band-to-band tunnelling and trap-assisted tunnelling are taken into account as a $\delta$ − function generation term at the location of the maximum electric field $F_m$ ($x = 0$, see fig. 2). The expression for $J_{bbt}$ (see table) is from ref. [3] while the expression for $J_{tat}$ is obtained by the integration of the corresponding generation rate, as given in refs. [2,4], over the depletion layer of a symmetrical step junction. Using $J_d = J_n + J_p$ for the diode current density, expression (1) can be solved exactly. Because this yields an expression which is far too complicated in a circuit simulation environment, the following approximations, which apply to a symmetrical step junction, are made:

© 1990 IOP Publishing Ltd

1) $R_{SRH}(x)$ is assumed to be constant in the depletion layer, having a value of $-n_i/2\tau$, except within a distance of $W_0/2$ from the depletion layer boundaries where $R_{SRH} = 0$ [5]. $W_0$ is the zero-bias depletion layer width.
2) The function $\phi(x) = \exp\left(-\int_x^{x_p}(\alpha_n - \alpha_p)\,dx'\right)$, occurring in the solution of (1), is approximated by $\phi(x) = \exp(-2\mu_{av})$ for $x < 0$ while $\phi(x) = 0$ for $x > 0$. At $x = 0$, $\phi(x) = \exp(-\mu_{av})$. The quantity $\mu_{av}$ is given by $\mu_{av} = c_{av} F_m^2 \exp(-b_n/F_m)$.
Now the solution of (1) is given by

$$J_d = \frac{(J_{bbt} + J_{tat})e^{-\mu_{av}} + (J_{SRH} + J_{is})(1 + e^{-2\mu_{av}})/2}{1 - 2\mu_{av}(1 + e^{-2\mu_{av}})}, \qquad (2)$$

where $J_{SRH}$ is due to SRH generation and $J_{is}$ is the constant ideal saturation current density. Breakdown occurs when the denominator of (2) becomes zero. This gives for $\mu_{av}$ at breakdown a value of 0.3295. Normalizing the above expression for $\mu_{av}$ to this value at breakdown yields the expression presented in the table below, where $F_{mbr}$ is the maximum field at breakdown.

| | |
|---|---|
| Band-to-band tunnelling | $J_{bbt} = -c_{bbt}\, V\, F_m^{3/2} \exp\left(\dfrac{-F_0}{F_m}\right)$ |
| Trap-assisted tunnelling | $J_{tat} = c_{tat}\, W\, \dfrac{\gamma}{F_m}\left[\exp\left(\left(\dfrac{F_m}{\gamma}\right)^2\right) - \exp\left(\left(\dfrac{F_m W_0}{\gamma W}\right)^2\right)\right]$ |
| SRH generation | $J_{SRH} = c_{SRH}(W - W_0)$ |
| Avalanche multiplication | $\mu_{av} = 0.3295\left(\dfrac{F_m}{F_{mbr}}\right)^2 \exp\left(b_n \dfrac{F_m - F_{mbr}}{F_m F_{mbr}}\right)$ |

Expression (2), together with the expressions given in the table, form the basis of the model. The value of $b_n$ from ref. [6] is used, while the parameters for tunnelling $c_{bbt}$, $F_0$ and $\gamma$ are obtained from refs. [2,3]. In the above expressions $W$ is the depletion layer width and $V$ is the junction voltage. The theoretical value of $c_{SRH}$ is $qn_i/2\tau$ and the theoretical value of $c_{tat}$ is $\sqrt{3\pi}\, qn_i/2\tau$. $\gamma = \sqrt{24m^*(kT)^{3/2}(q\hbar)^{-1}}$.

## 3 RESULTS

To show the validity of the mathematical simplifications involved, fig. 1 shows a comparison of results from a 1D numerical device simulation with compact model results. The corresponding doping profile is given in fig. 2. In the device simulation program the recombination model for tunnelling as described in ref. [2] is used, together with the conventional field-dependent model for avalanche multiplication. To obtain the compact model results, the numerically calculated values of $F_m$ and $W$ at each bias point are used. Also the quantities $F_{mbr}$ and $n_i/2\tau$ are taken from the numerical simulations. From fig. 1 we can observe that the agreement between numerical results and compact model results is good.

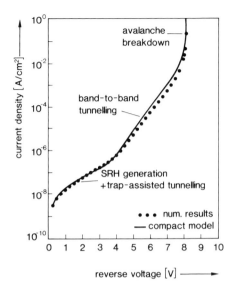 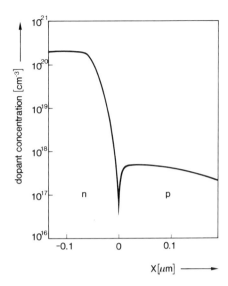

Fig. 1 Comparison of model results with numerical simulations.

Fig. 2 Doping profile of the diode in fig.1.

For practical use the problem arises that due to dopant inhomogeneities the electric field will generally not be exactly constant along the junction area. Because tunnelling effects and avalanche multiplication are extremely sensitive functions of the electric field, the current due to these effects is strongly confined to the area where the electric field is highest. This implies that the effective junction area for the reverse current is usually not known. For this reason the following strategy is followed for accurate fitting: The diffusion voltage $V_{diff}$ and the grading coefficient $p$ are obtained from fitting the depletion capacitance. The depletion layer width $W$ follows from the relation $W = W_0 (1 - V/V_{diff})^p$, while the zero bias depletion layer width $W_0$ (hence the effective junction area) is obtained by treating $F_{mbr}$ as a parameter. The relation between the electric field and the depletion layer width is obtained by using Gauss' law [7]. The model contains the following six parameters: $F_{mbr}$, $c_{SRH}$, $c_{tat}$, $c_{bbt}$, $V_{br}$ (all extracted from the reverse I-V characteristic) and $J_{is}$ (from the ideal forward characteristic). Notice that in the case that the current and not the current density is fitted, the prefactors also comprise the effective junction area.

Fig. 3 shows a comparison of model results with measurements on three diodes having different doping profiles. From the temperature dependence of the currents we have observed that for diode A band-to-band tunnelling dominates while for diode C both Shockley-Read-Hall generation and trap-assisted tunnelling are important. For diode B we observe two distinct regions, as can be seen in fig. 4, where the I-V curves are plotted for three temperatures. The parameters are fitted for the curve at $T = 338K$, while the other curves are obtained without any fitting. The temperature rules for the tunnelling parameters are obtained from refs. [2 – 4] while for the parameters $c_{SRH}$ and $c_{tat}$ the temperature dependence of $n_i$ is taken with a bandgap $E_g = 1.1eV$. Furthermore, $c_{bbt}$ is assumed to be temperature independent while $b_n$ and $V_{br}$ have the weak linear temperature dependence as given in ref. [8]. Notice the strong temperature dependence of the current at low bias due to the strong temperature dependence of $n_i$ and the weak temperature dependence at high bias conditions due to band-to-band tunnelling and avalanche multiplication. The resulting values of the parameters $c_{SRH}$ and $c_{tat}$ yield a value of $\tau$ of around $5\mu s$, for both diodes B and C. The resulting values of the maximum field at breakdown are $1.62\ 10^6$ V/cm, $1.53\ 10^6$ V/cm and $7.38\ 10^5$ V/cm for diodes A,B and C respectively. These are realistic values for such highly-doped diodes [9]. At the considered temperatures the ideal saturation current density $J_{is}$ is found to be unimportant in reverse bias.

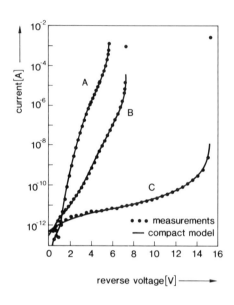 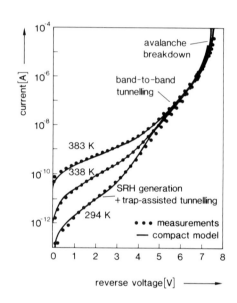

Fig. 3 Comparison of model results with measurements on three diodes.

Fig. 4 Measurements and model results for diode B at three temperatures.

## 4 CONCLUSIONS

From comparison with both numerical simulations and measurements it is found that the model provides a good description of reverse-biased I-V characteristics. Also the temperature dependence of these characteristics is described well by the model.

REFERENCES

1 E. Hackbarth and D.D. Tang 1988 IEEE Trans. Electron Devices 35 p 2108
2 G.A.M. Hurkx, D.B.M. Klaassen, M.P.G. Knuvers and F.G. O'Hara 1989 IEDM Techn. Dig. p 307
3 G.A.M. Hurkx 1989 Solid-St. Electron. 32 p 665
4 G. Vincent, A. Chantre and D. Bois 1979 J. Appl. Phys. 50 p 5484
5 P.U. Calzolari and S. Graffi 1972 Solid-St. Electron. 15 p 1003
6 R.J. van Overstraeten and H.J. de Man 1970 Solid-St. Electron. 13 p 583
7 H.C. de Graaff and F.M. Klaassen 1990 *Compact Transistor Modelling for Circuit Design* (Wien: Springer) p 74
8 R. Hall 1967 Int. J. Electronics 22 p 513
9 S.M. Sze 1981 *Physics of Semiconductor Devices* (New York: Wiley) p 100

# Compositional analysis of submicron silicon in one, two and three dimensions

Chris Hill

Plessey Research Caswell Limited, Towcester, Northants, England NN12 8EQ

Abstract  Shrinking VLSI geometries have resulted in increasingly non-planar structures and current flows, necessitating lateral as well as vertical information on dopant distribution for device design, modelling and optimisation. Continuing demand for improved vertical resolution (1-10nm) has led to improvements in existing 1D dopant profiling techniques: new 2D techniques of reasonable resolution (20-25nm) are beginning to meet the demand for lateral information; 3D techniques at present are too low in resolution (0.5 micron) or sensitivity ($10^{20}$/cc). The present status of dimensional compositional analysis techniques as applied to VLSI submicron structures is reviewed.

## Introduction

The electrical characteristics of the active components of silicon integrated circuits are sensitively dependent on the concentration and spatial distribution of the doping elements within them. Thus, right from the start of IC technology, there has been an associated analysis activity aimed at measuring dopant distributions in silicon planar structures. For many years, determination of the variation of concentration with depth on laterally uniformly doped silicon regions (so-called 1 dimension profiles) sufficed to adequately predict the electrical properties. However, the shrinkage of lateral dimensions to one micron and below created a need for analysis techniques capable of measuring dopant distributions as a function of two spatial directions in the device (two-dimensional distributions) or even all three spatial dimensions (three- dimensional distributions). The reasons for this can be understood in terms of the effect of lateral shrinkage on the basic components, the MOS and Bipolar transistor, shown schematically in vertical section in Fig 1.

In large geometry MOS, the transistor characteristics are essentially determined by the vertical dopant profile in the current-carrying channel under the gate dielectric, since this determines the field required to create an inversion layer of sufficient width to carry significant current.  In submicron devices, the electrical fields at the edges of source and drain adjacent to the channel become high enough to cause hot electron generation and subsequent degradation of the gate dielectric by charge injection.  This effect is strongly influenced by the lateral doping gradients in source and drain:  thus two-dimensional dopant distributions are now needed  to specify and optimise the total device characteristics.

In large geometry bipolar, the vertical current flow is wholly controlled by the change in material properties (including doping type and (concentration) in a vertical direction, as shown in Fig 1. In submicron bipolar, the technology uses self alignment techniques to reduce lateral separation of the emitter and base contact regions, which results in a very close approach by lateral diffusion of the emitter dopant (usually arsenic) and base contact dopant (usually boron). Achieving the optimum combination of sufficiently high breakdown voltage at the lateral junction with sufficiently low base resistance requires a knowledge of the two-dimensional dopant distributions in both regions.

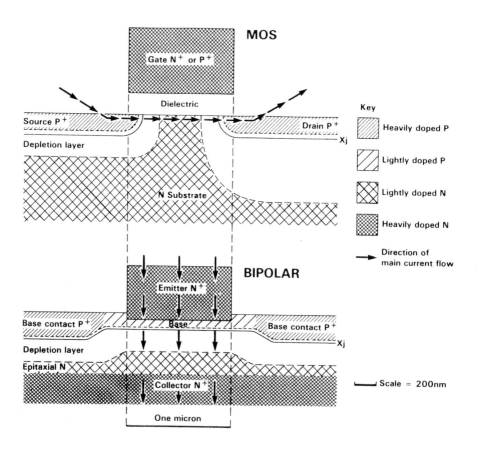

Fig. 1 Schematic vertical sections through the active regions of one-micron geometry MOS and Bipolar VLSI transistor structures, approximately to scale, showing positions of the p-n junctions, associated depletion layers, and current flows. The depletion layer positions shown are for the OFF state of the devices, the current flows shown are for the ON state of the devices. Lateral shrinkage of these critical device regions to submicron dimensions progressively increases the contribution of the lateral doping gradients to the electrical properties of the transistors. After Hill 1987.

In both MOS and Bipolar, the resolution required is of the order of 5% of feature size, about 20-50nm over the compositional range $10^{17}$ - $10^{21}$/cc. In addition, the very shallow emitter base regions of bipolar high switching speed transistors require high vertical resolution, typically 10% of the depth of region. For emitters (20-100nm) this requires 2-10nm resolution, and for base regions (50-200nm) about 5-20nm resolution. Polysilicon based emitters have a special analysis problem, in that the electron and hole current flows across the emitter-base junctions are strongly influenced by the presence of thin interfacial oxide (and its morphology) and a segregated As layer, both at the polysilicon/single crystal interface, and probably not more than 0.5nm thick, as reviewed by Ashburn 1989. Analysis of such regions require vertical resolution of 0.1nm or better. The fabrication techniques developed for submicron technology necessarily create photoengraved structures (and hence doped regions) with very sharp corners in plan view. This results in increasing electric field (due to junction curvature) from planar (1D) to edge (2D) to corner (3D)regions of the same doped region. For full specification of this field, analysis techniques with a 3D spatial resolution of about 20 x 20 x 20nm and compositional sensitivity ~$10^{17}$/cc are required.

The compositional analysis techniques currently available do not meet the above requirements. 1D techniques are nearest to meeting requirement and 3D the furthest away. One reason for this is the inverse relationship between spatial resolution and sensitivity described in the next section.

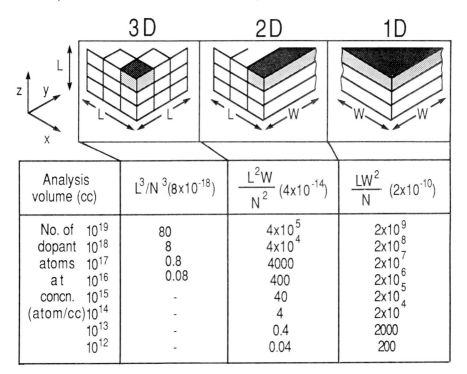

| Analysis volume (cc) | $L^3/N^3$ ($8 \times 10^{-18}$) | $\dfrac{L^2 W}{N^2}$ ($4 \times 10^{-14}$) | $\dfrac{LW^2}{N}$ ($2 \times 10^{-10}$) |
|---|---|---|---|
| No. of dopant atoms at concn. (atom/cc) $10^{19}$ | 80 | $4 \times 10^5$ | $2 \times 10^9$ |
| $10^{18}$ | 8 | $4 \times 10^4$ | $2 \times 10^8$ |
| $10^{17}$ | 0.8 | 4000 | $2 \times 10^7$ |
| $10^{16}$ | 0.08 | 400 | $2 \times 10^6$ |
| $10^{15}$ | - | 40 | $2 \times 10^5$ |
| $10^{14}$ | - | 4 | $2 \times 10^4$ |
| $10^{13}$ | - | 0.4 | 2000 |
| $10^{12}$ | - | 0.04 | 200 |

Fig 2  Relationship between analysis mode (1,2 or 3 dimensional), analysis volume, dopant concentration, and no of atoms. The table has been calculated for dimensions  L=400nm, W=100 microns, L/N=20nm. After Hill 1990, with the 3D column values corrected from arithmetical error in original.

## Spatial Resolution and Compositional Sensitivity

The ultimate sensitivity of many analytical techniques is determined by a machine background signal, ultimately equivalent to a definite number of dopant atoms detected. For a given dopant concentration, therefore, the larger the volume of material analysed, the higher above background will be the detected signal. This volume is necessarily reduced by improving spatial resolution in 1D, and reduced drastically by the square and cube of that dimension in 2D and 3D respectively, as illustrated in Fig 2, from Hill 1990. For instance, for a concentration of $10^{17}$ dopant atoms/cc, the analysis volume in 1D profiling can be made large simply by increasing surface area, so that even for an acceptable resolution of 20nm, there are $2 \times 10^7$ atoms to be detected (column 4); in a 2D determination, one lateral dimension is shrunk to 20nm, reducing the number of atoms to 4000; in a 3D determination all three dimensions must be 20nm, resulting in only 0.8 atoms for detection! This can be compared to the realistic detection limits for SIMS (secondary ion mass spectroscopy), the most sensitive of the direct analysis techniques, of about 2000 atoms where there is no interfering species of similar mass/charge ratio. Direct 3D analysis using SIMS is thus limited to much coarser spatial resolution. Bryan et al 1985 achieved a resolution of 4 microns x 4 microns x 20nm at an arsenic sensitivity of about $2 \times 10^{20}$/cc, and predicted an improvement to about $10^{18}$/cc if a primary ion beam with higher arsenic ion yield (Cs rather than O) was used. If, in addition, the third dimension was coarsened to match the other two, a limiting combination of $(700nm)^3$ and $10^{18}$ atoms/cc might be achieved.

It is worth noting that analysis techniques which detect atomic emissions, rather than the atoms themselves, are more limited by sample background (which is a function of analysis volume) than by instrumental background. Thus, although Auger Electron Spectroscopy (AES) has a relatively poor sensitivity for dopants (e.g. about $10^{20}$/cc for As), this sensitivity is maintained down to very small volumes: this means that using the above figures for SIMS sensitivity, at 3D resolutions below $(150nm)^3$, AES can be more sensitive for As than SIMS. Similar considerations apply to X-ray spectroscopy (XRS).

## One Dimensional Analysis Techniques

A wide range of 1D techniques have been applied to analysis of VLSI planar structures, including SIMS, AES, RBS, ERD, XPS, Spreading Resistance (SR) Profiling, Differential Sheet Resistance (DSR) Profiling and Capacitance Voltage Profiling (CV). These have been reviewed by Vandervorst and Bender 1989 and will not be described in detail here. Of these, only RBS and CV are inherently absolute techniques, covering the dopant ranges $10^{22} - 10^{19}$, and $10^{19} - 10^{14}$ respectively. All the other techniques require calibration using external standards as similar to the VLSI matrix material as possible. Resolution is limited by the thickness of the altered surface layer produced by the bombarding stripping ions in SIMS and AES profiling (typically 5-20nm), (Augustus et al 1989), and by carrier spilling and depletion layer effects in the electrical techniques SR, DSR, CV (typically 10-100nm), (Vandervorst and Bender 1989).

The majority of 1D dopant profiling is carried out using either SIMS (for atomic profiles) or SR (for carrier profiles). This is because both have a wide dynamic range ($\geqslant 10^{20}$/cc down to $\leqslant 10^{16}$/cc) and depth resolution (10-20nm) adequate for many VLSI characterisation requirements. Single atomic layer depth resolution has been achieved in field ion microscope profiling of Si - SiO$_2$ interface regions (Grovenor et al 1985); this is, however, a very specialised time-consuming technique, not applicable to samples in their planar form.

General Two Dimensional Analysis Techniques

A number of techniques have been used for the analysis of lateral dopant distributions in silicon VLSI device structures. Imaging Auger electron spectroscopy and imaging SIMS are direct machine developments of the well-established one-dimensional versions to produce surface maps of dopant for successively deeper layers exposed by planar ion sputtering. They are, thus, essentially 3D techniques; so although considerable improvements in resolution are being developed, (Levi-Setti et al 1985, Umbach et al 1989), the concomitant degradation in sensitivity for SIMS, and the intrinsic insensitivity of AES, limits analysis at 20nm resolution to concentrations greater than $10^{20}$ atom/cc. Quantification also poses problems in 2D analysis, (Walker et al 1988).

A closer match to VLSI requirements is achieved by indirect techniques, which process the sample before analysis, and map a resulting feature of the sample which is related to the underlying dopant concentration. Electron-beam-generated defects during TEM observation can delineate differently doped regions (Romano et al 1989). Selective oxidation TEM, (Hill et al 1985), uses the property that arsenic-doped silicon oxidises thermally at a rate determined by the surface arsenic concentration, to produce an oxide replica of the doped material which can be examined in TEM to reveal the two-dimensional surface arsenic distribution. The technique is relatively rapid, allows individual mapping of hundreds of doped windows in one TEM preparation and has good resolution ($\sim$ 10nm); the sensitivity is, however, poor ($\sim 10^{19}$/cc). More quantitative are techniques which are developments of the selective etch method described by Marcus & Sheng 1981, in which the etch rate of the silicon in an HF/HNO$_3$ etch solution is determined by the (active) dopant concentration at the surface. Selective etch SEM reveals up to 3 isoconcentration contours on a cleaved section through the device; the sensitivity is good, and resolution adequate for many applications. (30 x 30nm and $\sim 10^{17}$/cc), as reported by Gong et al 1989. Selective etch XTEM, Roberts et al 1985, examines selectively etched cross-section TEM samples of the device structure, which show a continuously varying thickness with dopant concentration. Thickness can be accurately measured at every point using many-beam thickness fringes, so that this technique reveals the full 2D dopant distribution at resolution of about 5-10nm and sensitivity about 5 x $10^{17}$/cc. Selective staining SEM has been used for delineation of a one concentration contour, usually the p-n junction position. By preparing vertical sections cleaved through the device structure at a small angle to the long edges of doped windows, a geometrical magnification of about 10 times in the lateral direction is obtained; after staining with a CuSO$_4$/HF/H$_2$O stain, the n type regions are delineated to an accuracy observable in SEM of about $\pm$ 50nm (vertical) and $\pm$ 30nm (lateral), (Subrahmanyan et al 1988, Ahn and Tiller 1988).

All these indirect techniques require calibration by matching a one-dimensional profile through the structure measured by a direct method (usually SIMS) to the 1D profile of the materials property being exploited. Quantification requires great care, being influenced by many variables in the sample preparation process (e.g. stress, temperature, etch dynamics, illumination level, presence of electrochemically different regions in the structure). Despite this, the etching and staining techniques, using either SEM or TEM observation, are at present the only methods that are sufficiently rapid, give adequate resolution, useful sensitivity, and can be applied to real VLSI device structures. For these reasons, and because they also reveal simultaneously 2D structural and dimensional information, they are likely to be the main tool of VLSI technology development engineers for 2D compositional mapping.

A direct 2D analysis technique has recently been reported by Renteln et al 1989 for analysis of Arsenic-doped source and drain regions in submicron MOS structures, with resolution of about 20nm x 20nm at sensitivity about 5 x $10^{18}$/cc. The method involves STEM-EDX mapping of XTEM samples containing the device structure and is a 2D development of the STEM-EDX technique used by Grovenor et al 1984 for very high resolution 1D profiles of arsenic across grain boundaries in polysilicon.

An important aspect of 2D analysis of VLSI device structures is the sectioning to reveal the 2D dopant distribution of interest. In simple structures which are repeated over a large array, this can be achieved by cleaving (for surface analysis) or by cleave and planar ion beam thinning (for transmission analysis). For individual submicron geometry transistors this approach has a low probability of success. A method appropriate to such cases is in-situ ion-beam imaging and the erosion of the desired structure, followed by SIMS analysis of the exposed sloping surface, as described by Bishop and Greenwood 1990.

Specialised Two-Dimensional Analysis Techniques

In order to obtain simultaneously adequate resolution and sensitivity (20nm x 20nm at $10^{17}$/cc) and full quantification, two-dimensional dopant distributions must be measured on specially fabricated samples with specialised analysis techniques. All three methods described below exploit the property of VLSI structures that the 2D section of most interest is at right angles to the long edge of a defined implant window (e.g. MOS source/drain, emitter stripe in a bipolar transistor), and the 2D dopant distribution across that section is the same wherever the long edge is sectioned. The analysis volume can thus be extended as far as desired parallel to the long edge, and by this means the sensitivity of the techniques are enhanced by orders of magnitude, without losing spatial resolution in the directions at right angles to the edge.

The angled-implant method is applicable to the determination of 2D lateral dopant distributions produced solely by ion-implantation. Furukawa and Matsumura 1973 showed that the lateral distribution function can be derived by combining information from 1D dopant depth profiles of samples implanted at different angles. In a fuller development of this technique, Oven et al 1988 have derived the depth-dependent lateral spread of implanted boron, and compared 2D Boltzman and Monte Carlo simulations of implants into amorphous silicon - excellent agreement is found, except in the near surface regions.

2D Carrier profiling, developed at Plessey Caswell, Hill et al 1983, 1988, Pearson and Hill 1988, Hill 1990, is based on the well-known planar anodic sectioning and differential conductance technique (Tannenbaum 1961), used from the beginning of silicon planar technology to establish conductivity depth profiles. The 2D carrier profiling technique has three major components (i) the special structures which contain the 2D distributions of interest (ii) an automatic anodic sectioning and conductance monitoring equipment (iii) a computer analysis programme for conversion of data to 2D dopant distributions. These are described in detail in the papers referenced, together with experimental 2D distributions of implanted and diffused boron at the edges of masked regions of silicon. A resolution of about ± (20nm x 20nm) and sensitivity about $10^{18}$/cc was obtained. At least one order of magnitude improvement in sensitivity is obtainable, but to achieve this the two-dimensional effects of carrier spilling, depletion layer position and oxide charge effects on the measured conductivity will need to be accurately modelled. Another limitation of the present technique is the spatial uncertainty (~ 50nm) in locating the 2D dopant distribution with reference to the original mask edge position.

2D SIMS, a technique developed in the last three years by Hill et al 1988, Dowsett et al 1989, has great promise for determining 2D atomic dopant profiles. Special samples are again required, but these are less complex than those needed for 2D carrier profiling. The high resolution of the technique (10nm - 50nm) is achieved by fabricating samples in which the implant windows are crossed by a regular array of plasma-etched trenches at a very shallow angle θ (0.03, 0.08 or 0.2°). A SIMS line scan across the array in raster mode then picks up dopant signals from successive portions of the lateral profile, each line scan effectively sampling an additional lateral strip w=stan θ wide, where s is the raster line separation. For a typical separation of 10 microns and θ = 0.8°, w = 14nm. The high sensitivity of the 2D SIMS technique ($10^{16}$/cc) is achieved through sampling a large number (300) of identical portions of 2D regions as the line scan crosses a 1.8mm square array of doped windows at 6 micron pitch. The main experimental difficulties of the technique have been in achieving planar removal of the surface during SIMS profiling; a technique for producing planar samples, and maintaining adequate planarity during profiling has now been developed, and the first 2D boron distributions have been measured by this technique, Cooke et al 1990.

Acknowledgements

This work has been funded by the E.E.C. and I.E.D., and Plessey Semiconductors Ltd and thanks are due to these bodies for permission to publish.

References

Ahn S.T.and Tiller W.A. 1988 J.Electrochem.Soc.135 No.9. 2370-2373.
Ashburn P. 1989 Proc.of 1989 Bipolar Circuits and Technology Meeting
  (Sept 18/19 1989 Minneapolis) IEEE Piscataway NJ 90-97.
Augustus P.D. Kightley P Hutchison J.L. Nicholson W.A.P. Clark E.A. Dowsett M.G. Spiller G.D.T. 1989 Microscopy of Semiconducting Materials 1989 Inst.Phys.Conf.Series 100 519-524.
Bishop H.E. and Greenwood S.J. 1990 Surf. and Int.Anal. 16 70-76.
Bryan S.R. Woodward W.S. Linton R.W. 1985 J.Vac.Sci.Techn. A 3(6) 2102-2107.
Cooke G. Dowsett M. Pearson P. Hill C. Goulding M. to be published.

Dowsett M.G. Cooke G. Hill C. Clark E.A. Pearson P. Snowden I. Lewis B. 1989 Proc. of SIMS VII (Monterey Ca Sept 4-8 1989) J Wiley and Sons N.Y.
Furukawa S. and Matsumura H. 1973 in Semiconductors and Other Materials (B.L. Crowder ed) Plenum Press 193-207.
Gong L. Barthel A. Lorenz J. Ryssel H. 1989 ESSDERC 89 (A. Heuberger Ryssel Large P. eds) Springer-Verlag Berlin 198-201.
Grovenor C.R.M. Batson P.E. Smith D.A. Wong C. 1984 Phil Mag A $\underline{50}$ 409.
Grovenor C.R.M. Cesezo A Smith C.D.W. 1985 in Microscopy of Semiconducting Materials 1985 (Cullis A.G. Holt D.B. eds) IOP Conf.Series 76 Adam Hilger Bristol 423-428.
Hill C. Holden A 1983 Two Dimensional Dopant Profile Measurement Res.Rpt RP9-275 MOCVD London.
Hill C. Augustus P.D. Ward A. 1985 in Microscopy of Semiconducting Materials 1985 (Cullis A.G. Holt D.B. eds) IOP Conf.Series 76 477-482.
Hill C. 1987 Nucl.Inst. and Methods in Phys.Res. B19/20 North Holland Amsterdam 348-358.
Hill C. Pearson P.J. Lewis B. Holden A.J. Allen R.W. 1988 Solid State Devices (Soncini G. Calzolari P.V. eds) North Holland Amsterdam 147-150.
Hill C. Dowsett M.G. Snowden I. Clark E.A. Lewis B. 1988 Paper O23 Proceedings Abstracts of QSA5 (November 1988 Teddington UK).
Hill C. 1990 in Analytical Techniques for Semiconductor Material and Process Characterisation (Kolbeson B.O, McCaughan D.V. Vanderworst W. eds) ECS Proc.Vol 90-11 ECS Pennington NJ 65-83.
Levi-Setti R. Gros G. Yang Y.L. 1985 Scanning Electron Microscopy 1985 535-551.
Oven R. Ashworth D.G. Hill C. 1988 in Simulation of Semiconductor Devices and Processes 3 (Baccarini G. Rudan M. eds) Technoprint Bologna 429-440.
Pearson P.J. and Hill C. 1988 J.de Physique $\underline{49}$ C4 No.9 515-518.
Renteln P. Ast D.G. Mele T.C. Krusius J.P. 1989 J.Electrochem.Soc.$\underline{136}$ No. 12 3828-3836.
Roberts M.C. Yallup K.J. Booker G.R. 1985 in Microscopy of Semiconducting Materials 1985 (Cullis A.G. Holt B.D. eds) IOP Conf.Series 76 483-488.
Romano A. Vanhellemont J, Morante J.R. Vanderworst W 1989 Micron and Microscopy Acta, $\underline{20}$ No 2 151-152.
Sheng T.T. and Marcus R.B. 1981 J.Electrochem.Soc. $\underline{128}$ 881-884.
Subrahmanyan R. Massoud H.Z. Fair R.B. 1988 Appl.Phys. Lett $\underline{52}$ 2145.
Tannenbaum E. 1961 Sol.State Electronics $\underline{2}$ 123-127.
Umbach A. Hoyer A. Brunger W.H. 1989 Surf. and Int.Analysis $\underline{14}$ 271-273.
Vanderworst W. and Bender H. 1989 in ESSEDERC 89 (Heuberger $\overline{A.}$ Ryssel H. Large P eds) Springer-Verlag Berlin 843-860.
Walker C.G.H. Peacock D.C. Prutton M. El-Gomati M.M. 1988 Surface and Interface Analysis $\underline{11}$ 266-278.

# TITAN-RTA: a 2D integrated equipment and process model for simulation of rapid thermal processing

S K Jones [1], A Gerodolle [2]

1 Plessey Research Caswell Ltd, Allen Clark Research Centre, Caswell, Towcester, Northants, UK

2 CNET/CNS, Chemin du Vieux-Chene, BP 98, F 38243 Meylan, France

ABSTRACT. An integrated model for the simulation of the exact thermal behaviour of a silicon wafer during optical rapid thermal processing coupled to the microscopic thermal processing effects of dopant diffusion and oxidation is presented.

## 1. Introduction

Rapid Thermal Processing (RTP) is now established for key silicon device processing steps such as implant activation, shallow dopant drive-ins from overlayers, and oxidation. Restrictions on process tolerances coupled with the high activation energies of many of the physical processes occuring during RTP mean that accurate control of absolute temperature and temperature non-uniformities during RTP is essential. It is known, however, that inherent features in the physics of operation of RTP systems can lead to such temperature non-uniformities and control errors (Hill et al 1989). These include wafer effects such as overlayer patterning, doping and geometry; and machine effects such as the lamp power spectrum and profile, chamber reflectance and ramp rates. Prediction of the resulting temperature control error and non-uniformity and their effect on the material processes occuring during RTP is therefore very important for successful process control. TITAN-RTA is a novel 2D simulator which integrates a sophisticated set of models for thermal simulation of RTP equipment into an existing comprehensive 2D process simulator TITAN-5 (Gerodolle et al 1989). The process models of TITAN-5 are then able to simulate dopant diffusion and oxidation for the realistic thermal conditions predicted to occur at specific locations of a wafer during RTP, rather than the idealised isothermal conditions conventionally used in process simulators. This allows study of the sensitivity of the processing both to machine variables and to wafer features such as proximity to different overlayer regions and to the wafer edge.

## 2. Model Description

The RTP equipment model within TITAN-RTA is suitable for high temperature applications such as oxidation and diffusion and is capable of simulating a wide class of machines. Commands have been introduced into TITAN-RTA which allow the user to define the important machine and wafer features, and the RTP operation variables. The description of the

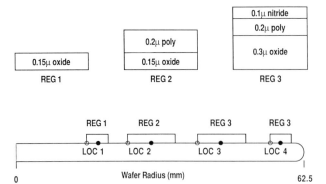

Fig 1

The layer structures and wafer patterning for the RTP simulation example. The wafer is modelled assuming radial symmetry. Temperature is monitored at the four locations shown (solid circles).

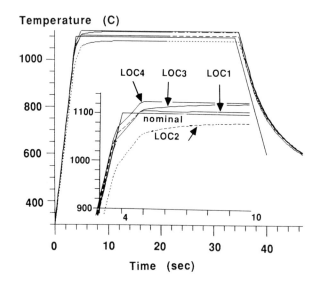

Fig 2

The predicted time-temperature profiles at the four locations in Fig 1 for a nominal 1100°C 30 sec anneal cycle. The inset shows the transient behaviour.

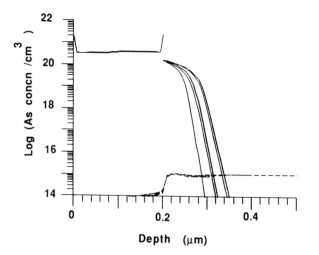

Fig 3

The predicted arsenic profiles following a 40keV $10^{16}/cm^2$ implant into a 0.2 μm polysilicon layer and drive-in for each of the thermal cycles of Fig 2. The profiles correspond to locations LOC2, nominal, LOC1, LOC3 and LOC4 in order of increasing depth.

RTP machine includes the wall temperature and reflectivity; the quartz isolation chamber; the spectral properties of the lamps; and the temperature controlling pyrometer. The wafer features include size, substrate doping and overlayer patterning. Operation variables include a general specification of the time temperature cycle in terms of a set of ramped and steady state components and the pyrometer target emissivity.

TITAN-RTA solves the coupled thermal diffusion and radiative emission and absorption for a 2D approximation to the whole wafer over the entire RTP cycle using a finite difference method. The radiative boundary conditions are calculated for each overlayer region of the wafer as a function of the material layer types and thicknesses and the temperatures of the wafer and lamps. Time-temperature profiles are stored internally at a number of user defined locations on the wafer. Because of the different length scales involved in the physical processes (O(10 mm) for thermal diffusion and O(1 μm) for dopant diffusion) the thermal processing effects of oxidation and diffusion are not simulated simultaneously with the thermal modelling. Dopant diffusion and oxidation are subsequently simulated over a typical device region using the stored time-temperature profiles to calculate temperature dependent coefficients at each stage of the RTP. The approximation of decoupling the thermal simulation from the induced process changes is valid for the high temperature processing considered here in which the growth of thin oxide layers has a negligible effect on the emission and absorption boundary conditions of the wafer.

## 3. RTP Thermal Simulation

Two options are available for thermal modelling within TITAN-RTA: either to simulate the temperature control error associated with incorrect emissivity programming of the pyrometer; or to simulate the temperature non-uniformity across the wafer associated with overlayer patterning and edge proximity assuming that the temperature is correctly controlled at the location monitored by the pyrometer. An example of temperature non-uniformity simulation is now given. Figure 1 shows the simulated silicon wafer and dominant overlayer pattern. Three types of region are modelled in this example and the regions are placed on the wafer as shown, occupying areas of width 5 - 10 mm. A simple thermal anneal cycle involving a ramp-up and down to a nominal steady state of 30 seconds at 1100°C is defined. Figure 2 displays the simulated time-temperature profiles at four different locations of the wafer shown in Figure 1, compared to the programmed cycle. Large transient non-uniformities are evident, also the steady state temperatures achieved at the different locations are significantly different. The effects of these non-uniformities on the processing outcome at different regions of the wafer is considered in the next section. TITAN-RTA can be used to minimise the transient temperature non-uniformity by, eg, reducing the ramp-rates. The steady state temperature differences are a function of layer thickness and region size: these can be optimised to minimise temperature difference.

## 4. RTP Process Simulation

An example of dopant out-diffusion from polysilicon is given for the RTP thermal conditions simulated above. Figure 3 shows the diffused arsenic profiles from a polysilicon emitter receiving each of the four predicted thermal treatments compared to the standard isothermal anneal. The effect on junction depth and sheet resistance is summarised in Table 1.

| Location | T °C | Xj μm | R Ω/□ | ΔT °C | ΔXj % |
|---|---|---|---|---|---|
| Nominal | 1100 | 0.112 | 167 | - | - |
| LOC 1 | 1107 | 0.117 | 154 | +7 | +4 |
| LOC 2 | 1083 | 0.088 | 239 | -17 | -21 |
| LOC 3 | 1123 | 0.139 | 118 | +23 | +24 |
| LOC 4 | 1125 | 0.143 | 115 | +25 | +28 |

Table 1
Simulated temperature, junction depth and sheet resistance for the polysilicon out-diffused arsenic shown in Figure 3 for the thermal conditions of Figure 2.

It can be seen that sensitivity to the location on the wafer is large in this case, with up to 28% variation in the junction depth, implying that modifications to the overlayer patterning or to the ramp conditions are necessary.

## 5. Conclusions

The example discussed above demonstrates the value of TITAN-RTA as a combined equipment and process simulator. It is shown that real effects occurring during RTP lead to significant variations in the processing outcome. TITAN-RTA is able to assist in process control and in the interpretation of physical effects occurring during RTP. It is believed that such simulators will play an increasingly important role in processing technology.

This work was partly funded under the European ESPRIT-2197 Project "STORM". The authors acknowledge the support of the Plessey and CNET Companies.

Gerodolle A, Corbex C, Poncet A, Pedron T, Martin S in Software Tools for Process, Device and Circuit Modelling (W Crans, ed) pp 56-67 Boole Press (1989)

Hill C, Jones S, Boys D in Reduced Thermal Processing for ULSI (R.A. Levy, ed) pp 143-180, Plenum Press New York (1989)

# Reflectivity measurements in a rapid thermal processor: application to silicide formation and solid phase regrowth

Dilhac J-M, Nolhier N, Ganibal C

Laboratoire d'Automatique et d'Analyse des Systèmes du CNRS, 7 avenue du colonel Roche, 31077 Toulouse CEDEX, FRANCE.

Abstract : Time resolved reflectivity is first applied to the measurement of the solid phase epitaxial growth rate of $As^+$ (60 keV, $4.10^{15} cm^{-2}$) implanted (100) Si wafers and then to the study of platinum silicide formation when samples of platinum films deposited on top of silicon wafers are annealed in a rapid thermal processor. The thermal cycles consist of a fast heating phase followed by an isothermal plateau. For both experiments, activation energies are calculated and in situ reflectivity monitoring is shown to be a powerful end point detection means in a rapid thermal processor.

## 1. Introduction

In situ process control is often cited as an advantage of single wafer processing (Hauser 1989), and hence of rapid thermal processing because real-time adaptive control can be performed if real-time process parameters are monitored (Saraswat 1989). For rapid thermal processing, the most commonly used parameter is wafer temperature which is measured either directly by thermocouples or by optical pyrometers. Unfortunately, these two sensors suffer such drawbacks as poor compatibility of the thermocouple materials with the processing environment, and the effect on optical pyrometer of unplanned wafer emissivity variations or parasitic light emitted by the heating lamps.

On the other hand, time resolved optical reflectivity is a noninvasive in situ measurement technique, which has already been used for direct monitoring of the real-time behaviour of silicon solid phase epitaxy (Olson 1988) and platinum silicide formation (Pan 1984). However, it had never been used in a rapid thermal processsor. These considerations prompted us to study the above phenomena using time resolved reflectivity in a rapid thermal processor.

## 2. Experimental methods

In our experiments the time resolved reflectivity technique utilizes either time-dependent optical interferences induced by change in thickness of a thin amorphous Si layer on top of a crystalline silicon substrate, or more simply, reflectivity changes associated with silicide formations. In the first case, if the interface and the surface are smooth compared to the light wavelength, optical interferences arise from the refractive index discontinuity between the a-Si thin film layer and the c-Si substrate. Interference conditions are as follow (see Fig.1) :

$$n_a t_1 \cos \theta_t = (m + \frac{1}{2}) \frac{\lambda}{2} \qquad (1)$$

$$n_a t_2 \cos \theta_t = m \frac{\lambda}{2} \qquad (2)$$

with $\qquad n_0 \sin \theta = n_a \sin \theta_t \qquad (3)$

where t is the thickness of the amorphous layer, $n_a$ and $n_0$ are the a-Si and nitrogen refractive indices respectively, $\theta$ is the incident angle, $\theta_t$ is the refraction angle in amorphous Si, $\lambda$ is the wavelength, and m is a positive integer. During solid phase epitaxy, the optical interferences induce oscillations of the reflected beam intensity. From equations 1 and 2, the thin film thickness variation $\Delta t$ corresponding to two consecutive reflectivity extrema (maximum and minimum) can be computed as :

$$\Delta t = t_1 - t_2 = \frac{\lambda}{4 n_a \cos \theta_t} \qquad (4)$$

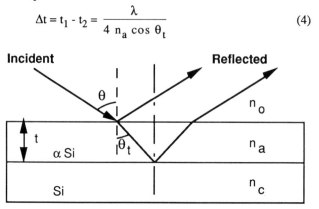

Fig. 1 Thin film interferences.

Since the optical constants of amorphous Si are temperature dependent, the use of equation (4) requires knowledge of the a-Si index of refraction $n_a$. It has been shown in the literature (Olson 1988) that at $\lambda = 0.6328$ µm, and over the 550°C to 750°C range, $n_a$ is given by :

$$n_a = 4.39 + 5.10^{-4} \, T \qquad (5)$$

with T in degrees Kelvin. This temperature dependence is the same as that of crystalline Si :

$$n_c = 3.72 + 4.88.10^{-4} T \qquad (6)$$

For nitrogen, we will take $n_0 = 1$. Therefore, at the nitrogen / amorphous silicon interface, the directly reflected laser beam undergoes a phase shift of $\pi$ because $n_a > n_0$. At the amorphous / crystalline interface, as $n_c < n_a$ (see eq. 5 and 6), no phase shift takes place. During solid phase epitaxy (SPE), the wafer reflectivity at a given wavelength consequently oscillates between extrema, and the last extremum to appear ($n_a t \cos \theta_t = \lambda/4$) when the amorphous crystalline interface reaches the near surface region, is therefore a reflectivity maximum. As consecutive extrema correspond to a known change in the amorphous layer thickness, it is therefore possible to assess the rate of interface movement. Unfortunately, during annealing, actual reflectivity may differ from the theoretical behaviour previously depicted. For example, random crystallization and formation of interface roughness during SPE reduce the interference contrast. The extrema amplitudes are also reduced for thick amorphous films because absorption in the film exponentially attenuates the beam reflected by the amorphous / crystalline interface.

With respect to platinum silicide formation, the fact that at $\lambda = 0.6328$ µm, platinum exhibits an absorption coefficient $\alpha$ of 190 µm$^{-1}$, and a refractive index of 2.3 (Palik 1985) has to be taken into account. Films thicker than $1/\alpha = 53$ Å will therefore exhibit the same reflectivity as an infinitely thick metallic sample, and films less than 53 Å will not produce interferences as their thickness will be much smaller than the value calculated from equation 4 (700 Å with $\theta = 30°$). Reflectivity will therefore be affected only when the initial metallic or silicide film has fully reacted to form a new phase.

The rapid thermal processor and the laser optical arrangement used in our experiments have already been detailed elsewhere (Dilhac 1989). Briefly, the rapid thermal processor consists of two rows of tungsten-halogen quartz lamps placed above and below a quartz processing chamber. In all the experiments reported below, the ambient is nitrogen, and the thermal cycles consist of a fast heating phase (between 100 and 200 °C/s) followed by an isothermal plateau ranging from 520 to 624 °C for SPE, and from 410 to 600 °C for silicide formation. The amorphous Si thin films used in our experiments were formed by ion implantation ($As^+$ ions, 60 keV, $4.10^{15}$ $cm^{-2}$) into (100) Si wafers. Amorphous layers approximately 900 Å thick were formed (Kirillov 1985). After implantation, the wafers were cleaved into small pieces which were then subjected to rapid thermal annealing. For silicide formation study, platinum films with a thickness of 120 nm were sputter deposited on (100) Si wafers (Dilhac 1989).

3. Results and discussion

We have first recorded the reflectivity variations in the case of SPE. A typical thermal cycle, together with the corresponding laser reflected light intensity as a function of the annealing time, is shown in Figure 2. The shape of the reflectivity oscillations is similar to the theoretical predictions : two extrema have been detected, the last one being a maximum. The reflected beam intensity then falls to a constant value corresponding to crystalline silicon reflectivity. It has been shown in the literature (Olson 1988) that, in intrinsic films and in those containing low impurity concentrations (less than 0.1 at.%), SPE is a thermally activated process characterized by an activation energy of 2.7 eV, and a pre-exponential factor of $3.1 \times 10^8$ cm/s over a large temperature range (470 - 1350 °C). However, for $As^+$ implanted a-Si with dopant impurity concentration greater than $10^{19}$ $cm^{-3}$, an intrinsic value rate enhancement is observed in the early stages of SPE, together with rate-retarding effects at a later stage, this enhancing/retarding effect being temperature dependent. It is therefore difficult to assign an accurate value to the regrowth rate of our samples, since the As concentration and hence the growth rate vary strongly with depth. Nevertheless, we have recorded the time intervals between the two measured extrema vs. annealing temperature. These time intervals are shown in an Arrhenius plot in Figure 3 : the measured points fall on a line.

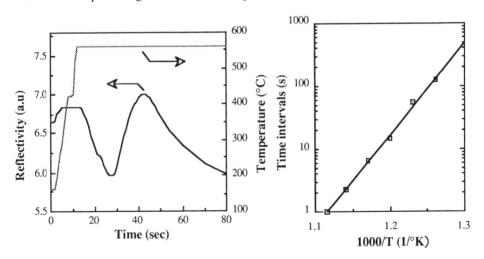

Fig. 2 Temperature and reflected light intensity as a function of time. Equation 4 gives $\Delta t = 327$ Å at 660°C ($\theta = 30°$).

Fig. 3 Arrhenius plot of the time intervals between extrema vs. annealing temperature. The activation energy is 2.8 eV.

The activation energy associated with these data is 2.8 eV. This energy is very close to the usual 2.7 eV. Rutherford backscattering investigations of the regrowth kinetics of $As^+$ implanted Si amorphous layer during RTP (Grob 1987) has also revealed a similar activation energy.

For platinum silicide formation, a typical thermal cycle together with the corresponding reflected light intensity are shown in Figure 4. A two-stage reflectivity drop is observed : these two drops are related to the formation of, first a $Pt_2Si$ and then a PtSi film (Dilhac 1989). It has also been shown (Pan 1984) that the time required to complete the phase formations corresponds to that taken by the reflectivity to drop by 50% during both the first and the second stage reflectivity changes. These silicide formation times are shown in Figure 5. As can be seen, there is a good correlation between the Arrhenius laws defined at low temperatures (Pan 1984) and our results.

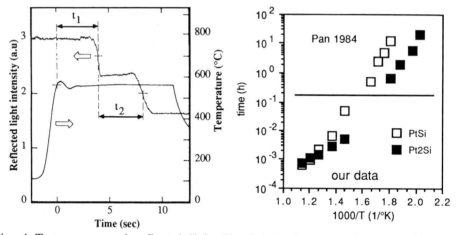

Fig. 4 Temperature and reflected light intensity during platinum silicide formation. $t_1$ and $t_2$ refers to $Pt_2Si$ and PtSi formations.

Fig. 5 Arrhenius plots of $t_1$ and $t_2$. Upper data are from Pan 1984, lower data are from our experiments.

4. Conclusion

We have shown that solid state reactions can be monitored in situ using reflectivity as a real-time process parameter. Time resolved reflectivity may therefore be used for end point detection in a rapid thermal processor, in addition to temperature measurement, and could automatically compensates for process variations occurring before or during thermal treatment.

References

Dilhac JM, Ganibal C, Castan T 1989, *Appl.Phys.Lett* 55(21), pp 2225-2226
Hauser J R, Masnari N A and Littlejohn M A 1989, *Materials Research Society Proceedings*, vol.146 ed Hodul D & all, pp 15-26
Olson G L and Roth J A 1988, *Materials Science Reports* 3, (Amsterdam : North-Holland) pp 1-78
Palik E D 1985, *Handbook of Optical Constants of Solids*, (Academic Press)
Pan J T and Blech I A 1984, *Thin Solid Films* 113, pp129-134
Saraswat K C, Len Booth, Grossman D D, Khuri-Yakub B T, Young Jin Lee, Moslehi M M and Wood S 1989, *Proc. SPIE 1189* ed Rajendra Singh, pp 2-14
Kirillov D, Powell R A and Hodul D T, *Materials Research Society Proceedings*, vol 52, pp 131-138

*Paper presented at ESSDERC 90, Nottingham, September 1990*
*Session 2A3*

## Temperature non-uniformities in a silicon test square during rapid thermal processing

A G O'Neill[1], D Boys[2], S K Jones[2] and C Hill[2]

[1] Department of Electrical and Electronic Engineering, University of Newcastle upon Tyne, NE1 7RU, UK

[2] Plessey Research Caswell Ltd, Towcester, Northants, UK

> Abstract. Rapid thermal processing of a small process square placed on a susceptor wafer of controlled emissivity is investigated. Under these conditions the process square is hotter than the susceptor wafer. Finite element modelling is compared with experimental data and used to predict the subsequently observed changes in process square temperature as its size and the condition of the susceptor wafer are changed.

1. Introduction

Rapid thermal processing (RTP) is now well established as a silicon processing technique in, for example, implant annealing, silicide formation and out-diffusion of dopants. The high activation energy of many of these technologically important physical processes makes very accurate temperature control and measurement important for both good process control and scientific investigation of the physical processes themselves. Because RTP machines are not black body cavities, and direct contact temperature measurement is usually precluded by contamination effects, sufficiently good temperature control and uniformity is only obtainable using susceptor wafers of controlled emissivity to support the process square (Hill et al 1989). It has been found that the temperature of process samples is higher than that of the susceptor wafer if the former are smaller than the latter. This situation commonly arises when part-wafers are processed or when small test samples are placed on whole wafers to monitor temperature.

In this paper the processing of silicon test squares of side 25mm and 10mm supported on a 125mm silicon susceptor is considered. Temperature distributions across the test square are compared with simulations using the finite element method. The extent of test square overheating as the susceptor wafer surface is changed, and as the test square side length is reduced, are investigated.

2. Experimental

Square samples of silicon, 625 microns thick, were cleaved from VLSI quality 125mm silicon single crystal CZ wafers of resistivity 7 ohm-cm n-type. The crystal orientation of the wafer surface was (100), cleavage directions were [110]. As supplied, the wafers had one face highly polished, the other ground and etched. Susceptor wafers, on

© 1990 IOP Publishing Ltd

which the test squares were supported during RTP, were from the same box of wafers.

RTP was carried out in an AG610 annealer, in an ambient of oxygen at a set temperature of 1100°C, the emissivity value in the pyrometer feedback loop being set to that appropriate for the etched wafer surface. The steady state temperature distributions across the test wafers were measured by ellipsometric determination of the thermally grown oxide thickness and comparison with previously established calibration graphs.

## 3. Theory

Typical processing temperatures range from 600°C to 1200°C so that heat loss due to black body radiation is significant. The system has been modelled by solving the thermal diffusion equation in two dimensions using the finite element method (O'Neill 1989):

$$\nabla \kappa \nabla T = (\alpha I - \epsilon \sigma (T^4 - T_0^4)) \, 2/z$$

where $\kappa$ is conductivity, $I$ is the intensity of incident lamp radiation, $\alpha$ and $\epsilon$ are the absorption and emissivity coefficients respectively, $\sigma$ is the Stefan-Boltzmann constant and $z$ is the thickness. Incomplete Cholesky decomposition with conjugate gradients (ICCG) is used to solve the set of equations resulting from the application of the finite element method to the original partial differential equation. The strongly non-linear source term is dealt with using Newton iteration. It is the ratio of $\alpha/\epsilon$, rather than $\alpha$ and $\epsilon$ independently, which influences the source term in this equation. Changes in the magnitude of $\alpha$ (or $\epsilon$) would be the same as introducing a small change in the conductivity.

The silicon thickness is such that variations in temperature between top and bottom surface are negligible and so a two-dimensional plan view of the susceptor wafer and test square is modelled. The test square is then a region of double thickness surrounded by a thin layer, of width equal to the silicon thickness, which represents the edge. The temperature variation across the susceptor and test square is small enough that a fixed value of thermal conductivity of 0.25 $Wcm^{-1}K^{-1}$ can be taken. Assuming a thermodynamic equilibrium, the absorption and emissivity coefficients for silicon can be determined from the refractive index, $n$, to be

$$\alpha = \epsilon = 1 - R = 1 - (n-1)^2/(n+1)^2 = 0.7$$

for the susceptor and test square, where $R$ is the reflectance. A fit with a set of experimental data is achieved by varying $\alpha$ and $\epsilon$ for the test square edge. The additional components of reflected lamp radiation (3000°C) and black body susceptor wafer radiation (1100°C) incident at the wafer edge are what give rise to the different values of $\alpha$ and $\epsilon$ here.

## 4. Results

Using $\alpha$ and $\epsilon$ as adjustable parameters for the test square edge, good agreement between theory and experiment can be obtained. Figure 1 shows a comparison of the simulated and measured temperature profiles

for a 25mm test square on a susceptor wafer.  Experimental data were
obtained moving away in two directions from the square centre, as shown
in the inset to figure 1.  A simulated temperature contour map for the
susceptor and test square is shown in figure 2.  It demonstrates that
the predicted temperature profiles are similar in both directions.  The
small differences observed in the experimental profiles in figure 1 may
be due to inaccuracy in the oxide measurement and to small asymmetries
in the test square geometry.  The test square temperature is around 2°C
above the susceptor wafer temperature in this case, where the test
square lies on the rough face of the susceptor wafer with the polished
side facing the pyrometer.

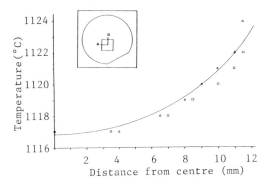
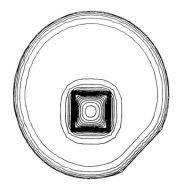

Fig. 1. Comparison of simulated
temperature profile with
experimental results for a 25mm
test square on a susceptor wafer

Fig. 2. Simulated contour plot
of temperature (1°C/division;
minimum temperature 1116°C)
for a 25mm test square

When the susceptor wafer is inverted it is found that the test square
temperature is around 10°C above the susceptor temperature.  Inverting
the susceptor wafer changes the emissivity of the wafer side facing the
pyrometer which in turn will decrease the lamp intensity of the RTP
machine.  A second effect is to alter the reflection and radiative
emission properties of the top surface which gives rise to a larger
secondary heating contribution to the test square through its edge.  It
is believed that this latter effect gives rise to the observed increase
in test square overheating.  Indeed, numerical integration shows that
an average of 4% of the radiation emitted per unit area of the polished
face of the silicon susceptor is intercepted by unit area of the test
square edge.  It is less easy to predict the lamp radiation which is
directly reflected into the test square edge, since this is dependent
on the exact chamber and reflector geometry.

The simulation of these two cases gives two sets of fitted test square
edge parameters: rough face up, ratio $\alpha/\epsilon$ = 1.34; polished face up,
ratio $\alpha/\epsilon$ = 1.5.  In both cases the ratio is greater than unity which
indicates an increase in the absorption and/or a decrease in
emissivity.  The increased absorption coefficient is expected because
of the additional lamp radiation reflected from the susceptor surface.
The re-absorption of black body radiation from the susceptor can be re-

interpreted as a decrease in the edge emissivity. The increased $\alpha/\epsilon$ ratio, for the case when the polished side of the susceptor is facing upwards, is consistent with the expected higher reflectivity of the polished surface.

When the test square is reduced in size, the model predicts that the degree of overheating will increase. The reason for this is that the test square edge, using a ratio of $\alpha/\epsilon$ greater than one, serves as a heat source which increases as the edge to volume ratio of the test square increases, i.e. as its dimensions decrease. Experiments using a test square of side 10mm reveal an overheating of 23°C when the susceptor polished face is up and 20°C when the rough side is up, as measured at the centre of the test square.

## 5. Summary

RTP has been investigated where small process test squares are placed on a supporting susceptor wafer. It is observed that the temperature of the process square is above that of the susceptor wafer, and so differs from the target temperature of the RTP machine. A simulation which solves the thermal diffusion equation in two dimensions using finite elements is able to give good agreement with experimental data and predict trends in behaviour if an adjustable parameter of $\alpha/\epsilon$ is used for the process square edge. To investigate the temperature control further, it will be necessary to introduce other changes to the system. For example, the addition of a thin oxide layer along one edge of the test square will serve to further increase the absorption coefficient of the edge surface, due to multiple reflection effects and because the oxide refractive index is intermediate between that of the ambient and that of silicon. In this way it is hoped to gain a better understanding and so control of the overheating mechanism.

Acknowledgements. This work was partly funded under the European ESPRIT-2197 project "STORM". The authors would acknowledge the support of the Plessey Company.

Hill C, Jones S and Boys D 1989 Reduced Thermal Processing
    for ULSI (R A Levy, Ed.) (New York: Plenum) pp 143-180
O'Neill A G 1989 Electron. Lett. 25, 1484

*Paper presented at ESSDERC 90, Nottingham, September 1990*
*Session 2A4*

# Prevention of boron penetration from $p^+$ poly gate by RTN produced thin gate oxide

T. Morimoto, H. S. Momose, K. Yamabe, and H. Iwai

ULSI Research Center, Toshiba Corporation
1, Komukai-Toshiba-cho, Saiwai-ku, Kawasaki, 210, Japan

## ABSTRACT

The effect of a nitrided oxide gate film produced by RTN on the boron penetration from a $p^+$ poly gate electrode was studied. It was found that the RTN oxide gate is very effective for the suppression of boron penetration, even when the nitrogen concentration is as low as a few percent.

## INTRODUCTION

Dual- (or symmetric-) gate CMOS structures have been developed mainly for application in lower-submicron CMOS-ULSIs. However, several problems have still to be overcome before the dual-gate structure can be introduced. The biggest is probably the effect of boron penetration from the $p^+$ poly Si gate into the substrate. This paper reports a study into a technique for suppressing boron penetration using a nitrided gate oxide produced by RTN (Rapid Thermal Nitridation). It was found that a nitrogen content of only a few percent in the gate oxide is sufficient to completely stop boron penetration.

## EXPERIMENTS AND RESULTS

$P^+$ poly gate p-MOSFETs with nitrided oxide gates and pure oxide gates were fabricated. In all cases, gate oxide was 6-7nm thick, a thickness expected to be used in actual 0.3μm devices. The process conditions are listed in Table 1. A nitrogen profile of the gate oxide, measured by AES, is shown in Fig.1.

Figure 2 shows the I-V characteristics of long-channel (10μm) p-MOSFETs with nitrided oxide gates and pure oxide gates. In this case, the process temperature after gate oxide formation was 850°C. In the pure gate oxide sample, boron in the poly-Si gate diffused into the substrate, and this resulted in punch-through I-V characteristics. In the nitrided oxide case, normal I-V characteristics were obtained, and boron diffusion from the poly-Si gate was suppressed. It should be noted that in these experiments, $BF_2$ was implanted instead of B as the gate poly-Si dopant. With $BF_2$, boron diffusion or penetration from the gate is very bad (Baker et al 1989, Sung et al 1989) compared with the B case. Table 2 summarizes the boron penetration for different process conditions. The nitrided oxide gate is very resistant to boron penetration, even when the peak nitrogen concentration is as low as 4 percent.

Figures 3 and 4 show the dependence of S factor (subthreshold slope) and $p^+$ poly sheet resistance on the $BF_2$ implantation dose. In the pure gate oxide case, only a $1 \times 10^{15}$ cm$^{-2}$ dose was necessary to suppress boron penetration (Fig.3), and this resulted in higher gate poly-resistance (Fig.4). Figure 5 shows the C-V characteristics of MOS diodes. For the pure gate oxide case, the active gate dopant concentration in the poly Si is insufficient. A depletion layer is formed (Chapman et al 1988, Wong et al 1988) in the gate poly-Si and the MOS capacitance becomes small (Fig.5), and hence the MOSFET transconductance becomes small. In the nitrided oxide gate case, even a $5 \times 10^{15}$ cm$^{-2}$ dose is enough to prevent boron penetration, achieving lower gate poly-resistance and higher transconductance. It should be noted that 1000°C RTA (Rapid Thermal Annealing) can be added to the nitrided oxide gate samples without penetration (Fig.3), resulting in further improvement of gate resistance and transconductance.

Figure 6 shows boron profiles measured by SIMS for the pure and nitrided oxide gate structures. It was confirmed that, in the case of pure gate oxide, high concentration boron penetrates into the Si substrate. Figure 7 shows hydrogen, fluorine, and nitrogen profiles. Regarding hydrogen and fluorine profiles, there were very few differences between the pure oxide gate and nitrided oxide gate samples. This suggests that nitrogen itself suppresses the boron penetration even when its concentration is as low as a few percent.

## CONCLUSION

A nitrided oxide gate produced by RTN was found to be very effective for the suppression of boron penetration, even when the nitrogen concentration is as low as a few percent. This RTN oxide gate will be a key technology in lower submicron dual-gate CMOS ULSIs.

## REFERENCES

Baker F K, Pfiester J R, Mele T C, Tseng H-H, Tobin P J, Hayden J D, Gunderson C D, and Parrillo L C  1989 *IEDM Tech. Dig.* pp 443-446

Sung J M, Lu C Y, Chen M L, and Hillenius S J  1989 *IEDM Tech. Dig.* pp 447-450

Chapman R A, Wei C C, Bell D A, Aur S, Brown G A, and Haken R A  1988 *IEDM Tech. Dig.* pp52-55

Wong C Y, Sun J Y-C, Taur J, Oh C S, Angelucci R, and Davari B  1988 *IEDM Tech. Dig.* pp238-241

Table 1. Sample fabrication conditions

| SAMPLE | PROCESS | | INTERFACE N conc (A.C%) | Tox nm |
|---|---|---|---|---|
| "PO" | FURNACE OXIDATION | | 0 | 6 |
| "NO" | FURNACE OXIDATION | →RTN→RTO | 4~10 | 6~7 |

Table 2. Dependence of boron penetration on process temperature and boron dose. In pure gate oxide case, only $1 \times 10^{15}$ / cm² boron dose is allowed.

| PROCESS | 850°C FURNACE | | 1000°C RTP | |
|---|---|---|---|---|
| BF₂ DOSE | PO | NO | PO | NO |
| $1 \times 10^{15}$/cm² | NO | NO | | |
| $3 \times 10^{15}$/cm² | YES | NO | | |
| $5 \times 10^{15}$/cm² | YES | NO | WEAK | NO |

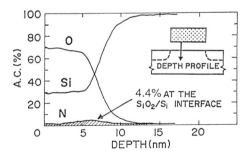

Fig 1. Si, O, N profiles measured by AES for NO sample.

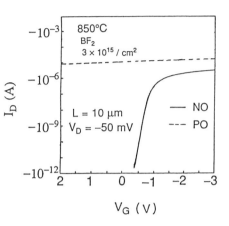

Fig 2. $I_d$ - $V_g$ characteristics of boron-penetrated and non-boron-penetrated p-MOSFETs.

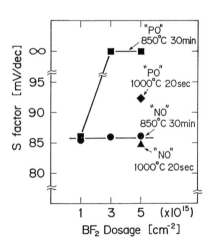

Fig 3. Dependence of S factor (or subthreshold slope) on boron dose and process temperature.

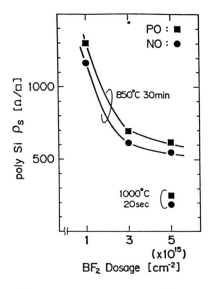

Fig 4. Dependence of p⁺ gate poly-Si sheet resistance on boron dose and process temperature.

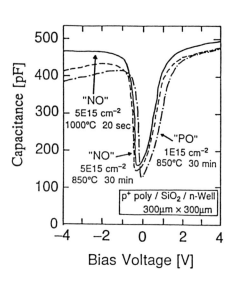

Fig 5. C-V characteristics of MOS diodes

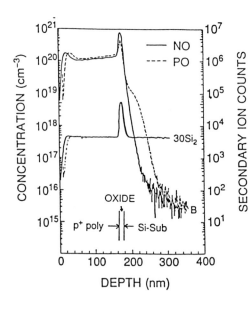

Fig 6. B profiles measured by SIMS for PO and NO samples. Process temperature is 850°C. Boron dose is $5 \times 10^{15}$ cm$^{-2}$.

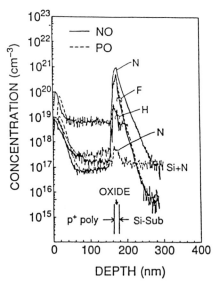

Fig 7. H, F, and N profiles measured by SIMS for PO and NO samples. Process temperature is 850°C. Boron dose is $5 \times 10^{15}$ cm$^{-2}$.

# Formation of $(Ti_xW_{1-x})Si_2/(Ti_xW_{1-x})N$ contacts by rapid thermal silicidation

H. Norström, K. Maex, J. Vanhellemont, G. Brijs, W. Vandervorst and U. Smith

*Interuniversity Microelectronics Center (IMEC v.z.w.), Kapeldreef 75, B-3030 Leuven, Belgium.*

*Royal Institute of Technology, S-16428 Kista, Sweden.*

## Abstract

The formation of $(Ti_xW_{1-x})Si_2/(Ti_xW_{1-x})N$ by rapid thermnal processing of $Ti_xW_{1-x}$ on Si in an $N_2$ ambient is investigated. A distinct snowploughing of As atoms is observed during silicide formation. The diffusion barrier properties of the $(Ti_xW_{1-x})Si_2/(Ti_xW_{1-x})N$ stack in contact with Al is investigated upon post-metal annealing.

## Introduction

Sputter deposited TiW has been widely used as a diffusion barrier between aluminium interconnection layers and silicided contacts in order to prevent spiking (1,2). More recently, TiW has also been applied as a direct contact material to shallow $n^+$ and $p^+$ areas (3-6) to avoid detrimental effects of high contact resistances caused by Si precipitation in narrow contact windows as well as increased junction leakage resulting from spiking.
The main disadvantage of TiW as a direct contact material is the frequently observed high contact resistance values to $p^+$ areas (5,6). The addition of nitrogen to the sputtering gas during deposition of the TiW layer further increases the problem of high contact resistance (5).
In an attempt to improve the quality of TiW contacts we have investigated the possibility of forming reacted contacts to silicon and simultaneously improving the barrier properties of the TiW film via rapid thermal annealing in nitrogen (7)

## Experimental

Si-wafers, of (100)-orientation, with a resistivity between 20 tot 40 ohm-cm were used as a starting material.
Boron doped bulk wafers were implanted with As to a dose of $5.E15/cm^2$ at 50 KeV. The phosphorus doped wafers were implanted with $BF_2$ at the same energy and with the same dose.
The As-implanted wafers were annealed in $N_2$ at 1000C for half an hour whereas the $BF_2$ implanted wafers were annealed at 925C for 40 minutes. Sputtering of TiW was done from a target containing 30 wt.% Ti. The as-deposited films were found, by means of RBS, to contain approximately 56 at.% Ti.
Rapid thermal processing (RTP) was carried out in an Heatpulse 610 flushed with $N_2$. The temperature of interest (700-900C) was monitored by a pyrometer. The processing temperature was fixed at 30 sec.
The composition of as-deposited films and subsequently reacted layers were determined by RBS.
Phase identification of the layers was made by means of X-ray (XRD) and electron diffraction (ED).
The thickness and morphology of the reacted layers were investigated by cross-sectional scanning and transmission electron microscopy (X-SEM and X-TEM).

The behaviour of the implanted dopants during the reaction of TiW and Si was investigated by RBS, SIMS and spreading resistance profiling (SRP).

© 1990 IOP Publishing Ltd

The thermal stability during post-metal annealing of the bi-layers, consisting of RTP treated TiW-films covered by Al, was investigated by sheet resistance measurements and RBS.

## Rapid thermal silicidation of TiW-layers on Si

$Ti_{.56}W_{.44}$ and $Ti_{.24}W_{.76}$ layers were RTP treated at temperatures between 700 and 900C. All structures were analyzed by RBS prior to and after removal of the unreacted TiW top layer by selective etching. Figure 1 shows RBS spectra of $Ti_{.56}W_{.44}$ coated wafers after RTP at 700C, 800C and 900C. At 700C the reaction has just started as verified by a small shift of the trailing edge of the Ti signal towards lower energies compared to an as deposited sample. At 800C and 900C, part of the TiW layer is converted into a ternary silicide, as evident by the plateau in the Si signal. The ratio of Ti to W was the same as in the as deposited films viz $Ti_{.56}W_{.44}Si_2$. The formation of a ternary silicide was confirmed both by X-ray and electron diffraction. A thin layer of unreacted TiW, which probably contains some nitrogen, is still present on the surface.

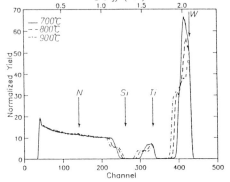

Fig. 1. RBS-spectra from TiW-coated wafers after RTP at 700, 800 and 900°C.

The X-TEM micrographs (fig.2) show that upon annealing at 700C, a silicide layer has formed with a smooth interface to the underlying Si. The silicide layer contains large grains (~ 150 nm laterally) arranged in a columnar structure with the grain boudaries perpendicular to the substrate surface. It is interesting to note , however, that two silicide layers appear upon silicidation of the TiW at 800C. Based on the results obtained by RBS, XRD and ED, it appears that the two silicide layers consist of the same phase and differ only with respect to grain size. No seperation of Ti and W has taken place in agreement with previous findings (8,9).

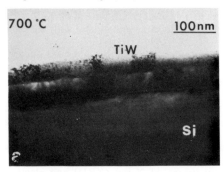
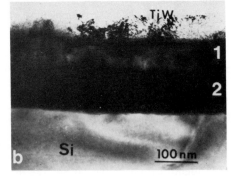

Fig. 2. a) Cross-sectional TEM micrograph showing the structure after the 700C anneal. In between the remaining TiW layer and the silicon substrate a thin silicide layer is formed.
b) The structure obtained after the 800C anneal. The silicide layer now consists of two sublayers (1,2) with different morphology.

The resistivity of the $Ti_{.56}W_{.44}Si_2$ layer was found to be 230 μOhm-cm. Films of $Ti_{.25}W_{.75}$ were similarly found to produce a ternary silicide, $Ti_{.25}W_{.75}Si_2$ upon silicidation.

## Movement of dopants during silicidation

Redistribution of the implanted arsenic during silicidation of the TiW layers was investigated by RBS. The RBS spectra recorded after silicidation at various temperatures are presented in fig.3. The normalized As signal at the silicide/Si interface, which is directly proportional to the As concentration, is clearly seen to increase for higher temperatures. Thus, part of the As, originally present in the mono-Si consumed during silicidation, has been pushed ahead of the reaction front. SRP measurements confirm a small but significant increase in the carrier concentration at the interface. The interface carrier concentration is highest for the samples annealed at 750C in agreement with the RBS data.
The redistribution of B during silicidation was also studied by SIMS and SRP. The B interface concentration is observed to decrease for increasing silicidation temperatures.

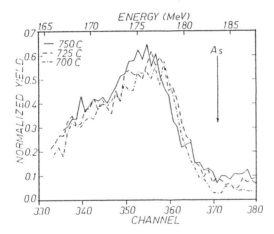

Fig. 3. Redistribution after $Ti_{.56}W_{.44}Si_2$ formation of As implanted in Si (5.E15/cm$^2$, 50KeV) as measured with RBS (2MeV He, Θ=170). Parameter is the RTP-temperature for silicide formation. The arrows indicate the position of As on the silicide surface.

## TiW films as a diffusion barrier

The thermal stability of Al coated $(Ti_xW_{1-x})Si_2/(Ti_xW_{1-x})N$ layers were compared to unsilicided TiW layers. To facilitate the interpretation of RBS spectra the Al thickness was kept below 250nm. RBS results for TiW structures silicided at 700C prior to Al deposition are shown in fig.4. The spectra for an unannealed sample (not shown) and a specimen annealed at 475C for 60 minutes are identical, indicating that no reaction has taken place. Annealing at 500C leads to a reaction between Al and the silicided TiW, as is evident from the shift of the leading edge of the W signal towards lower energies. For comparison the RBS spectra for samples with unsilicided TiW are shown in fig. 5. It is readily seen that an interaction between TiW and Al occurs already at 475C. The improved thermal stability of TiW layers silicided in an $N_2$ ambient is likely to have its origin in nitrogen blocking of the grain boudaries.

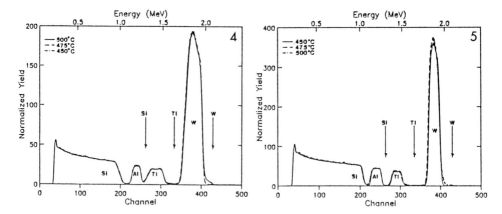

Fig. 4. RBS-spectra (2MeV He, Θ=170) of Al coated TiW-layers which had been RTP-treated at 700C and 725C for 30 sec. as a function of the post-metallization anneal temperature. The arrows indicate the channel NR. for the element present at the surface. The tail at the high energy side of the W-peak for an anneal of 500C are indicative of a Al-W-interreaction.

Fig. 5. RBS-spectra (2MeV He, Θ=170) of Al covered TiW layers as a function of annealing temperature. The arrows indicate the channel NR. for the element present at the surface. The tails on the high energy side of the W-peak for the anneals of 475C and 500C, is indicative of an Al-W-reaction.

## Conclusion

Rapid thermal silicidation of $Ti_xW_{1-x}$ has been evaluated for use as a contact material on $N^+$ and $P^+$ Si and as a diffusion barrier for Al. A smooth silicide interface was obtained. Snow ploughing of As occurs during silicidation. No increase in the reverse leakage current of diodes contacted via silicidation of the TiW was observed.
Due to a decoration of the grain boundaries with nitrogen, an improved thermal stability in contact with Al is obtained.

## References

1. J.M. Harris, S.S. Lau, M.-A. Nicolet and R.S. Nowicki, J. Electrochem. Soc. 123, pp. 120, 1976.
2. P.B. Ghate, J.C. Blair, C.R. Fuller and G.E. McGuire, Thin Solid Films, 53, pp. 117, 1978.
3. S.S. Cohen, M.J. Kim, B. Gorowitz, R. Saia and T.F. McNelly, Appl.Phys. Lett. 45, pp. 414, 1984.
4. M.J. Kim, D.M. Brown, S.S. Cohen, P. Picante and B. Gorowitz, IEEE Trans. Electron Dev. ED-32, pp. 1328, 1985.
5. R.A.M. Wolters, and A.J.M. Mellissen, Solid State Technol. 29, pp. 131, 1986.
6. A. Lindberg, M. Ostling, H. Norstrom and U. Wennstrom, Proc. 17th ESSDERC-87, pp. 191, 1987.
7. B.K. Mueller and T.S. Kalkur, Proc. Symp. on Rel. Semicond. Dev., Electrochem. Soc. Proc. 89-6, Eds. H.S. Rathore and G.C. Schwartz, pp. 289, 1989.
8. P. Gas, F.J. Tardy and F.M. d'Heurle, J. Appl. Phys. 60, pp. 193, 1986.
9. S.E. Babcock and K.N.Tu, J. Appl. Phys. 53, pp. 6898, 1982.

# Contact-related effects on junction behaviour

M. L. Polignano and N. Circelli

SGS-Thomson Microelectronics, 20041 Agrate MI, Italy

> The effect of a titanium-titanium nitride (Ti/TiN) barrier metallization on junctions characteristics was studied by comparing diodes of different geometries. An increase of the reverse current of junctions is observed in the presence of contacts when Ti/TiN/Al:Si metallization is used. It has been shown that the reverse current increase is related to contact perimeter, that it is independent of many relevant process steps and it is reduced by thermal treatments after metal deposition.

## 1. Introduction

Junction leakage is a very sensitive tool to evaluate junction quality, it can be affected by random contamination or by process variations and it is very sensitive to the fabrication process. It is well known that specific process steps such as the processes of denuding and precipitation of oxygen [1], the incomplete recristallization of ion-implanted layers [2] and silicide formation on the junction surface [3] can induce increase of junction leakage current. The impact of new metallization technologies on device performances is still under investigation. In this work we present an experimental study of the electrical behaviour of junctions with titanium-titanium nitride (Ti/TiN) metallization barrier. In order to identify the effect of the metallization scheme on junctions characteristics, diodes obtained by different metallization layers (Ti/TiN/Al:Si and standard Al:Si) have been compared. The study has been carried out both on $n^+$-p and $p^+$-n junctions. In order to separate the variuos contribution of junction geometry to current-voltage ($I$-$V$) characteristics, several different structures have been tested.

## 2. Experimental procedure

$n^+$-p and $p^+$-n junctions have been produced by three different processes giving junction depths below the contact in the range $1\mu m - 0.25\mu m$ [4] to test the dependance of the electrical results on junction depth. Gettering of metal impurities has been obtained by heavy phosphorus doping and a moderate temperature annealing at the end of the process. Contacts holes were usually opened by a standard reactive ion etching (RIE), however specific tests were carried out in order to compare RIE and wet etching. Just prior to metallization the samples were chemically etched in order to remove native $SiO_2$. A standard Al:Si(1%) [$0.9\mu m$] metallization has been compared to a Ti[50nm]/TiN[60nm]/Al:Si (1%) [$0.9\mu m$] metallization with TiN deposited by reactive sputtering. Metal patterns were defined by a multistep RIE process.

In order to investigate how contacts affect the electrical behaviour of junctions and to separate the various contributions to current-voltage characteristics, several different structures have been tested (fig. 1): "*area diodes*", whose leakage current is almost completely due to the junction area, "*contact arrays*" consisting of the parallel connection of a large number of small diodes and "*perimeter structures*", consisting of one large perimeter diode with just a few contacts.

The junctions have been tested under reverse bias conditions. The most frequent value of leakage current is assumed as the typical value of the process and a diode is defined to be defective when its leakage current exceeds the typical value by more than one order of magnitude.

## 3. Experimental results

The distribution of reverse current in Al:Si contact arrays (fig. 2) is typical of well-gettered perimeter junctions [5]. With respect to this distribution, in Ti/TiN/Al:Si contact arrays (fig. 2) the typical value of leakage current is increased by about a factor of 4, and the reverse current distribution is significantly wider than in Al:Si contact arrays. The leakage current in perimeter structures is independent of metallization (fig. 3); this fact ensures that the observed variations can be ascribed to the contacts.

By comparing structures with different contact width, it has been found that, to a first approximation, the reverse current excess $\Delta I$ observed in Ti/TiN/Al:Si contact arrays increases in proportion to contact perimeter $p_c$ (fig. 4). No such correlation is observed for contact area.

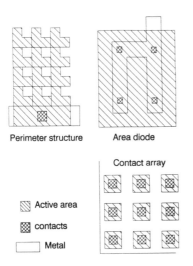
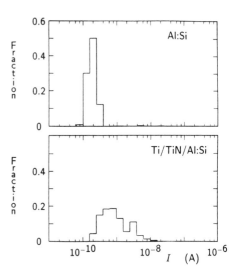

Fig. 1: Schematic representation of test structures.

Fig. 2: Leakage current distribution at 15 V reverse bias and 22°C in $n^+$-p contact arrays.

The percentage of defective junctions is independent of metallization, and can be entirely ascribed to a perimeter defect density (about $10^3$ defects/cm, [5]).

No influence on this phenomenon has been observed due to many relevant process variations, such as doping sign ($n^+$-p and $p^+$-n contact arrays show the same behaviour), junction depth below the contact (in the range 0.25–1$\mu$m), contact etching (whether wet or reactive ion etching), thermal treatments between contact etching and metal deposition (in the temperature range 950°–1050°C), TiN layer thickness (in the range 600 Å – 1200 Å)

The leakage current of contact arrays with Ti/TiN/Al:Si metallization is found to be sensitive to thermal treatments following metal deposition. Fig. 5 shows the reverse characteristics of Ti/TiN/Al:Si contact arrays and of Al:Si contact arrays after subsequent alloy treatments. It is observed that the leakage current excess induced by Ti/TiN contacts is gradually reduced, but not vanished, by these thermal treatments, while for Al:Si contact arrays no significant change was observed.

This annealing behaviour might suggest that the observed increment in the reverse current could be ascribed to surface generation-recombination, due to the presence of unannealed surface states at the Si/SiO$_2$ interface below the Ti/TiN layer. However, this interpretation has been excluded by a test carried out on suitably prepared capacitors. A reverse current increment of $0.5 - 2 \cdot 10^{-10}$ A/cm (per unity junction perimeter, typical values) would correspond to a surface state density of $1 - 5 \cdot 10^{12} \text{cm}^{-2} \text{eV}^{-1}$, which should produce large deviations in the

$C - V$ characteristics too (e.g., the threshold voltage should vary by a few volts). However, no significant change has been observed in the $C - V$ characteristics of polysilicon gate capacitors (150 Å oxide thickness) with the gate region fully covered by a metal plate. This result indicates that the reverse current increment in Ti/TiN contact arrays cannot be explained by a retarded annealing of surface states below the Ti/TiN layer. The mechanism responsible for this reverse current increment must be more complex and really related to contacts.

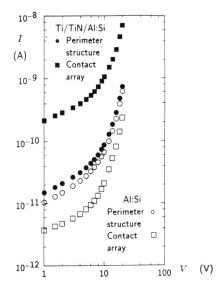
Fig. 3: Reverse characteristics at 22°C of $n^+$-p contact arrays and perimeter structures

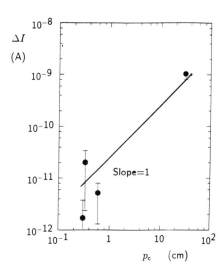
Fig. 4: Increment in the reverse current at 10 V reverse bias and 22°C induced by Ti/TiN metallization vs. contact perimeter.

The annealing behaviour of the reverse current excess in Ti/TiN contact arrays also shows that this current excess cannot be ascribed to starting formation of alloy spikes (which of course could not be annealed out). We have assumed that this contribution to the reverse current can be described in terms of generation-recombination phenomena.
The previous results suggests a few simplifying hypotheses:
• The current in a contact array can be written as the sum of a perimeter term and of a term increasing in proportion to contact perimeter, $I = pJ_p + p_c J_c$, where $J_p$ is the perimeter current density, $p$ is the perimeter of the junction, $J_c$ is a current density per unity contact perimeter.
• Ti/TiN/Al:Si and Al:Si contact arrays are assumed to the same perimeter current density, so that the difference between them is entirely ascribed to contacts.
• We assume $J_c = 0$ for Al:Si contact arrays.
Both $J_p$ and $J_c$ have been assumed to be due to one, single-level generation-recombination center. According to the procedure stated in ref. [6], $J_p$ and $J_c$ have been written as the sum of a diffusion term and of a generation-recombination term. This second term depends exponentially on $\Delta E_p/k_B T$ and $\Delta E_c/k_B T$, where $\Delta E_p = | E_p - E_i |$, $\Delta E_c = | E_c - E_i |$, $E_i$ is the intrinsic Fermi level, and $E_c$ and $E_p$ are the energy levels of the generation-recombination centers responsible for perimeter and contact current, respectively, $k_B$ is Boltzmann constant and $T$ is measurement temperature. A best fit to measured characteristics in the temperature range 22 – 200°C gives the energy differences $\Delta E_p$ and $\Delta E_c$. This analysis has been applied to measurements performed both on $n^+$-p and on $p^+$-n contact arrays. Fig. 6 shows the experimental characteristics and the fitting curves for Ti/TiN/Al:Si and Al:Si $p^+$-n contact arrays under 1 V reverse bias vs. $1/T$.

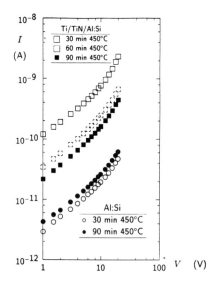

Table 1. Values of some fitting parameters.

|  | $n^+$-p | $p^+$-n |
|---|---|---|
| $\Delta E_p$ (eV) | 0.15 | 0.095 |
| $\Delta E_c$ (eV) | 0.06 | 0.098 |

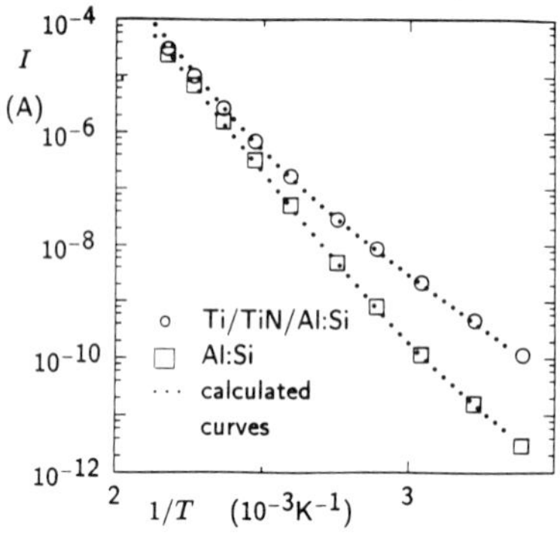

Fig. 5: Reverse characteristics at 22°C of $p^+$-n contact arrays, before and after additional alloy treatments.

Fig. 6: Current at 1 V reverse bias vs. $1/T$ in $p^+$-n contact arrays.

These results suggest a few remarks. Both in $n^+$-p and in $p^+$-n contact arrays $\Delta E_p$ is very close to the value obtained in area diodes [6], indicating that the same mechanisms are responsible for area and perimeter current densities. The contribution induced by Ti/TiN contacts is satisfactorily described by a generation-recombination model. This fact allows us to exclude that this contribution could be identified with a "soft" reverse characteristic: in fact, in such a case the reverse current should be independent or weakly dependent on temperature [7].

## 4. Conclusions

A contribution to junction reverse current induced by Ti/TiN/Al:Si metallization has been observed. This contribution is surely related to the presence of contacts, and it is observed both in $n^+$-p and in $p^+$-n junctions. In addition, the contribution to the reverse current induced by Ti/TiN contacts shows the following features:
- it approximately increases in proportion to contact perimeter, while it is not related to contact area;
- it is not affected by many relevant process variations, such as the technology of contact etching, thermal treatments following contact etching, junction depth variations etc.;
- it can be partially reduced by thermal treatments after metal deposition;
- it can be satisfactorily described in terms of a generation-recombination center induced by Ti/TiN/Al:Si metallization.

## References

[1] J. M. Hwang and D. K. Schroder, J. Appl. Phys. **59**, 2476 (1986)
[2] E. Landi and S. Solmi, Solid St. Electron. **29**, 1181 (1986)
[3] R. Liu, D. S. Williams and W. T. Lynch, J. Appl. Phys. **63**, 1990 (1988)
[4] M. L. Polignano and N. Circelli, J. Appl. Phys., (1990) to be published
[5] M. L. Polignano, G. F. Cerofolini, H. Bender, and C. Claeys, J. Appl. Phys. **64**, 869 (1988)
[6] G. F. Cerofolini and M. L. Polignano, J. Appl. Phys. **64**, 6349 (1988)
[7] H. H. Busta and H. A. Waggener, J. Electrochem. Soc. **124**, 1424 (1977)

Paper presented at ESSDERC 90, Nottingham, September 1990
Session 2A7

# A new improved electrical vernier to measure mask misalignment

D. Morrow†, A.J. Walton, W.R. Gammie, M. Fallon, J.T.M Stevenson, R.J. Holwill

Edinburgh Microfabrication Facility, University of Edinburgh, Edinburgh, EH9 3JL

**Abstract.** A new interconnect scheme is proposed which reduces the pad to tooth ratio for passive electrical verniers. This design is based upon the maximum theoretical number of direct connections between $N$ pads which is $N(N-1)/2$. This concept is developed further and it is demonstrated that the use of diodes can reduce the ratio to $N(N-1) : N$. Some experimental results are also presented.

## 1. Introduction

Figure 1 gives the operating principle of the electrical vernier which works by testing for continuity. Its advantages for measuring misalignment have previously been presented at ESSDERC 89 (Walton 89) and the diode vernier which was proposed in this paper had a tooth to pad ratio of $(N/2)^2 : N$ (where $N$ is the number of pads). By employing an appropriate measurement scheme it has been demonstrated that the diodes can be dispensed with producing a passive device which can be processed in exactly the same manner as analogue misalignment structures (Walton 90). This paper proposes an interconnect scheme that can be used to further improve the pad to tooth ratio of electrical verniers.

## 2. The Reduced Pad Count Passive Vernier

The maximum theoretical number of direct connections between $N$ pads is $N(N-1)/2$ (Kreyszig 1972) and this sets the limit on the pad to tooth ratio for a passive vernier. For example, four pads give 6 direct connections as indicated by the resistors in figure 2. An example of using such an approach in a manner more amenable to IC layout is given in Figure 3. This vernier has 6 pads which results in 15 tests for connectivity (ie a vernier with 15 teeth). In this schematic the resistors represent the interconnect resistance and the method of measurement is highlighted. If connectivity is being tested between pads 1 and 2, our design ensures that there is only one resistor (tooth) that directly connects these two pads. However, current will obviously flow via other parallel paths. This can be prevented by earthing all the pads not involved in the measurement. In this case current flows to earth as illustrated in figure 3 but, because the potential difference between these pads and the sensing ammeter is negligible, no significant current will flow through the ammeter via the parallel resistor paths. Connectivity measurements made on the structure shown in figure 3 gave the nominal value of 100 Ω when there was connection. When two or three teeth were

---

† Now with Digital, South Queensferry, West Lothian, EH30 9SH

not connected to their opposite numbers the open circuit measurement was about 25 kΩ. This degree of discrimination makes the automatic measurement of these designs a simple software exercise.

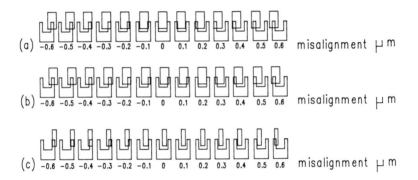

Figure 1. An electrical vernier which compensates for over-etch. (a) No over-etch with zero misalignment. (b) No over-etch with +0.2μm misalignment. (c) 0.3μm over-etch with +0.2μm misalignment.

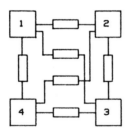

Figure 2. The maximum number of direct connections (6) when four pads are used.

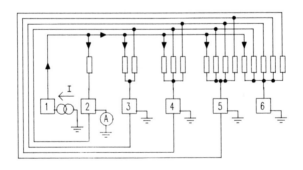

Figure 3. Schematic of the vernier interconnect for the theoretical maximum number of direct connections between pads. (The location of the vernier teeth is at the bottom of each resistor.)

## 3. Diode Vernier

Since it is possible to measure the resistance by forcing current in either direction the number of connectivity tests can be doubled by the use of diodes. This is illustrated in figure 4 where a single resistor has been replaced by two resistors and two diodes. In this case the connectivity of each path may be tested by forcing current in different directions. Figure 5 gives a schematic layout of this improved diode vernier which has 4 pads and tests the connectivity of 12 teeth. It can be observed that this approach doubles the number of teeth for a given pad count. The increase to $N(N-1)$ teeth is at the expense of extra processing which is required for the fabrication of the diodes. A comparison of the pad to tooth ratio of existing verniers and the two new designs presented in this paper are summarised in table 1.

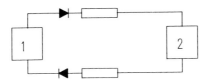

**Figure 4.** The replacement of a bi-directional connection with two uni-directional connections.

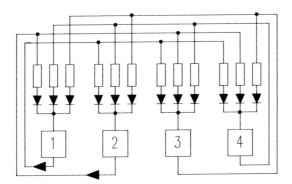

**Figure 5.** Schematic of the diode vernier interconnect for the theoretical maximum number of uni-directional direct connections between pads. (The location of the vernier teeth is at the bottom of each diode.)

## 4. Measurements

Figure 6 gives a comparison of $x$ axis misalignment as measured by a passive $0.1\mu m$ resolution vernier and optical measurement of an adjacent 'box-in-box' structure using a Bio-Rad Nanoquest Quaestor System. It can be observed that there is a good correlation between the two measurements which confirms that the measurement scheme is quite robust in its discrimination between connected and open circuit teeth.

| Number of Pads | Number of Vernier Teeth | | | | |
|---|---|---|---|---|---|
| | Standard Vernier | Old Passive Vernier | Old Diode Vernier | New Passive Vernier | New Diode Vernier |
| 4 | 3 | 4 | 4 | 6 | 12 |
| 5 | 4 | 6 | 6 | 10 | 20 |
| 6 | 5 | 9 | 9 | 15 | 30 |
| 7 | 6 | 12 | 12 | 21 | 42 |
| 8 | 7 | 16 | 16 | 28 | 56 |
| 9 | 8 | 20 | 20 | 36 | 72 |
| 10 | 9 | 25 | 25 | 45 | 90 |

Table 1. Comparison of the number of vernier teeth which can be checked for a set number of pads for different designs. (The standard design is one side of all vernier teeth connected to a single pad with the other sides all connected to individual pads)

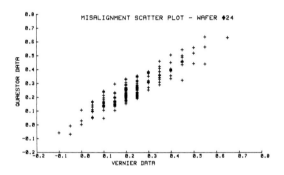

Figure 6. A comparison of the misalignment given by the passive vernier compared with that from an optical method.

## 5. Conclusions

In conclusion a new interconnect scheme has been presented which significantly reduces the number of pads required on electrical verniers to measure mask misalignment. As a result it becomes feasible to design an $x$ and $y$ axis vernier with diodes (8 pads) or passive vernier (10 pads) with a similar number of pads to the potentiometer test structure (10 pads) which is widely employed for measuring misregistration (Dikeman 1988). The vernier designs which have been presented give all the many advantages of a digital measurement without the overhead of a high pad count.

## References

Dikeman J M Roenker K P 1988 IEEE Trans. Electron Devices **35** 2419
Kreyszig E 1972 Advanced Engineering Mathematics (New York: Wiley) pp 719
Walton A J Ward D Robertson J M 1989 Proc. ESSDERC 89 ed A Heuberger H Ryssel P. Lange (Berlin: Springer-Verlag) pp 950-53
Walton A J Gammie W Fallon M Ward D Holwill R J 1990 Submitted for publication.

Paper presented at ESSDERC 90, Nottingham, September 1990
Session 2A8

# New aspects of intrinsic gettering for CCD imagers

N A Sobolev    I Yu Shapiro    V I Sokolov    E D Vasilyeva

A F Ioffe Physico-Technical Institute Academy of Sciences of the USSR, Leningrad, USSR

Abstract. The influence of as-grown microdefects on the CCD-Imager parameters is observed. It's shown that the defect structure of the initial silicon wafers defines directly both denuded zone/gettering zone formaition and $Si/SiO_2$ interface behavior in MIS-structures under intrinsic gettering annealing. For the defect structure investigation the gamma-diffraction method was used side by side with some traditional methods.

## 1. Introduction

Functional yield of CCD-Imagers is often limited both by impurities and defects in the bulk of silicon wafer and by electrophysical parameters of $Si/SiO_2$ interface. Impurities and defects can create electrically active generation-recombination sites within an active region. These sites result in cosmetic defects of CCD-Imagers ( so called "white spots") and in a local inhomogeniety of dark current. $Si/SiO_2$ interface states influence on the complete value of dark current. Thus, it's necessary to regulate a bulk defect structure of the wafers and to decrease $Si/SiO_2$ interface state density in order to increase functional yield and to improve CCD-Imager parameters. We consider that both these problems can be resolved in frames of intrinsic gettering process. The working out of the non-destructive methods of a defect structure examination is a very urgent problem. Gamma-diffraction method was used. A basis of this method and its ability to study the swirl-defect crystals were considered.

The aim of present work was to investigate a defect structure of the initial silicon wafers and to understand the features of intrinsic getter formation and their influence on the $Si/SiO_2$ interface state behavior under three-step annealing gettering cycle.

## 2. Experiments

Wafers of dislocation-free single crystal CZ-grown commercial silicon (n-type, 15 Ohm cm, 76 mm diameter with (100) orientation of the surface) were used in experiments. Two groups of wafers were used: there were as-grown swirl-defects (with a density about $5 \cdot 10^5 - 10^6$ cm$^{-3}$) in the wafers of the first group whereas such defects were absent in the wafers of the second group. Tree-step gettering cycle ( Craven, 1981) was carried out: first step- 1100°C , 4 hours; second step- 650°C , 16 hours; third step- 1000°C , 4 hours.

© 1990 IOP Publishing Ltd

We used three types of gate dielectrics in MIS-structures: oxide formed at the first step of the gettering cycle ( this step was carried out in oxidizing ambient) and preliminary formed one- ($SiO_2$) or two-layered ($SiO_2 + Si_3N_4$) dielectric ( in this case the first gettering step was carried out in nitrogen). The second and the third steps were carried out in nitrogen in all cases. Aluminium contacts were formed by thermal sputtering. High frequency C-V curves were measured at 78K and 300K after each annealing step and the interface state density ($N_{ss}$) was calculated emploing the Gray-Brown (1966) and the Goetzberger (1968 ) techniques.

The silicon structural perfection was investigated by selective chemical etching and gamma-diffraction (Kurbakov,1986). Diffraction experiments have been carried out on a computer-controlled gamma-diffractometer (Kurbakov,1987). A neutron activated $^{198}Au$ plate was the gamma-ray source. Rocking curves have been measured in transmission geometry $\omega$-scanning. We used the 412 keV gamma-line corresponding to a wavelength of 0.003 nm. The energy width of the line is only $\Delta E/E=10^{-6}$ so that the instrumental resolution is solely determined by the angular divergence of the incident beam (the horizontal divergence is 10" and the vertical divergence is 30').

The wafers coated with two-layered dielectric films were used for CCD-Imager fabrication.

3. Results and Discussion

Fig.1 and 2 show the integrated reflectivities and the rocking curves measured for silicon plates 26 mm in diameter and 3 mm thick with(a) and without (b) swirl-defects. The (111) oriented wafers were used. Selective chemical etching of samples indicated that the point defect clusters in the first crystal were distributed nonuniformly in the cross-

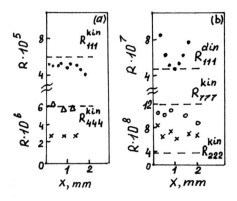
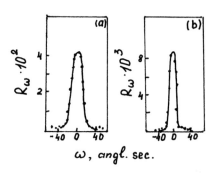

Fig. 1 Integrated reflectivity distribution along crystals with (a) and without (b) swirl-defects. Black circles correspond (111) reflection, triangles-(444), open circles-(777), crosses-(222). Dashed lines represent kinematic ($R^{kin}$) and dinamic ($R^{dyn}$) limits

Fig. 2 Diffraction ray intensity versus crystal rotation angle ($\omega$=0 corresponds the ideal Bragg angle) for crystals with (a) and without (b) swirl-defects

section and the swirl-free areas were 0.5 mm thick and clusters in the
second crystal were absent. For the swirl-defect crystal the (111) and
(444) integrated reflectivities coincide sufficiently well with the
kinematic estimates, the (222) forbidden integrated reflectivity is
essentialy higher than the kinematic value and the half-width of the
rocking curve is 20". For the swirl-free crystal the (111) integrated
reflectivity decreases considerably approaching the dynamic limit,
only the (777) integrated reflectivity is equal the kinematic estimate,
the (222) forbidden integrated reflectivity drops drastically
to the kinematic limit, the half-width of the rocking curve is
practically equal that of the instrumental response (10"). Thus,
correlation between the structural perfection of silicon crystals
and gamma-diffraction parameters was observed. The increase of the
concentration, size and inhomogeneous distribution of clusters is
accompanied by the rise of the level of local strains and leads to
essential increase of the integrated reflectivities and the half-width
of the rocking curves. Note that diffraction of hard gamma-radiation is
non-destructive method.

The optical analysis of a wafer splitting after a complete gettering
cycle with following selective etching was carried out. It was found
that the differences in the denuded zone (DZ) width and in the defect
density in gettering zone (GZ) depend on the as-grown defect
structure (Table 1).

Table 1

| Type of the defect structure | with as-grown swirl-defects | without as-grown swirl-defects |
|---|---|---|
| DZ width (microns) | 31.2 | 24.0 |
| Defect density in GZ ($10^{-6}$, cm$^{-2}$) | 4.7 | 1.7 |

It´s seen that the DZ width is larger and the defect density in the GZ
is higher in the wafers with as-grown swirl-defects. There isn´t a
difference in the concentration of opical active interstitial oxygen in
the both groups of the wafers. Thus, the distinctions in the DZ and GZ
formation are determined mainly by the as-grown defect structure. There
aren´t also the essential differences in the concentration of other
impurities.

Fig.3 demonstrates the change of $N_{ss}$ value after each gettering cycle
step (the oxide was formed at the first step). One can see the
difference in the behavior of $N_{ss}$ values between two groups of the
wafers. In the swirl-contained wafers a sharp increase of $N_{ss}$ value
after second gettering step takes place whereas this increase isn´t
observed in the swirl-free wafers of the second group. After the
complete gettering cycle there is the appreciable decrease of the $N_{ss}$
value in comparison with the initial one.

In that way the as-grown defect structure of the wafer defines
essentially both the DZ and GZ formation and the $Si/SiO_2$ interface
behavior under gettering annealing cycle. Optical microscopy data and
$N_{ss}$ value change ones confirm each other.

Fig.4 demonstrates $N_{ss}$ value change at the $Si/SiO_2$ interface for the
wafers preliminary coated with one- or two-layered dielectric film (all

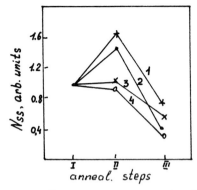
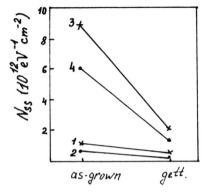

Fig. 3 $N_{ss}^n$ value change at the Si/SiO$_2$ interface under three-step gettering annealing.
1,2 - swirl-contaned samples;
3,4 - swirl-free samples.
$N_{ss}$ value was calculated according to the Gray-Brown (curves 1,3) and Goetzberger (curves 2,4) techniques

Fig. 4 The change of the Si/SiO$_2$ interface state density due to gettering annealing in MOS (1,2) and MNOS (3,4) structures.

gettering steps were carried out in nitrogen). It's seen, firstly, that the deposition of Si$_3$N$_4$ onto SiO$_2$ results in an increase of N$_{ss}$ value about an order of magnitude (because of additional mechanical stress induced by a Si$_3$N$_4$ film at the Si/SiO$_2$ interface) and, secondly, that the N$_{ss}$ value decreases after a complete gettering cycle for both types of the structures.

The sequence of two processes - dielectric film formation and gettering annealing - is significant for improvement of Si/SiO$_2$ interface. We carried out a technological chain "SiO$_2$ formation - gettering cycle - oxide removal - "new oxide" formation" and obtained the same N$_{ss}$ value for "new oxide" that the initial one was (before gettering cycle). We observed the same behavior for two-layered dielectric film. Thus, it's necessary to carry out gettering annealing only after gate dielectric formation to achieve considerable improvement of the Si/SiO$_2$ interface parameters.

The investigations carried out allowed us to modify the technology of CCD-Imager fabrication. We used swirl-contained silicon wafers coated with two-layered dielectric, then we carried out the gettering cycle. In such devices we observed a significant decrease of a "white spots" failure. Dark current decreases strongly (approximately by an order of magnitude) and charge transport efficiency increases. We also obtained a considerable growth of the functional yield (by a factor of 5). Thus, it's established that the presence of swirl-defects in the initial wafers increases gettering efficiency.

Craven R A and Korb H W 1981 Solid St. Technol. 24 55
Goetzberger A I and Irvin J C 1968 IEEE Trans. Electron. Dev. ED-15 1009
Gray P W and Brown D M 1966 Appl. Phys. Lett. 8 31
Kurbakov A I et al. 1986 Crystallography 31 979
Kurbakov A I et al. 1987 Preprint N 1307 (Leningrad: Nucl. Phys. Inst.)

Paper presented at ESSDERC 90, Nottingham, September 1990
Session 2A9

# Direct observation of the mask edge effect in a boron implantation

L. Gong, J. Lorenz, and H. Ryssel[*)]

Fraunhofer Arbeitsgruppe für Integrierte Schaltungen
Artilleriestr.12, D-8520 Erlangen, West-Germany
*) also: Lehrstuhl für Elektronische Bauelemente
Artilleriestr.12, D-8520 Erlangen, West-Germany

**Summary:** With a novel delineation technique, the increase in dopant concentration due to ions which have been scattered out of a vertical mask edge during ion implantation was demonstrated. The equiconcentration lines show that the implanted concentration near the mask edge is greater than that in the middle of the implantation window. This effect could be simulated qualitatively with a Monte Carlo simulation program.

## Introduction

The exact knowledge and control of dopant profiles in two and three dimensions is very important for device design and performance. With shrinking device dimensions and diffusion steps being frequently replaced by RTA, the importance of as-implanted profiles is growing. Various influences on ion implantation profiles have been measured and simulated, such as channeling, multilayer implantation and lateral spread. In this paper we will demonstrate the local increase in the implanted dopant concentration near a perpendicular mask edge using the novel delineation technique reported at ESSDERC '89 [1]. A comparison between the experiment and a Monte Carlo simulation will be given.

## Experiment

With the delineation technique reported in [1], several samples were etched. The etching solution is a mixture of $HNO_3$(70%), HF(50%) and $CH_3COOH$(100%). The samples were implanted with boron at energies of 200, 170 and 140 keV and a dose of $5\times10^{15}$ cm$^{-2}$ and annealed for 10 seconds at 1000°C. To investigate the effect of the mask edge, both implantations with and without 7° tilt were processed. To recognize the mask edge in a SEM micrograph, a LTO layer with a thickness of 400 Å was deposited before depositing the polysilicon passivation layer. The height of the

© 1990 IOP Publishing Ltd

polysilicon mask was 1.6 μm.

**Results**

Fig.1a) shows a SEM-micrograph of two equiconcentration lines delineated in one sample which was implanted without tilt while Fig.2a) and Fig.3a) show three and two equiconcentration lines after an implantation with 7° tilt. The equiconcentration lines correspond to the concentrations of about $5 \cdot 10^{19}$ and $1 \cdot 10^{19} cm^{-3}$ in Fig.1a), $2 \cdot 10^{20}$, $2 \cdot 10^{19}$ and $1 \cdot 10^{19} cm^{-3}$ in Fig.2a) and $1.7 \cdot 10^{20}$ and $1.5 \cdot 10^{19} cm^{-3}$ in Fig.3a), respectively. Because of the thin LTO-layer, the implantation mask can be clearly seen. The equiconcentraion lines with a concentration of $1 \cdot 10^{19} cm^{-3}$ in Fig.1 a) and Fig.2a) and $1.5 \cdot 10^{19} cm^{-3}$ in Fig.3a) are much closer to the surface near the mask edge than in the middle of the implantation window, which means that the dopant concentration near the mask edge is higher than in the middle of the implantation window. This additional dopant concentration near the mask edge is due to ions which have entered the polysilicon mask near its edge, have been scattered out of the silicon into the trench and have reentered the silicon target in the bulk region near the mask edge. In both figures one can see that the equiconcentration line indicating a high concentration level (higher than $2 \cdot 10^{19} cm^{-3}$) does not show this effect. The reason for this is that the additional concentration near the mask edge due to the scattered and reentered ions is about 10% of the peak concentration which is too low to modify the highly doped areas.

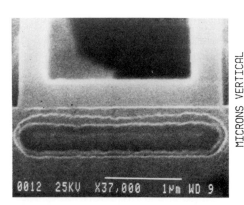
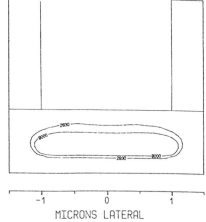

Fig.1a) SEM-micrograph of a boron implantation (energy 140keV, dose $5 \cdot 10^{15} cm^{-2}$) without tilt. The concentration of the equiconcentration lines are about $5 \cdot 10^{19}$ and $1 \cdot 10^{19} cm^{-3}$, respectively. The window width is 3μm.

Fig.1b) Corresponding Monte Carlo simulation of a boron implantation using 40,000 particles

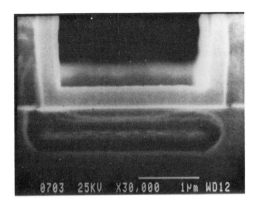

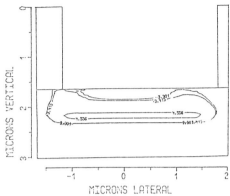

Fig.2a) SEM-micrograph of boron implantation (energie 170keV, dose $5 \cdot 10^{15}$ cm$^{-2}$) with 7° tilt. The concentration of the equiconcentration lines are about $2 \cdot 10^{20}$, $2 \cdot 10^{19}$ and $1 \cdot 10^{19}$ cm$^{-3}$, respectively. The window width is 3 μm.

Fig.2b) Corresponding Monte Carlo simulation of a boron implantation using 40,000 particles

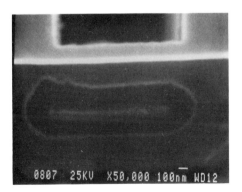

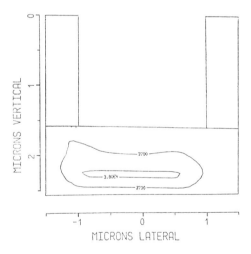

Fig.3a) SEM-micrograph of a boron implantation (energy 200 keV, dose $5 \times 10^{15}$ cm$^{-2}$). The concentration of the equiconcentration lines are about $1.7 \cdot 10^{20}$ and $1.5 \cdot 10^{19}$ cm$^{-3}$, respectively. The window width is 1.8 μm

Fig.3b) Corresponding Monte Carlo simulation of a boron implantation using 40,000 particles.

## Simulation

Many analytical models [2,3,4] have been developed to simulate the distribution of implanted ions in two dimensions. However, there is no analytical model which could give a description of the mask edge effect discussed above. The CPU expensive Monte Carlo simulation is the only suitable method to simulate this effect. We have used a Monte Carlo program [5] which uses the same approach and physical models as an earlier TRIM version [6]. Fig.1b), 2b) and 3b) show Monte Carlo simulations which correspond to the measurements shown in Fig.1a), 2a) and 3a). One can see that the simulations show the same effect as the etched equiconcentration lines. The effect discussed above influences the distribution of implanted ions only considerably at high energies because at low energies the concentrations closer to the surface are already too large. For boron, the effect can be seen for energies above 100 keV and for phosphorus above 200 keV, respectively. Both the experiment and Monte Carlo simulation show that the two dimensional distribution of the implanted ions depends not only on the implantation energy, but also because of the mask edge effect on aspect ratio of the implantation window.

## Conclusions

The increase of dopant concentration due to ions which have been scattered out of a vertical mask edge during ion implantation and have been reimplanted was directly observed. Monte Carlo results agree well with the experimental observations. Due to this mask edge effect, the dopant concentration at the bottom of a trench depends on the aspect ratio of the trench.

## References

[1] L. Gong, A. Barthel, J. Lorenz, and H. Ryssel, Simulation of the Lateral Spread of Implanted Ions: Experiments, in Proc. ESSDERC'89 (eds: Heuberger, Ryssel and lange), pp.198-202, Springer Verlag Berlin (1989).

[2] S. Furukawa, H. Matsumura and H. Ishiwara, Theoretical Considerations on Lateral Spread of Implanted Ions, Japan. J. Appl. Phys.11(2) 134 (1972).

[3] H. Runge, Distribution of Implanted Ions under Arbitrarily Shaped Mask Edges, phys. Stat. sol.(a)39, 595 (1977).

[4] J.Lorenz, W.Krüger and A.Barthel, Simulation of the Lateral Spread of Implantated Ions: Theory, in proc. NASECODE 89 (eds: Miller), Ireland (1989)

[5] G. Hobler, Dissertationsarbeit, TU Wien, 1988

[6] J.P. Biersack and L.G. Haggmark, A Monte Carlo Computer Program for the Transport of Energetic Ions in Amorphous Targets, Nucl. Instr. and Meth.174 pp 257-269 (1980)

Paper presented at ESSDERC 90, Nottingham, September 1990
Session 2A10

# *In situ* cleaning of silicon wafers for selective epitaxial growth (SEG)

M R Goulding, P Kightley, P D Augustus and C Hill.

Plessey Research Caswell Ltd, Caswell, Towcester, Northants. NN12 8EQ.

Abstract: The effects of pre-bake temperature, pressure and duration have been studied for silicon SEG substrates. Silicon cusp and trough formation adjacent to the oxide sidewalls in association with the undercutting of the oxide features is reported. Activation energies of $\Delta E_u=3.35eV$ for the undercut reaction and $\Delta E_t=1.98eV$ for the trough forming reaction have been calculated from measurements of the variation of undercut length and trough depth with temperature.

Introduction

The removal of surface oxide from a silicon wafer is one of the most important factors determining the quality of epitaxial layer growth. The thermal pre-treatment (pre-bake) of the substrate is determined to a large extent by both solid state out diffusion and autodoping considerations, but more fundamentally by the stability of the surface native (or chemically regrown) oxide to be removed, in-situ, immediately prior to growth. SEG has been highlighted for use in many advanced silicon process technologies [1]. It is neccessary to ensure that the quality of SEG layers is comparable to those grown on planar substrates. When using patterned oxide wafers for SEG, in addition to the considerations outlined above for conventional epitaxy, the disproportionation reaction between the silicon substrate and the oxide features must also be considered: ie, depending upon the the pre-bake conditions, such a reaction can result in the undercutting of the $Si/SiO_2$ interface. This undercut has been proposed as a source for the generation of planar crystallographic defects during growth [2]. In this paper the dependance of the undercutting upon the temperature (850°C to 1050°C), pressure (10 to 30τ) and duration (5 to 30 minutes) of the $H_2$ pre-bake is studied. In addition to the oxide undercutting reaction a second surface reaction has been identified which etches the silicon wafer close to the oxide edges to form a trough. The activation energies for both these reactions have been determined and a mechanism for the undercutting is proposed.

Experimental

The SEG substrates were prepared by growing ~0.6 μm of thermal oxide (1000°C : $O_2/H_2/HCl$) on 25Ωcm p-type <001> wafers, followed by ~325Å of nitride deposition. The SEG 'seed windows' were dry etched using a two stage process followed by the growth of a ~200Å 'sacraficial oxide' for etch damage removal [3]. Prior to loading, the sacraficial oxide was removed using buffered HF, followed by an RCA type clean [4]. All wafers were handled using back-surface vacuum pickup and VESPEL handling tools. The $H_2$ pre-bakes were performed using a modified (operational pressure $\leq 8\tau$) Applied Materials AMC 7811 epitaxial reactor, with the $H_2$ supplied by a Johnson Matthey HM2 diffuser.

Transmission Electron Microscopy was used to determine the extent of the undercutting reactions. Samples from these wafers were prepared in [110] cross-section, revealing the undercut regions. A JEOL 120CX was used for the TEM observations, operated at an accelerating voltage of 120keV. Images were mostly obtained from the (000) undiffracted beam at the [110] zone axis. SEM plan views were obtained using a JEOL IC 845 microscope.

Results and Discussion

Figure 1 shows a [110] TEM cross-section of a long oxide stripe after a $H_2$ (20 l/min) pre-bake at 1050°C/8$\tau$ for 5 minutes; the undercut distance is described by $x_u$. Also present is a cusp of silicon labelled C and an etched trough adjacent to the oxide feature; the maximum depth of the trough is described by $x_t$. To obtain the real value of the trough depth the thickness of silicon consumed by the formation of the sacraficial oxide is subtracted to give the value $x_t'$. The distance the oxide sidewall edge was set-back from the nitride edge is designated $x_{eb}$. The oxide edge is etched back from the nitride edge during the HF etch of the sacraficial oxide. The distance of the etch-back was determined experimentally and is subtracted from $x_{eb}$ to yield the value $x_{eb}'$.

Arrhenius behaviour was noted for the variation of both $x_u$ and $x_t'$ with temperature (figure 2). Activation energies of $\Delta E_u = 3.35$ eV for the undercut reaction and $\Delta E_t = 1.98$ eV for the formation of the trough were calculated. A strong linear pressure dependance was also noted for $x_u$ which is shown in figure 3. The activation energy for the undercutting process is in excellent agreement with the value of 3.3eV calculated by Liu et al [5]. In agreement with these authors it is proposed that the rate limiting undercutting reaction is described by

$$Si (s) + SiO_2 (s) \rightarrow 2SiO (g) \qquad (1).$$

Measurements of $x_{eb}'$ show no variation with temperature, pressure or duration of the pre-bake thus it is concluded that the rate of the reduction reaction,

$$SiO_2 (s) + H_2 (g) \rightarrow SiO (g) + H_2O (g) \qquad (2)$$

is negligible under these conditions and can be neglected as a cause of undercutting. It is considered that the trough formation occurs by the surface diffusion of silicon to the sidewall [6]. The migrating silicon atoms form the observed cusps which enables reaction (1) to proceed. A possible driving force could be the stress induced between the $SiO_2$ and the Si. The trough depth, $x_t$, and the cusp size have been observed by SEM to vary along the length of the oxide stripe; however, the undercut length ($x_u$) is constant. The variation of the cusp size is shown in figure 4. Figure 4a shows a grid of oxide stripes after a pre-bake for 5 minutes at 1050°C/10τ. Figure 4b shows the same sample after the oxide has been etched away. The cusps and the troughs are clearly visible. These results suggest that the cusp itself is mobile over the range of pre-bake temperatures used, maintaining the undercutting reaction along the length of the oxide stripe. Experiments have shown that the 4270Å of undercut observed for a 5 minute 1050°C/10τ $H_2$ pre-bake is reduced to ~50Å at 900°C/10τ and becomes essentially zero at 850°C/10τ for pre-bakes of up to 30 minutes duration. Experiments have also demonstrated that that the 850°C/10τ pre-bake does not fully remove an RCA re-grown surface oxide, with the subsequent growth of highly defective or polycrystalline material. Alternatively, SIMS analysis has shown that when using a 5 minute 900°C/10τ pre-bake, no interfacial $O_2$ is observable above the background detection limit of ~$10^{17}$ cm$^{-3}$, as shown in figure 5. It would appear from these results that to successfully remove the surface oxide from the SEG seed windows it will be neccessary to incurr a small undercut.

Conclusions

At present a 5 minute 900°C/10τ pre-bake is being used for SEG resulting in a small, ~50Å, undercut being present during the subsequent growth. TEM observation of SEG structures has shown that, possibly due to the surface kinetic control mechanism operating under the 850°C/10τ growth conditions, the filling of the undercut regions may not necessarily generate defects. This represents a considerable advantage when using higher pre-bake and/or growth temperatures within the kinetic regime to, for example, minimise As autodoping in thin epitaxial layers [7]. Autodoping and defect investigations are in hand.

Acknowledgements

This work was conducted as a part of ESPRIT TIP BASE workpackage 5. The authors would like to acknowledge Mr Alan Whitehead and Mrs G B Davies for their expert assistance. This paper is published with the permission of the Directors of Plessey Research Caswell Ltd and Plessey Semiconductors Ltd.

## References

1. IEDM 1987, Session 2 *'Selective Epitaxy and BICMOS Technology'*. Washington DC.
2. A Ishitani, H Kitajima, N Endo and N Kasai. Jpn.J.Appl. Phys. (1989) **28** 841
3. H Liaw, D Weston, B Reuss, M Brittella and J Rose *'Chemical Vapour Deposition 1984'*, The Electrochemical Society PV84-6 463
4. W Kern and D A Puotinen, RCA Review (1987) 187
5. S T Liu, L Chan and J O Borland. *'Chemical Vapour Deposition'* 1987 Ed. G W Cullen. Proc.10th Int.Conf. on CVD Honolulu p428
6. R Tromp, G W Rubloff, P Balk, F K LeGoues and E J van Loenen. Phys.Rev.Lett. (1985) **55** 2332
7. M R Goulding and J O Borland *'Semiconductor International'* May 1988, p56.

Figure 1  TEM cross section of an oxide stripe after a 5 minute 1050°C/10t pre-bake.

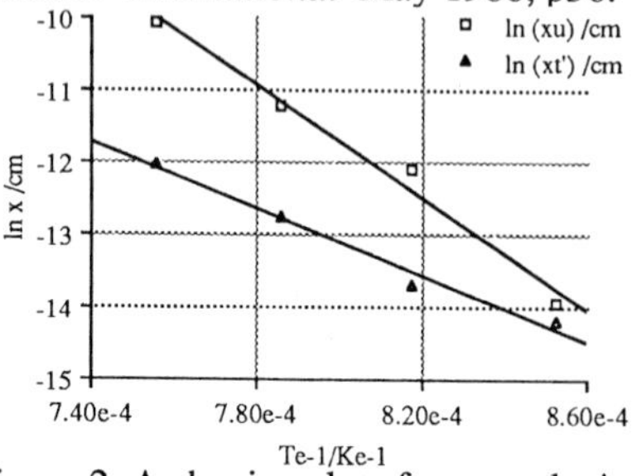

Figure 2  Arrhenius plots for $x_u$ and $x_t'$ as a function of temperature.

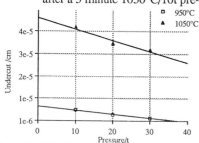

Figure 3  The linear variation of $x_u$ with pre-bake pressure.

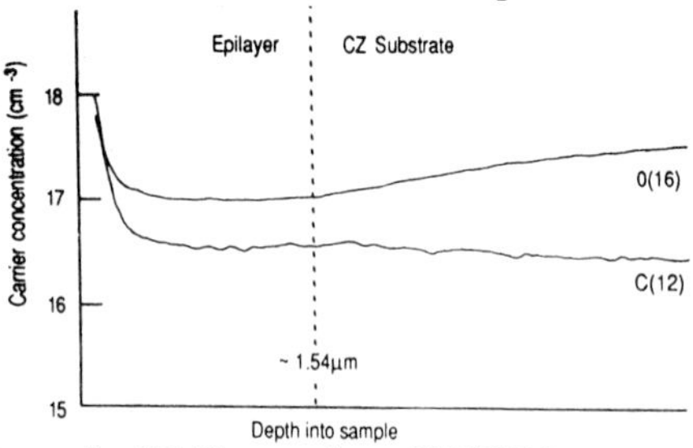

Figure 5  SIMS trace for a 900°C/10τ pre-bake showing no interfacial $O_2$ or Carbon.

 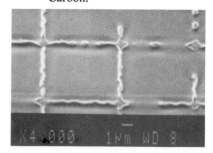

Figure 4  SEM micrographs of a grid of oxide stripes a) directly after a 1050°C/10τ/5 minute pre-bake and b) with the oxide removed.

Paper presented at ESSDERC 90, Nottingham, September 1990
Session 2A12

# Complementary silicon JFETs using novel ultra-shallow gate junctions

Ö. Grelsson, A. Söderbärg and U. Magnusson

Uppsala University, Dept. of Technology, P.O. Box 534, S-751 21 Uppsala, Sweden

Abstract. Measurements on normally-off type junction field effect transistors (JFET's) of both n- and p-type for low power applications are presented. A new method, using ultra-shallow gate junctions, when integrating p-and n-type JFET's to form a complementary JFET (CJFET) structure is proposed and a suitable technology is presented. The ultra-shallow gate junctions used in the devices are formed by low temperature diffusion of boron and antimony in amorphous silicon. Electrical characterisations and SIMS measurements on these ultra-shallow gate junctions are also presented.

1. Introduction

One of the major problems in manufacturing normally off JFET's is to control the threshold voltage. This factor is strongly dependent on several other parameters such as the location of the gate-channel junction, the buried junction and the channel doping. To improve the control of the threshold voltage it would be of great value if one of the above mentioned three parameters could be neglected. One way of doing this would be to use low temperature epitaxy or to deposit a polysilicon gate-structure in order to create an ultra shallow $p^+n$ or $n^+p$ gate junction for the n- and p-channel JFET respectively.

2. Experimental and results

In order to investigate the ultra-shallow gate junctions, different samples for characterisation of the junctions were made. To start with the $p^+n$ junction, a 2500Å thick amorphous silicon film was evaporated on an n-type silicon wafer. After the evaporation the film was ion implanted with $^{49}BF_2^+$ to a dose of $1 \cdot 10^{15}$ cm$^{-2}$ at 100 keV. The samples were then heat treated in an argon ambient to activate the dopants and to crystallize the silicon film. Different times and temperatures between 600 and 850°C were used in order to optimize the process. SIMS profiling and electrical measurements showed that the best results were obtained at an annealing temperature of 850°C for 30 minutes.(figure 1.)

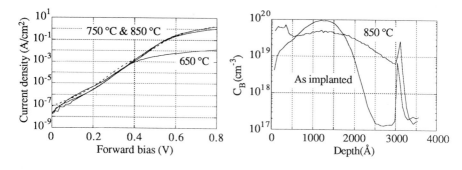

Figure 1a.

IV characteristics of the p⁺n junction.

Figure 1b.

SIMS profile of the boron concentration versus depth.

The same technique was used with Arsenic (As) and Phosphorus (P) in order to create a good n⁺p junction, but the results were not satisfying. To end up with results comparable with the p⁺n junction, a method described by Söderbärg et. al. (1990), where a thin Antimony (Sb) film was evaporated on the substrate prior to the silicon evaporation was used. A similar method, using Sb to enhance silicon crystallization have earlier been reported by Gong et. al. (1987). The two evaporations in our experiment were done without breaking the vacuum in between. During the following heat treatment the dopants are redistributed and activated, the presence of Sb also enhances the crystallization of the silicon film. Also in this experiment, different annealing times and temperatures were used and both electrical characterization and SIMS profiling gave at hand that a proper annealing temperature would be 900°C for 90 minutes. (figure 2.)

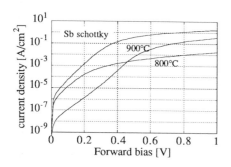
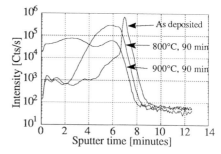

Figure 2a.

IV characteristics of the n⁺p junction.

Figure 2b.

SIMS profile of the antimony intensity versus sputtering time.

The SIMS analysis and theoretical calculations show that both of the above mentioned $p^+n$ or $n^+p$ junctions are located less than 300Å beneath the interface.

Both n- and p-type normally off type JFET's were made using these structures as gate-junctions. The transistors were manufactured on n-type <100> silicon substrates with sheet resistivites of 6-9 Ωcm. A p-well was formed when fabricating the n-type transistor, in order to make the process as comparable as possible to the proposed complementary process. The IV characteristics of the two transistor types with gate dimensions of 5x100μm are shown in figure 3a and 3b.

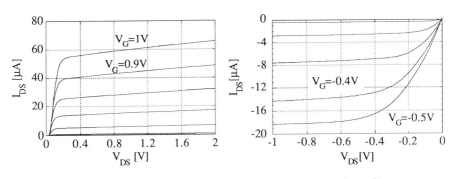

Figure 3a.
IV characteristics of the n-type JFET.

Figure 3b.
IV characteristics of the p-type JFET.

For the p-type transistor the threshold voltage was measured to $V_T = -0.04$ V and the transconductance was measured to be $g_m = 0.7$ mS/mm at $V_{GS} = -0.35$ V, which is very good for a device of these dimensions. The same measurements on the n-channel device gave the following results, $V_T = 0.4$ V and $g_m = 0.7$ mS/mm at $V_{GS} = 0.7$ V.

To make these transistors work as complementary components, the threshold voltages have to be adjusted to about 0.1 V off for both types of transistors. This is readily done by adjusting the channel implantation dose. A fabrication process for the complementary JFET's is proposed as follows:

1. p-well implantation and activation.
2. oxidation.
3. masking for source and drain.
4. source and drain implantations.
5. oxidation and activation.
6. masking for channel implantation.
7. channel implantation.
8. oxidation and activation.
9. masking for $p^+$ gate.
10. silicon evaporation.
11. gate implantation and activation.
12. masking for $n^+$ gate.
13. $n^+$ gate deposition and activation.
14. etching of contact holes.
15. metallization.

In order to predict the static power consumption of the device, subthreshold measurements were carried out. (figure 4.)

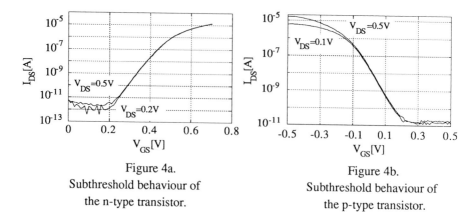

Figure 4a.
Subthreshold behaviour of
the n-type transistor.

Figure 4b.
Subthreshold behaviour of
the p-type transistor.

From these curves the static power consumption from the "off" transistor could be predicted to be in the order of 0.3-0.4 nW at a supply voltage of 0.5 V. The subthreshold swing for the p-type transistor was measured to be 65 mV/decade which is quite comparable to small scale MOS. The n-type transistor showed a subthreshold swing of 50 mV/decade.

3. Acknowledgment

We want to thank B. Mohadjeri at the Royal Institute of Technology for his work with the SIMS analysis. This work has been financially supported by the Swedish Board for Technical Development (STU).

4. References

1. Gong S. F. et. al. (1987), J. Appl. Phys. 62 (9), 3726
2. Söderbärg A. et. al. (1990), accepted for publication, J. Appl. Phys.

Paper presented at ESSDERC 90, Nottingham, September 1990
Session 2B1

# Submicron pseudomorphic $Al_{0.2}Ga_{0.8}As/In_{0.25}Ga_{0.75}As$-HFET made by conventional optical lithography for microwave circuit applications above 100 GHz

H. Meschede, J. Kraus, R. Bertenburg, W. Brockerhoff, W. Prost, K. Heime[*], H. Nickel[+], W. Schlapp[+], R. Lösch[+]

Universität -GH- Duisburg, Halbleitertechnik/Halbleitertechnologie, SFB 254, Kommandantenstr. 60, D-4100 Duisburg 1

[*]RWTH-Aachen, Institut für Halbleitertechnik, Sommerfeldstr., D-5100 Aachen
[+]Deutsche Bundespost Telekom, Am Kavalleriesand 3, D-6100 Darmstadt

> **Abstract:** In this work the device performance of a submicrometer pseudomorphic FET made by conventional optical lithography using a single-layer photoresist will be presented. With this technique a gate-length of 0.6µm can be reproducible achieved. Using a multiple finger structure with air-bridge technology a maximum transconductance of 470mS/mm and a cutoff-frequency of 111GHz was obtained.

## Introduction

The microwave properties of $Al_xGa_{1-x}As$/GaAs heterostructure-field effect transistors can be drastically improved by inserting an $In_yGa_{1-y}As$ channel /1/. Due to the higher conduction-band discontinuity between the $In_yGa_{1-y}As$ channel and the doped $Al_xGa_{1-x}As$ layer the carrier concentration and velocity as well as the carrier mobility will be increased. In order to achieve cutoff-frequencies above 100GHz very short gate-lengths are required. Gate-lengths far below 1µm with a sophisticated recessed-gate profile generally are obtained by electron beam lithography with a complicate multilayer photoresist technology. In this contribution we will show that even conventional optical lithography with a simple single layer photoresist structure is able to provide cutoff-frequencies above 100GHz, if an optimized multiple finger layout is used.

## Layout and Device Technology

One of the most important limiting factors of submicron FET performance are the ohmic losses in the input circuit of the FET. Especially the gate-resistance increases drastically and thus the microwave performance degrades with decreasing gate-length. Mushroom gates with a complicate photoresist technology may be used to reduce this gate-resistance. A more simple way is the reduction of gate-width down to typically less than 20µm. However for microwave circuit applications a sufficient available power can be achieved, only if a gate-width above 100µm is chosen. Fig.1 shows the dependence of cutoff-frequency ($f_{MAG=1}$) on the gate-width $W_G$ for different gate-length $L_G$. Reducing the gate-length beyond 0.5µm leads to an increased MAG only in combination with simultaneously shortening the gate-width.

© 1990 IOP Publishing Ltd

Therefore in this work a multiple finger FET (fig.2) has been developed which has a reduced gate-resistance and simultaneously a high available power. It consists of 8 gate-fingers with a total gate-width of 160μm. The source pads are connected via an air-bridge.

A photoresist (AZ1505) with a thickness of 0.5μm suitable for exposure in the 300nm-400nm wavelength region was used to define the gate structure. With a nominal length of 0.5μm in the mask and a resist development optimized for high contrast a standard gate-length of 0.6μm can be achieved. After a wet chemical gate-recess the Cr/Au metallization was evaporated into the recess groove directly. Ohmic contacts were made of AuGe/Ni/Au layers. The air-bridge process utilizes a different photoresist (AZ4330) with a height of 3μm and a galvanic gold metallization of 3μm, too.

The layer sequence shown in fig.3 was designed for high carrier concentration in the 2 DEG channel at room temperature. To avoid problems with the lattice strain a channel thickness of 9nm was used /2/. The sample was grown by molecular beam epitaxy. Due to the highly doped cap layer the magnetotransconductance method was chosen to characterize the transport behavior of the 2 DEG /3/. For the sheet carrier concentration at T=300K a value of $n_s=2.5 \cdot 10^{12} cm^{-2}$ and a mobility of $\mu=5900 cm^2/Vs$ can be measured using special processed test FET.

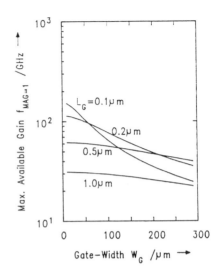

Fig.1: Dependence of cutoff-frequency $f_{MAG=1}$ on gate-width $W_G$ (gate-length $L_G$ as parameter)

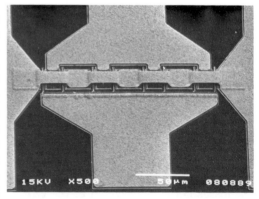

Fig.2: Multiple finger FET with a total gate-width of 160μm

| | |
|---|---|
| GaAs : Si<br>n=3.8·10$^{18}$ cm$^{-3}$ | Cap Layer<br>40 nm |
| Al$_x$Ga$_{1-x}$As : Si<br>x=0.2->0 | Grading<br>20 nm |
| Al$_{0.2}$Ga$_{0.8}$As : Si<br>n=3.0·10$^{18}$ cm$^{-3}$ | Doping Layer<br>20 nm |
| Al$_{0.2}$Ga$_{0.8}$As | Spacer<br>2.2 nm |
| In$_{0.25}$Ga$_{0.75}$As | 2 DEG<br>9 nm |
| GaAs | Buffer<br>4 μm |
| s. i. GaAs | Substrate |

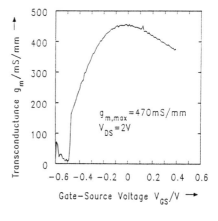

Fig.3: Cross section of the investigated device

Fig.4: Transconductance $g_m$ versus gate-source voltage $V_{GS}$

## Experimental Results

Depletion type FETs with a gate-length of 0.6μm and a drain-source spacing of $L_{DS}$=1.6μm were processed by this structure on a pseudomorphic layer with 25% In-content. The devices show a maximum extrinsic transconductance of 470mS/mm (fig.4) and excellent IV-characteristics with a drain current of $I_D$=400mA/mm at $V_{GS}$=0.4V and $V_{DS}$=2V. The output power is suitable for applications in a microwave circuit. Fig.5 shows the measured s-parameters from which the equivalent circuit elements shown in table 1 can be calculated.

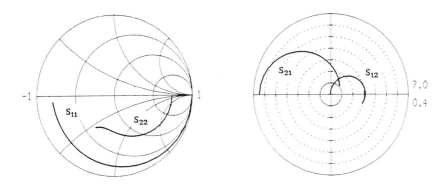

Fig.5: S-parameters of a multiple gate FET measured from 0.045 to 26.5 GHz at $V_{GS}$=-0.1V and $V_{DS}$=1.5V ($L_G$=0.6μm, $W_G$=160μm)

The input resistance of the device is about 5Ω, only. The gate-source capacitance is $C_{gs}$=0.28pF and the intrinsic transconductance $g_m$ is about 77mS ($\hat{=}$480mS/mm).

| $R_{in}$ | $C_{gs}$ | $g_m$ | $C_{gd}$ | $R_d$ |
|---|---|---|---|---|
| 5Ω | 0.28pF | 77mS | 71fF | 252Ω |

Table 1: Equivalent circuit elements calculated from s-parameters ($W_G=160\mu m$)

In fig.6 the unilateral gain (GU) versus frequency is shown. For the frequency limit of the unilateral gain a value of $f_{max}$=111GHz can be extracted. The improved transport properties of the $In_{0.25}Ga_{0.75}As$ channel should lead to an improved saturation velocity in the channel. In a first approximation the current gain cutoff-frequency $f_T$ can be used to extract a value for the over-all saturation velocity. Using the following formula:

$$f_T \approx \frac{g_m}{2\pi \cdot C_{gs}} = \frac{\bar{v}}{2\pi \cdot L_G}$$

together with the data of tab.1 a $f_T$ of 43GHz and an over-all saturation velocity can be calculated to $\bar{v}=1.62\cdot 10^7$ cm/s for a gate-length of $L_G=0.6\mu m$.

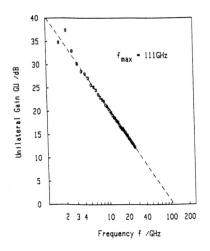

Fig.6: Unilateral gain (GU) versus frequency

## Conclusions

A submicrometer pseudomorphic HFET made by conventional optical contact lithography was presented. With a single-layer photoresist gate-lengths of 0.6μm can be reproducible achieved. Using a multiple finger structure with a gate-width $W_G$=160μm a maximum extrinsic transconductance of $g_m$=470mS/mm and a current of $I_D$=400mA/mm can be obtained. For the unilateral gain a frequency limit of $f_{max}$=111GHz was extracted from the measured s-parameters. The use of $In_yGa_{1-y}As$ in the channel of a HFET leading to an improved over-all saturation velocity is pointed out. These results show the possible applications of the introduced HFET especially for high frequency oszillator circuits.

## Acknowledgement

The authors are very thankful to A. Osinski for device preparations.

## References

[1] P.M. Smith, L.F. Lester, P.C. Chao, R.P. Smith, J.M. Ballingall; IEEE Electron Device Lett., Vol.10, No.10, Oct.1989, pp.437-439.

[2] B. Elman, E.S. Koteles, P. Melman, C. Jagannath, J. Lee, D. Dugger; Appl. Phys. Lett., 55(16), 16.Oct. 1989, pp.1659-1661.

[3] W. Prost, W. Bettermann, K. Heime, I. Gyuro, H. Dämbkes, M. Heuken, W. Schlapp, G. Weimann; Proceedings of the 16th Int. Conf. on GaAs & Related Compounds, Japan, 1989, pp.501-506.

# High electron density and mobility InGaAs/InAlAs modulation doped structures grown on InP

F. Gueissaz, R. Houdré, M. Ilegems.
Institut de Micro- et Optoélectronique. Ecole Polytechnique Fédérale.
CH 1015 Lausanne, Switzerland.

Abstract. Results of Two-dimensional Electron Gas Field Effect Transistors (TEGFETs) processed on InGaAs/InAlAs modulation doped heterostructures grown by Molecular Beam Epitaxy (MBE) are presented. We obtain extrinsic transconductances of 424mS/mm for 1X50μm$^2$ depletion mode TEGFETs. A detailed static characterization of the devices and energy band profile calculation reveal that a more efficient channel modulation at high current densities can be achieved.

## 1. Introduction.

Lattice matched InGaAs/InAlAs modulation doped structures have been recognized as potentially excellent candidates for high performance microwave TEGFETs. The large conduction band discontinuity at the heterointerface and the high low-field electron mobility in InGaAs lead to high electron sheet densities and mobilities in the two-dimensional electron gas. This combined with the high electron saturation velocity in InGaAs results in high current densities and low access resistances in the TEGFETs. However, the growth of these structures and the processing of high performance microwave devices present new challenges.

## 2. MBE growth.

Atomic plane modulation doped heterostructures on InP with different spacer thicknesses between the doping plane and the heterointerface have been prepared. Lattice matching to better than $0.5 \cdot 10^{-3}$ was reproducibly achieved as measured by four crystal x-ray diffractometry.

We present results on two different samples labeled A and B. Starting from the InP:Fe substrate interface, the sample A (resp. B) consists of a 4000Å InAlAs buffer layer, a 500Å InGaAs channel layer, a 70Å (resp. 50Å) InAlAs channel spacer layer, an atomic doping plane containing $6.0 \cdot 10^{12} cm^{-2}$ Si atoms, a 200Å InAlAs top spacer layer and a 50Å n$^+$ InGaAs cap layer which is fully depleted.

The growth temperature was held at 530°C and the typical growth rate was 0.67µm/h.

The two samples exhibit 300K electron sheet densities ($n_S$) and mobilities ($\mu_H$) of $2.7 \cdot 10^{12} cm^{-2}$ (resp. $3.1 \cdot 10^{12} cm^{-2}$) and $10'600 cm^2/Vs$ (resp. $7'900 cm^2/Vs$) (figures 1 and 2). The resulting sheet resistances are 220 Ω/square (resp. 260 Ω/square). A relatively weak (< 5% change in $n_S$) persistent photoconductivity effect (PPC) was observed at 13K.

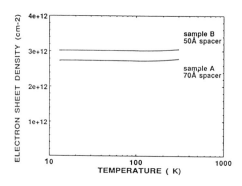

Fig. 1 Electron sheet density vs temperature.

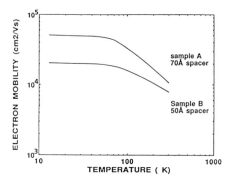

Fig. 2 Hall electron mobility vs temperature.

### 3. Device processing and results.

TEGFETs have been fabricated by a standard mesa etch / ohmic contact deposition, alloy / gate recess, metal deposition and overlay metal deposition process using optical lithography with the i and h spectral lines of a mercury vapour UV lamp. A high definition image reversal photoresist was used to pattern the metal layers by lift-off. Preliminary results have been obtained from a sample identical to sample A with lower electron mobility ($8500 cm^2/Vs$ at 300K).

The ohmic contact characterization yielded 0.6Ωmm. The gate Schottky barrier height of Titanium over InAlAs as calculated from I(V) measurements is found to be near 500meV. The resulting leakage currents are thus quite high but low enough to allow capacitance measurements ($C_{GS}(V_{GS})$) at 1MHz.

Extrinsic transconductances of 424mS/mm were obtained with 1X50µm² TEGFETs at $V_{GS}$= -0.5V and 175mA/mm drain current density (figure 3).

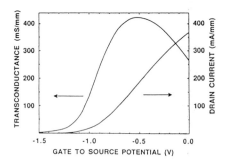

Fig. 3 Transfer characteristic of a 1x50µm² TEGFET.

The extrinsic transconductance can be significantly increased by reducing the resistance of the ohmic contacts. W.P. Hong et al. (1986) and L. Palmateer (1989) obtained values as low as 0.1Ωmm on similar structures. The transconductance to output conductance ratio is typically 12 for a drain to source potential up to 2V.

## 4. Calculation of the energy band profile.

A detailed analysis of the $C_{GS}(V_{GS})$ and low field $I_D(V_{GS})$ characteristics provides the necessary data to establish the equilibrium energy band profile across the heterojunction at different gate to source potentials (figure 4 and 5). We apply the triangular quantum well approach, described by D. Delagebeaudeuf and N.T. Linh (1982) or T. Ando and further by B. Vinter (1985) for the GaAs/AlGaAs heterostructure, with modified material parameters.

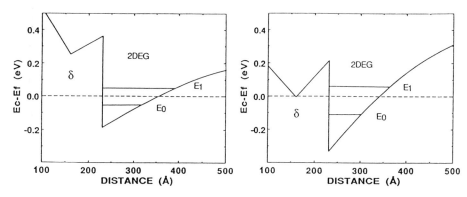

Fig. 4 Energy band profile vs distance from the surface at $V_{GS} = -0.53V$.

Fig. 5 Energy band profile vs distance from the surface at $V_{GS} = 0V$.

For a given 2DEG electron sheet density ($n_{2DEG}$) experimentally determined by $I_D(V_{GS})$ and $C_{GS}(V_{GS})$ measurements, we calculated the longitudinal quantized electron subband energies and the position of the Fermi energy in the InGaAs channel; with a $\Delta E_C$ of 0,55eV and a given spacer thickness, we can then determine the position of the conduction band dip above the Fermi energy at the atomic doping plane in the InAlAs ($E_{C\delta}-E_F$). For ease, we first assume that no free electrons are present in the atomic doping plane, thus yielding a constant slope of the conduction band on both sides of the latter. The peak transconductance occurs at $V_{GS}=-0.53V$, where the $n_{2DEG}$ is found to be $1.1 \cdot 10^{12} cm^{-2}$ and we calculate 0.25eV for $E_{C\delta}-E_F$ (figure 4). At $V_{GS}=0V$, $n_{2DEG}$ should reach $2.2 \cdot 10^{12} cm^{-2}$ and therefore $E_{C\delta}-E_F$ becomes negative which clearly indicates that free electrons will appear in the atomic doping plane (figure 5), creating a parasitic parallel conduction and reducing the modulation efficiency. At this gate bias, the measured transconductance is decreased by 30%.

A 50Å thick channel spacer (as in sample B) would result in a $E_{C\delta}-E_F$ of approximately 0.06eV at $n_{2DEG} = 2.2 \cdot 10^{12} cm^{-2}$, decreasing the parallel conduction by approximately one order of magnitude compared to the case of sample A and allowing the transconductance to reach much higher values at zero gate bias.

## 5. Conclusion

We succeeded to grow high performance modulation doped heterosructures and gained useful information for further optimization by evaluating the performance of processed TEGFETs. The fairly thick channel spacer allowed high electron mobilities at still high electron sheet densities but the modulation efficiency decreases already at $n_{2DEG}$ above $1.1 \cdot 10^{12} cm^{-2}$. The spacer thickness has to be optimized in order to reduce the parallel conduction while still maintaining high electron mobilities for low access resistances.

## Aknowledgment

This work was supported by THOMSON-CSF (France).

## References

T. Ando, J. Phys. Soc. Jap., Vol 51, No 12, pp.3893-3907, 1982.

D. Delagebeaudeuf and N.T. Linh, IEEE Trans. El. Dev., Vol. ED-29, No 6, pp.955-960, 1982.

W.P. Hong, S.S. Kwang, P.K. Bhattacharya and H. Lee, IEEE El.Dev. Lett., Vol. EDL-7, No 5, pp.320-323 , 1986.

L.Palmateer, PhD thesis work at Cornell University, Ithaca, New York, 1989 (unpublished).

B. Vinter, Appl. Phys. Lett. 44(3), pp.307-309, 1984.

B. Vinter, Proceedings of the winter school Les Houches, France; G. Allan, G. Bastard, N. Boccara, M. Lannoo and M. Voos editors, Springer Verlag (1986).

*Paper presented at ESSDERC 90, Nottingham, September 1990*
*Session 2B3*

# Surface characteristics of plasma treated WN$_x$/GaAs contacts from *C–V* measurements

P.E.Bagnoli [a], A.Paccagnella [b], A.Callegari [c], F.Fantini [d]

a) Istituto di Elettronica e Telecomunicazioni, University of Pisa, Pisa, Italy
b) Dipartimento di Elettronica e Informatica, University of Padova, Padova, Italy
c) Thomas J. Watson Research Center I.B.M., Yorktown Heigths, N.Y., U.S.A.
d) Scuola Superiore di Studi Universitari e di Perfezionamento S. Anna, SSSUP, Pisa, Italy

> Abstract . The density and the average penetration depth of acceptors near the semiconductor surface were calculated from C-V and J-V measurements on p+/n Shannon structures. The WN$_X$ / GaAs diodes were fabricated using chemically and plasma cleaned GaAs surfaces and annealed at several temperatures. It was found that the semiconductor surface cleaning before metal deposition is a key factor to control the rectifying properties of this type of metal / semiconductor contact.

1. Introduction

Tungsten nitride (WN$_X$) is a promising gate material for self-aligned GaAs metal-semiconductor field-effect transistor (MESFET) technology [Yamagishi 1987]. This material is suitable for microelectronic processing, owing to the easy and reproducible deposition by sputtering of pure metal in a nitrogen/argon atmosphere and steep side-wall definition by reactive ion etching (RIE). Moreover WN$_x$ shows lower sheet resistance and higher thermal stability in comparison with the widely used tungsten silicide.
The rectifying properties of the WN$_X$ / GaAs junction depend not only on the film composition and post-metallization annealing temperature, but also on the surface preparation procedure before the metal deposition [Paccagnella 1989]. It was found [Zhang 1987], for instance, that the barrier height of the contact is increased by the presence of nitrogen atoms or nitrogen/defect complexes acting as acceptor-like impurities near the semiconductor surface. This last property is also affected by the technological procedure used for the diode fabrication.
In this paper we compare the electrical characteristics of WN$_X$-GaAs diodes fabricated using three different surface cleaning procedures and three different temperatures of the post-metallization annealing. The conventional current-voltage (J-V) and capacitance-voltage (C-V) measurements were interpreted in terms of the model of the enhanced barrier Shannon contact [Eglash 1987], from which an evaluation of the effective surface acceptor density, the average penetration depth and the total amount of the acceptors per unit area was obtained.

2. Theory

For the analysis of the experimental results, we assume that our samples have a p+/n Shannon contact structure, where the donors are uniformly distributed through the semiconductor substrate and the acceptors are fully depleted and have a constant step-like distribution in depth from the surface up to the depth L. It follows that the $C^{-2}$-V curve is a straight-line whose slope depends on the donor density (N$_D$) in the semiconductor bulk . The voltage intercept (V$_{INT}$), as determined by $C^{-2}$-V measurements, is an increasing function of both the acceptor density (N$_A$) and the surface layer thickness [Zhang 1987, Eglash 1987], as shown by the following relation:

$$V_{INT} = V_O + \frac{q N_A L^2}{2 \varepsilon_S} - V_n - \frac{kT}{q} \qquad (1)$$

where $V_O$ is the Fermi level position at the semiconductor surface, $V_n$ is the energy difference between the conduction band and the bulk Fermi level and $\varepsilon_S$ is the GaAs dielectric constant. However, if the acceptor density is large enough, the voltage intercept is always higher than the actual value of the potential maximum which is located at the depth $x_m$ ($0 \leq x_m < L$).
On the other hand, if the top of the barrier is located under the semiconductor surface and it is consequently a weak function of the applied bias, the ideality factor (n) of the forward thermionic current-voltage characteristic is $> 1$ and also depends on the parameters L an $N_A$. Neglecting second order effects, such as the tunnel current across the barrier, its expression is given by [Eglash 1987]:

$$\frac{1}{n} = \left(1 + \frac{N_D}{N_A - N_D}\right)\left[1 - \sqrt{\frac{q N_D L^2}{q N_A L^2 + 2 \varepsilon_S (V_O - V_n - V_a)}}\right] \qquad (2)$$

where $V_a$ is the applied bias.
Note that, from (2), n is an increasing function of the penetration depth L but becomes closer to unit as the acceptor density increases.
The relations (1) and (2) provide a couple of equations from which the unknown parameters L and $N_A$ can be obtained using experimental values of $V_{INT}$ and n, by assuming a proper value for the parameter $V_O$ and calculating $N_D$ from the slope of the $C^{-2}$-V curve.
The values of L obtained using this method must be retained only as average values of the penetration depths, because the continuous decreasing profile of the acceptors has been approximated by an abrupt step distribution. However this assumption is consistent with the nearly-abrupt profiles of the acceptor density shown by Fourkas and Cheung (1988) for similar experiments.

## 3. Experiment

The $WN_x$ / GaAs diodes were fabricated on bulk GaAs <100> wafers ($N_D$= 2-4 x $10^{17}$ cm$^{-3}$). Three different surface preparation procedures were used: a) chemical etching in dilute HCl (CHEM samples), b) chemical etching plus hydrogen plasma at 200 °C (H2 samples) and c) chemical etching plus hydrogen plasma and plus nitrogen plasma also at 200 °C (N2 samples). The 200 nm thick $WN_x$ layer was deposited using a DC magnetron sputtering in an argon/nitrogen mixture. Details of the plasma treatments and metal nitride deposition procedures are reported elsewhere [Paccagnella 1989]. Standard photolythography and RIE were used for the definition of the $WN_x$ dots. Then different groups of diodes for every preparation sequence were annealed for 10 minutes at three different temperatures (700, 800, 850 °C) in arsine overpressure. Indium was eventually deposited on the wafer back side as an ohmic contact. Capacitance (1MHz) vs. reverse voltage and d.c. current vs. forward voltage measurements were performed at room temperature by means of HP 4280A and HP 4140 computer controlled capacitance and current meters respectively.

## 4. Results and Discussion

The figures 1 and 2 show the current and the capacitance characteristics of H2 samples annealed at several temperatures. The characteristics of all the diodes show good linearity as a consequence of the complete recovering of sputter-induced damage.
The voltage intercepts and the ideality factors (calculated at $V_a$=0.2V) are shown in the figure 3 and 4. The behaviour of CHEM samples completely differs from that of plasma treated samples: the voltage intercept is nearly temperature independent and the ideality factor slightly increases with the annealing temperature.
On the contrary, the rectifying properties of plasma treated diodes improve (i.e. the voltage intercept increases and the ideality factor decreases) as the annealing temperature increases.

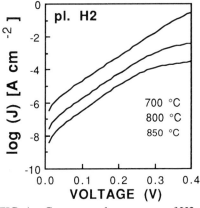

FIG. 1 : Current - voltage curves of H2 samples at several annealing temperatures

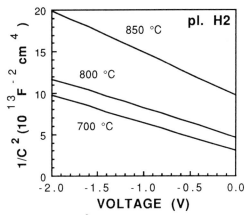

FIG. 2 : $C^{-2}$ vs. reverse voltage curves of H2 samples at several annealing temperatures

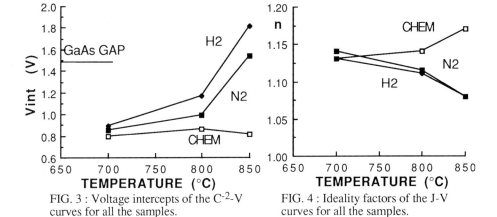

FIG. 3 : Voltage intercepts of the $C^{-2}$-V curves for all the samples.

FIG. 4 : Ideality factors of the J-V curves for all the samples.

The thermionic barrier height of all the samples, calculated from the J-V curves, lie in the range 0.65-0.91 eV and is always lower than the corresponding value obtained from the voltage intercept. The discrepancy increases with increasing annealing temperatures.

In the calculations of L and $N_A$ by means of equations (1) and (2), the parameter $V_O$ was kept constant for all the diodes because of the Fermi level pinning caused by the high density of surface states in GaAs. The value assumed for $V_O$ was 0.65 eV below the conduction band, which is lower than the commonly used 0.8 eV [Eglash 1987] and takes into account both the low barrier height of the W-GaAs contact and the effect of the image force lowering on the energy potential near the semiconductor surface.

The calculated values of the average penetration depth and the acceptor density near the surface are shown in figures 5 and 6 respectively and are summarized in figure 7.

The average penetration depth L increases with the annealing temperature in each sample type, while the effective acceptor density $N_A$ decreases in the chemically cleaned samples and increases in the plasma treated ones.

According to the hypothesis that the acceptor-like impurities are due to nitrogen atoms and/or nitrogen-defect complexes [Zhang 1987], the behaviour shown by CHEM samples corresponds to the anneal-induced diffusion of those nitrogen atoms implanted during the sputtering process. However the presence of a residual chemisorbed oxygen layer at the interface acts as a barrier for further diffusion of nitrogen from the $WN_x$ layer. Therefore the total amount of the acceptors per unit area, which is given by the product $N_A L$, remains rather constant. On the contrary, since the plasma treatment produces a cleaner and oxigen-free semiconductor surface, in H2 and N2 samples there are no obstacles for the diffusion of nitrogen atoms from the metal-nitride layer during the thermal treatment. The increase of both L and $N_A$ (and therefore of the total amount of the acceptors) is responsible for the higher voltage intercepts and barrier heights shown by these diodes at the higher temperatures.

As far as the comparison between the H2 and N2 samples concerns, it is interesting to notice that the additional nitrogen plasma, which was used with the purpose of increasing the nitrogen content of the surface, actually yields slightly worse rectifying characteristics with respect to the H2 diodes. This phenomenon can be due to localized microreactions during the annealing treatment.

5. Acknowledgement

This work was partially supported by CNR - Progetto Finalizzato Materiali e Dispositivi per l'Elettronica a Stato Solido.

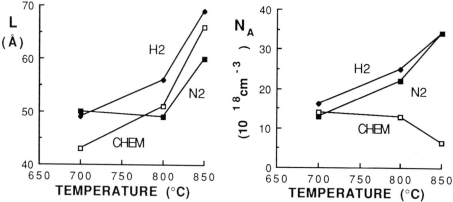

FIG. 5 : Calculated penetration depths of the acceptors under the GaAs surface

FIG. 6 : Calculated acceptor densities in the heavily doped surface layer

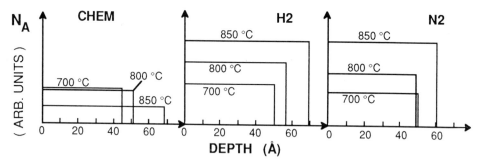

FIG. 7 : Idealized acceptor density profiles for all the samples.

References
Yamagishi H. et al. 1987 *Jap. J. Appl. Phys.* **26** 122
Paccagnella A. et al. 1989 *IEEE Trans. Electron Devices* **36** 2595
Zhang L.C. et al. 1987 *J.Vac. Sci. Technol.* ,**B5** 1716
Eglash S.J. et al. 1987 *J. Appl. Phys.* **61** 5159
Fourkas R.M., Cheung N.W. 1988 *IEEE Trans. Electron Devices* **35** 1384

*Paper presented at ESSDERC 90, Nottingham, September 1990*
*Session 2B4*

# Gate technologies for AlInAs/InGaAs HEMTs

T.D. Hunt, J. Urquhart, J. Thompson, R.A. Davies, and R.H. Wallis

Plessey Research Caswell Ltd., Caswell, Towcester, Northants. NN12 8EQ, U.K.

Abstract: The properties of Ti and Pt Schottky barriers on AlInAs have been investigated to assess their suitability as gate metals for AlInAs/InGaAs HEMTs. Pt was found to give a slightly higher Schottky barrier than Ti and, more significantly, gave a reverse leakage current about two orders of magnitude lower than Ti at typical gate biases. AlInAs/InGaAs HEMTs using PtAu gates have been fabricated, and a peak d.c. transconductance of 400 mS/mm achieved. Devices with 0.7μm gate lengths gave extrapolated values of $f_T$ up to 52 GHz.

1. Introduction

HEMTs based on the InGaAs/AlInAs material system, grown lattice-matched to InP substrates, have demonstrated microwave performance superior to any other type of transistor. Extrapolated values of $f_T$ of 186 GHz and $f_{max}$ of 405 GHz, together with noise figures as low as 0.3 dB at 18 GHz, have been achieved for very short gate length (0.15μm) devices (Chao et al 1990). Even for 1.15μm gate length devices, Hikosaka et al (1988) have achieved $f_T$s of 22.7 GHz, and shown that the $f_T$-gate length produce should be nearly 50% greater than for a conventional AlGaAs/GaAs HEMT. In addition, the InP substrate makes the InGaAs/AlInAs HEMT compatible with InP-based optoelectronic components, and therefore a candidate for use in optoelectronic integrated circuits such as optical receivers.

The performance advantages of the InGaAs/AlInAs HEMT compared to the GaAs/AlGaAs HEMT result from the superior electron transport properties of the $In_{0.53}Ga_{0.47}As$ channel compared to GaAs. Additionally, the $In_{0.53}Ga_{0.47}As/Al_{0.48}In_{0.52}As$ conduction-band discontinuity (~0.5 eV) is about twice that of a typical GaAs/AlGaAs heterojunction, allowing a higher density of electrons to be confined in the channel. The technology of InGaAs/AlInAs HEMTs is similar to, and has been derived from, that for GaAs/AlGaAs HEMTs, which in turn has been derived from that for MESFETs. In particular the Schottky gate metallisation is frequently TiPtAu, as generally employed in AlGaAs/GaAs HEMTs and MESFETs. However it is commonly observed (e.g. Chao et al 1990) that gate leakage currents in InGaAs/AlInAs HEMTs are high. A preliminary study of Schottky barrier metals on AlInAs by Hodson et al (1988) showed that Ti and Au gave lower barrier heights than Pt and Al, thereby suggesting that the TiPtAu gate metallisation is not optimum for AlInAs/InGaAs HEMTs.

In this paper we report the results of a comparative study of Pt and Ti Schottky barriers on AlInAs and show that Pt offers clear advantages for low leakage current gates. We also report the results of the first InGaAs/AlInAs HEMTs fabricated using PtAu as the gate metallisation.

© 1990 IOP Publishing Ltd

## 2. Fabrication and Characterisation of Schottky diodes on AlInAs

The Schottky diodes used in this study were fabricated on doped $Al_{0.48}In_{0.52}As$ grown epitaxially by low-pressure MOVPE on 2" $n^+$ InP substrates. The growth conditions were similar to those previously described by Davies et al (1988). The layers consisted of an initial 0.3μm of heavily doped ($3 \times 10^{18}$ cm$^{-3}$) AlInAs, to avoid series resistance effects, followed by ~1.5μm of nominally undoped AlInAs, for which the background carrier density was in the mid $10^{-16}$ cm$^{-3}$ range. The first stage of processing was to evaporate an ohmic contact on to the back face of the substrate. The wafers were then cut in several pieces so that different metallisation schemes and surface treatments could be compared on the same material. Circular Pt and Ti Schottky diodes were then prepared by e-beam evaporation and standard photolithographic and lift-off techniques. Additional Au and PtAu metallisations were also deposited on the Pt and Ti respectively to facilitate device probing and wire bonding. Immediately prior to evaporation the surface of the AlInAs was lightly etched in $H_3PO_4/H_2O_2/H_2O$. The devices were characterised by both forward and reverse current-voltage (I-V) measurements and reverse-bias capacitance-voltage (C-V) measurements. Most measurements were carried out on-wafer, although in some cases the devices were diced and bonded into headers, with no significant degradation in their characteristics.

Fig. 1 shows the forward I-V characteristic of a 110μm diameter Pt Schottky diode. For biases greater than 100 mV the characteristics of all devices obeyed the standard expression

$$I = I_0 \exp(qV/nkT)$$

over more than five decades of current, with ideality factors n in the range 1.08 to 1.10. Extrapolating the linear region to V = 0 to obtain $I_0$ and using the standard expression

$$I_0 = SA^{**}T^2 \exp(-\phi_B/kT)$$

(where $A^{**}$ is the effective Richardson constant and S the diode area), barrier heights $\phi_B$ in the range 0.72 eV to 0.75 eV were obtained. For the Ti diodes, similar values of ideality were obtained, but the values of $I_0$ were typically an order of magnitude greater than for Pt, corresponding to barrier heights about 60 meV lower, in the range 0.66 eV to 0.69 eV. Fig. 2 shows a 1/$C^2$-V plot for a 110μm diameter Pt diode, measured at 100 kHz. The good linearity,

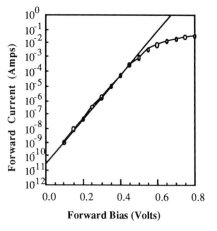

Fig. 1 Forward current-voltage plot for a 110μm diameter Pt Schottky diode on undoped AlInAs

resulting from the uniform background donor density in the epitaxial AlInAs, allows an accurate determination of the built-in voltage $V_{bi}$, (0.66V for the diode shown in Fig. 2.

From this value and the position of the Fermi level (obtained from the donor density determined from the slope of the 1/$C^2$-V plot) a barrier height of 0.76 eV may be deduced. Similar results were obtained from other Pt diodes, whereas Ti diodes consistently gave values which were about 40-50 meV smaller. The two techniques for barrier height determination are therefore in good agreement both in their absolute values and in showing that Pt gives a consistently higher barrier than Ti. The barrier heights for the Pt diodes are slightly higher than the value of 0.69 ± 0.05 eV obtained by Chu et al (1988) for Au and Al barriers on MBE-grown semiconducting AlInAs.

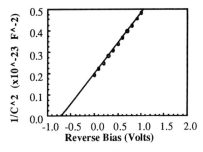
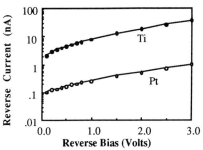

Fig. 2 1/C^2-V plot for a 110μm diameter Pt Schottky diode on undoped AlInAs

Fig. 3 Comparison of the reverse leakage currents for Pt and Ti Schottky barriers on the same undoped AlInAs layer

A more significant difference between the Pt and Ti diodes is seen in their reverse I-V characteristics, the Ti diodes having a leakage current between twenty and fifty times greater than the Pt diodes, as shown in Fig. 3. In both cases the reverse leakage currents were higher than expected from the values of $I_o$ extracted from the forward characteristics. To investigate whether this non-ideal behaviour was related to an interfacial layer beneath the Schottky metal, other pre-deposition surface treatments were examined, and found to give slightly higher reverse leakage current. No apparent change in the reverse current density was detected for diodes of diameters ranging from 50μm to 150μm, thereby indicating that edge effects were not important.

## 3. AlInAs/InGaAs HEMTs with Pt Gates

In view of the superiority of Pt over Ti as a Schottky barrier on AlInAs, HEMTs have been fabricated using PtAu gate metallisation in place of the more normal TiPtAu metallisation. The layer structure used for these devices, which was again grown by low-pressure MOVPE, is shown schematically in Fig. 4. The use of a very thin (20 nm) AlInAs buffer has been described elsewhere (Hunt et al, 1990). The doping in the AlInAs supply layer was confined to a region 10 nm thick above the 5 nm AlInAs spacer layer, with a further 40 nm of undoped AlInAs above this. The heavily doped InGaAs cap layer served to facilitate the fabrication of low resistance ohmic contacts.

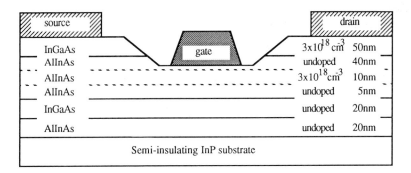

Fig. 4 Schematic diagram of the structure used for AlInAs/InGaAs HEMTs with PtAu gates

Device fabrication began with mesa etching followed by ohmic contact fabrication, using AuGeNi metallisation and rapid thermal annealing. Next gates were defined by optical lithography, and the gate recess etched using $H_3PO_4/H_2O_2/H_2O$. PtAu gates (100 nm Pt, 390 nm Au) were deposited by e-beam evaporation. The process was completed by plating up the bonding pads with thick Au to assist in device probing and bonding. No problems of adhesion of the PtAu gate metal were experienced.

The $I_D$-$V_{DS}$ characteristics of a typical device with a 150μm gate width are shown in Fig. 5. A maximum transconductance of 400 mS/mm was obtained for a slight forward bias on the gate. Also notable are the low output conductance and excellent pinch-off characteristics. Gate leakage currents at pinch-off were only ~10μA for $V_{DS}$ = 1.0V. This is a very low value for AlInAs/InGaAs HEMTs. It is, however, much higher than would be expected from the reverse I-V characteristics for Pt diodes of Fig. 3. Whether this difference results from the higher fields under the gate, due to the high doping of the AlInAs, or surface leakage between the gate and drain contact is not yet clear.

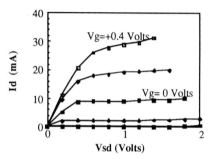
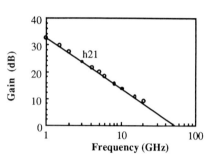

Fig. 5 Id-Vds plots for an AlInAs/InGaAs HEMT with a 1.1μm x 150μm PtAu gate

Fig. 6 Forward current gain h21 versus frequency for a 0.7μm gate length AlInAs/InGaAs HEMT

Fig. 6 shows a plot of the forward current gain $h_{21}$ as a function of frequency, derived from on-wafer S-parameter measurements over the range 0.1-20 GHz. An extrapolated value of $f_T$ of 52 GHz is obtained. This is a very high value for a 0.7μm gate length device, and exceeds that expected on the basis of $f_T$-gate length product of 26 GHz.μm given by Hikosaka et al (1988).

Conclusions

We have demonstrated that Pt forms a Schottky barrier on AlInAs which is slightly higher than Ti and more significantly, shows lower reverse leakage current. AlInAs/InGaAs HEMTs using PtAu gates have shown low gate leakage currents. A value of 52 GHz for $f_T$ has been obtained for a 0.7μm gate length devices.

Acknowledgements

We would like to thank C.G. Eddison and D.M. Brookbanks for RF assessment of the devices. This work was partially supported by the EEC under ESPRIT Project 2035.

References

Chao P C, Tessmer A J, Duh K-H G, Ho P, Kao M-Y, Smith P M, Ballingall J M, Liu S-M J and Jabra A A 1990 IEEE Electron Device Lett. **11** 59
Chu P, Lin C L and Wieder H H 1988 Appl Phys Lett **51** 3 2423
Davies J I, Hodson P D, Marshall A C, Scott M D and Griffiths R J M 1988 Semicond Sci Technol **3** 223
Hikosaka K, Sasa S, Harada N and Kuroda S 1988 IEEE Electron Device Lett **9** 241
Hodson P D, Wallis R H, Davies J I, Riffat J R and Marshall A C 1988 Semicond Sci Technol **3** 1136
Hunt T D, Davies R A, Thompson J, Eddison C G, Jessup M and Wallis R H 1990 to be published.

Paper presented at ESSDERC 90, Nottingham, September 1990
Session 2B5

# A parametric investigation of the reactive ion etching of InP in $CH_4/H_2$ plasmas using response surface methodology

D J Thomas and S J Clements

STC Technology Ltd, London Road, Harlow, Essex, CM17 9NA

> Abstract. Response surface methodology (RSM) is applied to study the reactive ion etching (RIE) of InP in $CH_4/H_2$ plasmas. Mechanistic information is inferred through careful interpretation of the contour plots derived from RSM. The data enable the process conditions required for specific device applications to be successfully selected.

1. Introduction

Plasma-assisted processing is commonly applied to etch III-V materials as part of the overall fabrication of optoelectronic devices. It is vital to understand how the plasma conditions affect the nature of the processed structure. Because of the diverse parameter space available to the operator and the probable interaction between these parameters, single-variable experiments are inappropriate for developing and studying plasma etching processes. Response Surface Methodology (RSM), however, allows a broad examination of the relationship between process variables (factors) and process outputs (responses) by using statistical methods both to design the experiments and to model the data.

Niggebrügge et al (1985) were the first to describe the RIE of InP and related compounds using $CH_4/H_2$ plasmas. This process chemistry offers several distinct advantages over chlorinated gases. One important feature is that surfaces which are not etched become passivated by the selective deposition of a hydrocarbon polymer. It becomes necessary to study both the etching of InP and the polymer deposition if the overall process is to be better understood.

2. Experimental

The InP(100) was reactive ion etched, through a silicon oxide mask, using a commercial parallel plate reactor (Plasma Technology Plasmalab μP) operating at 13.56 MHz. The samples were placed upon a silicon wafer to prevent back-sputtering of the electrode material.

A five-factor matrix was statistically designed according to the conditions of Table 1. Six responses were chosen to characterise the etching process. The InP etch rate was measured by stylus profilometry. The deposition rate of polymer onto a silicon sample and the polymer refractive index were determined by ellipsometry. The InP wall angle and the lateral etch rate of the $SiO_2$ mask (see Figure 1) were derived from measurements on scanning electron micrographs. Finally the DC self-bias potential was measured directly during plasma operation.

© 1990 IOP Publishing Ltd

| Factor | Range |
| --- | --- |
| $CH_4$ content | 10-25 % |
| RF power | 50-160 W |
| Pressure | 25-70 mtorr |
| Flow Rate | 30-60 Sccm |
| Temperature | 0-25 °C |

Table 1 Factors and appropriate ranges chosen for the RSM experimental design.

## 3. Results

Figure 2 shows six process responses as a function of the RF power and the process pressure and at a constant $CH_4$ content, total flow rate and electrode temperature. Contour plots of this type allow quantitative exploration of the effects of various factor combinations and represent a qualitative method of comparing different responses to the same factor pairs.

Figure 2a demonstrates that raising the RF power at a constant pressure has a significantly greater effect on the etch rate if the process pressure is high. Many of the contours exhibit this type of behaviour which could easily be overlooked by single-variable experiments.

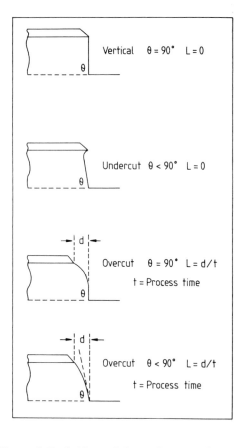

Figure 1 Definition of the wall angle, $\theta$ and the lateral etch rate of the $SiO_2$ mask, L for vertical, undercut and overcut profiles of InP.

The similarity between the InP etch rate and the deposition rate of polymer (Figures 2a and 2b) suggest that the same types of species are involved in the etching and deposition mechanisms. This is consistent with observations by other workers that $CH_3$ is the predominant polymer pre-cursor in RF plasmas of pure $CH_4$ (Kline et al 1989) and is the most likely reactant in removing In from surfaces of InP (Hayes et al 1989). Almost identical trends to those of Figures 2a-2f were observed for contours involving RF power and $CH_4$ content (not shown) confirming that the partial pressure of $CH_4$ is more critical than either the total pressure or the $CH_4$ fraction of the parent gas (Niggebrűgge et al 1985).

The wall angle contours (Figure 2e) exhibit opposing trends to those of the polymer deposition rate (Figure 2b) suggesting that the predominantly anisotropic etching which we observe is unlikely to be due to the passivation of sidewalls by thick layers of

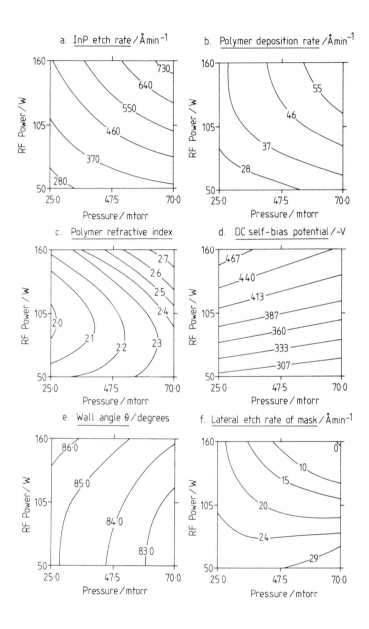

Figure 2 Contour plots derived from RSM as a function of RF power and process pressure. 17.5% $CH_4$, 45 Sccm flow rate and 25 °C electrode temperature. 160 W is equivalent to 0.7 $Wcm^{-2}$ of lower electrode area.

polymer. Figures 2d and 2e indicate that the energy of ions striking the InP surface is important in controlling the anisotropy. However, this appears to be only part of the explanation. The wall angle is weakly dependent on RF power in the range 50-105 W when the pressure (or the $CH_4$ content) is held constant. In contrast, the DC self-bias potential is almost unaffected by pressure (or $CH_4$ content), implying that the average ion energy remains approximately unchanged. The increased rate of lateral etching is therefore more likely the result of an increase in the neutral reactant supply [ie. $CH_x$ (X=1-3) radicals]. This explanation necessarily assumes that any differences in the ion scattering cross-sections, for the various plasma constituents within the plasma sheath, are negligible in influencing the wall angle. It appears, therefore, that lateral etching of InP in our apparatus is the result of a low rate predominantly chemical mechanism.

Figure 1 shows schematically the overcutting which can occur in our apparatus. Because the edge of the $SiO_2$ mask is tapered (due to limitations in the photolithography and the oxide etching) it can recede sideways during the RIE of InP if it is insufficiently protected by hydrocarbon polymer (Figures 2b and 2f). Clearly, steep walls (>85°) are only achievable at the expense of significant lateral $SiO_2$ etching (Figures 2e and 2f), which has a tendency to "round-off" the InP closest to the mask.

## 4. Conclusions

Response surface methodology is ideally suited to studying plasma etching processes and has allowed us to derive, from the contour plots, mechanistic information regarding the RIE of InP in $CH_4/H_2$ plasmas. We are now in a position to successfully select the process conditions for the fabrication of specific optoelectronic devices.

## 5. Acknowledgments

We wish to thank Dr. Suresh Ojha for his help in running the RSM program and David Spear for instructive discussions concerning this work. The technical assistance of Julie Champelovier, Dave Moule and Sue Wheeler is also gratefully appreciated.

## 6. References

Hayes T R, Dreisbach M A, Thomas P M, Dautremont-Smith W C and
    Heimbrook L A 1989 J. Vac. Sci. Technol. **B7** 1130
Kline L E, Partlow W D and Bies W E 1989 J. Appl. Phys. **65** 70
Niggebrügge U, Klug M and Garus G 1985 GaAs and Related Compounds, Inst. Phys.
    Conf. Ser. Karuiza, Japan 1985 **79** 367

# Effects of the passivation process on the electrical characteristics of GaInAs planar photodiodes

F. Ducroquet, G. Guillot, J.C. Renaud*, A. Nouailhat**

Laboratoire de Physique de la Matière, INSA-Lyon, 69621 Villeurbanne cedex (France)
* CNET- Laboratoire de Bagneux, 196 av. H. Ravera, 92220 Bagneux (France)[1]
** CNET-CNS, Chemin du vieux chêne, B.P. 98, 38243 Meylan cedex (France)

Abstract: The electrical properties of GaInAs PIN photodiodes passivated by a silicon nitride film have been compared for different deposition techniques: CVD, PECVD, UVCVD. Passivation induced defects have been observed by admittance spectroscopy and DLTS measurements. The nature, density and profile of these defects are found to be strongly dependent on the deposition process. The latter also largely influences the dark current and its temporal stability.

## 1. Introduction

The dark current of GaInAs PIN photodiodes, used as detectors for optical fibre telecommunications systems working at 1.3-1.55 µm, is often dominated by a surface current, source of noise and temporal instabilities (Ripoche et al 1985, Bauer et al 1988). This current can be reduced by an efficient passivation which thus appears as one of the key technological parameters to achieve highly sensitive and reliable devices. At the present time, the most widely used technique is the Plasma Enhanced Chemical Vapour Deposition (PECVD) which gives reliable and reproducible results (Ripoche et al 1985). However, this process damages the electrical surface properties and leads to the creation of defects not only at the insulator/semiconductor interface but also in depth in the bulk material (Ota et al 1987). Therefore, it is important to develop new deposition methods which alllow an efficient surface passivation without degrading the device performances. Thus, the UVCVD (Ultra-Violet CVD) technique appears yet very promising in particular for the passivation at the end of the technological process of mesa type components (Le Bellégo et al 1989).

A comparative study of different passivation techniques has been performed by investigating the electrical characteristics of GaInAs PIN photodiodes. The experimental results are based firstly on the analysis of the dark current and its temporal stability and secondly on the characterization of the passivation induced defects in the GaInAs layer by admittance and Deep Level Transient Spectroscopy (DLTS) measurements.

## 2. Experimental

The investigated planar GaInAs/InP photodiodes consist of a n-$Ga_{0.47}In_{0.53}As$ absorbing layer and a buffer layer grown on an InP substrate as shown in Figure 1. The thickness of the active layer is about 3 to 4 µm and the carrier concentration about $2\text{-}5\ 10^{15}$ cm$^{-3}$. The junction is formed by a selective zinc diffusion through 40 and 70 µm diameter windows opened in a $SiN_x$ film. This film is kept to passivate the junction. AuGeNi and Ti-Au alloys are evaporated

---

[1] New address: Thomson-CSF, Laboratoire Central de Recherches, Domaine de Corbeville, 91401 Orsay

for n and p contacts respectively. Three dielectric deposition techniques have been compared:

- the Chemical Vapour Deposition (CVD) for which the sample is placed on a graphite support carried at 750°C in a few seconds
- the conventional PECVD at 350°C in which energetic particles are involved
- the UVCVD performed at low temperature ($\approx 180°C$) and without bombardment of the surface.

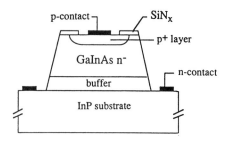

Fig 1: Cross-sectional view of a PIN photodiode

## 3. Experimental results

### 3.1 Temporal stability

The experimental results show important surface related effects which appear specific for each deposition method. If the temporal stability of the dark current is confirmed in the case of the PECVD technique (Ripoche et al 1985), slow drift phenomena take place for both other processes as previously reported (Ducroquet et al 1989a). By recording the current evolution under a fixed reverse voltage, a decrease quasi-logarithmical with time is observed, indicating a charge transfer between the semiconductor and dielectric layer. At high voltages (>10V), the large variations indicate that the dark current is influenced by the surface potential evolution resulting from the charge transfert into the insulator (Bauer et al 1989). Moreover, on some UVCVD photodiodes, a current increase under reverse bias voltage occurs at room temperature. This can be explained by the accumulation of mobile ionic charges at the junction perimeter (Kuhara et al 1985). These charges should be present in the native oxide layer remaining between the GaInAs surface and the silicon nitride film after deposition.

### 3.2 Current analysis

From the analysis of the current-voltage curves at different temperatures, we have identified several contributions to the dark current. At low bias voltages, the dominant contribution results from a generation-recombination mechanism. On the UVCVD samples, this current varies linearly with the depleted region width and its activation energy is close to that of the

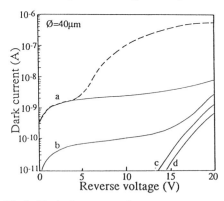

Fig 2: Typical current-voltage curves at different temperatures obtained on UVCVD passivated diodes: a - 300K, b - 250K, c - 200K, d - 150K
- - - : additional resistive-type conduction

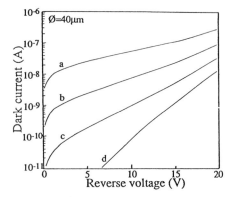

Fig 3: Typical current-voltage curves at different temperatures obtained on PECVD passivated diodes: a - 300K, b - 250K, c - 200K, d - 150K

mid-gap, indicating that the generation mainly takes place in the GaInAs bulk layer (Figure 2). On the PECVD and CVD photodiodes, the current is one order of magnitude higher and increases quasi- exponentially with the reverse voltage (Figure 3). As reported by Philippe *et al* (1986), this variation can be explained by the electrical field assisted generation process. In agreement with an ideality factor close to 2 and taking into account the temporal current evolution noticed on the CVD samples, we think that this process is localized near the surface and occurs through mid-gap interface states. At higher voltages, a surface tunneling current is involved as shown by its weak dependence on temperature, its marked increase with bias and its large drift induced by the surface potential variations with time as observed on the UVCVD samples. At room temperature, the ionic charge migration into the surface oxide layer leads to an important additional leakage current which can be attributed to a resistive-like conduction at the junction periphery (Figure 2). This contribution only appears on some UVCVD photodiodes, for which the oxide layer is not completely eliminated during the $SiN_x$ deposition.

### 3.3 Deep level characterization

Passivation induced defects have been characterized by admittance and capacitance transient spectroscopy measurements. On the UVCVD and PECVD passivated diodes, DLTS spectra reveal the presence of one main level positioned at $E_c - 0.35\pm0.03eV$. This trap has previously been located near the $SiN_x$/GaInAs interface, from a correlation established on the UVCVD samples between the current drift and the temporal evolution of the DLTS peak amplitude (Ducroquet *et al* 1989b). The origin of this defect should be related to the surface degradation induced by the passivation process. Indeed, this trap is observed in greater concentration in the case of the PECVD technique for which such a degradation is expected (Ota *et al* 1987). Concerning the UVCVD deposition, it is only detected significantly on the diodes for which the surface oxide is entirely removed, corresponding moreover to a reduction in the current drift phenomena. When the surface remains protected by a native oxide layer, this trap is almost undetectable, thus illustrating that the region damaged by the UVCVD process is minimized. On the CVD passivated diodes, an electron trap is observed in great concentration. The GaInAs non intentionally doped active layer ($n \approx 2\text{-}5.10^{15}cm^{-3}$ at room temperature) becomes semi-insulating at low temperatures, indicating that this defect is an acceptor-type one. It is located at $E_c - 0.19+0.03eV$ from admittance spectroscopy measurements. This level is not detected on the PECVD and CVD diodes and thus appears to be specific to the CVD technique, possibly induced by the high temperature thermal treatment which is characteristic of this process. Taking into account the V element volatility on such III-V compounds, the origin of this defect could be related to the surface arsenic evaporation.

## 4. Discussion

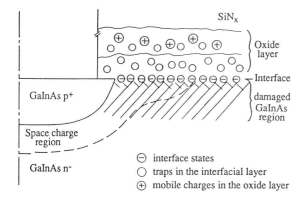

Fig 4: Proposed surface representation at the junction periphery

This comparative study points out the specific influence of each passivation technique on the electrical characteristics of these planar photodiodes. The interpretation of the experi- -mental results has led us to make some assumptions about about the nature and role of the traps and charges which can exist at the insulator/ semiconductor interface. These are summarized on the junction periphery modelling propo- -sed in Figure 4. On this schematic representation, we can discern three regions: the interface itself, the insulator layer and the surface GaInAs region. On GaInAs,it is known that the interface states act as efficient generation centers (Ota *et al* 1987). This is in agreement with the higher generation-recombination current we have observed on both the more energetic depositions (PECVD and CVD), the larger surface degradation resulting in a higher interface state density. We can also note in this case the electrical field effect on the generation process. In contrast, this is not observed on the UVCVD diodes for which the interface perturbation and consequently generation current are reduced. It is also known that the existence of traps in the interfacial layer are responsible for temporal instabilities. Indeed, if these traps are located very near the interface, charge transfer by tunneling can occur from the semiconductor. The drift analysis shows that this transfer can result in large variations in the current as observed at high reverse voltages on the UVCVD and CVD samples when the current becomes largely dependent on the surface electric field. Another cause of instability is attributed to the charge migration inside the native oxide layer (Kuhara *et al* 1985). This can give rise to the formation of a surface "conduction channel", resulting in a large leakage current as observed on some UVCVD samples. This illustrates, for weakly energetic processes such as UVCVD, the critical aspect of the surface preparation, especially the deoxidation step before deposition. On the other hand, the stability of the PECVD diodes is in agreement with the excellent reliability of this technique (Ripoche *et al* 1985). The DLTS measurements also confirm that the PECVD deposition leads to a more extended GaInAs damaged region than that induced by the UVCVD deposition. Moreover, the very large defect concentration observed on the CVD diodes suggests that a high temperature treatment induces the creation of defects in depth in the material bulk. Such effects have already been reported on InP (Krawczyk *et al* 1988).

## 5. Conclusion

Three passivation techniques (CVD, PECVD, UVCVD) have been compared by analyzing the dark current and deposition induced defects on GaInAs PIN photodiodes passivated by silicon nitride film. The specific influence of each deposition method is shown. We confirm that the UVCVD process induces much less damage than the conventional PECVD and CVD techniques. In particular, when the surface preparation before deposition is perfectly controlled, a very low and stable dark current can be obtained.

## Acknowledgements

The authors are grateful to P. Blanconnier and J.P. Praseuth for the epitaxial layer growth, to P. Dimitriou for deposition of the UVCVD $SiN_x$ films and to A. Scavennec for fruitful discussions. This study was supported by CNET.

## References

Bauer J G and Trommer R 1988 *IEEE Trans. Electron. Dev.* **35** p.2349-2353
Ducroquet F, Guillot G, Nouailhat A and Renaud J C 1989a *Rev. Phys. Appl.* **24** p.57-63
Ducroquet F, Brémond G, Guillot G, Renaud J C , Nouailhat A 1989b *Proc. Int. Conf. on the Science and Technol. of Defect Control in Semicond. Yokohama (Jpn)* to be published
Krawczyk S, Schohe K, Longère J Y, Leyral P, Hartnagel H L and Schutz R Z 1988 *Le Vide, Les couches Minces* **43** 241 p.193-194
Kuhara Y, Terauchi H and Nishizawa H 1985 *Proc. IOOC-EOOC* p.533-537
Le Bellégo Y, Renaud J C, Blanconnier P and Praseuth J P 1989 *Appl. Surf. Sci.* **39** p.168-177
Ota Y, Hu P H-S, Seabury C W and Brown M G 1987 *J. Appl. Phys.* **61** 1 p.404-410
Philippe P, Poulain P, Kazmierski K and de Crémoux B 1986 *J. Appl. Phys.* **59** p.1771-1773
Ripoche G, Decor Ph, Blanjot C, Bourdon B, Salsac P and Duda E 1985 *IEEE Electron. Dev. Lett.* **ED-6** 12 p.631-633

*Paper presented at ESSDERC 90, Nottingham, September 1990*
*Session 2B7*

# The application of selective electroless plating for microelectronics applications

C.NiDheasuna° T.Spalding°, A.Mathewson[†]

° Department of Chemistry University College Cork, Ireland
† National Microelectronics Research Centre, Cork, Ireland

### Abstract

The feasibility of using electroless Nickel deposition for contact hole filling in Si VLSI technology has been demonstrated. The fundamentals of nickel plating are observed in the effects of such variables as temperature, nickel concentration and pH. Three types of electroless nickel solutions have been examined. In the first bath, the reducing agent was Nickel Hypophosphite, in the second, DMAB and in the third, Hydrazine. Pretreatment of the wafers involved immersing the wafers in an organic solvent. Surface topography and planarisation has been investigated by scanning electron microscopy (SEM) while the plating solution composition was investigated using nuclear magnetic resonance spectroscopy (NMR).

### Introduction

The reduction of device geometries to the submicron regime that has taken place over the last few years has brought with it new problems which require extensive investment in new technology and expensive process chemicals. One of the most significant problems to be encountered in the current generation of CMOS technology is that of contact hole planarisation. This is necessary to provide reliable metal contacts to device terminals through the high aspect ratio contact holes that are now being produced in current fine geometry CMOS processes. This can be performed in a number of ways including the LPCVD of refractory metals and the laser melting of Al films to provide planarised contact holes and vias [1].

This work concentrates on a technique that has been applied to power device technology for a long time but has only recently become a potential candidate for high throughput VLSI technology metalisation schemes. The selective electroless plating of Ni onto silicon substrates provides an inexpensive and very high throughput contact fill and total metalisation scheme which has been reported in the recent literature [2]

### Experimental Work

The purpose of this work is to evaluate different electroless Ni plating baths in relation to their selectivity for the deposition of Ni onto silicon surfaces.

Three different reducing agents have been investigated

i] Boron based ($BH_3$-amine adducts) e.g. DMAB and Tert-butyl Amine Borane

ii] Hypophosphite based $[H_2PO_2]^-$ and

iii] Hydrazine Based, [ $NH_2NH_2H_2SO_4$] and [$NH_2NH_2H_2O$]

© 1990 IOP Publishing Ltd

The investigation focused on the determination of the conditions for good plating including an investigation of pH, temperature, time and concentration of various reagents in the various baths.
pH control is very important for the boron based and hypophosphite baths. The boron based baths require slightly acidic environments i.e. a pH of ~6, whereas the hypophosphite baths require alkaline conditions in order to achieve a measure of uniform plating. Plating is enhanced at elevated temperatures and this is particularly important for the boron based baths. The rate of plating increases linearly with an increase of temperature from 60°C to 80°C. All baths which involved hypophosphite or hydrazine reagents required temperatures in the range 90-95°C

For all three systems the concentration of various reagents in the bath were seen to be vital, The concentrations used in the Hypophosphite bath were varied as shown in Table 1 Solution 1 produced a coating with poor adhesion which was easily removed in a scotch tape test with only 10-12% of the film remaining. Solution 2 provided a film with good adhesion and uniformity. However, when hypophosphite was used as a reducing agent the deposit was seen to be doped with phosphorus [3] and these films have been studied using EDX analysis. Figure 1 shows an EDX spectrum of the deposit which shows the level of phosporous included in the film. The peaks which indicate the presence of Zn, Cu and Co are arrtifacts of the sample holder and can be neglected in this analysis.

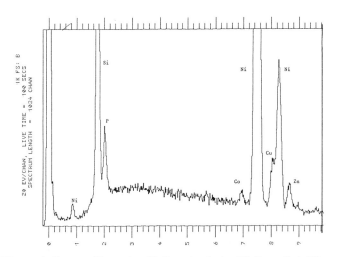

**Figure 1 Energy Dispersive X Ray Analysis Of Deposited Film**

| Solution | [NiSO4] mole.L$^{-1}$(x10) | Ni[H$_2$PO$_2$)$_2$.2H$_2$O mole.L$^{-1}$ (x1E-3) |
|---|---|---|
| 1 | 21.44 | 7.93 |
| 2 | 10.72 | 3.96 |
| 3 | 5.36 | 1.98 |
| 4 | 2.68 | 0.99 |

**Table 1 - Concentrations In Plating Baths Studied**

Adhesion, selectivity and plating rate were also investigated. Adhesion was seen to improve dramatically when a rinse in an organic solvent such as methanol, acetonitrile or chloroform was performed prior to deposition. The SEM studies shown in Figure 2a,b shows planarised and unplanarised contact holes which illustrate the selectivity and usefulness of the technique.

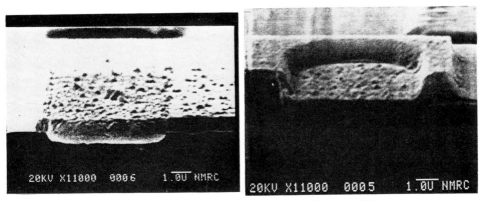

Figure 2 a,b Planarised And Normal Contact Hole

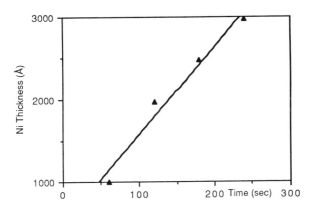

Figure 3 - Film Thickness vs Time

Figure 3 shows that a linear dependence exists between film thickness and elapsed time. The plating rate also shows a dependence on the composition of the nickel bath. When Boric Acid and glycine are added the plating rates increase from 8.3Å/sec to 21.64Å/sec and 12.79Å/sec respectively. It is believed that Boric Acid [4] acts as both a catalyst and buffer in the reaction. However, very few reports of the species present in the reducing baths have been studied.

We have initiated $B^{11}$ and $P^{31}$ NMR spectroscopic studies to elucidate some of the species present in the reaction. From these investigations two of the species in the nickel bath have been positively identified as DMAB and Boric Acid at -16ppm and 20 ppm respectively in the NMR spectrum shown in Figure 3. Further studies are required to correctly identify the other species in the solution but it is believed that DMAB forms a complex with lactic acid shown by the two peaks in the spectrum at 8 and 10 ppm

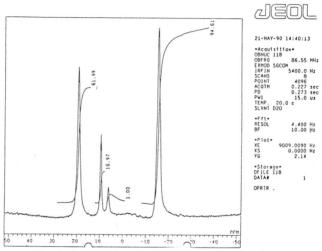

Figure 4 NMR Spectrum Of DMAB Plating Bath

## Conclusion

From our studies the bath which uses the hypophosphite salt as the reducing agent provides the most promising results in terms of adhesion, uniformity and selectivity, although a greater understanding of the chemistry of the reaction and its effect on Si devices is required before it can be used in a production environment. The presence of phosphorous in the depoited film has been observed but it is felt that this can be avoided by the use of the other systems under investigateion to provide equivalent results. The impact of this process on CMOS technology will also be presented.

## References

1] **Characterisation of Laser Planarised Aluminium For Submicron Double Level Metal CMOS Circuits**
D.Pramanik and S.Chen, IEDM89 P673

2] **Selective Electroless Metal Deposition For Integrated Circuit Fabrication**
C.H.Ting and M.Paunovic J.Electrochem.Soc. Vol 136 No 2 Feb 1989 P456

3] **The crystallisation of an Electroless Ni-P Deposit**
K.L.Lin and P.J.Lai J.Electrochem.Soc. Vol 136 No 12 Dec 1989 P3803

4] **Boric Acid as a Catalyst in Ni Plating Solutions**
J.P.Hoare J.Electrochem.Soc. Vol 134 No 12 Dec 1987 P3102

Paper presented at ESSDERC 90, Nottingham, September 1990
Session 2B8

## p-Type AuMn ohmic contact on GaAs: integration in a HBT processing technology

C.Dubon-Chevallier, A.M.Duchenois, A.C.Papadopoulo, L.Bricard, F.Héliot, P.Launay

Centre National d'Etudes des Télécommunications,
Laboratoire de Bagneux,
196 avenue Henri Ravéra, 92220 Bagneux, FRANCE

### Abstract

The influence of the p-type doping level in GaAs on the AuMn specific contact resistivity has been investigated, showing that AuMn could be used as an ohmic contact on epitaxial layers with a very large range of doping levels. The integration of this contact in the processing technology of heterojunction bipolar integrated circuits has been analysed, showing the influence of the processing steps preceding the contact evaporation. The contact resistivity has been found to be very sensitive to defects created by ion beam etching, making necessary a light chemical etch before depositing the contact.

### I- Introduction

The reliability of a component is directly related to the integrity of its ohmic contacts. It is particularly the case for Heterojunction Bipolar Transistors (HBTs) which require both n-type and p-type ohmic contacts. Low resistivity n-type ohmic contacts, either classical AuGeNi or refractory metallizations like GeMoW (C.Dubon-Chevallier 1989), can be achieved rather easily. Reliable low resistivity p-type ohmic contact is much more a problem. For this purpose, we have developed a p-type ohmic contact using AuMn metallization (C.Dubon-Chevallier 1985), which is among the most performant p-type ohmic contacts. The ohmic contact metallurgy as well as the annealing cycle has already been optimized and published. The main results of this previous study are that the specific contact resistivity obtained on GaAs layer with a doping level of $10^{19}$ at.cm$^{-3}$ is very low, $2 \cdot 10^{-7}$ $\Omega$.cm$^2$, the morphology remains very smooth after annealing and the Mn penetration is very shallow, less than 500 Å.
This AuMn contact has been extensively used for the fabrication of circuits implemented with HBTs. This experience revealed that the resistivity measured on real devices was often higher than what was usually measured during the optimization study. In this paper, after reporting complementary data on AuMn contact, we will investigate the reasons of this resistivity increase.

### II- Dependence of the resistivity with the p-type doping level

To complete the study on the AuMn p-type ohmic contact, we have investigated the variation of the resistivity with the p-type doping level. For this purposdÇ ohmic contact specimens were formed on Be-doped p-type GaAs layers grown by molecular beam epitaxy on semi-insulating substrates. Doping levels were in the range $10^{15}$-$10^{19}$ cm$^{-3}$. Prior to deposition, the samples were deoxidised in diluted HCl solutions, blown in dry nitrogen and transferred into the evaporation chamber through a preparation chamber. The metals were deposited using e-beam evaporation, Mn was deposited first followed by Au. The samples were annealed in a classical furnace at 300°C. The specific contact resistivity was deduced from TLM measurements. The curve giving the contact resistivity as a function of the p-type doping level is presented in figure 1. Several remarks can be pointed out. First, ohmic contacts are achieved with p-type doping level as low

as $10^{15}$ cm$^{-3}$; secondly there is a great dependence of the resistivity on the doping level. When the doping level is lower than $10^{18}$ cm$^{-3}$, the contact resistivity follows the law :
$$\rho\ (\Omega.\text{cm}^2) = 10^{18.66}/N_A^{1.38}$$

When the p-type doping level is higher than $10^{18}$ cm$^{-3}$, there is a much weaker dependence of the resistivity with the doping level.

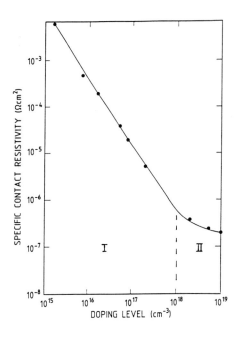

Fig.1 Dependence of the contact resistivity with the doping level.

These results are in good agreement with the model proposed by Dingfen (1986), which explains the $N_A^{-1}$ dependence of the resistivity when the doping level is lower than the average carrier concentration due to the contact formation, and the independence when the doping level is higher. The hole transport in domain I is controlled by a barrier between the highly and low doped semiconductor regions, while the hole transport in the domain II is controlled by the barrier between the metal and the highly doped semiconductor region induced by the contact formation.

However, if the AuMn contact presents these two domains according to the model, it is characterized by some differences as compared to conventional AuZn or AgZn contacts. The dependence of the resistivity with the concentration in the first domain is stronger: there is a variation of the resistivity as $N^{-1.38}$ against $N^{-1}$ for the other contacts; the average carrier concentration due to the contact formation is rather low, $10^{18}$ cm$^{-3}$ versus $10^{19}$ cm$^{-3}$ usually achieved. These results show the quality of this contact which permits to reach on a very large range of doping levels lower resistivities than what is generally achieved.

This contact has been used extensively in the fabrication technology of heterojunction bipolar integrated circuits. This experience revealed that the resistivity measured on real devices was often higher than what was usually measured during the optimization work. Study have then been carried out to investigate the reasons of this increase in resistivity. The first to be investigated was the influence of the damages induced by the etching process.

## III- Damages induced by etching

In the double mesa technology frequently used to process HBTs, the p-type ohmic contact is deposited on the GaAs base layer which has been exposed by etching. Chemical etching or dry etch techniques can be used for that purpose. We are using the ion beam etching technique, since it allows to obtain the uniformity and the reproducibility necessary for the realization of integrated circuits. It is also possible to control the isotropy of the etch by changing the incidence angle. However, this technique induces surface defects, making the realization of p-type ohmic contact rather difficult to achieve. We have then investigated by cathodoluminescence intensity measurements, the damages induced by ion beam etching for different ion beam energies. TLM measurements have also been realized to deduce the specific contact resistivity of AuMn ohmic contact deposited on etched layers.

We have investigated the amount of damage created by etching 500nm of GaAs at 250 and 400 eV. The samples were 5 $10^{18}$ cm$^{-3}$ doped, 1 µm thick GaAs layers grown by MBE. After each etch, cathodoluminescence intensity measurements have been realized on etched and protected regions of the same sample.

When the ion beam energy is 400 eV, the amount of damage is very important, chiefly close to the surface, giving a cathodoluminescence intensity which is merely 30% of the initial intensity. When the etch is achieved at 250 eV, it induces a lower decrease of the cathodoluminescence intensity which is about 60% of the initial intensity. Moreover, thermal treatments were found to be unable to insure the recovery of the material qualities. In order to eliminate the damaged surface layer, chemical etching was performed after the ion beam etching. We have then processed the layers in two steps, the first 500 nm were etched by ion beam etching either at 250 or at 400 eV, it was followed by a light chemical etch using the $H_3PO_4/H_2O_2/H_2O$, 3/1/40 solution, which etch rate is 100 nm/min. The etch time was varied for both samples between 20 and 75 s.

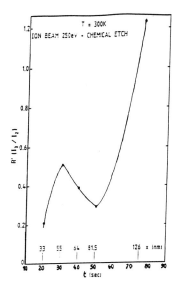 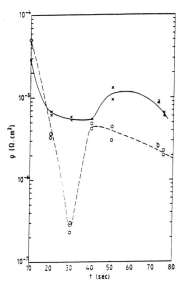

Fig.2: Evolution of the ratio $I_3/I_2$ with chemical etching duration, the ion beam energy being 250 eV.

Fig.3: Variation of AuMn contact resistivity with chemical etching duration. Chemical etching follows ion beam etching performed at a) 250 eV, b) 400 eV.

When the sample was etched at 400 eV, the chemical etch did not permit to recover the initial intensity, even with an etching time of 75 s. On the other hand, for the etching at 250 eV, it was almost possible to recover the quality of the etched region. In figure 2, we have plotted the ratio of the cathodoluminescence intensity after ion and chemical etching ($I_3$) to the intensity after ion beam etching only ($I_2$), as a function of the etching time. It appears that a chemical etch of 75 s, corresponding to an etch depth of 125 nm, permits to eliminate the damaged region and to recover about 75% of the initial intensity. TLM measurements have been realized on the samples, which present a doping level of $5 \cdot 10^{18}$ cm$^{-3}$, after different etching durations. In figure 3, the specific contact resistivity deduced from TLM measurements, is plotted as a function of the duration of chemical etch performed after the ion beam etching at 250 or 400 eV. After ion beam etching, there is an increase of at least two decades of the contact resistivity as compared to the result obtained on a sample without etching. For the sample etched at 400 eV, the chemical etch permits to improve the resistivity. However, the effect of chemical treatment is much more sensible in the case of the sample etched at 250 eV, since it permits to obtain contact resistivity as low as $10^{-6}$ $\Omega.cm^2$, within one decade from the value for unetched samples.

## IV- AlGaAs emitter layer etch

Another reason for the increase of contact resistivity is related to the presence of a remaining thin AlGaAs layer on top of the GaAs base layer. The $H_3PO_4$ solution used to contact the GaAs base layer does not exhibit any selectivity between AlGaAs and GaAs, the etch stop is then determined by electrical means. However, if this method is rather satisfying, it can happen that a thin AlGaAs layer remains on top of the GaAs layer. This point was put in evidence by electrical measurements achieved on processed devices. Among our test patterns, there is a TLM pattern realized on the etched base region, which permits to deduce the extrinsic base resistance after etching. There is another TLM pattern where only the regions on which the contacts will be deposited, are etched down to the base, the regions between the contacts being protected during the etch.

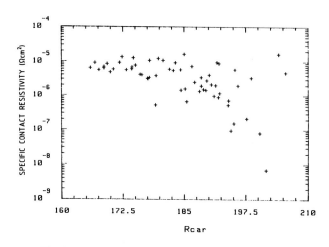

Fig.4: correlation curve between the contact resistivity and the extrinsic base resistance, the mean intrinsic base resistance is 185 $\Omega/\square$.

This pattern permits to deduce the base resistance intrinsic to the epitaxial structure. The comparison between these two values allows to detect whether or not the AlGaAs emitter layer has been totally removed. Moreover, the presence of a correlation between the contact resistivity and the extrinsic base resistance, as presented in figure 4, permits to show the drastic influence of a remaining AlGaAs layer. When the extrinsic resistance is lower than the intrinsic one, the contact resistivity is high ($10^{-5}$ $\Omega.cm^2$); as soon as the intrinsic base resistance is lower than the extrinsic one, meaning that the AlGaAs emitter layer has been totally removed, there is a decrease of the resistivity ($<10^{-6}$ $\Omega.cm^2$).

## V- Conclusion

We have presented a study on the integration of the AuMn p-type ohmic contact in the heterojunction bipolar circuits fabrication technology. We have pointed out two reasons which lead to measured resistivities higher than what was obtained during the optimization phase. The first reason is related to the defects created during the etching process, the second one is due to an imperfect etch of the AlGaAs layer. When the emitter etch is performed under good conditions, low resistivities are achieved, which permit to obtain on discrete devices with a non self-aligned technology, cut-off frequencies higher than 30 GHz.

## Acknowledgment

The authors will like to thank F.Alexandre and J.Riou for supplying the samples, J.Dangla, A.Scavennec and M.Bon for fruitful discussions.

## References

Dubon-Chevallier C, Henoc P, Glas F, Gao Y, Bresse J F, Blanconnier P, Besombes C 1989 Proc. ESSDERC.
Dubon-Chevallier C, Duchenois A M, Bresse J F, Ankri D, 1985 Proc. ESSDERC.
Dingfen W, Dening W, Heime K, 1986 Solid-St Electron. **29**(5) 487.

## A fast method of parameter extraction for MOS transistors

P R Karlsson and K O Jeppson

Dept of Solid State Electronics, Chalmers University of Technology, S-412 96 Göteborg, Sweden

> Abstract. A fast method of parameter extraction using a limited number of data points is developed for the SPICE level 3 MOS transistor model. Analytical expressions or numerical equations that converge fast are used to calculate the parameters and all interactions between parameters are taken into account. Proper selection of data points ensures physically reasonable values for most extracted parameters.

### 1. Introduction

The problem of extracting transistor parameters is as old as the transistor itself and many different methods have been proposed in the literature. Traditionally they can be classified into two different groups. In the first, special measurements are used for each parameter and interaction between different parameters is ignored. In the second, optimization techniques are used.

In this paper a fast method of parameter extraction using a limited number of data points is presented. The basic idea is to use only one data point for each transistor parameter as suggested by Tuinhout et al (1988). To achieve fast parameter extraction, analytical expressions (or numerical equations that converge fast) are used on a small number of data points. Proper selection of data points and taking account of interaction between different parameters ensures physically reasonable values on most extracted parameters. The extraction algorithm is developed for the SPICE level-3 MOS transistor model.

### 2. Extraction algorithm

The proposed extraction algorithm starts by determining initial values for the threshold voltage, $V_T$, the surface mobility, $\mu_0$, and the mobility modulation, $\theta$, due to a transverse electrical field. This is done in the linear region where the following equation is used in SPICE for the drain current, $I_{DS}$:

$$I_{DS} = \frac{\mu_0 C_{ox} \frac{W}{L} \left(V_{GS} - V_T - \frac{1+F_B}{2} V_{DS}\right) V_{DS}}{1 + \theta(V_{GS} - V_T) + \frac{\mu_0 V_{DS}}{v_{max} L}} \qquad (1)$$

The drain current is measured for three values of the gate source voltage, $V_{GS}$, with a small voltage applied between drain and source ($V_{DS}=50$mV). Three data points, ($V_{GS1}$, $I_{DS1}$), ($V_{GS2}$, $I_{DS2}$) and ($V_{GS3}$, $I_{DS3}$), (Figure 1) are enough to determine $V_T$, $\mu_0$ and $\theta$ from equations similar to Hamer's (1986):

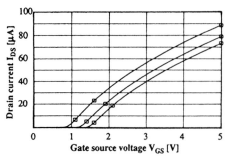
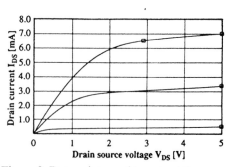

Figure 1. Data points used to extract the threshold voltage, the surface mobility and the mobility modulation in the linear region for three values of $V_{BS}$.

Figure 2. Data points used to extract the threshold voltage, the maximum drift velocity and the saturation field factor in the saturation region.

$$V_T = V'_T - \frac{1+F_B}{2} V_{DS} \qquad (2)$$

$$\theta = \frac{C_1}{1-aC_1-bC_2} \qquad (3)$$

$$\mu_0 = \frac{C_2}{1-aC_1-bC_2} \qquad (4)$$

where

$$V'_T = \frac{I_{DS1}I_{DS2}V_{GS3}(V_{GS2}-V_{GS1})-I_{DS1}I_{DS3}V_{GS2}(V_{GS3}-V_{GS1})+I_{DS2}I_{DS3}V_{GS1}(V_{GS3}-V_{GS2})}{I_{DS1}I_{DS2}(V_{GS2}-V_{GS1})-I_{DS1}I_{DS3}(V_{GS3}-V_{GS1})+I_{DS2}I_{DS3}(V_{GS3}-V_{GS2})} \qquad (5)$$

$$C_1 = [V_{GS3}(I_{DS2}-I_{DS1})+V_{GS2}(I_{DS1}-I_{DS3})+V_{GS1}(I_{DS3}-I_{DS2})] * $$
$$* \frac{V_{GS3}I_{DS3}(I_{DS2}-I_{DS1})+V_{GS2}I_{DS2}(I_{DS1}-I_{DS3})+V_{GS1}I_{DS1}(I_{DS3}-I_{DS2})}{(V_{GS3}-V_{GS1})(V_{GS3}-V_{GS2})(V_{GS2}-V_{GS1})(I_{DS3}-I_{DS1})(I_{DS3}-I_{DS2})(I_{DS2}-I_{DS1})} \qquad (6)$$

$$C_2 = \frac{L[V_{GS1}I_{DS1}(I_{DS3}-I_{DS2})+V_{GS2}I_{DS2}(I_{DS1}-I_{DS3})+V_{GS3}I_{DS3}(I_{DS2}-I_{DS1})]^2}{C_{ox}WV_{DS}(V_{GS3}-V_{GS1})(V_{GS3}-V_{GS2})(V_{GS2}-V_{GS1})(I_{DS3}-I_{DS1})(I_{DS3}-I_{DS2})(I_{DS2}-I_{DS1})} \qquad (7)$$

$a=(1+F_B)V_{DS}/2$ and $b=V_{DS}/(v_{max}L)$. $F_B=0$ and $v_{max}=\infty$ are used as starting approximations. Later on as $F_B$ and $v_{max}$ are determined, new values of $V_T$, $\mu_0$ and $\theta$ are calculated to improve accuracy.

To achieve proper selection of data points, the values of $V_{GS}$ are automatically adjusted with respect to the threshold voltage. For $V_{GS1}$ smaller than $V_T$ the extraction algorithm extracts a value of $V_T$ close to $V_{GS1}$. $V_{GS1}$ is then increased until $|V_T|+0.2 \leq V_{GS1} \leq |V_T|+0.4$. The other gate voltages are chosen as $V_{GS2}=V_{GS1}+0.5$ and $V_{GS3}=V_{DD}$, typically 5V.

These measurements are performed for three different values of the bulk-source voltage, $V_{BSi}$ (i=1, 2 and 3), and the threshold voltage, $V_{Ti}$, is determined for each bulk-source voltage. The threshold voltage is given by

$$V_{Ti}=V_{FB}+2\psi_F+\gamma F_{Si}\sqrt{2\psi_F+V_{BSi}}+F_N(2\psi_F+V_{BSi})-\sigma V_{DS} \quad i=1, 2 \text{ and } 3 \qquad (8)$$

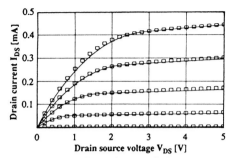
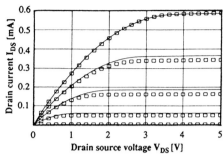

Figure 3. Plot of I-V characteristics comparing measured (squares) with simulated (lines) characteristics for an n-type transistor with L=2μm and W=3μm. $V_{GS}$ varies from 0 to 5V.

Figure 4. Plot of I-V characteristics comparing measured (squares) with simulated (lines) characteristics for an n-type transistor with L=30μm and W=50μm. $V_{GS}$ varies from 0 to 5V.

From this system of three linear equations the following parameters can be determined: the flatband voltage, $V_{FB}$ (neglecting $\sigma V_{DS}$), the bulk threshold parameter, $\gamma$, and the narrow-channel factor, $F_N$. To solve this system of equations the short-channel factor, $F_S$, and the surface potential, $2\psi_F$, are regarded as constants, calculated with $N_{SUB}=10^{22}m^{-3}$. Once $\gamma$ is determined a new value of $N_{SUB}$ can be calculated and the system of equations is solved again. This is repeated until all parameters are stable. Usually, only a few iterations are necessary. The SPICE level 3 model parameters $V_{TO}$ (zero-bias threshold voltage), $N_{SUB}$ (substrate doping) and $\delta$ (width effect on threshold voltage) are determined from the parameters $V_{FB}$, $\gamma$ and $F_N$.

In our measurements nine data points are used in the linear region to determine the five SPICE-parameters $\theta$, $\mu_0$, $V_{TO}$, $N_{SUB}$ and $\delta$. These nine data points make it possible to determine the $V_{BS}$ dependence of $\mu_0$ and $\theta$ but such dependencies are neglected in the SPICE level 3 model. Hence the price of four extra data points is paid to standardize the method of threshold voltage determination. Also, the threshold must be determined very carefully since the system of equations above (eq (8)) is very sensitive to variations in the threshold voltage (its condition number is around 40).

So far all parameters have been determined in the linear region, but the remaining parameters will be determined in the saturation region. First $\sigma$ is determined as $\sigma = - \Delta V_T / \Delta V_{DS}$ where $\Delta V_T$ is the decrease in the threshold voltage when $V_{DS}$ is increased $\Delta V_{DS}$ from 50mV to 5V. From $\sigma$ the SPICE-parameter $\eta$, known as the static feedback, is determined as

$$\eta = \sigma \frac{C_{ox} L^3}{8.15 \cdot 10^{-22}} \quad (9)$$

Once $\sigma$ is known the value of $V_{FB}$ can be improved by adding the initially neglected term $\sigma V_{DS}$.

The threshold voltage, $V_{T5}$, for $V_{DS}=5V$ in the saturation region can be determined using the same method as in the linear region if the square root of the saturation current is approximated by

$$\sqrt{I_{DSi}} = \frac{A(V_{GSi} - V_{T5})}{1 + B(V_{GSi} - V_{T5})} \quad i=1, 2 \text{ and } 3 \quad (10)$$

Finally the maximum drift velocity, $v_{max}$, and the saturation field factor, $\kappa$, are calculated from two values of $I_{DS}$ in the saturation region. Both drain current values are measured for the same gate voltage but with different drain voltages (Figure 2). One value of $V_{DS}$ is chosen close to the saturation voltage, $V_{DSAT}$. In short, one can say that $\kappa$ is determined from the slope of $I_{DS}$ vs $V_{DS}$ and $v_{max}$ is determined from the absolute current level (for $V_{DS}$ close to $V_{DSAT}$).

Using this method, the eight DC parameters in the linear region and saturation region are determined from only 13 data points. In the first round, interaction between parameters is partly neglected since each parameter is determined without knowledge of the parameters to be determined later. In the following rounds extraction can be improved by accounting for the interaction between parameters. In the linear region, the same equations as before are used to determine improved values of $\theta$, $\mu_0$, $V_{TO}$, $N_{SUB}$ and $\delta$. In the saturation region, approximations are no longer needed to determine $\sigma$ (or $v_{max}$ and $\kappa$ for that matter), these three parameters now being determined using three of the four data points in the saturation region.

The density of fast surface states, $N_{FS}$, is determined from the slope of the drain current versus the gate voltage in the subthreshold region. Two data points are used to calculate the slope.

## 3. Implementation

The extraction algorithm has been implemented on a HP-9000/340 work station and is written in HP-BASIC running under UNIX. The drain current is measured with a parameter analyzer (HP-4145A) and all set-ups are made automatically from the computer. Total time to extract all parameters is around half a minute. Only a minor part of the time is used by the computer to calculate the parameter values.

## 4. Results

The algorithm was tested on several different sized transistors of both N- and P-type available from Norchip with drawn lengths ranging from 2µm to 30µm and drawn widths ranging from 3µm to 50µm.

Figure 3 shows good agreement between measured and simulated characteristics for an N-channel transistor with L=2µm and W=3µm. In this graph, lines represent simulated curves and the squares represent measured data points. It is important to point out that these data points are measured for comparison with simulated curves only. During the parameter extraction process only a total of 13 data points were used. In this plot the average error is 2.3% and the peak error is 7.0%. These errors compare well with those determined by global optimization techniques using a large number of data points. The proposed extraction algorithm reaches comparable accuracy using only 13 data points.

The results from an extraction of a large N-channel transistor with L=30µm and W=50µm is shown in figure 4. In this graph of $I_{DS}$ versus $V_{DS}$ the average error is 10.5% and the peak error is 27.3%.

## 5. References

Hamer M F 1986, "First-order parameter extraction on enhancement silicon MOS transistors", *IEE PROCEEDINGS*, Vol. 133, Pt. 1, No 2, pp 49-54.

Tuinhout H P, Swaving S, Joosten J J M 1988, "A Fully Analytical MOSFET Model Parameter Extraction Approach", *1988 IEEE Proceedings on Microelectronic Test Structures*, Vol 1, No 1, pp 79-84.

## A charge and capacitance model for modern MOSFETs

T Smedes and F M Klaassen[1]

Eindhoven University of Technology, Faculty of Electrical Engineering,
Electron Devices Group, P.O.Box 513, 5600 MB Eindhoven, the Netherlands.

Abstract. A new physical, compact charge and capacitance model for long to submicron size MOSFETs is presented. It includes the important short channel effects and the effects of parasitic elements present in modern MOSFETs. The model is compared with numerical device simulations and measurements on actual devices. Good agreement is found between modelling, simulation and measurement.

### 1. Introduction

For a correct prediction of circuit behaviour, eg. in analogue or time critical digital applications, an accurate charge and capacitance model for the MOSFET is essential. Most previously published models either lack a physical basis or neglect effects that become increasingly important in modern MOSFETs. This paper will describe a model that includes such effects as a non-uniform doping profile, carrier velocity saturation, overlap of gate over source and drain and series resistances. The discussion will be divided in a section on the intrinsic part of the transistor (inversion channel) and a section on the extrinsic part with the parasitic elements.

### 2. The intrinsic part of the transistor

For application in a circuit simulator it is necessary to attribute the charges in the intrinsic transistor to its terminals. Thus the charge in the inversion channel has to be partitioned between the source and the drain terminal. Using a physical partitioning scheme (Oh 1980), all charges in the semiconductor have been calculated. The source charge is given as:

$$Q_s = W \int_0^{V_{DS}} \left(1 - \frac{x}{L}\right) q\, n(x)\, \frac{\partial x}{\partial V}\, dV,$$

where x denotes the lateral position in the channel relative to the source junction and V the quasi fermi level for the electrons. The other symbols take their usual meaning. A similar expression can be written for the drain charge.

The change of integration variable represented by the last factor of the integrand can be found from the accompanying DC model (De Graaff and Klaassen 1990). In this model carrier mobility degradation has been taken into account by:

$$\mu(x) = \frac{\mu_0}{\{1+\theta_A(V_{GS}-V_T)\}(1 + E_C^{-1}\partial V/\partial x)}$$

---
[1] also with PHILIPS Research Laboratories Eindhoven.

The first factor in the denominator describes the transversal field degradation, whereas the second factor describes the velocity saturation effect, where $E_c$ denotes the critical lateral electric field.

The effects of a threshold adjustment implant and the bulk charge have been calculated by a box integration method to obtain a bias dependent body effect factor.

Finally the gate charge is found from the charge neutrality condition. The resulting expressions are quite complicated, but of a closed form.

The capacitances are now found as the partial derivatives of the charges with respect to the terminal potentials:

$$C_{IJ} = \frac{\partial Q_I}{\partial V_J}, \qquad \text{with } I,J \in \{G,D,S,B\}.$$

As an example the expression for the drain-gate capacitance, as resulting from our analysis, is presented here:

$$C_{DG} = WLC_{OX}\left\{\frac{1}{2} + \frac{(1+\theta_c V_{DS})\varphi^2}{6(2V_{GT}-\varphi)^2}\left[1 + 2\theta_c V_{DS} - \frac{4}{5}\frac{(5V_{GT}-2\varphi)(1+\theta_c V_{DS})}{2V_{GT}-\varphi}\right]\right\}$$

where $V_{GT} = V_{GS}-V_T$, $\varphi = (1+\delta)V_{DS}$, $\theta_c = (LE_c)^{-1}$ and $\delta$ is the body effect factor.

The other symbols take their usual meaning. The first term represents the symmetric situation (no drain-source bias), while the second term describes the charge distribution in the non-symmetric situation. The influence of the velocity saturation is clearly seen. The transversal field degradation does not appear in the capacitance expressions, since this does not affect the charge distribution.

The definition of a complete set of capacitances, together with the previously mentioned condition for the gate charge ensures charge conservation within the circuit model. Since the charges are nonlinear functions of several variables, in general the capacitances will be non-reciprocal,

i.e. $C_{IJ} \neq C_{JI}$,

and even may be negative. This is illustrated by figure 1, showing several measured and modelled capacitance curves for a long transistor. The measurements were done with a PC controlled system based on the HP 4284 LCR meter.

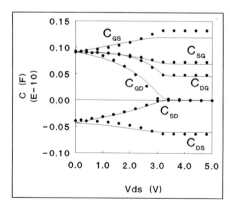 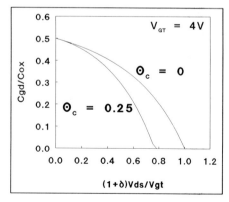

Figure 1: Measured (—) and modelled (•) capacitances vs. drain-source bias for a long transistor (W=L=100μm, $t_{ox}$=17.5nm).

Figure 2: Comparison of long channel ($\theta_c = 0$) and short channel ($\theta_c = 0.25$) capacitance curves.

The influence of velocity saturation on the intrinsic capacitances is illustrated by figure 2. It shows the gate-drain capacitance as a function of the drain-source bias normalized to the long channel saturation voltage. Clearly the short channel curve saturates before the long channel saturation voltage. A mere adaptation of the long channel expressions by using a corrected saturation voltage will result in a discontinuity in the capacitance curves, leading to erroneous circuit simulation results. The present model shows a continuous transition from the linear to the saturation region.

## 3. The extrinsic parts of the device

Since most modern MOSFETs are realized in an lightly doped drain (LDD) process, they may have considerable drain and source series resistances. As described by Klaassen (1988) the effects of the series resistances on the drain current can be expressed in an explicit form by use of generalized mobility reduction parameters. With this procedure the bias of the intrinsic transistor is known and therefore the intrinsic charges and capacitances can be calculated in an explicit form. Figure 3 shows the effect of a series resistance on the gate-drain capacitance. Note the much smaller slope of the curve at lower biases if the resistance is taken into account.

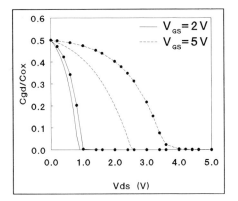
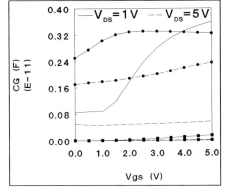

Figure 3: modelled intrinsic gate-drain capacitance with (•) and without (—) an LDD resistance.

Figure 4: Components of total gate-drain capacitance: intrinsic gate (—), source overlap (■) and drain overlap (•).

Another noticeable side effect of the LDD structure is the overlap of the gate over the source and drain implants. These overlaps cause capacitances in addition to the intrinsic capacitances. Numerical AC device simulations (CURRY 1990) were carried out to study this effect. In order to separate the intrinsic part from the extrinsic parts of the gate a small insulating gap was placed between these parts of the gate. All parts were kept at the same potential. Therefore it is possible to calculate the small signal currents through those separate parts. Figure 4 shows the resulting three components of the gate-drain capacitance for several bias situations. Clearly in the saturation mode of operation the drain overlap component is essential.

It can be shown (Cetner 1988) that these capacitances can be described as parallel plate capacitances, with a geometry dependent fringing field correction factor. However due to the relatively low doping concentrations of the LDD structure the above capacitance becomes bias-dependent since part of the region may become depleted or even inverted when the device is biased in the saturation mode (Figure 5). Therefore the effective overlap length is modulated. The resulting decrease of the capacitance can be seen in figure 4.

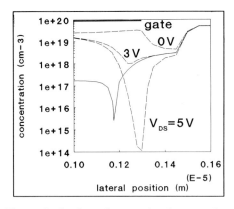
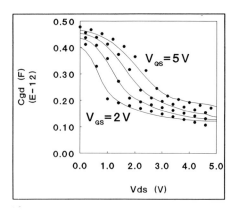

Figure 5: Surface electron (- -) and doping (—) concentrations at the drain side for 3 bias conditions with $V_{gs}$=3V.

Figure 6: Measured (—) and modelled (•) drain-gate capacitance of a submicron transistor (L=0.7μm, $t_{ox}$=17.5nm).

## 4. Results for a submicron transistor

As an example of the combined intrinsic and extrinsic effects figure 6 shows the measured and modelled gate-drain capacitance curves of a transistor with a drawn gate length of 0.7 μm (effective length ≈ 0.5 μm) and a gate width of 500 μm. Clearly the described smoother saturation behaviour (compared with a long channel transistor) can be seen. The error in the modelling is caused by the still rather simple approximation of the lateral doping profile of the LDD structure.

## 5. Conclusions

A compact charge and capacitance model for long to submicron MOSFETs was presented. It was shown by simulations and measurements that several parasitic elements of modern transistors have to be taken into account. The presented model included the effects of these elements in an explicit form. Good agreement is found between modelling, simulation and measurements for long as well as short channel MOSFETs.

## 6. Acknowledgement

This work is sponsored by the Dutch Foundation for Fundamental Research on Matter (F.O.M.) under project nr. EEL 46.017 of the Innovatively Oriented Research Project for IC Technology (IOP-IC).
We would like to thank the Philips Research Laboratories for providing the test samples and offering the opportunity to work with CURRY.

## 7. References

Cetner A, Iniewski K and Jakubowski A 1988 *Solid State Devices* **Vol 31** pp 973-4
CURRY *User reference manual V8.3* 1990 PHILIPS propriety
De Graaff H C and Klaassen F M 1990 *Compact Transistor Modelling for Circuit Design*.
 (Vienna: Springer-Verlag) pp 213-23
Klaassen F M, Biermans P T J and Velghe R M D 1988 *Proc. ESSDERC*
 (Les Ulis Cedex: Les Editions de Physique) pp 257-60
Oh S-Y, Ward D E and Dutton R W 1980 *IEEE J. of Solid-State Circuits* **SC-15**
 pp 636-43

*Paper presented at ESSDERC 90, Nottingham, September 1990*
*Session 2C3*

# Electron velocity overshoot in sub-micron silicon MOS transistors

P J H Elias, Th G van de Roer and F M Klaassen[1]

Eindhoven University of Technology, Faculty of Electrical Engineering,
Electron Devices Group, P.O.Box 513, 5600 MB Eindhoven, the Netherlands.

Abstract. A 1-D non-stationary hydrodynamic transport model is presented. Simulation results of $n^+$-n-$n^+$ drift devices using this model are compared with results using a Monte Carlo model, and it is shown that it is better not to include the heat flow in the model. Simulations of deep sub-micron NMOS transistors show velocity overshoot in the channel. These results are compared with experimental data and a good agreement is found.

## 1. Introduction

Electron velocity overshoot is due to the non-equivalence of electron momentum and energy relaxation times. In sub-micron silicon MOS transistors large field gradients give rise to this effect and may play an important role in device behavior. This has been shown experimentally by Shahidi et al (1988) and by Sai-Halasz et al (1988). To model velocity overshoot effects various different efforts have been made: Hänsch (1989) has modified the electron mobility in a drift diffusion approach in such a way that the electron velocity effect is taken care of. Probably a more realistic approach is obtained if energy relaxation is included explicitly in the model; this may be done by using the Monte Carlo methods such as used by Laux and Fischetti (1988), or by solving the hydrodynamic set of transport equations including the third moment of the Boltzmann equation (e.g. Kobayashi (1985), Forghieri (1988) and Feng (1988)). The latter method can not deal with some aspects (e.g. ballistic electrons) that a Monte Carlo method can handle, but its main advantage is that it needs considerably less computational effort.

## 2. Transport Parameters

In the model proposed the various relevant transport parameters (electron momentum relaxation time, electron energy relaxation time, and averaged electron effective mass), are considered to be functions of the local electron energy. These functions are obtained from Ensemble Monte Carlo simulations of bulk silicon. In the Monte Carlo model the physical parameters needed are obtained from Tang and Hess (1983). This includes non-parabolic X- and L-valleys, acoustic mode intra-valley scattering and optical mode intra-valley and inter-valley scattering. The Brooks-Herring model is used to include ionized impurity scattering. To model impact ionization scattering the model developed by Keldysh (1963) is used. Simulation results are given in Figures 1a and 1b. The effects of doping concentration (ionized impurity scattering) and weak avalanche (impact ionization scattering) on the transport parameters are taken into account. It is shown that the momentum relaxation time is strongly affected by the first, but hardly by the second; in contrast, the energy relaxation time is

---

[1] Also PHILIPS Research Laboratories Eindhoven.

hardly affected by the first, but strongly by the second; and the effective mass is not affected by either. Analytical expressions for the transport parameters are obtained by fitting procedures, that are valid within the energy range .04 eV to 1.5 eV.

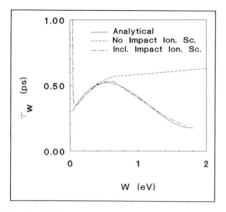 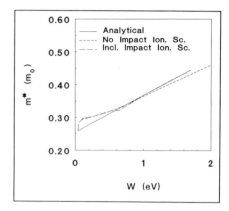

Fig.1. Silicon electron energy relaxation time (1a) and averaged electron effective mass (1b) as functions of average electron energy ($T_{lattice}$=300 K). The effect of the inclusion of impact ionization scattering in the simulations is shown.

## 3. Validity of the Hydrodynamic Model

Most authors include the heat flow in their hydrodynamic models, but others (Feng (1988)) neglect the heat flow completely. For this reason two different models are being compared: one that does include the heat flow (to be called the full hydrodynamic model), and one that does not (to be called the semi-hydrodynamic model). The full hydrodynamic model is a combination of Feng's approach (which allows also the inclusion of non-stationary effects) and Forghieri's (which allows the inclusion of the heat flow term, using the Wiedemann-Franz law). In both models also the velocity convective term $(v \cdot \nabla) \cdot v$ is included.

To check the validity of the hydrodynamic models a 1-D Ensemble Monte Carlo device model has been developed and all models are compared with each other and also with a drift-diffusion model. Only electron transport is considered. Simulations of several 1-D $n^+$-n-$n^+$-silicon drift devices with various values of drift region length and dope, show that the Monte Carlo and the hydrodynamic models are in good agreement with regard to the terminal device currents, giving substantially higher values than obtained from the drift-diffusion model. This proves the hydrodynamic model to be an adequate substitute for the Monte Carlo model. Some typical results are plotted in Figure 2a. The effects of the heat flow term on the terminal device currents appear to be rather small. However, if the internal charge, field, energy and electron velocity distributions are considered, clearly some differences occur between the two hydrodynamic models. The most dramatic one is plotted in Figure 2b: inclusion of the heat flow term causes a pronounced velocity overshoot at the drain side. This is not found if the semi-hydrodynamic model is used, and is also not found in the Monte Carlo simulations. This leads us to the conclusion that the way the heat flow is defined does not improve the model, and it is better to leave it out.

Simulations also show the effects of the velocity convective term to be negligible in the steady state, although it plays an important role in transient behaviour.

Some authors e.g. Jacoboni et al (1989) have simplified the Monte Carlo transport model by fixing a priori the device potential distribution to values calculated in a drift-diffusion simulation. Our simulations show a significant difference in the potential distribution for different models at high fields, and this makes it doubtful whether the method described can be justified.

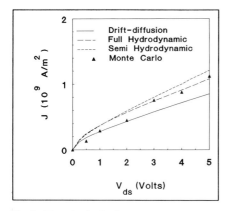
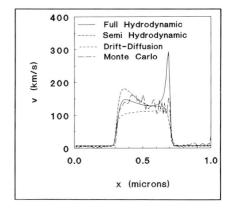

Fig.2. Terminal device current per unit area as a function of the applied voltages (2a) and electron velocity distribution as a function of the position in the device (2b) calculated by various models of a silicon $n^+$-n-$n^+$ drift device at T= 300 K.

## 4. Hydrodynamic NMOST Model

A 1-D silicon NMOST model has been developed that includes the hydrodynamic transport equations. Neither the heat flow term, nor the velocity convective term are included for reasons stated before. The surface potential $\psi_s$ is found by solving the Poisson equation, taking Gauss law at the interface directly below the gate contact; there

$$\psi_s = V_G - \psi_{ms} + C_{ox}(-q\,N_{Ad}W_d + \rho_s W_{inv} + \tfrac{1}{2}\rho_{ox}W_{ox} + \sigma_{sur})$$

$$W_d = 2\left[\frac{\psi_B \varepsilon_s}{q\,N_{Ad}}\right]^{1/2} \qquad \psi_B = \frac{k_B T}{q}\ln\left(\frac{N_{Ad}}{n_i}\right) \qquad \rho_s = q\,(p_s - n_s - N_{As})$$

$\rho_s$, $n_s$, $p_s$ represent the space charge, the electron and hole densities in the inversion layer; $N_{As}$, $N_{Ad}$ represent the acceptor concentration in the inversion layer and depletion layer, respectively; $W_{ox}$, $W_d$, $W_{inv}$, the thickness of the oxide, the depletion and inversion layer. All other parameters are defined as usual.
The inversion layer is assumed to be uniform in the vertical direction.

## 5. Results and Discussion

Calculations have been performed on deep sub-micron NMOS devices. The device sizes and doping profiles were based on the data given by Shahidi (1988). The thickness of the inversion layer is chosen to be 5 nm for a best fit with the characteristics of the 90 nm device given by Shahidi. This also agrees well with 2-D drift diffusion simulations using MINIMOS. To include surface scattering effects the ohmic channel mobility is taken to be 300 cm$^2$/Vs.
The simulations show that if the devices are operating in saturation, at the drain side velocity overshoot occurs up to two times the bulk saturation velocity. This is shown in Figure 3. From the results the average channel velocity is calculated in a similar way as was done by Shahidi and co-workers from their experimental results. A good agreement between experiment and simulations is obtained (Figure 4). Simulations show that the increase in average channel velocity if the channel length decreases is not in the main due to the fact that the electron velocity increases at shorter

channels, but to the fact that a high velocity occurs in a relatively larger part of the channel.

Similar results were obtained by Laux (1988) from Monte Carlo simulations based on other devices. But our model is of course much simpler and faster.

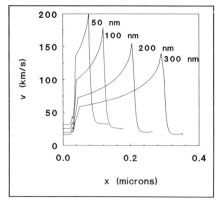
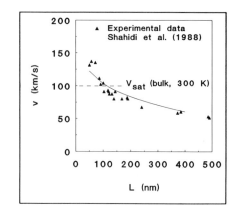

Fig.3. Electron velocity distribution in NMOSTs at various channel lengths ($V_{gs}$ = 2 Volts, $V_{ds}$ = 3 Volts).

Fig.4. Average channel velocity in NMOSTs as a function of channel length ($V_{gs}$ = 2 Volts, $V_{ds}$ = 3 Volts).

## 6. References

Feng Y and Hintz A 1988 *IEEE* **ED-35** 1419
Forghieri A, Guerrieri R, Ciampolini P, Gnudi A, Rudan M and Baccarani G 1988 *IEEE* **CAD-7** 231
Hänsch W and Jacobs H 1989 *Proc.ESSDERC'89* pp 583
Jacoboni C and Lugli P 1989 *The Monte Carlo Method for Semiconductor Device Simulation* (Springer-Verlag) pp 301
Keldysh L V 1965 *Sov.Phys.JETP.* **21** 1135
Kobayashi T and Saito K 1985 *IEEE* **ED-32** 788
Laux S E and Fischetti M V 1988 *IEEE* **EDL-9** 467
Sai-Halasz G A et al 1988 *IEEE* **EDL-9** 464
Shahidi G G, Antoniadis D A and Smith H I 1988 *J.Vac.Sci.Techn.B.* **6** 137
Tang J Y and Hess K 1983 *J.Appl.Phys.* **54** 5139

# New short-channel effects on nitrided oxide gate MOSFETs

H. S. Momose, T. Morimoto, S. Takagi, K. Yamabe, S. Onga, and H. Iwai

ULSI Research Center, Toshiba Corporation
1, Komukai-Toshiba-cho, Saiwai-ku, Kawasaki, 210, Japan,

## ABSTRACT

New short channel effects with nitride-oxide gate MOSFETs were found, where threshold voltage reduction occurs in a relatively long channel region. These effects would be explained by trapped charges or interface states induced by the mechanical stress at the Si and the nitride-oxide gate film in the course of the heat process.

## INTRODUCTION

Nitrided oxide gate films, produced by rapid thermal nitridation (RTN), have been investigated for use in lower submicron MOSFETs, because of their high TDDB (Time Dependent Dielectric Breakdown) reliability. Recently, an interesting effect of nitrided oxide films on mobility in n- and p-MOSFETs was reported by Momose et al (1989) and Wu et al (1989), where the mobility of the nitrided oxide gate n-MOS under high gate bias is larger than that of the pure oxide gate, while the mobility of the nitrided gate p-MOS is lower. The reason for this strange mobility phenomenon was first attributed to donor layer formation caused by nitrogen diffusion from the gate insulator into the substrate, which turns the nMOSFET into a buried channel type of device, and the pMOSFET into more of a surface type (Wu et al 1989). Recently, another reason has been proposed by Iwai et al (1990), suggesting that nitrided oxide induces tensile stress compared with pure gate oxide, which increases n-MOSFET mobility and makes p-MOSFET mobility lower. This paper reports a new MOS short channel effect, which is peculiar to nitrided oxide gate MOSFETs, is reported for the n-MOSFET case. The authors also show that this phenomenon is not explained by donor layer formation and that it is probably caused by mechanical stress.

## EXPERIMENTS AND RESULTS

N-MOSFETs with nitride-oxide gates and pure oxide gates were fabricated. In some samples, nitridation was accomplished by RTN (Rapid Thermal Nitridation). A stacked

© 1990 IOP Publishing Ltd

**Table 1.** Sample fabrication conditions

| | SAMPLE FABRICATION PROCESS | | | $T_{OX}$ (C-V) | N CONC. AT Si/SiO$_2$ INTERFACE |
|---|---|---|---|---|---|
| | FURNACE OXIDATION | NITRIDATION | RE-OXIDATION | | |
| PURE OXIDE | 10 nm | – | – | 10.8 nm | 0% |
| RAPID NITRIDATION a | 10 nm | 1200°C RTN | 1200°C RTO | 10.9 nm | 8.3% |
| b | | 1000 | 1000 | 10.9 | 2.8 |
| c | | 900 | 900 | 11.1 | 2.0 |
| d | | 1200 | 1000 | 10.5 | 8.4 |
| e | | 1200 | 900 | 10.4 | 9.8 |
| STACKED NITRIDATION f | – | LPCVD Si$_3$N$_4$ 6 nm | – | 4.9 nm | ? % |
| g | 5 nm | | – | 7.6 | 1.0 |
| h | – | | FURNACE OXIDATION | 6.0 | ? |
| i | 5 nm | | | 8.9 | 4.0 |

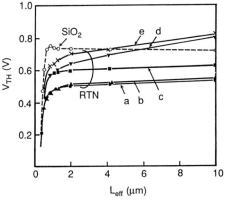

**Fig 1.** $V_{th}$ dependence on $L_{eff}$ for several surface channel n-MOSFETs with RTN gate oxide.

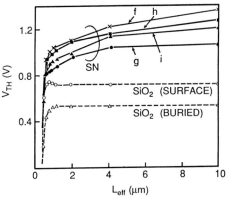

**Fig 2.** $V_{th}$ dependence on $L_{eff}$ for several surface channel n-MOSFETs with stacked nitride gates, in comparison with surface and buried pure gate oxide samples.

nitride film, deposited by LPCVD onto the oxide film, was also used as a gate insulator. Sample fabrication conditions are listed in Table 1. Figures 1 and 2 show the threshold voltage dependence on the effective channel length. In both cases, some of the nitride-oxide n-MOSFETs show reduced threshold voltage even when $L_{eff}$ is 4 µm. The threshold voltage reduction in nitrided gate n-MOSFETs may be explained by donor layer formation. P$^+$ poly-gate buried-channel n-MOSFETs with an arsenic implanted layer were fabricated. The short channel effects were measured and plotted, as shown in Fig.2. It should be noted that, when $L_{eff}$ was 4 µm, no threshold voltage reduction was observed. In general, in the pure gate oxide case, when the source and drain junction depths are about 0.2 μm, as in this case, no threshold voltage reduction at $L_{eff}$ = 4 μm is observed, even in a buried channel device. Thus, the significant threshold voltage reduction in nitrided oxide gate n-MOSFETs cannot be explained by the formation of a donor layer.

Figure 3 shows an SIMS profile of the nitrogen in the silicon substrate. This shows that there is very little difference between the nitrided oxide gate sample (Fig.3 (a)) and a plain Si sample (Fig.3 (b)), so it cannot be stated that donor layer formation occurs preferentially in the nitride-oxide gate samples.

Figure 4 shows the threshold voltage reduction dependence on the re-oxidation temperature. It should be noted that, when the re-oxidation temperature is higher, the threshold voltage reduction becomes small. Donor layer formation due to nitrogen diffusion cannot explain this phenomenon, which is probably related to the mechanical stress as in the mobility case. High temperature re-oxidation will relax stress at the silicon oxide interface.

Figure 5 compares the threshold voltage dependence on the channel length with the stress dependence when the wafer is bent, reported by Hamada et al (1990). Similar dependence on threshold voltage reduction was observed, which further suggests that stress is the cause for the new short-channel effects. Figure 6 shows the I-V characteristics for n-MOSFETs, where $L_{eff}$ values are 10 μm and 4 μm. The long-channel n-MOSFET with nitrided oxide gate seems to have more negative charges in the oxide due to the mechanical stress.

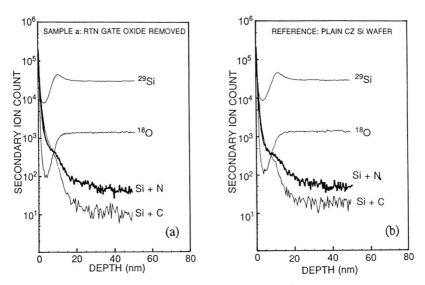

**Fig 3.** Si, O, N, C profiles measured by SIMS ($^{18}O_2^+$ source)
(a) RTN sample: gate oxide removed
(b) Reference: plain Cz-Si Wafer

## SUMMARY AND CONCLUSION

Short channel effects were compared between pure oxide and nitrided oxide gate n-MOSFETs.

Some of the nitrided oxide gate samples showed significant threshold voltage reduction at relatively long channel region.

The cause for this short-channel effect was investigated in detail.

These effects are probably explained by the mechanical stress in the nitride-oxide gate film in the course of heat process.

## REFERENCES

Hamada A, Furusawa T and Takeda E  1990 *Dig. of Tech. Papers VLSI Symp. on Tech. (Honolulu)* pp 113-114

Iwai H, Momose H S, Takagi S, Morimoto T, Kitagawa S, Kambayashi S, Yamabe K and Onga S  1990 *Dig. of Tech. Papers VLSI Symp. on Tech. (Honolulu)* pp 131-132

Momose H S, Kitagawa S, Yamabe K and Iwai H  1989 *IEDM Tech. Dig.* pp 267-270

Wu A T, Chan T Y, Murali V, Lee S W, Nulman J and Garner M  1989 *IEDM Tech Dig.* pp 271-274

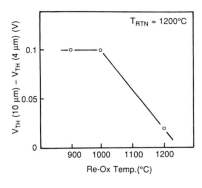

**Fig 4.** $V_{th}$ shift dependence, between $L_{eff}$ = 10 μm and 4 μm, on re-oxidation temperature.

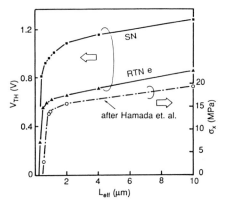

**Fig 5.** Comparison between $V_{th}$ and $\sigma_x$ (mechanical strain) dependences on $L_{eff}$.

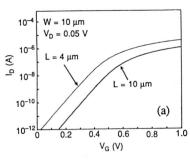

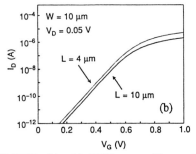

**Fig 6.** $I_d$ - $V_g$ characteristics for n-MOSFETs (a) with RTN gate oxides, and (b) with pure gate oxides.

Paper presented at ESSDERC 90, Nottingham, September 1990
Session 2C5

## Analytical model for circuit simulation with quarter micron MOSFETs: subthreshold characteristics

M. Miura-Mattausch and H. Jacobs

Siemens AG, Corporate Research and Development, Otto-Hahn-Ring 6, Munich 83

Abstract: We show a new simple model which includes the gradient of the lateral electric field analytically. The model describes the subthreshold characteristics relating to short-channel effects correctly down to $0.1\mu m$ effective channel length $L_{eff}$ with physical parameters $(N_{sub}, C_{ox}, V_{fb}, \phi_f)$ taken from the long-channel device.

## 1 Introduction

For deep submicron MOSFETs short-channel effects dominate the transistor characteristics. For transistors with reduced channel lengths, the drain current $I_D$ in the subthreshold region is no more independent of the drain voltage $V_D$ but increases in absolute value. As a result, three important phenomena are observed (DeGraaff 1989). One is a reduction of the threshold voltage $V_{th}$. Second is a reduction of the body effect. Third is an increase of the subthreshold swing for increased $V_D$. The widely used charge-sharing model (Yau 1974) is no longer appropriate for deep submicron MOSFETs.

By reducing the channel length but keeping applied voltages and other parameters as they are, only one physical quantity is changed, that is, the lateral electric field which becomes larger. The contribution of the lateral electric field causes the 2D current flow in the channel. There are several publications showing analytical solutions of the 2D Poisson equation (Toyabe 1979, Ratnakumar 1982). However, either the solutions are too complicated for CAD applications or rather crude approximations are made. We show here that a simple analytical model can be given by combining theory and experiment. The model describes all short-channel effects in a self-consistent way.

## 2 Theory and Results

### 2.1 Lateral electric field

A theory can be developed by applying the Gauss law to a narrow polygon in the depletion region under the gate. Under the assumption that the lateral electric field is independent of the vertical position, we get a simple relation between the vertical electric field at the surface $E_x$ and the gradient of the lateral electric field $E_y$

$$E_x(y) + \sqrt{B(\phi_s(y) - V_{sub})}E_{yy}(y) = \sqrt{2\epsilon_{Si}qN_{sub}\phi_s(y)} + Q_m, \quad (1)$$

$$B = 2\epsilon_{Si}/(qN_{sub}), \qquad E_{yy}(y) = dE_y(y)/dy$$

where $N_{sub}$ is the substrate doping, $\phi_s$ is the surface potential, and $Q_m$ is the mobile charge density. Fig.1 shows the vertical field $E_x$ at the surface calculated by the 2D simulator MINIMOS and by Eq. 1. In the analytical calculation the lateral field gradient is taken from the MINIMOS result. The reduction of the maximum $E_x$ seen in the MINIMOS result can be reproduced by including the lateral electric field. The reduction is known as the drain-induced barrier lowering, which corresponds to the reduction of $V_{th}$ for short-channel devices.

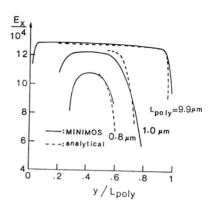

Fig.1. $E_x$ distribution along the channel calculated by MINIMOS. Broken curves are results obtained by our theory. Applied voltages are the respective $V_G = V_{th}$ for each $L_{poly}$, $V_D = 4V$, and $V_{sub} = 0$.

## 2.2 The dependence of $\Delta V_{th}$ on $L_{eff}$

The threshold voltage shift from the long-channel device $\Delta V_{th}$ is written

$$\Delta V_{th} = A\sqrt{B\phi_s} E_{yy}, \quad (2)$$

$$A = \epsilon_{Si}/C_{ox} \quad , \quad \phi_s = \phi_{s0} - V_{sub}$$

where $\phi_{s0}$ is the surface potential at threshold for a long-channel transistor. By giving an approximation that the surface potential is a linear function of $V_G$ at threshold, we can eliminate the surface potential from Eq. 2. The equation can be further simplified according to the magnitude of $E_{yy}$

$$\Delta V_{th} \simeq \sqrt{c} A \sqrt{B} E_{yy} \quad \text{for} \quad E_{yy} \leq 10^9 \frac{V}{cm^2}, \quad (3)$$

$$\Delta V_{th} \simeq a A^2 B E_{yy}^2 \quad \text{for} \quad E_{yy} \gg 10^{10} \frac{V}{cm^2}. \quad (4)$$

The parameters $a$ and $c$ define the gradient and the intersection of $\phi_{s0}$ as a function of $V_G$ around threshold. These parameters are calculated by solving the Poisson equation incorporating the contribution of $E_y$ (cf. Eq. 1). Values for $N_{sub} = 6 \times 10^{16} cm^{-3}$ are about 0.3 and 0.55V, respectively.

For an analytical model description we need an expression for the lateral field gradient. We assume a quadratic function for the lateral potential. Figure 2 shows $\Delta V_{th}$ values as a function of $L_{eff}$ calculated using this assumption. In the calculation no fitting parameter is used. The value of $N_{sub}$ is determined from the measured $V_{th}$ for the long-channel transistor. The result of the charge-sharing model is depicted as well. The MINIMOS result shows that the magnitude of $E_{yy}$ for short-channel devices is around $1 \times 10^9 V cm^{-2}$. In this case Eq. 3 is valid. Since the potential distribution is assumed to be a quadratic function, the dependence of $\Delta V_{th}$ is $L_{eff}^{-2}$. If $E_{yy}$ becomes larger than $10^{10} V cm^{-2}$, the dependence becomes $L_{eff}^{-4}$. By comparing the calculated and measured $\Delta V_{th}$, we can estimate $L_{eff}$ which is equal to $L_{poly} - 0.5 \mu m$ for transistors studied here.

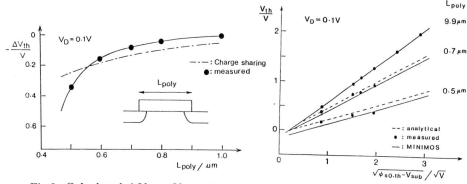

Fig.2. Calculated $\Delta V_{th}$ at $V_{sub} = 0$. The result by the charge-sharing model is shown by a stippled curve.

Fig.3. Measured and calculated $V_{th}$ as a function of $\sqrt{\phi_{s0} - V_{sub}}$, where $\phi_{s0}$ is the surface potential.

This is in good agreement with 2D doping profile simulations. For comparison measured $\Delta V_{th}$ values are also shown in Fig.2.

## 2.3 The dependence of $\Delta V_{th}$ on $V_{sub}$

Figure 3 shows the $V_{th}$ dependence on $\sqrt{\phi_{s0} - V_{sub}}$. The agreement with measurements is very good. Thus the reduction of the body coefficient can be described correctly by the contribution of the lateral field gradient. A square root dependence on $V_{sub}$ can be seen as well as in the MINIMOS result. This suggests that the lateral field gradient is independent of $V_{sub}$. On the other hand, the charge-sharing model shows a linear dependence of $\Delta V_{th}$ on $V_{sub}$.

## 2.4 The dependence of the subthreshold swing on $V_D$

The theory can be extended to get an equation to evaluate the gate voltage swing S

$$S/\log 10 = (1 + \frac{\gamma}{2\sqrt{\phi_s}} - \frac{A\sqrt{B}}{2\sqrt{\phi_s}}E_{yy} - A\sqrt{B\phi_s}\frac{dE_{yy}}{d\phi_{s0}})/\beta, \qquad (5)$$

where $\gamma$ is the body coefficient and $\beta$ is the inverse of the thermal voltage. The potential distribution in the inversion layer is approximated by the quadratic function. As $V_D$ increases, pinch-off occurs at the drain side. This is commonly treated as a reduction of $L_{eff}$ by $\Delta L$. Figure 4 shows the measured dependence of $\Delta V_{th}$ on $V_D$. This measurement is used to calculate $\Delta L$. With increasing $V_D$ values the subthreshold swing increases for short-channel transistors. This can be calculated with estimated $\Delta L$. The result in Fig. 5 shows good agreement with measurement.

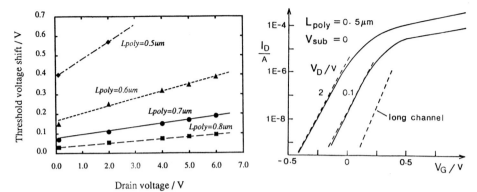

Fig.4. Measured $\Delta V_{th}$ as a function of $V_D$ at $V_{sub} = 0$.

Fig.5. Subthreshold characteristics for the short-channel device. Solid curves are measurements and broken lines are calculated gradients. The gradient of the long-channel device is shown for comparison.

## 3 Conclusion

Our model which includes the lateral electric field can describe all subthreshold phenomena in a self-consistent way. The potential distribution along the channel is approximated by a quadratic function, which seems suitable. Calculated reduction of the body coefficient and calculated increase of the gate voltage swing agree well with measurements with $L_{eff}$ and its reduction as a function of $V_D$ evaluated by combining the theory and the experiment.

**Acknowledgements:** We would like to thank W. Neumüller and D. Takacs for the measurements.

DeGraaff H C and Klaassen F M 1989 *Compact Transistor Modelling for Circuit Design* (Wien: Springer-Verlag)
Ratnakumar K N and Meindl J D 1982 *IEEE J. Solid-State Circuits* SC **17** pp 937-47
Toyabe T and Asai S 1979 *IEEE Trans. Electron Devices* ED **26** pp 453-61
Yau L D 1974 *Solid-State Electron.* **17** pp 1059-63

*Paper presented at ESSDERC 90, Nottingham, September 1990*
*Session 2C6*

# Small-signal modelling of MOSFET for circuit design applications

## V. Altschul, E. Finkman, D. Lubzens

Technion – Israel Institute of Technology, Haifa 32 000, Israel

### Abstract

We propose simple, explicit expressions for the gate–bulk incremental capacitance and terminal conductances of a long–channel MOS transistor. The new expressions are valid in the moderate inversion region and consequently are superior to the traditional circuit simulation MOS models. We compare results to a numerical charge–sheet model and experimental measurements. Because of the explicit dependence on applied voltages, the new expressions are computationally efficient. They allow a semi-empirical inclusion of the second–order effects similar to the traditional models.

### 1. Introduction

Most MOS transistor models currently used in circuit design were originally created for digital application. Consequently, they provide accurate description of the device behavior well above and far below the threshold voltage. This is not adequate for analog design where the intermediate range of gate voltages, known as the moderate inversion region (Tsividis 1982), must be accurately modeled as well. In SPICE (Antognetti at al 1988) and BSIM (Sheu 1985) the moderate inversion region is characterized by numerical splicing of the above- and sub-threshold characteristics. A particularly large error appears in the in the small-signal modeling where equivalent circuit parameters are derived by differentiation of inaccurate large-signal expressions. A numerical large–signal device simulation model of Pao and Sah (1966) relies on double integration of Poisson's equation. An iterative solution was used in the charge sheet approach, Brews (1978), Baccarani et al (1978) for example. A comprehensive small–signal model based on a charge sheet approach was suggested by Turchetti et al (1983). Selberherr et al (1980) applied advanced numerical techniques in a two-dimensional MOS model. All these methods are too inefficient computationally to be useful in circuit design. Also, they do not allow easy parameter extraction and empirical inclusion of the second order effects.

Thus, the goal of this work is to derive simple analytical expressions for MOS gate-bulk capacitance $C_{GB}$, drain and source conductances $g_d$ and $g_s$, and transconductance $g_m$. The models should be valid in moderate inversion and allow easy inclusion of the second-order effects.

© 1990 IOP Publishing Ltd

Consider a long-channel, uniformly doped, idealized NMOS structure. We define all potentials in respect to the bulk which reflects the physical symmetry of the device. Inversion charge density $Q_I$ can be expressed from the charge and potential balance as

$$Q_I = C_{ox}(V_G - V_{FB} - \psi_s) - Q_B. \tag{1}$$

Here $C_{ox}$ is the oxide capacitance and $V_{FB}$ is the flat-band voltage. The ionized impurity charge density $Q_B$ is expressed using the depletion approximation as $Q_B \simeq C_{ox}\gamma\sqrt{\psi_s - \phi_t}$ where $\gamma$ is the body effect coefficient and $\phi_t = KT/q$ is the thermal potential. The onset of moderate inversion was defined by Tsividis (1982) for $\psi_s$ greater than $2\phi_f + \phi_n$. Here $\phi_f$ is the Fermi potential and $\phi_n$ is the electron quasi–Fermi potential, varying along the channel between source voltage $V_S$ and drain voltage $V_D$. In traditional circuit simulation models $\psi_s$ is assumed constant in moderate and strong inversion pinned to some value $\phi_B \geq 2\phi_f + \phi_n$. Tsividis (1987) demonstrated that variation of the surface potential with applied voltage strongly affects device behavior in moderate inversion. Using Boltzmann's approximation of the semiconductor statistics and with the help of Poisson's equation, $\psi_s$ has been approximated in moderate inversion by Altschul et al (1990) as

$$\psi_s \simeq 2\phi_f + \phi_n + \ln\left[\xi_1(V_G - V_{TO})^2 + \xi_2(V_G - V_{TO}) + 1\right]. \tag{2}$$

$V_{TO}$ is defined as the gate voltage corresponding to the onset of moderate inversion $V_{TO} = V_{FB} + 2\phi_f + \phi_n + \gamma\sqrt{2\phi_f + \phi_n}$. Coefficients $\xi_1 = 1/(\gamma^2\phi_t)$ and $\xi_2 = (1 + 0.5\gamma/\sqrt{2\phi_f + \phi_n})^{-1}$ were obtained analytically from boundary conditions in the weak and strong inversion limits (Altschul et al 1990). We compare $Q_I$ from (1) and (2) with an iterative charge sheet model in Fig.1; a traditional approximation for $Q_I$ (Tsividis 1987, p.59) was used in weak inversion.

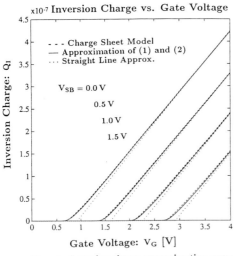
Fig. 1. Inversion charge approximation compared to a charge sheet model and a traditional strong inversion model for different substrate biases $V_{SB}$.

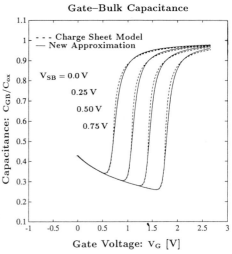
Fig. 2. Comparison of the new model for $C_{GB}$ with a charge sheet model for different substrate biases $V_{SB}$.

Equations (1) and (2) provide explicit expressions for $\psi_s$ and $Q_I$ in terms of applied voltages and process parameters. These equations are now used in the derivation of the small-signal parameters in moderate and strong inversion.

## 2. Incremental Gate–Bulk Capacitance

An analytical expression for the incremental gate–bulk capacitance $C_{GB} \equiv \partial Q_G/\partial V_G$ is highly desirable in small–signal MOS modeling. Unfortunately, such an expression exists only in the depletion and weak inversion (Antognetti et al 1988, p. 152). Assuming that parasitic charges do not depend on $V_G$ we can write from the charge balance $\Delta Q_G = \Delta Q_I + \Delta Q_B$. Thus, the behavior of $C_{GB}$ is controlled by the variations of the bulk and inversion charges with $V_G$. Using absolute values of all charges $C_{GB}$ can be written with the help of (1) as

$$C_{GB} = \frac{\partial (Q_I + Q_B)}{\partial V_G} = C_{ox}\left(1 - \frac{\partial \psi_s}{\partial V_G}\right). \qquad (3a)$$

Now by using (2) we obtain an analytical approximation for $C_{GB}$ in moderate and strong inversion:

$$C_{GB} \simeq C_{ox}\left[1 - \phi_t \frac{2\xi_1(V_G - V_{TO}) + \xi_2}{\xi_1(V_G - V_{TO})^2 + \xi_2(V_G - V_{TO}) + 1}\right]. \qquad (3b)$$

The new model for $C_{GB}$ is compared to a charge sheet model in all regions of inversion in Fig. 2.

## 3. Transconductance and Drain Conductance

We define the small–signal terminal conductances as derivatives of the drain current $I_D$ in respect to applied voltages. An analytical expression for $I_D$ is derived usually by an approximation of the physical model that includes both the drift and diffusion current components:

$$I_D = \mu^* \frac{W}{L} \int_{V_S}^{V_D} Q_I \, d\phi_n . \qquad (4)$$

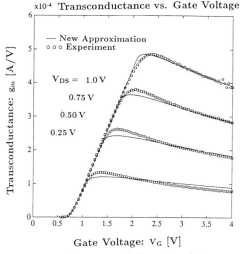

Fig. 3. The new model for transconductance compared to the experimental measurements of MOS transistor. $N_A \simeq 3.8 \times 10^{16}$, $D_{ox} = 250$ Å, $V_{FB} \simeq -0.75$ V.

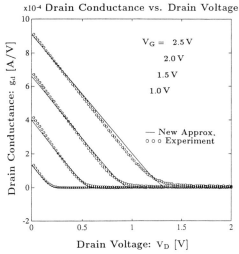

Fig. 4. The new model for drain conductance compared to experimental measurements of the same device as in Fig.3.

Here $\mu^*$ is the effective surface mobility, $W$ and $L$ are the length and width of the device. An approximate model for $I_D$ using (1), (2), and (4) was presented by Altschul et al (1990). A derivative of a function usually exhibits a larger error than the function itself. Consequently, expressions for conductances, derived by differentiation of an approximate model for $I_D$, will demonstrate a larger error than $I_D$. We try to use analytical models for terminal conductances obtained without differentiation of $I_D$ and apply then $Q_I$ and $\psi_s$ from (1) and (2). The drain small–signal conductance is derived directly from (4), as in Baccarani et al (1978), resulting in a precise formulation in all regions of inversion:

$$g_d \equiv \frac{\partial I_D}{\partial V_D} = \mu^* \frac{W}{L} Q_I(V_D) . \tag{5}$$

The final expression for $g_d$ is obtained from (5) using $Q_I$ of (1) and (2) evaluated at $\phi_n = V_D$. A model for $g_s$ can be similarly obtained. An expression for transconductance $g_m$ is derived with the help of (2) using an approximation suggested by Brews (1978)

$$g_m \equiv \frac{\partial I_D}{\partial V_G} \simeq \mu^* \frac{W}{L} C_{ox} [\psi_s(V_S) - \psi_s(V_D)] . \tag{6}$$

A comparison of the new conductances models with experimental measurements of a production quality NMOS transistor is presented in Figures 3 and 4. The field dependent mobility was modeled empirically (Sheu 1985) and non-uniform substrate doping was included in the description of $Q_B$.

## 4. Conclusions

Thus, we propose new simple models for the gate–bulk incremental capacitance and terminal conductances. The models are derived with the help of analytical expressions for the inversion charge density and surface potential. Differentiation of inaccurate large–signal approximations was avoided. The expressions are valid in moderate and strong inversion and can be matched with known approximations in the subthreshold regime of operation. The second order effects can be included empirically similar to the traditional approach. Therefore, the new models seem suitable for simulation of the small-signal MOS parameters in analog circuit design.

## References

V. Altschul, Y. Shacham-Diamand, to be published in *IEEE Trans. on Electron Devices*, July 1990.

P. Antognetti, G. Massobrio "Semiconductor Device Modeling With SPICE", McGraw-Hill, 1988.

G. Baccarani, M. Rudan, G. Spadini, *Solid–St. and Electron Dev.*, vol. 2, pp. 62, 1978.

J. R. Brews, *Solid–State Electron.*, vol. 21, pp. 345-355, 1978.

H. C. Pao and C. T. Sah, *Solid–State Electron.*, vol.9, pp.927–937, 1966.

S. Selberherr, A. Shutz, H.W. Potzl, *IEEE Trans. on Electron Devices*, vol. ED-27, p.1540, 1980.

B.J. Sheu, *Memorandum UCB/ERL M85/85*, UC Berkeley, 1985.

Y. Tsividis, *Solid–State Electron.*, vol. 25, p. 1099, 1982.

Y. Tsividis, "Operation and Modeling of the MOS Transistor", McGraw-Hill, 1987.

C. Turchetti, G. Masetti, and Y. Tsividis, *Solid–State Electron.*, vol. 26, pp. 941, 1983.

*Paper presented at ESSDERC 90, Nottingham, September 1990*
*Session 2C7*

# Numerical simulation of MOS devices with non-degenerate gate

Predrag Habaš and Siegfried Selberherr
Institute for Microelectronics
Gußhausstrasse 27-29, 1040 Vienna, Austria

> **Abstract.** In order to analyze implanted polysilicon-gate devices our MOS-device simulator MINIMOS has been extended to solve the basic semiconductor equations also in the poly-gate area self-consistently. Heavy doping effects in the gate as well as surface charge at the gate/oxide interface have been taken into account. The impact of some technological parameters related to the poly-gate effect on MOS-device performance is studied.

## 1. Introduction

Implanted gate MOS-devices have become common in submicron technologies. Due to higher diffusivity of the impurities along the grain boundaries in polysilicon compared to the single crystal it is possible to achieve high chemical concentration of the impurities near the gate/oxide interface, although the poly-gate thickness is larger than source/drain junctions depth. However, a significant part of the impurities (phosphorus or arsenic) segregate at grain boundaries and remain there non-activated after annealing (Mandurah et al. 1980). Therefore, the activated impurity concentration at the gate/oxide interface in N-type gates can be significantly lower than the chemical concentration (Sun et al. 1988), which depends on many parameters of the technological process (e.g. type of impurity, grain size, annealing cycle). The chemical concentration of impurity in P-type gates is usually lower than in N-type. Schwalke et al. (1988) report a saturation of boron chemical concentration at $1 \div 2 \cdot 10^{19} cm^{-3}$ in silicide/polysilicon gate structures. Therefore, in spite of high activation of boron (the absence of boron segregation at grain boundaries has been reported by Mandurah et al. 1980), the final activated impurity concentration in P-type gate can be low, too.

Shift of the high-frequency C-V curve (Wong et al. 1988) as well as the degradation of the quasi-static C-V curve (Chapman et al. 1988, Lu et al. 1989) has been experimentally observed in implanted poly-gate devices. The latter effect suggests a reduction of the driving capabilities of implanted gate devices in comparison with their degenerate-gate counterparts. These experimental findings have been related to shift of the Fermi level in poly-gate and depletion in the poly-gate near the oxide due to the penetration of the electric field into the gate (Chapman et al. 1988, Wong et al. 1988) or/and the existence of acceptor type interface traps at the gate/oxide interface (Lifshitz et al. 1983). An additional effect may be boron penetration (e.g. Tseng et al. 1990).

Consequently, the implanted (non-degenerate) poly-gate can no longer be assumed an equipotential area, especially in the modeling of thin oxide devices. We have published a 1D analytical model of thin oxide devices accounting for the potential drop in the poly-gate elsewhere (Habaš et al. 1990). In this paper, the numerical modeling of the poly-gate effect is presented, and this enables us to account for realistic doping profiles and 2D effects in submicron devices.

## 2. Model

We have extended MINIMOS to solve self-consistently the basic semiconductor equations also in the poly-gate area (in fully non-planar geometry). Poisson's equation is solved usually in the whole simulation area (from $y_t$ until $y_B$ – Figure 1). For the continuity equations two (in steady-state equivalent) approaches have been implemented:
1) The solution of both discretized continuity equations in the poly-gate simultaneously with the bulk area (from $y_G$ until $y_B$ – Figure 1). Such an approach is interesting for the transient simulation.
2) In steady-state the poly-gate is in thermodynamic equilibrium (the leakage currents are negligible and net recombination vanishes) and the assumption of a constant Fermi level holds. As a consequence, the carrier concentrations in the poly-gate $n,p$ can be calculated analytically as a function of the local potential $\Psi$. This approach performs the calculation in a significantly shorter computer-time than the first. It permits that the band gap narrowing and Fermi-Dirac statistics may be implemented in a simpler way. Assuming a rigid parabolic band model it follows for N-type gate:

$$n(\Psi) = N_c F_{1/2}\left[(\Psi - \Psi_G + \Phi_{fc} + \delta E_c - \delta E_{cG})/U_T\right] \quad ; \quad \Phi_{fc} = U_T F_{1/2}^{-1}\left(N_{gG}/N_c\right)$$

where $N_c$ is the effective density of states for conduction band, $\delta E_c$ is the local shift of the conduction band due to band gap narrowing, and $N_g$ is the activated impurity concentration in the gate. The index $G$ denotes the quantities at the gate–polysilicon contact ($y_G$ at Figure 1). For $p(\Psi)$ an analogous relation holds. The top gate potential (boundary condition) is given by

$$\Psi_G = \Phi_{fc} - \delta E_{cG} + U_{GS} + (E_{co} - E_{io})$$

with respect to the Fermi level in the source, where $U_{GS}$ is the terminal voltage and $E_{co}, E_{io}$ denote the conduction band edge and the intrinsic level in the ideal silicon band. These equations account properly for a position dependent band gap narrowing, and ensure that the potential $\Psi$ is continuous in the total simulation area. The analogous model has been implemented for a P-type gate. Fermi integrals $F_{1/2}$ can be calculated efficiently by analytical approximations (e.g. Blakemore 1982).

The charge at the polysilicon-gate/oxide interface affects the field in the gate at the gate/oxide interface, and changes remarkably the potential drop in the gate (Habaš et al. 1990). Therefore, a fixed oxide charge and interface trapped charge at the polysilicon/oxide interface have been incorporated in simulation. If $N_g$ is several times higher than the equivalent volume trap density in polysilicon, the trapped charge is negligible compared to the space charge due to impurity ions. Since we restrict ourselves to $N_g > 10^{18} cm^{-3}$, the traps at the grain boundaries have not been taken into account in the present model.

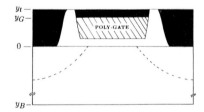

Figure 1: Simulation area

## 3. Simulation results and discussion

Figure 2 shows the distribution of the potential and the electron and hole concentrations in the gate of N-channel/N-gate device. Due to thin oxide (8nm), low ionized impurity concentration at the gate/oxide interface ($5 \cdot 10^{18} cm^{-3}$) and high gate bias ($U_{GS} = 5V$), a significant potential drop appears in the gate (inversion in the gate). In spite of high drain bias ($U_{DS} = 5V$) (which reduces the gate–channel potential difference), there is inversion in the gate along a significant part of the channel, due to the convex shape of the channel potential. The threshold voltage and the potential drop in the poly-gate versus ionized impurity concentration $N_g$ are shown for a P-gate/P-channel device in Figure 3. The charge at the gate/oxide interface $Q_{go}$ (here assumed as fixed) has a strong influence on the threshold voltage. There is not much information about the nature of $Q_{go}$ in literature. Yaron et al. (1980) obtained experimentally a positive total (fixed and trapped) interface charge of order $\sim 10^{12} cm^{-2}$ for the polysilicon deposited over thermally grown oxide. Such a large positive charge lowers significantly the field penetration into the gate in N-channel devices, but it increases the potential drop in the gate in P-gate/P-channel devices (Habaš et al. 1990). It is necessary to account for this charge in the analysis of the threshold voltage instability of thin oxide devices.

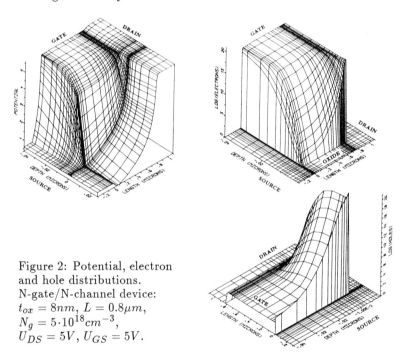

Figure 2: Potential, electron and hole distributions. N-gate/N-channel device: $t_{ox} = 8nm$, $L = 0.8\mu m$, $N_g = 5 \cdot 10^{18} cm^{-3}$, $U_{DS} = 5V$, $U_{GS} = 5V$.

The fall-off of the drain current in the linear region with $N_g$ as parameter is shown in Figure 4. The charge $Q_{go}$ has a minor influence, and $N_g$ is the main parameter in determination of the drain current degradation. The inversion in the gate causes a kink in the curve $10^{18}$, leading to the recovery of the transconductance (experimental finding by Lu et al. 1989). There is a recovery of the quasi-static C-V curve, too (obtained experimentally by Chapman et al. 1988, Lu et al. 1989). In order to suppress the reduction of the gate drive, the activated impurity concentration in the gate at the gate/oxide interface must be higher than $10^{19} cm^{-3}$ for $10nm$-oxide devices. Note that the Fermi-Dirac statistics has a small influence on the analysis performed.

The poly-gate effect depends approximately on $t_{ox}^2 N_g / U_{GS}$. It seems that it will not become more severe in deep submicron devices, because of the restricted reduction of the oxide thickness (down to $\sim 5nm$) and significant reduction of the supply voltage established in literature. However, Okazaki et al. (1990) have recently reported a $3.5nm$-thick oxide subquarter-$\mu m$ CMOS technology with $2V$ proposed supply voltage.

The $N_g$ necessary to suppress the gate effect at the supply voltage is for such a device at least four times higher than for the device at Figure 4. The gate effect will be also important regarding the possible application of $Ta_2O_5$ as gate-insulator (see e.g. Nishioka et al. 1987), because the gate effect depends on the square of the insulator permittivity.

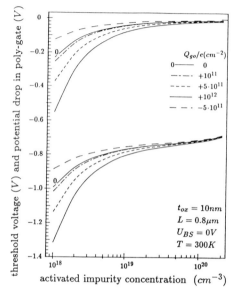

Figure 3: Threshold voltage dependence. Parameter $Q_{go}$ is fixed charge density at gate/oxide interface.

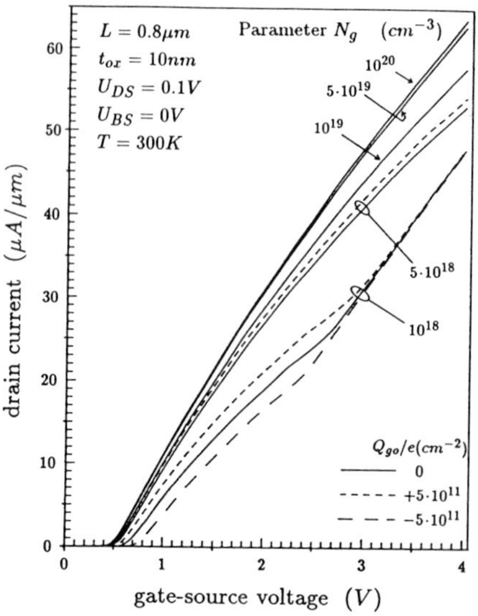

Figure 4: Linear region characteristics

*Acknowledgement* — Our work is considerably supported by the research laboratories of Digital Equipment Corporation at Hudson U.S.A.

**References**
Blakemore J S 1982 *Solid-State Electronics* **25**(11) p.1067
Chapman R A, Wei C C, Bell D A, Aur S, Brown G A and Haken R A 1988
  in *IEDM-88 Tech. Dig.* p.52
Habaš P and Selberherr S 1990 accepted for publication in *Solid-State Electronics*
Lifshitz N and Luryi S 1983 *IEEE Trans. on Electron Devices* **30**(7) p. 833
Lu C-Y, Sung J M, Kirsch H C, Hillenius S J, Smith T E and Manchanda L 1989
  *IEEE Electron Device Letters* **10**(5) p.192
Mandurah M M, Saraswat K C and Helms C R 1980 *J. Appl. Phys.* **51**(11) p.5755
Nishioka Y, Shinriki H and Mukai K 1987 *J. Appl. Phys.* **61**(6) p.2335
Okazaki Y, Kobayashi T, Miyake M, Matsuda T, Sakuma K, Kawai Y, Takahashi M
  and Kanisawa K 1990 *IEEE Electron Device Letters* **11**(4) p.134
Schwalke U, Mazure C and Neppl F 1988 *Mat. Res. Soc. Symp. Proc.* **106** p.187
Sun J Y-C, Angelucci R, Wong C Y, Scilla G and Landi E 1988
  *Journal de physique* **C4**(9) p.401
Tseng H-H, Tobin Ph J, Baker F K, Pfiester J R, Evans K and Fejes P 1990
  *IEEE Proc. of 1990 Symp. on VLSI Technology* p.111
Wong C Y, Sun J Y-C, Taur Y, Oh C S, Angelucci R and Davari B 1988
  in *IEDM-88 Tech. Dig.* p.238
Yaron G and Frohman-Bentchkowsky D 1980 *Solid-State Electronics* **23** p.433

Paper presented at ESSDERC 90, Nottingham, September 1990
Session 2C8

## Series resistance effects on EPROM programming

R.Bez, D.Cantarelli, P.Cappelletti, A.Maurelli and L.Ravazzi

SGS-Thomson Microelectronics, 20041 Agrate MI, Italy

**Abstract.** An extensive characterization of the influence of series resistance on programming of $1.0\mu m$ technology EPROM cells is presented. The different effects of series resistances on the source or on the drain have been pointed out. Furthermore a simple analytical model has been developed to simulate the influences on the programming.

### 1. Introduction

As well known, an Electrically Programmable Read Only Memory (EPROM) cell is programmed by injection of channel hot electrons into a floating–gate (FG). The influence of device parameters and applied voltages on programming behavior or, equivalently, on the gate current, has been extensively studied both on memory cells (Eitan and Frohman–Bentchkowsky - EFB - 1981) and on transistors (Takeda et al 1982, Ng and Taylor 1983).
Due to new designs for high density EPROM cell array (Mitchell et al 1987, Bellezza et al 1989, Yoshikawa et al 1990), series resistances, either on the source (S) or on the drain (D), are no more negligible and their effects can drastically affect the programming characteristics of the cell.
In this work an extensive characterization of the impact of series resistances has been performed. Due to the intrinsic feedback behaviors, the influence of each effect of the series resistance on the programming characteristics has been pointed out measuring the cell in particular bias configurations.
Furthermore a simple analytical model, able to calculate the voltage drop in programming conditions as a function of the series resistance, has been developed to improve the understanding of these effects.

### 2. Experimental

Programming curves, defined as the cell threshold voltage shift as a function of the programming time, have been measured on an EPROM cell with a series resistance ($R_S$) alternatively on S and on D (Figure 1), using a box waveform programming pulse. The measurements have been performed on the same cell, whose main geometrical and technological features are reported in Table 1. As can be observed, the effects of an increase in the $R_S$ are: i) a decrease in the programming speed, defined as the programming time to obtain a fixed threshold shift; ii) a decrease in the final threshold voltage ($V_{Tf}$) when the $R_S$ is applied on S; iii) an increasing difference in the programming speed between the two cases ($R_S$ on S and $R_S$ on D).
Considering that the voltage drop on $R_S$, due to the channel current, strongly depends on the FG voltage, to better understand the influences of $R_S$, the cell has been measured with a constant voltage shift, $\Delta V$, applied on S or on D (Figure 2).
Furthermore a lot of information can be obtained analyzing the programming currents, i.e. the gate current $I_G$, as a function of the FG voltage, $V_{FG}$. This is accomplished considering the time derivative of the programming curve and the total capacitance of the cell to determine the

© 1990 IOP Publishing Ltd

programming current, and the cell coupling ratios to determine the $V_{FG}$ (EFB 1981). In Figure 3 the gate currents, as derived from the programming curves of **Figure 2**, are reported.

| Gate length | 1.2 μm |
| Channel width | 1.0 μm |
| Gate oxide thickness | 28 nm |
| Interpoly oxide thickness | 30 nm |
| $n^+$ junction depth | 0.3 μm |
| Cell size | 4.4 × 4.3 μm² |

Table 1: Main technological and geometrical features of the measured EPROM cell.

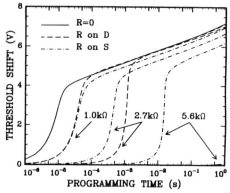

Figure 1: Comparison between programming curves measured without (solid line) and with series resistance (dashed line) applied respectively on D and on S. The programming voltages have been $V_{CG} = 12.5$V, $V_D = 7$V and $V_S = V_B = 0$. Three different $R_S$ have been considered.

## 3. Discussion

The observations i)–iii), referred to Figure 1, hold true for the programming curves of Figure 2 too. In this case: i) increasing $\Delta V$, the programming speed decreases; ii) a shift in the $V_S$, and then in the $V_{GS}$, produces an equivalent shift in the $V_{Tf}$ and iii) increasing $\Delta V$, the difference in the programming speed between the two cases ($\Delta V$ on S or $\Delta V$ on D) increases.

The $I_G$ of Figure 3 can be divided in two distinct regions: for $V_{FG} < V_D$, $I_G$ is an exponential function of $V_{FG}$ ( the so-called by EFB "electrode–limited" region) and for $V_{FG} > V_D$, $I_G$ is a weak function of $V_{FG}$ (the so-called by EFB "injection–limited" region). As can be observed:

a) the $I_G$ curves obtained with different $V_D$ and $V_S$, but with the same $V_{DS}$, are superimposed in the "injected–limited" region. This means that the programming speed depends on the $V_{DS}$ and not only on the $V_D$;

b) the two regions are separated by the condition $V_{FG} = V_D$. Hence $I_G$ curves, obtained with the same $V_{DS}$ but with different $V_D$ and $V_S$, show a different dependence on $V_{FG}$ in the "electrode–limited" region.

A simple model has been developed in order to calculate the impact of the $R_S$ on the programming characteristics. Since $V_{DS}$ is the most important parameter for $I_G$ in the "injected–limited" region, it has been related to the $R_S$ or to the threshold shift. The model is based upon the following items.

• The channel current $I_D$ is calculated in the saturation region for velocity saturation regime and without Early effect, i.e. in programming conditions:

$$I_D = \beta(V_{FG} - V_S - V_T)$$

where $\beta$ is the dummy cell transconductance and $V_T$ is the FG threshold voltage.

• The voltage drop due to the $R_S$ is calculated to obtain $V_{DS}$:

$$V_{DS} = V_D^{ext} - R_S I_D = V_D^{ext} - R_S \beta(V_{FG} - V_S - V_T)$$

where $V_D^{ext}$ is the applied bias to the drain node.
- Obviously when the $R_S$ is applied on S, the body effect has been taken into account to evaluate the $V_T$:

$$V_T = V_{T0} + \gamma(\sqrt{R_S I_D + 2\phi_F} - \sqrt{2\phi_F})$$

where $V_{T0}$ is the reference threshold voltage, $\gamma$ is the body factor and $\phi_F$ is the Fermi potential in the channel.
- The programming parameters and the coupling ratios have been considered to determine the $V_{FG}$:

$$V_{FG} = \alpha_G V_{CG} + \alpha_D V_D + \alpha_S V_S - \alpha_G \Delta V_T$$

where $\alpha_G$, $\alpha_D$ and $\alpha_S$ are respectively the gate, the drain and the source coupling ratio and $\Delta V_T$ is the threshold shift.

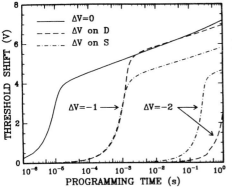

Figure 2: Comparison between programming curves without (solid line) and with a voltage shift $\Delta V$ (dashed line) in the $V_{DS}$. Standard programming voltages have been: $V_{CG} = 12.5V$, $V_D = 7V$ and $V_S = V_B = 0$. Two different $\Delta V$ have been considered.

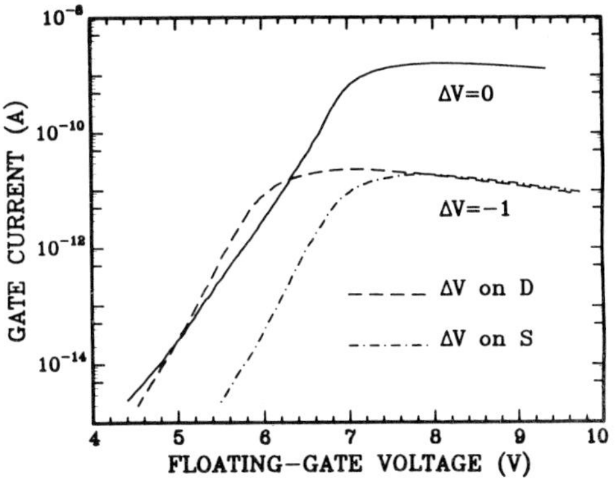

Figure 3: Gate currents as a function of the floating gate voltage are derived with the method reported by Eitan and Frohman-Bentchkowsky (1981), considering the programming curves of Figure 2.

Figure 4 shows the $V_{DS}$ vs. $R_S$ when the cell is virgin, i.e. for $\Delta V_T=0$. When the $R_S$ increases, the $V_{DS}$, and then the programming speed, for $R_S$ on S is higher than for $R_S$ on D; this explains the great difference in programming speeds for high $R_S$ values in Figure 1. The higher programming speed for $R_S$ on S is due to the lowering of the channel current because of the body effect.
Figure 5 shows the $V_{DS}$ vs. the $\Delta V_T$ of the cell for different $R_S$. For low $R_S$, the voltage drop is small compared with the applied biases and the difference between $R_S$ on S and on D is not appreciable. This influence becomes more relevant when the $R_S$ value is increased. The delayed programming for the $R_S$ on D is easily explained considering the higher voltage drop, due to the higher channel current. A reduction of the channel current, due to a $\Delta V_T$ increases, enhances the programming speed.

## 4. Conclusions

The programming curves with series resistances on S or on D have been measured. A simplification of the problem has been obtained considering a constant voltage shift in the $V_{DS}$. The main results have been:
- a decrease in the programming speed when $R_S$ increases;

- a decrease in the $V_{Tf}$ when $R_S$ is applied on S;
- the programming speed reduction is more severe when $R_S$ is on D than on S.

Furthermore a simple analytical model has been developed in order to simulate the impact of the $R_S$ on the programming characteristics.

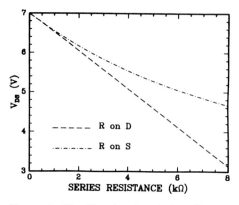

Figure 4: The $V_{DS}$ for the virgin cell as a function of the series resistance $R$ is calculated considering the analytical model described in the text.

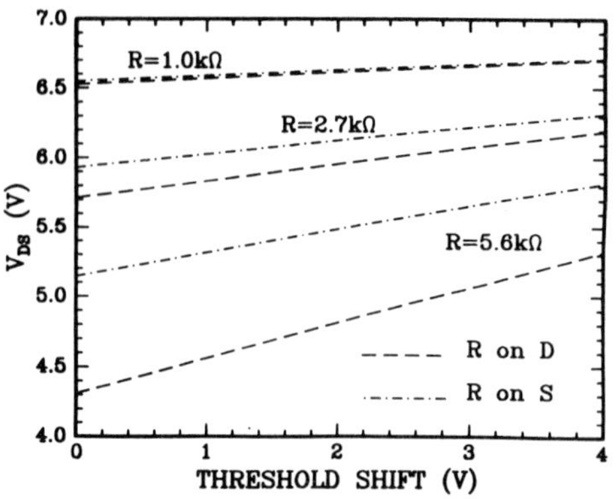

Figure 5: The $V_{DS}$ as a function of the threshold shift of the cell, considering the series resistance as a parameter.

Acknowledgments

The authors would like to thank G.Corda for helpful discussions.
This work was partially supported by the ESPRIT Project 2039–APBB.

References

Bellezza O, Laurenzi D and Malanotte M 1989 *IEDM Tech. Dig.* **25.1** 579
Eitan B and Frohman-Bentchkowsky D 1981 *IEEE Trans. El. Dev.* **28** 328
Mitchell A T, Huffman C and Esquivel A L 1987 *IEDM Tech. Dig.* **25.5** 548
Ng K K and Taylor G 1983 *IEEE Trans. El. Dev.* **29** 611
Takeda E, Kume H, Toyabe T and Asai S 1982 *IEEE Trans. El. Dev.* **30** 871
Yoshikawa K, Sato M and Ohshima Y 1990 *IEEE Trans. El. Dev.* **37** 999

*Paper presented at ESSDERC 90, Nottingham, September 1990*
*Session 2C9*

# A flash EEPROM cell scaling including tunnel oxide limitations

Kuniyoshi Yoshikawa, Seiichi Mori, Eiji Sakagami,
Norihisa Arai*, Yukio Kaneko, and Yoichi Ohshima

Semiconductor Device Engineering Laboratory, Toshiba Corp.
* Toshiba Microelectronics Corporation.
1,Komukai-Toshiba, Saiwai-ku, Kawasaki 210, JAPAN

**Abstract.** A new flash EEPROM cell scaling scenario is proposed, which takes tunnel oxide thinning limitation into consideration. The derived scaling scenario, performance, and reliability are discussed, in comparison with EPROM scaling.

## 1. Introduction

Currently, flash EEPROM technology is receiving a lot of attention, since it has a great potential for replacing disk memories in future. Among the various proposed cell structures, an EPROM with a tunnel oxide (ETOX) cell (Tam 1987) is very similar to the standard EPROM cell, except for its thin tunnel cell gate oxide.

This paper describes an ETOX cell scaling scenario which considers tunnel oxide scaling limitations. Operation voltages are also derived from this guideline. Performance and reliability (including various disturb resistances) are evaluated using simple analytical equations and compared with a reported EPROM scaling scenario (Yoshikawa 1989).

## 2. Tunnel oxide limitations

Besides direct tunneling, two other main tunnel oxide limitations for flash memory applications are high-temperature bake retention degradation, and stress induced leakage current increase after Write/Erase cycling.

Fig.1 shows high-temperature bake retention as a function of cell gate oxide thickness. EPROM cells with a 20nm gate oxide show 30% stored charge decrease after 3000H of $300^{\circ}C$ baking. This slow decay rate is considered a mass produced EPROM standard criteria for high reliability. This decay means that positively charged contaminants introduced during processing are gradually attracted by the electric field of charged cells. Note that, by decreasing the tunnel oxide thickness to 7.3nm, the cell retention capability can be preserved. However, for 5.3nm oxide samples, more data destruction was observed, showing that electrons leak through the thin tunnel oxide.

The increase in stress-induced leakage current (Naruke 1988) must be the most serious constraint for flash tunnel oxide scaling, especially in realizing 1E6 write/erase cycles. Fig.2 shows the decrease in leakage field strength after charge injection as a function of tunnel oxide thickness. Initially, in 9.8nm to 6.8nm samples, the leakage field is almost equal, while after stressing, the tunnel oxide thickness dependence of the field becomes prominent. In particular, below the $10nA/cm^2$ region, this dependence becomes more sever. Since the leakage criteria become stricter as the memory density increases, an 8nm-9nm tunnel oxide thickness is the minimum to ensure charge retention during program-disturb duration in the 1Gbit era.

Therefore, we assumed a value of 8nm for the tunnel oxide limitation in 1Gbit memories. Any new scaling scenario should be capable of this 8nm limitation, without resulting in performance degradation.

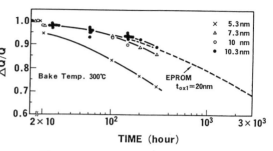

Fig.1. High temperature bake retention as a function of tunnel oxide thickness.

## 3. Flash Cell Scaling

We have already derived an EPROM cell scaling scenario (Yoshikawa 1989) by satisfying reliability and performance constraints simultaneously. These constraints require that dielectric (gate oxide and interpoly ONO) field strengths and the maximum channel fields in program/read operations should be maintained compared with previous generation devices.

In order to derive an ETOX scaling rule, the EPROM scaling scenario has been modified to include tunnel oxide and erase voltage reduction. The starting parameters for 4M flash are given in Table 1. A 4M flash tunnel oxide is assumed to be 12nm, considering the current 12V program/erase voltage standards. Scaling ratios for gate length/width, interpoly insulator thickness, and tunnel oxide thickness are obtained as $1/k$, $1/k^{0.5}$ and $1/k^{0.25}$, respectively. Here, k is a lateral dimension scaling factor. Note that, for ETOX scaling, tunnel oxide and interpoly insulator scaling factors are different, while the EPROM scenario requires the same scaling factor for both insulators.

Assuming the same programming speed for EPROM and ETOX cells in the 4Mbit 0.9um regime, the same maximum channel electric field (Em.p) will arise, because the hot electron generation rate is exponentially determined by $-1/Em.p$. Since Em.p is proportional to $(Vdp-Vdp,sat)/(Tox1*Xj)^{0.33}$ (Kakumu 1990), the program drain voltage (Vdp) is $5.9/k^{0.42}$(V).

For read-disturb (soft-write) suppression, the read drain voltage (Vdr) should be close to the saturation drain voltage (Vdr.sat), because the

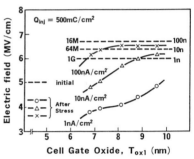

Fig.2. Stress-induced leakage field strength as a function of tunnel oxide thickness, Tox1.

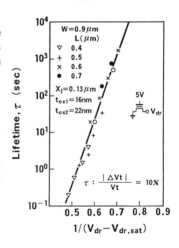

Fig.3. Read-disturb (Soft-write) lifetime vs. $1/(Vdr-Vdr,sat)$

soft-write lifetime is proportional to $1/\{(Vdr-Vdr,sat)^2+C\}^{0.5}$ as shown in Fig.3. Thus, $Vdr=0.96/k^{0.42}$ is obtained.

The cell read current (Icell), which determines the bit line delay, is maximized when Vdr equals Vdr.sat, since Icell reaches saturation. Icell is approximated as 113 $k^{-0.15}$. This is 1.4 times higher than the value of EPROMs ($-80k^{-0.2}$). Furthermore, even if an "intelligent" erasing scheme is used (erased Vt is held 1V higher than intrinsic Vt to prevent over-erasing), Icell is still comparable to that for EPROMs as shown in Fig.4. The bit-line capacitance per cell is decreased as $k^{-0.3}$ (mainly determined by the circumference length of the drain diffusion area), however, the total bit-line delay increases as $k^{1.4}$. Therefore, an appropriate divided bit-line technique should be introduced to ensure the device retains high speed sensing.

For program-disturb resistance, the disturb due to band-to-band tunneling induced hot hole injection is less important, since the lateral electric field in the depletion layer (which generates hot holes) is weaken due to Vdp scaling, compared with the thick-oxide EPROM case. Thus, program-disturb resistance can be maintained as shown in Fig.5.

Stress induced leakage current limits tunnel oxide scaling, because the data retention capability is degraded after W/E cycling. The allowable electric field after W/E cycling rapidly decreases with device scaling as shown in Fig.6. This is due to the reduced stored charge in each cell and increased stress time as memory density rises. Note that oxide field conditions in program-disturb are severer than in the DC-programming condition. For the 1Gbit regime, 8nm-9nm will be the minimum, again requiring a divided bit-line scheme for reliability.

In order to prevent punchthrough and drain turn-on, a DSA p-pocket structure (Yoshikawa 1984) is suitable for high punchthrough resistance without increasing cell threshold voltage. Because DSA structure can increase channel doping locally adjacent to the drain, as shown in Fig.7.

For erasure, the erase voltage should be reduced as $1/k^{0.25}$ to keep a constant 11MV/cm electric field. Since the erase voltage will be limited by

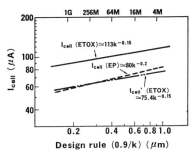

Fig.4. Cell read currents for ETOX and EPROM for each generation rule. Icell' is defined as the current where cell Vt = (intrinsic Vt)+1 V.

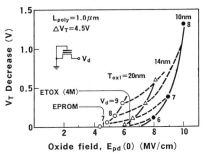

Fig.5. Cell Vt decrease vs. oxide field Epd(0), in program-disturb condition.

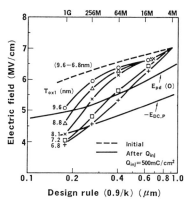

Fig.6. Allowable electric field strengths after stressing for various oxide thicknesses. Epd(0) and -Edc,p are field strength for program-disturb and DC-program conditions, respectively.

junction breakdown, the source N⁻ doping must be optimized and hot-hole induced degradation must be reduced.

TDDB lifetime predictions for intrinsic and $10mm^2$ oxide do not limit the scaling as shown in Fig.8, because oxide fields for Epd(0) and -Edc,p become slightly weaker with scaling according to the difference in scaling factors for interpoly ONO and tunnel oxide. ETOX scaling predictions are summarized in Table 1.

### 4.Summary

A new scaling scenario for ETOX flash cells is proposed. The performance is comparable or superior to that of EPROMs. The various electric fields in the cell are slightly weaker with scaling, preserving the reliability and disturb resistance. This kind of scaling consideration for flash cells will become more important in future applications.

### References

Kakumu, M., et al., 1990, IEEE ED. p1334
Hokari, Y., 1989, VLSI Sym. Workshop, p265
Naruke, K., et al., 1988, IEDM, p424
Tam, S., et al. 1988, VLSI Sym. Tech. p31
Yoshikawa, K., et al., 1984, IEDM, p456
Yoshikawa, K., et al., 1989, IEDM, p587

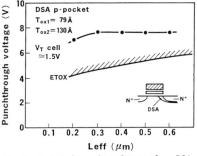

Fig.7. Punchthrough voltage for DSA ETOX cells.

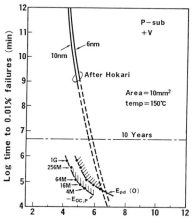

Fig.8. Oxide TDDB lifetime vs. stress fields for program-disturb and DC-program conditions.
Data are taken after Hokari Y, 1989.

Table 1. ETOX scaling predictions from 16Mbit to 1Gbit densities.

| Parameters/Voltages | | | Scaling | 4M(Input) (0.9um) | 16M (0.6um) | 64M (0.4um) | 256M (0.27um) | 1G (0.18um) | Expression | EPROM (Ref.) |
|---|---|---|---|---|---|---|---|---|---|---|
| Chanell Length | Leff | (um) | $1/k$ | 0.58 | 0.39 | 0.26 | 0.17 | 0.11 | $0.58/k$ | $0.58/k$ |
| Gate Width | W | (um) | $1/k$ | 0.9 | 0.6 | 0.4 | 0.27 | 0.18 | $0.9/k$ | $0.9/k$ |
| Junction Depth | Xj | (um) | $1/k$ | 0.2 | 0.13 | 0.09 | 0.06 | 0.04 | $0.2/k$ | $0.2/k$ |
| Cell Gate Oxide. | Tox1 | (um) | $1/k^{0.25}$ | 12 | 10.8 | 9.8 | 8.85 | 8.0 | $12/k^{0.25}$ | $20/\sqrt{k}$ |
| Interpoly ONO. | Tox2 | (um) | $1/\sqrt{k}$ | 30 | 24.5 | 20 | 16.3 | 13.3 | $30/\sqrt{k}$ | $30/\sqrt{k}$ |
| Prog. Gate Volt. | Vpp | (V) | $1/\sqrt{k}$ | 12.0 | 9.8 | 8.0 | 6.53 | 5.3 | $12/\sqrt{k}$ | $12.5/\sqrt{k}$ |
| Prog. Drain Volt. | Vdp | (V) | $1/k^{0.42}$ | 5.9 | 4.98 | 4.21 | 3.55 | 3.0 | $5.9/k^{0.42}$ | $7/\sqrt{k}$ |
| Read Gate Volt. | Vcc | (V) | 1 | 5 | 5 | 5 | 5 | 5 | 5 | 5 |
| Read Drain Volt. | Vdr | (V) | $1/k^{0.42}$ | 0.96 | 0.81 | 0.68 | 0.58 | 0.49 | $0.96/k^{0.42}$ | $1.2/\sqrt{k}$ |
| Erase Source Volt. | Ve | (V) | $1/k^{0.25}$ | 12 | 10.8 | 9.8 | 8.85 | 8.0 | $12/k^{0.25}$ | — |
| Cell Vt | Vtcell | (V) | 1 | 2 | 2 | 2 | 2 | 2 | 2 | 2 |

Paper presented at ESSDERC 90, Nottingham, September 1990
Session 2C10

# A new 0·5 µm² DRAM cell with internal charge gain investigated by 2D transient device simulation

R Richter, K E Ehwald, B Heinemann, W-E Matzke, H Gajewski[*], W Winkler

Institute for Physics of Semiconductors
Academy of Sciences of the GDR
Walter-Korsing-Str. 2, 1200 Frankfurt (Oder), GDR

[*] Karl-Weierstrass-Institute of Mathematics
Academy of Sciences of the GDR
Mohrenstr. 39, 1086 Berlin, GDR

Abstract. The Vertically Integrated Gain (VIG) cell is a new DRAM structure for 64/256 Mbit DRAMs. By using a trench structure and 0.25 µm design rules a cell size of $0.55 \mu m^2$ is attainable. The cell function bases on merging 2 bipolar junction transistors (BJT), 1 junction field effect transistor (JFET) and 2 capacitors. 2D transient device simulation is used to investigate the electrical behaviour of the VIG cell. A high read out signal, 2 control lines, operation voltages between 0 and 5 volts only and the capability to maintain the stored information during read operation are the main features of the proposed DRAM cell.

## 1. Introduction

The scaling of DRAMs up to 64/256 Mbit storage capability requires a vertical configuration of the access transistor and the storage capacitor. Richardson et al. (1985) proposed this concept and Sunouchi et al. (1989) further developed it with the Surrounding Gate Transistor (SGT) cell by use of matrix-like trenches. A size of $1.2 \mu m^2$ using a design rule of 0.5 µm was estimated for the SGT cell. However in a SGT cell array neighbouring cells can not share their word lines and the trenches have to be broad enough to arrange 2 word lines and their intermediate isolation.
The VIG cell functional principles base on amplifying the stored charge during reading the "1" state by a JFET and a BJT. Opposite to the 1 T-Cells the control regime of VIG cells allows shared word lines between neighbouring cells. Contrary to previous gain cells the vertical arrangement of the storage capacitors ensure an extremely small cell size and a higher storage charge adjustable by the trench depth.

## 2. Cell Structure

Fig. 1 shows the cross section of the VIG cell (a), the equivalent circuit (b) and the arrangement of VIG cells in a storage array (c). The isolation between the cells consists of

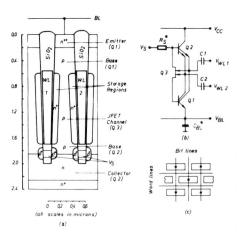

Fig 1. (a) Schematic cross section of the VIG cell
(b) Equivalent circuit
(c) Array organisation scheme

matrix-like trenches. The trenches contain a first conducting layer contacting shallow $p^+$ - layers near by the bottom of the trench and a second conducting layer forming buried word lines. The isolation between the conducting layers and the filling of the trenches perpendiculary arranged to the word line are made by CVD oxide. The word lines and the storage region are isolated by a thin ONO layer and form the capacitors C1 and C2. Simultaneously the storage regions act as collectors for a first BJT Q1, as emitters for a second BJT Q2 and as gates of a JFET Q3 as shown in Fig. 1 (b). The base of Q1 and the base of Q2 form drain and source of Q3, respectively. The emitter of Q1 is connected to the bit line (BL). The supply voltage $V_{DD}$ of 5 V is applied to the buried $n^+/n$ -well acting as collector of Q2. The array organisation scheme proposed in Fig. 1(c) leads to a cell size of 8 $F^2$ (F-feature size). By using 2 metallization levels for the bit lines a reduction to 4 $F^2$ could be achieved. $R_S^*$ and $C_{BL}^*$ (Fig. 1 (b)) are not components of the VIG cell. $R_S^*$ regards the sheet resistance of the first conducting layer connected to the constant voltage $V_S$ (2.5 V) and $C_{BL}^*$ is the bit line capacitance. Both elements are included into the simulation to form an idea of the electrical behaviour of the cell under array conditions.

## 3. Simulation of Cell Operation

For the cell simulation a minimum feature size of 0.25 µm for the trench structures was assumed. The length and the width of the emitter of Q1 were 0.3 µm and 0.75 µm, respectively. This results to a cell size of 0.55 µm$^2$, a word line pitch of 0.55 µm and a bit line pitch of 0.5 µm. The depth and the obliquity of the trenches were 2 µm and 87.5°, respectively and the equivalent oxide thickness for the dielectric of C1 and C2 was chosen to 5 nm. For the doping concentrations of the storage regions, the base and the emitter of Q1 and the $p^+$-layer Gaussian profiles were assumed. Table 1 summerizes the pn junction depths $x_j$ and the maximum concentrations $N_{max}$. In order to simplify the simulated structure the $p^+$-layers were arranged on the bottoms of the trenches. We used the 2D device simulator TOSCA (Gajewski et al. 1986) to calculate complete read-write cycles for both of the storage states. The automatically generated mesh of the discretized cell structure containing 1172 vertices and 2164 triangles is shown in Fig. 2. Under these conditions the transient simulation took a VAX station 3100 21 hours CPU time. The cell operation is described on the help

Table 1 : Doping profile characteristics

|  | $x_j$ (μm) | $N_{max}$ (cm$^{-3}$) |
|---|---|---|
| storage region | 0.1 | $1.1\ 10^{19}$ |
| emitter (Q1) | 0.28 | $8\ 10^{20}$ |
| base (Q1) | 0.6 | $2\ 10^{17}$ |
| $p^+$-layer | 0.14 | $2.5\ 10^{19}$ |
| channel (Q3) | const. | $1.4\ 10^{17}$ |
| collector (Q2) | const. | $3\ 10^{17}$ |

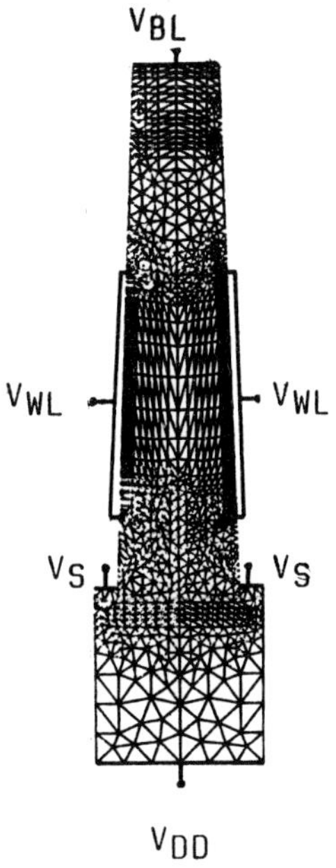

of the waveforms and the time dependent development of the storage region potentials shown in Fig. 3.

WRITING The write operation is realized by grounding of word lines (WL1 and WL2). In consequence of the capacitive coupling between the word lines and the storage regions their potential falls until Q1 and Q2 turn on. If a "0" is written the bit line voltage $V_{BL}$ is held at 5 V and Q1 operates in the reverse and Q2 in the forward mode injecting the electrons of the storage regions into the bases. As a result of the electron deficiency in the storage regions their potentials are increased after returning the word lines to 5 V. To write an "1" the bit line precharge level $V_{BL}^{pr}$ of about 1 V is used to fill the storage regions with electrons, because Q1 injects them from the bit line to the storage regions. While the first WL returns to 5 V the injection has to continue holding the storage region potentials down. After that the injection is stopped due to switching $V_{BL}$ to 5 V. The return of the second WL to 5 V turns off Q3 suppressing any current in an "1" storing cell. The "0" and "1" states of the VIG cell are characterized by different charges in the storage regions, resulting in mean storage potentials of about 6.5 V and 4 V, respectively Fig. 4 a and b show the 2D potential distributions of "0" and "1" storing cells. It should be noted, that in order to reduce the parasitic word line capacitance the storage regions are not connected to each other but they interact by punch through in the channel of Q3. The punch through voltage $V_{PT}$ was adjusted to 2 V. This measure reduces the mean word

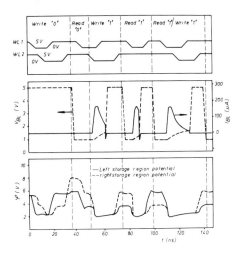

Fig 2. Mesh of the simulated cell

Fig. 3 Operation waveforms, calculated read out signals and internal storage region potentials ($R_S$=10kΩ, $C_{BL}$=2pF)

line displacement current by the factor $(\Delta V_{WL}-V_{PT}) / \Delta V_{WL}$.
READING To read the VIG cell one word line has to be activated (0V) decreasing the storage region potential (Fig. 3). In the case of reading "0" the JFET current is suppressed because the gate source voltage ($V_{GS}$) of Q3 is greater than its pinch off voltage (about 1V). During a read "1" operation Q3 turns on charging the base of Q1. Q1 amplifies the JFET current. The collector current discharges the storage regions causing a positive feedback due to the decreasing of $V_{GS}$ of Q3. This process is restrained by turning on Q2 which transports the electrons to the n-well. A read out signal of 300 mV has developed at a 2 pF bit line capacitance 9 ns after the begin of word line activation (Fig. 3). A read out current $I_{BL}$ of 170 μA points out the high driving capability of the VIG cell. $I_{BL}$ is adjustable by the cell geometry and by $V_S$ and $V_{BL}{}^{Pr}$, respectively. Because of the non destroying read operation a rewrite cycle after reading should not be necessary. The option of a completely selective read access to single cells allows a very small power dissipation of a VIG cell array.

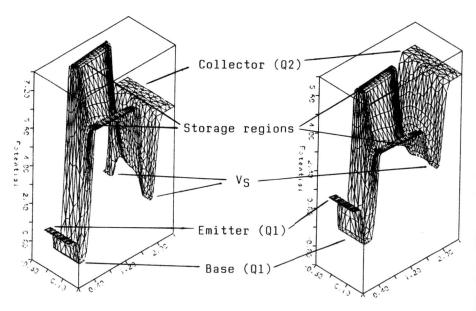

Fig 4 (a) and (b)  Potential distributions of storing "0" and "1" cells (all scales in microns)

References

Richardson W et al., "A Trench Transistor cross point cell", Techn. Dig., IEDM pp. 714-717, 1985
Sunouchi K et al.,"Surrounding Gate Transistor (SGT) cell for 64/256 Mbit DRAMS", Techn. Dig., IEDM pp. 23-26, 1989
Gajewski H et al., "TOSCA user`s guide", Karl-Weierstrass-Institut for Mathematics Academy of Sciences of the GDR, Berlin 1986

# A novel flash erase EEPROM memory cell with asperities aided erase

Alaaeldin A. M. Amin [1]

Computer Eng. Dept., King Fahd Univ. of Petroleum & Minerals,
Dhahran, Saudi Arabia

**Abstract.** A novel[2] flash erase EEPROM memory cell structure is presented. The cell uses triple poly layers and two independent $N^+$ implants. The first poly is used as the control gate, the second poly as the floating gate and the third poly as an erase electrode. Cell programming is by avalanche injection of hot electrons into the floating gate, while erasure is performed by asperities-aided Fowler Nordheim tunneling of electrons. Asperities introduced at the top surface of the floating poly gate allow using thicker interpoly oxide at lower erase voltage and less than 0.1 second erase time.

## 1. Introduction

Several Flash EEPROM (**FEEPROM**) memory cell structures have been recently reported [1] - [6]. Contrary to UV-light erase, electrical erase is not a self-limiting process. This has resulted in over-erase problems in single transistor FEEPROM memory cells. This problem is avoided [5] through the use of a merged transistor memory cell structure. Merged transistor FEEPROM memory cells use an integral select transistor as part of the memory cell. Such structure avoids over-erase problems since the cell threshold voltage during erasure will be limited by the series select transistor. For such structure, however, both programming and erasing will have to be performed at the drain junction, which necessitate careful optimization of its profile to meet the conflicting requirements of the program and erase operations. A reported triple poly structure [6] avoids these problems by using a special erase electrode. However, this structure suffers from low programming efficiency since the drain junction does not overlap the floating gate. The first poly layer is used as the erase electrode while the second poly is used as a floating gate. A high integrity oxide is used between the first and the second poly layers which requires fairly high voltages for both programming and erasing. In addition, the control gate critical dimensions, being the third poly layer, is poorly controlled. This paper presents a new triple poly FEEPROM memory cell structure which overcomes these problems and enjoys a smaller cell area.

---

[1] The author was with National Semiconductor Corp.
[2] Filed patent application

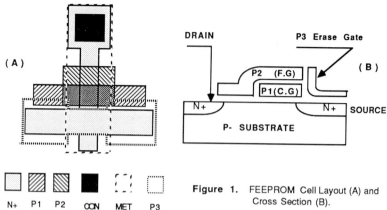

Figure 1. FEEPROM Cell Layout (A) and Cross Section (B).

## 2. Cell Structure and Process

The cell uses a merged transistor structure with triple poly layers, the first poly is used as a control gate, the second as a floating gate while the third poly layer is used as an erase electrode. The cell layout and cross section are shown in figure 1. Using modified $1.5\mu$ EPROM design rules, the cell has an area of 24.6 $\mu^2$.

Standard LOCOS process is used to define the active area. A thin gate oxide of 250Å is grown and covered with the first polysilicon layer. After doping, the first poly layer is defined and etched to form the control gate. This is followed by the first $N^+$ implant to form the source junction. A subsequent high temperature interpoly oxide is grown (400 Å) and covered with the second poly layer. This layer is then doped and defined to form the floating gate. A second $N^+$ implant is then used to form the drain junction followed by the second interpoly oxidation step. The third poly layer is then deposited and defined to form the erase electrode. The third poly layer is located on top of the $N^+$ Vss source line and overlaps the floating gate.

The second interpoly oxide, grown at low temperature (900 $C^o$), has higher conductivity than the first interpoly oxide. With the lower oxidation temperature, the top surface of the second poly gets texturized and poly asperities are formed. Such asperities cause locally enhanced oxide electric fields under erase condition. This leads to a lowering of the effective oxide electron energy barrier [7] near the injecting point and thus enhancing the Fowler Nordheim tunneling current. This barrier lowering allows the use of thicker interpoly oxides (600Å- 800Å) without requiring excessively high erase voltages ($\approx$ 12 volts). Using thicker interpoly oxide between the second and third poly layers improves the erase and the programming coupling ratios to 65% and 5% respectively. A SEM picture of the cell cross section is shown in figure 2. The programming coupling ratio of the new cell is independent of both the first gate oxide thickness as well as the interpoly oxide thickness. This is due to the fact that both the floating gate to substrate oxide thickness and the floating gate to the control gate oxide thickness will have a fixed ratio since both are grown at the same oxidation step. Thus, the programming coupling ratio of this cell is mainly dependant on the cell geometry. The second and the third poly layers, typ-

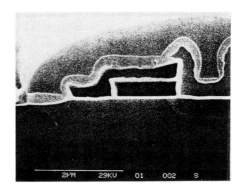

Figure 2. SEM Picture of the Triple Poly Cell.

ically not used for interconnect, can tolerate higher sheet resistances. This allows reducing their thicknesses to 3500Å - 4000Å for better metal step coverage.

## 3. Operation and Performance

**3.1 Reading:** The control gate voltage is raised to 5 volts, a drain bias of 1.7 volts is applied while the source and the erase gate are grounded. Typical read current is 70-80 $\mu$A for an erased cell and less than 100 pico-Amps for a programmed cell.

**3.2 Programming:** The cell is programmed into the high threshold state by avalanche injection of hot electrons into the floating gate. The control gate voltage is raised to 14 volts, the drain is pulsed to 10 volts while the source and erase gate are kept at 0 volts. The programming characteristics can be optimized by proper choice of the drain junction implant profile and adequate channel implant. The DC programming characteristics of the cell are shown in figure 3. A 100 $\mu$ sec programming pulse is enough to shift the threshold voltage by over 5 volts.

**3.3 Erasing:** Electrical erasure to the low threshold state is accomplished by Fowler Nordheim tunneling of electrons from the floating gate to the erase electrode. This is done by grounding the control gate, the drain and the source junctions to 0 volts while raising the erase gate voltage to 12 volts. The erase voltage is dependent on both the type and thickness of the second interpoly oxide and on the top surface texturization of the floating gate. Over-erase condition will result in a threshold voltage which is less than the native threshold. However, a net positive threshold will be maintained because of the merged series select transistor. Erase time is typically less than 0.1 second. The erase characteristics of the cell is shown in figure 4.

## 4. Conclusion

A novel flash EEPROM cell structure is described. The new cell allows the independent optimization of process steps to satisfy the cell program and erase requirements. The cell uses a non-self-aligned poly etch process with two independent $N^+$ implants and three poly layers. The cell is programmed by avalanche injection of hot electrons into the floating gate but is erased by asperities-aided Fowler Nordheim tunneling of electrons from the floating gate to the third poly erase electrode.

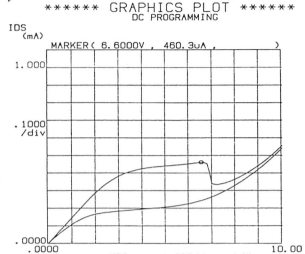

Figure 3. DC-Programming Characteristics. (Vcg = 14 volts)

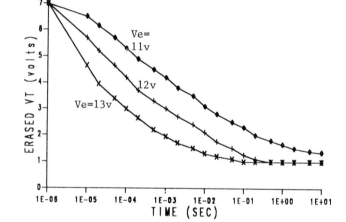

Figure 4. Electrical Erase Characteristics (Ve = 11, 12 and 13 volts)

## 5. Acknowledgements

The auothor acknowledges the MOS memory R & D group of National Semiconductor Corp. and in particular the FEEPROM development group. The author also acknowledges King Fahd University of Petroleum and Minerals for support.

## 6. References

1. S. Mukherjee, et. al., IEDM Tech. Digest of Papers 1985, 26.1, pp. 616-619.
2. V. Kynett, et. al., IEEE JSSC, Vol. 23, Oct. 1988, pp. 1157-1163.
3. V. Kynett, et. al., IEEE JSSC, Vol. 24, No. 5, Oct. 1989, pp. 1259-1264.
4. H. Kume, et. al, IEDM Tech. Digest of Paper 1987, 25.4, pp. 560-563.
5. G. Samachisa, et. al, ISSCC Dig. of Tech Papers 1987, pp.76-77.
6. Fujio Masouka, et. al, IEEE JSSC, Vol. SC-22, No. 4, Aug. 1987, pp. 548-552.
7. L. Faroane, IEEE Trans. Electron Dev., Vol. ED-33, Nov. 1986, pp. 1785-1794.

# Circuit level models for VLSI components

F.M. Klaassen

Philips Research Laboratories, Eindhoven, The Netherlands

**Abstract**

Physics-based, analytical device models intended for use in circuit simulation are the subject of this review. We discuss in turn the modelling of the MOSFET, the modelling of the bipolar transistor, and the acquisition, process dependence and statistics of parameters.

## Introduction

Essentially compact or circuit-level models give a description of the currents and charges of electronic devices in terms of the applied terminal voltages. Generally three different types can be distinguished: physics-based models, empirical models and table look-up models. In the first type the above relations are analytical expressions, which have been derived via an in-depth understanding of the physical mechanisms underlying current transport, etc. In the second type the relations are of a curve fitting nature and in the last type the characteristics are reconstructed via tables of measured or simulated data.

Usually in physics-based models the relations are of a type $I_D = I_D(V_i, P_j)$, etc., where $V_i$ is a terminal voltage and $P_j$ is a parameter. Owing to its physical base, the parameters obey geometrical scaling rules, when the device dimensions are altered. Furthermore realistic statistical modelling of circuit properties is feasible. Naturally, in order to avoid excessive CPU-time, a trade-off between complexity and accuracy has to be made.

Despite their short development time, owing to the huge data storage required for the description of small devices, until now table models have not realized a breakthrough. In addition this approach has no predicting, scaling nor statistics capability. Self evidently the latter facts also apply to empirical models.

This review is limited to physics-based modelling. The modelling of the MOSFET, the modelling of the bipolar transistor and the acquisition, process dependence and statistics of parameters are discussed in turn. In addition mainly basic approaches are given. For detailed formulae we refer to [1].

## 1 MOSFET modelling

### 1.1 Threshold voltage

Since the threshold voltage has a pivot function in MOSFET characteristics, its description is discussed first. In order to understand the approach in fig. 1, the cross section of a small $n$-type MOSFET is given in a direction parallel and perpendicular to current flow. Practically the onset of inversion occurs in this structure, if a voltage drop across

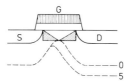

Fig. 1a. Shape of the depletion region in a short-channel MOSFET at zero and high drain bias. The grey areas indicate the charge shared by the gate and the junctions at zero drain bias.

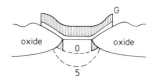

Fig. 1b. Shape of the depletion region in a narrow-width MOSFET at zero and high back-bias. The grey areas indicate excess charge, which is formed under the LOCOS region.

the depleted substrate layer is induced, which approximately equals the diffusion voltage $\phi_F(N_B)$. Under this condition the threshold voltage $V_T$ is determined from a balance of the applied gate charge

$$Q_G = C_{ox} Z L (V_T - V_{FB} - \phi_F) \tag{1}$$

and the depletion charge

$$Q_B = C_{ox}(Z + 2\delta Z)(L - 2\delta L) q \int_0^{y_d} N_B(y) dy. \tag{2}$$

In these equations $V_{FB}$ is the flatband voltage, $C_{ox}$ is the unit area gate capacitance, $N_B$ is the substrate doping, $y_d(\phi_F)$ is the depletion width and $L$, $Z$ are the actual channel length and width, respectively.

The $Q_B$ expression can be evaluated using several simplifying assumptions. Since $V_T$ is the result of an integration, the implanted Gaussian-type doping profile $N_B(y)$ is replaced by an equivalent box profile[2]. The increase $\delta Z$ of the depletion layer width is attributed to its spreading into the channel stop area under the field insulator. Finally the decrease in length $\delta L$ is associated with charge sharing by the junctions and the channel area. The latter effects have been calculated either by assuming trapezoidal forms of the depletion region[3] or by making use of pseudo-2D solutions of Poisson's equation[4,5].

Naturally according to the above approach $V_T$ is expressed in terms of the substrate doping[2]. Since however the latter quantity cannot be extracted directly from the I-V characteristics, it is more practical[1] to rewrite the result in terms of the substrate coefficients $\gamma_i$ and $\gamma$, which are the asymptotic values of $\partial V_T / \partial V_S$ (compare fig. 2)

$$V_T = V_{T0} + \Delta V_T(\gamma_i, \gamma, V_{SX}, V_S). \tag{3}$$

Fig. 2 gives a comparison between measured and modelled results. For the $\gamma$-parameters the following scaling relations apply

$$\gamma = \gamma_\infty - \gamma_L/L + \gamma_Z/Z. \tag{4}$$

Although a buried channel causes the evaluation of the dope integral (2) to be more complicated for $p$-channel transistors, it has been shown[6] that eqs (3) and (4) still apply.

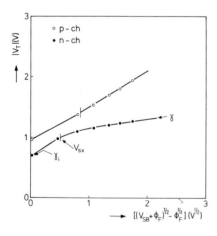
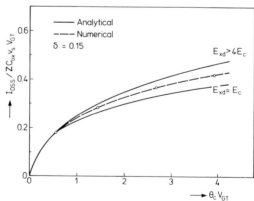

Fig. 2. Comparison of measured values of the threshold voltage with calculated data (fully drawn lines).

Fig. 3. Modelled value of saturated drain current as a function of normalized gate driving voltage at several values of short-channel parameter $E_{xd}/E_c$. The broken line represents values calculated numerically.

## 1.2 Drain current at strong inversion

In the strong inversion regime current transport is mainly by drift. Consequently the channel current (at position $x$ with a potential $V$) is given by

$$I_D = Z\mu_s C_{ox}\{(V_G - V_{FB} - V) - q_B(V)\}\frac{dV}{dx}, \tag{5}$$

where $\mu_s$ is the local surface mobility, the first term between brackets represents the applied gate charge and $q_B$ is the charge in the depletion layer. Generally the mobility is affected by the normal and lateral field[7]. In practice this can be satisfactorily modelled writing[6, 8, 9]

$$\mu_s = \mu_{so}[1 + \theta_A(V_G - V_T) + \theta_B V_S]^{-1}(1 + E_x/E_c)^{-1}, \tag{6}$$

where $\theta_A$, $\theta_B$ and $E_c$ have to be considered as parameters and the last term represents velocity saturation. For gate insulators thinner than 15 nm, owing to surface roughness an additional term $\theta_D V_{GS}^2$ becomes necessary[1]. By calculating $q_B$ for a box-type doping profile the drain current is obtained from an integration of eq (5). Since however for an implanted substrate the result is complicated[2] and in the case of short-channel devices leads to a quartic equation for the saturation voltage, it is useful to approximate the depletion charge

$$q_B(V) \simeq q_B(V_S) + \delta C_{ox}(V - V_S). \tag{7}$$

If $\gamma_i = \gamma$, $\delta \simeq 0.3\gamma(V_S + \phi_F)^{-1/2}$; however for $\gamma_i > \gamma$, $\delta(\gamma_i, \gamma)$ becomes more complicated[1].

Since for higher drain bias current saturation is induced by reversal of the normal field at the drain end of the channel, formally the saturated drain current can be written

$$I_{DSAT} = ZC_{ox}[V_G - V_T - (1 + \delta)V_{DSAT}]\frac{\mu_{so}E_{xd}}{1 + E_{xd}/E_c}, \tag{8}$$

where $E_{xd}$ is the actual lateral field and $V_{DSAT}$ is the saturation voltage. By equating eqs (5) and (8) for this saturation condition, $I_{DSAT}$ and $V_{DSAT}$ are obtained from a square-law equation. Since $E_{xd}$ is not known exactly, MOSFET models have been based on the assumption $E_{xd} = E_c$ [10, 11] or $E_{xd} \gg E_c$ [12, 13]. In reality as shown by fig. 3 the truth lies somewhere in between. In this figure the normalized saturation current according to the above approaches has been compared with results of 2-D numerical device simulation[14]. Since in practice $\theta_c V_{GT} < 1.5$ and $\theta_c = (LE_c)^{-1}$ is a measurable parameter, the error of both models can be made sufficiently small. Finally fig. 4 gives a

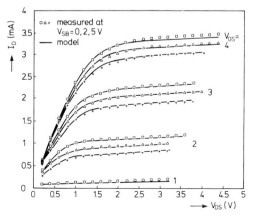

Fig. 4. Measured and modelled characteristics of 0.5 $\mu$m n-type MOSFET.

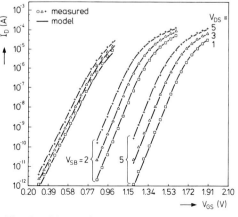

Fig. 5. Measured and modelled subthreshold characteristics for 0.7 $\mu$m n-type MOSFET.

comparison between the modelled and measured characteristics of a submicron $n$-channel MOSFET. In order to obtain a good fit of the drain conductance in the saturation region, the above model has been extended. This is discussed in section 1.4.

## 1.3 Subthreshold current

When a gate bias well below threshold is applied, a potential barrier exists between the channel area and the source. In this case current flow is possible by means of carriers which are able to pass the above barrier, and transport takes place by diffusion rather than by drift. Therefore the current can be written in the simple form $I_D = ZqD_n n(0)/L$, where $D_n$ is a diffusion constant and $n(0)$ is the number of carriers that can pass the above barrier. By solving the charge balance equation discussed in 1.1., it can be proved that

$$I_D = \frac{Z}{L} I_0 \exp\left\{\frac{q(V_G - V_T)}{M k_B T}\right\}, \tag{9}$$

in which $I_0$ is a current constant and the slope factor M is given by $M = 1 + m(V_S + \phi_F)^{-1/2}$.

Unfortunately for short-channel devices the value of the underlying potential barrier is affected by an increase of drain bias. Since an accurate analysis of this so-called DIBL effect requires a 2-D solution of Poisson's equation, several approximations have been proposed[15, 16]. Following the latter one, in which the effect of the drain-induced field is transformed into an apparent decrease of the substrate doping, it can be shown[1] that the DIBL effect can be expressed as a decrease of threshold voltage

$$\Delta V_T \simeq -s(V_S)V_{DS}. \tag{10}$$

In the case of a uniformly doped substrate $s(V_S) \simeq s(V_S + \phi_F)^{1/2}$.

For the weak inversion region around $V_T$ transport by drift and diffusion equally contribute to the current[17]. Therefore presently no satisfactory analytical results are available. However the expressions for strong inversion and subthreshold mode can be successfully merged either by adding eq (9) after multiplication with a limiting function to eq (5)[12, 13] or by defining a generalized gate driving voltage[9]. Fig. 5 gives a comparison of the measured and modelled characteristics below inversion. Note the increase of the DIBL effect at higher values of back bias.

## 1.4 Saturation mode

As already shown in fig. 4, the drain current increases slightly in the saturation region. Generally this is caused by three physical mechanisms. First an increase of drain bias causes the point of normal field reversal to shift towards the source, next for shorter channels near the same point excess mobile charge is induced and finally for very short channels weak avalanche multiplication will occur. The first effect, which can be interpreted as modulation of effective channel length, has to be calculated via a 2-D solution of Poisson's equation, including mobile carrier space charge (compare fig. 6). Therefore few published analytical models sustain a comparison with experimental results[1]. After a slight adaptation satisfactory results are obtained[16] for a model[15], which expresses the effective length by[16].

$$L_{eff} \simeq L\left[1 + \alpha \ln\left(1 + \frac{V_D - V_{DSAT}}{\alpha V_p}\right)\right]^{-1}, \tag{11}$$

where $\alpha$ and $V_p$ are parameters.

The second effect, which is known as static feedback, can be successfully taken into account by adding a term to the gate drive[12]

$$V_{GT} = V_G - V_T + f(V_{DS}). \tag{12}$$

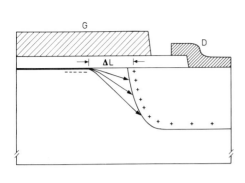
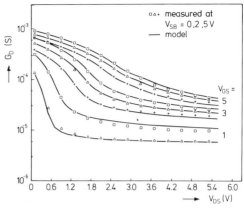

Fig. 6. Illustration of physical mechanisms affecting the value of the drain conductance.

Fig. 7. Measured and modelled drain conductance characteristics of 0.8 μm p-type MOSFET.

For digital applications a first-order approach $f(V_{DS}) = fV_{DS}$ is sufficient, but for an accurate description of the drain conductance in analog applications a more refined approach is necessary[16].

Fig. 7 gives a comparison of the modelled and measured characteristics of the drain conductance of a submicron p-channel MOSFET.

## 1.5 Charges and capacitances

Compared with the notice given to the MOSFET current, the modelling of charges and capacitances lags behind. For a long time models have been used taking only a few small signal capacitances derived for a long channel device. This approach not only leads to errors in some capacitances of short-channel devices, but generally the charge conservation principle

$$Q_G + Q_S + Q_D + Q_B = 0 \qquad (13)$$

is violated[17, 18]. In order to maintain the latter a model has to be based on three independent terminal charges and, since such charges depend on other node voltages too, nine independent capacitances have to be taken into account.

For the calculation of the charges $Q_S$ and $Q_D$ a physical partitioning of the charge in the channel is required. Such a split has been derived along two different lines of approach[19, 20]. For $Q_D$ then we have

$$Q_D = -qZ \int_0^L n(x,t)\frac{x}{L}dx. \qquad (14)$$

Similarly for $Q_S$ a weighting factor $(1 - x/L)$ has to be used. In order to simplify the evaluation of (14) quasi-static operation is assumed, which means that $n(x,t)$ only depends on the instantaneous value of node voltages. In practice this implies that the application is limited to transients in a time step larger than twice the channel transit time[1]. Since n is known only as a function of the potential V, the variable x has to be replaced by V. This can be achieved via the current relation (5). For a short-channel device the expression for $Q_S$ and $Q_D$, although complicated, can be given in a closed form[21]. In addition, effects of velocity saturation, static feedback, etc. can be taken into account and the continuity of the resulting capacitances across boundaries between possible modes of operation is assured. As an example fig. 8 gives a plot of $C_{DG} = -\partial Q_D/\partial V_G$ for all possible modes. Generally, the capacitances are non-reciprocal.

Finally similar to the $V_T$ calculation, the bulk charge is obtained from a solution of Poisson's equation for a box-type doping profile.

## 1.6 Parasitics and model extensions

In addition to the intrinsic properties discussed previously, modelling of parasitics becomes increasingly important. Within the limited scope of this review, only a few remarks are made. Owing to the necessity to reduce hot carrier effects via graded junction or off-set gate structures, series resistances have a large impact on the characteristics of submicron MOSFETS. Although these resistances can be added to a circuit model, it is more practical to avoid the use of additional nodes and to include the resistance effect implicitly in the model equations. Generally this is achieved by generalizing the $\theta$-parameters[22].

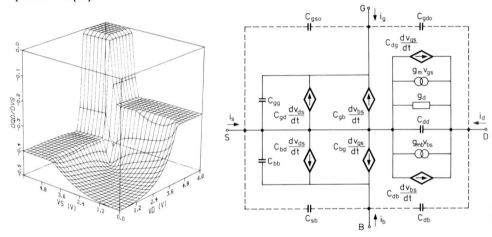

Fig. 8. The normalized drain-gate capacitance vs. drain and source bias at a fixed value of gate bias ($V_G = 6$ V, $V_T = 1$ V).

Fig. 9. Equivalent circuit of a MOSFET, showing all intrinsic capacitances (with fully drawn interconnections) and parasitic capacitances (with dashed interconnections).

Since the gate-junction and the bulk-junction capacitance do not scale in the same manner as the intrinsic parts and additionally their voltage dependence becomes more complicated, an accurate description is required[1]. A complete MOSFET model is shown in fig. 9.

For wide applications such a model has to be extended with a description of the geometry and temperature dependence of parameters, the weak-avalanche induced substrate current and the noise sources[1].

## 2 Bipolar modelling

### 2.1 Carrier transport

In addition to a diffusion current, owing to the built-in field associated with doping gradients, in bipolar devices transport by drift has to be taken into account as well. Consequently the electron current density is given by

$$J_n = q\mu_n n E + q D_n \nabla_r n \tag{15}$$

with a similar equation for holes.

Furthermore under transient conditions the continuity equation applies:

$$\frac{\partial n}{\partial t} - q^{-1} \nabla_r . J_n = G - R, \tag{16}$$

where $G$ and $R$ are the generation and recombination rate, respectively.

Generally the above equations can only be solved numerically. However with almost no loss in accuracy, analytic solutions are possible under the following assumptions: current flow is 1-D and for the base region exclusively controlled by minority carriers; outside depleted regions quasi-neutrality exists; the right-hand side of eq (16) is determined by recombination with a single time constant. Using these assumptions the field is eliminated from eq (15) yielding the result

$$J_n = \frac{qD_n}{p}\frac{d(pn)}{dx}. \tag{17a}$$

Finally at the boundaries of depleted regions the condition applies

$$pn = n_i^2 \exp(V_J/U_T), \tag{17b}$$

where $n_i$ is the intrinsic carrier density, $V_J$ is the applied junction voltage and $U_T = k_B T/q$.

## 2.2 Collector current

### 2.2.1 Charge control approach

By integrating eq (17a) from emitter to collector and using the boundary condition (17b), the electron current density in an $npn$ transistor (compare fig. 10) is given by[23]

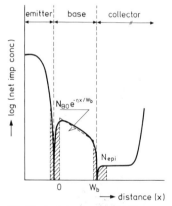

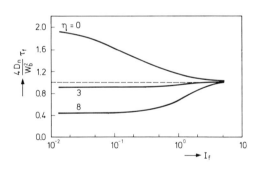

Fig. 10. Typical doping profile in an $npn$ transistor. Hatched areas denote depletion regions. The dashed line gives the approximation of Eq. (21).

Fig. 11. Current dependence of the forward ($\tau_f$) base transit time.

$$J_n = qn_{io}^2 G_b^{-1}\{\exp(V_{BE}/U_T) - \exp(V_{BC}/U_T)\}, \tag{18}$$

in which the Gummel number

$$G_b = \int_E^C \frac{p(x)}{D_n(x)}\left(\frac{n_{io}}{n_i(x)}\right)^2 dx.$$

In these equations $V_{BE}, V_{BC}$ are the applied junction voltages and $n_i$ is considered to be subject to bandgap narrowing[24]. Fortunately in the latter integral $D_n(n_i/n_{io})^2 \simeq 25 cm^2 s^{-1}$ in a wide range of doping densities[1]. Therefore the collector current is inversely proportional to the total hole charge $Q_b$ between emitter and collector. Consequently the above current is written as

$$I_c = I_s(Q_{bo}/Q_b)\{\exp(V_{BE}/U_T) - \exp(V_{BC}/U_T)\}, \tag{19a}$$

where $I_s$ is a process-related current constant.

In principle $Q_b$ can be divided into parts:

$$Q_b = Q_{bo} + Q_{Te} + Q_{Tc} + Q_{be} + Q_{bc}. \tag{19b}$$

$Q_{bo}$ is the zero bias base charge, $Q_{Te}$ and $Q_{Tc}$ are the depletion charges of the junctions, and $Q_{be}$ and $Q_{bc}$ are the stored charges of the minority carriers, injected from the emitter and collector, respectively.

Disregarding all charges except $Q_{bo}$ in the denominator, eq (19a) forms the base for the well-known Ebers-Moll model[25]. In order to describe better the characteristics at higher currents, $Q_b$ is further evaluated in the current Gummel Poon model[26]. Then the depletion charges are calculated by integration of a bias-dependent depletion capacitance (see 2.4) and the stored charges are related to the injected currents by means of the charge control principle, e.g.

$$I_f = I_s[\exp(V_{BE}/U_T) - 1] = Q_{be}\tau_f^{-1}, \qquad (20)$$

where $\tau_f$ is called the forward base transit time. For $Q_{bc}$ a similar relation can be given. In the above concept stored charge moving across the base is considered as the driving force of current. Since all charges in (19b) are now modelled in terms of junction voltages, $Q_b$ can be eliminated from eq (19a) and a closed form expression

$$I_C = I_C(V_{BE}, V_{BC}, Q_{bo}, I_s, \tau_f, \tau_r)$$

is obtained, which is claimed to be valid for all operation modes.

In spite of its wide use, the G-P model poorly describes the characteristics under high current, low bias conditions (compare figs. 14 and 15). Generally this is caused by the neglect of quasi-saturation and the fact that $\tau_f$ and $\tau_r$ are not constants.
Although the latter can be cured by making them bias dependent[27], a basic new approach is preferred. This is discussed next.

### 2.2.2 Carrier control approach

An alternative approach to the description of currents and charges is to relate them to the injected minority carrier densities in the base, which in turn depend on the applied junction voltage[28, 29, 30]. In this approach the hole density $p$ is eliminated from eq (17a) by approximating the base doping profile (see fig. 10)

$$N_B(x) = N_{BO} \exp(-\eta x/W_B), \qquad (21)$$

where the parameter also is a measure for the built-in field. In this way for the neutral base region from eq (17a) the basic equation

$$I_n = qD_n \left\{ \frac{2n + N_B}{n + N_B} \frac{dn}{dx} + \frac{n}{n + N_B} \frac{dN_B}{dx} \right\} \qquad (22)$$

is obtained. Generally this equation can be solved analytically for the low injection case ($n \ll N_B$) and the high injection case ($n \gg N_B$). By writing $I_c = I_f(V_{BE}) - I_r(V_{BC})$, for instance in the latter case we have

$$I_f = (I_s I_k)^{1/2}[\exp(V_{BE}/2U_T) - 1], \qquad (23)$$

where the parameter $I_k$ (onset current value of high injection) is determined by $I_k = 4D_n W_B^{-2} Q_{bo}$.

For the general case, where no closed form solution is possible, an accurate fit formula is obtained by smoothly merging both asymptotic solutions[28]. In this way the bias dependence of the transit time parameters is obtained naturally. Fig. 11 gives the normalized value of $\tau_f$ as a function of injected current, clearly showing the asymptotic values.

## 2.3 Base current

By integrating the continuity equation (16) between emitter and collector and interpreting the resulting terms, the base current is expressed in three parts

$$I_B(t) = I_{EP} + \frac{Q_b - Q_{bo}}{\tau_p} + \frac{dQ_b}{dt},$$

where $\tau_p$ is the hole lifetime and the hole current $I_{EP}$ arriving at the emitter contact has a form similar to eq (18) with a characteristic Gummel number ($G_e$) for the emitter. Owing to surface recombination in the depletion regions, usually the second right-hand side term dominates at low forward bias. On the other hand $I_{EP}$ becomes dominant at higher bias conditions. Therefore, introducing a current gain parameter $\beta_f$ and a non-ideality factor $m_f$, the dc base current is satisfactorily described by

$$I_B(V_{BE}) = \frac{I_s}{\beta_f}\left\{\exp(\frac{V_{BE}}{U_T}) - 1\right\} + I_{bf}\left\{\exp(\frac{V_{BE}}{m_f U_T}) - 1\right\} \tag{24}$$

with a similar equation for the collector bias dependent part.

## 2.4 Charges and capacitances

Since the total stored base charge is proportional to the integrated minority carrier density $n(x)$ and the latter is obtained by solving eq (22), the charge $Q_{BE}$ and $Q_{BC}$ can be calculated in principle [28]. Here we give only the high injection result

$$Q_{BE} = \tau_f I_f = (W_B^2/4D_n)I_f.$$

Similar to the current, the general charge stores are obtained by merging the low-injection and high-injection solutions. As $\tau_f$, $\tau_r$ are not suitable parameters (compare fig. 11), in the practical model $I_C$, $Q_{BE}$ and $Q_{BC}$ are expressed in terms of the current constants $I_s$, $I_k$, the zero-bias base charge $Q_{bo}$ and the parameter $\eta$, e.g.

$$Q_{BE} = Q_{BE}(V_{BE}, I_s, I_k, Q_{bo}, \eta).$$

Usually the depletion charges are calculated by integrating the junction depletion capacitance expression $C_T = C_o[1-(V_J/\phi_F)^p]$, where the parameter $p$ is the grading coefficient. Since the resulting expression

$$Q_T = \frac{C_o \phi_F}{1-p}[1 - (1 - V_J/\phi_F)^{1-p}] \tag{25}$$

has a singularity at $V_J = \phi_F$, in practice a slight modification is applied[1].

Under high forward injected current conditions a complication arises at the collector junction. First the moving minority carrier space charge modulates $Q_{Tc}$ and secondly charge may be injected in the collector. This is further discussed in the next section.

## 2.5 Quasi-saturation

Due to an internal voltage drop in the collector region, the base-collector junction may become forward biased at high currents although the external voltage constitutes a reverse bias. This effect, which occurs mainly under the emitter, is called quasi-saturation. The locally forward biased junction causes an increase of reverse current $I_r$ and the injection of an excess charge $Q_{epi}$ in the collector epilayer (base push-out effect). As a result the current gain and the cut-off frequency will decrease. However, depending on the combination of current injected and applied collector bias, a number of different cases has to be considered[31]. This is illustrated in fig. 12, where the typical field and electron density distribution are given.

The main effect on the model described previously is that the boundary condition (17b) at the collector-side is changed and that the charge $Q_{epi}$ has to be taken into account. Using the same assumptions as in section 2.1, the current in the quasi-neutral collector region is controlled by

$$J_n = qD_n\left(1 + \frac{n}{p}\right)\frac{dn}{dx} = qD_n\left(2 + \frac{N_{epi}}{p}\right)\frac{dp}{dx} \tag{26a}$$

with the approximate boundary condition (see fig. 12b)

$$p(x_i) \simeq n_i. \tag{26b}$$

Solving this equation, the charge $Q_{epi}$ can be expressed explicitly in terms of the injected carrier density $p(o)$ at the junction[31]. Since however $p(o)$ according to (26b) depends on $x_i$ and the latter via the voltage drop on $I_C$ and $V_{CB}$, finally $I_C$ and the product $Q_{epi} * I_C$ become an implicit function of the external $V_{CB}$ voltage. In practice $Q_{epi}$ and $I_C$ are solved in a few steps using first an estimated value of $I_C$.

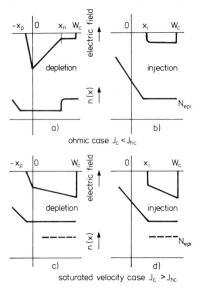

Fig. 12. The field and electron density in the collector epilayer ($n$-type) for the various depletion and injection conditions.

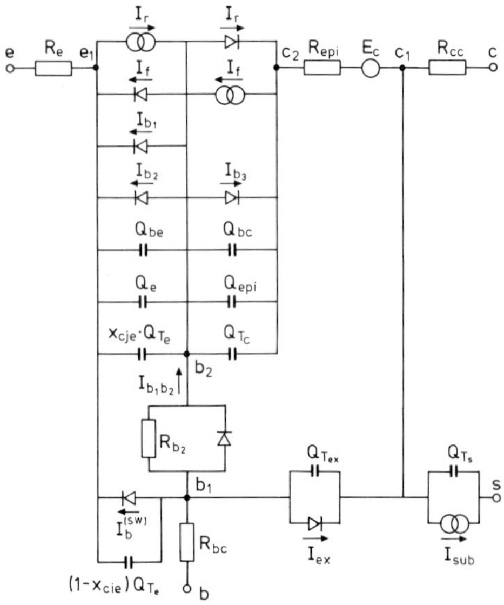

Fig. 13. Equivalent circuit diagram of the MEXTRAM model.

## 2.6 Parasitics and model extensions

Since an increase of depletion charge occurs at the cost of the neutral base width, the currents $I_f$, etc. are affected via a modulation of the zero bias base charge $Q_{bo}$ (Early effect). Usually this effect is taken into account by putting $Q_{bo} + Q_{Te} + Q_{Tc} = Q_{bo}(1+\xi)$ instead of $Q_{bo}$. It is easily shown that in the forward mode

$$\xi \simeq (V_{BE}/V_{ear}) - (V_{CB}/V_{eaf}), \tag{27}$$

where the parameters $V_{ear}$, $V_{eaf}$ are called an Early voltage.

As with the above correction the most important intrinsic properties have been discussed, next an equivalent circuit[32] is given in fig. 13. However in this scheme already parasitic elements such as series resistances and the substrate transistor have been added. Usually the base resistance is split in a part below and adjacent to the emitter. To account for current crowding a diode has to be put in parallel to the first resistor[33, 34].

In contrast to the Gummel-Poon model the above MEXTRAM model describes well the characteristics under high current–low voltage conditions. This is illustrated in fig. 14 for the common emitter dc characteristics and in fig. 15 for the cut-off frequency.

Naturally for wider applications the above model has been extended[1]. Without further discussion we mention here temperature effects, charge storage in the emitter, avalanche multiplication, non-linear Early effect, noise sources and non-quasi static charge behaviour[35, 36].

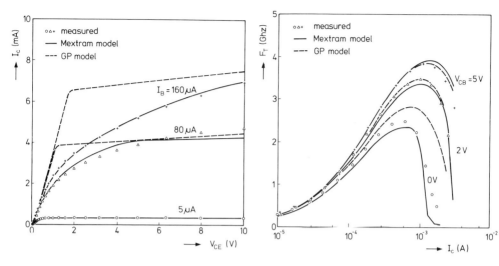

Fig. 14. Measured and modelled dc characteristics of bipolar transistor.

Fig. 15. Measured and modelled cut-off frequency of bipolar transistor.

## 3 Parameters

Although the number of electrical parameters of the above models is large (typically > 20), usually their effect on the characteristics is limited to a specific range. Therefore it is practical to determine the parameters in characteristic groups. For MOSFETS these groups are characteristic for the channel conductance in strong inversion, the saturated $I_D$-$V_{DS}$ characteristics, the subthreshold characteristics, the weak avalanche induced substrate current and the intrinsic capacitances, successively. Similarly, for bipolar transistors the groups have to be determined from the depletion capacitances, the Early effects, the forward and reverse Gummel plots, the quasi-saturation region and the cut-off frequency.

Owing to junction sidewall effects or MOSFET short channel effects most electrical parameters are geometry dependent. Usually simple geometrical rules apply such as eq 4. In addition to the geometry, the parameters also may depend on specific process quantities. For instance, the bipolar current constant $I_s$ will depend on the active base sheet resistance $\rho_{b1}$. For other reasons the MOSFET threshold voltage may depend on the gate insulator thickness and the threshold implantation dose. Using the above dependences on geometry and process quantities, the electrical parameters can be calculated from process data[1].

Owing to their relation with process parameters, compact model parameters can be regarded as statistical variables. For bipolar devices the major process variables to characterize parameter spread are the geometry corrections ($\Delta L$, $\Delta Z$), the active and non-active base sheet resistances ($\rho_{b1}$, $\rho_{b2}$) and the breakdown voltage $V_{cbo}$ [1]. For MOSFETS in addition to $\Delta L$, $\Delta Z$, the gain factor $\beta_0$ and the threshold voltage $V_{TO}$ of a large area device and the unit area interconnect capacitance $C_{int}$ have been used[38]. This can be further extended to predict the spread of a designed electronic function via circuit simulation[37, 38].

## Acknowledgement

The author acknowledges the advice and help of his colleagues H.C. de Graaff, W.J. Kloosterman and R.M.D. Velghe.

# References

1. H.C. de Graaff and F.M. Klaassen, *Compact transistor modelling for circuit design*, Springer Verlag, Vienna (1990).
2. E. Demoulin, F. v.d. Wiele, in *Process and Device Modelling for IC Design*, Noordhoff, Leyden, 617-675 (1977).
3. L.D. Yau, *Solid State Electronics* **17**, 1059–1063 (1974).
4. K.N. Ratnakumar et al., *IEEE Journ.* SSC-17, 937–947 (1982).
5. T. Skotnicki et al., *Proc. ESSDERC 87*, North Holland, Amsterdam, 543–546 (1987).
6. .M. Klaassen et al., *Solid State Electronics* **28**, 359–373 (1985).
7. T. Ando et al., *Rev. Modern Physics* **54**, 437–672 (1982).
8. G. Merckel et al., *IEEE Trans.* ED-19, 681–690 (1972).
9. G.T. Wright, *IEEE Trans.* ED-34, 823–833 (1987).
10. B. Hoeflinger et al., *IEEE Trans.* ED-26, 513–520 (1979).
11. T. Poorter et al., *Solid State Electronics*, **23**, 765–772 (1980).
12. F.M. Klaassen et al., *Solid State Electronics* **23**, 237–242 (1980).
13. B.J. Sheu et al., *IEEE Journ.* SSC-22, 558–566 (1987).
14. CURRY, Proprietary Philips 2-D Simulation Program.
15. P.K. Ko et al., *Technical Digest IEDM* **81**, 600–603 (1981).
16. F.M. Klaassen et al., *Proc. ESSDERC '89*, Springer Verlag, 418–422 (1989).
17. P. Yang et al., *IEEE Journ*, SCC-18, 128–138 (1983).
18. D.E. Ward et al., *IEEE Journ.*, SSC-13, 703–707 (1978).
19. S.Y. Oh et al., *IEEE Journ.*, SSC-15, 636–643 (1980).
20. M.F. Sevat, *Digest ICCAD87*, 208–210 (1987).
21. T. Smedes et al., *Proc. ESSDERC 90*, present issue.
22. F.M. Klaassen et al., Proc. ESSDERC 88, *Journ. Physique* **C4**, 257–260 (1988).
23. H.K. Gummel, *Bell Syst. Techn. J.* **49**, 115–122 (1970).
24. J.W. Slotboom et al., Solid State Electronics **19**, 857–862 (1976).
25. J.J. Ebers, J.L. Moll, *Proc. IRE* **42**, 1761–1771 (1954).
26. H.K. Gummel, H.C. Poon, *Bell Syst. Techn. J.* **49**, 827–844 (1970).
27. H.M. Rein et al., *IEEE Trans.* ED-34, 1741–1761 (1987).
28. H.C. de Graaff et al., *IEEE Trans.* ED-32, 2415–2419 (1985).
29. J.G. Fossum, *Proc. IEEE* Bip. Circuits and Technology Meeting, 234–241 (1989).
30. K.W. Michaels, A.J. Strojwas, *Ibid.* Bip. Circuits and Technology Meeting, 242–244 (1989).
31. H.C. de Graaff, in *Process and Device Modelling for IC Design*, Noordhoff, Leyden, 419–442 (1977).
32. H.C. de Graaff et al., *18th Conf. Solid State Devices*, Tokyo, 287 (1986).
33. G. Rey, *Solid State Electronics* **12**, 645–655 (1969).
34. H. Groendijk, *IEEE Trans.* ED-20, 329–333 (1973).
35. J. te Winkel, *Adv. Electronics/Electron Physics* **39**, 253–289 (1976).
36. J.G. Fossum et al., *IEEE* EDL-7, 652–654 (1986).
37. P. Yang et al., *Techn. Digest IEDM* 82, 286–289 (1982).
38. M.J.B. Bolt et al. *ISMSS 90*, to be published (1989).

Paper presented at ESSDERC 90, Nottingham, September 1990
Session 3A1

# Nondestructive 2D doping profiling by the numerical inversion of CV measurement

G J L Ouwerling[*] and M Kleefstra

Delft University of Technology, Faculty of Electrical Engineering, Electrical Materials Laboratory, P.O. Box 5031, 2600 GA Delft, the Netherlands.
[*] Now with: Philips Research Laboratories, P.O. Box 80.000, 5600 JA Eindhoven, the Netherlands.

**Abstract:** Capacitance-voltage measurements provide an attractive nondestructive method to determine the doping profile in not too highly doped semiconductor layers. However, traditionally the $CV$ method is limited to one space dimension. Also, some error due to the abrupt depletion approximation can occur for steep profiles. In this paper, a numerical measurement data inversion algorithm is presented that circumvents the above limitations of the $CV$ method. It uses a discretization of the doping profile on a one- or two- dimensional grid and involves the repeated solution of a linear least squares system.

## 1 Introduction

To extend the possibilities of $CV$ profiling, we regard the determination of doping profiles from $CV$ measurement data as an *inverse problem*. With a *known* doping profile, a capacitance measurement can be simulated by the *forward* solution of Poisson's equation in the semiconductor device. Consequently, the reconstruction of an *unknown* doping profile from a capacitance measurement requires the *inverse* solution of that equation. This approach was inspired by similar problems in geophysic and biophysics as e.g. summarized by Lines and Treitel (1984).

Figure 1 symbolically shows the extension of 1D CV profiling to 2D problems. It is intuitively clear that excitation by two independently varied gate voltages $V_1$ and $V_2$ is required to obtain sufficient measurement information $C(V_1, V_2)$. Also, only (numerical) inverse methods are capable of reconstructing the 2D doping profile, whereas in the 1D case the $CV$ formula as first derived by Schottky (1942) can be used. The use of inverse methods allows for the nondestructive determination of the 2D doping profile, in contrast with such methods as 2D spreading resistance measurements (Hill et al. 1987), chemical junction delineation (Godfrey 1986) or 2D reconstructions from multiple angle SIMS profiles (van Schie et al. 1989).

Often, inverse solutions are obtained by iteratively solving the forward problem to minimize the difference between measurement and simulation as expressed in a least squares sum

$$\xi = \frac{1}{N} \sum_{j=1}^{N} w_j^2 \left[ C_j^{meas} - C_j^{model} \right]^2 \qquad (1)$$

where $w_j$ is a suitable weight function. As clearly described by Tijhuis (1987), in iterative least squares inversion the available algorithms can be divided into two categories.

In the first place, the forward model can be *parametrized* as $\vec{C}^{model} = \vec{C}^{model}(\vec{p})$ with $\vec{p}$ a vector of parameters with a clear physical interpretation. Starting from a "guestimate" $\vec{p}^0$, the error sum (1) is minimized by a sequence of parameter estimates $\vec{p}^k$ such that $\xi(\vec{p}^{k+1}) < \xi(\vec{p}^k)$, using standard nonlinear parameter estimation methods. For doping profiling an implanted profile might be described by parameters like the projected range and straggle. Results with

© 1990 IOP Publishing Ltd

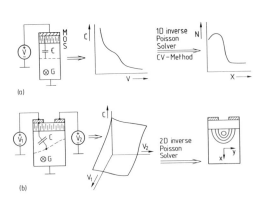
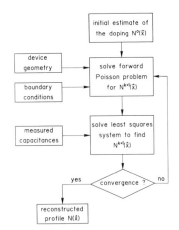

**Fig. 1:** Principle of 1D (a) and 2D (b) doping profiling by inverse methods. The first arrow symbolizes measurement data acquisition, the second data interpretation. The dashed lines indicate depletion boundaries.

**Fig. 2:** Flow diagram of the reconstruction of a doping profile $N(\vec{x})$ from capacitance measurements by the iterative solution of a linear least squares system. $k$ is the iteration index.

this method were reported by us in earlier work (Ouwerling et al. 1988, Ouwerling 1989a, 1990).

In the second place, the unknown quantity — in our case the doping profile — can be *discretized* on a geometrical grid in the device, yielding the forward model as $\vec{C}^{\,model} = \vec{C}^{\,model}(\vec{n}\,)$, where $\vec{n}$ is the vector of geometrically discretized doping values. Some internal knowledge about the forward model is used to define an iterative linear least squares problem $A^k \vec{n}^{k+1} = \vec{C}^{\,meas}$ with $k$ the iteration counter as indicated in the flowchart of Fig. 2. This second method is the subject of this paper. A more extensive discussion can be found in Ouwerling (1989a).

## 2   The iterative linear least squares algorithm

Fig. 3 shows a MOS device on which a sequence of $N$ voltages $V_j$ is applied to measure a sequence of small signal capacitances $C_j$. A linearization is obtained by the definition of the depletion rate $d_E$ (Fig. 3c) and the *differential* depletion rate $d_D$ (Fig. 3d) as

$$d_D^k(x;V_j) = d_E^k(x;V_j+\Delta V_j) - d_E^k(x;V_j) = \frac{\rho^k(x;V_j+\Delta V_j) - \rho^k(x;V_j)}{qN_N^k(x)} \qquad (2)$$

With $d_D(x,V_j)$, the measured incremental depletion charge $\Delta Q = C \cdot \Delta V$ can be written as

$$\Delta Q_j^k = C_j \cdot \Delta V_j = A_d \int_0^L \left[\rho^k(x;V_j+\Delta V_j) - \rho^k(x;V_j)\right] dx = qA_d \int_0^L N_N^k(x) \cdot d_D^k(x;V_j)\, dx, \quad (3)$$

for all measurements $j \in [1,N]$, and with $A_d$ the device area. From this set of equations the desired linear least squares system can be found by 3 steps: **1.** substitution of the measured $\Delta Q_j^M$ for the calculated $\Delta Q_j^k$, **2.** substitution of the unknown $N_N^{k+1}(x)$ for the known $N_N^k(x)$, and finally **3.** discretization of the unknown doping $N_N^{k+1}(x)$ on a geometrical grid with $M$ elements $n_i$ as indicated in Figs. 4a and b. This yields the (iterative, see index $k$) linear least squares system

$$C_j^M = \frac{1}{\Delta V_j} qA_d \sum_{i=1}^M n_i^{k+1} \int_{x_i}^{x_{i+1}} d_D^k(x;V_j)\, dx, \qquad j \in [1,N] \qquad (4)$$

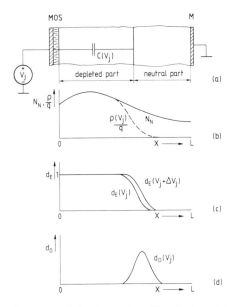
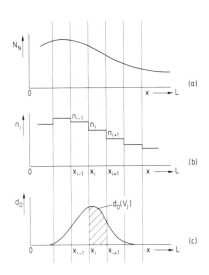

**Fig. 3:** Depletion in a 1D profiling device. (a) partially depleted MOS device (b) Doping concentration $N_N$ and space charge $\rho$ (scaled to $q$) (c) Depletion rate $d_E = \rho/qN_N$ (d) Differential depletion rate $d_D$

**Fig. 4:** Discretization of the unknown doping profile. (a) Doping profile update $N_N^{k+1}$ to be determined. (b) Zero-th order discretization of $N_N^{k+1}$. (c) Differential depletion rate $d_D(x; V_j)$. The hatched area gives matrix element $a_{ji}$.

or in matrix form

$$\left(A\right)^k \left(\vec{n}\right)^{k+1} = \left(\vec{C}\right), \quad \text{with} \quad a_{ji}^k = \frac{1}{\Delta V_j} q A_d \int_{x_i}^{x_{i+1}} d_D^k(x; V_j)\, dx \qquad (5)$$

where $\vec{n}$ is the vector of $M$ unknown discrete doping values $n_i$, $\vec{C}$ the vector of $N$ measured capacitances $C_j$, and $a_{ji}$ one element of the matrix $A$, given by the hatched area of Fig. 4c.

To resolve possible ill-posedness of the inverse problem, the linear least squares system is solved by the singular value decomposition of the design matrix $A$ as advocated by Press et al. (1986). The problem is easily extended to the two-dimensional case by using the 2D capacitances as shown in Fig. 1, a 2D dope grid, and a 2D definition of the differential depletion rate as $d_D(x, y)$.

## 3 Experimental

The iterative linear least squares method was succesfully applied to the inversion of 1D capacitance measurement data. The reconstruction capability for 2D profiles was studied using synthetic measurement data. Evidently, this does not provide experimental information. On the other hand, it does provide an excellent way to assess the measurement error sensitivity of the method, because the original doping profile is exactly known. E.g., a perturbation analysis showed that a measurement noise of 0.5 RMS % did not impair the convergence of the algorithm to the correct profile shape (Ouwerling 1989a).

Two typical results are shown in Figs. 5 and 6. Fig. 5 shows the first 8 iterations of the reconstruction of a "camelback" profile from synthetic measurement data, including convergence behaviour. Fig. 6 shows the reconstruction of a 2D implantation profile discretized on a grid as shown. The 80 required iterations took approximately 4 hours of CPU time on a Convex C210 vector machine with the forward Poisson solver according to Ouwerling (1989b) on a 41 × 61 solution mesh.

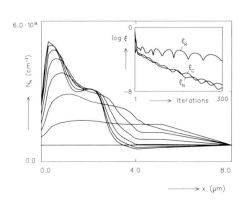

**Fig. 5:** First 8 iterations of the reconstruction of the camelback profile with the iterative linear least squares method using noise-free data. The insert shows the convergence behaviour for 300 iterations. Here $\xi_C$ is the error as given by eq. 1, and $\xi_R$ the residue of system (5).

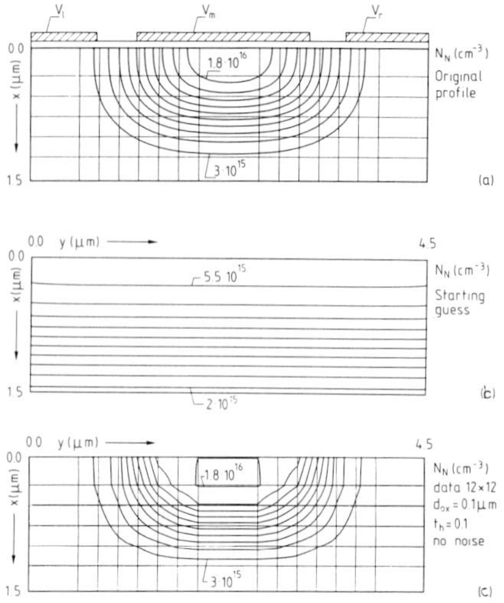

**Fig. 6:** Reconstruction of a 2D doping profile from CV measurements (a) Trimos device with independently varied gate voltages $V_l$ and $V_r$. (b) Starting guess $N^0$ of the 2D doping profile. (c) Reconstruction $N^{80}$ from 12 × 12 synthetic capacitance measurements.

## Acknowledgements

The authors are much indebted to H. Blok, P.W. Hemker and A. Wexler for discussion on numerical methods, and to W. Crans for stimulating the use of inverse methods

## References

Hill C, Pearson P J, Lewis B, Holden A J and Allen R W 1987 *Proc. ESSDERC-87* 923
Godfrey D J 1986 *GEC Journal of Research* **4** 57
Lines L R and Treitel S 1984 *Geophysical Prospecting* **32** 159
Ouwerling G J L, Rijs F van, Wentinck H M, Staalenburg J C and Crans W 1988 *Proc. SISDEP-3* ed Baccarani G and Rudan M (Bologna: Tecnoprint)
Ouwerling G J L 1989a *Nondestructive one- and two-dimensional doping profiling by inverse methods* Thesis, Delft University, available from University Microfilms Int., Ann Arbor, Michigan USA
Ouwerling G J L 1989b *Journal of Applied Physics* **66** 6144
Ouwerling G J L 1990 *To be published in Solid-St. Electron.* (Spec. Issue on Proc. and Mat. Char.)
Press W H, Flannery B P, Teukolsky S A and Vetterling W T 1986 *Numerical Recipes* (Cambridge UP)
Schie van E, Middelhoek J and Zalm P 1989 *Nucl. Inst. Meth. Phys. Res.* **B42**, 109
Schottky W 1942 *Zeitschrift für Physik* **118** 539
Tijhuis A G 1987 *Electromagnetic inverse profiling* (Utrecht: VNU Science Press)

# Low leakage current evaluations for process characterizations

P. GIRARD, B. PISTOULET and P. NOUET

Laboratoire d'Automatique et de Microélectronique de Montpellier (U.A. D03710 CNRS)
Université de Montpellier II : Sciences et Techniques du Languedoc
Pl. E. Bataillon, 34095 MONTPELLIER Cedex 5, FRANCE

Abstract. In this paper the principle of operation of a structure allowing accurate determination of currents in the fA range is given. A device has been implemented on silicon and experimentally tested. The capability of 0.1 fA range measurements is demonstrated.

## 1. Introduction

In today submicron technologies, elementary device leakage currents become very small. In order to characterize technological processes, microelectronic test structures are commonly used (Carver 1980). In this paper, we propose a test structure which allows the determination of currents smaller than 1 fA using classical experimental apparatus such as Transistor Parameter Analyzer and a Scanning Electron Microscope (SEM). In the first part of the paper the principle of the method is given, then experimental data arising from the implementation of a test structure on a 2 μm CMOS process are reported, and finally discussed.

## 2. Principle of operation

In order to provide the required amplification, the leaky device under test is connected to the gate of a MOS transistor (see Figure 1). The device bias $V_{GS}$ is obtained with an electron-beam charge deposition on an unpassivated aluminum target, so, extra capacitances and supplementary leakage currents are eliminated. Once the various capacitors are charged, the electron beam irradiation is stopped and the charge evacuation occurs via the leaky device.

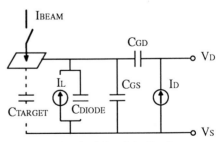

Figure 1: Principle of the Leakage Current Test Structure

The leakage current $I_L$ is given by:

$$I_L = dQ_{stored} / dt = d(\Sigma C \cdot V_{GS}) / dt \qquad (1)$$

Assuming that total capacitance ($\Sigma C$) variations with device bias are negligible, and that the main contribution to the total capacitance results from the MOS transistor gate, equation 1 becomes:

$$I_L = C_{ox} \cdot W \cdot L \cdot (dV_{GS}/dI_{DS}) \cdot (dI_{DS}/dt) \qquad (2)$$

Assuming simple laws for drain–source currents variations versus gate source bias (Sze 1981), it finally comes:

$$I_L = K \cdot dI_{DS}/dt \quad \text{with } K = L^2/(\mu \cdot V_{DS}) \qquad (3)$$

where $C_{ox}$, $W$, $L$, $\mu$, $I_{DS}$ are respectively the gate capacitance per unit area, the width and length of the MOS transistor, the carrier mobility and the drain–source current.
Consequently, with typical values corresponding to the test structure implemented ($L = 5$ μm, $\mu = 175$ cm$^2$/(V.s), $V_{DS} = 0.1$ V) the leakage current remains about $10^{-8}$ times smaller than the drain–source current variations with time.

Since $I_{DS} = f(t)$ is an experimental data, low currents could be evaluated. The bias of the device results from the knowledge of the $I_{DS} = f(V_{GS})_{V_{DS}}$ characteristics of the MOS transistor used for amplification. In order to avoid errors due to approximations, these relations are experimentally obtained with a classical MOS transistor used as reference. This transistor has the same dimensions as the amplificying one.

Consequently, the method for experimental evaluation of very low leakage currents is as follow: i) the electron beam is used in order to reach a sufficient bias of the gate, ii) drain–source current is measured versus time during all the discharge, iii) drain–source currents variations versus gate bias are measured for the reference MOS transistor, iv) gate voltage variations are numerically deduced from these characteristics, v) leakage currents are evaluated using equation 1.

3. Experimental results

The test structure has been implemented on silicon using a 2 μm industrial CMOS technology (see Figure 2). Since a negative charge deposition is obtained on aluminum at 1 keV primary energy, a p channel MOS transistor has been chosen in order to obtain an initial off state and a switching on after irradiation.

In order to ensure a very easy charging by the electron-beam, the aluminum target size is 100 x 140 μm$^2$. The dimensions of the MOS transistor used for amplification are 16 μm wide and 5 μm long. The leaky device to evaluate corresponds here to a 4 x 4 μm$^2$ reversely biased pn junction. The device is placed, under vacuum, inside an ISI SS40 Scanning Electron Microscope chamber, and current measurements are achieved with an HP 4145B Transistor Parameter Analyzer.

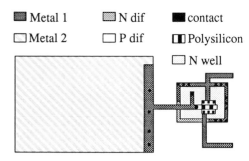

Figure 2: Layout of the implemented test structure.

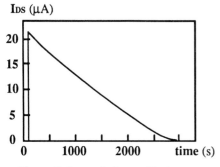
Figure 3: $I_{DS}$ variations with time.

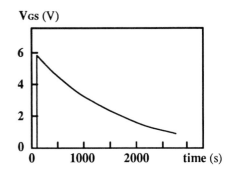
Figure 4: Deduced $V_{GS}$ variations with time.

On Figure 3, we have reported experimental variations of the drain–source current versus time ($I_{DS} = f(t)$). This current allows the determination of $V_{GS}$ gate bias variations with time. Using an electron beam pointed for a few seconds on the target a sufficient bias of the gate is obtained. As shown on Figure 4, it is higher than five volts. When the electron-beam is stopped, the charge evacuation via the pn junction slowly decreases the gate–source bias down to zero.

Using equation 1, leakage current is deduced from the decay rate $dV_{GS}/dt$. This current is reported versus gate bias on Figure 5. Observed magnitudes are consistent with the technological parameters (CMOS 1987) and the capability of 0.2 fA currents measurements is clearly shown. All capacitances of the structure (Figure 1 & 2) have been taken into account, as we will see below .

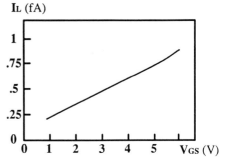
Figure 5: Leakage current variations versus reverse bias voltage.

### 4. Discussions

Due to important dimensions of the target, the total capacitance can not be strictly reduced to the gate one. An evaluation of this capacitance (CMOS 1987) is made in order to obtain the leakage current with equation 1. Consequently, the amplification term ($K^{-1}$) is different from the expected one in equation 3. Essentially due to the real value of the total capacitance which is nearly an order of magnitude higher than the gate one, the experimental value of the amplification term is about $10^7$ s$^{-1}$. This enables the leakage current evaluation in the fA range, measuring decay rate ($dI_{DS}/dt$) of about 10 nA / s.

Additional experimental measurements done with MOS bias $V_{DS}$ up to 1 V have given the same final results, i.e. the same variations of the leakage current versus the reverse voltage applied to the junction. These results agree with equation 3.

Connecting devices with different leakage currents to our structure results in variation of the decay rate ($dI_{DS}/dt$). Limitations of our device are given by an upper and a lower range of leakage currents.
For good operation conditions the leakage current must be lower than the effective charging current: a typical value for beam current is about 1 nA, so, the upper range for leakage currents

lies around 0.1 nA. Besides, the maximum time allowed by the user for the discharge and, the sensitivity of current measurements versus time, may drive to a limitation in the lower range.

The capabilities in lower range leakage current measurements of our device could be determined by the implementation of test structure in submicron technologies in order to reach the $10^{-17}$A range.

## 5. Conclusions

In this paper we have given the principle of a test structure allowing the evaluation of low currents, i.e. in the fA range. This test structure has been implemented on silicon in a 2 µm CMOS process. On the one hand, experimental results correspond to expected capabilities of the device, i.e evaluation of currents in the $10^{-16}$A range. On the other hand, determined values of currents in a reversely biased junction are consistent with those obtained from technological parameters.

Acknowledgements. the authors want to acknowledge the French "Groupement Circuits Intégrés Silicium" (GCIS) for financial support.

## References

Carver GP, Linholm LW and Russel TJ, sept. 1980, "Use of microelectronic test structures to characterize IC Materials, Processes, and Processing Equipment", Solid State Technology, pp 85–92.

Sze SM 1981, "Physics of semiconductor devices", 2$^{nd}$ edition, pp 440–442, John Wiley, New–York.

CMOS 1987, CMOS 2 µm Industrial Design Rules.

Paper presented at ESSDERC 90, Nottingham, September 1990
Session 3A3

# Computer simulation of oxygen precipitation in CZ-silicon during rapid thermal anneals

M.Schrems[1], P.Pongratz[3], M.Budil[1], H.W.Pötzl[1],[4]
J.Hage [2], E.Guerrero [2], D.Huber[2]

[1] Institut für Allgemeine Elektrotechnik u. Elektronik, Technische Universität Wien, Gußhausstraße 27–29, A-1040 Vienna, AUSTRIA

[2] Wacker-Chemitronic GmbH, D-8263 Burghausen, FRG

[3] Institut für Angewandte Physik, Technische Universität Wien, Wiedner Hauptstraße 8-10, A-1040 Vienna, AUSTRIA

[4] Ludwig Boltzmann Institut für Festkörperphysik, Kopernikusgasse 15, A-1060 Vienna, AUSTRIA

**Abstract.** A new computer model for the simulation of oxygen precipitation in conventional as well as rapid thermal anneals is presented. The kinetic part of the model combines chemical rate equations and a Fokker-Planck equation. This allows for an adequate description of both small and larger precipitate sizes. The model can be fitted to experimental data reported by Hawkins and Lavine [1] concerning reduction of interstitial oxygen after a $950°C/1h$ - $1200°C/10h$ annealing cycle. If the annealing cycle is preceded by a short thermal pulse ($1200°C/2s$) retardation of oxygen precipitation is observed experimentally. In our simulations this can be explained by the dissolution of small precipitates, which have formed during crystal growth.

## 1. Introduction

The paper is concerned with the computer simulation of oxygen precipitation during RTA anneals as well as conventional anneals in a diffusion furnace. This requires the simultaneous description of growth and dissolution of both very small and of larger precipitates. Previous simulation models using a Fokker-Planck equation only (e.g. [3]) are not adequate for this purpose, since they assume that the concentration of the smallest precipitates follows an equilibrium distribution. This assumption does not hold in case of very short annealing times (e.g. $1200°C/2s$), or at lower temperatures (e.g. $450°C/10h$).

## 2. Theory

### 2.1 Kinetic Model for Oxygen Precipitation

In the following we present a kinetic model allowing for a non-equilibrium description even of the smallest precipitate sizes. It is based on the concept of linking a rate equation description for small precipitates to a Fokker-Planck equation, which describes larger precipitates more efficiently [2]. The elemantary process for the formation of oxygen precipitates containing n O-atoms follows the reaction equations

$$(n) + (1) \xrightarrow{g(n,t)} (n+1) \qquad (n) \xrightarrow{d(n,t)} (n-1) + (1) \qquad (1)$$

© 1990 IOP Publishing Ltd

with the rates $g(n,t)$ for precipitate growth and $d(n,t)$ for precipitate dissolution. Mathematically this corresponds to a set of rate equations

$$\frac{\partial f_n}{\partial t} = I_n - I_{n+1} \qquad (n = 2, 3, ...) \qquad (2)$$

$$I_n(t) = g(n-1,t) \cdot f_{n-1} - d(n,t) \cdot f_n \qquad (3)$$

with $f_n(t)$ denoting the concentration of precipitates consisting of n O-atoms. For the single O-atoms ($n=1$) the conservation of particles within a small given volume inside a sample leads to an additional term $Q_1$:

$$\frac{\partial f_1}{\partial t} = -I_2 + Q_1 \qquad Q_1(t) = -\sum_{n=2}^{n_{max}} I_n \qquad (4)$$

The maximum number of atoms $n_{max}$ in a precipitate is equal to the number of ordinary differential equations required. If precipitates with a radius of about 500nm occur, the number of equations required will exceed $n_{max} = 1 \cdot 10^{10}$. Therefore the set of Eqns. 2,3 is usally approximated by a single partial differential equation called "Fokker-Planck equation" (FPE) :

$$\frac{\partial}{\partial t} f(n,t) = -\frac{\partial}{\partial n} s \qquad s(n,t) = -B \cdot \frac{\partial f}{\partial n} + A \cdot f \qquad (5)$$

$$B(n,t) = \frac{g(n,t) + d(n,t)}{2} \qquad A(n,t) = g(n,t) - d(n,t) - \frac{\partial}{\partial n} B \qquad (6)$$

The quantities $f(n,t)$, $g(n,t)$ and $d(n,t)$ denote the concentration, the growth rate and the dissolution rate for precipitates containing $n$ O-atoms. The FPE approximating Eqns.2,3 is based on a Taylor series expansion, which gives inaccurate results for the smallest precipitate sizes n. Therefore we use rate equations for the smallest values of n up to a value $n_0$ in combination and a FPE for $n \geq n_0 + 1$. Rate equations are linked to the FPE by a boundary condition for the precipitation flux

$$s(n_0 + 1, t) = I_{n_0+1}(t) = g(n_0, t)f_{n_0}(t) - d(n_0 + 1, t)f(n_0 + 1, t) \qquad (7)$$

2.2 Growth and Dissolution Rates
The growth rate g(n,t) for a precipitate containing n O-atoms is modelled by

$$g(n,t) = 4\pi r^2 \delta \cdot C_O(r) \nu \exp\left(\frac{-\Delta G_{n \to n+1}}{kT}\right) \qquad (8)$$

r denotes the radius of a spherical precipitate and is related to the number n of O-atoms in the precipitate by

$$n = \frac{4\pi}{3v_O} \cdot r^3 \qquad v_O = \frac{v_{SiO_2}}{2} \cdot (1 - e_C)^3 \qquad (9)$$

with $v_O$ denoting the average volume of an O-atom in a strained precipitate (stochiometric composition $\approx SiO_2$), $v_{SiO_2}$ the average volume of $SiO_2$ and $e_C$ the constrained strain. The thickness $\delta$ of the interface precipitate matrix in Eqn. 8 is chosen to be roughly equal to the distance between two Si atoms in the lattice (0.235 nm). $C_O(r)$

is the concentration of O-atoms at the interface precipitate-matrix. $\nu = D_O/\delta^2$ ($D_O$... diffusion coefficient of oxygen) denotes the frequency of jumps of the O-atoms in order to surmount the energy barrier $\Delta G_{n \to n+1}$ for incorporation into the precipitate.

$$\Delta G_{n \to n+1} = \Delta G(n+1,t) - \Delta G(n,t) + \Delta G_{activation} \tag{10}$$

The activation energy for the reaction $\Delta G_{activation} \approx 0.25eV$ is taken from Turnbull and Fisher [4], while the Gibbs energy $\Delta G(n,t)$ of precipitate containing n O-atoms, is modelled as the sum of volume energy ($\Delta G_O$), energy for emission of Si self interstitials ($\Delta G_I$) and absorption of vacancies ($\Delta G_V$), stress energy ($\Delta G_s$), and interfacial energy ($\Delta G_{if}$) in the same manner as in our previous work [2]. An expression for the dissolution rate d(n,t) can be obtained from Eqn.8 assuming the growth law

$$\frac{dn}{dt} = g(n,t) - d(n,t) = h \cdot (C_O(r) - C_{Oo}(r)) \tag{11}$$

for an individual precipitate containing n O-atoms.

$$d(n,t) = h \cdot C_{Oo}(r) = 4\pi r^2 \delta \nu \exp\left(\frac{-\Delta G_{n \to n+1}}{kT}\right) \cdot C_{Oo}(r) \tag{12}$$

Solving the diffusion equation outside a spherical precipitate with radius r yields

$$\frac{dn}{dt} = \frac{dn}{dr} \cdot \frac{dr}{dt} = \frac{dn}{dr} v_O \cdot \frac{D_O}{r}(C_O(t) - C_O(r)) = g \cdot (C_O(t) - C_O(r)) \tag{13}$$

$D_O$ denotes the diffusion coefficient for O in Si, $C_O(t) = f_1(t)$ (see Eqn.4) is the bulk concentration of interstitial oxygen. The equilibrium concentration of O-atoms at the interface precipitate-matrix $C_{Oo}(r)$ can be calculated by setting $\partial \Delta G/\partial r = 0$ and solving for $C_O = C_{Oo}$ [2]. From Eqns.11,8,12 the non-equilibrium concentration $C_O(r)$ of O-atoms at the interface precipitate-matrix is obtained:

$$C_O(r) = \frac{g \cdot C_O + h \cdot C_{Oo}(r)}{g + h} \tag{14}$$

2.3. Additional Model Equations and Numerical Tools

The model also considers the emission of Si-selfinterstitials (I) and absorption of vacancies (V) by $SiO_2$ precipitates, the formation of stacking fault loops, Frenkel pair recombination and diffusion of O,I,V by additional equations [2]. The Fokker-Planck equation has been solved using ZOMBIE [5].

3. Results

The above model has been fitted to experimental results for the reduction in interstitial oxygen obtained by Hawkins and Lavine [1]. We used the initial concentrations of point defects and the Gibbs energies for the smallest O-precipitates (2 O-atoms) as parameters. Precipitate formation during cooling in CZ crystal-growth is roughly estimated using cooling rates extracted from theoretical calculations [6]. Fig.1 shows the oxygen loss as a function of initial oxygen for a 950°C/1h+1200°C/10h furnace cycle preceeded by a 1200°C/2s rapid thermal anneal ("RTA"). Significant retardation of oxygen precipitation compared to similar data without the preceeding RTA step can be seen. In

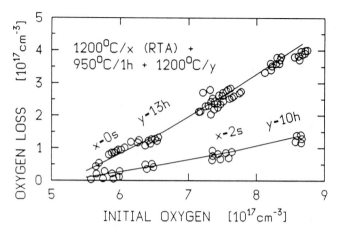

Figure 1: Oxygen loss vs. initial oxygen content, experimental results by Hawkins and Lavine [1] (circles), our simulations (full lines)

agreement with Hawkins and Lavine [1] we find, that this can be explained by the dissolution of precipitates existing previous to the thermal anneals. In our simulations these precipitates form during cooling in CZ crystal-growth.

### 4. Conclusion

A new computer model has been applied to studying retardation effects in oxygen precipitation caused by rapid thermal annealing. In our simulation model the retardation effect is explained by the dissolution of precipitates originating from crystal growth during the RTA step.

*Acknowledgement* — This work has been supported by the Fonds zur Förderung der wissenschaftlichen Forschung, project no. P7495-PHY. Stimulating discussions with Dr.Hobler are appreciated.

### 5. References

[1] G.A. Hawkins and J.P.Lavine, J.Appl.Phys. 65 (1989)3644-3654.

[2] M.Schrems,T.Brabec,P.Pongratz,M.Budil,H.W.Poetzl,E.Guerrero, D.Huber, Semiconductor Silicon (1990)144-155.

[3] M.Schrems,P.Pongratz,M.Budil,H.W.Poetzl,J.Hage,E.Guerrero,D.Huber, Materials Science and Engineering B4 (1989)393-399.

[4] D.Turnbull and J.C.Fisher,J.Chem.Phys.17,71 (1949)

[5] W.Juengling,P.Pichler,S.Selberherr,E.Guerrero,H.W.Poetzl, IEEE J.Solid State Circuits Vol.SC-20 (1985)76-87.

[6] W.Zulehner, Semiconductor Silicon (1990)30-44.

*Paper presented at ESSDERC 90, Nottingham, September 1990*
*Session 3A4*

# Simulation of arsenic and boron diffusion during rapid thermal annealing in silicon

Michael HEINRICH[a], Matthias BUDIL[a] and Hans W. PÖTZL[a,b]

[a] TU-Wien, Institut für allgemeine Elektrotechnik und Elektronik,
Abteilung physikalische Elektronik, Gußhausstraße 27-29, A-1040 Wien,
Austria, [b] Ludwig Boltzmann Institut für Festkörperphysik, Wien

ABSTRACT

Rapid thermal annealing for boron was simulated by a recently developed pair diffusion model. Boron is assumed to reside on interstitial sites after implantation. Decay starts as the temperature rises due to the reaction $BI \rightleftharpoons B + I$. Damage has been taken into account. The model accounts for the temperature dependencey of the transient diffusion effect. The fact that arsenic yields a much less pronounced transient effect is adressed and discussed in our theoretical framework.

INTRODUCTION

Increasing interest in the use of rapid thermal annealing (RTA) as a promising processing step in the fabrication of shallow junctions for VLSI circuits can be found. The desired effect would be to anneal out the defects while causing almost no motion of the dopants. A transient diffusion enhancement has been reported for boron[1], phosphorous[2], antimony[3], and arsenic[4]. While considerable effort has been spent on experimental verification of diffusion enhancement for arsenic, boron, phosphorus and antimony, modeling of RTA still provides challenge.

DIFFUSION MODEL

We simulated rapid thermal annealing by using the pair diffusion model developed by Budil[5], which accounts for the concentration of the dopant, the dopant-interstitial and dopant-vacancy pairs and the interstitials and vacancies. Diffusion equations are solved for all particles. The dopant resides on substitutional sites and is not allowed to diffuse by itself. Diffusion takes place solely via dopant-point defect pairs. Although the model initially has been developed to describe the "kink and tail" diffusion of phosphorus, it can be used for other dopants, too. Boron is assumed to diffuse mainly via an interstitial mechanism. It is allowed to react with interstials:

$$BI \rightleftharpoons B + I \qquad (1)$$

The point defects and impurity-point defect pairs have different charge states and react with electrons or holes.

$$I^{i-1} \rightleftharpoons I^i + e^-, I^{i+1} \rightleftharpoons I^i + h^+ \qquad (2)$$

$$BI^{i-1} \rightleftharpoons BI^i + e^-, BI^{i+1} \rightleftharpoons BI^i + h^+ \qquad (3)$$

For all these species a diffusion equation describes the motion through the bulk with respect to the electrical potential $\psi$. No diffusion may take place via the substitutional dopants. The model yields equations with a mean diffusion coefficient $\overline{D}$ and a mean

© 1990 IOP Publishing Ltd

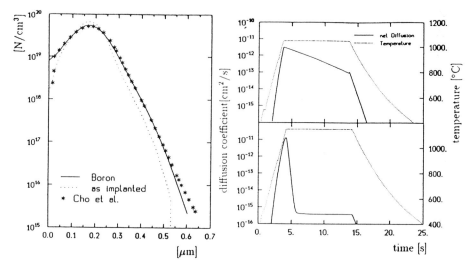

Figure 1: Simulation of rapid thermal annealing of $^{11}$B at 1050°C (solid line) compared with the experimental values (taken from the experiments of Cho[12] and represented by stars). The dashed line represents the implanted profile that has been used as initial value for our simulation.

Figure 2: 'Transient diffusion effect': The net diffusion coefficient obtained by our model and temperature used during our simulation are plotted over the simulation time. Rapid thermal annealing of $^{11}$B both at 1150°C and at 1050°C was simulated (experimental values taken from Cho[12]). It can be seen, that the diffusion process lasts for about 1 second at 1150°C, while it needs about 10 seconds to saturate at 1050°C.

electrical charge $\overline{Q}$ depending on the electron concentration.

$$\overline{D}_{BI} = \frac{\sum_i D_{BI}^i k_{BI}^i \left(\frac{n}{n_i}\right)^{-i}}{\sum_i k_{BI}^i \left(\frac{n}{n_i}\right)^{-i}} \quad , \quad \overline{D}_I = \frac{\sum_i D_I^i k_I^i \left(\frac{n}{n_i}\right)^{-i}}{\sum_i k_I^i \left(\frac{n}{n_i}\right)^{-i}} \tag{4}$$

$$\overline{Q}_{BI} = \frac{\sum_i (-1+i) k_{BI}^i \left(\frac{n}{n_i}\right)^{-i}}{\sum_i k_{BI}^i \left(\frac{n}{n_i}\right)^{-i}} \quad , \quad \overline{Q}_I = \frac{\sum_i i k_I^i \left(\frac{n}{n_i}\right)^{-i}}{\sum_i k_I^i \left(\frac{n}{n_i}\right)^{-i}} \tag{5}$$

The flux J must be equal to the diffusion current plus the field current:

$$J_{BI} = -\overline{D}_{BI} \left(\frac{\partial c_{BI}}{\partial x} + \overline{Q}_{BI} c_{BI} \frac{\partial \psi}{\partial x}\right) \quad , \quad J_I = -\overline{D}_I \left(\frac{\partial c_I}{\partial x} + \overline{Q}_I c_I \frac{\partial \psi}{\partial x}\right) \tag{6}$$

Formation and decay of BI pairs is expressed by a generation term $G_{BI}$:

$$\frac{\partial c_{BI}}{\partial t} + \frac{\partial J_{BI}}{\partial x} = -G_{BI} \quad , \quad \frac{\partial c_B}{\partial t} = G_{BI} \quad , \quad \frac{\partial c_I}{\partial t} + \frac{\partial J_I}{\partial x} = G_{BI} \tag{7}$$

$$G_{BI} = k_{decay} c_{BI} \sum_i k_I^i \left(\frac{n}{n_i}\right)^{-i} -$$

$$-k_{generation} c_B c_I \sum_i k_{BI}^i \left(\frac{n}{n_i}\right)^{-i} \quad (8)$$

The equations for boron, boron-interstitial pairs and for interstitials need to be solved. The equations are solved by the program package ZOMBIE [6], a one dimensional solver for systems of coupled parabolic, elliptic, and ordinary differential equations. Parameter extraction by error minimization has been done by PROFILE [7], a general purpose measurement data processor. Interstitial equilibrium concentration, as has been calculated by Bronner[8], has been taken into account as a boundary condition on both sides of the wafer. For substitutional and interstitial boron the flux at the surface was assumed to be 0. The diffusivity of interstitials was taken from the work of Bronner[8].

Implantation damage from the work of Hobler[9] has been included. We assume that after implantation boron is primarily found on interstitial sites, as has been shown by the experiments of Fink[10]. Diffusion enhancement takes place due to the significantly higher diffusion coefficient of boron-interstitial pairs (compared with the 'standard' diffusion coefficient for boron) and the implantation induced oversaturation of boron interstitials.

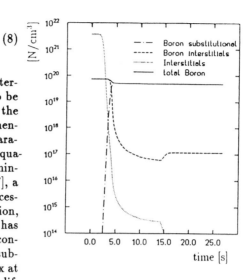

Figure 3: Maximum peak concentration for substitutional boron $B_s$, interstitial boron $B_i$, interstitials and the sum of $B_s + B_i$ over time. Boron is assumed to reside mainly on interstitial sites after ion implantation. After the temperature has risen to significant values, interstitial boron starts to decay into substitutional boron and interstitials. 'Normal' equilibrium conditions are reached as the decay comes to an end. The 'transient diffusion' has normalized by this time (compare with figure [2]).

The anomalous effect comes to an end, when the supersaturation of boron-interstials has dissolved via the reaction $BI \rightleftharpoons B + I$ and reached the equilibrium level. At this point the transient effect has vanished and the model describes the 'normal' diffusion behaviour. In our model the dissolution rate of boron-interstitial pairs can be controlled via the reaction term of the abovementioned equations.

Michel[11] stated that the interstitial model fails to explain the well known temperature dependence of the displacement of the dopant and the increase in the time necessary for the effect to saturate. While saturation of the anomalous diffusion enhancement lasts for about 1 second at 1000°C it can take several hundreds of seconds at 800°C. An activation energy of the order of 4eV should be required for the boron-interstitial pair. Generation and recombination terms are $k_{generation} = k_{velocity} * k_{equilibrium}$, $k_{recombination} = k_{velocity}$ with $k_{velocity} = const * e^{-4.3eV/kT}$ (4.3eV being the activation energy for the anomalous displacement[11]).

It can be shown, that the interstitial model is able to account for the temperature dependence of the anomalous displacement as can be seen from figure[2]. $k_{equilibrium}$ is used to control the equilibrium $BI$-pair concentration. $const$ must be provided as a fitting parameter. We simulated the experimental data gained from the experiments of Cho[12]. Cho carried out implantation of $10^{15}$, 50keV $^{11}$B at room temperature into Czochralski grown n-type <100> Si wafers. Annealing was done under $N_2$ atmosphere and temperatures at 1050°C and 1150°C. In these experiments annealing took place for 10s. Profiles were measured with SIMS. The simulation is shown in figure[1] for 1050°C. In figure[2] the net diffusion coefficient for boron at 1150°C can be compared with the net diffusion coefficient at 1050°C. A short diffusion enhancement of about 1 second at 1150°C can be compared with the long diffusion enhancement at 1050°C (about 10 seconds). The model clearly yields the temperature dependence of the 'transient diffusion'.

In the case of arsenic the transient diffusion effect has been much less pronounced than in the case of boron. We have shown[13] that arsenic-point defect pairs have rather low diffusivities and therefore we find a high percentage (in the order of 20%) of $As(I, V)$ pairs during normal diffusion conditions. In our framework a significant transient diffusion can only occur if (1) the equilibrium concentration of AsI pairs is significantly lower than As concentration and, consequently, (2) the AsI equilibrium diffusivity is significantly higher than As equilibrium diffusivity. In the case of arsenic, high supersaturation of arsenic interstitial pairs after ion implantation cannot be achieved. The high equilibrium $As$ point defect pair concentration will obviously diminish any transient effect for arsenic.

REFERENCES

[1] M.Miyake, S.Aoyama, *J.Appl.Phys.***63**(5), p.1754,1988

[2] N.E.B. Cowern, D.J.Godfrey, D.E.Sykes, *Appl.Phys.Lett.***49**(25), p.1711, 1986

[3] V.E.Borisenko, V.A.Labunov, *Phys. stat. solidi***A72**, K173, 1982

[4] R.Kalish, T.O.Sedgwick, S.Mader, S.Shatas, *Appl. Phys. Lett.***42**, p.107, 1984

[5] M. Budil, H. Pötzl, G. Stingeder, M. Grasserbauer, K. Goser, Materials Science Forum (Proc. 15. ICDS), Vol. 38-41, Part 2, pp. 719-724, 1989

[6] W. Jüngling, IEEE Trans **ED-32**, 1985, p. 156

[7] G. Ouwerling, F.van Rijs, B. Jansen, W. Crans, Proc. NASECODE VI(1989) p.78

[8] G.B.Bronner, J.D. Plummer, *J.Appl. Phys.* **61**(12), p. 5286, 1987

[9] G. Hobler, S. Selberherr, *IEEE Transactions on* **CAD-7** , p. 175

[10] D. Fink, J.P. Biersack, H.D. Carstanjen, F. Jahnel, K. Muller, H. Ryssel, A. Osei, Radiation Effects **77**, 1983, p.11

[11] A.E. Michel, *Nucl. Inst. and Meth. in Phys. Res.* **B37/38** , p.379

[12] K. Cho, M. Numan, T.G. Finstad, W.K. Chu, *Appl. Phys. Lett.* **47**(12), 1321

[13] M. Heinrich, M. Budil, H.W. Pötzl, Proc. of the MRS-fall meeting, Boston 1989

*Paper presented at ESSDERC 90, Nottingham, September 1990*
*Session 3A5*

# An improved model of plasma etching including temperature dependence: comparison between simulation and experimental results

A Gérodolle, J Pelletier, S Drouot

Centre National d'Etudes des Télécommunications, BP 98, 38243 Meylan cedex, France

Abstract. The temperature dependence of the plasma etching of silicon by $SF_6$ has been studied. A purely reactive model involving a few parameters is presented. Parameter values are extracted from experimental data, and results of the simulations carried out with these parameters are compared with the experimental profiles.

1. Introduction

Most of the simulation studies of plasma etching published up to now mainly show geometric and size effects (trench-bowing, dove-tails) and sputtering effects (mask erosion, redeposition). Generally, these effects are the results of high energy ion bombardment. In contrast, simulation results on the etching of Si at low ion energy have also been presented [1]; although the model [2] involves only a few parameters and excludes sputtering effects, it allows the simulation of most geometric effects. This paper shows how this model takes into account the evolution with temperature of the etching characteristics of Si in fluorine-based plasma. The results obtained by the simulation are then compared with the experimental kinetics and profiles [3].

2. Model description

The physical model used in this study has been developed from experimental results on $SF_6$ plasma etching of Si [2]. It accounts for most of the observed etching charateristics such as isotropy/anisotropy transition and etch rate. The main hypotheses are : large lateral repulsive interactions between chemisorbed fluorine adatoms in nearest neighbour positions, fluorine diffusion and multilayer adsorption of fluorine on the silicon surface. The expression of the balance between the F-atom adsorption flux from the gas phase, the spontaneous and/or induced desorption flux of $SiF_4$ and the surface diffusion is a partial derivative equation, the solving of which allows determination of the adatom coverage and thus of the etching profiles.

Solving the balance equation analytically is not possible in the general case. To predict the shapes of two-dimensional trench structures, simulation is necessary. Only the surface reactions are considered, very simple assumptions being made on the flux distribution as described in [1]. Briefly, the etching time being divided into time-steps, at each time-step the balance equation is numerically solved and the surface is then moved according to the computed local adatom coverage. This allows the TITAN program [4] to take into account variations in adatom coverage versus location, or its evolution with etching time. These effects cannot be computed analytically.

## 3. Temperature dependence

Besides the main plasma parameters involved in the etching process (flux of fluorine atoms, current density and energy of ion bombardment), temperature must be considered. Indeed the spontaneous desorption rate is linked with $\tau$ which is the lifetime of the fluorine adatoms on the silicon surface. $\tau$ obeys an Arrhenius law :

$$\tau = \tau_0 \exp(R/kT) \qquad (1)$$

where $\tau_0$ is a pre-exponential factor, R the activation energy for the associative desorption, k the Boltzmann's constant and T the absolute temperature. Likewise, the surface diffusion coefficient is typically observed to depend on temperature:

$$D = D_0 \exp(-E/kT) \qquad (2)$$

where E is the activation energy for diffusion and $D_0$ is a pre-exponential factor. Other parameters are temperature independent.

Although the balance equation cannot be solved analytically in the general case, it is possible when the diffusion length and trench width are large with respect to the depth to be etched. In this case, the ratio of lateral to vertical etch rate [5] is expressed as:

$$\frac{V_1}{V_v} = \frac{\kappa p_F - \eta j \sigma_0 \theta c}{\kappa p_F (1 + \eta j \tau)} \qquad (3)$$

where j is the current density, $p_F$ the partial pressure of atomic fluorine, $\sigma_0$ the number of adsorption sites on the surface, and $\kappa$ en $\eta$ are constants. Two new practical parameters $\alpha$ and $\beta$, both temperature independent, are introduced in order to simplify the expression of $V_1/V_v$:

$$\alpha = \frac{\kappa p_F - \eta j \sigma_0 \theta_c}{\kappa p_F} \quad , \quad \beta = \eta j \tau_0 \qquad (4)$$

and thus :

$$\frac{V_1}{V_v} = \alpha \left(1 + \beta \exp \frac{R}{kT}\right)^{-1} \qquad (5)$$

The values of $\alpha$, $\beta$ and R are required in order to determine the temperature evolution of the etching profiles. The study of $V_1/V_v$ in terms of temperature shows that more information is obtained when this ratio is plotted in terms of $1/T$. This curve allows determination of the numerical values of $\alpha$, $\beta$ and R, as reported in Figure 1.

Figure 1. $V_l/V_v$ versus 1/T curve. Experimental points from [3] and the corresponding scale are reported.

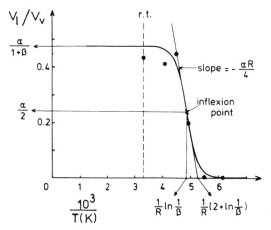

## 4. Comparison with experiments

The determination method illustrated in Figure 1 has been applied to the experimental results described in [3]. The calculated values are 0.475 for $\alpha$, $3 \times 10^{-11}$ for $\beta$ and 5000 K for R/k (i.e. 0.43 eV for R). The set of parameters required for the simulation can be merely deduced from $\alpha$, $\beta$ and R. The values of the diffusion constants are rather well assessed thanks to other experiments which have been performed. Thus, E and $L^2$, where L is the diffusion length in ion bombardment free areas at room temperature, are estimated to be 0.12 eV and 10 $\mu m^2$, respectively.

The simulations of the experiments described in [3] were performed with the set of parameter values indicated above. The results were qualitatively satisfactory, but the lateral etching was underestimated. This was due to the fact that the method used to determine the parameters is precise only if the trench depth is negligible with respect to the width, which is obviously not the case here. The parameters were therefore adjusted and the simulations were performed with this new set of parameter values where $\beta$ was divided by two ($1.5 \times 10^{-11}$ instead of $3 \times 10^{-11}$). This led to an excellent agreement between simulation results and the experiments (Figure 2). Note that the surface diffusion length increases with decreasing temperature, although the diffusion coefficient decreases. This prevents the appearance of the dove-tail effect, as seen in [1].

## 5. Conclusion

A model for silicon etching by $SF_6$-based plasmas, which is based on the surface diffusion of reactive species from the trench sidewalls toward the areas under ion bombardment, has been presented. The temperature dependence of the anisotropy is shown to be accurately described by this model. Simulation results also exhibit the influence of trench size on the anisotropy. Experiments are therefore being carried out for a more exhaustive study.

## References
[1] A. GERODOLLE, proc. ESSDERC' 89, 19th European Solid State Device Research Conference, Berlin 1989, Ed. Springer-Verlag, p. 206
[2] B. PETIT and J. PELLETIER, Revue Phys. Appl. 21, 377 (1988)

[3] S. TACHI, K. TSUJIMOTO and S. OKODAIRA, Appl. Phys. Lett. 52, 616 (1988)
[4] A. GERODOLLE, C. CORBEX, A. PONCET, T. PEDRON and S. MARTIN, NASECODE VI conference, Short course lecture notes, Dublin 1989, Ed. Miller, Boole press
[5] J. PELLETIER, Appl. Phys. Lett. 53, 1655 (1988)

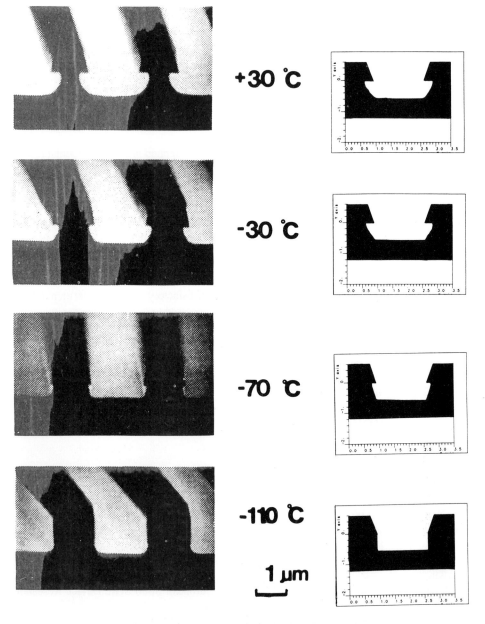

Figure 2. Comparison between temperature evolution of experimental [3] (left) and simulated (right) profiles.

# Simulation of a polysilicon LPCVD reactor

# Fluid-dynamics and Error Analysis

Ch. Hopfmann, J.I. Ulacia F., Ch. Werner

ZFE SPT 33, SIEMENS AG, Otto-Hahn-Ring 6,
D-8000 Munich 83, Germany, Tel. (+4989) 636-44646.

## Abstract

The deposition of polysilicon is numerically simulated with a model that computes the fluid flow, transport coefficients and surface chemistry inside the LPCVD reactor. The results exhibit relatively small gas recirculations as a result of the temperature gradients in the empty inlet area and in the heating zone. Comparing simulations with and without reactive surfaces, inter-wafer recirculations near non-reactive waferedges are eliminated at reactive wafer. A comparison of four deposition models and an error analysis leads to the conclusion, that no complex chemical models are necessary as long as the uncertainty of the activation energy is by far the largest source of error.

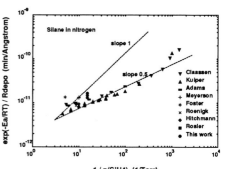

Figure 1. Overview of experimental data

## 1 Introduction

Polysilicon is a widely used material in semiconductor technology for interconnects, transistor gates, capacitor electrodes, and other thin-film applications. For its importance, there is a large experimental and theoretical knowledge on this process; however, the models developed so far [Kuiper 1982, Brekel 1981, Jensen 1983, Roenigk 1985] predict different influences under equivalent experimental conditions. It is the purpose of this work to provide a general view between the different models, to demonstrate a 3D numerical-simulation methodology using fluid-dynamics, and to discuss the influences of silane pressure, temperature and activation energy on the deposition rate. As illustrated in experimental data of undoped polysilicon in Figure 1, the deposition rate follows a square root dependence at silane-partial pressures lower than $50 mTorr$ as $P_{SiH_4}^{1/2}$. Deposition rates proportional to $P_{SiH_4}$ at higher pressures could be explained by the fact that the processes might be limited by mass transport. Increasing temperature leads to higher deposition rates, which are described by the Arrhenius plot in Figure 2. The wide variation in activation energy derived from the Arrhenius plot as illustrated in Figure 3 is a result of experimental error found under different working conditions and processing equipment used. Based on the experimental data presented, all the deposition models proposed can be described by a general deposition rate equation as

$$R_{depo} = \frac{k_0 \, exp(-E_a/(R\,T)) \, P_{SiH_4}^n}{1 + A\, P_{H_2}^m + B\, P_{SiH_4}^n}$$

where $k_0$, $A$, and $B$ are constants, $T$ is the temperature in $K$, $E_a$ is the activation energy in $Kcal/mole$, $P_{H_2}^m$ is the hydrogen partial pressure in $mTorr$, and $n$ and $m$ are the factors for pressure that describe the kinetics of the reaction and can be obtained from the previous figures. For the different models, the reported values that fit the general deposition equation can be found in Tab. 1.

© 1990 IOP Publishing Ltd

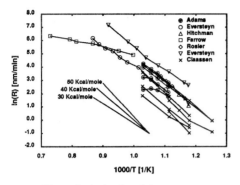

Figure 2. Arrhenius plotFigure 3. Activation energy versus silane pressure

## 2 Mathematical model

The mathematical description of a LPCVD reactor is based on partial differential equations for the continuity of mass, momentum, energy and chemical species [Bird 1960]. These equations are coupled through transport coefficients by suitable models that are a function of temperature and particle concentrations [Bird 1960, Reid 1986]. For the low pressure regime used in the calculations the particle transport is between the fluid and free particle flow as observed from a Knudsen number of 0.12; in these conditions gas phase reactions can be neglected and simple surface kinetics are sufficient. The resulting overall chemical reaction $SiH_4 \longrightarrow Si + 2 H_2$ requires a flux boundary condition in the species-continuity equations for silane and hydrogen at the reactive surfaces, and an effective mass-loss for the mass-continuity equation. These fluxes are derived from the general deposition rate equation corrected by the stoichiometry of the reaction. The experimental results presented are for an ASM horizontal reactor with 96 wafers in the batch under typical working conditions.

## 3 Results and Discussion

### 3.1 Internal distributions

In typical working conditions, the gas mixture is incompressible and the total pressure can be treated as constant, because pressure variation caused by volume expansion in the inlet area and in chemical reactive regions is only about one percent from inlet to outlet. A small recirculation flow formed in the inlet area does not influence the wafer batch and a laminar flow with a maximum velocity of 5 $m/s$ surrounds the wafer; this is a result of the temperature distribution and the low-pressure conditions, reflected in a Reynolds number of 67 in the middle the reactor between batch and wall. The cold inlet gas is heated up in the first 30 $cm$ of the heating zone and the remaining temperature gradient between batch and reactor wall is 3 $K$. If no chemical reaction takes place on the surfaces, a recirculation flow between the wafer with a penetration depth of the same magnitude as the wafer spacing can be seen in Figure 4, whereas the consideration of an effective-mass flow into the wafer-surfaces overlaps the recirculation flow almost completly, as demonstrated in Figure 5. The species concentration contours, on the other hand, exhibit a silane decrease and hydrogen increase in reactive areas, and as a result of the diffusion controlled behaviour no flow influences occur, reflected by Peclet numbers for each species of less than 0.01 between the wafer, and of approx. 30 in main flow direction. A black box approach of the silane conversion with an average deposition rate of 8 $nm/min$ predicts that 50% of the silane inflow is converted into silicon and hydrogen; therefore the mass density decreases about 16% in the reactive area at constant temperature. The following results have been obtained by the calculation of the transport coefficients: the laminar kinematic viscosity contours are similar to the temperature contour, and variations with gas composition are negligible. On the other hand no influence of temperature variations on the heat capacity occur although it is a function of temperature, because hydrogen with its 29fold higher specific heat capacity covers the influence of argon and silane, and its variation with temperature is small. The strong changes of heat capacity with gas composition determine the thermal conductivity in the first reactor half, but in the second half the temperature dependence dominates because of a almost constant heat capacity. The mutual dependence of the species continuity equations and the diffusion coefficients leads in a stationary system to a fix species distribution; so in the first reactor half, temperature and silane concentration together determine the silane diffusivity, whereas temperature is important in the second half, and the silane diffusion coefficient is constant at constant temperature. The hydrogen diffusion coefficient is only dependent on temperature ($\sim T^{3/2}$), the argon diffusion coefficient on the other hand is inversely proportional to the silane concentration at constant temperature.

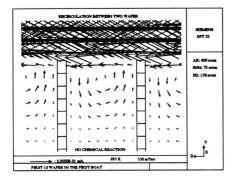
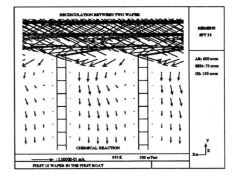

Figure 4. Recirculation flow between wafer without chemical reactions

Figure 5. Mass flow between wafer with surface reactions

## 3.2 Deposition rates

The four deposition models are compared against each other and with experimental data for axial deposition uniformitiy as illustrated in Figure 6. The deposition rate contours are different in intercept and slope, but they all have a similar behaviour with a retarded growth rate from inlet to outlet caused by silane consumption; however, the experimental observation of decreasing growth rates at both ends of the batch is not seen, because the effect of wafer cooled by radiative heat loss is neglected. The radial growth rate variation of 1% inhomogeneity on a single wafer from center to edge demonstrated in Figure 7 is an effect of silane consumption caused by chemical reaction, and is in good agreement with experiment. The Roenigk model has been adjusted in activation energy to experimental data of this reactor by extrapolation of the growth rate to the first wafer, and the new activation energy with 37.45 $Kcal/mole$ is only 2% larger than Roenigks value.

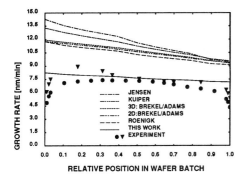
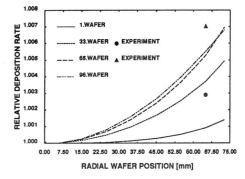

Figure 6. Deposition rates over the whole batch

Figure 7. Radial wafer deposition uniformity

## 3.3 Error analysis

To evaluate the experimental uncertainty on activation energy, a linear regression of Claassens experimental data [Claassen 1982] was made, with an average value of 36.95 $Kcal/mole$ and a relative average error of $\sigma_{E_a}/E_a = 6.93\%$. Almost any adjustment to experimental data by the activation energy is in the range of error, demonstrated in Figure 3 with the calculated average value and its 99% confidence limits. An error propagation of all four models, which reflects the influence of uncertainty in every part of the deposition models, leads to the conclusion, that errors in temperature and activation energy mainly influence the deposition rate (Tab. 2), and that the other coefficients do not have significant influence on the deposition rate error.

# 4 Conclusions

Under typical working conditions in this reactor no mass transport limitations occur, and transport by diffusion dominates between the wafer; the flow is therefore laminar around the wafer batch and the empty inlet area does not influence the wafer batch. For nonreacting gases a remaining recirculation flow between the wafers is identified, but this is totally covered by the net inflow of reaction gas into the wafer spacing; therefore deposition inhomogeneities at the edge of the wafers are not expect to be caused by this effect. The deposition rate contours are influenced in axial direction by silane consumption and temperature effects together, in radial direction silane depletion and hydrogen inhibition dominates. As a result of the large uncertainty in the experimental determination of the activation energy, deposition rates cannot be predicted *a priori* with exact values; however, wafer to wafer homogeneities and radial deposition profiles show the correct qualitative behaviour. For the low pressure regime without *in situ* doping as it was investigated in this work, a simple model for the mass flow into a surface was fully sufficient and no need for more complex gasphase and surface chemistry models is found.

Table 1: Coefficients for the general deposition rate

|  | n | m | $k_0$ | $E_a$ | A | B |
|---|---|---|---|---|---|---|
| [Kuiper 1982] | 0.5 | - | 2.81 | 0.0 | 0.0 | 6.25E-2 |
| [Brekel 1981] | 1.0 | - | 0.95 | 0.0 | 0.0 | 5.9E-2 |
| [Jensen 1983] | 1.0 | 1.0 | 1.19E9 | 36.73 | 2.3E-3 | 5.26E-2 |
| [Roenigk 1985] | 1.0 | 0.5 | 1.52E9 | 36.73 | 6.88E-2 | 9.21E-2 |

Table 2: Relative variation of $R_{depo}$ in percent with $\pm 5\ K$ and in the 99% confidence limits of $E_a$

|  | 615 $C$ | 620 $C$ | 625 $C$ |
|---|---|---|---|
| 32.74 $Kcal/mole$ | -868.4 | -97.5 | -1091 |
| 36.95 $Kcal/mole$ | 11.1 | 0.0 | -12.3 |
| 41.16 $Kcal/mole$ | 91.8 | 90.7 | 89.4 |

## References

R.B. Bird, W.E. Stewart, E.N. Lightfoot. *TRANSPORT PHENOMENA*. Wiley Int. Edition, 1960
C.H.J. van den Brekel, L.J.M. Bollen. *J. Cryst. Growth*, **54**, 310 (1981)
W.A.P. Claassen, J. Bloem, W.G.J.N. Valkenburg, C.H.J. v. d. Brekel.
　*J. Cryst. Growth*, **57**, 259 (1982).
K.F. Jensen, D.B. Graves. *J. Electrochem. Soc.*, **30**, 1950 (1983)
A.E.T. Kuiper, C.J.H. v. d. Brekel, J. de Groot, G.W. Veltkamp.
　*J. Electrochem. Soc.*, **129**, 2288 (1982)
R.C. Reid, J.M. Prausnitz, B.E. Poling. *THE PROPERTIES OF GASES & LIQUIDS*.
　McGraw-Hill Book Company, Fourth Edition 1986
K.F. Roenigk, K.F. Jensen. *J. Electrochem. Soc.*, **132**, 448 (1985)

Paper presented at ESSDERC 90, Nottingham, September 1990
Session 3A7

# Channeling of boron in silicon: experiments and simulation

G. Hobler[1], H. Pötzl[1,4], R. Schork[2], J. Lorenz[2], S. Gara[3], G. Stingeder[3]

[1] Institut für Allgemeine Elektrotechnik u. Elektronik, Technische Universität Wien, Gußhausstraße 27–29, A-1040 Vienna, AUSTRIA

[2] Fraunhofer-Arbeitsgruppe für Integrierte Schaltungen, Artilleriestraße 12, D-8520 Erlangen, GERMANY

[3] Institut für Analytische Chemie, Technische Universität Wien, Getreidemarkt 9, A-1060 Vienna, AUSTRIA

[4] Ludwig Boltzmann Institut für Festkörperphysik, Kopernikusgasse 15, A-1060 Vienna, AUSTRIA

**Abstract.** In order to establish a rigorous test for the various models used to simulate ion implantation in crystalline silicon, we have implanted 17–150 keV boron ions with low doses into <111> and <100> silicon in channeling direction. Simulations using the program MARLOWE show that impact parameter dependent electronic stopping is essential. The binary collision approximation is fully justified. Moliere and ZBL potential are almost equivalent. Although simulation results are qualitatively good, some systematic disagreement with experiments can be observed.

## 1. Introduction

Ion implantation of boron into crystalline silicon is strongly affected by channeling. Although several Monte Carlo codes exist which qualitatively predict channeling tails, physical models are not thoroughly tested. Usual 7° implants are not ideally suited to investigate this question, as statistical fluctuations of the Monte Carlo results at low concentrations (i. e. in the tail) and experimental uncertainties make it difficult to draw unambiguous conclusions. The channeling effect can best be studied by implantations in channeling direction. In the cases investigated (17–150 keV boron in <100> and <111> silicon) a channeling peak forms in place of the channeling tail. This means that the majority of the ions are channeled and thus experience the effect of interest. For this reason experimental profiles of channeled implants are used in this study in order to investigate the physical models contained in the simulator MARLOWE [1].

## 2. Experimental

Boron implantations into <100> and <111> silicon were performed using a medium current implanter with electrostatically scanned beam (Varian 350 DF). The boron dose ($10^{13}$ cm$^{-2}$) was low enough to preclude any influence of damage on the profile. The distance between vertical (horizontal) deflection plates and wafer is 1.80 m (2.10 m).

© 1990 IOP Publishing Ltd

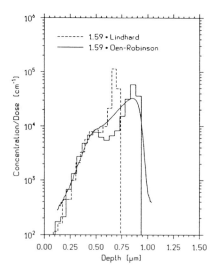

Figure 1: $B \rightarrow <111>$ Si, 150 keV, no tilt. Histograms are MARLOWE results

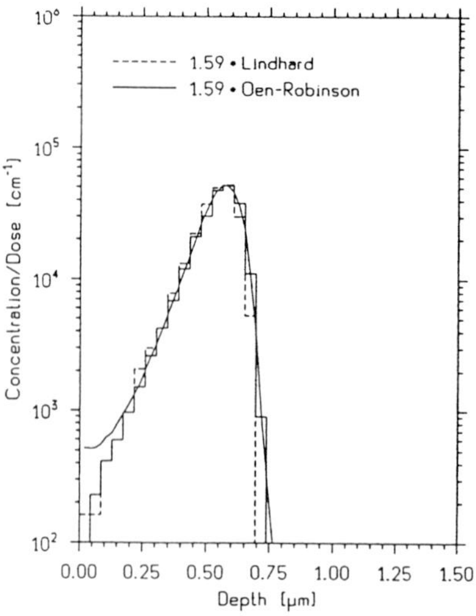

Figure 2: $B \rightarrow \alpha$-Si, 200 keV. Experiments (full line) are from Hofker [9]. Histograms are MARLOWE results

Thus, the actual incidence angle $\theta$ of the implanted ions with respect to the surface normal varies between 0° and $\approx 1°$ on a 3" wafer. In this way profiles with tilt angles $\theta \leq 1°$ under otherwise identical conditions can be obtained.

In order to determine the point of 0° incidence SIMS measurements were taken at different positions of the wafer. First the diameter perpendicular to the flat was scanned. As the dependence on $\theta$ is more pronounced for large $\theta$, the position with maximum channeling was determined as the center of two points with identical profiles near the edge of the diameter. This procedure was repeated on a line parallel to the flat crossing the diameter at the point with maximum channeling. Considering the geometry of the implanter, $\theta$ can be determined with an accuracy of about 0.1°.

The SIMS measurements were performed using a Cameca-IMS 3f ion microanalyzer. An $O_2^+$-primary ion beam with an impact energy of 5.5 keV and a current intensity of 2 $\mu$A was rastered over an area of 500 × 500 $\mu m^2$. Positive secondary ions were detected from an analyzed area with 150 $\mu$m in diameter. The $^{14}Si^{++}$-signal was used as a reference.

### 3. Simulation

The simulations were performed with MARLOWE, Version 12. A layer of 2 nm native oxide was considered. Thermal vibrations of the lattice atoms were included using the Debye model. Without thermal vibrations no reasonable results can be obtained. The electronic stopping power according to Lindhard [2] and Oen and Robinson [3] was corrected by a factor of 1.59 as proposed by Eisen [4]. The ZBL potential [5] was used to describe the interatomic forces. It was found that the maximum distance of target atoms included in the crystal description needs to be taken at least 1.5 times the unit cell edge in order to obtain results which are independent of this parameter.

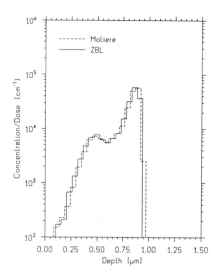
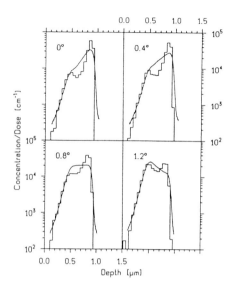

Figure 3: $B \to <111>$ Si, 150 keV, no tilt. Comparison of Moliere and ZBL potential

Figure 4: $B \to <111>$ Si, 150 keV, various tilt angles. Histograms are MARLOWE results

## 4. Results

The first question we have investigated is the validity of the binary collision approximation, i. e. the assumption that the ion interacts with only one target atom at the same time. It was claimed [6,7] that simultaneous interaction with more than one atom plays an important role. In contrast, we did not find the slightest difference between profiles obtained with and without "simultaneous" collisions. This is important as all models for many-body interaction are only approximate.

Comparing the results using non-local Lindhard electronic stopping with experiments, one finds that the penetration of channeled ions is much too low (Fig. 1). As this approach works well for amorphous targets (Fig. 2), it can be concluded that electronic stopping is impact parameter dependent with an average value given by Lindhard. Channeled ions will then experience less stopping, as they are never traveling close to target atoms. Using the impact parameter dependence $\exp\{-0.3p/a\}$ proposed by Oen and Robinson [3] the channeling peak is shifted to the correct position (Fig. 1). Although this can also be achieved by other models (e. g. 80% Lindhard + 20% p-dependent with $\exp\{-p/a\}$), we prefer the Oen-Robinson model, as it is also theoretically supported [8].

In the cases investigated, the channeling effect is not sensitive to whether the Moliere or ZBL potential is used (Fig. 3).

Finally, we investigated the implant angle and energy dependence of the profiles. It can be seen from Figs. 4-6 that the general trend is well predicted by the simulations. However, the dechanneling mechanism seems to be underestimated in all cases, as the channeling peak is too pronounced, there is a "valley" between random peak and channeling peak, and there is a little too much channeling for the larger incidence angles.

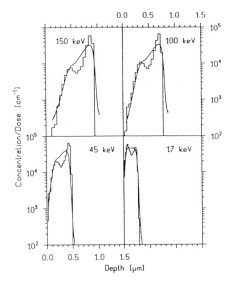
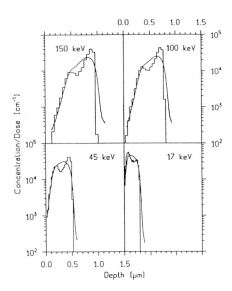

Figure 5: $B \rightarrow$ <111> Si, no tilt, various implant energies. Histograms are MARLOWE results

Figure 6: $B \rightarrow$ <100> Si, no tilt, various implant energies. Histograms are MARLOWE results

Possible explanations could be the neglection of the non-spherical distribution of the valence electrons in the electronic stopping power or a shortcoming in the thermal vibration model.

*Acknowledgement* — This work has been supported by the Fonds zur Förderung der wissenschaftlichen Forschung, project no. P7495-PHY.

## 5. References

[1] M. T. Robinson, *MARLOWE Version 12, User's Guide*, Oak Ridge National Laboratory, Oak Ridge, Tennessee 37831.

[2] J. Lindhard, M. Scharff, *Phys. Rev.* 124, 128, 1961.

[3] O. S. Oen, M. T. Robinson, *Nucl. Instr. Meth.* 132, 641, 1976.

[4] F. H. Eisen et al, *Atomic Collision Processes in Solids* (Pergamon Press. London), p. 111, 1970.

[5] J. P. Biersack, J. F. Ziegler, *Springer Series in Electrophysics.* Vol. 10 (Springer, Berlin), p. 122, 1982.

[6] M. Hane, M. Fukuma, *Proc. IEDM-88*, 648, 1988.

[7] B. J. Mulvaney, W. B. Richardson, T. L. Crandle, *IEEE Trans. CAD*, 8, 336, 1989.

[8] N. Azziz, P. C. Murley, *phys. stat. sol.* (b) 151, K105, 1989.

[9] W. K. Hofker et al., *Rad. Eff.* 24, 223, 1975.

# Towards the limit of ion implantation and rapid thermal annealing as a technique for shallow junction formation

J.L.Altrip, A.G.R.Evans, J.R.Logan[1] and C.Jeynes[2]

Department of Electronics and Computer Science, University of Southampton, Highfield, Southampton S09 5NH, England.
[1] Lucas Automotive Ltd., Advanced Engineering Centre, Dogkennel Lane, Shirley, Solihull, W. Midlands B90 4JJ, England.
[2] Department of Electronic and Electrical Engineering, University of Surrey, Guildford, Surrey, GU2 5XH, England.

Abstract
High temperature, very short time annealing techniques have been used to study dopant activation during and immediately after solid phase epitaxial regrowth of amorphous layers produced by ion implantation of As into Si. Short annealing timescales have revealed electrically inactive As tails, correlated with a region of implant-induced excess point defects, indicating the formation of stable dopant-interstitial complexes which are not removed during the timescales of these anneals.

Introduction
With the requirement of VLSI for continually smaller geometries, significant research efforts have been directed towards ion implantation and rapid thermal annealing (RTA) as a viable combination for shallow junction formation. High temperature, short time annealing schedules are now routinely used to remove lattice damage and simultaneously achieve high electrical activation of the implanted dopant. However, the choice of a suitable temperature-time window to satisfy the conflicting requirements of defect removal and restriction of dopant diffusion is, of necessity, a compromise.

The complex annealing behaviour of an implanted layer is related to the initial distribution of damage within the layer and its changing nature during the anneal. Even on conventional RTA timescales, processes occurring during the later stages of the anneal often mask the dopant activation and diffusion behaviour occurring during the initial anneal stages. In this work, shallow $n^+$ layers of the type required for source and drain regions in CMOS technology, have been fabricated by ion implantation and RTA techniques. High temperature very short time annealing has allowed a study of the kinetics of As activation and diffusion during and immediately after high temperature solid phase epitaxial (SPE) regrowth.

Experimental Details and Results
Amorphous layers produced by room temperature 80KeV As implants ($5 \times 10^{14}$-$2 \times 10^{16}$ As cm$^{-2}$) into p-type <100> 17.5-33.0 $\Omega$cm substrates were subsequently capped with low temperature CVD oxide and annealed using a high power flashlamp system described previously[1]. A single anneal schedule consisted of a 5 second background anneal at 600°C, followed by a 30J cm$^{-2}$ flashlamp pulse to raise the surface of the sample to high temperature in very short time (~0.3ms). Multiple anneal schedules were used to create a series of samples in various stages of partial regrowth. During the background anneal, chosen to enable a peak temperature of ~1100°C to be achieved, some regrowth of amorphous material will have occurred but this is small compared to that regrown during the high temperature anneal.

SPE regrowth of a high dose ($2 \times 10^{16}$ As cm$^{-2}$) amorphous layer is shown as a function of annealing time (number of flash anneals) in Figure 1. The as-implanted amorphous layer, initially 1450Å deep, is seen to regrow from the original amorphous/crystalline ($\alpha$/c) interface towards the surface with increasing annealing time. The regrowth rate is non-uniform, slowing down in the region corresponding to the implanted profile peak. In all the partially regrown samples, a region of excess non-substitutional Si atoms can be identified, located at the position of the original $\alpha$/c interface. The corresponding random and channelled As spectra shown in Figure 2 illustrate that as the $\alpha$/c interface proceeds towards the surface, As atoms are incorporated onto substitutional sites within the regrown material. Similar amorphous layer regrowth behaviour was seen for medium ($5\times10^{15}$ As cm$^{-2}$) and low dose ($5\times10^{14}$ As cm$^{-2}$) implants. Both were completely regrown by six flash anneals, despite different initial amorphous layer thicknesses of 1450Å and 1090Å respectively, but no similar regrowth retardation was observed.

Changes in As substitutionality as a function of depth and annealing time extracted from this data are shown in Figure 3 for the high dose sample. Although As atoms are incorporated onto substitutional sites during regrowth, additional increases in substitutionality within the regrown material occur with subsequent annealing. After 4 flash anneals, the substitutional concentration within the regrown material has an approximately uniform value of ~$5\times10^{20}$ cm$^{-3}$. Following 6 flash anneals, when the α/c interface position is ~590Å from the surface, increasing As substitutionality is seen with increasing depth within the regrown material because atoms deeper in the profile have had longer time in the crystalline state to reorder. The substitutional As profile shows no discontinuity across the 4-flash α/c interface position (866Å) and its peak is slightly displaced from that of the total As curve. After 10 flash anneals, when the amorphous layer has almost completely regrown, high substitutionality (>90%) is seen across the profile, falling slightly in the near surface region.

A comparison of chemical (SIMS) and electrical (SRP) profiles for a high dose sample after 6 flash anneals is shown in figure 4, revealing an electrically inactive As tail region at the as-implanted α/c interface. The maximum electrically active concentration ~$2\times10^{20}$ cm$^{-3}$ is in approximate agreement with equilibrium electrical solubility derived from furnace annealing experiments. No evidence of metastable excess electrically active As is seen even for these very short anneal schedules, and electrical activity is observed in the amorphous material ahead of the moving interface, in agreement with that reported after low temperature annealing[2]. Also shown is the SRP profile for a low dose sample, fully regrown after 6 flash anneals, together with the smaller as-implanted α/c interface.

Discussion
A knowledge of the depth distribution of implant-induced defects is required to interpret activation and diffusion mechanisms occurring at very short timescales. Monte Carlo simulations based on point defect kinetics have predicted excess vacancy formation in a region between the surface and $0.8R_p$, and excess interstitial formation between $R_p$ and $2R_p$[3]. Depth profiles of lattice strain measured by x-ray diffraction techniques have been shown to correlate with these regions of vacancy and interstitial excess[4], which occur in addition to the more commonly observed α/c dislocation loops evidenced by TEM. The position of excess Si interstitials revealed by Rutherford Backscattering (RBS) analysis (Figure 1) correlates directly with the electrically inactive tail region of the As profile (figure 4), indicating the formation of inactive dopant-interstitial complexes, which are not removed by these very short annealing timescales. Diffusion length calculations based on crystalline interstitial diffusivities[5] indicate that within the timescale of a single flash anneal, Si interstitials are sufficiently mobile to form dopant-interstitial complexes within the crystalline material beyond the initial α/c interface.

Electrically inactive dopant is also seen in the initially amorphous material close to the interface, regrown by the first few flash anneals. Dopant-impurity complex formation in the amorphous material ahead of the regrowing interface may either be due to interstitial diffusion similar to that within crystalline material, or to trapping of interstitials by dopant atoms during implantation. Projected range defects identified by TEM for group V implants into Si[6] have been attributed to self interstitials trapped by dopant atoms during implantation, which survive SPE regrowth but break up during subsequent annealing, causing transient enhanced diffusion. The increase in As substitutionality observed within the regrown material with additional annealing (Figure 3) may correspond to the dissociation of some of these dopant-interstitial complexes. In contrast, the end of range damage defects revealed by RBS (Figure 1) are not removed after 10 flash anneals. Within the resolution of the SIMS technique, no evidence of As diffusion processes have been observed, consistent with the idea that these anneal schedules are not sufficiently long to release excess Si interstitials trapped in dopant-interstitial complexes. However the extent of any enhanced diffusion has been shown to be determined both by the relative position of dopant and point defect excesses, and the diffusion mechanism of the

Silicon processing II    223

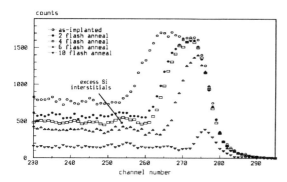

Figure 1. RBS spectra showing SPE regrowth of a $2 \times 10^{16}$ As/cm$^2$ implanted amorphous layer. Also seen is a region of excess Si interstitials coincident with the amorphous/crystalline interface.

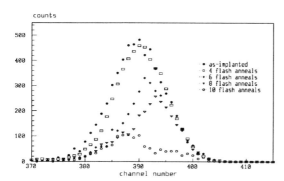

Figure 2. Channelled and random RBS spectra showing incorporation of As atoms onto substitutional sites during SPE regrowth of a $2 \times 10^{16}$ As/cm$^2$ implanted amorphous layer.

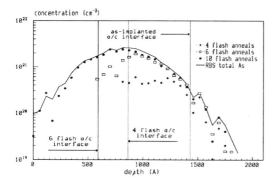

Figure 3. As substitutionality profiles as a function of depth and annealing time, showing an increase in substitutionality within regrown material following additional annealing.

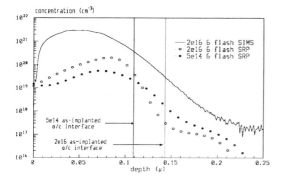

Figure 4. Comparison of chemical and electrical As profiles after 6 flash anneals, revealing inactive As in the tail region of the implant, coincident with the original amorphous/crystalline interface.

dopant[4]. Transient enhanced diffusion effects have been reported predominantly for elements which diffuse by an interstitial mechanism, although small diffusion enhancements have also been reported for As[7]. Despite the smaller initial amorphous layer thickness of the low dose sample, the position of the SRP shoulder after 6 anneal flashes is in approximate agreement with the higher dose implant (Figure 4). Additionally, higher electrical activity is observed in the tail of the low dose sample. The depth error on both profiles is ~10%. If excess interstitials are produced by implanted As knocking Si atoms into interstitial positions in the tail region of the profile, a similar depth distribution of excess interstitials would be produced for equal energy implants. The excess Si concentration would however be significantly greater for the higher dose implant since each implanted As atom produces more than one Si interstitial. This, together with the increased average As atom separation, decreases the probability of dopant-interstitial complex formation in the low doped sample. Electrically inactive As tails, similarly reported following furnace annealing (725-1000°C) of As implants into <111> Si were attributed to enhanced diffusion during implantation by an interstitial mechanism[8].

Conclusions
High temperature, short time annealing techniques in combination with ion implantation have produced shallow virtually diffusion-free layers whose characteristics are determined by the initial implant-induced damage distribution and its changing nature during the anneal. Chemical and electrical characterisation of partially and fully regrown layers indicates that full activation does not coincide with SPE regrowth or complete As substitutionality. Although As atoms are incorporated onto substitutional sites during SPE regrowth in concentrations exceeding equilibrium electrical solubility, additional annealing causes further increases in substitutionality within the regrown material. Short annealing timescales have revealed electrically inactive As tails, correlated with excess point defects (Si interstitials), and electrical activation within amorphous material. These results indicate that for current implant technology, the limit on shallow junction formation achievable by ion implantation and RTA is determined by implant induced damage distributions. Although shallower implanted junctions are achievable for As using lower energy implants, this becomes increasingly difficult for boron which has higher diffusivity and larger projected range, and for which the effects of transient enhanced diffusion become more significant since it diffuses predominantly via an interstitial mechanism. Whilst preamorphization techniques can be used to suppress axial channelling and enhanced diffusion, the relative position of implanted dopant and preamorphization-induced damage critically determines the electrical behaviour of the junction[9] and ultimately limits the junction depth achievable[10]. To retain the advantages of ion implantation (uniformity, controllability and reproducibility) for future generation devices will require separating the junction and implant damage by, for example, implantation into surface layers which act as diffusion sources[11].

Acknowledgements The authors would like to thank L. Nanu for useful discussions during this work and to acknowledge the help of Dr.M.Dowsett and E.A.Clark (Warwick University) in obtaining SIMS profiles and Dr.M.Pawlik (SAS Ltd.) for spreading resistance analysis. John Altrip would also like to acknowledge postgraduate student funding by the Science and Engineering Research Council.

References
1. J.L.Altrip, A.G.R.Evans, J.R.Logan and C.Jeynes, Sol Stat Electron (in press).
2. J.Said, H.Jaouen, G.Ghibaudo, I.Stoemenos and P.Zaumseil, 19$^{th}$ Sol Stat Dev Res Conf Proc, Berlin 1989, p225, Springer-Verlag 1989.
3. A.M.Mazzone, Phys Stat Sol (a)95, 149 (1986).
4. M.Servidori, R.Angeluci, F.Cembali, P.Negrini, S.Solmi, P.Zaumseil and U.Winter, J Appl Phys, 61(5), March 1987.
5. F.F.Morehead, Mat Res Soc Symp Proc, 104, 1983.
6. S.J.Pennycook, R.J.Culbertson and J.Narayan, J Mat Res, 1(3), 476, 1986.
7. R.Angelucci, F.Cembali, P.Negrini, M.Servidori, S.Solmi, J Electrochem Soc, 134, 12, 1987.
8. F.N.Schwettmann, Appl Phys Lett, 22(11), 570, 1973.
9. S.D.Brotherton, J.P.Gowers, N.D.Young, J.B.Clegg, J.R.Ayres, J Appl Phys, 60, 3567, 1986.
10. S.N.Hong, G.A.Ruggles, J.J.Paulos and J.J.Wortman, Electron Lett, 25, 16, 1100, 1989.
11. H.J.Bohm, H.Wendt, H.Oppolzer, K.Masseli and R.Kassing, J Appl Phys, 62, 7, 2784, 1987.

*Paper presented at ESSDERC 90, Nottingham, September 1990*
*Session 3B1*

# A two-dimensional approach to the noise simulation of GaAs MESFETs

Giovanni Ghione
Dipartimento di Elettronica, Politecnico di Milano, Piazza Leonardo da Vinci 32, Milano, Italy

**Abstract.** A two-dimensional technique for the physical noise majority-carrier modeling of GaAs MESFETs is presented, based upon a computationally efficient 2D implementation of the impedance-field method. Preliminary results concerning an epitaxial device on a semi-insulating substrate are discussed.

## 1. Introduction

Noise in electronic devices is caused by microscopic fluctuations, e.g. of the current density $\underline{J}(\underline{r})$. While Montecarlo techniques evaluate both the average and the fluctuations of the external voltages or currents, classical methods such as the impedance-field method (IFM) [Shockley 66] or the transfer-impedance method (TIM) [van Vliet 75] follow a two-step strategy. (Although the matter is somewhat controversial, we shall mantain that the equivalence between the IFM and TIM in two or three dimension, as discussed in [van Vliet 75], holds true, and shall base the following discussion on the IFM. ) Firstly, the statistical properties of the microscopic sources are established (e.g. the power spectrum $S_{\delta J \delta J}$ of the current density fluctuations). Secondly, the effect of the microscopic noise sources on external voltage or current fluctuations is evaluated in linearity and expressed in terms of a proper Green's function. Moreover, if microscopic sources are supposed to be spatially uncorrelated, the fluctuation spectra of external variables can be simply recovered through power summation, thereby yielding the well-known IFM expression for the power spectra ($i = j$) and cross spectra ($i \neq j$) of the open-circuit voltage fluctuations $\delta\phi$ on the electrodes $i$ and $j$:

$$S_{\delta\phi_i \delta\phi_j}(\omega) = \int_\Omega \nabla Z_i(\underline{r},\omega) \cdot S_{\delta J \delta J}(\underline{r},\omega) \cdot \nabla Z_j^*(\underline{r},\omega) d\underline{r} \qquad (1)$$

where $Z$ is the scalar impedance field, $\Omega$ the device cross section. Since MESFETs can be modeled as 2D structures invariant for translation along the gate, all functions are evaluated on a per-unit-length basis and then properly denormalized for the actual gate width.

## 2. IFM implementation

The scalar impedance field $Z_i = G_\chi(\underline{r}_i, \underline{r}', \omega)$ is the Green's function of the problem: inject into point $\underline{r}'$ a scalar current source $\delta I$ and evaluate the voltage on electrode $i$ as $\delta\phi(\underline{r}_i,\omega) = G(\underline{r}_i, \underline{r}', \omega)\delta I(\underline{r}',\omega)$; $\chi$ denotes a set of boundary conditions, here: electrode $i$ is *open*, all others are *grounded*. The evaluation of $G_\chi(\underline{r}_i, \underline{r}', \omega)$ can be readily carried out within the framework of a 2D frequency-domain small-signal model. However, the 2D evaluation of IF is cumbersome, since it amounts to solving as many linear problems as the number of the discretization points, and only one-dimensional implementations have been proposed so far [Nougier 85]. In the present work the IF evaluation is addressed through an efficient technique [Ghione 89], akin to the so-called *adjoint* approach to the noise analysis of lumped networks [Rohrer 71].

© 1990 IOP Publishing Ltd

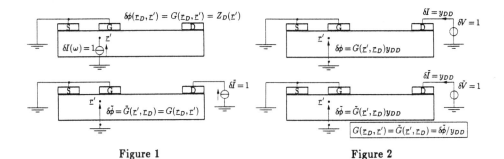

Figure 1        Figure 2

The starting point is a small-signal 2D frequency-domain monopolar drift-diffusion model:

$$j\omega \delta n = \nabla \cdot [\delta n \mu_0 E_0 - n_0 \underline{\mu} \nabla \delta \phi + D_0 \nabla \delta n] \quad (2)$$

$$\nabla^2 \delta \phi = \alpha \delta n \quad (3)$$

where $\delta n$ and $\delta \phi$ are the s.s. frequency-domain charge density and potentials, $E$ is the electric field, the subscript 0 refers to the working point, $D$ is the diffusivity, $\mu$ the mobility, $\alpha = q/\epsilon$ and $\underline{\mu}$ the small-signal mobility tensor accounting for the field dependency of mobility. On substituting (3) into (2) a fourth-order equation in $\delta \phi$ only can be derived:

$$\nabla \cdot \left[ D \nabla \nabla^2 \delta \phi + \mu E_0 \nabla^2 \delta \phi - (n_0 \alpha \underline{\mu} + j\omega I) \nabla \delta \phi \right] = 0 \quad (4)$$

or, shortly, $\mathcal{L}_\chi \delta \phi = 0$. The boundary condition set $\chi$ is conventional; charge neutrality on ohmic contacts reads $\nabla^2 \delta \phi = 0$. Instead of finding the Green's function of (4) (i.e. the gate or drain voltage response to a unity current source injected into a point internal to the device, see Fig.1, above, for the drain case), let us derive an *adjoint problem* such as to be *interreciprocal* to the original one. The adjoint problem $\tilde{\mathcal{L}}_{\tilde{\chi}} \delta \tilde{\phi} = 0$, where $\delta \tilde{\phi}$ is the adjoint potential and $\tilde{\chi}$ the adjoint set of boundary conditions, must satisfy the classical definition of adjoint operator, i.e. $< \delta \tilde{\phi}, \mathcal{L}_\chi \delta \phi > = < \tilde{\mathcal{L}}_{\tilde{\chi}} \delta \tilde{\phi}, \delta \phi >$, where the bracket pair denotes the symmetric integral scalar product. By repeatedly applying Gauss theorem and setting the residual contour integral to zero the adjoint problem can be shown to take the form:

$$\nabla \cdot \left[ \nabla \nabla \cdot (D_0 \nabla \delta \tilde{\phi}) - \nabla (\mu_0 E_0 \cdot \nabla \delta \tilde{\phi}) - (n_0 \alpha \underline{\mu}^T + j\omega I) \nabla \delta \tilde{\phi} \right] = 0. \quad (5)$$

The adjoint boundary conditions are the same as the direct one, apart from the ohmic contacts, where they read $\nabla \delta \tilde{\phi} \cdot \mu E_0 - \nabla \cdot (D \nabla \delta \tilde{\phi}) = 0$. It can be shown that the adjoint problem actually is *interreciprocal* to the direct one, i.e. that $G_\chi(\underline{r}_i, \underline{r}_j, \omega) = \tilde{G}_{\tilde{\chi}}(\underline{r}_j, \underline{r}_i, \omega)$, where $\tilde{G}$ is the Green's function of the adjoint problem. Interreciprocity can be exploited as depicted in Fig. 1 (below), i.e. the evaluation of the gate and drain impedance fields can be simply performed as $Z_D(\underline{r}) = \tilde{G}(\underline{r}, \underline{r}_D)$ and $Z_G(\underline{r}) = \tilde{G}(\underline{r}, \underline{r}_G)$, by solving *two* adjoint linear problems at each frequency, instead of twice as many as the discretization nodes (i.e. 2000-4000), as needed by the direct approach.

Rather than directly discretizing the adjoint problem, an easier and more consistent way is to discretize the direct problem and take its adjoint. Since the discretized direct problem leads to an admittance-matrix formulation akin to the one used in network theory, the adjoint problem is connected to the related adjoint network [Rohrer 71]; however, attention must be paid to the adjoint boundary conditions, whose discretized treatment goes beyond the simple network analog. Moreover, instead of solving a direct and adjoint I-driven pair, it is computationally more convenient to solve the E-driven pair shown in Fig.2. The I-driven adjoint problem can be easily recovered from the device admittance matrix. For more details, see [Ghione 89].

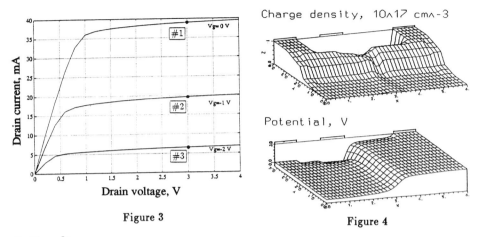

Figure 3                                  Figure 4

## 3. Results

The technique outlined in the last section has been implemented within the framework of the MESFET simulator MESS [Ghione 88], which makes use of the well-known Scharfetter-Gummel discretization scheme on a triangular grid. Only *diffusion noise* has been considered in the results shown here, i.e. $\mathbf{S}_{\delta J \delta J}(\underline{r},\omega) = 2q^2[\mathbf{D}(\omega) + \mathbf{D}^\dagger(\omega)] < n(\underline{r}) >$. Moreover, although the IFM can handle anisotropic diffusivity, a scalar $D$ was assumed. RG noise and intervalley scattering noise have been implemented as well, though the latter can be self-consistently simulated only through an energy-transport model. As a case study, an epitaxial one-micron MESFET, with 300 $\mu$m gate periphery, active layer thickness 0.2 $\mu$m, $N_D = 10^{17}$ cm$^{-3}$ and semi-insulating buffer layer has been considered. The DC curves are shown in Fig.3; small signal and noise analysis was performed for f=0 ... 14 GHz on the working points marked in Fig.3. The charge density and potential for the working point #1 are shown in Fig.4, while, for the same working point and $f = 4$ GHz, Fig.5 shows the short-circuit gate and drain AC currents deriving from unit current injection in a point internal to the device. Notice that these are the gate and drain scalar impedance fields, normalized with respect to the gate and drain short-circuit AC admittance. The level curves of the IF are almost perpendicular to the device surface, thereby yielding a vector IF almost parallel to the AC current lines, as 1D analyses

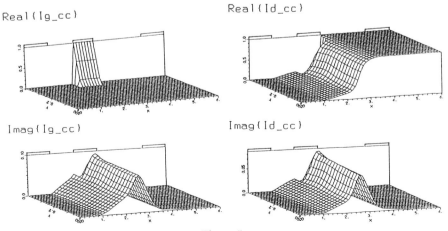

Figure 5

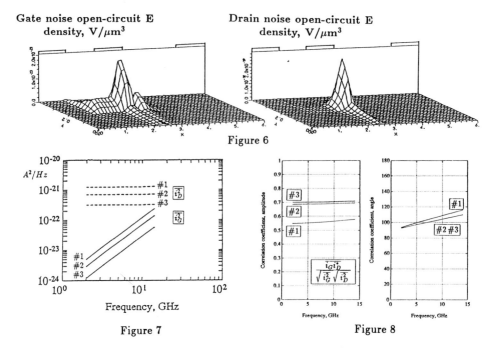

Figure 6

Figure 7

Figure 8

assume. (Generally speaking, in two or three dimensions this exactly occurs only in a *reciprocal medium*). Fig.6 shows the integrand of the IFM formula (1) i.e. what could be defined as the open-circuit gate and drain noise "spatial density" referred to 1 Hz bandwidth. The major noise contributions are seen to come from both the ohmic and saturated part of the channel. Finally, Fig.7 and Fig.8 show the gate and drain noise short-circuit current spectral densities and the correlation coefficient, respectively, for all working points considered. Notice the $\omega^2$ behaviour of the gate generator and that the correlation coefficient is almost imaginary at low frequency. The qualitative behaviour and the values obtained are in agreement with analytical or 1D noise models [Cappy 88].

## 4. Conclusions

An efficient technique for the 2D noise simulation of MESFETs, based on the impedance-field method, has been demonstrated. The adjoint approach allows the noise parameters to be estimated with very little computational overhead with respect to the frequency-domain small-signal physical simulation. Preliminary results agree with those derived from other approaches.

## References

[Cappy 88]  Cappy A *IEEE Trans. on Microwave Theory Tech.*, Vol. MTT-36, No.1, pp 1-10.

[Ghione 88]  Ghione G et al. *Alta Frequenza*, Vol. LVII, N.7, pp 295-309.

[Ghione 89]  Ghione G, Bellotti E, Filicori F *Proceedings of Int. El. Dev. Meeting 89*, pp 550-554

[Nougier 85]  Nougier J P et al. *Proc. of ESSDERC 1985* (Amsterdam: North Holland) pp 260-263

[Rohrer 71]  Rohrer D et al. *IEEE Journal of Solid-State Circuits*, Vol. SC-6, No.4, pp 204-212.

[Shockley 66]  Shockley W, Copeland J A, James P, in *Quantum theory of atoms, molecules and solid state*, ed P O Lowdin (New York: Academic Press) pp 537-563.

[van Vliet 75]  van Vliet K M et al. *Journal of Appl. Physics*, Vol. 48, No.4, pp 1804-1813.

Paper presented at ESSDERC 90, Nottingham, September 1990
Session 3B2

# 3D Integration of GaAs MESFET and varactor diode for a VCO-MMIC

M. Joseph, B. Roth[1], F. Scheffer, H. Meschede, A. Beyer[1], K. Heime[2]

Halbleitertechnik/Halbleitertechnologie
[1] Allgemeine und Theoretische Elektrotechnik
Sonderforschungsbereich 254, Universität-GH-Duisburg
[2] Institut für Halbleitertechnik, RWTH Aachen

**Abstract.** For the optimization of MESFET and varactor diode in a monolithic microwave integrated circuit (MMIC) different epitaxial layer structures are required. We report on the 3-d integration of the different devices by stacking the specific layers upon each other. DC and RF performances of the integrated FET are improved introducing an insulating layer sequence. A voltage controlled oscillator (VCO) has been realized with the first design for a frequency of oscillation of 9 GHz.

## 1. Introduction

Integrating different kind of devices, which each have the demand for a specially designed layer structure, leads either to selective epitaxy or to 3-dimensional integration. Selective area epitaxy has been used to fulfill these requirements especially for the combination of optical and electrical components /1/. However, additional technological steps like groove etching and silicon nitride deposition are needed and the growth process is difficult. The growth rate strongly depends on the size of the individual growth islands. Stacking the different epitaxial layers has been demonstrated /2/. But for an optimal FET performance all layers have been grown on top of the FET layer. Therefore complicated technologies had to be employed for the fine line lithography of the gate structure of the FET. In our case the MESFET is realized on top of the varactor. Then the key problem for a production process for 3-dimensional integrated circuits is an insulating layer between the different kind of devices.

## 2. Experimental

The first example we realized was a varactor tuned voltage controlled oscillator (VCO). Usually varactor diodes are fabricated with the FET within the same epitaxial layer with a high doping level which is constant in vertical direction and has a thickness of a few hundred nm. This technology results in a poor realization of varactor diodes /3/,/4/,/5/. For a better frequency voltage dependence and a higher tuning range of the oscillator we calculated a doping profile with maximum capaci-

© 1990 IOP Publishing Ltd

voltage $V_T$ as a function of backgate voltage $V_{BS}$ were measured and compared. The saturation current $I_{DSS}$ of the FET was measured at $V_{DS}$ = 1.5 V and $V_{GS}$ = 0 V. The influence of the backgate voltage $V_{BS}$ upon the threshold voltage $V_T$ was measured at $I_{DS}$= 0.01 $I_{DSS}$. The backgate voltage $V_{BS}$ was varied from -15 V to +3 V. Tab. 1 shows the variation of saturation current $I_{DSS}$ as well as threshold voltage $V_T$. $\Delta I_{DSS}$ is expressed in percent of the value at 0V $V_{GS}$, $\Delta V_T$ in mV.

| | | sample: | a | b | c | d | e |
|---|---|---|---|---|---|---|---|
| | | buffer: | AlGaAs (x=0.5) | AlGaAs (x=0.3) | GaAs p⁻ | GaAs undoped | GaAs )* |
| $\dfrac{\Delta I_{DSS}}{\%}$ at | $V_{BS}$ = -15V : | | 1 | 5.4 | <0.1 | <0.1 | 5.9 |
| | $V_{BS}$ = +3V : | | <0.1 | 1.8 | <0.1 | 0.42 | 1.6 |
| $\dfrac{\Delta V_T}{mV}$ at | $V_{BS}$ = -15V : | | 14.3 | 66 | 0.3 | 0.7 | 110 |
| | $V_{BS}$ = +3V : | | 11 | 34 | 0.3 | 0.7 | 53 |

)* undoped GaAs buffer and superlattice on top of a varactor layer sequence

Tab.1: Variation of $I_{DSS}$ and $V_T$ os GaAs MESFET for $V_{BS}$ = -15V and $V_{BS}$ = +3V Different buffer layers ( sample a-d) and a layer sequence for the integration of MESFET and Varactor (sample e) are investigated.

The undoped and the p-type doped buffer layer show the best performances in all investigations. The variation of $I_{DSS}$ is below 1 % for both samples and $\Delta V_T$ reaches a maximum value of 0.7 mV with the nominally undoped buffer and 0.3 mV for the p⁻ only. For further investigations a layer sequence for the integration of MESFET and Varactor was grown. An insulating layer sequence consisting of undoped layer and superlattice as described above leads to a good saturation and pinch off behaviour of the MESFET. Backgating effects are not yet negligible, variation of threshold voltage $V_T$ rises up to 110 mV and $\Delta I_{DSS}$ is as high as 6 % (cf. Tab. 1). But DC performance degradation is totally avoided, as a result a transconductance of 250 mS/mm and a saturation current of 400 mA/mm were achieved with 1.2 μm gate length and 140 μm gate width (cf. fig. 3).

For the MESFET 200 nm GaAs with a carrier concentration of $n=3 \cdot 10^{17}$ cm⁻³ is choosen resulting in a maximum frequency of power gain of $f_{MAG=1}$ = 41GHz as grown on semiinsulating substrate. Stacking the MESFET on top of the varactor $f_{MAG=1}$ is reduced to 25 GHz. However a voltage controlled oscillator has been realized with a frequency of oscillation of 9 GHz with an output power of 2 dBm.

Comparing the elements of the FET equivalent circuit extracted from s-parameter measurements with those from a non integrated planar FET, an increase of the drain capacitance $C_D$ and gate-drain capacitance $C_{gd}$ can be observed. Fig. 4 shows the

tance ratio for a hyperabrupt varactor. This profile ranging from n = $1 \cdot 10^{15}$cm$^{-3}$ to n=$1 \cdot 10^{17}$ cm$^{-3}$ with a thickness $d_v$ = 2 µm for a capacitance ratio $C_{max}/C_{min}$ = 20 was grown by MOVPE /6/. To obtain a high Q-factor, a n$^+$layer (n= $1 \cdot 10^{18}$cm$^{-3}$) is used for the ohmic contact. In this way the series resistance of the varactor diode is reduced. The MESFET layer (n=$3 \cdot 10^{17}$ cm$^{-3}$, d=200nm) is grown on top of the varactor and separated by an insulating layer sequence which consists of an 1.5 µm undoped buffer and a 10 period GaAs/AlAs superlattice (SL). The complete layer sequence is shown in fig. 1.

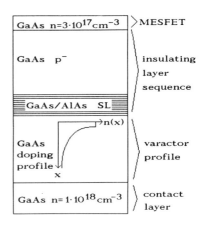

Fig. 1: layer sequence for 3-d integration of MESFET and varactor diode

The AlAs/GaAs superlattice (cf. fig.1) provides an excellent etch stop for a highly selective dry etch process. In this way the Schottky contact of the varactor can be reproducibly placed in the same depth. The interconnection between the FET, the varactor diode and additional passive elements is carried out by air bridges to pass the mesa edges without contacting the underlying layers.

Fig. 2:

SEM photograph of the connection of the FET via an airbridge passing the mesa edge

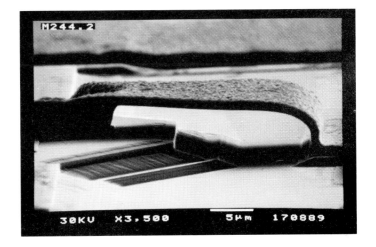

## 3. Results

At first different kind of MOVPE-grown buffer layers were investigated without underlying varactor to optimize the FET performance. A planar MESFET with a backgate-electrode was used. Variation of saturation current $I_{DSS}$ and threshold

dependence of the capacitances from the thickness $d_i$ of the insulating layer sequence between the varactor layer and the source and drain pads of the FET. A simple dependence of the additional parasitic capacitances on the geometric factor $1/d_i$ is obtained.

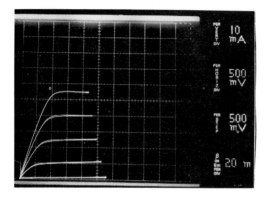
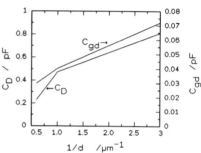

*Fig. 3:*
DC characteristic of a MESFET
$L_g = 1.2 \mu m$, $W_g = 50 \mu m$, $n = 3 \cdot 10^{17} cm^{-3}$

*Fig. 4:*
$C_d$ and $C_{gd}$ as a function of $1/d_i$
$d_i$ = thickness of insulating layers

## 4. Conclusion

We have shown, that 3-d integration is a good solution for a high performance VCO. We developed a reproducible process, compatible to standard MMIC production technique. The influence of the varactor layers on the FET parameters has been studied. The DC performance degradation of the FET is totally avoided. The cause for the reduction of maximum frequency $f_{MAG=1}$ is characterized. With the first design, a VCO with an output of 2 dBm at 9 GHz has been realized.

## 5. References

/1/ see for example : W.S. Lee, D.H. Spear, P.J. Dawe, S.W. Bland, J. Mun; Proc. of first international meeting on "Advanced Processing and Characterization Technologies", Tokyo (1989)

/2/ N. Suzuki, H. Furuyama, Y. Hirayama, M. Morinaga, K. Eguchi, M. Kushibe, M. Furinazu, M. Nakamura; Electron. Lett. 24, 467 (1989)

/3/ T. Ohira, T. Hiraoka, H, Kato; IEEE Trans. Microwave Theory and Techn., Vol. MTT-35, No 7, 657 (1987)

/4/ A.M. Boifot, I. Melhus, and O. Pedersen; Proc. 19th European Microwave Conference, London (1989)

/5/ B.N. Scott, G.E. Brehm, IEEE Trans. Microwave Theory and Techn. Vol. MTT-30, No 12, 2172 (1982)

/6/ I. Gyuro, F. Scheffer, M. Joseph, B. Roth, M. Heuken, K. Heime; Int. Conf. on MOVPE, Aachen (1990), to be publ. in J. of Chrystal Growth

# High performance 0·5 μm GaAs MESFET for MMIC applications

Areski BELACHE, Serge GOURRIER

L.E.P. : Laboratoires d'Electronique Philips,
22 avenue Descartes, B.P. 15, 94453 LIMEIL-BREVANNES CEDEX, FRANCE

Abstract :
Epitaxial and implanted GaAs MESFET devices fabricated with 0.5 µm gate lengths are presented. From d.c. measurements, very high transconductances, respectively equal to 500 mS/mm and 354 mS/mm, are achieved. In each cases, average current gain cut-off frequency Ft of 42.6 GHz and 35 GHz are calculated from S-parameters measurements. The best device exhibit a Ft value of 47 GHz associated with a maximum available cut-off frequency Fmax greater than 98 GHz. These results demonstrate the high performance of 0.5 µm GaAs MESFET for millimeter integrated circuits.

Very high performance MESFETs have been recently reported (Ft = 48 GHz, Gm = 350 mS/mm for a 0.5 µm gate device) (1). Theoretical and experimental analysis show that this can be achieved using a proper design of the active layer and an appropriate technological process in order to minimise the distance between the source end of the recess and the gate edge (Lrg) of the MESFET (2, 3). So parasitic phenomena which occur at the entrance of submicrometer devices are minimised. In this paper, we report the DC and microwave results obtained with half-micrometer gate length MESFETs, for microwave monolithic integrated circuit (MMIC). An optimised recess-gate provides, after Ti/Pt/Au evaporation, a very small lateral etched region (close to 50 nm) and yields improved DC and microwave performance.

Most of the results are obtained with MESFET fabricated on epitaxial layers. The structure was grown by metalorganic vapour phase epitaxy. It consists of a (40 nm Si doped, $2.10^{18}$ cm$^{-3}$) n+ GaAs layer, a (100 nm Si doped, $1.10^{18}$ cm$^{-3}$) n-type GaAs active layer and an undoped GaAs layer grown on a (100) semi-insulating substrate. The sheet resistance is 116 ohm/□ , which is low enough to ensure low source access resitance Rs and velocity overshoot beneath the gate. Protection of the active device layer is achieved by depositing a silicon nitride layer ($Si_3N_4$, 150 nm). Source-to-drain spacing is equal to 2.5 µm, with a gate shifted toward the source contact (Lsg = 0.5 µm). Finally, to reduce the gate access resistance, a six finger structure has been chosen (6 x 33 µm).

© 1990 IOP Publishing Ltd

The average drain current and transconductance versus gate voltage, across the wafer, are shown figure 1. A maximum extrinsic Gm as high as 500 mS/mm is measured when the depleted zone reaches its minimum thickness. The small deviation of this parameter (figure 2) shows the uniformity of the material and the technological process. The intrinsic Gm value is 570 mS/mm (Rs is close to 0.25 ohm.mm). A gate-to-drain breakdown voltage close to - 3.6 V is measured.

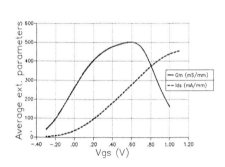
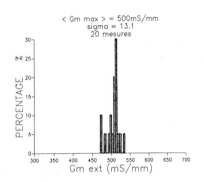

Figure 1
Average transconductance and drain current versus gate voltage (Vds = 2.5 V) for 0.5 μm gate devices

Figure 2
Gm max distribution on the wafer (Vgs = 0.7 V)

In a second part, we describe the millimeter-wave performance extracted from S-parameters measured on-wafer. The maximum current gain (H21) and the unilateral power gain (U) are extrapolated to unity for more than 20 devices (figures 3a & 3b). The average value of Ft and Fmax are respectively 42.6 GHz and 92 GHz with standard deviations of 2.1 GHz and 12 GHz., measured for gate bias voltages approximatively equal to 0 (Ids = Idss), and Vds = 2 V. The best device exhibited extrinsic Ft and Fmax of 47 GHz and 98 GHz, respectively (figure 4).

These values confirm recently published results (1). It is interesting to note that the surface passivation has no significant effect on the device performance. Simulated with SUPERCOMPACT, the small signal equivalent circuit is extracted over 10 devices and an intrinsic Ft (Gm/$2\pi$ Cgs) = 52 GHz is predicted for a 0.5 μm. This indicates a very high average electron velocity under the gate of $1.65 \cdot 10^5$ m/S (calculated with Ft = $<v>/2 \pi$ Lg), caused by an increase of overshoot effect in the channel and hence, a minimised hot carriers energy injected under the gate. According to device simulations, this high electron velocity corresponds to a short Lrg spacing in the transistors.

Preliminary results are also obtained with ion-implanted devices of the same geometry. The MESFET active channel is formed by silicon ion implantation into a semi-insulating GaAs substrate. After an optimised capless annealing, a peak carrier concentration of approximatively $1.10^{18}$ cm$^{-3}$ is realised. The sheet resistance $R_\square$ is closed to $340\ \Omega/\square$.

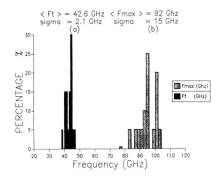
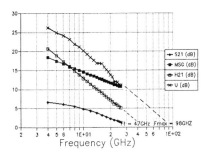

Figure 3
Ft (a) and Fmax (b) dispersions at Ids = Idss (Vds = 2 V)

Figure 4
Current gain (H21), maximum stage gain (MSG), unilateral gain (U) and (S21) as a function of frequency (0.5 x 200 μm)

In figure 5, we give dc performance typically obtained for ion-implanted MESFET's. An average extrinsic Gm of 354 mS/mm, associated with a drain current of 315 mA/mm is measured. For these devices, higher gate to drain breakdown voltages, close to (-6,1 V) are obtained.

The maximum transconductance is lower than for the MESFET on an epitaxial layer. In fact, the lower sheet resistance of the epitaxial layer corresponds to a lower source resistance (better access and contact resistances) which has 2 beneficial effects : (i) improving the extrinsic gm for a given intrinsic gm ; (ii) improving intrinsic gm by reducing the transfert of electrons to the lower velocity L valley under the gate.

Microwave results are also given. The average value and standard deviation of Ft and Fmax, which are respectively equal to (35 ± 4.6) GHz and (69 ± 6 GHz) are shown on figure 6.

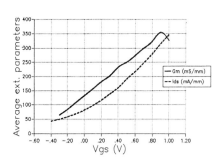

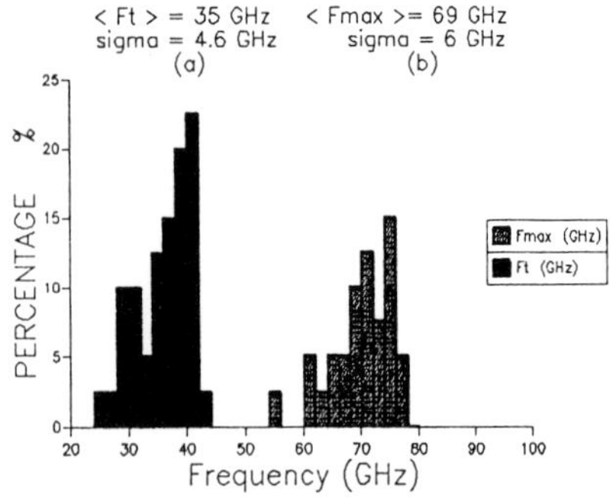

Figure 5
Average transconductance and drain current versus gate voltage for 0.5 μm gate devices

Figure 6
Dispersions of the maximum Ft (a) and Fmax (b) at Vds = 2 V

## Conclusion
Half-micrometer GaAs MESFETs have been realised with high performance in gain and cut-off frequency (Gm, Ft, Fmax) with low deviation across the wafer. This is in agreement with theoretical analysis and experimental results and confirms that the distance between the source edge of the recess and the gate-edge must be minimised to obtain the proper electron dynamics under the gate.

## Acknowledgements
The authors thank Peter Frijlink and José Maluenda for the epitaxial layer supply, José Bellaiche and Patrick Chambéry for technological assistance, and Isabelle Lecuru and Pascal Talbot for measurements.

## References
(1)  M Feng et al., Inst. Phys., GaAs and Related Compounds, Chapter 7, p. 513, Altanta, Georgia, 1988
(2)  F. Heliodore et al., IEEE Trans. on Electron Devices, vol. 35, n° 7, July 1988
(3)  P. Godts et al., Electr. Letters, 23rd June 1988, vol. 24, n°13

## GaAs FET and HFET on InP substrate
### A.CLEI, R.AZOULAY, N.DRAIDA, S.BIBLEMONT, C.JOLY

CENTRE NATIONAL D'ETUDES DES TELECOMMUNICATIONS
LABORATOIRE DE BAGNEUX
196 RUE HENRY RAVERA 92220 BAGNEUX ( FRANCE )
TELEX : CNET BAGNEUX 202 266 F   FAX : (1) 45295405

Abstract : GaAs FET and HFET structures have been grown by mismatched MOVPE epitaxy on InP substrates and processed using conventionnal technology. These transistors exhibit characteristics comparable to those measured on GaAs substrate. Ft and Fmax values of 12 and 30 GHz are measured on 1µm MESFET. Intrinsic transconductance greater than 250 mS/mm have been extracted for 1µm gatelength DMT.

## I. INTRODUCTION

Optoelectronic Integrated Circuits ( OEICs ) are very promising for high performance and low cost optical fiber communications. Many realisations have been demonstrated throughout the world and particularly in our laboratory at 0.85µm ( GaAs/GaAlAs ) and 1.3µm ( InP/InGaAs ).[1-3]
1.3 or 1.55µm optical communications (i.e. lasers and photodiodes lattice matched to InP) are generally preferred due to the minimum attenuation and dispersion of silica fibers at these wavelengths. However, InP microelectronics suffer from the lack of a good Schottky barrier precluding the fabrication of MESFET. Thus, it is necessary to develop other types of transistors (Junction-FET, MIS-FET, Barrier Enhanced-FET, Bipolar Transistors...) but the maturity of these devices is not yet sufficient. On the other hand, the good maturity of GaAs microelectronics has led to high performance devices and circuits of large complexity have been demonstrated. Rapid progress in mismatched hetero-epitaxy makes monolithic integration of lasers and photodiodes on InP with mismatched GaAs microelectronics a good solution for OEICs.[4]
The great difference in lattice parameters ( 3.8% ) between GaAs and InP assigns severe constraints on epitaxy. However, no antiphase problems and less severe difference in thermal expansion coefficient makes the problem less difficult than the intensively explored GaAs on silicon.

© 1990 IOP Publishing Ltd

## II. M.O.V.P.E. GROWTH.

GaAs growth has been made by atmospheric pressure M.O.V.P.E. on (100) Fe doped InP substrate. Growth parameters optimization, described elsewhere [5], led to mirror-like surfaces for layers grown on 3µm thick GaAs buffer. The epitaxy temperature sequence is shown in figure 3. After the growth of a 100 A nucleation layer at 450°C, the substrate temperature is raised to 700°C. After a 5 minutes anneal under arsine pressure, the growth of the structure is started (no phosphine is used). The FWHM of the ( 004 ) peak of the double X-Ray diffraction spectra measured on a 3 µm layer thickness is better than 220 arcsec. The surface quality is improved when the substrate is slightly misoriented toward (110) planes. Hall mobility measurements indicate $3500 cm^2/V.s.$ at 300K for $2.10^{17} cm^{-3}$ doping.

## III. TRANSISTORS FABRICATION AND RESULTS.

Field Effect Transistors have been processed on these GaAs layers. A conventional technology was used. A 2000 A thick channel doped to $2.10^{17} cm^{-3}$ was grown on a 3µm buffer. A 1000 A contact layer doped to $2.10^{18} cm^{-3}$ was added. Transistor areas were isolated by mesa etching or boron implantation. Metallizations were defined by contact optical lithography using polarity inversion resist for easy lift off, and deposited by E-Beam evaporation. AuGeNi ohmic contacts were alloyed by RTA, leading to a contact resistivity better than $10^{-6}$ $Ohm.cm^2$. TiPdAu gates were deposited after a recess of the channel by a citric acid solution. Schottky contacts show usual forward and reverse characteristics. A TiAu interconnection level was added to connect pads for on wafer static and dynamic measurements.
I-V characteristics display good pinch off and curve tracer examination shows no hysteresis, indicating the absence of deep levels (Fig.1). Maximum transconductance of 200 mS/mm was measured for 1µm gatelength and 3 µm channel length. A unity current-gain frequency of 12GHz and a power-gain cutoff frequency greater than 30GHz were measured by on wafer Cascade probing (Fig.2).

Doped channel MIS-like FETs have also been studied. Their structure (Fig.4) consist of a 3µm GaAs buffer, a 25nm GaAs channel doped to $2.10^{18} cm^{-3}$, a 30nm $Ga_{1-x}Al_xAs$ ( x=0.3 ) non intentionaly doped and a 10nm GaAs cap layer. This cap layer was added to prevent GaAlAs from oxydation and to make contact fabrication easy. It appears that doping of this layer does not change significantly the value of ohmic contact resistance measured in the $10^{-6}$ $Ohm.cm^2$ range. However, implantation of the contact regions is required to improve significantly the access resistance values, due to the 300 A GaAlAs undoped layer. A self aligned structure is even more desirable due to the high sheet resistance of the channel layer and our work was oriented in this direction.

Conventional FET technology as described above was used to make the transistors. Hall mobility of 2000cm$^2$/V.s was measured at room temperature. Without any implantation of the contact regions, a maximum transconductance of 145mS/mm was obtained(fig.5). Taking into account the high sheet resistance of the channel layer (2000 Ohm per square), an intrinsic transconductance greater than 250 mS/mm is infered.

N+ doping of the contacts was done by silicon implantation. Activation annealing under arsine overpressure gave good results in terms of surface quality while close contact annealing led to degraded surfaces. Access resistances of these implanted structures were reduced and a transconductance of 180 mS/mm was obtained. However, the annealing was accompanied by a degradation of the channel. This is illustrated by a shift of the threshold voltage of the transistors. Neither outdiffusion of Fe from InP substrate, nor silicon diffusion from the channel can fully explain this behaviour. Interaction of donors with traps seems to be a more likely explanation. Work is in progress to understand the mecanisms of degradation and optimise the self aligned structures.

## IV. CONCLUSION

In conclusion, mismatched GaAs FETs on InP substrates show performances close to those obtained with lattice-matched components. Doped Channel MIS-like FETs have been made whith results at the best published level. Improvements are still needed if annealing at high temperature must be used. Based on present results, it appears that hetero-epitaxy is a very promising way for the realization of OEICs working at long wavelengths.

Aknowledgements: The authors thank M.Bon, A.Mircea, A.Scavennec for fruitful discussions, L.Dugrand for technical assistance, B.Descouts and P.Krauz for ion implantation, M.Laporte for frequency measurements

[1] T.Horimatsu and M.Sasaki J.Lightwave Techno. 7/11/1989/1612
[2] F.Brillouet, A.Clei, A.Kampfer, S.Biblemont, R.Azoulay, N.Duhamel Electron.Lett. 22/1988/1259
[3] J.C.Renaud, L.Nguyen, M.Allovon, P.Blanconnier, S.Vuye, A.Scavennec J.Lightwave Techno. 6/1988/1507
[4] T.Suzaki et al. Electron.Lett. 24/1988/1283
[5] R.Azoulay et al. to be published in J.of Crystal Growth

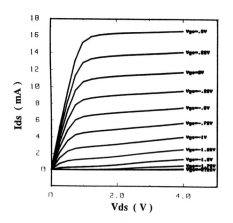

Fig. 1

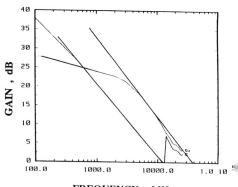

Fig. 2

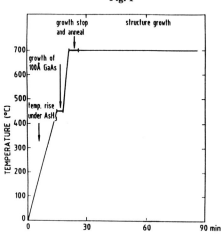

Fig. 3

Fig. 1 : MESFET I-V characteristics
( Lg = 1 µm , Wg = 50 µm )

Fig. 2 : Frequency dependance of current gain $h_{21}$ and unilateral gain

Fig. 3 : Epitaxy substrate temperature profile

Fig. 4 : DMT structure

Fig. 5 : DMT I-V characteristics
( Lg = 1 µm , Wg = 150 µm )

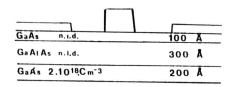

Fig.4

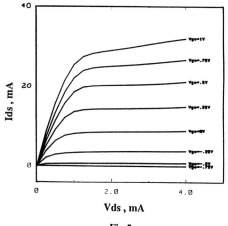

Fig.5

Paper presented at ESSDERC 90, Nottingham, September 1990
Session 3B5

# Analysis of the breakdown phenomena in GaAs MESFETs

J. Ashworth and P. Lindorfer*

Siemens Research Laboratories, Otto-Hahn-Ring 6, D-8000 Munich 83
*TU Vienna, Gußhausstr. 27-29, A-1040 Vienna

Abstract - Experimental investigations and 2-D numerical simulations have been performed to obtain a better physical understanding of the breakdown phenomena in GaAs MESFETs. The nature and location of carrier generation in a typical device have been determined and used to qualitatively explain experimentally observed breakdown features. It has been found that holes play an important role in determining breakdown behaviour.

I. Introduction

GaAs MESFETs find application in power amplification circuits which are key elements in microwave radar and communication systems. To obtain maximum output power from such devices great attention must be paid to the design of the structure and choice of technological parameters so that both a high drain current and high breakdown voltage are achieved. In order to fulfill this second requirement an understanding of the physical processes responsible for the breakdown phenomena in GaAs MESFETs and their dependence upon the parameters of a given FET is needed, which was the aim of this study.

Novel to previous considerations of the breakdown phenomena in GaAs MESFETs is the treatment of both before and after pinch-off voltage breakdown behaviour, analysed from experimental investigations and the numerical simulation program MINIMOS. Superior to the majority of contemporary programs used for breakdown simulation MINIMOS performs a self-consistent solution of carrier generation due to impact ionisation in full 2-D, considering both carrier types and the surface properties of GaAs MESFET's. Not only does a self-consistent solution of carrier generation provide the most accurate simulation of breakdown behaviour, it allows the observation of breakdown current in the drain characteristics directly and realises the calculation of the distribution of the generation rate of carriers in the device due to impact ionisation. This distribution has been used, for the first time, to determine the nature and location of carrier generation in a device under various bias conditions.

II. Summary of Experimental Results

Salient features of breakdown behaviour can be observed in the D.C. drain characteristics of fig. 1.

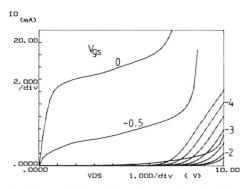

Fig. 1 Typical D.C. breakdown characteristics for FET's with undepleted sheet concentration $> 2.6 \times 10^{12} cm^{-2}$.

Initiation of 'excess current' on the curves is seen to have a specific gate-bias dependence. For gate bias from 0 V to -1 V ($\approx$ pinch-off), the breakdown voltage ($V_{ds}$ at which excess current is appreciable) increases with increasing gate bias magnitude whereas beyond pinch-off the breakdown voltage decreases with increasing gate bias magnitude. The breakdown characteristics as shown in fig. 1 were typical for devices with relatively large undepleted sheet charge concentration ($> 2.6 \times 10^{12}$ cm$^{-2}$, as defined by Wemple (1980)). Measurements were also taken on transistors with sheet concentrations of below $2 \times 10^{12}$ cm$^{-2}$. These devices exhibited a gate bias dependent breakdown behaviour where the breakdown

© 1990 IOP Publishing Ltd

voltage increased with increasing gate bias magnitude from $V_{gs}=0$ V to well beyond the pinch-off voltage. From these unique measurements of the complete breakdown characteristics i.e. before and after channel pinch-off, one can conjecture that for the case in fig. 1 there is a transition in the physical mechanism governing charge generation, between open and closed channel bias conditions, whereas for devices with low sheet concentration the breakdown mechanism remains unchanged for all gate biases. Experimental measurements however reveal little about the physical mechanisms taking place in a device and thus numerical simulations were turned to.

## III. Simulation Method

The breakdown behaviour as shown in fig. 1 is generally attributed to impact ionisation. This physical mechanism is such that when the electric field is increased in the saturated region of a device characteristic that at some point the charge carriers have sufficient energy to generate an electron-hole pair when they collide with the lattice. The total generation rate of carriers G due to this process can be expressed as:

$$G = \alpha_n \cdot J_n/q + \alpha_p \cdot J_p/q \tag{1}$$

where $J_n$, $J_p$ are the electron and hole current densities respectively and $\alpha_n$, $\alpha_p$ are the ionisation rates. The ionisation rates for electrons and holes are defined as expressing the number of generated electron-hole pairs per unit distance travelled by the charge carrier. $\alpha_n$ and $\alpha_p$ are very dependent upon the energy of the charge carrier as determined by the electric field which is usually expressed by:

$$\alpha_{n,p}(E) = A_{n,p} \exp(-B_{n,p}/E)^\beta \tag{2}$$

where E is the component of the electric field in the direction of the current vector. The coefficients $A_{n,p} = 3.5 \times 10^5$ cm$^{-1}$, $B_{n,p} = 6.85 \times 10^5$ Vcm$^{-1}$ and $\beta = 2$ were taken from Sze et al (1966). Equations (1) and (2) are solved self-consistently in the framework of the 2-D numerical simulation program MINIMOS which solves the basic semiconductor equations as given by Lindorfer (1989).

The Poisson Equation predicts that the electric field distribution, and thus $\alpha$ and G (see equations (1) and (2)), is dependent upon the charge density distribution in a device. It is of paramount importance therefore that the doping profile, surface and substrate deep-level trap effects present in the real device be correctly modelled for the accurate simulation of breakdown behaviour. The surface state properties of GaAs MESFET's were taken into account by using acceptor impurities on the surface between the device electrodes to affect the boundary conditions on the displacement vectors normal to the interface GaAs to Air (Selberherr (1984)). This was used to give rise to a surface depletion region and surface potential of -0.6 V with respect to that of the bulk material beneath. The effect of substrate deep level traps on the potential distribution of the device simulated, as given in fig. 2, is assumed to be negligible due to the choice of a p-type substrate which strongly determines the properties of the active channel-substrate interface anyhow.

## IV. Results

Geometry, doping and parameters of the device simulated are given in fig. 2 and Table I. Drain characteristics with and without generation mechanisms allowed for are presented in fig. 3. The simulated gate-bias dependence of the 'excess current' initiation both before and after pinch-off voltage are found to

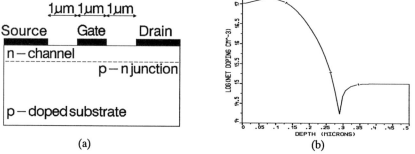

Fig. 2 Simulated MESFET structure (a) and net doping profile (b), showing p-n junction at $\approx 0.29$ µm.

| | |
|---|---|
| Surface acceptor density | $1.0 \times 10^{12}$ cm$^{-2}$ |
| Schottky barrier height | 0.8 V |
| Gate width | 100 μm |
| Si implant energy | 80 keV |
| Si dose | $2.2 \times 10^{12}$ cm$^{-2}$ |

Table 1 Device parameters for the structure simulated.

be in excellent qualitative agreement with that observed experimentally for low doped structures by Yamamoto (1978) and Wemple(1980). Simulations of devices with large sheet concentration were performed, to attempt to simulate the behaviour as in fig. 1, but convergence problems at high current levels were encountered.

In order to understand the gate-bias dependent breakdown behaviour shown in fig. 3, the carrier generation rate distribution in the device and its component factors have been examined for the biases of $V_{gs}=0$ V $V_{ds}=8$ V and $V_{gs}=-2$ V $V_{ds}=15$ V. Figs. 4 (a) and (b) reveal that the location and degree of

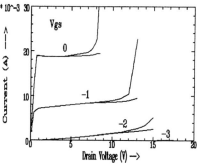

Fig. 3 Simulated drain characteristics of the FET in fig. 2. The upper curves portray the currents obtained with carrier generation taken into account and the lower curves with no generation mechanisms allowed for.

localisation of charge carrier generation in a device is strongly dependent upon bias conditions. For a fully open channel, carrier generation is sited at the edge of the drain contact and can be attributed to electron impact ionisation. This can be deduced from figs 5(a) and 6(a) showing that the maximum of the electric field (and hence that of α) is below the drain contact where there is a negligible hole current density (not shown) but appreciable electron current density. On the other hand with a gate voltage of -2 V (≈ pinch-off voltage) carrier generation is located at the edge of the gate contact (see fig. 4(b)) and arises due to hole impact ionisation, as can be deduced from figs. 5(b) and 6(b) (electron current density is not shown as it is negligible below the gate contact for $V_{gs}=-2$ V).

In an n-doped device the hole concentration is many magnitudes lower than the electron concentration. This means that the situation of a large number of mobile electrons per unit area moving below the drain contact is easier to achieve than a large number of mobile holes moving below the gate contact. In other words, the requisite for 'noticeable excess current' in the drain characteristics for $V_{gs}=0$ V is significantly easier to fulfill than that for $V_{gs}=-2$ V. Therefore, near pinch-off a higher electric field (or a larger $V_{ds}$) is required for breakdown to occur compared to that for open channel, qualitatively explaining the gate-bias dependence of breakdown behaviour. Only the application of a higher field enables the low concentration of mobile holes to generate a number of electron-hole pairs large enough such that the contribution from these secondary charge carriers be at all observable in I-V curves.

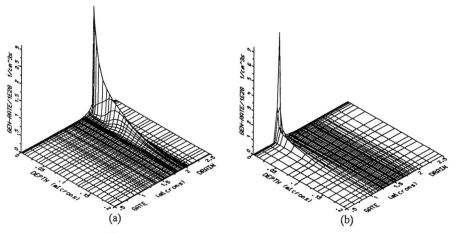

Fig. 4 Avalanche Generation rate in the drain-gate region for
(a) $V_{gs}=0$ V, $V_{ds}=8$ V (b) $V_{gs}=-2$ V, $V_{ds}=15$ V

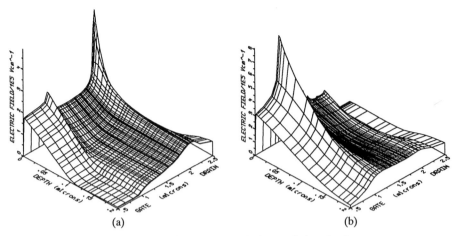

Fig. 5 Electric field distribution in the gate-drain region for
(a) $V_{gs}=0\,V$, $V_{ds}=8\,V$ (b) $V_{gs}=-2\,V$, $V_{ds}=15\,V$

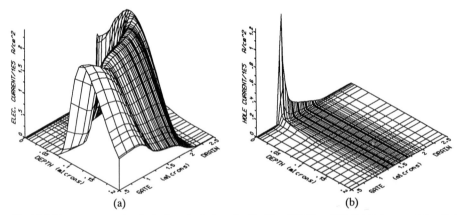

Fig. 6 (a) Electron current density in the gate-drain region for $V_{gs}=0\,V$, $V_{ds}=8\,V$. (b) Hole current density in the gate-drain region for $V_{gs}=-2\,V$, $V_{ds}=15\,V$.

## V. Conclusion

Simulations of impact ionisation in a simple device structure and doping indicate that carrier generation for $V_{gs}=0$ V takes place below the drain contact due to electron collision with the lattice whereby for $V_{gs}$ near pinch-off carrier generation was found to take place by the gate contact and stem from hole impact ionisation. It is concluded that the reliance of near pinch-off voltage breakdown phenomena on hole behaviour determines the gate-bias dependence of breakdown current initiation, as is observed in both measured and simulated drain characteristics.

## References

Lindorfer P et al 1989 GaAs IC Symposium Technical Digest
Selberherr S 1984 Analysis and Simulation of Semiconductor Devices (Vienna:Springer)
Son I et al 1989 IEEE Trans ED Vol 36 No 4
Sze S M et al 1966 Appl. Phys. Lett. Vol 8
Wemple S H et al 1980 IEEE Trans ED Vol 27 No 6
Yamamoto R et al 1978 IEEE Trans. ED Vol 25 No 6

Paper presented at ESSDERC 90, Nottingham, September 1990
Session 3C1

# Dynamic hot-carrier stress on submicron n and p-channel transistors

C Bergonzoni, G Dalla Libera and A Nannini
SGS-THOMSON Microelectronics - Central R & D
v. Olivetti 2 - 20041 Agrate Brianza (MI) - ITALY

Abstract: MOSFET hot carrier degradation under dynamic stress conditions is analyzed on submicron N- and P-channel transistors manufactured with up to date CMOS processes. Comparison of dynamic data to static stressing results shows the inadequacy of conventional methods in life-time prediction. Novel effects, mainly for P-channel devices, are observed and discussed.

## 1. Introduction

The effects of hot carrier degradation on MOS devices have been so far mainly analyzed with regard to static stressing procedures. The most common life-time prediction methods assume as a worst case the maximum degradation static bias configuration (which, in general, for given Vd, is located at the maximum of the Ib vs. Vg curve). Attempts to correlate SRAM circuit life-time (in terms of access time) to static aging results were also reported (Van der Pol, Koomen 1990). Only in recent works (Weber 1988; Hansch and Weber 1989), the study has been focused on the effects of fully dynamic inverter-like operation on the MOSFET electric characteristics, simulating the actual CMOS operation cycles. In this work we analyze the effects of realistic inverter-like operation on submicron CMOS MOSFETs, comparing them to the results of conventional static stressing at the substrate current maximum; the effect of gate and drain rise and fall times, simulating the actual loading condition of circuit nodes, is also analyzed. Novel degradation modes, pointing definitely out the inadequacy of conventional static procedures in transistor life-time prediction, are discussed for the P-channel device case.

## 2. Experimental and discussion

Two sets of devices, manufactured with two different CMOS processes, were tested. Set A includes N- and P-channel 0.8 $\mu$m gate length transistors, manufactured with a twin-well CMOS process. Dry gate oxide thickness is 20 nm and both N- and P-channel devices feature LDD drain engineering. Set B includes P-channel 1.0 $\mu$m gate length devices manufactured with a N-well CMOS process; dry gate oxide thickness is 28 nm, and the P+ implantation was performed after the sidewall spacers formation, without LDD (offset type P-channel transistors). The stress equipment was set up in order to simulate the actual CMOS operation, that is square pulses shifted by 180 degrees were applied to gate and drain; we call this stress configuration an 'inverter-like' operation (Figure 1). The effect of such an operation cycle in the Vd-Vg space is shown in Figure 2 for a N-channel

© 1990 IOP Publishing Ltd

transistor, basing on a simple linear charge/discharge model. The paths reported in Figure 2 show that during dynamic operation the transistor runs through several bias conditions; we consider particularly relevant for hot carrier injection the saturation region which, in the Vd-Vg space, lies below the Vg = Vd line. The importance of the Td/Tg ratio on the degradation rate can be argued from Figure 2; it may be seen that increasing the time ratio leads to a widening of the operation cycle, which reflects

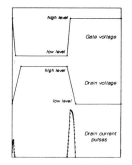

Figure 1 Voltage and drain current pulses in inverter-like configuration

in higher drain voltages applied during the saturation period of the cycle and subsequent higher degradation rates. This is experimentally confirmed in Figure 3, that reports a comparison of static Vg = Vd/2 stressing to various Td/Tg rates inverter-like dynamic configurations for a 0.8 μm gate length LDD N-channel transistor. In each case the current decrease depends on stress time according to the power law (Takeda and Suzuki 1983):

$$\Delta I_d/I_{d0} = At^n \qquad (1)$$

All dynamic stress cases can be fit by the same exponent n, testifying for the same physical damage mechanisms, but it can be noticed how the coefficient A increases as Td/Tg ratio increases. The importance of properly choosing the Td/Tg ratio is then evident. During dynamic stresses of both N- and P-channel MOSFETs we set Td/Tg = 8/3, a value compatible to actual loading condition seen during real operation in the 4 Megabit EPROM circuit where these transistors are used.

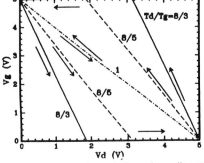

Figure 2 Vd-Vg paths variations depending on drain rise/fall time to gate rise/fall time ratio

## 2.1 N-channel LDD transistors

Several aging measurements were performed at different stress voltages on set "A" N-channel devices, both in dynamic "inverter-like" operation with 10 kHz pulse frequency and in static Vg = Vd/2 stress conditions. In all cases the linear zone drain current degrades according to (1) with slope n about 0.2-0.25 in the

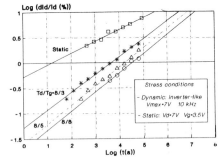

Figure 3 N-channel MOSFETs drain current degradation after static and dynamic stress

static stress case and about 0.35-0.4 in the dynamic stress case. This enhanced degradation efficiency in the dynamic mode turns out from the presence of different bias condition, that is different damage mechanisms, during the inverter cycle: hole injection phase (low Vg, high Vd) originates oxide electron traps subsequently filled during electron injection phase (Vg = Vd) (Doyle et al. 1990); the presence of the bias condition Vg = Vd/2 during the cycle testifies also for interface states creation. On the contrary, in static hot carrier stress, only one bias condition is present. The complete qualitative and quantitative inadequacy of static stress measurements in predicting device lifetimes during real operation follows from our experiments and considerations.

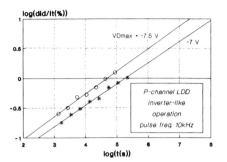

Figure 4 LDD P-channel MOSFETs drain current degradation after dynamic stress

### 2.2 P-channel LDD transistors

Dynamic stress measurements were performed on set "A" LDD P-channel transistors in inverter-like configuration with 10 kHz pulse frequency. Typical results are shown in figure 4. A decrease in drain current, depending on stress time according to a power law (1), is observed. This is not consistent with what is usually reported (Mittl and Hargrove 1989) for static degradation, which suggests electron trapping in the gate oxide. Dynamic stress results can only be explained in terms of positive charge localization in the gate oxide or at the oxide/silicon interface. Again static stress results appear completely inadequate in predicting device lifetime in real operation conditions.

### 2.3 P-channel offset transistors

Dynamic "inverter-like" stress applied to set B offset type P-channel transistors turned out into complex degradation phenomena (Figure 5): an initial drain current decrease converts after some stressing time into a trend inversion, where the drain current begins to increase towards the original value. Stresses were performed at 9.0, 9.25, 9.35, 9.5 and 10.0 V maximum voltage, showing a decrease of the inversion point time with

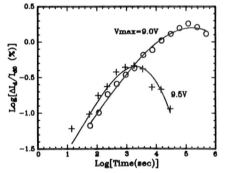

Figure 5 Offset type P-channel MOSFETs drain current variation after dynamic stress

increasing maximum voltage; the aging performed at Vmax = 10.0 V turned out in increasing drain current right from the start. This behaviour may be explained in terms of two parallel damage processes charachterized by different kinetics: initial current degradation is due to positive charge localization in the gate oxide and/or to the formation of donor states at the $Si/SiO_2$ interface; the following current increase is due to negative charge localization in the gate oxide.

The drain current trend may be reproduced by an extention of the formula (1) currently used to model electrical parameters degradation, that is by:

$$\Delta I_d/I_{d0} = A_h\ t^{nh} + A_e\ t^{ne} \qquad (2)$$

the observed trend may be explained if $A_e < 0$, $|A_e| < |A_h|$ and $n_e > n_h$. Arguing from analogy to LDD static and dynamic stress data, we suppose that the exponents $n_e$ and $n_h$

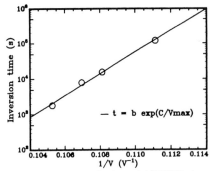

Figure 6 Offset type P-channel MOSFETs inversion time as a function of stress high level voltage

do not depend on the maximum stress voltage. The dependance from $V_{max}$ in (2) appears in the coefficents $A_e$ and $A_h$. Applying the Takeda model (Takeda and Suzuki 1983) to each degradation mechanism separately, the dependance of the inversion time ($t^*$) on high level voltage can be found in the form (C. Bergonzoni et al. 1990):

$$t^* = b \; \exp(C/V_{max}) \qquad (3)$$

where b and C are positive constants; both can be taken as fitting parameters. Figure 6 shows the excellent accordance between (3) and experimental data. Moreover, the extrapolation of (3) to $V_{max} = 10.0$ V predicts an inversion time of the order of 10 seconds; it is explained why we were not able to observe any initial decrease at $V_{max} = 10.0$ V.

3. Conclusions

The inadequacy of static hot carrier stress measurements in predicting transistors degradation during real operation, and the consequent necessity to reproduce dynamic operating conditions, is pointed out for N- and P-channel devices.
In performing dynamic stress tests in inverter-like configuration, strong care must be taken in simulating real loading conditions.
Dynamic stresses on P-channel LDD devices suggest positive charge localization in the gate oxide or at the Si/SiO$_2$ interface.
Dynamic stresses on offset type P-channel transistors reveal the presence of two parallel damage processes involving opposite sign charges localization.

4. References

- Bergonzoni C, Dalla Libera G and Nannini A, 1990, not yet published
- Doyle B S, Bourcerie M, Bergonzoni C, Benecchi R, Bravais A, Mistry K R and Boudou A 1990, to be published in IEEE Trans. El. Dev., July issue
- Hansch H and Weber W, 1989, IEEE El. Dev. Letters, vol.10, p.252
- Mittl S W and Hargrove M J, 1989, IRPS Conf. Proc., p.98
- van der Pol J A and Koomen J J, 1990, IRPS Conf. Proc., p.178
- Takeda E and Suzuki N, 1983, IEEE El. Dev. Letters, vol EDL 4, n.4
- Weber W, 1988, IEEE Trans. El. Dev., Vol.35, n.9

Paper presented at ESSDERC 90, Nottingham, September 1990
Session 3C2

# Hot-carrier experiments on scaled NMOS transistors

R. Woltjer, G.M. Paulzen, P.H. Woerlee, C.A.H. Juffermans and H. Lifka

Philips Research Laboratories, 5600 JA Eindhoven, The Netherlands
Tel: 31-40-743551   Fax: 31-40-743390   Telex: 35000 phtc nl nlwtfau

**Abstract.** Hot-carrier reliability limits the operation voltage for downscaled NMOS transistors. Transistors made according to Quasi Constant Voltage Scaling with design rules between 2.5 $\mu$m and 0.25 $\mu$m are tested for reliability and performance. We show that this scaling is in accordance with hot-carrier reliability. The maximal output power ( per unit width ) is independent of the design rule, but a factor of two more output is possible for LDD transistors. The gate delay appears to be proportional to the design rule, independent of the drain structure.

## 1. Introduction

Hot-carrier degradation of NMOS transistors is a limiting factor in ULSI scaling. For scaling of the dimensions with a factor of $\lambda$ three scenarios ( Chatterjee et al 1980 ) are proposed: Constant Voltage Scaling ( $V_D$ constant ), Constant Field Scaling ( $V_D \propto \lambda$ ) and Quasi Constant Voltage Scaling ( $V_D \propto \sqrt{\lambda}$ ). The Constant Voltage scaling ( $V_D = 5$ V ) is preferrable because different generations of ICs can use one power supply and ICs may be coupled to each other ( e.g. TTL ) without voltage conversion. For gate lengths shorter than 2 $\mu$m, the conventional drain in the NMOS transistors is replaced by an LDD ( Lightly Doped Drain ) to avoid excessive hot-carrier degradation due to the peaked electric fields.

Only recently, a lower operating voltage has been introduced for logic applications to avoid unacceptable power dissipation. For the half-micron generation, a lively discussion continues whether the 5 volt standard can still be maintained or not. The most critical points are oxide integrity and hot-carrier degradation. Simulations have been used for extrapolations to the future generations ( Kakumu et al 1986 ), leading to the conclusion that the Quasi Constant Voltage Scaling ( QCV ) appears to be optimal when both performance and reliability are taken into account. The 5 Volt standard should, when it is no more feasible, be replaced by the QCV scaling according to these simulations.

## 2. Device scaling

To study the hot-carrier hardness and the driving capability of scaled NMOS transistors experimentally, we prepared a set of transistors with decreasing design rules ranging from $L_{poly} = 2.5\,\mu$m ( $t_{ox} = 50$ nm ) down to $L_{poly} = 0.25\,\mu$m ( $t_{ox} = 7.5$ nm ) according to QCV scaling. Scaling of the design rules with a factor of $\lambda$ reduces the gate oxide thickness $t_{ox} \propto \lambda^{0.8}$ and channel doping is scaled to ensure 0.5 V < $V_T$ < 0.8

© 1990 IOP Publishing Ltd

V for all transistors. We use n⁺ polysilicon gates defined by optical lithography or ( below 0.7 μm ) electron beam lithography. Two series of transistors are made with either conventional drain or LDD of approximately $4 \times 10^{13}$ cm$^{-2}$ dosis.

## 3. Hot-carrier degradation experiments

We present accelerated hot-carrier degradation experiments at room temperature on wide transistors ( $W/L > 10$ ). We measure the worst-case d.c. degradation for various drain voltages and for various transistor lengths for each transistor type. The maximal degradation is generally found at $V_G \approx V_D / 2 - 0.5$ V. These experiments yield the time dependence of the transconductance degradation ( measured at $V_D = 0.1$ V ). Interpolation or extrapolation of the measured powerlaw dependence yields the lifetime $\tau_{life}$ defined by a 10 % decrease in transconductance.

We give our results for transistors with minimal gate lengths according to the design rules. Note that our gate length is not the effective length, but the length of the polysilicon gate. In figure 1, we show our measurements for conventional drain transistors with 0.25 μm and 0.5 μm design rule. Extrapolation in plots of the logarithm of the lifetime versus the reciprocal drain voltage yields the maximal operating voltage $V_{D,max}$ that guarantees a lifetime longer than 10 years under the worst-case degradation condition. This results in figure 1 for 0.25 μm and 0.5 μm in maximal operating voltages $V_{D,max} = 2.2$ V and $V_{D,max} = 2.7$ V, respectively. No important new degradation mechanism is observed in these experiments down to 0.25 μm design rule.

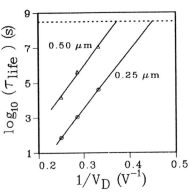
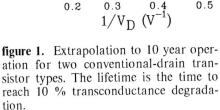

**figure 1.** Extrapolation to 10 year operation for two conventional-drain transistor types. The lifetime is the time to reach 10 % transconductance degradation.

**figure 2.** Maximal operating voltage that avoids 10 % change in transconductance during 10 years of worst-case d.c. operation. $L_{poly}$ is the minimal gate length according to the design rules.

Figure 2 shows that for ( QCV ) scaled devices the maximal operating voltage decreases approximately with the square root of the gate length for both conventional and LDD transistors. This is consistent with the quasi-constant-voltage scaling introduced by Chatterjee et al ( 1980 ) and found in simulations by Kakumu et al ( 1986 ).

## 4. Discussion

Figure 2 yields important information: 5 V ($\pm$ 10 %) operation for transistors shorter than 2 $\mu$m is difficult with conventional drain transistors when quasi-constant-voltage scaling is used. When the power dissipation is not the limiting factor, an optimized LDD will reduce the peak electric field sufficiently to allow 5 volt operation down to 0.8 $\mu$m design rules. For gate lengths below 0.8 $\mu$m the operation of LDD transistors at 5 volt is questionable.

For 3.3 volt operation, half micron transistors can be obtained with optimized LDD structures, but the next generation (0.35 $\mu$m) must either operate at a lower drain voltage or have a more optimal transistor design like LATID (Large Angle Tilted Implanted Drain) (Hori et al 1988), inverted T (Huang et al 1986) or GOLD (Gate Overlapping Drain) (Izawa et al 1987).

## 5. Output current, power and speed of scaled transistors

The output current per unit width $I_{D,max}$ at $V_G = V_D = V_{D,max}$ is given in figure 3 for the transistors of minimal length according to the design rules. We clearly see that the current drive increases approximately with $L_{poly}^{-1/2}$. This results in a constant maximal output power (per micron width) $P_{max} = I_{D,max} \times V_{D,max}$ for all design rules, as shown in figure 4. The maximal output power of LDD transistors is a factor of two higher than the output power of conventional-drain transistors.

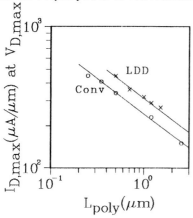
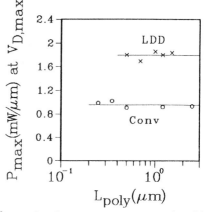

**figure 3.** Drain current per unit width at $V_G = V_D = V_{D,max}$ for the transistor types in figure 2.

**figure 4.** Output power per unit width at $V_G = V_D = V_{D,max}$ for the transistor types in figure 2.

The gate delay is approximately proportional to $V_{D,max}/I_{D,max}$ according to Sodini et al (1989). In figure 5, the gate delay appears to scale linearly with the design rule. The advantage of LDD with respect to the conventional transistors appears to be negligible, but optimization of the LDD transistors may improve their speed.

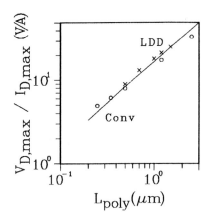

**figure 5.** Gate delay ( proportional to $V_{D,max}/I_{D,max}$ according to Sodini et al ( 1989 ) ) for the transistor types in figure 2. The advantage of LDD with respect to the conventional transistors appears to be negligible.

## 6. Concluding remarks

We studied the hot-carrier degradation for scaled NMOS transistors experimentally. Figure 2 presents the maximal operating voltage with respect to hot carrier degradation, but other effects might impose extra constraints on the power supply. Figure 4 implies that the power per unit IC area increases approximately proportionally to the reciprocal design rules ( the transistor length scales with the design rule ), making clear that the QCV scaling might still result in too much power dissipation when a large number of transistors is switched simultaneously e.g. in logic circuits like microprocessors. Excessive power dissipation may be a reason to operate transistors at a lower voltage than the maximal voltage given in figure 2.

For decreasing design rules, the maximal electric field in the gate oxide ( $V_{D,max}/t_{ox}$ ) increases significantly: For conventional drain transistors we find 1 MV/cm for 2.5 $\mu$m and 3 MV/cm for 0.25 $\mu$m design rule, respectively. This reliability hazards ( oxide degradation ) is not studied for our scaled transistors and may impose another constraint on the operating voltage for downscaled NMOS transistors. Finally, the sensitivity for degradation of single transistors in an IC may differ significantly. Design optimization with respect to duty cycle may reduce the effective time of worst-case degradation. Design optimization with respect to the sensitivity for variations in transistor parameters allows more ( than 10 % ) degradation for single transistors. Both approaches may result in operating voltages exceeding those given in figure 2.

Chatterjee P B, Hinter W R, Holloway T C and Lin Y T 1980
    *IEEE El.Dev.Lett. EDL-1 220*
Hori T, Kurimoto K, Yabu T and Fuse G 1988
    *Tech. Digest for Symp. on VLSI Technol. 15*
Huang T, Yao W, Martin R, Lewis A, Ko M and Chen J 1986
    *IEDM Tech. Dig. 742*
Izawa R, Kure T, Iijima S and Takeda E 1987 *IEDM Tech. Dig. 38*
Kakumu M, Kinugawa M, Hashimoto K and Matsunaga J 1986
    *IEDM Tech. Dig. 399*
Sodini C G, Wong S S and Ko P 1989 *IEEE Solid-State Circuits SC-24 118*

# Noise characterisation of silicon MOSFETs degraded by F–N injection

C. Nguyen–Duc, G. Ghibaudo and F. Balestra
Laboratoire de Physique des Composants à Semiconducteurs,
ENSERG, 23 rue des martyrs, B.P. 257, 38016 Grenoble, France.

**Abstract** : MOS transistors are investigated by low frequency noise measurements before and after Fowler-Nordheim (F-N) injection. The comparison with other techniques such as static I(V) characterisation and charge pumping method is also done in order to correlate the noise level to the MOSFET interface properties. The results are discussed in term of number and mobility fluctuations noise model.

## 1. INTRODUCTION

The 1/f noise in MOSFETs being very sensitive to interface properties, it can be successfully used for the analysis of hot-carrier-induced aging (Fang et al 1986) as well as for the extraction of the oxide trap density near the conduction band edge of silicon (Jayaraman and Sodini 1989).

In this work, MOS transistors are studied by low frequency noise measurements before and after Fowler-Nordheim (F-N) injection. The comparison with other techniques such as static I(V) characterisation and charge pumping method is also done in order to correlate the noise level to the MOSFET interface properties.

The devices used are conventional n-channel MOSFETs with gate width $W = 25$ $\mu$m, channel length $L = 1$ to $5$ $\mu$m, gate oxide thickness $t_{ox} = 28$ nm and channel doping $N_a \simeq 10^{16}/cm^3$. The devices were subjected to F-N injection with constant current density of $10^{-4} A/cm^2$ for different period of time : the dose ranging from $5\times10^{17}$ to $5\times10^{18}$ $/cm^2$.

## 2. RESULTS AND DISCUSSION

Fig. 1 shows typical equivalent input gate voltage spectral densities $S_{vg}(V_g-V_t)$ characteristics measured at 10 Hz and $V_d$=50mV for two MOSFETs, $L=1\mu$m (a) and $L=5\mu$m (b), with injection dose as a parameter. The figures show a considerable increase in low frequency noise induced by F–N injection. The gate voltage dependence in the exponent $\gamma$ of the $1/f^\gamma$ noise spectrum is presented in Fig. 2 for the 1-$\mu$m-long device. Note that the frequency exponent $\gamma$ is taking values closer to 1 after degradation. The data also indicate that $\gamma$ does not increase monotonically with gate drive. It first decreases reaching a minimum and then increases approaching the unity at large $V_g$ for all the cases. After F-N injection of $3\times10^{18}$charges/cm$^2$, the threshold voltage shift $\Delta V_t$ is about 0.26 V and the mobility degradation $\Delta\mu/\mu$ around 23 % (520 to 400 cm$^2$/Vs) as determined from static I(V) measurement shown in Fig. 3 where the device transconductance is plotted against gate voltage. Moreover, by using the charge pumping current $I_{cp}$ (Fig. 4), we have found the interface state density to be around $9 \times 10^9/eVcm^2$ for unstressed devices, and $10^{11}$ and $2.4 \times 10^{11}/eVcm^2$ for devices stressed with dose of $8 \times 10^{17}$ and $3 \times 10^{18}$ charges/cm², respectively.

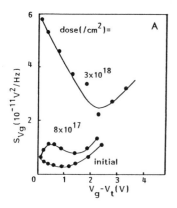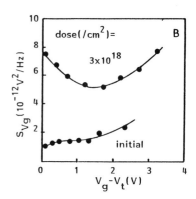

Fig. 1 : $S_{vg}$ variations with gate voltage after F-N degradation at various doses and for two device lengths.

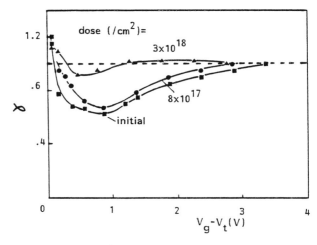

Fig. 2 : Frequency exponent $\gamma$ dependence with gate voltage for several injection doses.

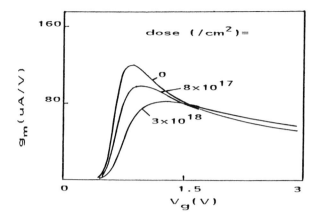

Fig. 3 : Transconductance characteristics versus gate voltage for initial and after F–N injection.

These results can be discussed in terms of mobility fluctuation model ($\delta\mu$) and carrier number fluctuation model ($\delta n$) with and without interface-charge-induced mobility fluctuations. Theoretically, Hooge's $\delta\mu$ model predicts $S_{vg}(V_g)$ characteristics with an abrupt decrease at weak inversion and a superlinear variation in strong inversion. In the pure $\delta n$ model $S_{vg}$ should not depend on gate voltage for constant interface trap density (Ghibaudo 1989). In contrast, $\delta n$ model taking into account the interface-charge-induced $\delta\mu$ provides a $S_{vg}(V_g)$ behaviour first independent of $V_g$ below threshold and followed then by a parabolic increase with $V_g$ above threshold (Ghibaudo 1990, Hung et al 1990). In our case, the noise level has a more complicated dependence on gate voltage than $\delta n$ or $\delta\mu$ theory predicts. Nevertheless, $\delta n$ model seems to prevail as supported by the strong correlation between the noise level and the interface state density deduced from $I_{cp}$ data and, by the fact that $S_{vg}$ is either independent of gate voltage or increases with $V_g$. The number fluctuation model is also supported by the overall strong correlation between the normalised drain current spectral density $S_{id}/I_d^2$ and transconductance–current ratio squared $(g_m/I_d)^2$ vs drain current as persented in Fig. 5 for the devices described above.

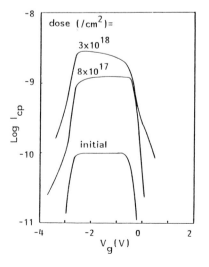

Fig. 4 : Typical charge pumping current $I_{cp}$ versus base gate voltage as obtained after F-N stress.

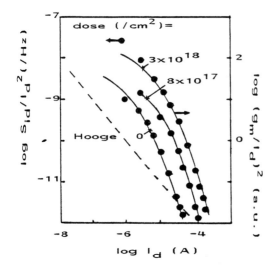

Fig. 5 : Normalised drain current spectral density and transconductance–drain current ratio squared as a function of drain current for initial and after stress.

Based on the relation,

$$\frac{S_{Id}}{I_d^2} = \left(\frac{g_m}{I_d}\right)^2 S_{Vg}$$

we have found the average values for $S_{Vg}$ of $5\times10^{-12}$, $1\times10^{-11}$ and $5\times10^{-11}$ V²/Hz for unstressed device and devives stressed with dose of $8\times10^{17}$ and $3\times10^{18}$ charges/cm², respectively.

As can also be seen from Fig. 5, the mobility fluctuation model represented by the dashed line is unable to provide a satisfactory fit of our results in the drain current range studied here ($10^{-8} - 10^{-3}$ A).

## 4. CONCLUSION.

In conclusion, F-N degradations can be used as an excellent way to change the noise behaviour of Silicon MOSFETs and, therefore, to discuss the ability of currently admitted noise models to interpret noise data. Furthermore, noise measurements can be regarded as a sensitive characterisation method for the assessment of slow interface trap densities in MOS transistors.

## REFERENCES

Z.H. Fang Z H Cristoloveanu S and Chovet A 1986 IEEE Electron Device Letters **EDL-7** 371.
Ghibaudo G 1989 Sol State Electron **32** 563.
Ghibaudo G 1990 unpublished.
Hung K K, Ko P K, Hu C and Cheng Y C 1990 IEEE Trans Electron Devices **ED-37** 654.
Jayaraman R and Sodini C 1989 IEEE Trans Electron Devices **ED-36** 1773.

Paper presented at ESSDERC 90, Nottingham, September 1990
Session 3C4

# Hot-carrier induced degradation in short-channel silicon-on-insulator MOSFETs

T Ouisse[1], S Cristoloveanu[2], G Reimbold[3], G Borel[1]

[1]Thomson–TMS, BP 123, 38521 Saint-Egrève Cedex, France.
[2]LPCS, INPG, ENSERG, BP 257, 38016 Grenoble Cedex, France.
[3]LETI, CENG, 85X, F 38041 Grenoble Cedex, France.

**Abstract**. Submicron MOSFET's fabricated on SIMOX substrates present a good tolerance to aging induced by hot carrier injection in the front gate oxide. Stressing the back channel reveals the formation of localized defects at the buried interface which are responsible for a large transconductance overshoot and an attenuation of the kink effect.

## 1. Introduction

Silicon On Insulator (SOI) materials have made great progress in the recent years, being now able to produce rather complex integrated circuits which are faster than bulk Si counterparts. A key argument for the wide application of SOI in the future might be the improved tolerance of SOI MOSFET's to hot carrier-induced degradation. However, this aspect has received only limited attention to date (Colinge 1987, Woerlee et al 1989), probably because the configuration of SOI–MOSFET's (two gates, two oxides and three interfaces) and the high number of experimental parameters make the problem even more complex than in bulk Si. Presented here are new results on the vulnerability of the front and back oxides and interfaces in SIMOX transistors.

The SIMOX substrates were fabricated by high current implantation of oxygen ($1.8 \times 10^{18}\ cm^{-2}$, $200 keV$, $600°C$) and annealing above $1300°C$ in argon ambient. The devices under inspection were LOCOS isolated n-channel MOSFET's with optimized LDD configuration. The basic channel length was 1 $\mu m$ and the thicknesses of the gate oxide, buried oxide and Si film were 24 $nm$, 400 $nm$ and 150 $nm$, respectively.

This work is focused on the influence of the stressing conditions (front gate $V_{G1}$, back gate $V_{G2}$ and drain $V_D$ biases, stress duration and channel-length) on the performance degradation. Several parameters were monitored : threshold voltage shift, transconductance peak $g_{max}$ and corresponding voltage $V_{gmax}$, subthreshold slope and leakage current, mobility attenuation coefficient $\theta$ which accounts for the influence of series resistances in LDD structures, drain current in linear and saturation regions.

Since the substrate current cannot be measured and used, as in bulk Si, to predict the device lifetime, we have instead considered some specific parameters of SOI, related to the bipolar transistor activation and kink effect. The $I(V_G, V_D)$ characteristics of the front and back channels were plotted also by exchanging the drain and source.

## 2. Front Channel Degradation

Relevant conclusions for the conventional mode of operation were derived by stressing the front channel only. The gate oxide and interface of SIMOX-MOSFET's present a

© 1990 IOP Publishing Ltd

good tolerance to aging. This is illustrated by an insignificant shift of the extrapolated threshold voltage which remains in the noise level. The degradation of the transconductance $g_{max}$ is also acceptable: 5% after more than 200 hours of stress. The change in $V_{gmax1}$ is more pronounced for stress at $V_{G1} \simeq V_D$ whereas $g_{max1}$ is more degraded for stress at $V_{G1} \simeq V_D/3$. The rate of change of the reversed saturation current stands as the most sensitive parameter to stress (Fig.1(a)). In general, the aging kinetics appear to be similar in SIMOX and bulk Si transistors.

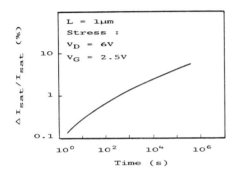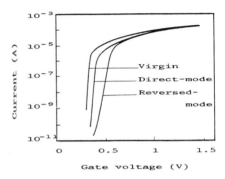

**Fig.1** *Front channel characteristics. (a) Relative variation of the saturation current (measured in the reversed mode for $V_D = V_{G1} = 5V$) versus stress time. (b) Subthreshold curves at $V_D = 5V$, before and after 168 hours of stress.*

Very interesting is the evolution of those properties which are related to the impact ionization current. The *blocking voltage* (defined as the gate voltage leading to current *transients* in weak inversion and for $V_D = 5V$) increases for both the direct and reversed modes of operation (Fig.1(b)). This means that the margin of this critical parameters improves during stress. Furthermore, the *kink effect* is attenuated after degradation as the activation of the parasitic bipolar transistor occurs at a higher voltage and the excess current is reduced. These two phenomena can be explained by a reduction of the ionization rate. As the optimized $N^-$ doping is relatively low, the lateral field is at its maximum under the spacer, leading to electron trapping and interface state generation. Depletion of the $N^-$ region occurs and causes the drain series resistance and related voltage drop to increase. For a fixed value of $V_D$, the high field region extends and the peak field is lowered, so that the impact ionization rate decreases in both direct and reverse modes. This model also accounts for the sublinear law of current degradation (Fig.1(a)). Probing the back channel demonstrates that absolutely no degradation of the buried oxide occurs after stressing the front gate. This result infirms recent speculations (Woerlee et al 1989).

## 3. Back Channel Degradation

Stressing the back channel indicates that the particular nature of the buried oxide makes it very sensitive and vulnerable to hot-carrier injection. In MOSFET's processed on wafers SIMOX–A, the application of an alternating sequence of substrate biasing resulted, even for $V_D = 0$, in substantial oscillations of $V_{T2}$ and $\mu_2$ (Fig.2(a)). The decrease of $V_{T2}$ and $\mu_2$, when applying a positive voltage $V_{G2}$, indicates the drift of positive charges from the lower interface towards the upper interface of the buried oxide. An opposite evolution is observed for $V_{G2} < 0$. This corresponds to the transfer, back and forth in the oxide, of a mobile positive charge $(2 \times 10^{11} cm^{-2})$, which probably originates from accidental contamination during annealing. Such an experiment offers, therefore, an easy way to inspect the material quality.

Although wafers SIMOX–B do not suffer from the above detrimental effects, very large shifts of $V_{T2}$ are induced by hot carrier injection at $V_D = 6V$ and $30V < V_{G2} < 60V$ (Fig.2(b)). They may reach 15 V after only 20 hours of stress at $V_{G2} = 60V$. The shift obeys the time dependence law $\Delta V_{T2} \sim t^n$, where the exponent ($n \simeq 0.16$) is much lower than for bulk Si MOSFET's. A reduction of $V_{G2}$ or a channel length increase clearly attenuates the degradation without modifying this power law. No defect formation was identified in very long channels ($10\mu m$). Also monitored was the threshold voltage shift of the parasitical sidewall transistor. Its degradation is more rapid with time and suggests that many defects are built-up not only at the back interface but also in the bottom of the Si island edges.

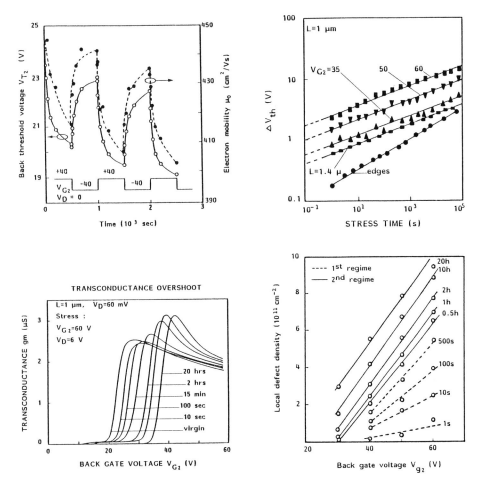

**Fig.2** *Back channel characteristics. (a) Periodic variations of threshold voltage and mobility with $V_{G2}$. (b) Stress-induced threshold voltage shifts. (c) Transconductance overshoot. (d) Defect density in the damaged region versus $V_{G2}$ and time.*

The back channel transconductance $g_{max2}$ (Fig.2(c)) presents a very interesting behaviour: after being reduced in the very first minutes of stress, it then improves substantially during the degradation. This *transconductance overshoot*, which may exceed 40%, has been anticipated earlier and explained by the coupling of the defective and non-defective regions of the channel (Haddara and Cristoloveanu 1988). It happens when the transistor is under the control of the damaged region which is just entering into strong inversion. This model allows extracting the position and density of interface defects and carrier mobility.

It is found that the extension of the damaged region increases rapidly with $V_D$ and can reach 40–50 % of the channel length for $V_D \geq 5.5V$. The defect density depends rather linearly on $V_D$. The variation of the amount of charged defects as a function of stressing time and voltage $V_{G2}$ (Fig.2(d)) gives evidence for a two-step process. Firstly, negative charges are injected, the density of which saturates to a value depending linearly on the oxide field or $V_{G2}$. After a short period, the electron injection becomes independent on $V_{G2}$ and probably results in a local generation of interface states. This second mechanism seems to govern the whole aging process for low values of $V_{G2}$.

The saturation of the first process may be explained by the gradual reduction of the vertical oxide field subsequent to the filling of slow oxide traps. This lowers the level of the highest trap which can be filled, until it equals the uppermost level already filled. Both the model and experiment show that (i) this saturation level depends on $V_{G2}$ and (ii) the initial process should be reversible with a long time constant.

As for the front interface, the kink effect occurs at a higher $V_D$ and the excess current is reduced after the stress. Since the defective region is localized near the drain terminal and the bipolar transistor is governed by the body potential close to the source, the kink effect is more substantially attenuated in the reverse mode of operation. In the reverse mode, all defects are situated above the channel while in the direct mode a portion of them are inefficient being located besides the pinch-off point.

## 4. Conclusions

Systematical results on the aging of SIMOX MOSFET's show that distinction must be made between the properties of the two channels. The front interface presents a good tolerance to hot carrier injection while the buried interface (if activated) is far less resistant. This is not critical for standard operation conditions and simply indicates that the hot carrier degradation of the buried oxide might be chosen as a sensitive criterion for optimizing the quality of SIMOX structures. Preliminary models have been introduced to explain the influence of localized defects on the kink effect.

**Acknowledgements.** This work was supported by various "SIMOX" Projects funded by Esprit, JESSI and GCIS. Many thanks are due to our colleagues, Drs. A-J. Auberton-Hervé, J. Margail, C. Jaussaud, M. Bruel and P. Chatagnon, who have made possible the fabrication of outstanding SIMOX material and devices.

## References

Colinge J-P 1987 *IEEE Trans. Electron Devices* **ED–34** *p 2173*
Haddara H and Cristoloveanu S 1988 *Solid-St. Electron.* **31** *p 1573*
Woerlee P H, van Ommen A H, Lifka H, Juffermans C 1989 *IEDM'89, Technical Digest p 821*

Paper presented at ESSDERC 90, Nottingham, September 1990
Session 3C5

# Study of the enhanced hot-electron injection in split-gate transistor structures

J. Van Houdt, P. Heremans[+], J.S. Witters[°], G. Groeseneken and H.E. Maes

IMEC v.z.w. - Kapeldreef 75 - B3030 Leuven - Belgium

> Abstract. When applying a high voltage to the floating gate of a split-gate device, enhanced hot-electron injection is observed. This phenomenon could be used for 5V-compatible EPROM or Flash EEPROM device operation. The gate current is proven to be equal to the total electron injection current. Charge-pumping measurements and simulations were performed in order to analyse the nature of the injection mechanism in this five-terminal transistor structure.

## 1. Introduction

The channel-hot-electron injection mechanism is widely used as a programming mechanism for EPROM and Flash EEPROM operation. However, from the point of view of programming efficiency, this mechanism is not at all favourable because the conditions for high hot-carrier generation and injection and those for optimal electron collection onto the (floating) gate are incompatible. Indeed, the simultaneous occurrence of a high vertical and a high lateral field can not be obtained in a conventional transistor. This usually results in high gate and drain voltages and a high power consumption, which imposes the use of an external supply voltage to program the device.

By using a configuration with an extra gate, it is however possible to accomplish a combination of a high lateral and a high vertical field (Wu *et al* 1986): a sufficiently high voltage is applied at the gate at the drain side of the device (further called the D-gate), while a low voltage is applied at the gate at the source side of the device (further called the S-gate). This results in an extension of the drain junction by an inversion layer underneath the D-gate. Consequently, the pinch-off condition appears in the region separating the two gates. Because of the high gate voltage at the D-gate, a high vertical field will appear at this "split point".

From our work, it is shown that the increased efficiency of the sidewall-gate concept (Wu *et al* 1986) can also be achieved with a double polysilicon split-gate transistor structure which is compatible with standard CMOS-processes, since no complex spacer technology is required.

## 2. The enhanced injection phenomenon

Fig.1a shows the device structure while fig.1b displays the substrate current ($I_{sub}$) and the gate current measured at the D-gate ($I_g$) as a function of the S-gate voltage ($V_{sg}$), for a D-gate voltage ($V_{dg}$) of 9V and a drain voltage of 5V. A gate current of 1nA/μm device width is observed at an S-gate voltage of only 1.5V. The gate-voltage dependence of the gate current is very similar to that of the substrate current. This is an indication that the gate current is equal to the hot-electron injection current, generated in the channel region. In conventional transistors

---

[*] part of this work has been carried out under the ESPRIT-APBB project 2039

[+] P. Heremans is a Senior Research Assistant of the Belgian National Fund for Scientific Research
[°] J.S. Witters has joined MIETEC, Oudenaarde, Belgium

© 1990 IOP Publishing Ltd

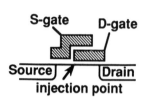

Fig.1a The split-gate transistor structure

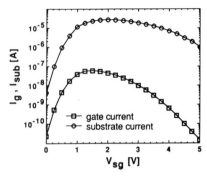

Fig.1b Gate and substrate currents as a function of the S-gate voltage for a D-gate voltage of 9V and a drain voltage of 5V

the gate current would be typically several orders of magnitude smaller and would reach its peak value at a gate voltage more or less equal to the drain voltage. This is due to the repulsive vertical field for gate voltages smaller than $V_d$, which repels the injected hot electrons back into the channel. In the present case however, the vertical field is always highly favourable for electron collection onto the D-gate. According to Hu *et al* (1985), the injection efficiency I and the multiplication factor M can be written respectively as:

$$I = \frac{I_g}{I_d} = C_1 \exp\left(\frac{-\Phi_b}{q\lambda E_m}\right) \quad (1)$$

$$M = \frac{I_{sub}}{I_d} = C_2 \exp\left(\frac{-\Phi_i}{q\lambda E_m}\right) \quad (2)$$

In these expressions, $\Phi_b$ is the effective energy-barrier height for electrons at the Si/SiO$_2$ interface, $\Phi_i$ is the impact-ionization energy, q is the electron charge, $E_m$ is the lateral electric field at the drain and $\lambda$ is the electron mean free path. Notice that the first relation does not include the influence of the vertical field. For conventional MOS-transistors, this means that equation (1) is only valid in the linear regime ($V_g > V_d$), since for $V_g < V_d$, the injection efficiency drops down to zero because of the repelling oxide field (Hu *et al* 1985). From (1) and (2), the injection efficiency can be written as a function of the multiplication factor:

$$I = C_3 M^{\frac{\Phi_b}{\Phi_i}} \quad (3)$$

When plotting this relationship on a bilogarithmic scale, the obtained slope is equal to the ratio of $\Phi_b$ and $\Phi_i$. Since $\Phi_b$ is equal to 3.2 eV for an oxide field equal to zero, a good estimate of $\Phi_i$ can be obtained from this slope (Hu *et al* 1985). Fig.2 shows this plot for the measurements on the split-gate device of fig.1a. A constant slope of 2.2 is now obtained for $V_{sg}$ going from 0 up to 6V. The corresponding $\Phi_i$-value is equal to 1.5 eV. Consequently, equation (3) now holds in the linear as well as in the saturation regime ($V_g < V_d$), and one can conclude that all carriers which are injected towards the D-gate, are also collected at the D-gate. The measured gate current of fig.1 is thus indeed equal to the electron injection current at the split point.

3. Charge-pumping measurements

Fig.3 shows the charge-pumping characteristics, measured by pulsing the D-gate, while keeping the S-gate grounded. In this way, only the part of the channel that is controlled by the D-gate is analysed. The measurements (which are shown for the fresh device and for stress

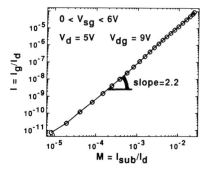
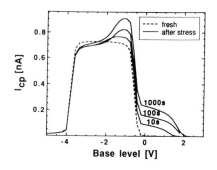

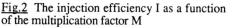

Fig.2 The injection efficiency I as a function of the multiplication factor M

Fig.3 Charge-pumping characteristics for the D-gate area before and after stress

conditions of $V_{sg}$=4V, $V_{dg}$=9V, $V_d$=6V after respectively 10s, 100s, and 1000s) reveal a local increase in the interface-trap density and a large amount of trapped negative charge, which is indicated by the increase of the charge-pumping current at high base-level voltages (Heremans *et al* 1989). This shows that electrons have been trapped underneath the D-gate. An analogous experiment to examine the channel area under the S-gate showed no such degradation.

More information on the location of the generated interface traps and trapped electrons can be obtained by performing CP-measurements using pulses at S-gate and D-gate simultaneously, and by disconnecting either source or drain. Fig.4a shows the CP-characteristics when both gates are pulsed simultaneously, while the source contact is left open. After stress, it is noticed that a portion of the interface traps, present in the fresh device, are not included in the CP-signal at low base levels ($V_{base}$<-2V). Indeed, by disconnecting the source, the only supplier of minority carriers is the drain. For low base levels, the trapped negative charge at the split point prohibits minority carriers to flow under the S-gate because it forms a potential barrier for the electrons which are flowing from the drain into the channel area. Only when the gate pulse exceeds the local threshold voltage in the region of trapped negative charge, the S-gate region starts to contribute to the CP-current. The longer the device has been stressed, the higher the amount of negative charge that has been trapped at the split point and consequently the higher the base-level voltage needed for the S-gate region to contribute to the CP-current (fig.4a).

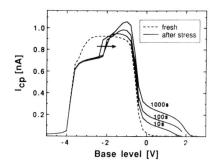
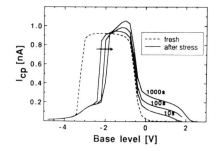

Fig.4a CP-characteristics when pulsing both gates while disconnecting the source, before and after stress ($V_{sg}$=4V, $V_{dg}$=9V, $V_d$=6V)

Fig.4b CP-characteristics when pulsing both gates while disconnecting the drain, before and after stress ($V_{sg}$=4V, $V_{dg}$=9V, $V_d$=6V)

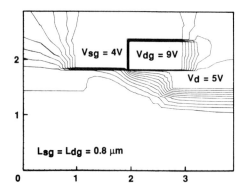

Fig.5 Equipotential lines for the split-gate device under the shown conditions. Scales are in μm. The length of the S-gate region $L_{sg}$ and the D-gate length $L_{dg}$ are both equal to 0.8 μm.

A similar result is obtained when leaving the drain contact open, as shown in fig.4b: now the contribution of the D-gate region to the CP-current is inhibited, since the carriers have to be provided by the source. At low base levels ($V_{base}$<-2V), the curve starts off as if only the S-gate were examined, but at a certain voltage (which is again higher when the stress time increases) the barrier at the split point is "exceeded" and electrons from the source are able to flow into the D-gate-controlled part of the channel.

We conclude that charge pumping allows to determine the nature and the position of the hot-carrier degradation: the electron injection occurs under the D-gate, close to the split point, and causes an increase of the interface-trap density and a negative charge build-up.

4. Simulation results

A series of device simulations have been carried out with a home-made device simulator in order to check the previous results. Fig.5 shows the equipotential lines inside a split-gate transistor for the bias conditions shown in the figure. An inversion layer appears underneath the D-gate: the equipotential lines are clearly bent towards the split point which means that pinch-off appears there instead of at the drain. This causes a high lateral field peak, which is combined with a high vertical injection field, caused by the presence of the high gate voltage at the drain side of the device.

5. Conclusion

It has been the main issue of this work to prove the existence of an enhanced injection phenomenon in split-gate transistors with contacted floating gate. The resulting gate current is shown to be equal to the electron injection current, which cannot be measured in conventional MOS-devices. A highly attractive EPROM or Flash EEPROM concept is suggested here since the maximum hot-carrier injection efficiency is obtained with a minimum of technological complications. Charge pumping is very effective to determine the nature and the position of the hot-carrier injection. Simulations have been shown to confirm the experimental results.

References

Heremans P et al 1989 *IEEE Trans. Electron Devices* **36** p1318
Hu C et al 1985 *IEEE Trans. Electron Devices* **32** p375
Wu A et al 1986 *IEDM Tech. Dig.* p584

# Electron conduction and charge trapping behaviour of $SiO_2$ prepared by plasma anodisation

J.F. Zhang, P. Watkinson, S. Taylor, W. Eccleston and N.D. Young*

Department of Electrical Engineering and Electronics,
University of Liverpool, P.O. Box 147, Liverpool, L69 3BX.

*Philips Research Laboratories, Cross Oak Lane, Redhill, Surrey.

<u>Abstract</u>  Fowler-Nordheim tunneling of electrons dominates the electron conduction mechanism in low temperature plasma grown silicon dioxide. Avalanche injection studies yield typical trap capture cross-sections of $10^{-15}$ and $10^{-17}$ $cm^2$ following anodisation. If a high temperature post-oxidation anneal is used then the only capture cross-section observed has a value $10^{-18}$ $cm^2$.

## 1. Introduction

The advantages of forming dielectrics at low temperatures for advanced VLSI has been recognised for many years. Recently efforts have been made to improve the quality of plasma grown SiO2 through optimisation of the dielectric breakdown strength and reduction of oxide fixed charge and interface trap density (Taylor et al 1988, Zhang et al 1989). Plasma anodisation is also of importance for structures like superlattices, and the low temperature oxidation of Si/Ge has been described in some detail (Hall et al 1989). Poly-Si TFT's and delta doped FET's can also benefit from the use of the technique (Mattey et al 1990, Young and Gill 1990). However there is little information available on the electron conduction and charge trapping behaviour in plasma grown SiO2. The properties of the oxides are compared with those of thermally grown SiO2, prepared on substrates from the same wafer batch.

## 2. Experimental

The silicon substrates used were Boron doped, orientation [100], resistivity 0.17-0.23 ohm-cm supplied by Wacker Ltd. The thermal oxide samples were grown at 900°C in dry oxygen to a thickness of 48 nm and the plasma oxide samples were grown at less than 100°C to a similar thickness in an oxygen plasma. MOS capacitors (MOSC) of various diameter were formed by evaporation of high purity aluminium through a shadow mask. I-V measurements were performed using an automated Keithley type 617 electrometer and voltage source. Avalanche injection was performed using a 50 kHz rectangular waveform with the amplitude automatically adjusted to maintain a constant average avalanche current density.

## 3. Results and Discussion

### 3.1 Conduction mechanism in plasma oxides

Figure 1 shows a typical current density versus electric field plot for a thermally grown and a plasma grown oxide. Shown also is the theoretical curve assuming Fowler-Nordheim (F-N) tunneling of electrons from the Aluminium gate electrode into the oxide, as described by Lenzlinger and Snow (1969).

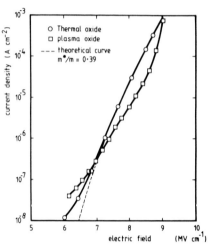

Fig. 1  J against E for thermal and plasma grown oxides

As can be seen the thermal oxide result is coincident with the theoretical prediction for much of the curve, whereas the plasma oxide shows some departure at both lower and high field strengths. To determine whether the electron conduction method in the case of the plasma oxide is due to F-N tunneling or some other mechanism such as Poole-Frenkel (P-F) conduction (Frenkel 1938), it was decided to investigate both the temperature dependence of the conduction and the effect of the electrode material. Figure 2 shows the ratio of J/E versus reciprocal temperature at a fixed field of E = 6.63 MV/cm. From this plot it can be readily seen that the conductivity varies only slightly with temperature in the range 20-250°C.

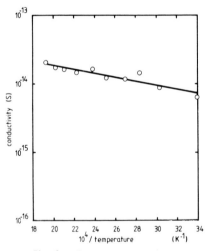

Fig. 2  Temperature variation of J/E for plasma grown oxide

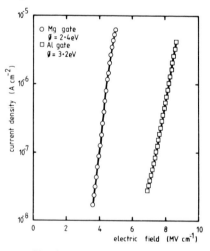

Fig. 3  Electrode dependence of conduction mechanism

This is in contrast to most other conduction mechanisms (Sze 1981) including P-F conduction. From the slope of the curve an effective trap barrier height, which is the depth of the trap potential well in the oxide bulk, may be calculated as $\phi$ = 0.059 eV, if a P-F process is assumed. This is so close to the oxide conduction band edge that at room temperature any trapped electrons would be re-emitted and the conduction process would not be limited by these bulk traps. Further evidence that the conduction mechanism is electrode rather than bulk limited is provided in figure 3 which shows the current density versus electric field curves for two different gate electrode materials deposited on the plasma oxide.

3.2 Avalanche injection

The question still remains as to why the experimental current density versus electric field curve shown in figure 1 departs from the theoretical curve for SiO2 grown by plasma anodisation. To address this problem we consider figure 4 which shows the results of avalanche injection (AI) for thermal and plasma oxide samples. In curve A the midgap voltage shift (dV) versus total injected charge density (Q) for plasma oxide is considerably higher than the corresponding curve for thermal oxide (curve B). However if the plasma oxide is subjected to a post-oxidation anneal (POA) in Nitrogen at 960°C then the midgap voltage shift during AI is considerably reduced (curve C). The AI curves may be fitted using a double exponential fit to the first order trapping equation of Young et al (1979):

$$dV = dV_t [1 - \exp(\sigma Q/q)]$$

where $dV_t$ = total midgap voltage shift if all the traps are filled, and $\sigma$ = trap cross section ($cm^2$). The effective trap density per unit area Neff is related to $dV_t$ by Neff = $(\varepsilon/qd).dV_t$, where d = oxide thickness. The values of Neff and $\sigma$ are listed in Table 1 for the various conditions.

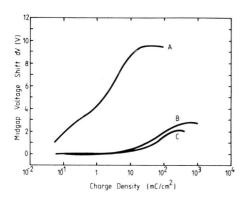

Fig. 4   Avalanche injection results:
A - plasma oxide, B - thermal oxide, C - plasma oxide following 30 min POA in $N_2$ at 960°C

Traps with cross sections $10^{-17}$ to $10^{-18}$ $cm^2$ are thought to be water related. The $10^{-15}$ trap which is eliminated by a short high temperature anneal is possibly due to unsaturated oxygen bonds through incomplete oxidation. Traps with similar cross sections are also observed in other low temperature oxides (Young and Gill 1990) which are also eliminated by a short high temperature anneal in a neutral ambient. It should be noted that the effective density of trapped electrons might be underestimated due to the anomalous positive charge generation during AI (Young et al 1979). Further work is in progress to address this problem.

Table 1  Comparison of trap capture cross sections ($\sigma$) and effective trap densities ($N_{eff}$) for thermal and plasma oxides.

| | $\sigma_1 (cm^2)$ | $N_{eff} (cm^{-2})$ | $\sigma_2 (cm^2)$ | $N_{eff} (cm^{-2})$ | $\sigma_3 (cm^2)$ | $N_{eff} (cm^{-2})$ |
|---|---|---|---|---|---|---|
| Thermal | - | - | $10^{-17}$ | $2.6 \times 10^{11}$ | $1.4 \times 10^{-18}$ | $10^{12}$ |
| Plasma (no POA) | $9 \times 10^{-16}$ | $1.2 \times 10^{12}$ | $3.3 \times 10^{-17}$ | $3.1 \times 10^{12}$ | - | - |
| Plasma (POA 960°C for 1 min in $N_2$) | - | - | - | - | $4.4 \times 10^{-18}$ | $2.7 \times 10^{12}$ |
| Plasma (POA 960°C for 30 min in $N_2$) | - | - | - | - | $1.7 \times 10^{-18}$ | $10^{12}$ |

## 4. Conclusion

The electron conduction mechanism for SiO2 grown by plasma anodisation at low temperatures is seen to be Fowler-Nordheim tunneling of electrons from the gate electrode into the oxide conduction band. This is confirmed by experiments showing the electrode dependence and correct temperature dependence of the conduction mechanism, and is in accord with the results of previous workers for thermal SiO2. The difference between the experimental and theoretical current density versus electric field curves may be explained by enhanced charge trapping in the plasma grown oxide as confirmed from the results of avalanche injection studies. The implication for solid state devices is that using the plasma anodisation technique gate oxides may be grown at less than 100°C with conduction, breakdown and electron trapping properties comparable to conventional thermally grown oxides.

## References

Taylor, S., Kennedy, G. and Eccleston, W., 1988, Vacuum, 38, (8-10), 643

Zhang, J.F., Taylor, S., Watkinson, P. and Eccleston, W., 1989, J. Appl. Surf. Sci., 39, 374

Hall, S., Zhang, J.F., Taylor, S., Eccleston, W., Beahan, P., Tatlock, G.T., Gibbings, C.J., Smith, C. and Tuppen, C., 1989, J. Appl. Surf. Sci., 39, 57

Mattey, N., Dowsett, M.G., Parker, E.H.C., Zhang, J.F., Taylor, S. and Whall, T.E., 1990, submitted to Appl. Phys. Lett.

Lenzlinger, M. and Snow, E.H., 1969, J. Appl. Phys., 40, (1), 278

Frenkel, J., 1938, Phys. Rev., 54, 647

Sze, S.M., 1981, Physics of Semiconductor Devices, (New York: Wiley), pp 402-7

Young, D.R., Irene, E.A., Di Maria, D.J. and De Keersmaecker, R.F., 1979 J.Appl. Phys., 50, 6366

Young, N.D. and Gill, A., 1990 this conference.

Paper presented at ESSDERC 90, Nottingham, September 1990
Session 3C7

# A study of multiplication-induced breakdown in buried-channel p-MOSFETs

T. Skotnicki, G. Merckel and A. Merrachi

CNET-CNS; B.P. 98; 28, chemin du Vieux Chêne; 38243 Meylan; France.

Abstract. The multiplication-induced breakdown (MIB) in buried-channel (BC) P-MOSFETs is studied by means of numerical simulation, thus leading to a better comprehension of its physical mechanism. On this basis an analytical model is derived and verified according to measured breakdown characteristics. The very good accuracy of the model (error of the order of 5%) is demonstrated.

## 1. Introduction

To date, MIB has been associated solely with surface-channel (SC) N-MOSFETs. In BC P-MOSFETs, MIB has been considered to occur beyond the practically used range ($0 \leq V_{DS} \leq 10V$) of drain biases, comp. Ong et.al.(1987), thus resulting in rather scant interest obtained by this phenomenon in the literature. As shown in Fig. 1, illustrating substrate current $I_{sub}$ in a BC P-MOSFET, MIB in the short channel case, may begin at $V_{DS}$ as low as 5V, which indicates a need for a complete study of the phenomenon.

The approach proposed in this paper is an extension of that of Skotnicki et.al. (1989) applied to SC N-MOSFETs. A common physical background of MIB in both types of transistors as well as some essential differences are pointed out and analysed. MIB in BC P-MOSFETs has been found to be composed of two consecutive phases: the first resulting from the multiplication equivalent threshold lowering (METL) and the second from the multiplication induced bipolar injection (MIBI), similarly as is the case for SC N-MOSFETs.

## 2. METL

At the beginning the up-bending of the $I_{source}$ versus $V_{DS}$ characteristic is almost exclusively due to METL, see Fig. 2. The electrons multiplied in the drain domain forward-bias the channel-bulk junction when passing through the bulk spreading resistance $R_{sub}$. This leads to an extra current $I_{METL}$ to appear in the channel, see Fig. 3a, and to add to the ordinary channel current $I_{ch}$, which would occur without multiplication at the same biases (comp. Fig. 2). This extra current can be easily accounted for in terms of a certain threshold voltage reduction (absolute value), being just the essence of METL. In spite of some similarities, METL cannot be replaced by the body effect, as might be desirable at first sight. The incompatibility between these two effects results from the fact that the body effect does not involve any electrical field, whereas METL has to account for the non-zero field in the bulk, necessary to drive the $I_{sub}$ current. As a result, an original model of METL has had to be developed.

© 1990 IOP Publishing Ltd

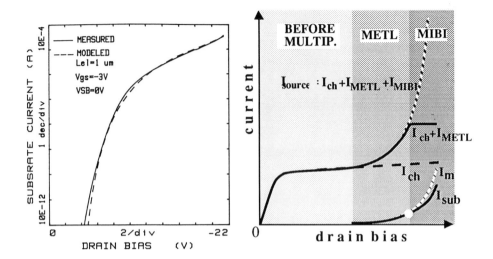

Fig. 1. Measured and analytically modeled $I_{sub}$ versus $V_{DS}$ characteristic for a BC P-MOSFET.

Fig. 2. Balance of currents in the structure of a BC P-MOSFET.

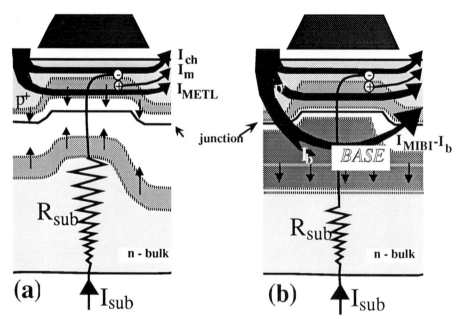

Fig. 3. Physics of METL-(a), and MIBI-(b) phases in a BC P-MOSFET.

3. MIBI

METL increases until the space-charge region practically disappears, which corresponds to the end of the METL phase. In other words the current handling capacity of the channel saturates (current flows through the entire channel opening), and any further increase in $V_{DS}$ inevitably leads to the expansion of the current flow into the bulk, comp. Fig. 3b.

The latter is attained through injection of holes and constitutes what we have called MIBI. In the MIBI phase $I_{METL}$ levels off, as illustrated in Fig. 2, and the further up-bending of the source current $I_{source}$ (broken line) is ensured by the rapidly growing $I_{MIBI}$ component alone.

MIBI is shown to be possible to treat in terms of a parasitic bipolar transistor but only on condition that the variable geometry and unconventional mode of biasing of the base are taken into account. Numerical simulation results obtained with the use of the TITAN5 program, Gerodolle et.al. (1989), show that the size of the base increases considerably with $V_{DS}$, see Fig. 4. Additionally, the base bias $V_{be}$ is found to be equal to $V_{sub}+\delta-V_{SB}$, rather than the base size to be constant and $V_{be}=V_{sub}-V_{SB}$, as might appear at first sight ($V_{sub}=R_{sub}I_{sub}$). Note that $V_{be}=V_{sub}-V_{SB}$ applied to the diode law (emitter diode) $I_e=I_0[\exp(V_{be}/U_T)-1]$ would result in $I_e$=const (no breakdown) since $V_{sub}$ saturates. Only the expanding base together with the $\delta$ term added to the $V_{be}$ expression allow the diode law to be reconciled with measurements. $\delta$ has been found explicitly to be a logarithmic function of $I_{sub}$.

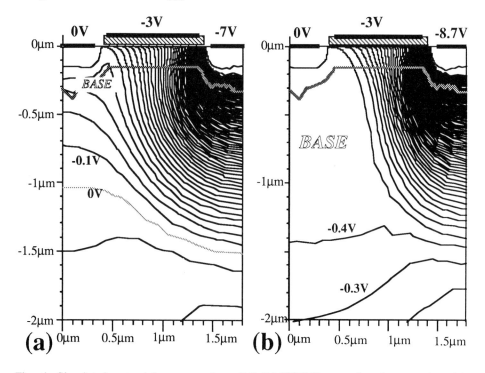

Fig. 4. Simulated potential contours in a BC P-MOSFET operating in a weak - (a), and heavy - (b) multiplication regime.

## 4. Results

The new image of the physics of MIB in BC P-MOSFETs leads to an accurate analytical model. Figs. 1 and 5 show exemplary results of comparison between the model and experiment for $I_{sub}$ versus $V_{DS}$, and $I_D$ versus $V_{DS}$ characteristics, respectively. The mean error over a wide range of channel lengths and biases is of the order of 5%. The latter has been obtained with a constant set of parameters, unique for a given technology.

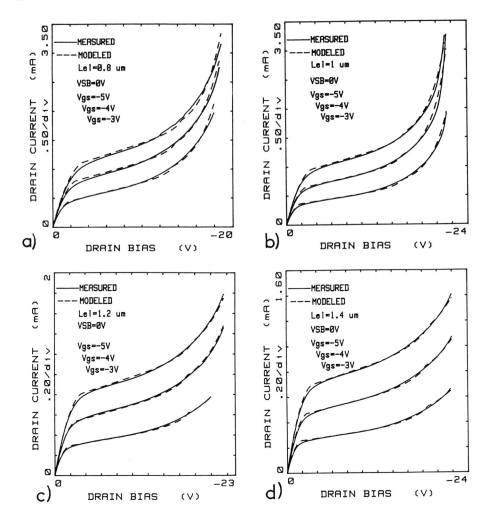

Fig. 5. Measured and analytically modeled $I_D$ versus $V_{DS}$ characteristics for a series of BC P-MOSFETs differing only in electrical channel length $L_{el}$ which reads: (a) 0.8μm, (b) 1.0μm, (c) 1.2μm, and (d) 1.4μm.

## 5. Conclusion

The reported good accuracy and the wide range of validity confirm the correctness of the model. Thanks to these features the model is expected to find applications in the design and optimization of BC P-MOSFETs which has to date, only been available using 2-dim. numerical simulation.

## References

Gerodolle A. et. al. 1989, Proc. NASECODE VI, pp. 56-67.
Ong T.-C. et. al. 1987, IEEE Electron Device Letters, pp. 413-416, vol. EDL-8, No.9.
Skotnicki T. et. al., 1989, IEDM Tech. Digest, pp.87-90.

# Electrical characterisation of ferroelectric thin films for integration into VLSI

D.M. Swanston, D.J. Johnson, D.T. Amm, E. Griswold and M.Sayer

Department of Physics, Queen's University,
Kingston, Ontario CANADA K7L 3N6

Abstract. A variety of electrical measurements for the characterization of ferroelectric thin films are presented. Particular attention is given to experimental details and sample properties that can lead to erroneous results. Ferroelectric fatigue is demonstrated as an end use of a fully characterized system.

## 1. Introduction

The benefits of ferroelectric thin films in non-volatile memory systems [Evans and Womack 1988], as well as in dynamic RAM [Parker and Tasch 1990], are apparent and the importance of accurately characterizing these films has become more crucial. Electrical characterization of these materials is particularly arduous due to their intrinsic non-linearity and hysteretic behaviour. The electrical data presented here are the results from a variety of PZT (lead zirconate titanate) thin films, which are being developed for non-volatile memory devices.

## 2. Electrical Measurements

There are three general types of measurements for characterizing ferroelectric thin films: standard hysteresis loop analysis, pulsed measurements and small signal dielectric response.

A typical PZT hysteresis shown in Figure 1(b) is recorded with a standard Sawyer-Tower circuit (Figure 1(a))[Burfoot and Taylor 1979]. The saturation (maximum) polarisation ($P_s$), remanent polarisation ($P_r$), and the coercive force ($E_c$) indicated in Figure 1(b) are easily extracted.

An example of pulse results and the associated measurement circuit are given in Figure 2. The sequence of four pulses in Figure 2(b) elicits the "switching" and "non-switching" response for both positive and negative polarisation. (The current peaks marked "S" are the switching response and all others are the non-switching response in Figure 2(b)).

The small signal analysis is usually performed with an impedance analyzer in order to give an equivalent parallel capacitance and resistance of the ferroelectric sample. An amplitude study of the dielectric response of the ferroelectric demonstrates the difference between this measurement and the hysteresis and pulse results. Figures 3 and 4 show the equivalent parallel capacitance (proportional to the dielectric permittivity) and the energy dissipated per cycle (the area of the hysteresis loop) as a function of applied voltage amplitude at a frequency of 1 kHz. For amplitudes below about 100 mV, the capacitance is constant and the dissipated energy is practically zero. In the region above 100 mV, domain switching begins, and both the effective capacitance and energy dissipation increase. Thus at

© 1990 IOP Publishing Ltd

low amplitudes, the dielectric response is due to the reversible component of the polarisation, whereas larger signal tests, such as the hysteresis and pulse measurements, are due to the irreversible domain switching.

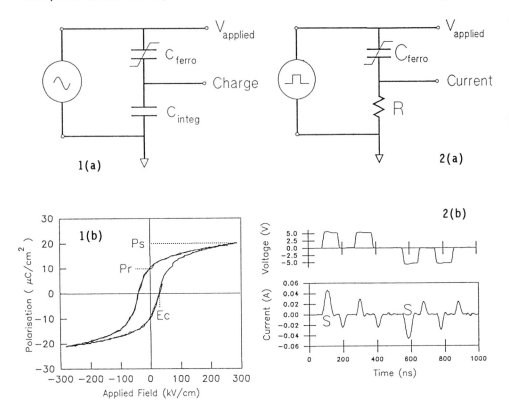

Fig. 1  Hysteresis Loop Measurement

Fig. 2  Pulse Measurement

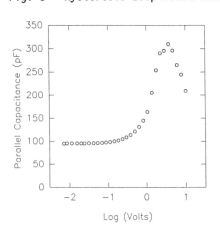

Fig. 3  Amplitude Study of Parallel Capacitance

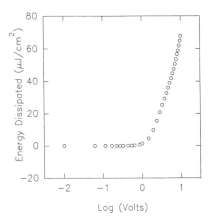

Fig. 4  Amplitude Study of Energy Dissipation

In general, hysteresis and pulse measurements can be classed as
measurements of the large signal, non-linear and irreversible nature of the
ferroelectric. Conversely, the dielectric response probes the small
signal, linear, reversible properties of the material. Thus, each of these
measurements can yield different information about the ferroelectric. For
material characterization, the continuous wave (CW) hysteresis loops give
good measures of $P_r$, $P_s$ and $E_c$. For device applications, pulse responses
better simulate end use. However, although pulse measurements yield good
values for $P_r$ and $P_s$, little information is given about $E_c$.

## 3. Erroneous Results

The possibility of erroneous results obtained from hysteresis loop analysis
is demonstrated in Figure 5. The ratio of the actual to measured remanent
polarisation is recorded as a function of frequency under a variety of test
and sample conditions. The dropoff in measured polarisation below 100 Hz
(curve A) is due to charge leakage from the integrating capacitor into the
measuring instrument (usually a 1 MΩ oscilloscope input). This problem can
be avoided by replacing the integrating capacitor with an active OPAMP
integrator (curve B), but this limits the upper operation frequency and can
be unstable under some conditions. Curve C displays the results obtained
for a "leaky" ferroelectric sample. The roll-off observed above 10 kHz
(curve D) can result from series resistance effects (resistive electrodes)
in the ferroelectric sample. Similar erroneous results are observed in the
coercive field. Ideally, the polarisation and coercive force should be
virtually independent of frequency up to the intrinsic switching time (10
to 100 MHz) but practically, measurements become difficult above 1 MHz due
to matching and noise problems.

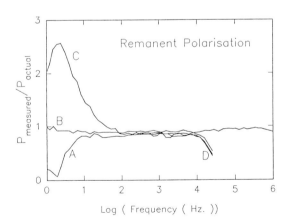

Fig. 5   Measurement Problems in Hysteresis Loop Measurements

Pulse measurements are also susceptible to errors, particularly with
respect to the switching time. For example, a pulse generator of 50Ω
output impedance can only supply 0.1 A current at 5 volts. Thus, a 100 μm
by 100 μm sample with a polarisation of 15 μC/cm² would have an effective
RC time constant of 30 ns, which could easily mask the "intrinsic"
switching time.

## 4. Fatigue

One of the principle reasons for studying the measurement system response is to be able to understand ferroelectric fatigue - the degradation of ferroelectric properties as the polarisation state is continuously cycled between its two states. Early results also indicate that this fatigue is not only dependent on the number of polarisation reversals, but the test system itself. An example of the degradation of various PZT films is indicated in figure 6. The $P_r$ was measured from hysteresis loop analysis. This figure shows a wide variety of material lifetimes, and in sample D, no degradation was measurable up to $10^{12}$ reversal cycles. Although there is no universal explanation for this degradation, it is usually related to the creation of defects, be it ionic conduction or physical defects (such as microcracking). It is plausible to relate the generation of defects to the amount of energy deposited into the films during polarisation cycling. This energy is proportional to the area enclosed in the hysteresis loop. For the sample in Figure 5, the energy dissipated initially was 2.1, 0.85, 0.73, and 0.06 mJ/cm² per cycle for samples S1, S2, S3 and S4 respectively. This correlates quite well with the observed lifetimes and the general trend for long lifetimes has been towards materials with lower polarisation, lower coercive force and as a result, lower internal stress and energy dissipation.

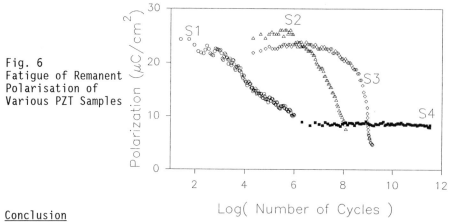

Fig. 6 Fatigue of Remanent Polarisation of Various PZT Samples

## 5. Conclusion

In order to study the mechanisms of ferroelectric fatigue it is essential to thoroughly understand the system response to a variety of sample properties. A number of electrical measurement techniques have been presented, indicating their limits and possible pitfalls during operation.

## 6. Acknowledgements

This work was supported in part by the Natural Science and Engineering Research Council of Canada.

## 7. References

Burfoot J C, Taylor G W, 1979 *Polar Dielectrics and Their Applications* (University of California Press, Berkeley)
Evans J T, Womack R, 1988 *IEEE Journal on Solid-State Circuits* **23** 1171
Parker L H, Tasch A F, 1990 *IEEE Circuits and Devices Mag.* Jan. pp 17-26

Paper presented at ESSDERC 90, Nottingham, September 1990
Session 4IP1

# Recent trends in multilayer process simulation for submicron technologies

A. Poncet, A. Gérodolle and S. Martin
CNET-Grenoble, BP 98, 38243 MEYLAN-Cedex

**Abstract** Modelling and algorithmic aspects of process simulation for submicron technologies are addressed, with a special emphasis on multi-dimensional and multilayer applications. A bibliography is attached, which is quite exhaustive over the last three years period.

## -1-Introduction

Process and device simulators are nowadays widely used to facilitate the development of new technologies. However, the development of such software tools is an endless task: as a result of decreasing dimensions and increasing complexity of processes, many physical and geometrical effects which were considered as second order effects few years ago, must be accurately taken into account in modern simulators. These new requirements concern both bulk substrate and overlayers, and a special care must be taken of interfaces between different materials, because they are the main locations of defects, and consequently anomalous dopant profiles and parasitic effects.

The main advances which will be expected in process/device modelling and simulation in the forecoming decade are:

-1-*physically sound models* for all process steps involved in submicron technologies, including a better knowledge of correlations between process conditions, generation of defects and traps, and electrical behaviour of devices; insofar as these effects depend drastically on equipments, progress on equipment modelling and simulation are becoming mandatory.

-2-*improvements in measurement techniques*: indirect measurements, such as SRP for dopant profiling for instance, can be significantly improved by implementing simulation tools in their built-in software. Similar progress is expected for in-line, in-situ measurements by coupling signal analysis and simulation [8,88].

-3-*facilities for optimal design*: Eventhough software environments to support device design have had a steady improvement in the last few years, the tuning of process parameters is still often left to process designers who, by their own, have to choose the trade-offs among many objectives in competition with one another. However, inverse modelling is improving rapidly [2-8], mainly because of the increasing capabilities of computers, and it is expected that general purpose optimization software tools become very helpful in a near future for both technology optimization and physical parameter fit.

-4-*general pupose process simulators:* the emergence of multilayer silicon technologies, together with needs in process simulation for III-V and II-VI compounds [9,10] will give a strong push to the development of simulators in which a wide variety of materials will be accounted for [11-16].

-5- *fully three-dimensional process-device simulators*: reductions in dimensions of both active and interconnect parts of integrated circuits lead to three dimensional effects. They are taken into account in many device simulators, but, despite few exceptions in Japan [17-19], process simulators did not reach up to now the same level of sophistication, especially for topography: delays between device, process and equipment modelling are usual, for any dimension, and this situation can be related to the facilities for academical research, and to the wide gap in geometrical problems between two and three dimensions. It is worthwhile to say that computers are more powerful for solving algebraic equations than geometrical problems!

---

* This work has been partially supported by the E.E.C. under ESPRIT regulation (E-2197 project)

In this paper, we will emphazise on multi-dimensional and multi-layer aspects of process simulation. Although they are on the critical path towards the development of modern simulation tools, these aspects are less developed than physical modelling or device simulation problems in the literature.

## 2- Process Modelling / Process Simulation : where is the limit ?

Process modelling activity consists in translating physical phenomena involved technology, into sets of equations, whereas process simulation consists in solving these equations on computers.

### 2.1 Models
There are three levels of models, each of them providing specific insights, but specific simulation problems as well:

-1- *Particle Models* are the most physically sound models for solid state devices; there are nomore limitations to the use of particle models in process/device simulation. They are used in traditional domains such as device modelling and ion implantation [20-29], but also in epitaxial growth [97], deposition [94-96], etching [112], and electron beam lithography [117]. Numerical solutions are computed following stochastic (Monte Carlo) or (less frequently) determinist approaches. Cellular algorithms for topography can be considered as "macro-particle" models, insofar as they do not involve continuous equations.

-2- *Partial Differential Equations* have been initially used for solving field problems (thermal transfer, electrostatic potential, structural mechanics,...) which remain valid at any range. Next, they have been applied successfully to mechanisms which are more specific to solid state devices, such as diffusion or light propagation in amorphous or crystalline solids,... but the atomic scale determines the limit of validity of PDEs in solids[1].

-3- *Analytical Expressions* are very helpful in process simulation, when models are either only phenomenological laws, or so expensive that they can be used for computing standard solutions, but not for optimizing complete processes. In that case simplified analytical models can be fitted once for all and implemented in a process simulator.

### 2.2 Algorithms
- *Particle Models*: Monte-Carlo simulations are usually limited by the power of computers, because the number of particles involved must be significant anywhere in the structure to investigate. Let take the simulation of implantation tails as an example: concentrations of interest are several orders of magnitude lower than mean concentration in the structure; it follows that, in some circumstances, millions of particules must be introduced! In order to overcome this requirement, deterministic particle models have been developed for solving Boltzmann transport equations in devices [119]; in our knowledge, they have never been applied to process simulation.

- *Partial Differential Equations:* Process simulations require time and space discretizations of PDEs involved, and the resolution of resulting algebraic systems. Such systems are usually non-linear, and then require iterative procedures to solve them. Stability and consistancy of dicretization schemes (Finite Elements, Finite Differences, time integration) and convergence properties of linearization procedures are the main qualities of numerical algorithms.

Many trends have been made to transform some of PDEs involved in process modelling into integral equations, by using Green's formulas, and to solve them whith Boundary Element Method (BEM) [47,48, 64,70, 73,74]. But, whereas avoiding mesh generation in 2D, this approach has severe limitations which make it unsuitable for non linear problems.

### 2.3 Software
The major challenge in process simulator development is to provide software tools with the following characteristics:

-1- *Robustness*: a simulator must be able to anticipate problems which may occur from unexpected situations: it must not stop with a fatal error message when a process can not been simulated, but it must say *why* this process can not been simulated!

*-2- Accuracy and Confidence*: process designers usually wish numerical errors to be negligible in comparison with errors coming from physical models, but, *in fine*, they wish both to be as small as possible!

*-3- Friendliness*: a careful attention must be payed to the design of user interfaces,

*-4- Open software architecture*: a program structure which leads to a straightforward implementation of new models is obviously recommended.

*-5- Efficiency*: process simulators are usually much less CPU time consuming than device simulators; however, the increasing number of process steps which require sophisticated physical models leads to take care of this problem. Parallel computers may lead to significant reduction of time consumption in Monte-Carlo simulators, whereas pipeline architectures are more suited to the discretization of partial differential equations.

## 2.4 Numerical accuracy

Although requirements on numerical accuracy do not make any difficulty in one dimension, they are more difficult to meet in multi-dimensional process simulation: the critical stage is the generation of suitable meshes, and several attemps have been made recently to improve simulation results without increasing drastically CPU time. For that purpose, adaptive mesh generators have been developed, which are able to automatically refine meshes in critical areas. This task becomes very complex when several mechanisms are involved at the same time: for instance, moving boundaries, mechanical stress relaxation and dopant diffusion produce conflicting constrains on meshes during oxidation; *a priori* error estimates are difficult to evaluate in such cases, and the translation of academic investigations into industrial tools is somehow hasardeous; in practice, *a posteriori* error estimates are much easier to implement in process simulators [54,118].

# 3- Ion Implantation

Analytical models are widely used in process simulators for computing as-implanted dopant profiles. However, Monte Carlo simulation is the basic tool for investigating ion implantation; it offers a very confident way of fitting analytical models. Recent progresses come from improvements in both characterization and simulation: 2D profiling is becoming more accurate [30,46], and Monte Carlo simulators tend to be:
- valid for both crystalline and amorphous targets, in a wide energy range [26,27],
- able to predict the lateral spread of dopants [28], even in non planar and multi-layer structures, as well as induced point defects [29];
- faster [20,25]: As mentionned in section 2, different attempts have been made to reduce huge CPU time consumption, for instance by anticipating on the trajectories, in order to determine more closely the area to investigate around each particle [20].

# 4- Dopant Diffusion

Dopant diffusion is the mechanism which has been the most deeply investigated in the past decade: diffusion models involving point defects are nowadays commonly used in process simulators, although some controversies still remain on the respective influences of vacancies and intersticials and the more appropriate way of simulating point defects [31-44]. However, dependencies versus initial conditions (i.e. as-implanted profiles) and non-equilibrium defects generated on interfaces become the dominant factors in rapid thermal processing.

On a physical point of view, it means that bulk diffusion can no longer be accurately predicted without getting more accurate data from ion implantation and oxidation. On the other hand, diffusion in and from overlayers (polysilicon, silicides, epitaxial layers,...) require new models to predict point defects generation along interfaces.

On a numerical point of view, it results from shorter annealing times that "smoothing" properties of diffusion operators are less effective and that more attention must be payed to initial and boundary conditions. Standard discretization schemes are Finite Element or Finite Difference methods; improvements to these classical methods [50-56] and alternative methods [47-49] have been recently investigated, with a special attention to adaptive mesh

generation problems[53-56]. However, "old" numerical problems in dopant diffusion must be kept in mind; they can be illustrated by some (non exhaustive) examples of this paradox:
*how to spend more CPU time and get worse results ?*
- by using a coarse mesh in an oxide layer, instead of appropriate boundary condition at the Si/SiO$_2$ interface without mesh in the oxide;
- by using high order quadrature rules instead of "mass lumping" on time derivatives, and then getting negative concentrations.

It comes out that process simulators for the nineties must become expert systems.

### 4.1- Bulk diffusion

Rather accurate results can be obtained in 1D dopant diffusion, when no precipitation occurs. In this case computer programs which solve coupled impurities-intersticials-vacancies diffusion can be used successfully, because initial and boundary values for point defects can be easily fitted [38,44]. The most illustrative exemple is the accurate simulation of phosphorus diffusion. However, recent progress in 2D profiling [46] reveal large discrepancies between measurements and 2D simulations [44] which result from lacks of accurate models for defects generated by ion implantation and oxidation.

High doses effects are accounted for by simulating precipitation mechanism [57].

### 4.2- Overlayers

-1-polysilicon Two mechanisms are involved in dopant diffusion in polysilicon layers: diffusion inside grains, and diffusion at grain boundaries. 2D local models have been developed [58,59], and further simplified to provide 1D macroscopic model, in which the following quantities are evaluated after homogenization: grain size, and, for each dopant, volume concentration and grain boundary concentration [59]. This model has been extended in 2D for non planar polysilicon layers [60], and completed with a model to predict grain growth during re-crystallization [61].

-2-silicides Diffusion modelling in silicides/silicon bilayers is at an earlier stage of development [62], because of the difficulty of problems involved: for instance, stability of metal-silicon-dopant systems requires thermodynamical studies, capping layers must be taken into account,...; moreover, the implementation of doped Silicon consumption requires careful attentionin 2D. TiSi2 and CoSi2 characterizations and Monte Carlo simulations have been reported in [63].

### 4.3- Thermal analysis

Thermal simulations are very helpful during rapid thermal annealing for studiing temperature discrepancies on wafers and resulting lacks in process reliability. Although thermal analysis is strongly related to equipment modelling [65,66], thermal simulations must be mentionned here, insofar as they are an efficient way of providing RTP simulators with input temperature profiles [67,68]. Moreover, they can be coupled with thermal stress simulations in order to predict defect onset [69].

## 5- Topography

### 5.1- Basic simulation problems

The main numerical difficulties in topography simulation arise from moving boundaries. In two dimensions, string algorithms have been developed for a long time for deposition, etching [89] and lithography [114]. They consist in moving points on boundaries or interfaces with a given velocity, and refreshing the topology at each time step accordingly. A rigorous analysis of this method has been performed in the case of sputtering, where the yield depends only versus incidence angle [91,92].

The extension of such an approach to three dimension simulation is tremendous; therefore, recent trends are to overcome geometrical problems by using alternative methods such as particle and cellular methods.

In many circumstances, kinetics are defined by physical models which involve partial differential equations to be solved together with moving boundaries. These equations apply either on the material itself (oxidation, light propagation), or in the surrounding sheath (plasma assisted processes); in both cases, many difficulties are met: different mechanisms are usually involved simultaneously, such as magneto-hydrodynamics together with diffusion/reaction equations; moreover, in the second case, problems depend drastically on equipment parameters, and must often be treated in 3D [113] or cylindrical 2D.

## 5.2- Oxidation of Silicon and Stress Analysis

Numerical simulations of local oxidation have been reported in the literature for a long time [70,71]; they allow realistic predictions of the shape of "bird's beaks" in non planar structures, that is impossible by using analytical profiles. However, accurate predictions cannot be achieved without accounting for mechanical stresses which reduce drastically oxidation kinetics under thick masks or at trench corners.

Despite a large number of publications on both modelling and numerical aspects, few simulation results have been presented up to now, which account accurately for stress effects on complex non planar structures [76,82]; this is probably due to early stage of development of physical models, as well as numerical unstabilities and geometry modelling problems which are somehow difficult to overcome:

-*modelling*: different expressions of diffusivity, reaction rate and viscosity versus mechanical stresses in oxide layers have been proposed [75,80]; some authors have introduced upper bounds in tensile stress enhancement [78], that is not rigorously motivated. More recently, the well known Kooi's effect has been schematically introduced to reduce more drastically oxidation velocity under nitride masks [81].

-*numerical schemes*: three kinds of numerical algorithms have been explored in 2D: -1- boundary elements [64,70, 73,74], -2- finite elements on irregular meshes [71], -3- finite elements (or finite differences) on regular grids obtained by conformal mapping [78,83,84], Introducing stress effects in the first approach is quite impossible, whereas stress relaxation requires numerous interpolations in the second one [82]. Recent trends with the third approach are to avoid complex geometrical manipulations [83,84]: $Si/SiO_2$ interfaces are not updated at each time step; Si and $SiO_2$ layers are traited as a unique unhomogeneous material. This approach looks promising in view of an extension to 3D.

Finally, the most advanced simulators are able to account for the elastic behaviour of silicon [78,84], which is mandatory in trench re-oxidation, Polysilicon-Buffered LOCOS, or SOI process modellingg. Simulations of non uniform thermal expansion in $Si_3N_4/SiO_2/Si$ trilayers are expected in a near future, in order to complete stress analysis and predict more accurately oxidation induced defects in silicon.

## 5.3- Planarization

Although recent attempts to planarize interconnexion layers at low temperatures by using sputtering mechanisms are successful, glass reflow of BPSGs is still widely used in CMOS technology in order to smooth contact holes and gate overlaps. Therefore, it is convenient to provide simulation facilities for both planarizing deposition (Section 5.4) and glass reflow.

The standard model used for glass reflow in 2D process simulators is an incompressive viscous flow, and the driving force is the surface tension, which is proportional to the inverse of curvature radius [85-87]

The poor reproducibility of highly doped glasses led to the development of an optical in-situ method to control of their flow rate, by analysing diffraction diagrams over test patterns [88]. Comparisons between experiments and simulations permit viscosity measurement.

## 5.4- Deposition and Epitaxy

Improvements from existing deposition models [89,90] are expected in two directions: towards kinetics which strongly depends on the equipment, and towards simulation of local geometrical effects. Simulations of 2D effects in deposition are usually based on purely geometrical models which determine velocity vectors to be applied for moving boundary points in suitable string algorithm. For investigating deposition kinetics, rather than 2D effects during PECVD, Monte Carlo simulations give insights on mechanisms involved at atomic scale, as reported in [93,94]. Similar approaches have been followed for balistic sputter deposition [95] and epitaxial growth [97].

## 5.5- Etching

Etching rate, anisotropy and selectivity are, besides uniformity, the main etching characteristics to be controlled in submicron technologies. In particular, processes with high aspect ratios and large selectivities are now developed, and several explanations can be found for the observed various two-dimensional effects (dove-tail, barelling, trench size effect...).

Since the first simulator for etching process [90], many simulation studies have been performed using different calculation methods and physical models [98-105]. Most of them, however, use a model of the sputter etching type, in which either the total etch rate, or the anisotropic part, is proportional to the ion flux or to the deposited energy flux. All of them introduce an angular distribution of energetic particles, either gaussian or resulting from the computation of ion-neutral collisions in the plasma sheath. Sidewall passivation is also taken into account [99,102,103,105]. Glancing ions and energy or flux threshold have been accounted for in [101]. It is generally assumed that etching results from the linear summation of up to three independent components, i.e. isotropic chemical etching with uniform distribution, anisotropic ion etching with angular distribution and anisotropic "radical" etching with uniform distribution. The concept of surface diffusion for chemisorbed reactive species has been introduced recently [102,105].

Although some models explain local effects by ion angular distribution, either analytically [109-111], or via Monte-Carlo simulation [112], another model making very simple assumptions on this distribution explains the same effects by the surface migration of adsorbed particles [105-108].

## 5.6- Lithography

*Optical lithography:*

Simulation of photo-lithography is usually splitted into three stages, which are the simulations of aerial image, light propagation into the resist and final resist profile after development [114,115].

Recent trends in modern simulators are to account for non planar substrates and non ideal objectives. The effects of defocusing can be simulated, and corresponding numerical results provide an explanation for asymmetries observed in the focus offset dependence of submicron resist image, when the depth of focus becomes comparable to the layer thickness.

Parallel algorithms have been developed very recently in order to speed up the simulation of light reflection and ligth propagation in resits [116].

*Electron Beam Lithography*

Although investigations have been reported on 3D Monte Carlo simulations [117], electron beam lithography is not a domain of intensive modelling activities. The emergence of new equipments will probably lead to further developments.

## 6- Conclusion

Recent advances in process modelling and simulation have been outlined; it has been shown that the simulation chain is now complete and rather accurate from lithography to thermal annealing. Enlarged investigations on physico-chemical modelling of all mechanisms involved in semiconductor processing, together with faster computers and self-adaptive numerical schemes are expected to make process simulation even more attractive in the future.

**Acknowlegdements.** The authors whish to thank all their co-workers in STORM project for stimulating discussions.

### BIBLIOGRAPHY
General
[1] W. Hänsch and F. Lau "On the physical problems in process and device simulation: Possibilities and limitations" Proc. NASECODE-V Short Course. Dublin, Boole Press Ed., June 1987
Optimization, Inverse Modelling
[2] Y. Aoki, H. Masuda S. Shimada and S. Sato "A New Design-Centering Methodology for VLSI Device Development" IEEE/CAD Vol. 6, No. 3, May 1987
[3] S. Onga and T. Wada "Process Diagnostic and Fluctuation Analysis Using a Two-Dimensional Simulation System with a New Linearized Iteration Method" IEDM'87 Tech. Digest, Washington, Dec. 1987

[4] K. Tanaka and M. Fukuma "Design methodology for deep submicron CMOS" IEDM'87 Tech. Digest, Washington, Dec. 1987
[5] G.J.L. Ouwerling, F. van Rijs, H.M. Wentinck, J.C. Staalenburg, and W. Crans "Physical Parameter Extraction by Inverse Modelling of Semiconductor Devices" Proc. SISDEP Vol. 3, Bologna, Sept. 26-28, 1988.
[6] G.J.L. Ouwerling " Non-Destructive Measurement of 2D Doping Profiles by Inverse Modelling" Proc. NASECODE-VI Conf. Dublin, Boole Press Ed., July 1989
[7] J. Brockmann and S. Director "A Macromodelling Approach to Process Simulator Tuning" NUPAD-III Workshop, Honolulu, June 1990
[8] J. Wenstrand, H. Iwai, M. Norishima, H. Tanimoto, T. Wada and R. Dutton "Intelligent Simulation for Optimization of Fabrication Processes" NUPAD-III Workshop, Honolulu, June 1990

Simulators

[9] E. Caquot, J. Dangla, M. Laporte, J.F. Palmier, A. Marrocco, F. Hecht and K. Souissi "Integration of process, device and circuit models for III-V devices" Proc. NASECODE-V Short Course. Dublin, Boole Press Ed., June 1987
[10] M.D. Deal, S.E. Hansen and T.W. Sigmon "SUPREM 3.5 - Process Modelling of GaAs Integrated Circuit Technology" IEEE/CAD Vol. 8 No. 9, Sept. 1989
[11] J. Lorenz, J. Pelka, H. Ryssel, A. Sachs, A. Seidl and M. Svoboda, "COMPOSITE, a complete modelling program of silicon technology" IEEE/ED Vol. ED-32 No. 10. 1985
[12] K. Lee and A. R. Neureuther "SIMPL-2 (SIMulated Profiles from Layout- Version 2) IEEE/CAD Vol. 7, No. 2, Feb. 1988
[13] B. Baccus, D. Collard, E. Dubois and D. Morel "IMPACT4 - A General Two-Dimensional Multilayer Process SImulator" Proc. SISDEP Vol. 3, Bologna, Sept. 26-28, 1988.
[14] K. Nishi, K. Sakamoto, S. Kuroda, J. Ueda, T. Miyoshi and S. Ushio " A General-Purpose Two-Dimensional Process Simulator - OPUS- for Arbitrary Structures" IEEE/CAD Vol. 8 No. 1, Jan. 1989
[15] B.J. Mulvaney, W.B. Richardson, and T.L. Crandle " PEPER - A Process Simulator for VLSI" IEEE/CAD Vol. 8 No. 4, April 1989
[16] A. Gérodolle, C. Corbex, A. Poncet, T. Pedron and S. Martin "TITAN-5 a two-dimensional process and device simulator" Proc. NASECODE-VI short course, Dublin, Boole Press Ed. July 1989
[17] K. Hane "Supercomputing for process/device Simulation" Proc. NASECODE-VI Conf. Dublin, Boole Press Ed., July 1989
[18] K. Nishi, S. Kuroda, K. Kai and J. Ueda "Efficient Three-Dimensional Process Simulation of MOS Devices" Proc. NASECODE-VI Conf. Dublin, Boole Press Ed., July 1989
[19] S. Odanaka, H. Umimoto, M. Wakabayashi and H. Esaki "SMART-P: Rigorous Three-Dimensional Process Simulator on a Supercomputer" IEEE/CAD Vol. 7 No. 6, June. 1988

Ion Implantation

[20] G. Hobler and S. Selberherr "Efficient Two-Dimensional Monte-Carlo Simulation of Ion Implantation " Proc. NASECODE-V Conf. Dublin, Boole Press Ed., June 1987
[21] P. Luigi "Monte-Carlo Simulation of Semiconductor Devices and Processes" Proc. SISDEP Vol. 3, Bologna, Sept. 26-28, 1988.
[22] G. Hobler and S. Selberherr "Monte-Carlo Simulation of Ion Implantation in Two- and Three-Dimensional Structures" IEEE/CAD Vol. 8 No. 12, Dec. 1989
[23] G. Hobler "Multiple Use of Trajectories in Monte-Carlo Ion Implantation Simulations" Proc. NASECODE-VI Conf. Dublin, Boole Press Ed., July 1989
[24] G. Srinivasan "Monte-Carlo Modelling of Ion Implantation for Silicon Devices" IEDM'89 Tech. Digest, Washington, Dec. 1989
[25] E. Van Schie and J. Middelhoek "Two Methods to Improve the Performances of Monte-Carlo Simulation of Ion Implantation in Amorphous Targets" IEEE/CAD Vol. 8 No. 2, Feb. 1989
[26] A.M Mazzone "Charge states of heavy ions with energy in the MeV range in crystalline semiconductor targets" MRS meeting, Boston, November 1989
[27] A.M Mazzone "Preliminary report on Monte Carlo simulation" STORM deliverable1.c.4.1 -Jan. 1990
[28] J. Lorenz, W. Krueger and A. Barthel " Simulation of the Lateral Spread of Implanted Ions: Theory" Proc. NASECODE-VI Conf. Dublin, Boole Press Ed., July 1989
[29] G. Hobler and S. Selberherr "Two-Dimensional Modelling of Ion Implantation Induced Point Defects" IEEE:CAD Vol. 7, No. 2, Feb. 1988
[30] R. Oven, D.G. Ashworth and C. Hill " Simulation and Measurements of the Lateral Spreading of Ion Implanted into Amorphous Targets" Proc. SISDEP Vol. 3, Bologna, Sept. 26-28, 1988.

Diffusion

[31] M. E. Law and R.W. Dutton "Verification of Analytical Point Defect Models Using Suprem-IV" IEEE/CAD Vol. 7, No. 2, Feb. 1988
[32] M. R. Kump and R.W. Dutton "The Efficient Simulation of Coupled Point Defect and Impurity Diffusion" Vol. 7, No. 2, Feb. 1988
[33] S. Martin, A. Gérodolle and D. Mathiot " 2D Diffusion Models and Nonequilibrium Point Defects" Proc. NASECODE-V Conf. Dublin, Boole Press Ed., June 1987

[34] M. Budil, E. Guerrero, T. Brabec, S. Selberherr and H. Potzl "A New Model for the determination of point defect equilibrium concentrations in silicon" COMPEL Vol. 6, No. 1, March 1987
[35] N.E.B. Cowern and D.J. Godfrey "A model for coupled dopant diffusion in silicon" COMPEL Vol. 6, No. 1, March 1987
[36] R.B Fair "Low-Thermal-Budget Process Modelling with the PREDICT$^{TM}$ Computer Program" IEEE/ED Vol. 35 No. 3, Mar. 1988
[37] M. Orlowski "Progress in Process Simulation for Submicron MOSFETs" Proc. SISDEP Vol. 3, Bologna, Sept. 26-28, 1988.
[38] R. Dürr and P. Pichler "Influence of Initial Conditions on Point Defect Diffusion: Impact on Models" Proc. SISDEP Vol. 3, Bologna, Sept. 26-28, 1988.
[39] M.D. Giles "Defect-Coupled Diffusion at High Concentrations" IEEE/CAD Vol. 8 No. 5, May. 1989
[40] M. Orlowski "A Novel Concise Physically Motivated Algorithm for the Evaluation of Multiphase Diffusion Including Dopant Redistribution at the Interfaces" Proc. NASECODE-VI Conf. Dublin, Boole Press Ed., July 1989
[41] M. Orlowski "A Rigorous model for Dopant-Dopant Pair Diffiusion in Silicon" NUPAD-III Workshop, Honolulu, June 1990
[42] M. Law "Parameters for Point Defect Diffusion and Recombination" NUPAD-III Workshop, Honolulu, June 1990
[43] G. Hobler, S. Halama, K. Wimmer, S. Selberherr and H. Potzl "RTA-Simulations with the 2-D Process Simlulator PROMIS" NUPAD-III Workshop, Honolulu, June 1990
[44] R. Subrahmanyan, H. Z. Massoud and R. B. Fair "Comparison of Measured and Simulated Two-Dimensional Phosphorus Diffusion Profiles in Silicon" J. Electrochem. Soc. Vol. 137, No. 5, May 1990
[45] A. De Keersgieter "Diffusion in Silicon: Boron" STORM Tech. Rep No.1-a.2.i-Jan. 1990
[46] P.J. Pearson and C. Hill "2D boron distribution after ion implant and transient anneal" Proc. ESSDERC'88 Journal de Physique, No. 9 (49) Sept. 1988
[47] M. Hane and K. Hane "Finite and Boundary Element Approach to Proces Simulation with Conjugate Gradient-Based Method" Proc. NASECODE-V Conf. Dublin, Boole Press Ed., June 1987
[48] X.Tian and A.J. Strojwas "A Numerical Integral Method for Diffusion Modelling" Proc. NASECODE-VI Conf. Dublin, Boole Press Ed., July 1989
[49] R.E. Lowther " A Discretization Scheme That Allows Coarse Grid-Spacing in Finite-Difference Process Simulation" Vol. 8 No. 8, Aug. 1989
[50] W. Joppich "A Multigrid Method for Solving the Nonlinear Diffusion Equation on a Time-Dependent Domain Using Rectangular Grid in Cartesian Coordinates" Proc. NASECODE-V Conf. Dublin, Boole Press Ed., June 1987
[51] J. Lorenz and M. Svoboda "ASWR-Method for the Simulation of Dopant Redistribution in Silicon" Proc. SISDEP Vol. 3, Bologna, Sept. 26-28, 1988.
[52] M. Paffrath, and K. Steger "Numerical Solution of Diffusion Equations on Time-variant Domains" NUPAD-III Workshop, Honolulu, June 1990
[53] G. Punz and K. Dienstl "Comparison of Grid Strategies for Point-Defect Diffusion Simulations" Proc. SISDEP Vol. 3, Bologna, Sept. 26-28, 1988.
[54] R. Ismail, and G. Amaratunga "Application of local Neumann Error Criteria for Remeshing in Dopant Diffusion Problems" Proc. SISDEP Vol. 3, Bologna, Sept. 26-28, 1988.
[55] R. Ismail and G. Amaratunga "Adaptive Meshing Schemes for Simulating Dopant Diffusion" IEEE/CAD Vol. 9 No. 2, March 1990
[56] S. Ushio, K. Nishi and J. Ueda "Numerical errors in impurity transport calculations through the boundary of finite-area cells, and a new practical model to reduce such errors" Solid State Electronics, Vol. 33, No. 1 1990
[57] S. Solmi, E. Landi and F. Baruffaldi "High concentration boron diffusion in Silicon. Simulation of the precipitation phenomena" STORM Tech. Rep. No. 1-a.2.g, Jan. 1990
[58] I. Yamamoto, K. Hane, H. Kuwano and T. Suzuki "Boundary Condition for Grain Boundary Segregation" Proc. NASECODE-V Conf. Dublin, Boole Press Ed., June 1987
[59] S.K. Jones and C. Hill "Modelling Dopant Diffusion in Polysilicon" Proc. SISDEP Vol. 3, Bologna, Sept. 26-28, 1988.
[60] S.K. Jones and A. Gérodolle "Implementation of planar Polysilicon model" STORM Tech. Rep No. 1-a.1.g July 1989
[61] S. K. Jones and C. Hill "Report on Recrystallisation of Polysilicon" STORM Tech. Rep. No. 1-a.1.k - Nov. 1989
[62] P.B. Moynagh, A.A. Brown and P.G. Rosser " Modelling diffusion in silicides" Proc. ESSDERC'88 Journal de Physique, No. 9 (49) Sept. 1988
[63] P.D. Cole, G.M. Crean, K. Maex and W Eichhammer " Report on Boron diffusion from Titanium silicide" STORM Tech. Rep. No. 1-a.1.l -Jan. 1990
[64] C.C. Lee, A.L. Palisoc, and J.M.W. Baynham "Thermal Analysis of Solid-State Devices Using the Boundary Element Method" IEEE/ED, Vol. 35, No. 7, July 1988

[65] T.-J. Shieh and R.L. Carter "RAPS-A Rapid Thermal Processor Simulation Program" IEEE/ED, Vol. 36, No. 1, Jan. 1989
[66] R. Kakoschke, E. Bussmann and H. Föll "Modelling of Wafer Heating Dring Rapid Thermal Processing" Appl Phys. A 50 pp141-150 (1990)
[67] S.K. Jones, C. Hill and D. Boys "Modelling temperature control and temperature uniformity for rapid thermal annealing" in Reduced Thermal Processing for ULSI (R.A. Levy ed.) Plenum Press,New-York (1989)
[68] S. K. Jones, and A. Gérodolle "TITAN-RTA, a 2D integrated equipment and process model for simulation rapid thermal processing" these Conference
[69] K. Mokuya and I. Matsuba "Prediction of Defects Onset Conditions in Heat Cyclic Based on a Thermoelastic Wafer Model" IEEE/ED, Vol. 36, No. 2, Feb. 1989

Oxidation/Stresses

[70] D.J. Chin, S.-Y. Oh, and R.W. Dutton "A General Solution Method for Two-Dimensional Non Planar Oxidation"IEEE/ED, Vol. ED-30n No. 9, Sept. 1983
[71] A. Poncet "Numerical Simulation of Locval Oxidation of Silicon" IEEE/CAD, Vol. 4, No. 1, Jan. 1985
[72] P. Sutardja, Y. Shacham-Diamand and W.G. Oldham "Simulation of Stress Effects on Reaction Kinetics and Oxidant Diffusion in Silicon Oxidation" IEDM'86 Tech. Digest, Los Angeles, Dec. 1986
[73] S. Isomae and S. Yamamoto "A New Two-dimensional Silicon Oxidation Model" IEEE/CAD, Vol. CAD-6, No. 3, May 1987
[74] T.-L. Tung, J. Connor and D. A. Antoniadis "A viscoelastic BEM for modelling oxidation" COMPEL Vol. 6, No. 2, June 1987
[75] D.-B. Kao, J. P. McVittie, W. D.Nix and K.C. Saraswat " Two-dimensional Thermal Oxidation of Silicon -II. Modelling Stress Effects in Wet Oxides" IEEE/ED Vol. ED-35, No. 1, Jan. 1988
[76] A. Seidl, V. Huber and E. Lorenz "Implementation of Models for Stress-Reduced Oxidation into 2-D Simulator" Proc. SISDEP Vol. 3, Bologna, Sept. 26-28, 1988.
[78] H. Umimoto, S. Odanaka, I. Nakao and H. Esaki " Numerical Modelling of Non-Planar Oxidation Coupled with Stress Effects" IEEE/CAD Vol. 8 No. 6, June. 1989
[79] T. Kato and K. Hane "Viscoelastic treatment of Two-Dimensional Oxidation of Silicon" Proc. NASECODE-VI Conf. Dublin, Boole Press Ed., July 1989
[80] P. Sturdja and W.G. Oldham. IEEE/ED, Vol ED-36, No. 11, Nov. 1989
[81] N. Saito, H. Miura, S. Sakata, M. Ikegawa and T. Shimizu "A two-Dimensional Thermal Oxidation Simulator using Visco-Elastic Stress Analysis" IEDM'89 Tech. Digest, Washington, Dec. 1989
[82] A. Poncet "Numerical oxidation module. First release" STORM Tech. Rep. No. 1-b.2.b -Jan. 1990
[83] E. Rank "A New Finite Element Approach to the Local Oxidation of Silicon" Proc. NASECODE-VI Conf. Dublin, Boole Press Ed., July 1989
[84] E. Rank and U. Weinert " A Simulation System for Diffusive Oxidation of Silicon: A Two-Dimensional Finite Element Approach" IEEE/CAD Vol. 9 No. 5, May. 1990

Planarisation, Deposition, Epitaxy, Etching.

[85] P. Sutardja, Y. Shacham-Diamond and W.G. Oldham "Two-Dimensional Simulation of Glass-Reflow and Silicon Oxidation" Proc. Symp. on VLSI Technology, San Diego, 1986
[86] A. Tissier, A. Poncet, and J.F. Teissier "Glass reflow modelling for process optimization" Proc. ESSDERC'87 Conf. Bologna, Sept. 14-17, 1987
[87] F. A. Leon "Numerical Modelling of Glass Flow and Spin-On Planarization" IEEE/CAD, Vol. 7, No. 2, Feb. 1988
[88] A. Tissier, A. Poncet, G. Giroult, D. Maystre and P. Vincent "A non-destructive method for in-line glass flow control" Proc. V-MIC Conf. June 13-14, 1988
[89] W.G. Oldham, A.R. Neureuther, C. Sung, J.L. Reynolds and S. Nandgaonkar"A general Simulator for VLSI lithography and Etching Processes: Part II - Application to Deposition and Etching " IEEE/ED, Vol. ED-27, No. 8, Aug. 1980
[90] A. R. Neureuther "Algorithms for Wafer Topography Simulation" Proc. NASECODE-IV Conf. Dublin, Boole Press Ed. June 1985
[91] R. Smith, S.J. Wilde, G. Carter, I.V. Katadjiev and M.J. Nobes "The simulation of two-dimensional surface erosion and deposition processes" J. Vac. Sci. Technol. B. Vol. 5, No. 2 Mar-Apr. 1987
[92] M.J. Nobes, I.V. Katadjiev, G. Carter and R. Smith " Analytic, geometric and computer techniques for the prediction of morphology evolution of solid surfaces for multiple processes" J. Phys. D: Appl. Phys. Vol. 20, pp 870-879 (1987)
[93] M.J. Kushner "A model for the discharge kinetics and plasma chemistry during plasma enhanced chemical vapor deposition of amorphous silicon" J. Appl. Phys. Vol. 63, No. 8, Apr. 1988
[94] M.J. McCaughey and M. Kusher "Simulation of the bulk and surface properties of amorphous hydrgenated silicon deposited from silane plasma" Appl. Phys. Vol. 65 (1989)
[95] T. Smy, K.L. Westra and M.J. Brett "Simulation of Density Variation and Step Coverage for a Variety of Via/Contact Geometries Using SIMBAD" IEEE/ED Vol. 37, No. 3, March 1990
[96§§] M. Gross and C.M. Horwitz "Modelling of sloped sidewalls formed by simultaneous etching and deposition" J. Vac. Sci. Technol. B. Vol. 7, No. 3 May-June 1989

[97] D. Estève, M. Djafari-Rouhani, V.V. Pham, A. Amarani and J.J. Simonne "Molecular Mechanics and Monte-Carlo Simulations, a Tool to Analyse Defects Formation in Heteroepitaxial Growth" Proc. SPIE-88 Conf. New Port Beach, March 1988

[98] S.Yamamoto, T.Kure, M.Ohgo, K.Matsuzama, S.Tachi and H.Sunami, IEEE/CAD, Vol. 6, 417 (1987)

[99] J. Pelka, K.P. Müller and H. Mader "Simulation of Dry Etch Processes by COMPOSITE" IEEE/CAD Vol. 7, No. 2, Feb. 1988

[100] J.I.Ulacia, C.J.Petit and J.P.Mc Vittie, J.Electrochem. Soc. 135,1521 (1988)

[101] J.I.Ulacia and J.P.Mc Vittie, J.Appl.Phys. 65, 1484 (1989)

[102] W.Pilz, J.Pelka and P.Banks, Microcircuit Ing.Conf.89

[103] W. Pilz, H. Hübner, F. Heinrich, P. Hoffmann and M. Franosch "Discussion in profile phenomena in sub-µm resist reactive ion etching" Microelectronic Engineering Vol. 9 pp 491-494 (1989)

[104] E.S.G.Shaqfeh and C.W.Jurgensen, J.Appl.Phys. 66, 4664 (1989)

[105] J.Pelka, M.Weiss, W.Hoppe and D.Mewes, J.Vac.Sci.Technol. B7, 14873 (1989)

[106] B.Petit and J.Pelletier, Rev. Phys. Appl. 21, 377 (1986)

[107] A. Gérodolle " Two-Dimensional aspects of ion enhanced reactive etching of silicon with SF6" Proc. ESSDERC'89 Conf. Berlin Sept. 1989

[108] A. Gérodolle, J. Pelletier and S. Drouot "An improved model of plasma etching including temperature dependence: Comparison between simulation and experimental results" these proceedings.

[109] T. Arikado, K. Horioka, M. Sekine, H. Okana and Y. Horiike "Single Silicon Etching Profile Simulation" Japanese J. of Appl. Physics Vol. 27, No. 1, Jan. 1988

[110] F. Itah, A. Shimase, and S. Haraichi "Two-Dimensional Profile Simulation of focused Ion-Beam Milling of LSI" J. Electrochem. Soc. Vol. 137, No. 3, March 1990

[111] T. Thurgate "Segment Based Etch Algorithm and Modelling" NUPAD-III Workshop, Honolulu, June 1990

[112] T.J. Cotler, M. S. Barnes and M. E. Elta "A Monte Carlo microtopography model for investigating plasma/reactive ion etch profile evolution" J. Vac. Sci. Technol. B. Vol. 6, No. 2 Mar-Apr. 1988

[113] J. Ignacio " A 3-D fluid simulation of dry-etching aluminium hexode reactor" IEDM'89 Tech. Digest, Washington, Dec. 1989

[Lithography

[114] W.G. Oldham, S. Nandgaonkar, A.R. Neureuther and M.M O'Toole "A general Simulator for VLSI lithography and Etching Processes: Part I - Application to Projection Lithography" IEEE/ED, Vol. ED-26, No. 4, Apr. 1979

[115] B.P. Mathur, N.N. Kundu and S.N. Gupta "Quantitative Evaluation of Shape of Image on Photoresist of Square Apertures" IEEE/ED Vol. 35 No. 3, Mar. 1988

[116] R. Guerrieri, J. Gamelin, K. Tadros and A. Neureuther "Massively Parallel Algorithms for Scattering in Optical Lithography" NUPAD-III Workshop, Honolulu, June 1990

[117] Y. Ogawa, T. Hidaka, S. Hasegawa, K. Nakajima and Y. IIda "Three-Dimensional Electron Beam Lithography Simulator - Electron Beam Exposure Process" NEC Res. and Develop. J., No. 94, July 1989

miscellaneous

[118] R.E. Bank and A. Weisser "Some *a posteriori* error estimates for elliptic partial differential equations" Math. of Comp. Vol. 44, No. 170, pp 283-301, April 1985

[119] F.J. Mustieles and P. Degong "A determinist particle method for semiconductor Boltzmann equation" Proc. NASECODE-VI Conf. Dublin, Boole Press Ed. July 1989

Paper presented at ESSDERC 90, Nottingham, September 1990
Session 4A1

# Comparison of hot-carrier degradation in n- and p-MOSFETs with various nitride–oxide gate films

H. Iwai, H. S. Momose, T. Morimoto, S. Takagi, and K. Yamabe

ULSI Research Center, Toshiba Corporation
1, Komukai-Toshiba-cho, Saiwai-ku, Kawasaki, 210, Japan

## ABSTRACT

The electrical characteristics of MOSFETs with three types of gate insulator, pure gate oxide, rapid thermal nitrided oxide, and stacked nitride oxide, were compared. While the stacked nitride oxide sample has been regarded as a highly reliable insulator for TDDB, it has lower hot carrier reliability in both threshold voltage shift and interface state generation. RTP samples have very small interface state generation during stress application, while the threshold voltage shift is comparable (nMOS case) to or even worse (pMOS case) than that for the pure oxide gate sample. The relationship between hot carrier degradation and substrate/gate current during stress is discussed.

## INTRODUCTION

Recently, nitride-oxide (ONO, ON, NO) films have been studied as highly reliable gate insulators for MOSFETs. There are several combinations of these nitride and oxide films and several methods for producing the films. In general, the transconductance of the stacked nitride-oxide films is good, but detailed discussion on the transconductance and mobility was given elsewhere by Momose et al (1989b) and Iwai et al (1990). This paper reports on studies into the hot carrier reliability of various nitride-oxide gate n- and p-MOSFETs, in comparison with pure gate oxide MOSFETs.

## EXPERIMENTS AND RESULTS

Three gate-film formations were used in these experiments. They were i) stacked nitride film deposition by LPCVD, ii) nitridation of the oxide film by rapid thermal nitridation (RTN), and iii) pure gate oxide formation in a furnace. Process parameters for the samples are listed in Table 1.

Figure 1 shows the threshold voltage shift ($\Delta V_{th}$) dependence on the hot carrier stress gate voltage for n-MOSFETs. The stress was applied for 1000 seconds. Figure 2 shows the charge pumping current or the generated interface states dependence on the stress gate bias for n-MOSFETs. Figures 3 and 4 show the substrate and gate currents during the stress application for n-MOSFETs.

In general, stacked nitride and RNO samples are more reliable in terms of TDDB. However, regarding MOSFET hot carrier reliability, the situation seems to be a little different. The RNO samples exhibit higher interface state generation reliability than pure gate oxide samples, while, when the nitride film is thick, the stacked nitride samples have lower interface state generation reliability than the pure gate oxide samples (Fig.2). Regarding threshold voltage shifts, the stacked nitride samples are less reliable than pure oxide samples (Fig.1). When the nitride film is thick, the stacked samples have extremely large threshold voltage shifts. The threshold

© 1990 IOP Publishing Ltd

Table 1. Sample fabrication conditions

| | | SAMPLE FABRICATION PROCESS | | | $T_{ox}$ (C-V) | N CONC. AT Si/SiO$_2$ INTERFACE |
|---|---|---|---|---|---|---|
| | | FURNACE OXIDATION | NITRIDATION | RE-OXIDATION | | |
| PURE OXIDE | PO | 10 nm / 6 | — | — | 10.8 nm / 6.5 | 0% / 0 |
| STACKED NITRIDATION | SN | 5 nm / 5 | 4 nm / 6 | — | 6.5 nm / 7.6 | ? % / ? |
| | SNO | 5 nm / 5 | LPCVD Si$_3$N$_4$ 4 nm / 6 | FURNACE OXIDATION | 16.7 nm / 8.9 | 1.7% / 4.0 |
| RAPID NITRIDATION | RNO | 10 nm / 6 | RTN 1200°C | RTO 1200°C | 10.9 nm / 6.5 | 8.3% / 8.9 |

voltage shifts of the RNO sample were comparable with those of the pure gate oxide. However, it is interesting to note that, under some stress conditions, the signs of the trapped charge for the RNO samples were opposite to that of the PO, SN and SNO samples. Comparing Fig.3 with Figs.1 and 2, the degradation (specifically for the interface state generation) corresponds to the substrate current amount. However, it was difficult to find a general relationship between the degradation and the gate current amount (Compare Fig.4 with Figs.1 and 2).

P-MOSFET hot carrier reliability was also investigated. Figures 5 and 6 show the threshold voltage shift and charge pumping current dependence on the stress gate bias for p-MOSFETs. Figures 7 and 8 show the substrate and gate currents during the stress application for p-MOSFETs. The threshold voltage shifts (Fig.5) were larger for RTN, SN, and SNO samples. The interface state generations were very small for the pure oxide and RTN samples. Only for the SN samples, was the interface state generation found to be extremely large. For p-MOSFET case, it is difficult to find a universal relationship between the degradation and the substrate/gate current for all gate insulator samples.

For both n- and p-MOSFET cases, some of the SN and SNO samples showed very large hot carrier degradation. When the stress drain voltage was reduced, it is true that the degradation of those SN and SNO samples became smaller. However, it was confirmed that the degradation is still larger than that of other samples (PO and RNO).

## CONCLUSION

The electrical characteristics for MOSFETs with three types of gate insulator were compared. The stacked nitride sample has lower hot carrier reliability in regard to both threshold voltage shift and interface state generation. RTP samples show very small interface state generation during the stress application, while the threshold voltage shift is comparable (nMOS case) to or even worse (pMOS case) than that for the pure oxide gate sample.

## REFERENCES

Iwai H, Momose H S, Takagi S, Morimoto T, Kitagawa S, Kambayashi S, Yamabe K and Onga S 1990 *Dig. of Tech. Papers VLSI Symp. on Tech. (Honolulu)* pp 131-132

Momose H S, Kitagawa S, Yamabe K and Iwai H 1989a *IEDM Tech. Dig.* pp 267-270

Momose H S, Takagi S, Kitagawa S, Yamabe K, and Iwai H 1989b *IEEE Semiconductor Interface Specialists Conference, (Ft. Lauderdale, FL)* pp I.5

# Hot carriers in MOS devices 289

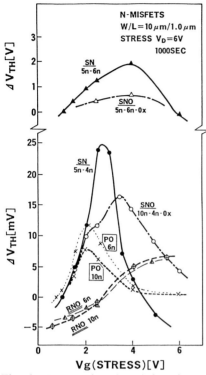

Fig 1. Hot carrier degradation for n-MOSFETs: Threshold voltage shift dependence on stress gate bias.

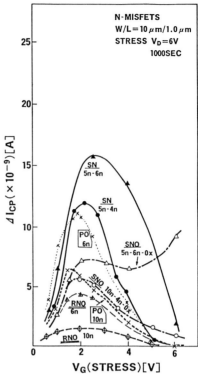

Fig 2. Hot carrier degradation for n-MOSFETs: Charge pumping current dependence on stress gate bias.

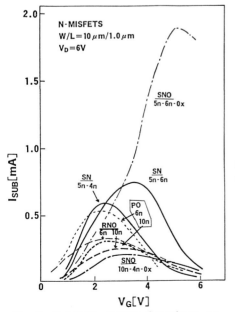

Fig 3. Substrate current dependence on stress gate bias for n-MOSFETs.

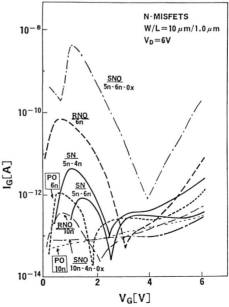

Fig 4. Gate current dependence on stress gate bias for for n-MOSFETs.

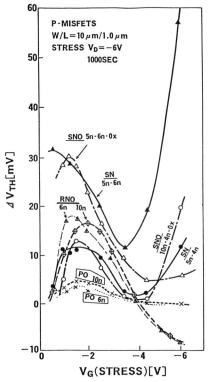

Fig 5. Hot carrier degradation for p-MOSFETs: Threshold voltage shift dependence on stress gate bias.

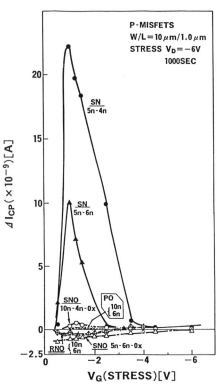

Fig 6. Hot carrier degradation for p-MOSFETs: Charge pumping current dependence on stress gate bias.

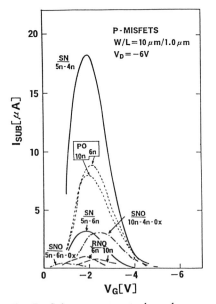

Fig 7. Substrate current dependence on stress gate bias for p-MOSFETs.

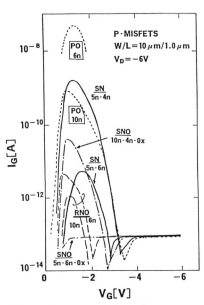

Fig 8. Gate current dependence on stress gate bias for for p-MOSFETs.

Paper presented at ESSDERC 90, Nottingham, September 1990
Session 4A2

# Duty cycles in digital logic applications: a realistic way of considering hot-carrier reliability

W. Weber, M. Brox, T. Künemund, D. Schmitt-Landsiedel, and Q. Wang

Siemens AG, Corporate Research and Development
Otto-Hahn-Ring 6, D8000 München 83, West Germany

Abstract. In this paper various stages as appearing in digital logic, like inverters, NANDs, NORs, and transfer gates are hot-carrier stressed. Transient effects and the one of voltage combinations are discussed and an estimation for a realistic lifetime criterion is given.

1. Introduction

As minimum feature sizes are progressively reduced in modern microelectronics, and technological measures are more or less exhausted, reliability assurance against hot-carrier degradation becomes an increasing problem. For this reason a quantitative knowledge of the degradation under realistic operating conditions is required. In this paper, different operating modes, as appearing in logic applications, are investigated with respect to their effect on hot-carrier degradation and a worst case duty cycle estimation for application in hot-carrier reliability assurance is given.

2. Transient effects

Transient effects were investigated for slopes down to 0.5ns. The n-MOS results are shown in Figure 1, proving clearly that by using a proper hot-carrier related duty cycle calculation (Weber 1988), the static and dynamic results agree and that no transient effects are present in the n-MOSFETs used here (cf. Hänsch 1989). A similar investigation in the p-MOSFET is not as simple, as the degradation alters the currents involved by a large amount. There are, however, clear hints that transient effects do not exist in p-MOSFETs either. These considerations allow duty cycles of MOSFETs in circuits to be calculated. They can vary within certain limits as shown in the results of Figure 2, where the varying duty cycle within an inverter chain causes changing degradation.
Within this framework, NAND and NOR-type stages were investigated. Their critical path was approximated by an inverter with dimensions adopted to the ones in the NAND (Figure 3). From a circuit simulation and the above-mentioned duty cycle calculation we expected an enhanced degradation of the n-channel device of the inverter stage following the NAND and of the p-channel device in the stage following the NOR. The reduced width of the p(n)-channel device in the NAND(NOR) causes a

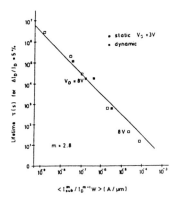

Fig. 1 Lifetime $\tau$ vs. time average of $I_{sub}^m/I_D^{m-1}$, the value of m being determined by a separate plot $\tau(I_{sub})$ for constant $I_D$ which is not shown here.

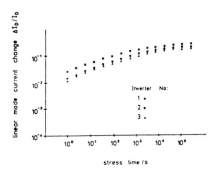

Fig. 2 Degradation vs. stress time for n-MOSFETs in an inverter chain with decreasing pulse overlaps, decreasing duty cycles and thus decreasing degradations

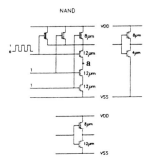

Fig. 3 Three-port NAND (above) and the related inverter-type circuit (below) which was used as a model of the NAND in the experiment (we assumed node a to be essentially at Vss). To the right the following inverter stage is drawn.

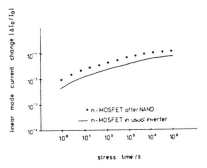

Fig. 4 Increased degradation of the n-channel MOSFET in the inverter following the NAND-type circuit

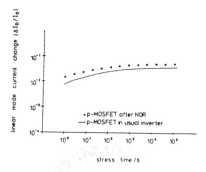

Fig. 5 Increased degradation of the p-channel MOSFET in the inverter following the NOR-type circuit

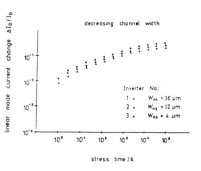

Fig. 6 Increased degradation due to overshoots in low-load inverter chains

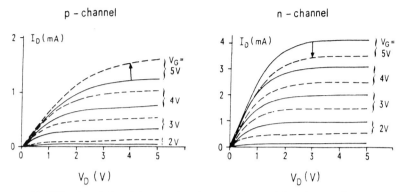

Fig. 7 The degradation of the normal-mode output characteristics after transfer gate stress in an n- and p-channel MOSFET. Due to the damage at drain <u>and</u> source, the change of the current is essentially homogeneous.

slower transient in the following stage. The experimental results are shown in Figures 4 and 5 proving the validity of the model.

In a further experiment the effect of overshoots was measured by using an inverter chain with decreasing width, decreasing capacitive load and thus increasing overshoots in the last stages. The results are shown in Figure 6. They show the increased degradation as expected.

In all cases an agreement with the above duty cycle calculation was confirmed. Furthermore it is emphasized that the differences between the degradation of an ordinary inverter and the specific applications is rather small (cf. Figures 4-6) leading to differences in lifetimes of about a factor of 4, only.

The effect of different combinations of gate and drain voltages in n-MOSFETs is described by Weber 1989. The combination of electron and hole injection can cause an increased degradation in n-MOSFETs which is not considered in the duty cycle calculation. In this way errors are introduced which may lead to a lifetime reduction of a factor of up to 4 if the duty cycle calculation is performed on the basis of static stresses alone. The calculations in this paper are, however, not precise enough to detect this deviation.

## 3. Bi-directional hot-carrier stress

The transfer gate is an important special case in logic applications. Source and drain are alternatingly switched into the high state and as a consequence the degradation takes place at both source <u>and</u> drain (cf. Mühlhoff 1987). The degradation of the output characteristics differs from the usual case of a stress on one side only (Figure 7). In the case of the p-MOSFET this greatly enhances effective channel length reduction and thus linear mode degradation. A $0.4\mu m$ longer p-channel is required in order to compensate for this effect (Figure 8). The linear-mode drain current degradation of the <u>n</u>-channel is, on the other hand, not increased as no degradation-enhancing effects are present here. Furthermore, this parameter is equally sensitive to damage at both source <u>and</u> drain. It thus represents a worst case (Figure 9).

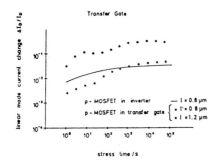 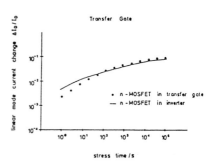

Fig. 8 Linear-mode drain current degradation of the p-MOSFET in a transfer gate is greatly increased as compared with an inverter-type stress.

Fig. 9 Linear-mode drain current degradation of the n-MOSFET in a transfer gate is very similar to that of an inverter-type stress.

## 4. Conclusion

Using experimental data a ratio between a typical dynamic (inverter case) and static lifetime can be evaluated to be 0.25 % (see 8V results in Figure 1) for a cycle time of 40ns. Including the deviations of specific applications like NANDs, NORs, and overshoots, a worst-case duty cycle of 1% can be estimated for the n-channel MOSFET and even less for the p-MOSFET. This holds good on the assumptions that subsequent stages should not differ by more than a factor of three in driving capability which is a common design rule.

Furthermore the p-channel degradation in the transfer gate must be investigated independently. Here it is important to consider reliability criteria which are adapted to the specific applications of transfer gates. Probably the transfer gates used in static latches with feedback can suffer much higher degradations than the ones used in dynamic circuits. These can lose their information if degradation-induced leakage currents (p-channel) appear.

Hänsch W and Weber W 1989 *IEEE Electron Device Lett.* **10** pp 252-4
Mühlhoff H-M, Murkin P, Orlowski M, Weber W, Küsters K H, Müller W, Rogers C M and Wendt H 1987 *Proc. VLSI Symp.* pp 57-8
Weber W 1988 IEEE *Trans. Electron Devices* **35**, pp 1476-86
Weber W and Borchert I 1989 *ESSDERC Tech. Dig.* pp 719-22

Paper presented at ESSDERC 90, Nottingham, September 1990
Session 4A3

# Annealing of fixed oxide charge induced by hot-carrier stressing

M. Brox and W. Weber
Siemens AG, Corporate Research and Development
Otto Hahn Ring 6, 8000 München 83, FRG

Abstract: A study on electric-field-assisted annealing of transistor degradation has been performed on hot-carrier degraded n- and p-channel Si-MOSFETs. Distinct recovery is found following stressing conditions which lead to accumulation of fixed charge in the gate oxide. Agreement is obtained with previous studies on hole-detrapping following ionizing radiation. Results on p-MOSFETs support the view that very thin oxides are resistant to degradation.

## 1. Introduction

Due to the continuous scaling down of MOSFET device dimensions, both the lateral and the transversal electric fields generated during device operation are steadily increasing. This enhances the injection of channel hot-carriers, thus aggravating the problem of oxide trapping and interface state generation. However, high-oxide electric fields are also able to influence the transistor degradation positively, as field-assisted detrapping of trapped charges can lead to a relaxation of the degradation problem. One example is the observation of Hiruta et al. (1989), who demonstrated p-MOSFETs with an ultra-thin gate oxide to be very resistant to hot-carrier degradation. They attributed the absence of the electron trapping usually observed to the tunneling of trapped electrons out of the oxide.

In this paper we aim to investigate relaxation effects following hot-carrier degradation of n- and p-channel MOSFETs where we focus on injection conditions which are known to cause accumulation of fixed charges in the gate oxide.

## 2. Experimental details

Unless otherwise noted, the experiments to be described in the following were carried out on conventional MOSFETs with an oxide thickness of 42nm (20nm) and a gate length of 1.7µm (1.0µm) for the n-channel (p-channel) device.

To study the annealing, the n-MOSFET was first stressed at a gate voltage $V_G = 1V$ and a drain voltage $V_D = 8V$ for 120sec, a condition which causes holes to be trapped in the oxide. Stress parameters for the p-MOSFET were $V_G = -3V$ and $V_D = -7V$ for a total of 100sec, which is usually observed to lead to the trapping of electrons. The time dependence of the hot-carrier-stress, characterized by the decrease of $I_{SUB}/I_D$ in the saturation region (Tsuchiya 1985), is given in Figure 1. We choose the multiplication factor as the characterization parameter, as it is both very sensitive to fixed oxide charges and, when measured at a low gate voltage, insensitive to generated interface traps above the drain. Therefore it allows these two types of damage to be distinguished which is difficult when examining linear mode parameter shifts alone.

© 1990 IOP Publishing Ltd

Trapped holes in n-MOSFETs and electrons in p-MOSFETs both lead to the appearance of a shoulder in the charge-pumping current (Heremans et al. 1988). Its diminution during the annealing was used to confirm that the reported effects originate from the discharging of oxide traps. However, charge-pumping was not in general carried out to minimize the application of bias during characterization in order to prevent unwanted detrapping.

During annealing the gate was biased, whereas the source, drain and substrate were grounded. The recovery characteristics are scaled linearly so that 0% refers to the post-degradation value and 100% to complete annealing.

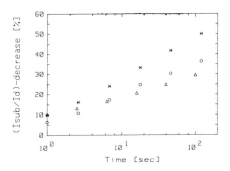

Fig. 1: Time dependence of $I_{sub}/I_d$ in the hot carrier stress (n-MOSFET (asterisk), n-MOSFET with SiN-passivation (circle), p-MOSFET (triangle)).

## 3. Detrapping of trapped holes in n-MOSFETs

The detrapping of holes out of $SiO_2$ after exposure to ionizing radiation was found to play a dominant role in the long-term recovery of irradiated MOS devices (for a review see Ma 1989). It was modelled as a tunneling transition of holes from an energetic level about 3eV above the $SiO_2$ valence band edge into the Si valence band. Manzini and Modelli (1983) derived an expression for the dependence of the time constant of the hole-discharge on the applied electric field F and the distance z between trapped hole and interface, which is given by:

$$\tau(z,F) = \tau_0 \exp\left( \frac{4}{3} \frac{\sqrt{2m^*}}{\hbar q F} \left( E_t^{1.5} - (E_t - qFz)^{1.5} \right) \right) \qquad (1)$$

where $E_t$ is the energetic level above the $SiO_2$ valence band, $\tau_0$ a characteristic tunneling time and $m^*$ the hole-tunneling effective mass. Figure 2 shows the tunneling time constant as calculated by (1).

The remaining flatband voltage $\Delta V(t)$ shift can be calculated by an integration over the un-detrapped fraction of the initially trapped holes, when the initial distribution of filled traps $n_0(z)$ is known:

$$\Delta V(t) = -\frac{q t_{ox}}{\varepsilon_{ox}\varepsilon_0} \int_0^{t_{ox}} \left(1 - \frac{z}{t_{ox}}\right) n_0(z) \exp\left(-\frac{t}{\tau(z,F)}\right) dz \qquad (2)$$

Figure 3 shows the detrapping response which we obtained after the injection of channel hot-holes into the Si MOSFET. In accordance with the model described above, hole detrapping proceeds faster with increasing positive gate bias. The experimental data was fitted to equation (2) by assuming that:

$$n_0(z) = \text{constant} \quad \text{for } z_1 < z < z_2; \quad n_0(z) = 0 \text{ otherwise} \qquad (3)$$

with the parameters as given in Figure 3. By the lower bound $z_1$ we take into account that the very near-interfacial trapped holes have tunneled out of the oxide before the first characterization after the hot-hole stress could be performed (Figure 2). The upper bound $z_2$ describes the fact that hole traps are found in a region much closer than 10nm to the interface only.

Charge pumping measurements showed that during detrapping only a small number of additional interface states built up. During the tunneling process therefore, interface states are not generated in a significant number.

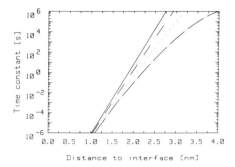
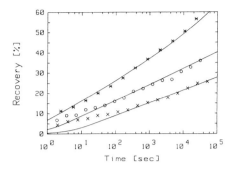

Fig. 2: Calculated time constant for the tunneling of trapped holes into the Si valence band as a function of the distance between the trapping site and the interface. Parameter is the oxide electric field: 0 (solid), 2 (dashed), 4 (dotted) and 6 MV/cm (dash-dotted). We used a value of $E_t$=2.5eV for the trap depth.

Fig. 3: Time dependence of the recovery after channel hot-hole stressing of n-MOSFETs. The oxide electric field during annealing was 1.9 (cross), 3.9 (circle) and 6.0 MV/cm (asterisk). The solid lines are calculations based on equations (1) and (2) with $z_1$=2.1nm and $z_2$=4.6nm. Recovery at negative gate bias is much smaller.

If the n-MOSFET is passivated by silicon nitride, the annealing behavior of the transistor changes (Figure 4). A merely quantitative difference is that the tunneling discharge proceeds much faster, which can be explained in terms of a trapped hole distribution located nearer to the interface. Additionally however, we observe that application of negative gate bias following the recovery at positive gate bias reverses the annealing. The recombination of the tunneling electrons with the trapped holes is no longer permanent.

A similar effect has recently been observed in radiation-hard devices after irradiation. Lelis et al. (1989) ascribed it to the strain at the interface, which is greater in hard devices as compared to soft ones. In their model, the two Si atoms of the E' center, constituting the trapped hole defect in the $SiO_2$ (Lenahan and Dressendorfer 1984), are driven apart from each other by the interfacial strain. An electron which tunnels to the strained defect site, can thus no longer reform the broken Si-Si bond. It only compensates the positive charge and can tunnel back into the silicon upon application of negative bias. In accordance with this model, we believe that the interfacial strain in SiN-passivated samples renders the structure of the E' center such that we observe reverse annealing.

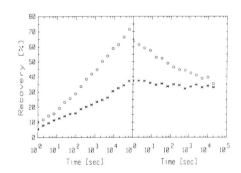

Fig. 4: Effect of SiN passivation on detrapping behavior. Following 60000sec of $E_{ox}$=3.9 MV/cm the field was switched to -2.5MV/cm. Pronounced reversal occurs for the SiN passivated sample only. Symbols as given in Figure 1.

## 4. Detrapping of trapped electrons in p-MOSFETs

The time dependence of the detrapping of trapped electrons which we obtained after the hot-carrier degradation of p-MOSFETs is given by Figure 5. It is clearly non-exponential and therefore cannot be explained by field-assisted thermal emission (this has been found for an electron trap generated by hot-hole injection

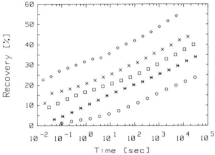
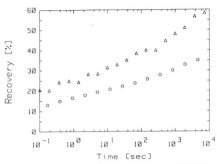

Fig. 5: Time dependence of the recovery after channel hot-electron stressing of p-MOSFETs. The oxide electric field during annealing was -7 (diamond), -5.5 (cross), -4.5 (square), -3.5 (asterisk) and -2.5 MV/cm (circle). In these samples the behavior is independent of the polarity of the oxide electric field.

Fig. 6: Dependence of the recovery in p-MOSFETs on the oxide thickness ($t_{ox}$=12.5nm (triangle), $t_{ox}$=20nm (circle)). The oxide electric field was -4.5 MV/cm. Annealing proceeds much faster in the device with the thinner oxide.

in n-MOSFETs (Bourcerie et al. 1989)) or trap band impact ionization by Fowler-Nordheim injected electrons (Nissan-Cohen 1985). Like in n-MOSFETs the logarithmic time dependence points to a tunneling discharge mechanism. However, the situation is not as clear as in the case of the trapped holes because we do not observe the anticipated asymmetry with respect to the polarity of the oxide electric field. According to a model in which trapped electrons tunnel into the Si conduction band, recovery at negative gate bias should proceed much faster than recovery at positive bias. In these samples, however, annealing at positive bias is as fast as that at negative bias, whereas in a second class of samples negative oxide fields indeed are more effective in removing the charge.

We assume that electrons which are trapped in the bulk of the oxide ($t_{ox}$=20nm) are not detrapped, which explains why only a recovery of approx. 50% could be achieved. Detrapping occurs in the vicinity of the electrodes only, which agrees with our observation that the recovery percentage increases with decreasing oxide thickness (Figure 6). These results give a hint as to why the ultra-thin oxides in the work of Hiruta et al. (1989) essentially showed no trapping, as in that case even the bulk of the oxide is within the detrapping region.

### References:

Bourcerie M., Marchetaux J., Boudou A. and Vuillaume D. 1989, Appl. Phys. Lett. 55, 2193

Heremans P., Witters J., Groeseneken G. and Maes H. E. 1989, IEEE Trans. Elec. Dev. TED-36, 1318

Hiruta Y., Oyamatsu H., Momose H. S., Iwai H. and Maeguchi K. 1989, ESSDERC Tech. Dig. 732

Lelis A. J., Oldham T. R., Boesch J. R. and McLean F. B. 1989, IEEE Trans. Nucl. Sci. NS-36, 1808

Lenahan P. and Dressendorfer P. V. 1984, J. Appl. Phys. 55, 3495

Ma T. P. and Dressendorfer P. V. (editors) 1989, "Ionizing Radiation Effects In MOS Devices and Circuits", John Wiley, New York

Manzini S. and Modelli A. 1983, INFOS Tech. Dig. 112

Nissan Cohen Y., Shappir J. and Frohmann-Bentchkowsky D. (1985), J. Appl. Phys. 58, 2252

Tsuchiya T. and Frey J. 1985, IEEE Elec. Dev. Lett. EDL-6, 8

[Paper presented at ESSDERC 90, Nottingham, September 1990]
[Session 4A4]

# Gate oxide integrity and hot-carrier degradation of $TaSi_2P^+$ polycide gate MOSFETs

U. Schwalke[*], W. Hänsch[*], M. Kerber, A. Lill and F. Neppl

Siemens AG, Corporate Research and Development
Otto-Hahn-Ring 6, D-8000 Munich 83, FR Germany
[*] Present Address: IBM, Essex-Junction, VT 05452, USA

Abstract. In this contribution we report on boron doped $TaSi_2$/polySi gates which provide low sheet resistance, the appropriate p+ workfunction and simultaneously suppress boron-penetration effects. Gate oxide integrity, hot-carrier degradation and performance of $TaSi_2$ p+ polycide NMOS and PMOS FETs are evaluated and compared to equivalent n+ polySi gate reference transistors.

## 1. Introduction

As CMOS devices have been scaled down, dual workfunction gates (i.e. n+ gate for NMOS and p+ gate for PMOS FETs) have been proposed in order to overcome the short channel effects of the buried channel n+ gate PMOS FETs (Sun et al. 1987). However, due to the interconnection of n+ and p+ gates, rapid dopand interdiffusion in dual workfunction gates has emerged as a serious problem since its limits the thermal budged (Hayashida et al. 1989). From this point of view, single workfunction p+ gates will be advantageous, since dopant interdiffusion is eliminated and the excellent short channel characteristics of the PMOS FET is preserved.

The purpose of this study is to evaluate the potential of a fully p+ polycide gate CMOS technology with respect to gate oxide integrity, hot-carrier susceptibility and transistor performance.

## 2. Experimental Procedures

A twin-well CMOS process with splitted well drive-in (Zeller et al. 1990), LOCOS isolation and 16nm gate oxide was used to realize CMOS devices with a minimum channel length of 0.4um. Except for the gate definition sequence, the processing was identical for both, p+ polycide gate and n+ polySi gate reference MOSFETs. For the n+ gate devices conventional phosphorous doped polySi was used. The p+ gate version consisted of $^{11}B$ doped $TaSi_2$/polySi bilayers (polycide) with low sheet resistance ($R_S$=2.9 Ohms/sq). The silicide films were prepared by cosputtering Ta and Si from separate targets. A LTO-cap was used to mask the gates against counterdoping from S/D implants. Conventional BPSG reflow, single level metallization and a forming gas anneal completed the process.

## 3. Results and Discussion

### 3.1 Gate Oxide Integrity

The influence of p+ polycide gate processing on gate oxide integrity is

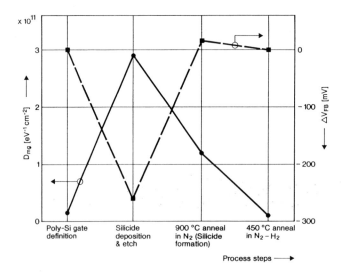

Fig.1 Relaxation of process-induced oxide damage by high and low temperature anneals.

examined at various process levels. Directly after gate structuring, the interface-trap density of p+ polycide gates is found to increase by an order of magnitude when compared to n+ polySi gates. In addition, negative $V_{FB}$-shifts indicate the presence of trapped oxide charge (Fig.1). Silicide sputter-deposition and to a minor degree reactive ion etching were identified to cause the observed radiation induced oxide damage. However, the oxide damage at the gate level is completely removed by subsequent high temperature (900°C, $N_2$) and low temperature (450°C, $N_2/H_2$) anneals as shown in Fig.1. Furthermore, charge-to-breakdown measurements confirmed the restored gate oxide integrity of p+ polycide gates and revealed $Q_{BD}$ values of 15 C/cm$^2$ at 25°C.

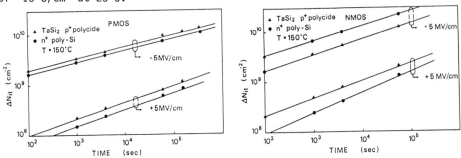

Fig.2 Low-field BT stress measurements performed on n+ polySi and p+ polycide MOSFETs. Because of its increased sensitivity the charge-pumping method was used to monitor BT degradation.

Boron-penetration induced gate oxide instabilities, which are of major concern in p+ polysilicon gates, are suppressed in silicided gates. Low-field bias-temperature (BT) stress measurements of MOSFETs (Fig.2) revealed no evidence for the dramatic +BT instability commonly observed for p+ polysilicon gates (Hiruta et al. 1987). The usual -BT instability ($E_A$=0.3eV) is the dominant mode for either gate type and transistor. The improved degradation hardness and $V_{th}$-stability of $TaSi_2$ p+ polycide gates is attributed to gettering of $H_2$ in the silicide as well as boron redistribution effects within the TaSi$_2$/polySi bilayer (Schwalke et al. 1989).

## 3.2 Hot-Carrier Degradation

Hot-carrier measurements revealed an improvement in liftime of more than two orders of magnitude (Fig.3) for the p+ polycide NMOS FET. This improvment is mainly due to the buried channel construction with the carrier generation center located deeper from the Si/SiO$_2$ interface. Consistently, the $I_{sub}/I_D$ values are decreased by 50% indicating a reduced lateral electric field. On the other hand, the peak $I_{sub}/I_D$ values of p+ polycide PMOS FETs are increased by 80% consistent with the change from buried channel (BC) to surface channel (SC) device. However, despite the larger hot-carrier generation rate of the SC PMOS FET, lifetime is enhanced by factor of five (Fig.4). The improved hot-carrier stability mainly results from the changed carrier injection conditions. A comparative study of hot-carrier degradation of MOSFETs with n+ and p+ gates (Schwalke et al. 1990) revealed an enhanced injection of holes combined with a reduced electron injection into the gate oxide. This results in a reduction of the gate current by 30%. Due to the enhanced injection of holes, negative charge near the interface is compensated. That moves the center of charge into the gate oxide. As a result it less affects the drain current.

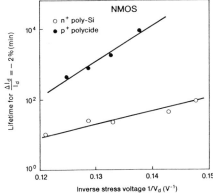

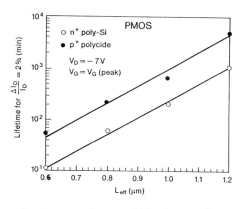

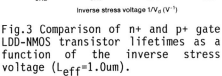

Fig.3 Comparison of n+ and p+ gate LDD-NMOS transistor lifetimes as a function of the inverse stress voltage ($L_{eff}$=1.0um).

Fig.4 Transistor lifetimes of n+ and p+ gate PMOS FETs as a function of effective channel length.

## 3.3 Performance and Scalability

Short channel effects due to the buried channel construction can be efficiently controlled in p+ gate NMOS FETs, since arsenic with its low diffusivity is used instead of boron as a buried-channel implant in p+ gate BC-NMOS FETs. This allows to realize shallow channel junctions which substantially improve threshold voltage roll-off (Fig.5) as well as subthreshold characteristics. In addition, device simulation results indicated the potential to scale down the p+ gate BC-NMOS into the deep submicron regime.

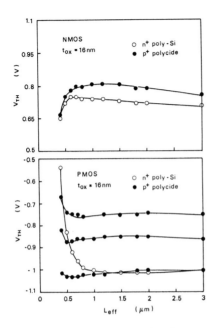
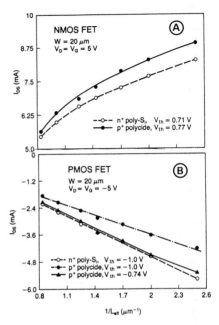

Fig.5 Plot of threshold voltage versus effective channel length.

Fig.6 Saturation drain current as a function of the inverse channel length of NMOS (A) and PMOS (B) devices

When compared to the equivalent n+ gate NMOS FET, an increase in saturation current of 10% was noted for the p+ polycide NMOS FET (Fig.6A). This improvement mainly results from the higher carrier mobility due to reduced surface scattering in the BC-NMOS FET. On the other hand, the p+ gate surface channel PMOS FET suffers from a loss in $I_{DS}$ of 20% at identical $V_{th}$. However, due to the improved surface channel characteristics, this loss could be compensated by a $V_{th}$-reduction as shown in Fig.6B.

## 4. Conclusion

In conclusion, the fully p+ polycide gate CMOS technology is a promising approach for future submicron CMOS devices. The enormous gain in hot-carrier lifetime is particularly attractive for 5V ULSI CMOS applications.

## References

Hayashida H, Toyoshima Y, Suizu Y, Mitsuhashi K, Iwai H, and Maegushi K 1989 Conf. Proc. VLSI Symp. pp 29-30
Hiruta Y, Matsuoka F, Hama K, Iwai H, Maeguchi K and K. Kanzaki 1987 IEDM Tech. Dig. p 578
Schwalke U, Mazure C and Neppl F 1989 J. Vac. Sci. Technol. B7 p 120
Schwalke U, Hänsch W, and Lill A 1990 Proc. Intern. Conf. on Solid State Devices and Materials, Sendai, Japan, in press
Sun JY-C, Taur Y, Dennard R H and Klepner S P 1987 IEEE Trans. Electron Dev. ED-34 p19
Zeller C, Mazure C and Lill A 1990 IEEE Electr. Dev. Lett. EDL-11(5) p 215

# The effects of gate and drain biases on the stability of low temperature poly-Si TFTs

N D Young and A Gill

Philips Research Laboratories, Redhill, Surrey, UK.

Abstract. Instabilities in low temperature poly-Si TFTs are shown to be due to the tunnelling of electrons into traps in the gate oxide, and to hot carrier induced interface acceptor state generation and trapping. Little evidence is seen for the creation of states in the poly-Si in the manner which has been reported for high temperature poly-Si TFTs. However, there is some evidence for the creation of mid gap acceptors and generation centres in our devices.

## 1. Introduction

Poly-Si TFTs are being studied with increasing interest as an alternative to αSi TFTs for electronics on glass applications, primarily because they demonstrate a high carrier mobility and better stability. They are generally fabricated at low temperatures, (<$650^0$C), for use with glass substrates, or high temperatures, (>$900^0$C), for use on quartz or silicon substrates. The important distinction is that high quality thermal oxide or fully densified LTO can only be used in the high temperature TFTs. In both cases plasma hydrogenation is used for defect passivation.

Investigations into the stability of both high and low temperature poly-Si TFTs have been made previously, and several different instability phenomena have been observed: (i) Threshold shifting attributed to mobile ions, (Young et al 1989), (ii) threshold shifting attributed to direct tunnelling into bulk oxide traps, (Young et al 1990a), and (iii) threshold shifting and subthreshold slope degradation attributed to state creation in the Poly-Si, (Wu et al 1990). The former two have been reported for low temperature TFTs, and the latter for high temperature TFTs, and the results clearly demonstrate a difference in the response of these devices to stress.

The present work reports upon further investigations into the stability of low temperature poly-Si TFTs.

## 2. Experiment

Low temperature autoregistered poly-Si TFTs were formed using 0.2μm columnar LPCVD poly-Si for the body and gate electrode, 0.15μm APCVD oxide for the gate dielectric, and phosphorus ion implantation for the source, drain and gate regions. Plasma hydrogenation was used for defect passivation, and typical device characteristics are given elsewhere, (Brotherton et al 1988). TFTs formed with recrystallised amorphous deposited silicon have also been investigated. The effects of gate bias and combined gate and drain biases have been studied with these devices at temperatures upto $200^0$C.

Detailed measurements of oxide charge trapping have been made on TFTs and mono-Si capacitors using DC Fowler-Nordheim, (FN), constant current injection, and on mono-Si capacitors using pulsed hot electron, (HE), injection, (Nicollian and Brews). In both the FN and the HE injection experiments the trapping is assessed by monitoring CV flat band shifts, and assuming that the trapping is spatially uniform.

## 3. Results and Discussion

### 3.1 Gate Bias

TFTs ,(W=50μm, L=10μm), subjected to gate bias stress, ($V_S=V_D=0V$, $V_G=20V$), at 50°C show $\Delta V_T < 1V$, and no degradation in $\mu_{fe}$, leakage current, $I_L$, or subthreshold slope, S, for times in excess of 8 months. Similar stresses at a temperature of 200°C reveal threshold shifting over much shorter time scales, as shown in figure 1, and again no significant degradation in $\mu_{fe}$ or $I_L$ is observed. It can be seen that a degradation in S occurs at around $I_D=10^{-10}$, though it is not yet clear whether this is due to near mid gap acceptor state creation or to a back channel effect. In view of the fact that both the negative and positive limbs of the transfer characteristic shift in the same direction, threshold shifts such as are shown in figure 1 have previously been attributed to the tunnelling of electrons into oxide traps, (Young et al 1990a).

Electron trap densities and cross sections in various oxides measured on mono-Si capacitors using HE and FN injection techniques are shown in Table I.

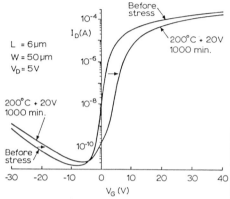

| Table I Oxide Trap Parameters from Electron Injection Experiments. | | | | |
|---|---|---|---|---|
| Oxide Type (Max Temp) | FN | | HE | |
|  | $N_T$ $10^{13}$ $cm^{-2}$ | $\sigma$ $10^{-15}$ $cm^2$ | $N_T$ $10^{13}$ $cm^{-2}$ | $\sigma$ $10^{-15}$ $cm^2$ |
| APCVD (600°C) | 0.5-1 | ⎫ | 2 | 0.3 |
| Plasma Anodized (600°C) | 0.2 | ⎬ 0.3-1 |  |  |
| Wet Thermal (700°C) | 0.3 | ⎭ |  |  |
| APCVD (1000°C) | 0.2 | 0.01 |  |  |
| Wet thermal (1000°C) | 0.4 | 0.010 | 0.6 | 0.005 |

Figure 1 TFT transfer characteristic before and after 20V gate biasing at 200°C.

The low temperature, ($\leq 700^\circ$C), processed oxides show a large cross section which is not present in the high temperature, ($\geq 1000^\circ$C), processed oxides, and this is most probably related to the oxide density. The smaller cross section found in the high temperature oxides is most likely associated with water, (Nicollian et al 1977). It is possible that these traps also exist in the low temperature oxides, where their detection could be hindered by the presence of the larger traps. Detailed annealing measurements show that the density of large traps cannot be significantly reduced at low temperatures, as shown in figure 2. It is apparent that equilibrium densities are achieved very rapidly, at least for the higher temperatures in the figure. Similar results have been obtained by Zhang et al (1990) for plasma anodized films.

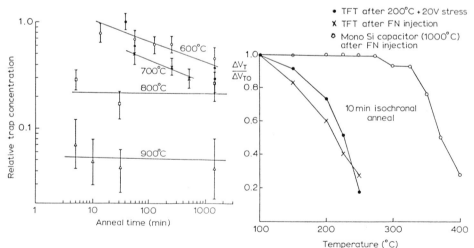

Figure 2 Density of the large trap in APCVD oxide and annealing.

Figure 3 The annealing of threshold shifts caused by various methods.

Results of experiments on post-stress annealing, shown in figure 3, lead us to believe that the larger of the two traps is responsible for the electron tunnelling instability in TFTs: Threshold shifts for TFTs stressed at $V_G$=20V for long times at 200°C, (traps filled by direct tunnelling), and for TFTs or mono-Si capacitors subjected to the constant current FN injection experiments are removed by annealing at 250°C, whereas annealing of threshold shifts for 1000°C mono-Si capacitors, (showing the small trap), does not occur until 350°C. It is also interesting to note that any CV stretchout generated during FN and HE injection was removed at 150-200°C, and this is taken to indicate that the associated interface states have been removed.

Finally, FN injection experiments on TFTs at high gate biases show that a given current density can be injected at lower biases than for mono-Si capacitors, and this is believed to be a result of field enhancement at asperities. SEM pictures confirmed the roughness of the poly-Si surface, and a field enhancement factor, $\eta$,= 1.5-2.5 is estimated, where $\eta$ is defined to be $E_I/E_{ox}$, where $E_I$ is the field at the interface and $E_{ox}=V_{ox}/t_{ox}$. The effect of this on the direct tunnelling instability in TFTs is demonstrated in figure 4, where the threshold shifting is shown for a columnar TFT, (rough interface), and a TFT formed using recrystallized deposited amorphous material, (smooth interface). Thus the roughness has a profound effect on stability. A detailed description of the effects of roughness and thermal assistance on the tunnelling is given by Young et al (1990b).

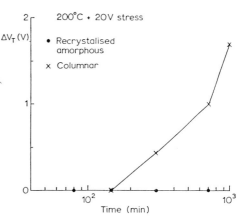

Figure 4 Threshold shifting for columnar poly-Si, (rough), and recrystalized αSi, (smooth), TFTs.

### 3.2 Gate and Drain Bias

TFTs, (W=50μm, L=6μm), have been stressed with gate and drain bias, ($V_G=V_D$=20V, $V_S$=0V), at temperatures of 50°C and 100°C for 6 months, and no degradation in $V_T$, S, $\mu_{fe}$ or $I_L$ has been seen. Previous work of this nature by Wu et al (1990), on high temperature TFTs, (W=50μm, L=15μm, $x_{ox}$=0.1μm), yielded appreciable threshold shifting and a degradation in S which was attributed to state creation in the poly-Si. Clearly this is not being seen in the devices studied here.

Application of higher drain biases gives rise to a field enhanced current, and this is shown for the off state, ($V_G$=-8V), in figure 5. The enhanced current is most likely due to avalanching near the drain contact. Walk out of the $I_D$-$V_D$ characteristic at high drain bias can be explained in terms of hot hole trapping in the gate oxide, which gives rise to field reduction near the drain.

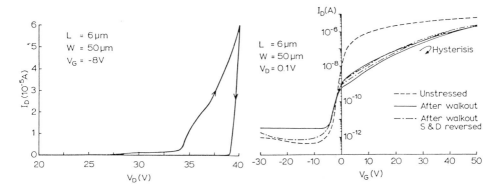

Figure 5 $I_D$-$V_D$ off state, ($V_G$=-8V), characteristic demonstrating walk out.

Figure 6 On and Off current degradation caused by walk out.

Further evidence for this is the fact that on subsequent identical sweeps the characteristic does not show the hysterisis loop, but follows the high voltage-low current limb of the curve. If the source and drain contacts are then reversed the effect can be recreated as carrier trapping occurs at the new drain end of the TFT. Moreover, degradation in the transfer characteristic is seen to occur during the drain bias sweep of figure 5, and this is shown in figure 6. The transfer characteristic after walk out shows a reduction in on current, and an increase in leakage current. Reversing the source and drain contacts has little effect on the on current measurement, but results in a much less degraded off current. The on current degradation is consistent with the generation of acceptor states in the upper half of the band gap near the drain contact, which affects the turn on of the TFTs in the same manner after contact reversal, (at least for low $V_D$). In view of the similarity to measurements in mono-Si MOSFETs, it is believed that the states are generated at the interface rather than in the bulk of the poly-Si. More detailed investigations into these effects are to be published elsewhere, (Young et al 1990c). The excess leakage current could be due to the creation of generation centres near the drain contact, and these will not of course contribute to the leakage current when the contacts are reversed.

Post-stress annealing experiments have been carried out on TFTs stressed at high drain bias, and it is found that both the walk out and the on current degradation can be removed at $150^0C$. Since the trapping of holes and interface state formation occur in the same experiments, and anneal at the same temperature, the two phenomena are no doubt intimately related. Furthermore, it should be noted that these effects anneal at much the same temperature as the interface states formed in mono-Si capacitors during FN injection above.

4. Conclusion

Carrier trapping instabilities in low temperature poly-Si TFTs are shown to arise on application of gate bias. Independent carrier injection experiments show the presence of at least two traps of cross section $0.3\text{-}1\text{x}10^{-15}cm^{-2}$ and $0.5\text{-}1\text{x}10^{-17}cm^{-2}$ in APCVD oxide. The larger trap is removed on annealing at temperatures $>900^0C$, and is believed to be related to the densification of the oxide. Thermal annealing experiments on trapped charges indicate that it is this larger trap which is responsible for the trapping instabilities in TFTs. TFTs fabricated with recrystallized $\alpha$Si show reduced trapping instability, and this is most likely due to a reduction in interface roughness. A small change in subthreshold characteristic could be due to the generation of near mid gap acceptor states, but further investigations are required.

The application of high drain bias into the avalanche region leads to drain current walk out and on and off state degradation. The walk out and on current degradation are consistent with hot hole injection and trapping accompanied by near band edge acceptor state generation. The trapped holes and acceptor states are formed in the same experiments, and also anneal at the same temperature indicating that there is an intimate relationship between the two phenomena. In view of similarities to results in mono-Si MOSFETs it is believed that the acceptors are created at the interface rather than in the bulk of the poly-Si. Further work is required to explain the off current degradation.

5. Acknowledgement

We would like to acknowledge S D Brotherton for useful discussions.

6. References

Brotherton S D  Young N D and Gill A 1988 Solid State Devices ed Soncini G
  and Cazoladi P (Amsterdam: Elsevier) p35.
Nicollian E H  Berglund C N  Schmidt P F and Andrews J M 1977 J Appl Phys 42 5654.
Nicollian E H and Brews J R MOS Physics and Technology (New York: Wiley) ch11.
Wu I-W  Jackson W B  Huang T Y  Lewis A G and Chiang A IEEE Electron Device Lett 11 4 167.
Young N D  Gill A and Clarence I R 1989 J Appl Phys 66 187.
Young N D and Gill A 1990a Semicond Sci Technol 5 72.
Young N D and Gill A 1990b to be published.
Young N D and Gill A 1990c accepted for publication in Semicond Sci Technol.
Zhang J F  Watkinson P  Taylor S  Eccleston W and Young N D  1990 This Conference.

Paper presented at ESSDERC 90, Nottingham, September 1990
Session 4A6

# Optimisation of a 5 nm ONO-multilayer-dielectric for 64 Mbit DRAMs

### A. Spitzer, H. Reisinger and W. Hönlein

### Siemens AG, Otto-Hahn-Ring 6, D-8000 München 83, F.R.G.

Abstract. The impact of the thicknesses of the nitride- and oxide layers on the performance of a 5nm ONO-layer was investigated. For an optimized 5nm ONO-dielectric $t_{bd}^{63\%}$ at 5MV/cm was found to be above $10^{12}$sec. In this dielectric the charge transport in the oxide layers is due to direct tunneling processes and the wear-out properties are dominated by the nitride layer. The limiting factors for the reduction of the individual layers of the stacked dielectric are discussed.

### 1. Introduction

Multilayer dielectrics composed of $SiO_2$-$Si_3N_4$-$SiO_2$ (ONO) are widely used as insulating films in the varactors of DRAMs. Recently, it was proposed that this concept is also applicable for the 64Mbit generation if the effective thickness can be scaled down to 5nm [1]. We investigated the scaling properties of ONO-layers and analysed different processes to realize 5nm ONO-dielectrics. Based on these investigations we present guidelines on how one can optimize these layers and what the wear-out mechanisms are.

### 2. Experimental

For the analysis, planar capacitors were fabricated on arsenic doped polysilicon. The ONO-layers were composed of thermal bottom oxide, LPCVD nitride, and thermal top oxide (dry oxidation). The planar capacitors were defined by structuring the $n^+$-doped polysilicon gate electrode with areas from $10^{-2}$ to 16mm$^2$. In these experiments 10 different samples were prepared whose thicknesses of the bottom oxide, nitride, and top oxide were varied while keeping the total oxide equivalent thickness between 5 and 6nm. For the purpose of comparison the same ONO-layers were available on p-doped (100)-silicon wafers. The thicknesses of the bottom oxide and the nitride layer were evaluated by ellipsometry and the top oxide thickness by combined measurements (ellipsometry and CV-measurements) of the complete ONO-layers. Some of the samples were also investigated by cross section TEM-analysis with respect to the layer thicknesses and the roughness of the polysilicon bottom electrode; the layer thicknesses of the different methods were in agreement within a few Angstrom.

### 3. Results

IV-characteristics of the different samples were measured for both gate polarities. From the measurements it is evident that the current density at a given electric field is a function of the thickness of the oxide layer, i.e. the current is lower for thicker oxides. This is shown in Fig.1 where the data are given for three samples with the same nitride thickness (4nm) but different bottom- and top oxide thicknesses (1.5 - 2nm).

© 1990 IOP Publishing Ltd

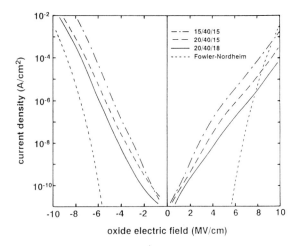

Fig.1: IV-curves of three ONO-layers between $n^+$-doped polysilicon electrodes; for each layer the thicknesses of the bottom oxide, nitride, and top oxide are given in Angstrom. A calculated FN-characteristic is shown for comparison.

For these plots the electric field was calculated from the electric charge collected on the plates since the calculation of the field from the applied voltage is difficult for very thin dielectrics due to the increased importance of the space charge regions in both electrodes. For negative gate polarity the IV-curves of the ONO-layers are always above a Fowler-Nordheim-characteristic calculated with the data of a silicon-oxide-polysilicon structure. The higher current densities of these thin multilayers are due to direct tunneling processes in the extremely thin oxide layers. For positive electric fields above 7MV/cm a crossing of the experimental curves with the calculated FN-characteristic is observed. This might be explained by the different field dependence of the nitride related Poole-Frenkel-conductivity; due to the different doping levels of the two polysilicon electrodes and due to the asymmetry of the oxide layers the relation of electron to hole conductivity depends on the gate polarity.

Therefore, in these 5nm ONO-layers the oxide layers are reducing the current compared to a single nitride layer. This is like a current reduction by series resistors where the resistors are realized by tunneling oxides. Since direct tunneling processes do not contribute significantly to the wear-out properties of these ONO-layers the time-to-breakdown values of the investigated samples should be nearly identical for stress conditions with the same current density and should be independent of the individual layer combinations. This is confirmed by time-to-breakdown measurements. In Fig.2 data are given for 5 different ONO-layers at a +12MV/cm stress as well as the data for one wafer with different electric fields. This plot shows that the time-to-breakdown is mainly a function of the current density and reflects only to a minor extent whether the different current densities are due to different electric fields at one particular layer or due to measurements of different samples at a fixed electric field of 12MV/cm. Regarding an optimization of a 5nm ONO-layer one should on the one hand minimize the current density at a given electric field by oxide layers as thick as possible and increase the dielectric lifetime in this way. On the other hand, the nitride layer must be so thick that it dominates the wear-out properties since additional investigations of the scaling properties of thicker ONO-layers have yielded that the reliability in constant voltage stress tests is increased if the composition is changed from a very thin nitride layer to nitride dominated multilayer. Furthermore, the nitride layer must prevent an increase of the bottom oxide during top oxide formation due to oxygen diffusion. For this reason the minimum nitride thickness is between 3 and 4nm depending on the conditions of the nitride deposition and the top oxide formation.

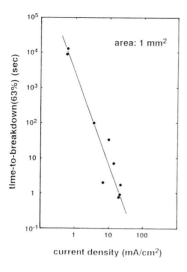

Fig.2: Time-to-breakdown (63% cumulative failure) versus current density. Data are given for 5 different ONO-layers measured at the same electric field and for one layer at different fields.

We have optimized an ONO-layer according to the above mentioned suggestions concerning current density and layer thicknesses; the effective thickness was 5.5nm and the bottom and top oxide was 2.0 and 1.8nm respectively. For this layer reliability relevant quantities like defect distribution, defect density, and dielectric lifetime as a function of the capacitor area and of the electric field were investigated. Breakdown distributions were measured at an electric field of -9MV/cm for three sets of 104 capacitors with an area of $3*10^{-2}$, 1, and 16 mm$^2$ respectively. The cumulative failure distribution of the 16 mm$^2$ samples are shown in a Weibull diagram in Fig.3. The data points are arranged in a straight line indicating that only one failure mode is found in these samples. Since one capacitor broke down in a weak spot related way the defect density is determined to be below 0.05cm$^{-2}$. The analysis of the breakdown distributions of the three areas showed that the single failure mode of the capacitors is not the intrinsic mode of the ONO-layer since the distributions are exactly scaled by the logarithm of the capacitor areas. Therefore, it is concluded that the observed failure mode is related to a layer property like the roughness of the nitride thickness.

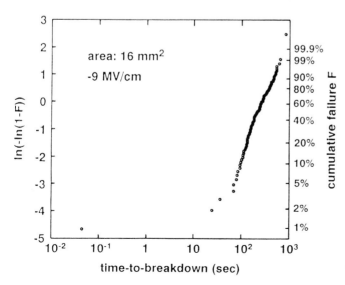

Fig.3: Cumulative failure versus the breakdown time in a Weibull diagram for a 5.5nm ONO-layer between polysilicon electrodes. The area of the planar capacitors was 16mm$^2$.

For each electric field of 8, 9, 10MV/cm for negative gate polarity and of 11, 12, 13MV/cm for a positive one more than 100 of the 1mm$^2$ capacitors were measured. The time-to breakdown values at a cumulative failure of 63% are given in Fig.4. The data in this figure are extrapolated to lower electric fields with the assumption that the logarithm of the breakdown times is a linear function of the electric field. From an experimental point of view this assumption cannot be proven due to the limitations in measuring time. Nevertheless it seems to be reasonable since in the range of breakdown times accessible by experiments the data can be nicely fitted by a straight line with this model. Besides that, the logarithm of the current density of the 5nm ONO-layers depends overall nearly linearly on the electric field and therefore supports this extrapolation. Furthermore, an extrapolation versus the electric field is more conservative than versus the reciprocal or the square root of the field as is usually done for nitride or oxide data. Even this extrapolation of $t_{bd}^{63\%}$ which is considering worst case yields a lifetime above $10^{12}$sec at 5MV/cm for the more critical polarity.

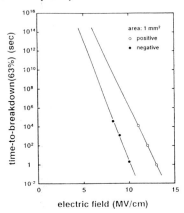

Fig.4: Time-to-breakdown (63% cumulative failure) of a 5.5nm ONO-layer as a function of the oxide equivalent electric field. Data of capacitors with 1mm$^2$ area are plotted for both polarities.

A comparison between the data of the 5nm ONO-layers on a (100)-silicon and on a polysilicon electrode showed no significant differences. The IV-characteristics and defect densities are identical whereas the time-to-breakdown values are a factor 2 higher for the layers on (100)-silicon. These results are remarkable since the polysilicon surface of the bottom electrode was fairly rough. It indicates that the properties of the described ONO-layers do not depend very much on the structure of the bottom electrode; this is confirmed by experiments on non-planar capacitors.

4. Conclusions

Our results clearly demonstrate the potential of 5nm ONO-layers as a reliable and promising dielectric for 64Mbit DRAMs. The contrary requirements on the thicknesses of the nitride- and oxide layers were shown and an optimized multilayer was successfully developed. On the other hand, the ONO concept reaches its limits at an effective thickness slightly below 5nm since thermal oxidation for the top oxide is inevitable for low defect densities and further reduction of the nitride thickness results in an anomalous oxidation of the nitride and therefore in an increase of the bottom oxide [2].

References

[1] J.Yugami, T.Mine, S.Iijima, and A.Hiraiwa, Extended Abstracts of the 20th Conference on Solid State Devices and Materials, Tokyo, 1988, 173.
[2] W.Hönlein and H.Reisinger, Appl. Surface Sci. 39 (1989) 178.

# Optimised and reliable drain structure for 0·5 μm n-channel devices

## G.GUEGAN, G.REIMBOLD, M.LERME
## D.LETI  CENG  85X  38041 GRENOBLE CEDEX  FRANCE

Abstract. Hot carrier induced degradation and short channel effect control are the most serious constraints on 0.5 μm NMOS device design. First, characteristics and device lifetime of various drain structures were compared. Then simulations were performed to achieve the best compromise between these previous constraints and drive capability. Based on these simulations, improved drain structure was processed and tested. This new device has both high current and improved lifetime with a better short channel effect control, than previously ones.

## 1. Introduction

Hot carriers and scaling laws have been recognized as major constraints on MOSFET device design, especially in 0.5 μm n-channel devices. A lightly doped drain (LDD) structure with spacer technology (Ogura et al 1981) has been used to reduce peak electric field. The main purpose of this work is to evaluate the best source/drain structure design of 0.5 μm and even 0.4 μm gate length n-channel devices in terms of current drive capability, short channel effect control and hot carrier degradation resistance. Although many studies have been carried out on reliability of submicron n-channel devices (Sanchez et al 1989 - Nagalingam et al 1989), no work has been reported on various drain structures to achieve both short channel effect control and reliability with a nominal and minimum gate lengths respectively equal to 0.5 μm and 0.4 μm .

The purpose of this work is to show the capability to properly design 0.5 μm n-channel device structure. Initially drain structure with various LDD designs and sidewall oxide spacer lengths were processed and compared. Device lifetime measurements show that a compromise must be made between reliability and short channel effects. Consequently, the drain structure was modified using 2D simulations, and the best compromise between all these various constraints was found.

## 2. Device Fabrication

N-channel LDD MOSFET's were fabricated with a 0.5 μm CMOS process (Guegan et al 1989 ). The transistors were designed with N+ polysilicon gate material, a nominal gate length of 0.5 μm, a gate oxide thickness of 12 nm and a power supply voltage of 3.3 V. The $TaSi_2$ gate was e-beam patterned while the other layers were printed using an optical stepper. After gate etching, various phosphorus doses ( $1.10^{13}$, $2.10^{13}$, $4.10^{13}$, $6.10^{13}$ cm$^{-2}$ ) with 65 keV accelerating voltage were used to form the n- region . Thereafter, various thicknesses of CVD $SiO_2$ layer were conformally deposited and anisotropically etched resulting in various sidewall oxide spacer lengths (0.10 - 0.15 - 0.20 μm ) . Following the sidewall spacer formation, arsenic ion implantation was used to form the n+ region. Then, Rapid Thermal Annealing was performed for both BPSG reflow and junction activation .

© 1990 IOP Publishing Ltd

## 3. Drain Structure Comparison

The electrical behaviour of these various drain structures was measured. Current drive capability in both triode and saturation regions were statistically measured for each variant. Results of structures formed by $6.10^{13}$ cm$^{-2}$ phosphorus implant, which exhibit drastic short channel effect, were not plotted. Figure 1 shows drain process parameter dependence of drain saturation current for the nominal gate length of 0.5 µm at VG=VD=3V. The current drive capability increases with increasing n- implantation dose and with decreasing spacer lengths. This behaviour is due to both effective channel length and parasitic series resistance reductions. Short channel effects, which are illustrated by the threshold voltage shift at 3V drain voltage (Figure 2), increase with increasing n- implantation dose. This is due to the increase of phosphorus junction depth with n- implantation dose.

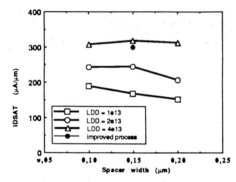

Fig.1 Saturation current versus spacer width at VD =VG = 3 V

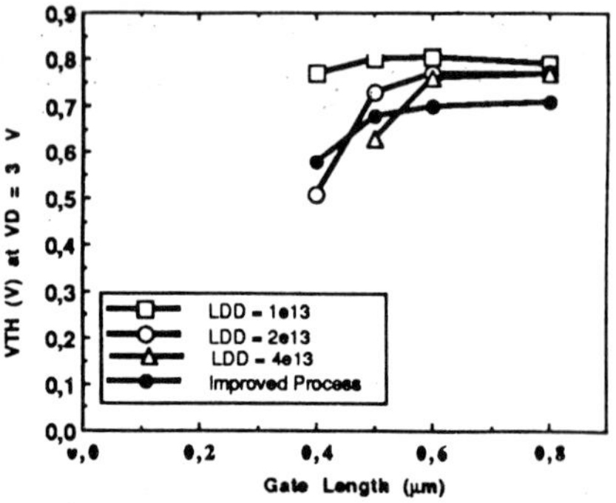

Fig.2 Threshold voltage versus gate length at VD = 3 V

Substrate current for nominal gate length was measured (Figure 3) with 4V of drain bias and a gate bias inducing the maximum substrate current. This current is quite independent of spacer length for the highest LDD dose ($4.10^{13}$cm$^{-2}$). In this case and even for the minimum spacer length (0.1µm), the n- region is not depleted at VD=VG=3 V. Consequently, an increase of spacer length over 0.1 µm will increase drain resistance without significantly modifying the position and the value of the peak electric field.

On the contrary, substrate current is strongly dependent of spacer length for the lowest LDD dose ($1.10^{13}$cm$^{-2}$). The n- region is depleted in this case, even for 0.2 µm spacer length. Electric field variations which spread all over the n- region, leads to a significant reduction of the field for 0.2 µm spacer length. A higher substrate current is observed for the smallest spacer length (0.1 µm) due to the field peak position which is determined by the transition between n- and n+ regions.

The substrate current variation with spacer length for a LDD dose of $2.10^{13}$ cm$^{-2}$ corresponds to an intermediate condition.

Accelerated hot electron life testing was done in addition to these statistical characterization.

The device lifetime is defined as the time during which the decrease of saturation current in the reverse mode is 10%. Figure 4 shows the overall comparison of 0.5 µm device lifetime measured at VG=1.5 V and VD=4 V. These electrical behaviours are correlated with 2D simulations. Simulated results show that the field peak position moves toward the inner gate region with increasing n- dose and consequently transistor degradation decreases. Both substrate current and peak electric field position are pratically independent of spacer length for the highest LDD dose. Consequently, dependence of device lifetime on spacer length is weak. On the contrary, substrate current which is strongly dependent of spacer length for the lowest LDD dose leads to a large increase in lifetime with spacer length increase according to the typical relation $\tau$=A Isub$^{-m}$ with m=3. A difference greater than two decades in device

Hot carriers in MOS devices    313

lifetime can be observed between the highest and the lowest doses for the shortest spacer length (0.1 µm ) in spite of quite similar substrate current. In the case of $1.10^{13}$ cm$^{-2}$ dose , the electric field peak is in the outer gate edge region and electron injection leads to a depletion of n- region. Thus, drain resistance increases and consequently device lifetime decreases. Lower degradation for higher dose ($4.10^{13}$ cm$^{-2}$) is due to the position of the maximum electric field under the gate associated with higher electron injection needed to deplete the n- region.

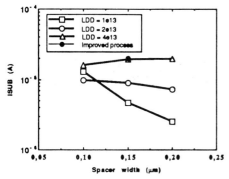 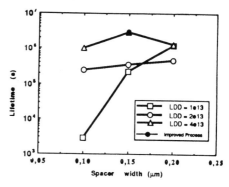

Fig.3 Substrate current versus spacer width at VG = 1.5 V and VD = 4 V (W=25 µm)   Fig.4 Device lifetime versus spacer width

The degradation decreases with n- dose increase under the same substrate current condition (Figure 5). This effect can be explained by the location of the peak electric field under the gate and lower degradation of drain resistance. Although high n- dose leads to higher substrate current, the overall device lifetime increase is significant.

The hot-electron-resistant device processed with $4.10^{13}$ cm$^{-2}$ LDD dose leads to the higher drive capability. Unfortunately this optimized dose leads to short channel effects and suffers from Drain Induced Barrier Lowering (DIBL) and leakage current. Consequently, extensive simulations were performed in order to improve the threshold voltage control.

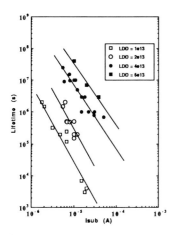 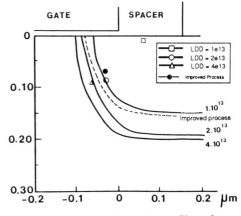

Fig.5 Device lifetime versus substrate current    Fig.6 Simulation of junction profile and location of peak electric field

## 4. Process Improvement

To assure both hot carrier lifetime and good short channel effect control, especially for the minimum gate length (0.4 µm), 2D device simulations were performed with MINIMOS to reach a good compromise, while keeping previous high reliability performance obtained on 0.15 µm spacer length and $4.10^{13}$ cm$^{-2}$ dose. A new drain structure processed with lowered LDD implant energy and optimized oxide sidewall thickness after gate etching was defined.

Figure 6 shows junction profiles, calculated by SUPRA, of the LDD formed by $1.10^{13}$-$2.10^{13}$-$4.10^{13}$ cm$^{-2}$ / 65 keV phosphorus implant and 0.15 µm spacer length. The position of the maximum lateral field calculated by MINIMOS at VG=1.5 V and VD=5 V occurs deeper under the gate with higher phosphorus dose and leads to higher device lifetime in spite of higher substrate current. In the case of lower dose, the peak electric field is in the outer gate edge region which leads to higher degradation of the parasitic transistor. Simulations of the new drain structure shows that the depth of the peak electric field position is about the same. By choosing both the proper oxide sidewall thickness processed before LDD implant and optimum phosphorus implant energy, effective channel length is increased and the junction depth is significantly decreased if compared with the previous LDD structure at the same dose ($4.10^{13}$cm$^{-2}$). Consequently, short channel effect control will be better. In order to maintain drive capability, channel doping implant was adjusted to slightly reduce the threshold voltage.

Devices with these minor modifications and this new drain structure were processed and tested. Measurement results show :
- a decreasing DIBL and consequently higher holding voltage (Figure 2)
- a quite similar drive capability if compared with the previous process at $4.10^{13}$ cm$^{-2}$ dose or a large increase in current if compared with $2.10^{13}$ cm$^{-2}$ (Figure 1) .
- assuming a duty cycle of 1/10 for CMOS technology, at least, a 10 years lifetime was obtained at 3.3 V supply voltage .

## 5. Conclusion

Drive capability, DIBL control, hot carrier degradation resistance are key points to optimize 0.5 µm n-channel devices. Extensive process and device simulations with statistical electrical characterization and device lifetime measurements were performed on devices with various drain structures. The n- doping level and spacer length dependence of drain current, substrate current, short channel effect and device lifetime were described and explained using 2D simulators.

Based on these analyses, an improved structure was proposed and processed. Results indicate that a light modification of the process can lead to a great device performance improvement. The capability to achieve the best compromise between the previously indicated constraints on a 0.5 µm and even on a 0.4 µm gate length device was proved.

Acknowledgements

This work has been supported by DRET contrat.
The authors would like to thank the different teams of LETI SMSC and SAME for device processing and testing.

References

G. Guegan et al "0.5 µm CMOS Devices and Circuit Characterization " ESSDERC 1989
S.J.S. Nagalingam et al "Optimization of submicron n-channel Devices for Performance and Reliability " ESSDERC 1989
S. Ogura et al "Elimination of hot electron gate by the lightly doped drain-source structure " IEDM 1981 pp 654
J.J. Sanchez et al "Drain -Engineered Hot-Electron Resistant Device Structures: A Review" IEEE Transactions On Electron Devices Vol 36 ,N° 6, June 1989

## Simulation of SOI-like kink effects in a 'horseshoe-drain' MOSFET for 16M and 64M DRAM applications

Ravi Subrahmanyan, Marius Orlowski, and Howard Kirsch
Motorola, Inc., APRDL
3501 Ed Bluestein Blvd., MS K10, Austin, TX 78721

### Abstract

The MOSFET structure of a scaled SCC trench cell with a buried drain has been studied using MINIMOS. In this cell the depletion regions surrounding the buried drain can pinch off the substrate, causing sharply increased avalanche carrier generation similar to, but more severe than, the kink effect in SOI MOSFETs. The mechanism for the enhanced avalanche generation and its dependence on bias conditions and geometry have been studied, and simple design rules for punchthrough and pinchoff by the buried drain have been established.

In the quest for ever-smaller DRAM cell structures, various stacked and trench capacitor cell concepts have been proposed, and applied at the 4Mb and 16Mb levels. This paper describes a simulation study of the scalability of the surrounding-high-capacitance trench cell (SCC), proposed by Inoue et. al. (Inoue 1988) for a 16Mb DRAM, to the 64Mb level. The MOSFET is made on a mesa completely surrounded by a trench which contains the capacitor. A highly doped n-type layer on the inner wall of the trench forms one plate of the capacitor, and this layer is connected to the conventional part of the drain of the MOSFET via a vertical $n^+$ connecting layer. The $n^+$ layer on the inner wall of the trench completely surrounds the MOSFET, and constitutes the buried part of the drain. Such an SCC structure provides a large storage capacitance with a small area and shallow trench depth.

The nominal transistor parameters used for the simulations were $L_G = 0.5\mu m$, $L_{eff} = 0.35\mu m$, $W = 1.0\mu m$, $T_{ox} = 12nm$, p-type substrate doping $N_{sub} = 2 \times 10^{17}$ cm$^{-3}$, and $V_{th,Vsub=0} = 0.66V$. The depth of the buried drain with respect to the source junction ($L_{punch}$ in Figure 1) was varied from $0.25\mu m$ to $0.75\mu m$ in the simulations.

A version of MINIMOS 4 (Hänsch 1987) modified to properly resolve the buried drain was used for the simulations. Doping profiles for the source, channel, and drain regions, and for the $n^+$ link region between the surface and buried drains were created using the process simulator PEPPER and input into MINIMOS. The buried drain was constructed by adding the buried layer to the channel profile.

This paper reports enhanced avalanche effects which severely lower the breakdown voltage of the MOSFET in the SCC structure when a sufficiently high drain bias causes the depletion regions around the buried drain to meet and pinch off the substrate (determined by $L_{cutoff}$ in Figure 1.) The avalanche effects are similar to the so called "kink"

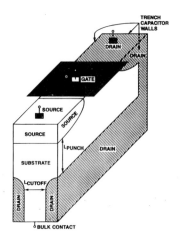

Figure 1: Perspective view of the SCC transistor with the horseshoe-drain structure.

effects found in SOI devices and MOSFETs (Colinge 1988a, Colinge 1988b, Hafez 1989) but are much more severe in this case. While in the case of the SOI device an increase in drain voltage beyond the kink leads to a moderate increase in the drain current Ids because the bulk potential is clamped by the forward-biased source-bulk diode, the pinched-off SCC transistor shows an exponential increase in Ids.

This is shown in Figure 2, which plots simulated I-V characteristics for the transistor with the "horseshoe-drain" and pinched-off substrate. The distance $L_{punch}$ between the source and buried drain was made $0.75\mu$m to avoid conventional punchthrough between the buried drain and the source. The two upper curves show the anomalous behaviour of Ids for Vgs = 2V and 3V, with the impact-ionisation model switched on. A dramatic increase in Ids can be seen above Vds = 1.5V. The two lower curves correspond to the

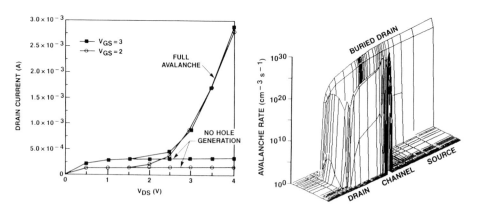

Figure 2: Ids vs. Vds curves for the pinched-off device with full avalanche generation and without hole generation.

Figure 3: Front view of avalanche generation rate in the structure.

same bias conditions, except in this case the hole generation due to impact ionisation is switched off while the electron generation is still on. The lower curves are nearly identical to I-V characteristics obtained for a conventional device with full avalanche (hole and electron generation) *and* a biased substrate.

This effect is explained along the same lines as the kink effect in SOI devices. Channel carriers (electrons) in high lateral fields generate hole-electron pairs near the drain region as shown in Figure 3. (Note that in this case avalanche multiplication takes place along the entire drain region including the buried drain, and this is explained below.) The generated holes accumulate under the source in the absence of a substrate contact, lowering the potential barrier between the source and bulk. This is clearly seen in Figure 4(a,b), which plots the potential distribution for a vertical cross-section under the source with (a) hole and electron generation, and (b) electron generation only. This barrier lowering, which is stronger in the vertical source-substrate junction than in the lateral source-channel junction, leads to an enhanced drain current, partly increased in the channel region due to a lowering of the threshold voltage by the increased bulk potential, and partly between the source and the buried drain. The enhanced channel (drain) current in turn produces more holes by increasing the avalanche multiplication process. However, the barrier lowering between the source and the bulk leads to a greatly increased vertical current flow from the source to the buried drain, which is instrumental in triggering breakdown along the buried drain as well, and this further increases the overall avalanche multiplication. It can be seen from Figure 3 that the avalanche rate along the buried drain is as much as the peak avalanche rate near the conventional drain. This mechanism is similar to that of collector-emitter breakdown in a bipolar transistor. These mechanisms are part of a positive-feedback cycle which results in the

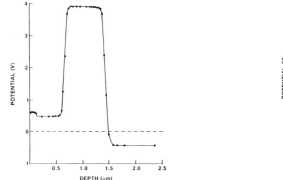
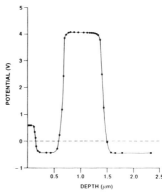

Figure 4: One-dimensional plot of potential under the source versus depth (a) with full avalanche generation and (b) without hole generation.

runaway increase of Ids shown in Figure 2. It should also be noted that because both holes and electrons accumulate in the substrate at high non-equilibrium concentrations, the enhanced avalanche effect depends strongly on the carrier recombination rate, as pointed out by Kato et. al. (Kato 1985) for SOI devices. The enhanced avalanche rate is also very sensitive to transistor geometry, as shown in Figure 5. The largest increase in Ids is observed for $L_{punch}=0.45\mu m$, and the lowest for $0.75\mu m$. Because the enhanced avalanche generation is caused by an accumulation of holes in the substrate, an increase in the volume of the substrate should decrease the hole concentration and thus decrease the avalanche effect, as confirmed by the simulation results. This is also consistent with the results of Hafez. et. al. (Hafez 1989) who have simulated kink effects in regular MOSFETs with unbiased substrates. Their results can be thought of as the limiting case of a very large $L_{punch}$ (implying a large volume available for holes to accumulate) where the hole accumulation in the substrate is not enough to induce the runaway avalanche multiplication.

Figure 6 shows the minimum distance $L_{cutoff}$ required to prevent pinchoff up to $V_{DB}$, as a function of doping. To reduce this distance further, a high concentration p-layer may be needed on the bulk side of the buried drain as reported by Inoue et. al. (Inoue 1988). For example, at the 64 Mb level, with a substrate doping of about $2\times10^{17}\text{cm}^{-3}$, the minimum distance is $0.7\mu m$ without an antipunch layer, which is too large. Similar considerations may limit the utility of these cells beyond the 16Mb level.

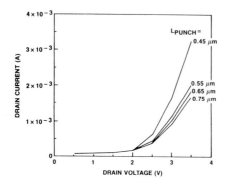 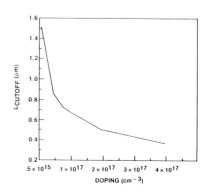

Figure 5: Ids vs. Vds characteristic for various buried layer depths at Vgs = 1.7 V

Figure 6: Minimum separation $L_{cutoff}$ required between ends of buried-drain layers to prevent cutoff of the bulk, at Vdb = 4V

# References

M. Inoue, T. Yamada, H. Kotani, H. Yamauchi, A. Fujiwara, J. Matsushima, H. Akamatsu, M. Fukumoto, M. Kubota, I. Nakao, N. Aoi, G. Fuse, S. Ogawa, S. Odanaka, A. Ueno, and H. Yamamoto 1988 *IEEE J. Solid State Circuits* **23** 1104.

W. Hänsch and S. Selberherr 1987 *IEEE Trans. on Electron Devices* **ED-34** 1074.

J.-P. Colinge 1988a *Microelectronic Engineering* **8** 127.

J.-P. Colinge 1988b *IEEE Electron Devices Letters* **EDL-9** 97.

I. M. Hafez, G. Ghibaudo, and F. Balestra 1989 *Proc. European Solid State Device Research Conf.* eds A. Heuberger, H. Ryssel, and P. Lange (Berlin: Springer-Verlag) pp 897-900.

K. Kato, T. Wada, and K. Taniguchi 1985 *IEEE Trans. on Electron Devices* **ED-32** 458.

# The future of epitaxy

**P. Balk**

DIMES, Delft University of Technology
P.O. Box 5053, 2600 GB DELFT, the Netherlands

Abstract. This paper reviews the present status and outlook for the epitaxy of III-V semiconductor structures. It focusses on the metalorganic approach and deals with MOCVD and MOMBE. After discussing some general features of these methods particular attention is given to the control of growth in the case of quantum well structures, selective deposition using dielectric masking and the development of new precursor materials.

1. Introduction

One of the most characteristic features of semiconductor device structures is the presence of potential barriers for electrons and/or holes. For compound semiconductors such barriers (or junctions) are not only located at the interface to a metal or at a p-n junction. They will also occur between different compound semiconductor materials (heterojunctions), where they permit the construction of device structures unattainable in the silicon technology. In contrast to the technology of silicon devices, where diffusion and ion implantation plays a major role in the formation of junctions, the dominant technique for preparing junctions in compound semiconductor structures is that of depositing them layer by layer by means of epitaxial growth. Since heterojunctions are of fundamental significance in modern compound semiconductor devices, like lasers, HEMTs and tunnel structures, mastering the epitaxial growth technique is an essential requirement for progress in this field. The present status of the epitaxial growth technique and the direction of its development is the topic of this review. For practical reasons it will focus on III-V materials.

An epitaxial technique used for the growth of device structures should fulfill a number of requirements: It should provide layers of high intrinsic purity, but, if necessary, also well-defined dopant profiles. The compositional transition at heterojunctions should be abrupt. An atomically smooth surface is obviously expected. Moreover, it is essential to exactly control the composition of ternary and quaternary compounds. This control of composition is necessary because it determines not only the electronic properties of the material, but also the barrier height (band offset) and the lattice fit at heterojunctions. Moreover, the growth technique should provide exact control of film thickness over an entire substrate wafer (2 or 3 inch diameter) for the fabrication of multilayer structures ranging from MeSFETs to superlattices and multiple quantum well lasers. To allow integration, growth on preselected areas should be possible.

Three basic techniques are available for epitaxial growth of III-V films. They distinguish themselves in the ease with which they fulfill some or all of the above requirements. A classical and convenient laboratory method is the growth of semiconductor films from a metal solution (for example, GaAs in Ga). This method (Liquid Phase Epitaxy, LPE [1]) generally yields high quality material, but the growth of very thin films and multilayer structures is difficult to control. Moreover, the surface morphology of the films often leaves to be desired. Excellent control of composition and growth rate is obtained in the deposition using molecular beams of the elements in a high vacuum apparatus (Molecular Beam Epitaxy, MBE [2]). However, here again the morphology of the films is not quite perfect. More importantly, the use of an elemental phosphorus source presents considerable difficulties, so that the use of this technique for P-containing compounds is impractical. The third method is chemical vapor deposition. Here two approaches are available. The older one (Vapor Phase Epitaxy, VPE [3]), starts from inorganic materials, for example, elemental Ga, $AsH_3$, HCl and the carrier gas $H_2$ for the deposition of GaAs. At reduced total pressure it produces high quality material at extremely high rates of deposition. For this reason it may be used to deposit thick layers to provide high quality substrates [4]. However, it is not suitable for the deposition of very thin film multilayer structures, like multiple quantum wells. Because of the low flow rate used in this method, it appears difficult to obtain abrupt transitions between the individual layers. The second chemical vapor deposition method uses metal organic group III precursors and group V hydrides, for example, $(CH_3)_3$ Ga (trimethyl Ga, TMG) and $AsH_3$ (arsine) for the deposition of GaAs (MetalOrganic Vapor Phase Epitaxy, MOVPE) [5]. Again $H_2$ is used as the carrier. It turned out that the starting compounds for MOVPE can also be used conveniently in MBE. This led to the "hybrid" technique MOMBE (MetalOrganic MBE) [6], also named CBE (Chemical Beam Epitaxy).

This multitude of methods is caused by the fact, that different material systems pose different requirements. It is also characteristic for a technology which is still developing. However, during the past few years there appears to be a strong focuss on MOVPE and the related technique MOMBE. These methods will be the topic of the rest of this review.

2. The metalorganic approach

MOVPE was developed because of difficulties experienced in the deposition of GaAlAs in the inorganic VPE. Still, also in MOVPE the progress in the late seventies was slow due to the difficulty to counteract the strong tendency of Al-containing semiconductors to incorporate oxygen and carbon. These contaminants are highly undesirable, since they negatively affect the electrical and luminescent behavior of the deposited GaAlAs films. Carbon is an intrinsic impurity introduced by the metalorganic precursor material. Here, the use of group III starting materials with a weak metal to carbon bond will reduce the carbon content of the solid. During the past decade MOVPE has developed to a technique which is capable of producing GaAlAs material useful for the preparation of lasers and high frequency devices. The MOVPE technique has been succesfully extended to combinations of the group III elements Ga, Al and In, using metal organic source materials, with the group V elements P and As, introduced in the form of gaseous hydrides, or Sb as $(C_2H_5)_3Sb$. A considerable amount of work has been done on the GaInAsP system, where the main interest arises from its use for devices for optoelectronic communication in the 1.3 to 1.55 $\mu$ wavelength region.

A major attraction of MOVPE is that the reactants are introduced as gases or vapors. For this reason MOVPE permits high flow rates, which is important for obtaining uniform films and abrupt transitions. This is a distinct advantage over the inorganic VPE method, where the necessity to obtain reaction equilibrium between HCl and the group III metal source permits only low flow rates. The convenient control of the injection of the gaseous precursors in MOVPE also led to their incorporation in the MBE technique and explains the name MOMBE. This approach not only allows avoiding the presence of hot effusion cells in the high vacuum environment, but also offers an attractive way to introduce phosphorus using the gaseous source material $PH_3$. However, to obtain deposition all group V hydrides have to be cracked at high temperatures before injecting them in the deposition chamber.

Even though MOVPE and MOMBE in principle use the same starting materials, there is one fundamental difference between the two techniques: Except for the cracking of the group V hydrides in MOMBE all reactions take place at the substrate surface. In contrast, in MOVPE the reactions leading to deposition take place both in the gas phase and at the substrate surface. These different reaction paths lead to differences in film characteristics (morphology, carbon incorporation) when using the same precursors in both methods. A further consequence of the different role of surface reactions in the two techniques becomes apparent in localized growth on partially masked surfaces, to be discussed later.

Finally, both methods permit n- and p-type doping during growth. In MOVPE gaseous sources are being used by necessity. On the other hand, in MOMBE effusion cells with elementary dopants still are frequently employed. However, there is considerable interest in employing gaseous dopants also in this technique. The feasibility of the latter approach has been demonstrated [7].

In the following we will discuss some examples of the recent achievements of the metalorganic approach.

## 3. Control of film thickness and abruptness of transitions

To achieve exact control in the deposition of heterostructures by MOVPE considerable sophistication in the design of the gas control system is required. Changing the contents of the gas phase should be executed by switching the gas flow from the vent to the reactor (vent-run system[8]) to avoid pressure overshoot. To obtain uniformity of film properties and abrupt transitions a high linear gas velocity is mandatory.

Quantum well structures are most suitable to demonstrate the level of growth control attainable. Particularly for very narrow wells small differences in the growth parameters lead to pronounced differences in the transition energies, as shown by photoluminescence (PL) studies. For GaInAs/InP wells of 1 to 3 monolayers width a change in the width (i.e. the thickness of the GaInAs layer) by one monolayer causes a shift in the PL peak of around 50 meV. Compositional changes in the well or in the barrier material (InP) at the interface will also affect the transition energies by affecting the shape of the well.

Fig. 1 shows PL spectra for 9 multiple quantum well (MQW) samples, each containing 10 GaInAs wells of the same nominal width, separated by 30 nm thick InP barriers and capped by a 60 nm InP top layer. The 9 samples differ only in the growth time of the wells, which was varied from 2.4 to 3.2 sec. in increments of 0.1 sec. At the conditions used in this experiment the bulk growth rate was 2.4 μm/h, which amounts to one monolayer per 0.4 sec. This yields a nominal well width of 6 monolayers for a growth time of 2.4 sec.

The spectrum of this MQW structures shows a single peak at 1024 meV. Increasing the growth time in four 0.1 sec. steps to 2.8 sec. causes this peak to gradually disappear. At the same time a new peak develops at 994 meV, which is attributed to a structure with 7 monolayer wells, as expected on the basis of the bulk growth rate.

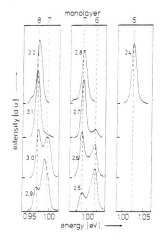

fig. 1  PL spectra of GaInAs-InP MQWs; parameter: growth time in sec.

The constant energy position of the two peaks gives an indication of the reproducibility of the composition of well and barrier from sample to sample and from well to well within each sample. A similar behavior is shown when increasing the growth time stepwise to 3.2 sec. The appearance of the peaks at distinct energies, i.e. the occurence of a discontinuous energy shift, indicates that the growth takes place monolayer after monolayer, i.e. in a 2-dimensional mode. This behavior, which is observed for wells down to one monolayer width (see fig. 2) is clear evidence of the level of control achievable in MOVPE [9]. MQWs prepared by MOMBE have also been reported [10]. However, in this case explicit reports on a discontinuous energy shift are not available as yet.

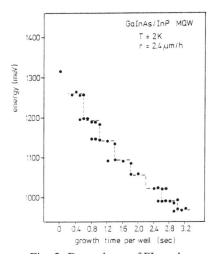

Fig. 2 Dependence of PL peak energies on growth time for GaInAs MQWs.

The samples of fig. 1 were deposited using an interruption of the growth of 5 sec. at the InP to GaInAs and of 3 sec. at the GaInAs to InP interface by stopping the flow of the group III precursors. During these interruptions the group V precursors were exchanged, so that the InP surface was exposed to $AsH_3$, the GaInAs surface to $PH_3$. Particularly during exposure of InP to $AsH_3$ substitution of P by As and formation of a graded InAsP layer takes place due to the high equilibrium vapor pressure of P over InP [9]. The extent of this graded layer is determined by the details of the switching process. However, since this grading will increase the effective well width, the PL peak will be shifted to lower energy. The magnitude of this shift for a nominal well width of 6 monolayers may amount up to 100 meV; it depends on the details of the switching process which require exact control.

The use of rotating substrates has been in use for many years in classical MBE. From there it has been introduced in MOMBE with considerable success [11]. In recent years this approach has also been adapted to MOVPE [12,13]. It enables deposition of layers of highly uniform thickness and composition on 2 or 3 inch substrates in both techniques.

4. Selective growth

The deposition of laterally localized semiconductor structures is particularly important for optoelectronic integration. Together with the growth on contoured substrates it opens the way to new device concepts. Using a masking layer, usually a dielectric film ($SiO_2$, $Si_3N_4$), growth can be confined to the exposed area of the semiconductor. This approach hinges on the sensitivity of the growth reactions on the nature of the surface. Since it is known from classical MBE that metals reaching the surface of the masked area can hardly be removed, it is important that conditions are avoided where the metalorganic precursor completely dissociates already in the gas phase. Instead, at least the final decomposition step should take place on the surface. This implies that low pressure conditions are favorable for obtaining selective growth. Indeed, the total absence of reactions in the gas phase, as realized in MOMBE, allows selective deposition at considerably lower temperatures than in MOVPE. It should be noted, that for a given set of growth conditions the tempreature region of selective growth shows a lower limit. This suggests that at the mask re-evaporation of arriving species rather than complete reaction is the dominant process at higher temperatures. This desorption is probably easier for undecomposed species (MOMBE) than for partially decomposed species (MOVPE).

In addition to the growth temperature the concentration of the reactants is important; it should not exceed a certain limit. In MOVPE only the pressure of the group III precursor has to be limited, in MOMBE also that of the group V component. However, in both techniques the conditions for selective growth are compatible with those for obtaining high quality material. For example, for InP and GaInAs the standard deposition temperatures can be used.

When growing in recesses in masks, the deposited structures will not only be bounded by a surface parallel to that of the mask, but also by sidewalls of different orientation. The different bonding situation and population with lattice atoms will again lead to different growth behavior, particularly when the dimensions of these facets are comparable to the migration lengths of the reactants on the surface. For InP structures oriented in (011) direction on a (100) InP substrate the evolving sidewalls are (III)B planes.

As shown schematically in fig. 3, in MOVPE at high partial pressures the migration length on (III)B and on (100) appears to be low, so that flat growth on both surfaces is obtained; the growth on the (III)B facets will lead to overgrowth on the mask. At lower $PH_3$ pressures the migration length on (III) B increases, so that enhanced growth at the edges of the (100) surface is observed. This is also the case for the higher $PH_3$ fluxes in MOMBE.

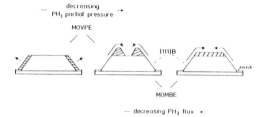

Fig. 3 Schematic of growth on partially masked substrates in MOVPE and MOMBE.

For lower $PH_3$ fluxes in the latter method all precursor species incident on the (III)B plane again reach the (100) surface. However, here the migration length on the (100) surface is also large enough to attain complete planarization without overgrowth on the mask [14].

Growth in the recesses of the masked area implies a nonuniform consumption of reactants across the surface. In MOMBE the growth rate in the recesses is the same as that on completely unmasked substrates. This indicates that surface migration of reactants does not play a role. However, in MOVPE the growth rate strongly depends on the dimensions of the masked and unmasked areas. Smaller unmasked areas and larger masked regions both lead to an increase in the growth rate, which may easily amount to a factor of 2 or more. It could be proven that this effect is caused by diffusion of reactants through the gas phase parallel to the surface.

The supply of material by diffusion will be diffierent for different precursors, i.e. species with different mass. This effect shows up in ternary alloys in the form of a dependence of the alloy composition on the size of the deposited patterns and their distance. In addition, the composition also varies laterally. These effects decrease for reduced overall pressure (fig. 4) [15] and are altogether absent in MOMBE. Thus, the use of MOMBE shows some clear advantages over that of MOVPE: In the latter case adjustment of the growth parameters to the structure to be deposited may be possible.

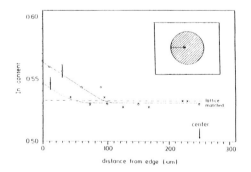

Fig. 4 Lateral variation of In content in GaInAs film, taken from Raman analysis; total pressure 20 mbar: circles; 10 mbar: crosses.

However, to correct for lateral compositional variations would require selection of group III precursors with the same diffusion constant; this does not seem to be a practical proposition.

## 5. Novel precursors

The classical precursors used in the metalorganic approach (trimethyl or triethyl group III compounds, group V hydrides) were primarily chosen while they were readily available, not because their use led to optimal film properties. In fact, these source materials have some distinct drawbacks: The hydrides combine a high vapor pressure with a high toxicity, which is extremely undesirable from a safety point of view. In addition, they have to be used in a large excess, which implies handling relatively large amounts of the hydrides, particularly in MOVPE. The trimethyl and triethyl group III compounds are pyrophoric. Moreover, they easily take up alkoxy groups, which may lead to oxygen contamination of the deposited films, particularly in the case of AlGaAs. Moreover, for AlGaAs, but also for GaAs there is a tendency for these group III compounds to induce carbon incorporation.

To replace $AsH_3$ and $PH_3$, completely substituted hydrides (like $(CH_3)_3As$) or partially substituted hydrides (like $(CH_3AsH_2)$ have been evaluated. It appears that the use of completely substituted hydrides interferes with the removal of carbon from the deposited layers. For example, the use of $(CH_3)_3$As combined with TMG in MOVPE results in GaAs films with high carbon background doping [16]. In MOMBE such problems do not arise, most likely because in this case thermal cracking of these completely substituted As precursors is a condition for growth [17]. It appears that in MOVPE As-bonded hydrogen facilitates reaction of the group III ligands to form stable and volatile hydrocarbons. Compounds like $C_6H_5AsH_2$ (phenylarsine) and $t(C_4H_9)AsH_2$ (tertiary butyl arsine) have been succesfully used in the deposition of GaAs [18,19]. Results reported for the deposition of other III-V compounds are in line with this concept.

A disadvantage of the partially substituted arsines is their strong Lewis base character. In combination with the standard group III precursors, which exhibit Lewis acid character, they tend to show strongly temperature dependent prereactions in the gas phase. This in turn causes a strong temperature dependence of the thickness and composition of the deposit; it thus presents a problem of process control. A promising approach to solve this problem is the use of coordinatively saturated group III compounds. Because of their low reactivity towards electron donors their use in film growth produces an extended region where the rate is only weakly dependent on the growth temperature. This contrasting behavior of the coordinatively unsaturated and saturated group III precursors is illustrated in fig. 5.

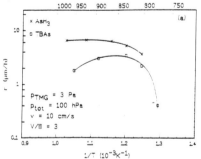
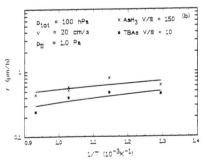

Fig. 5 Arrhenius plot of GaAs growth rates using TMG (a) and 1-3 dimethyl-aminopropyl-1 galla cyclohexane (b) for two different As sources.

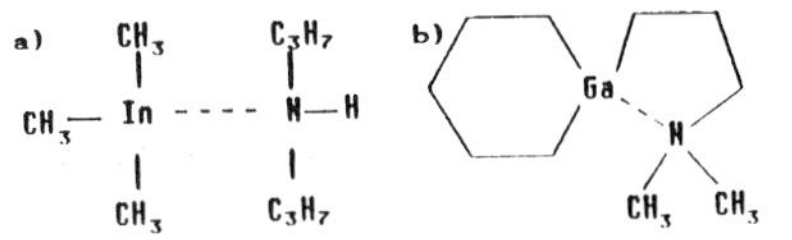

Fig. 6 Examples of inter- and intramolecularly coordinatively saturated group III precursors:
(a) trimethyl In-di-isopropylamine adduct; (b) 1-3 dimethyl-aminopropyl-1 galla cyclohexane

Fig. 6 presents two examples of compounds showing intra- or intermolecular coordinative saturation by means of amino groups. Comparable In compounds produce very high quality InP layers [20].

Coordinatively saturated compounds exhibit a distinctly reduced reactivity towards oxygen. This not only eliminates the pyrophoric characteristics found in the classical group III precursors. It is also important to avoid oxygen contamination in the growth of Al compounds. Specifically, the deposition of exceptionally high quality AlGaAs films was recently reported using the Ga precursor 1-3 dimethyl-aminopropyl-1 galla cyclohexane (fig. 6) and the comparable Al compound [21].

Similar concepts have also found increased interest for application in MOMBE [22]. Partially substituted group V hydrides can be used without precracking. However, in this technique the highest film quality is still obtained using the group V hydrides. The safety problems connected with their use may eventually be solved by in situ production from solid sources using electrochemical methods or by low pressure storage by means of adsorption. These methods and the in situ generation of the hydride from the group V elements with activated hydrogen in a glow discharge [23] have also been attempted for use in MOVPE.

## 6. Outlook

In the development of the III-V epitaxy in the past decade MOVPE has acquired a dominant position. The fast rise of MOMBE since the mid-eighties is certainly related to the broad acceptance of MOVPE. The combined effort of the research on the metalorganic approach has produced a number of important achievements:

* production of fundamental insight in the mechanisms of the deposition processes and deeper understanding of the critical process parameters

* deliberate and successful search for new precursors providing greater operating safety and improved material characteristics

* commercial availability of optimized equipment along with operating procedures and with a convergence of the technical aspects of the equipment, at least for MOVPE

The results presented at the latest international conferences on MOVPE and MOMBE show that we are beyond the time of revolutionary breakthroughs and big surprises. The field of epitaxy is clearly reaching maturity. This technology thus provides an ideal basis for application in devices and circuits, where the demands of modern communicaton technology will offer a strong impetus to further development. In the growth of this field the availability of sophisticated epitaxial techniques does not appear to be the limiting factor at the present time.

Acknowledgement: The author is indebted to A. Brauers, D. Grützmacher and M. Weyers for valuable support in writing this review.

## References

(1) J.C. Brice, in "Current Topics in Materials Science", Volume 2, eds E. Kaldis and K.G. Scheel (North-Holland, Amsterdam, 1977) p. 571
(2) K. Ploog, in "Crystals: Growth, Properties and Applications", Volume 3 (Springer, Berlin, 1980) p. 73
(3) M. Heyen and P. Balk, Prog. Crystal Growth Charact., **6**, 265 (1983)
(4) K. Grüter, M. Dechsler, H. Jürgensen, R. Beccard and P. Balk, J. Cryst. Growth, **94**, 607 (1989)
(5) G.B. Stringfellow, "Organometallic Vapor-phase Epitaxy" (Academic Press, Boston, 1989)
(6) H. Lüth, in "Solid State Devices 1986", ed D.F. Moore, Inst. of Physics Conf. Series Nr. 82 (Inst. of Phys., Bristol, 1987) p. 135
(7) M. Weyers, J. Musolf, D. Marx, A. Kohl and P. Balk, J. Crystal Growth, in press
(8) H. Haspeklo, U. König, M. Heyen and H. Jürgensen, J. Cryst. Growth, **77**, 79 (1986)
(9) D. Grützmacher, J. Hergeth, F. Reinhardt, K. Wolter and P. Balk, J. Electron. Mat., **19**, 471 (1990)
(10) R. Sauer, T.D. Harris and W.T. Tsang, J. Appl. Phys., **62**, 3374 (1987)
(11) H. Heinecke, H. Baur, R. Höger and A. Miklis, to appear in J. Cryst. Growth
(12) P.M. Frijlink, J. Cryst. Growth, **93**, 207 (1988)
(13) E. Woelk and H. Beneking, J. Cryst. Growth, **93**, 216 (1988)
(14) O. Kayer, to appear in J. Cryst. Growth
(15) J. Finders, J. Geurts, A. Kohl, M. Weyers, B. Opitz, O. Kayser and P. Balk, to appear in J. Cryst. Growth
(16) R.M. Lum, J.K. Klingert, D.W. Kisker, S.M. Aleys and F.A. Stevie, J. Cryst. Growth, **93**, 151 (1988)
(17) J. Musolf, M. Weyers, P. Balk, M. Zimmer and H. Hofmann, J. Cryst. Growth, in press
(18) A. Brauers, O. Kayser, R. Kall, H. Heinecke, P. Balk and H. Hofmann, J. Cryst. Growth, **93**, 7 (1988)
(19) G. Hanke, S.P. Watkins and H. Burkhard, Appl. Phys. Lett., **54**, 2029 (1989)
(20) F. Scholz, A. Molasiotti, M. Moser, B. Notheisen and G. Streubel, to be published in J. Cryst. Growth
(21) V. Frese, G.K. Regel, H. Hardtdegen, A. Brauers, P. Balk, M. Hostalek, M. Lokai, L. Pohl, A. Miklis and K. Werner, J. Electron. Mat., **19**, 305 (1990)
(22) M. Weyers, to be published in J. Cryst. Growth
(23) M. Naito, T. Soga, T. Jimbo and M. Umeno, J. Cryst. Growth, **93**, 32 (1988)

# High packing density techniques for cost effective multifunction GaAs MMICs

A.A. Lane and F.A. Myers

Plessey Research Caswell Limited, Caswell, Towcester, Northants. NN12 8EQ, England.

Abstract: Gallium Arsenide MMIC technology must now be considered mature with many proven circuit designs. However, the majority of these are "conventional" designs realised on the GaAs medium and as such do not take maximum advantage of the process. By utilising a sophisticated modern MMIC process considerable size, and hence cost, reductions can be achieved by using advanced high packing density techniques. These have led to the first cost effective "systems on a chip".

1. Introduction

Gallium Arsenide (GaAs) Monolithic Microwave Integrated Circuits (MMICs) can trace their history back almost thirty years now to the commencement of work on basic GaAs material growth in 1962. Key milestones were the demonstration of the world's first GaAs FET by Plessey in 1966 and of the world's first FET based MMIC by Plessey in 1976. Since that time the processes and design techniques have matured and the GaAs MMIC is a genuinely commercially available process, from dozens of manufacturers world wide.

The purpose of this paper is to describe a sophisticated MMIC process and to describe the design approach taken at Caswell in the application of the technology to a major new system - the active phased array radar.

2. GaAs MMIC Technology

Although MMICs are becoming available above 20 GHz the majority of the World's manufacturers have processes optimised for 1 to 20 GHz operation where the majority of the market exists. The process described below is the Plessey F20 0.5 micron gate process and is typical of the sophisticated MMIC processes.

In the process schematic of Fig. 1 the second level metal is created at the same mask level as the gate metal resulting in all the critical dimensional components being manufactured with the same sub-micron accuracy techniques inherent to the gate process. This metal is used for capacitor bottom plates, gates, all underpass interconnections as well as some transmission lines and other passive components. Interlayer dielectrics are then placed on the wafer to separate the metal layers, planarise the wafer surface, provide the overlay capacitor dielectric and to passivate the active devices. Via holes are etched through the dielectric layers and filled with metal at those points where connections have to be made between the various metal layers. At this stage a thick top level metal is added which contains the low-loss transmission lines, inductors, capacitor top plates, bond pads and other interconnection features. The front-face wafer processing is then completed by adding a covering dielectric to act as a surface protection/passivation layer.

The most widely used MMIC dielectrics are $Si_3N_4$, $SiO_2$ and polyimide although $Ta_2O_5$ has also been reported. The choice of which particular dielectric(s) to employ is dependent on many factors. For capacitor dielectrics the capacitance per unit area, thickness control and

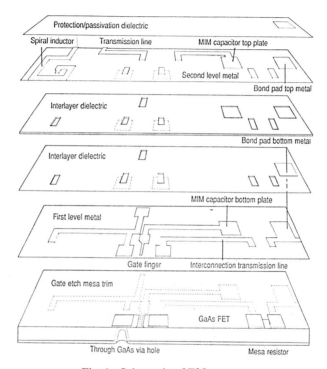

Fig. 1: Schematic of F20 process

freedom from pinhole defects are the main considerations. For interlayer dielectrics which separate the first and second level metals, the step coverage capability of the dielectric film is important in order to avoid short circuits at cross overs. Here polyimide is especially useful since it has a low dielectric constant (3.5) and can be deposited on the wafer at typically 1 micron thicknesses (c.f. $Si_3N_4$ at 0.2 micron).

Having completed the front-face processing the IC wafer is then thinned and through-GaAs via holes are etched before final back-face metallisation and chip separation. Thinning the wafer and backing with metal provides both a ground plane for the microstrip transmission lines and a controlled parasitic image plane for the lumped elements. The accurate control of the final wafer thickness is essential to maintain accurate transmission line characteristic impedances. Now, although GaAs can be thinned to $100 \pm 5$ microns, the mechanical fragility of such a thin wafer precludes its use in genuine production processes and 200 microns is that chosen by Caswell. Through GaAs vias are required to provide a low inductance path to ground and as such they are essential for high frequency ($\geq 12GHz$) and/or compact area circuits. The simple concept of "drilling" holes through the semiconductor is however far from straightforward technologically and a great deal of process R&D has been invested in the development of a high yield technique for the vias. Wet etchant chemistry can create the vias in thin GaAs wafers but its poor directional control causes difficulties in 200 micron thick wafers. Reactive ion etching (RIE) is a much more controllable technique and it is consequently becoming the common approach to vias.

Following the completion of the back surface processing the MMIC circuits are usually probed, e.g. by r.f.-on-wafer probes, before the chips are separated from the wafer by diamond sawing.

## 3. Phased Array Radar Circuits

The complex ceramic based hybrid T/R module (Green et al (1987)) shown in Fig. 2 was produced in 1986 and significant quantities of this type of module have been built for an important phased array radar programme.

The 140 mm long module of Fig. 2, although using the latest techniques available in the early 1980s is, however, far too complex and expensive to be suitable for the volumes required in production radars, where many thousands of modules/radar will be required. Accordingly, Caswell has for the last few years been investigating the increased use of GaAs MMICs and particularly developing low cost pacakging techniques and the use of higher integration levels in the MMICs.

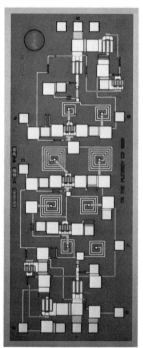

Fig. 2: Prototype T/R module (1986)   Fig. 3: Low noise amplifier (3-6 GHz)

A typical modern phased array radar circuit is shown in Fig. 3. This particular amplifier uses reactive matching techniques to obtain 20 dB of gain with 3 dB NFs over the 3-6 GHz frequency range. The particular circuit uses the F20 process and standard circuit design concepts and occupies ~5 sq.mms of GaAs. Similarly, phase shifters, T/R switches, medium power (~300 mW) amplifiers have all been realised as MMICs and used to produce the small signal module shown in Fig. 4. This module is mounted in a 17 x 17 mm high performance ceramic package with a CMOS controller chip in the centre and is dramatically simpler and cheaper to realise than the module shown in Fig. 2.

However, although a significant improvement, this module approach is still predicted to be too expensive and further improvements are required. The module shown in Fig. 4 requires

~25 sq.mm. of GaAs real estate and, because of its multichip approach, requires a large amount of chip and wire bonding. Caswell has accordingly been investigating advanced CAD techniques to allow components to be placed closer together and thus reduce the area of GaAs used. Also, the "traditional" design approaches used in the circuit of Fig. 3 are now being questioned and novel approaches resulting in the creation of new circuit elements are being investigated, Lane et al 1989.

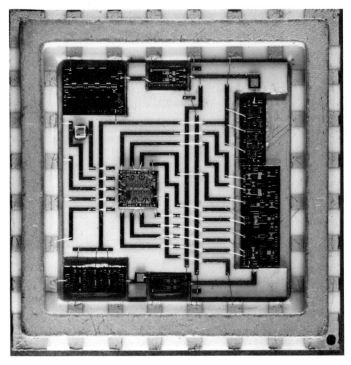

Fig. 4: Packaged module

The result of these approaches is that the area of GaAs required for a radar module is now dramatically reducing. As the F20 process is very high yielding (typically ≥50% of fully functional circuits/wafer) the possibiity now exists of creating multifunction chips, leading to a single chip T/R module.

Fig. 5 shows the functional block diagram of the latest Caswell development in MMIC based T/R modules. The realisation of the C-band concept is shown in Fig. 6, which is believed to be the world's most complex MMIC chip. The significant feature here is that this chip produces at least all of the functions of the module shown in Fig. 4 but in 10 sq.mm. of GaAs and in one chip. The design concepts used in this chip are believed to represent the future design approach for GaAs MMICs.

It should be noted that the MMIC based modules shown in Fig. 4 and Fig. 6 are small signal only, i.e. they produce only about 300 mW of r.f. power. As the full T/R module requires 5-10W of power, discrete GaAs power FETs have to be used as the output stages. Caswell is developing a power derivative of the F20 process and this will allow up to at least 5W of power to be achieved from a MMIC up to at least 10 GHz.

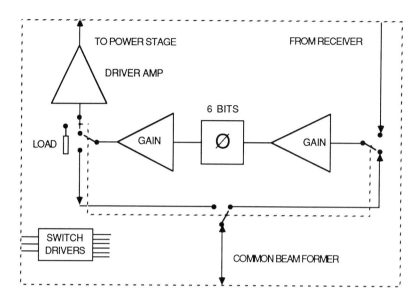

Fig. 5: Functional block diagram of latest multifunction chip (chip area 9.5 sq.mm)

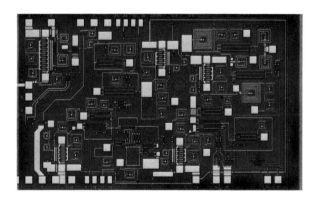

Fig. 6: Multifunction T/R module chip

4. Conclusions

The technological aspects of a MMIC process have been described and the use of this process in the realisation of cost effective T/R modules discussed. The design approaches being developed at Caswell are considered to represent the future applications of MMICs in complex subsystems.

References

Green C R, Lane A A, Tombs P N, Shukla R, Suffolk J R, Sparrow J A, Cooper P D, 'A 2 watt gallium arsenide Tx/Rx module with integral control circuitry for S-band phased array radars' IEEE MTT-S Las Vegas June 1987.

Lane A A, Jenkins J A, Green C R, Myers F A, 'S and C-band multifunction MMICs for phased array radar" 11th Annual GaAs IC Symposium San Diego October 1989.

*Paper presented at ESSDERC 90, Nottingham, September 1990*
*Session 4C1*

# 2D computer simulation of emitter resistance in presence of interfacial oxide break-up in polysilicon emitter bipolar transistors

J.S. Hamel[*], D.J. Roulston[*]
P. Ashburn[**], D. Gold[+], C.R. Selvakumar[*]

[*] *Department of Electrical and Computer Engineering*
*University of Waterloo, Waterloo, Ontario, Canada*

[**] *Department of Electronics and Computer Science*
*University of Southampton, England*

[+] *Department of Metallurgy and Science of Materials*
*University of Oxford, England*

*Abstract* Two-dimensional computer simulations of the emitter resistance and majority carrier current flow in the presence of interfacial oxide break-up in polysilicon emitter bipolar transistors are shown and compared with published experimental results. The simulations reveal that the creation of a large number of extremely small gaps in the oxide layer can result in a substantial reduction in the emitter resistance even though most of the oxide layer remains intact.

## 1. Introduction

A thin interfacial oxide layer at the polysilicon/monosilicon interface, while enhancing the gain by suppressing minority carrier flow in the emitter, also impedes the majority carrier flow giving rise to an emitter series resistance which has been shown [Yamaguchi et al. 1988] to seriously degrade the performance of bipolar VLSI circuits that employ small geometry devices. In order to yield small geometry devices with acceptable emitter resistances and reproducible gains, it is generally considered necessary to completely break up the oxide layer by thermal annealing at the expense of losing its gain enhancement properties. Since enhanced gain may be traded off for a higher base doping resulting in a reduced base resistance and higher maximum oscillation frequency, it is desirable to find optimum processing conditions which will yield the best trade-off between gain enhancement and emitter resistance by inducing only partial break-up of the oxide layer while still retaining reproducibility in gain and emitter resistance.

In this work, results from two-dimensional computer simulations of the emitter resistance in the presence of interfacial oxide layer break-up in polysilicon emitter bipolar transistors, which includes lateral majority carrier current flow, are presented. Examination of these results and comparison with experimental evidence suggests that a relatively short pre-anneal suffices to reduce all of the emitter resistance arising from the interfacial layer while leaving most of the interfacial layer intact.

© 1990 IOP Publishing Ltd

## 2. Details of Computer Simulation

Two-dimensional computer simulations of the emitter resistance are based on a typical polysilicon emitter structure as shown in Figure 1, where the contributions of metal contact resistance, the polysilicon resistance, the interface resistance, and the mono-crystalline emitter resistance, to the overall emitter resistance are included.

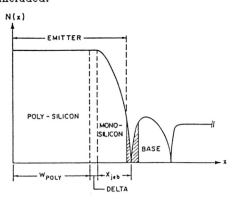

**Figure 1** Polysilicon Emitter Structure

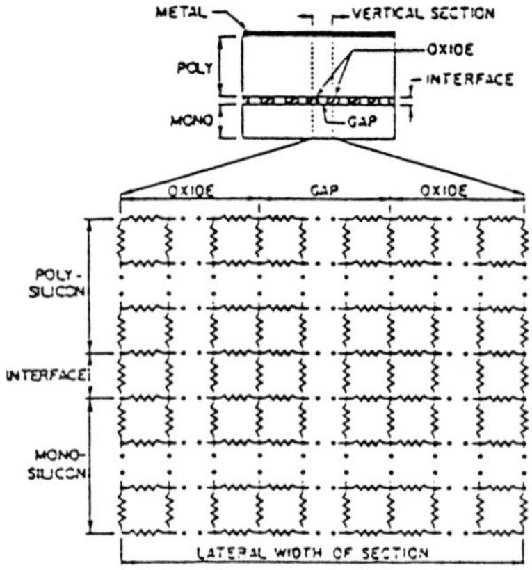

**Figure 2** Vertical section of emitter represented by a 2-D resistor network.

Current-dependent nonlinear oxide layer resistance is calculated using the tunneling expression of [Ashburn et al. 1987], where realistic elliptically-shaped oxide segments are assumed, consistent with High Resolution Transmission Electron microscopic (HRTEM) observations [Wolstenholme et al. 1987], to maintain a constant volume of oxide. All regions of the emitter - Gaussian-doped mono-silicon region with position dependent mobility, oxide layer with gaps, and a uniformly-doped polysilicon region, are represented by a two-dimensional resistor grid network to compute majority carrier current flow. HRTEM investigations [Wolstenholme et al. 1987] reveal that several gaps develop along the oxide interface during its progressive break-up. To model this phenomenon, several uniformly distributed gaps of equal size are assumed whose size depend upon the pre-anneal conditions. The simulation entails the calculation of the specific emitter resistance of a single vertical section of the emitter structure which contains one of the gaps surrounded symmetrically on two sides by elliptically-shaped oxide particles (Figure 2). The emitter resistance and the 2-D majority carrier current flow were studied for various emitter-base junction depths, emitter lateral dimensions, doping profiles, and varying degrees of oxide break-up.

## 3. Results and Discussion

In order to verify the computer simulations, calculations of the specific emitter resistance as a function of the percentage of interface with oxide remaining were compared with experimental resistance measurements of [Wolstenholme et al. 1989] and correlated with HRTEM observations of the percentage of oxide remaining along the length of the interface and the number of gaps in the oxide layer [Wolstenholme et al. 1987, Gold 1989] (Figure 3). The transistors used in the study received pre-anneals, ranging from 900° to 1100° $C$ for 10 minutes, to break up the 10 to 14 $A^o$ thick oxide interface by varying amounts, after the deposition of 0.4

µm of polycrystalline silicon at 610° C but before the emitter implant and drive-in, except for a control sample which received all processing steps without the pre-anneal.

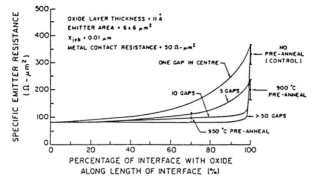

**Figure 3** Specific emitter resistance as a function of oxide layer break-up for different numbers of gaps. Comparison between experiment (•) and 2-D simulations.

The fact that epitaxial regrowth was observed by HRTEM [Gold 1989] to proceed up polysilicon grain boundaries at those locations where they intersected the interfacial layer (Figure 4), and that epitaxial regrowth cannot occur unless the oxide layer is broken, suggests that these intersection points provide energetically favorable gap formation sites. Higher mechanical stress in the oxide layer at these intersection points, caused by the presence of incident grain boundaries, may be responsible for gap formation at these locations.

For the devices in this study, it was estimated that there existed approximately 50 gaps along the length of a given cross section of the interface corresponding to HRTEM observations of an average distance of 1000 $A°$ between gaps. The computer simulations (Figure 3) show that as the *number* of gaps are increased, for a given amount of oxide remaining intact, the emitter resistance decreases since the total lateral majority carrier flow in the mono-emitter region is reduced.

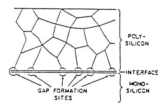

**Figure 4** Gap formation sites at interfacial layer / grain boundary intersection points.

Figure 5 shows a typical 2-D majority carrier current flow in a vertical section in the emitter containing one of many gaps in the oxide layer which comprises 30% of the length of the section.

Even though the oxide layer was observed [Wolstenholme et al. 1989] to be continuous (i.e. percentage of oxide remaining = 100%) for both the control (no pre-anneal) and 900° C pre-anneal samples, a substantial reduction in the emitter resistance between these two cases is evident in Figure 3. No explanation for this substantial reduction in emitter resistance is given in [Wolstenholme et al. 1989]. Since the series resistance of the heavily-doped polysilicon layer was found to be negligible compared to the metal contact and interface resistance, a reduction in the polysilicon resistance due to the 900° C pre-anneal cannot account for the dramatic reduction in observed emitter resistance.

The simulation results suggest, however, that a reduction of this magnitude can occur if a sufficient number of widely distributed gaps form, even though nearly 99% of the oxide layer remains intact along the length of a vertical cross-section (which would correspond to a total gap area of approximately 0.01% of the interface area). The HRTEM observations would not be able to detect such small gaps and the oxide layer would appear continuous. Experimental studies of the effects of pre-annealing on the base saturation current of similar devices [Gold 1989], however, reveal that with only 70% of the oxide layer remaining along the length of the interface there is still a 5 times enhancement in gain.

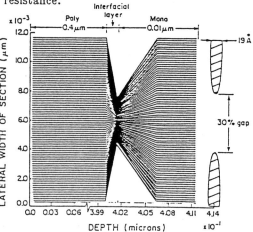

Figure 5 Representative 2-D vertical majority carrier current flow in presence of interfacial layer break-up through one of a large number of uniformly distributed gaps.

## Conclusions

Two-dimensional simulations of majority carrier flow in the emitter reveal that most of the emitter resistance arising from an oxide interfacial layer can be eliminated by the formation of a large number of widely distributed gaps in the oxide layer which, in turn, can be formed by a relatively short pre-anneal of fine-grained polysilicon such that most of the oxide interface remains intact. These results, combined with experimental studies of the effect of oxide break-up on the gain, suggest that the emitter resistance arising from the oxide layer can be eliminated while retaining much of its gain enhancing properties.

## References

Ashburn P, Roulston D J, and Selvakumar C R 1987 *IEEE Trans. Electron. Dev.* ED-34 1346

Gold D 1989 DPhil Thesis, Dept. of Metallurgy, Oxford University

Wolstenholme G R, Jorgensen N, Ashburn P, and Booker G R 1987 *J. Appl. Phys.* 61 225

Wolstenholme G R, Ashburn P, Jorgensen N, Gold D, and Booker G R 1988 *IEEE BCTM* 55

Yamaguchi T, Yu Y -C S, Drobny V, and Witkowski A 1988 *IEEE BCTM* 59

# Tunnelling in implanted emitter-base junctions in a low-power UHF process

B. Schlicht and L. Strobel

Philips GmbH, Röhren- und Halbleiterwerke, D-2000 Hamburg 54, FRG

Abstract
Tunneling currents in reverse-biased emitter-base junctions were investigated for different implanted base profiles in a low-power UHF process. A trade-off had to be found between improvement of the reverse I-V characteristic of the NPN's EB junction and deterioration of the device's HF performance. A semi-theoretical description of the experimental findings is given.

Introduction

High-frequency processes require thin base widths of the order of a few hundreds of nanometers. The narrower the base width the higher the base doping concentration has to be in order to avoid punch-through in the base. These requirements lead to shallow and steep emitter profiles and high peak concentrations in the base, resulting in considerable emitter-base junction electric fields. These in turn enhance the tunneling probability greatly. Recent studies on heavily doped junctions with doping levels between 1E18 and 1E19 $cm^{-3}$ showed that under reverse bias the current through the junction is dominated by tunneling (Stork and Isaac 1983). It should be emphasized that phonon-assisted band-to-band tunneling is an intrinsic mechanism in highly doped junctions, posing a lower bound for the current. The current may be further increased by surface states at the Si-SiO$_2$ interface (trap-assisted tunneling, Hackbarth and Tang 1988).

The transistors investigated in this study consisted of a boron base and a washed arsenic emitter each being implanted into an n-type epitaxial layer. In this concept, the emitter faces the highest boron concentration on its sidewalls. As long as the emitter is not walled against some dielectric, measures to lower the tunneling component of the reverse current must therefore aim at lowering the base peak concentration close to the surface without affecting the intrinsic base resistance too much. Of course, this measure leads to an increased extrinsic base resistance and a correspondingly deteriorated $f_{tmax}$ and noise figure of the device. A reasonable trade-off has to be found between low reverse currents and good HF performance.

Experimental Results

Fig. 1 shows a typical I-V characteristic of a reverse-biased emitter-base junction of an NPN transistor measured at 22°C and 130°C. Striking features are the strong voltage dependence of the current and the small positive temperature coefficient at medium reverse bias. The diffusion component is also apparent at low reverse bias and high temperature. The onset of avalanche breakdown with its negative temperature coefficient is visible near the highest measured reverse bias.

© 1990 IOP Publishing Ltd

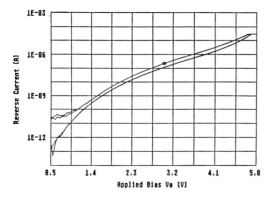

**Fig. 1:** Emitter-base reverse characteristics for two temperatures: lower curve 22°C, upper curve 130°C.

As mentioned in the introduction, the main goal of the present investigation was to study the dependence of the emitter-base reverse current on base profile while keeping the intrinsic base resistance at a constant value. To accomplish this, the implanted boron dose was successively lowered from 1.2E14 cm$^{-2}$ to 0.5E14 cm$^{-2}$ while correspondingly increasing the implantation energy from 30 keV to 44 keV. The base and emitter diffusion processes remained unchanged. Table 1 lists some relevant data.

| case | base dose [E14cm$^{-2}$] | II energy [keV] | sheet res. extr. base [Ohms/sq] | sheet res. intr. base [kOhms/sq] | $I_{eb,rev}$ @ $V_{eb}$=2.7V [nA] |
|---|---|---|---|---|---|
| 1 | 1.2 | 30 | 586 ± 3 | 10.0 ± 1.0 | 65 ± 3 |
| 2 | 1.0 | 33 | 634 ± 6 | 9.9 ± 0.8 | 27 ± 1 |
| 3 | 0.8 | 36 | 710 ± 4 | 10.0 ± 1.0 | 8.8 ± 0.6 |
| 4 | 0.7 | 38 | 767 ± 3 | 9.6 ± 0.4 | 3.7 ± 0.3 |
| 5 | 0.6 | 41 | 831 ± 4 | 9.3 ± 0.2 | 1.1 ± 0.2 |
| 6 | 0.5 | 44 | 920 ± 4 | 11.1 ± 1.4 | < 0.5 |

**Table 1:** Base implantation parameters and corresponding electrical parameters (mean values ± standard deviation)

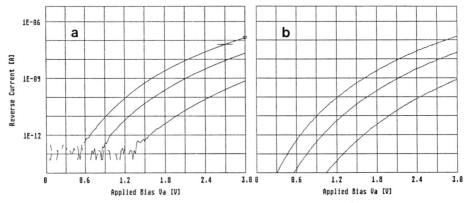

**Fig. 2:** Emitter-base reverse characteristics for cases 1 (upper curve), 3, and 6 (lower curve) (cf. Table 1), a) measured, b) calculated band-to-band tunneling component of the current

In Fig. 2a, typical reverse I-V characteristics for cases 1, 3, and 6 from Table 1 are displayed in the same frame for easy comparison. As can be seen, the absolute current level decreases by more than two orders of magnitude when going from case 1 to case 6, whereas the shape of the characteristic remains almost unchanged. On the other hand, the extrinsic base resistance increases by more than 50% going from case 1 to case 6. The latter fact leads to a decrease in $f_{tmax}$ by some 10 % and an increase in noise voltage by a similar amount. The final process version was chosen such that it provides a reasonable trade-off between low reverse currents and good HF performance.

Measurements on different device geometries on the same wafer revealed that the reverse current is predominantly a perimeter component.

Analysis and Discussion

As mentioned in the introduction, the doping conditions in the investigated devices produce electric fields in the junctions which are high enough for tunneling to become appreciable. Further evidence that this mechanism is responsible for the observed characteristics can be obtained from the temperature dependence and the extremely strong bias sensitivity of the current.

SUPREM3 calculations for the base doping profiles revealed that the peak boron concentration decreases from $4.8E18$ cm$^{-3}$ in case 1 to $1.7E18$ cm$^{-3}$ in case 6 while the peak position slightly moves away from the surface. The intrinsic base resistance is kept at 10 kOhms in all cases.

According to Sze (1981), the band-to-band phonon-assisted tunneling current density is given by:

$$J_t = K_1 \frac{E_p * V_a}{\sqrt{E_g}} \exp\left( -K_2 \frac{E_g^{3/2}}{E_p} \right) \tag{1}$$

where $V_a$ is the applied reverse bias, $E_p$ is the peak electric field at the junction, and $E_g$ is the temperature and doping dependent bandgap energy. The constants $K_1$ and $K_2$ are:

$$K_1 = \frac{\sqrt{2m^*}\, q^3}{4\pi^2 h^2} = 316 \frac{\sqrt{m^*}}{\sqrt{m_0}} \frac{A\sqrt{eV}}{V^2 cm}, \quad K_2 = \frac{4\sqrt{2m^*}}{3qh} = 6.83E7 \frac{\sqrt{m^*}}{\sqrt{m_0}} \frac{V}{(eV)^{3/2} cm}$$

where the symbols have their usual meaning (cf. Sze (1981)). m* is the reduced effective mass in the tunneling direction.

To get an idea about the electric fields occurring at the junctions, additional SUPREM3 calculations of spacial potential distributions were carried out. From these calculations, an empirical formula describing the bias and doping concentration dependence of the peak electric field at the junction could be deduced:

$$E_p = E_o * \ln(n/n_o) * (V_a + V_{bi})^a \tag{2}$$

where n is the boron concentration, $V_{bi}$ the built-in potential (1.12V), and $E_o$, $n_o$, and a are fitting parameters for the special process flow investigated in this study.

Eq. (2) can be plugged into eq. (1) to yield a formula describing (for a given temperature) the tunneling current density as a function of boron doping concentration and applied bias alone. The tunneling current for a given device geometry can now be calculated using:

$$I_t = P \ast \int_s J_t(n(x))dx + A \ast J_t(n(x_{je})) \tag{3}$$

where P is the emitter perimeter, A is the emitter area, and $x_{je}$ is the emitter junction depth. The integration has to be carried out along a vertical cut s through the sidewall of the emitter. For the present study, the emitter sidewall regions were approximated by cylindrical sections of radius $x_{je}$.

A calculation according to eq. (3) was performed for a specific device and for the cases 1, 3, and 6 of Table 1. Bandgap narrowing due to high doping concentrations was taken into account. $K_1$ and $K_2$ in eq. (1) were treated as free parameters. They were determined such that the measured currents for an applied bias of 2.7 V were reproduced correctly. The values found are:

$$K_1 = 0.357 \frac{A\sqrt{eV}}{V^2 cm} \quad \text{and} \quad K_2 = 1.75E7 \frac{V}{(eV)^{3/2} cm} .$$

The calculated results are shown in Fig. 2b for a direct comparison with the corresponding measured results in Fig. 2a. As can be seen from this figure, the agreement between measured and calculated characteristics is excellent. Doping dependence as well as bias dependence are modeled correctly. Temperature dependence can be modeled equally well (not shown here). This means that the mathematical description of the reverse current by eqs. (1) to (3) is at worst an excellent phenomenological one. At first glance, the agreement between theoretically and experimentally determined values for $K_1$ and $K_2$ is unsatisfactory. However, this finding is not unusual and has been encountered by others too (e.g. Hackbarth and Tang 1988). The discrepancies are usually explained by invoking a tunneling mechanism via bands or states within the bandgap (e.g. trap-assisted tunneling). The trap states are most probably confined to the $Si-SiO_2$ interface. The trap density, however, hardly is as closely related to the boron doping concentration as is necessary for an unambiguous explanation of our experimental results. It is far beyond the scope of this paper to attempt a profound theoretical explanation of the experiments. Some qualitative arguments should be given, however. It is well known that in highly doped semiconductors in a strong electric field, the band edges are not "sharp" but have exponential tails reaching into the bandgap (Kane 1963). Tunneling between the tails of valence and conduction band can be characterized qualitatively by a smaller effective tunneling density of states and a higher tunneling probability. This is in agreement with our fitting results.

Conclusion

The present investigation has shown that the I-V characteristics of low-power UHF emitter-base junctions under reverse bias conditions are determined by some kind of tunneling mechanism. It seems plausible that tunneling between band tails is responsible for the observed behaviour.

References

Hackbarth E and Tang D D 1988 IEEE Trans. Electron Devices ED-35
    pp. 2108-2118
Kane E O 1963 Phys. Rev. 131 79
Stork J M C and Isaac R D 1983 IEEE Trans. Electron Devices ED-30
    pp. 1527-1534
Sze S M 1981 "Physics of Semiconductor Devices" 2nd ed. (New York: Wiley)

# Low frequency noise of npn/pnp polysilicon emitter bipolar transistors

N. Siabi-Shahrivar, W. Redman-White, P. Ashburn and I. Post

Deptartment of Electronics & Computer Science,
University of Southampton, Highfield,
Southampton SO9 5NH, England.

*Abstract:* In this paper we will be presenting experimental and theoretical results on *Low Frequency* noise of polysilicon emitter transistors. The results will show that the noise is predominantly generated at the polysilicon/silicon interface. It will also be shown that the *fluorine* segregation at the same interface can cause a large reduction in the *Low Frequency* noise.

## Introduction:

Polysilicon Emitter Bipolar Transistor Technology is increasingly being used in modern VLSI, for high speed applications. The important advantages of *Polysilicon Emitter* technology is it's suitability for producing very shallow emitter/base junctions, thus allowing a co-ordinated scaling of both vertical and lateral device dimensions. The significantly higher gains obtainable also make this a useful technology for analogue applications. For these applications the noise performance of the device, especially the *Low Frequency* (*flicker*) noise is a very important parameter.

It is well known that the *LF* noise of semiconductor devices is strongly related to surface defects, more importantly those defects which are situated near the Emitter/Base junction. In this paper we will present experimental and theoretical results on the noise performance of *npn / pnp* polysilicon emitter bipolar transistors and will discuss the physical origins of the noise, particularly the *1/f* noise in these devices. Possible ways of reducing the noise will also be suggested.

## Experimental Results:

The emitter structure of an *npn* polysilicon emitter bipolar transistor comprises a shallow $n^+$ single crystal region and an $n^+$ polycrystalline region. *TEM* studies of the emitter structure have shown [*GR Wolstenholm etal.-1987*] that at the polysilicon/single silicon interface a thin layer of oxide is always present. The structure and thickness of this oxide is dependent on the chemical treatment given to the silicon surface prior to polysilicon deposition. Slices given an **HF** etch have been observed to have a thin and non-uniform layer of native oxide averaging ≈4Å at the interface, which may be discontinuous in places. By comparison, the **RCA** surface treatment has been observed to produce 10-14Å of relatively uniform layer of oxide at the interface.

Figure 1 shows the noise comparison between the two types of polysilicon emitter transistor discussed above, and a standard *npn* ( BC182L ) trnsistor [DC Murray etal.-1989]. The results clearly identify the interfacial oxide as the main source of *1/f* noise in these new devices.
Although the interfacial oxide is an inherent feature of all polysilicon emitter transistors, it is possible to minimize its influence by reducing its effective area. This reduction in the oxide area can be achieved through high temperature heat treatments. The heat cycle involves carrying out a high temperature preanneal to thermally stress the oxide after polysilicon deposition,

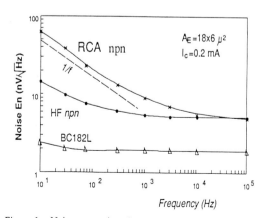

*Figure* 1 - Noise comparison between polysilicon emitter and conventional bipolar transistors

but before the emitter implant. This causes the interfacial oxide to form itself oxide balls during the emitter drive-in. The polysilicon also begins to regrow epitaxially with complete epitaxial regrowth occurring if the temperature is sufficiently high.

The noise comparison of the latter two structures is shown in *figure 2*. This result clearly shows that the polysilicon itself makes a very small contribution to the noise of these devices. Ordinarily, polysilicon is expected to exhibit significant levels of $1/f$ noise due to its highly disordered nature ( ie, high densities of grain boundaries ). However, since in these structures the polysilicon is heavily doped and degenerate, the grain boundaries do not actively participate in the noise process.

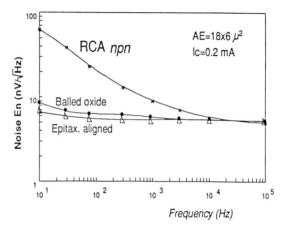

*Figure 2* - The effect of breaking-up the interfacial oxide on the noise of polysilicon emitter transistors.

## Physical Origins of Low Frequency Noise

Having established the interfacial oxide as the main source of *Low Frequency* noise in polysilicon emitter transistors, it is worth studying the physical origins and mechanisms invovled in the generation of noise. Defects in the $SiO_2$ gate dielectric of *MOSFETs* are widely believed to generate $1/f$ noise in these devices [McWhorter-1957]. This being the case, it is conceivable that similar noise generating mechanisms will also be present in polysilicon emitter transistors. The three common types of defects present in the $Si-SiO_2$ are; 1) $Si-Si$ stretched bonds or $O$-vacancy, 2) $Si-O$ stretched bonds, and 3) $Si$ dangling bonds. These defects come from incomplete bonding between the silicon and oxygen atoms during the oxidation stage and introduce states within the band gap of the silicon. The first and second groups of defects introduce states in the top ( *acceptor* state ) and bottom half ( *donor* state ) of the $Si$ bandgap respectively. The latter group of imperfections generally introduce states in the midgap ( *generation / recombination* centres). The effect of such imperfections is to produce a large and localised perturbation potential which captures or binds an electron or a hole.

The capture and subsequent emission of a carrier by a defect ( *Trap* ) centre, alternates the charge state of that centre. The fluctuating occupancy of the centre causes perturbations in the surface potential which in turn modulates the flow of current ( and hence giving rise to conductance fluctuations ) in the device. The modulation of the device current through fluactuation of occupancy of states in the oxide close to the silicon/silicon dioxide interface, can therefore lead to excess noise. It can be shown mathematically [ *McWhorter-1957*] that the noise will have a $1/f$ spectral shape if this process has a wide distribution of relaxation times.

## Noise reduction by annealing of defects:

Defect annealing by *hydrogen passivation* has long been practiced in the semiconductor industry. The effect of *hydrogenation* is to move the energy of the defect generated states outside the silicon forbidden gap ( 0.5 eV above and below the *valence* and *conduction* bands respectively) thereby effectively deactivating these traps. This process results in a dramatic improvement in the electrical characteristics; particularly the *Low Frequency* noise of the device.

Defect annealing can also be achieved with *Fluorine*. Like *Hydrogen*, the effect of *fluorine* is to shift the energy of the interface states outside the band gap of silicon, with the difference that the binding energy of *Si-F* is higher than a *Si-H* bonding and thus the *fluorine* annealed devices are more immune to electrical stress and hot electron injection [PJ Wright-1989].

Evidence of *fluorine* annealing comes from results on noise in *pnp* polysilicon emitter transistors, as shown in *figure 4*. In these devices, *fluorine* is introduced into the device through a $BF_2$ emitter implant. As shown in *figure 3*, the *SIMS* profile of the emitter shows a peak in the concentration of *fluorine* at the polysilicon/monosilicon interface. The lower *1/f* noise observed in the **RCA** *pnp* transistors could therefore be explained by the annealing of interface defects by the *fluorine*.

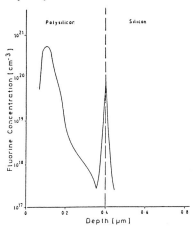
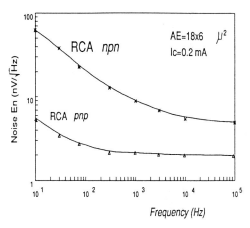

*Figure 3* - SIMS profile showing the *fluorine* concentration in the $BF_2$ doped emitter of *pnp* polysilicon emitter transistor

*Figure 4* - Noise comparison between *npn* and *pnp* polysilicon emitter transistors (effect of *fluorine* on low frequency noise)

## Theoretical model:

The fluctuations in the occupancy of the states within the bandgap causes fluctuations in the surface potential which in turn modulates of the surface recombination current $\overline{i_s}$. This process gives rise to a noise current with a pure *1/f* spectrum. Following Jäntsch's (1987) *random walk* model, the predicted magnitude of this noise current is defined by the expression:

$$i_n^2 = \frac{0.1\ (\overline{i_s})^2}{f\ A_E N_{it}} \tag{1}$$

Where,

$\overline{i_s}$ = Surface recombination current ( $A$ ), $A_E$ = Emitter area ( $cm^2$ ), $N_{it}$ = Number of interface States ( $cm^{-2}$ ) and $f$ = Frequency ( $Hz$ ).

Translating this noise current into an equivalent input referred noise voltage, and using $N_{it}$ as the fitting parameter, we can see from *figure 5a,b* that a good fit to the experimental data is possible by using a value of $N_{it}$ =2.5x$10^{12} cm^{-2}$. This value of $N_{it}$ agrees well with that obtained from the high frequency *CV* measurements of *thin-oxide MIS* structures [ HC Card-1979 ].

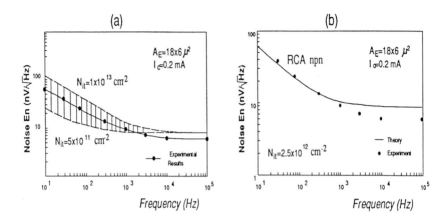

*Figure 5-* **a** ) Comparison between the experimental and theoretical results of noise in an *npn* **RCA** polysilicon emitter transistor fo differnt values of $N_{it}$. **b** ) Using the fitting parameter $N_{it}=2.5 \times 10^{12}$ $cm^{-2}$.

## Conclusions:

Polysilicon emitter bipolar transistors have been found to exhibit significantly higher *1/f* noise than the equivalent metal contacted devices. The increased noise level was seen to be as a direct result of the interfacial oxide. Significant improvement in the noise has been realised by breaking-up the interfacial oxide. It was also found that in some *pnp* polysilicon emitter devices, the *1/f* noise was considerably lower than the equivalent *npn* transistors. The reason for this difference is thought to be due to *fluorine* segregation at the polysilicon/monosilicon interface, thus suggesting a possible method for noise reduction in other types of polysilicon emitter devices. A theoretical model has also been proposed that accurately predicts the noise levels in these devices.

## Acknowledgements

The authors are indebted to Prof. HA Kemhadjian ( Southampton University ) and DC Murray for many helpful discussions. The authors also wish to thank the technical staff at the Sothampton University Microelectronics Centre for their help in preparation of the devices.

## References

[1] HC Card, *Solid State Electronics*, vol.22, 809-817, (1979)

[2] O Jäntsch, *IEEE Trans. on Elec. Devices*, vol.34, 1100-1113, (1987)

[3] AL McWhorter, *Semiconductor surface physics*, RH Kingston Ed., Philadelphia, PA: Penn. Press, 207-228, (1957)

[4] DC Murray, N Siabi-Shahrivar, AGR Evans, W Redman-White, JC Carter and JL Altrip, *Materials Science and Engineering*, B4, 367-372, (1989)

[5] GR Wolstenholme, N Jorgensen, P Ashburn and GR Booker, *J. Appl. Phys.*, vol.61, 225-233, (1987)

[6] PJ Wright and KC Saraswat, *IEEE Trans. on Elec. Devices*, Vol.36, 879-889, (1898)

# Small geometry effects in CMOS compatible self-aligned 'etched-polysilicon' emitter bipolar transistors

G. Giroult-Matlakowski *, A. Marty *, N. Degors, A. Chantre and A. Nouailhat

CNET/CNS Chemin du Vieux Chêne , BP98 , F-38243 Meylan Cedex , France.

* MOTOROLA Semiconductor SA , Toulouse , France

## Abstract.

This paper presents a detailed study of the small geometry effects of advanced CMOS compatible polysilicon emitter bipolar transistors. The current gain increase associated with a reduction in device width is a LOCOS proximity effect, leading to a reduction in base doping concentration under the bird's beak. The current gain modification for small emitter lengths has two major causes.: first, a local variation in the emitter-base dopant concentrations introduced by the oxidation step before spacer formation, and the L.D.E.B.(Lightly Doped Extrinsic Base) implantation; second, a geometrical effect due to the recess of the silicon substrate around the emitter.

## Introduction.

The polysilicon self-aligned bipolar transistor studied in this work has an emitter region defined by polysilicon dry etching (A. Nouailhat 1988). It is thus intrinsically different from the double polysilicon transistor which has been the subject of numerous publications, in particular concerning narrow emitter effects (G.P. Li 1988). Indeed, the short and narrow emitter effects discussed in this paper are very specific to the new transistor structure. These effects have not been reported so far, although a related device structure has been proposed recently (M.P. Brassington 1989 ).

## Device fabrication.

The "etched-polysilicon" emitter bipolar transistor is similar to a pMOS transistor without gate oxide (Fig. 1a). Its fabrication requires the addition of a single mask to the CMOS process mask set, for the implantation of the intrinsic base and the etch of the gate oxide before polysilicon deposition.

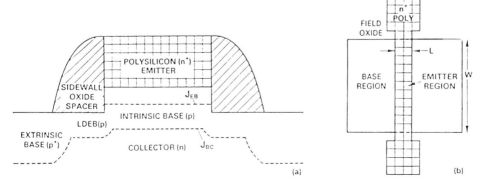

Figure 1:  (a) Schematic cross section of the "etched-polysilicon" emitter bipolar transistor.
(b) Top view of the device layout.

| Steps | Standard parameters Process splits |
|---|---|
| N substrate (collector) | |
| LOCOS isolation | |
| Gate oxidation | |
| Intrinsic base B implantation | 1 x 10¹³ cm⁻², 30 keV |
| Gate oxide etching | |
| RCA cleaning | |
| Polysilicon deposition | |
| As implantation | 1 x 10¹⁶ cm⁻², 100 keV |
| Polysilicon definition (RIE) | *overetch:* 500 / 1100 Å |
| Pedestal thermal oxidation | *thickness:* 0 / 280 Å |
| LDEB B implantation | no LDEB / LDEB = B₁ |
| Oxide spacer formation | |
| Extrinsic base B implantation | 2 x 10¹⁵ cm⁻², 25 keV |
| BPSG deposition | |
| Contact opening | |
| Rapid thermal annealing | 1060 °C, 20 sec |
| Metallization | |

The monocrystalline emitter region, which is driven-in during a final RTA treatment, is bounded by oxide-sidewall spacers along one dimension (width, W) and by the field oxide (LOCOS) along the other (length, L) (Fig. 1b). The devices used for this study were fabricated using a process derived from our 0.7 μm CMOS technology (Fig 2). Transistors of various geometries were available for the analyses. The experimental work includes site-by-site acquisition of Gummel and $h_{FE}(I_C)$ characteristics, and technological and device parameter extraction using statistical tools.

**Figure 2:** Simplified process used to study the emitter-base system

### Experimental results and interpretation.

The small geometry effects in this transistor structure are demonstrated by an increase in the maximum current gain on the $h_{FE}(I_C)$ curves, for narrow (small W) (Fig. 3a) and short (small L) (Fig. 3b) emitters, as the device dimensions approach 1μm.

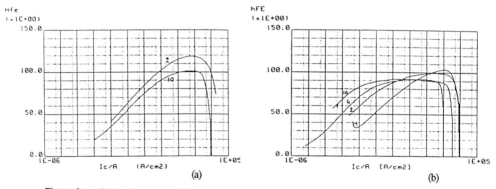

**Figure 3:** Plots of current gain ($h_{FE}$) versus collector current density ($I_C/A$) for transistors of:
 (a) same length L (2 μm) and different widths W (2, 10 μm) (dimensions on mask).
 (b) different lengths L (1, 2, 4, 10 μm) and large widths W (> 10 μm).

It should be pointed out that, in contrast, narrow emitter effects in double polysilicon transistors correspond to a current gain reduction for small dimension devices (G.P. Li,1988). By monitoring the individual changes in the base and collector currents with dimensions and process parameters, we have been able to understand the physical nature of these effects.

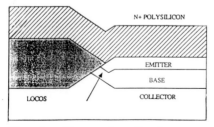

- **narrow emitter effect.** The current gain enhancement for small values of W can be interpreted as a decrease in the base Gummel number, due to a LOCOS edge masking effect during intinsic base implantation and to the gate oxide removal step prior polysilicon deposition, leading to a more lightly doped base under the bird's beak (Fig 4).

**Figure 4:** Narrow emitter effect: Locos proximity effect

### - short emitter effects.

The increase in $h_{FE}$ at small L is more complex due to the combined effects of two independent phenomena, and the role played by the L.D.E.B. implantation.

.The first effect is associated with the oxidation step prior to spacer oxide deposition, used to improve the peripheral emitter-base junction properties (A. Chantre 1989). The oxide growth leads to a local decrease in boron concentration in the base (segregation factor B(SiO2)/B(Si)=3 (S.M. Sze 1983)) and, on the contrary, an increase in arsenic concentration (As(SiO2)/As(Si)=.1 (S.M. Sze 1983))This results in a local variation of the emitter-base system at the spacer edge: decrease in the base Gummel number (increase of Ic) and increase in the emitter Gummel number (decrease of Ib). These effects depend on the duration of the oxidation step, i.e oxide thickness, as illustrated in Figure 5.

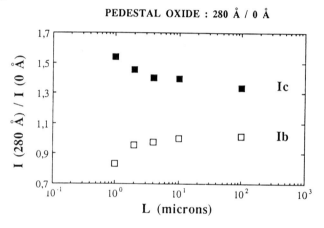

Figure 5: Base and collector current ratios for devices with and without 280 Å thick pedestal thermal oxide.

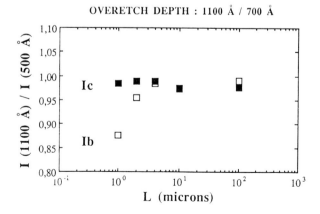

Figure 6: Base and collector current ratios for devices with 1100Å and 700Å silicon overetch.

.A second effect is related to the recess of the silicon substrate around the emitter during polysilicon patterning, which causes a base current reduction (Figure 6).Our interpretation is geometrical: the lateral diffusion of minority carriers (electrons) leads to a small, not visible, loss of collector current,but significant for the base current at small values of L. Of course, this exra current will decrease with overetch depth, as illustrated in Figure 6.

.These two effects, for given technological parameters, can be almost compensated by the L.D.E.B. implantation. With low or no L.D.E.B., they are increased dramatically, leading to a strong increase in current gain (up to 30% under our standard conditions).With optimized L.D.E.B.implant, the small geometry effects on L can be almost perfectly controlled.

## Conclusion.

We have identified the physical mechanisms behind both the short and narrow emitter effects in "etched-polysilicon" emitter bipolar transistors. Interestingly enough, the short emitter effect can be compensated by proper control of the lateral encroachment of the lightly doped extrinsic base region (the LDD region of the pMOS). Thus, this CMOS compatible bipolar transistor structure is believed to be scalable to deep submicron dimensions.

## References

[1]  A. Nouailhat , G. Giroult, P. Delpech and A. Gerodolle *Electron. Lett.* 24, 1581 (1988)

[2]  G.P. Li , E. Hackbarth and T.C. Chen.*IEEE Trans. Electron Dev.* ED-35, 1942 (1988)

[3]  M.P. Brassington, M.H. El-diwany, R.R. Razouk, M. Thomas and P.T. Tuntasood.*IEEE Trans. Electron Dev.* ED-36, 712 (1989)

[4]  A. Chantre, G. Festes, G. Giroult and A. Nouailhat. *ICVC '89 Technical Digest*, p.36

[5]  S.M. Sze (ed),*VLSI Technology,1983,McGraw-Hill Book Company.* pp158

## Acknowledgments

The authors wish to thank the staff of the CNET/CNS/APF pilot line for device processing and electrical test operating.

*Paper presented at ESSDERC 90, Nottingham, September 1990*
*Session 4C5*

# The application of a selective implanted collector to an advanced bipolar process

M C Wilson

Plessey Research (Caswell) Ltd., Caswell, Towcester, Northants NN12 8EQ

> Abstract. This report outlines a method to shallow an implanted base profile in a high speed bipolar process by selectively implanting phosphorus into the vicinity of the collector-base junction. The correct choice of implant conditions can give rise to reduced base width and increased cut-off frequency.

## 1. INTRODUCTION

In recent years the improvement in silicon bipolar device performance has been dramatic. This is best illustrated by the now regular reports of sub- 100ps/gate logic and sub- ns RAM access times. Such performance has been achieved by development of self aligned double polysilicon technology in which the device area has been shrunk and the parasitic capacitances substantially reduced. Similarly the device junction depths have been scaled down vertically. Further vertical scaling of an implanted boron base is limited by boron ion channelling. One solution, applicable to double polysilicon processes and existing emitter-base technology, is that of the selective implanted collector (SIC), or pedestal collector first reported by Tang and Solomon (1979) and later by Konaka et al. (1987).

In this case an n-type region is formed immediately below the base region of the device by implanting high energy phosphorus ions tailored to over-dope the channelled tail region of the base. A schematic diagram illustrating the location of the SIC region is reproduced in figure 1. This results in a narrower base, and reduced transit time. Furthermore, the inclusion of SIC can be combined with a reduction in doping concentration of the epi layer to give rise to reduced external

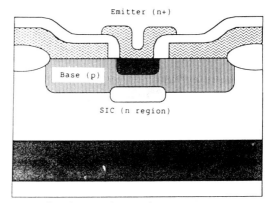

Fig 1. Schematic

collector-base capacitance (Ehinger et al. 1989). Konaka et al. working with the SST process + SIC report cut-off frequencies in the range 21.1 to 25.7GHz and fast ECL switching delays of

34.1ps/G.

In this report a range of phosphorus SIC implants have been included into an advanced Plessey double layer polysilicon bipolar technology known as Process HE. Transistors have been assessed by DC characterisation and cut-off frequency measurements. The effect on circuit performance has been assessed by measuring the performance of ECL prescalers. A marked improvement in cut-off frequency is reported.

## 2. EXPERIMENTAL

Devices have been processed using standard Process HE techniques details of which have been documented fully elsewhere (Hunt and Cooke 1988). High energy, singly charged phosphorus P31+ ions were selectively implanted into the emitter window immediately after the base pad oxidation and prior to base implantation. Phosphorus doses in the range $1E12 cm^{-2}$ to $1E13 cm^{-2}$ were used at an energy of 150kV. Apart from the high energy P31+ implants no additional processing operations were included. Slices were assessed using standard DC characterisation and s-parameter measurements. Cut-off frequency (FT) results were obtained from s-parameter measurements made on nominal 1x15um s-parameter structures at 4GHz with the collector current swept from 0.04 to 25mA at Vcb=0 and 2V. Circuit performance was determined by measuring the maximum frequency of oscillation on ECL prescalers.

## 3. RESULTS

DC measurements were made on all slices. Shown for example in figures 2 and 3 is the effect of the SIC implant on the forward current gain and Early voltage of nominal 1x5um devices.

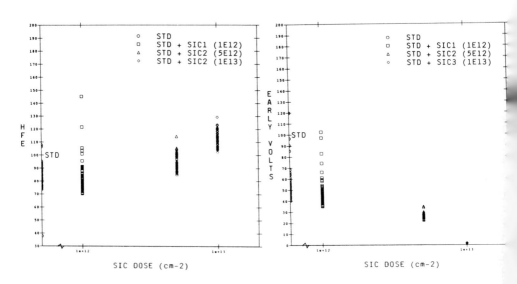

Fig 2.  HFE vs SIC dose.          Fig 3.  Early voltage vs SIC dose.

HFE increased and Early voltage and decreased for SIC doses greater than 1E12cm-2. In figures 2 and 3 values have been included from slices having no SIC implant for reference. Generally a there was a reduction in spread of HFE and Early voltage when a SIC implant was included and is due to the control of the vertical base and SIC profiles resulting from implantation at the same reference plane. Cut-off frequency data at Vcb=2.0V is shown in figure 4. A significant increase in peak FT was observed with the inclusion of a SIC implant, in this case from ~12GHz to ~20GHz. The collector current at which FTmax was obtained increases with SIC dose due to suppression of base push-out (ie suppression of the Kirk effect). The steeper roll-off was probably a result of reduced current spreading. Prescaler self oscillation frequency is shown in figure 5 and was observed to peak with a SIC dose of 5E12cm-2. The self oscillation of 6.5GHz compares to 5.5GHz for devices with no SIC implant.

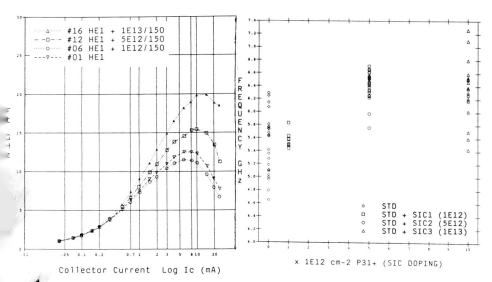

4. FT vs Ic (Vcb=2V)   Fig 5. Prescaler self oscillation

## DISCUSSION

Inclusion of a selective implanted collector in process HE can ilt in a higher cut-off frequency and improved ECL prescaler ormance. Though far from optimised the implant conditions sen were sufficient to highlight the main effects.

in important advantage of the SIC collector is that the intrinsic ase resistance (Rbb) can be reduced by including a higher base concentration whilst maintaining a base width less than the original. Thus with a SIC collector present it should be possible o achieve high cut-off frequency and lower base resistance at the same time. High base resistance is usually a consequence when attempting very shallow base profiles. In figure 6 very shallow base profiles modelled by SUPREM show that higher peak base doping and reduced base width can be obtained with the inclusion of SIC.

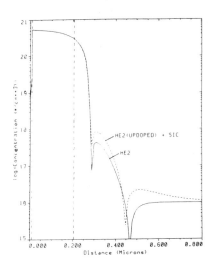

Fig 6. SUPREM modelled shallow base profiles.

In the initial assessment made here the collector capacitance not been measured though it is expected to increase with SIC Konaka et al (1989) report collector capacitance increased ~25%. For the case where base dose has been increased to rec base resistance it should be noted that emitter capacitance wi also increase.

5. SUMMARY

The inclusion of a selective implanted collector in proce resulted in significant improvement in cut-off freque prescaler performance. Forward current gain was obser increase slightly and Early voltage decreased with increas dose. A marked reduction in the spread of these paramete noticed.

The author gratefully acknowledges the support of The Company Ltd.

6. REFERENCES

Ehinger et al (1989) 19th ESSDERC Berlin, p.797
Hunt P C and Cooke M (1988) IEEE CICC Rochester, New York
Konaka et al (1987) 19th ICSSDM Tokyo p.331
Tang D D and Solomon P M (1979) IEEE J Sol Stat Circ SC-14, p.679

*Paper presented at ESSDERC 90, Nottingham, September 1990*
*Session 4C6*

# SiC pn structures grown by container-free LPE (GF LPE) and semiconductor devices based on these structures

V.A.Dmitriev, Ya.V.Morozenko, A.M.Strel'chuk, V.E.Chelnokov, and A.E.Cherenkov

Ioffe Insitute, Polytechnicheskaya st.26, Leningrad, USSR

Abstract. In this paper we consider the growth of SiC pn structures by COF LPE, properties of pn junctions and, lastly, parameters of the devices that were designed.

## 1. Introduction

For many years SiC has been attracting attention of researchers engaged in developing electronic devices. However, only in the last decade technological methods have been developed that brought real success in this field and formed the basis for commercial production of devices using hexagonal SiC. Among these techniques is the COF LPE of SiC (Dmitriev et al 1985a).

The COF LPE utilizes the idea of holding a molten metal suspended in an RF electromagnetic field. Application of this technique to epitaxial growth of SiC pn structures from Si melt makes it possible to exclude a physical contact between the chemically active melt and parts of technological equipment.

As a starting material we used commercial 6H-SiC and 4H-SiC single crystals grown by the Lely method. The epitaxial growth was carried out on (0001)Si face. The growth temperature was varied from 1440 to 1620°C. Epitaxial layers with thicknesses from .5 to 20 μm were grown at a rate in the 0,1-3 μm/min range. As dopants we used aluminum (acceptor) and nitrogen (donor). The doping level was controled in the range $N_d - N_a = (10^{16} - 10^{19})$ cm$^{-3}$ and $N_a - N_d = (10^{18} - 10^{20})$ cm$^{-3}$ for n-type layers and p-type layers, respectivly.

## 2. SiC pn junction

In this section we consider the properties of the $n^+$(substrate) $-n$(layer)$-p^+$(layer) structures grown by COF LPE in a single epitaxial process. The properties of SiC p-n junctions were studied on ⌀ 0.3-1.0 mm mesas fabricated by reactive ion-plasma etching (Popov et al 1986).

### 2.1. C-V characteristics

These were measured by the bridge technique using parallel substitution scheme over the temperature range from 300 to

© 1990 IOP Publishing Ltd

800 K. Measurements were carried out on the pn structures in which the measured was independent of the frequency in the 10-100 kHz range, and which produced C-V characteristics that were linear, when plotted in the C-V coordinates, over the entire range of voltage and temperature used.

Since the measured differential capacity in the srtuctures under investigation is frequency-independent, it represents the true differential capacity of the pn junction.
Furthermore, because the true differential capacity is linear in the C-V coordinates we may base our determinations of the electrostatic pn junction parameter from the C-V characteristics on the theory suggested by Shockley (1949).

Measurements show that the cutoff voltage $V_C$ was generally in the range 2.6-2.7 V for 6H-structures and 2.8-2.9 V for 4H-structures at room temperature, decreasing with increasing temperature. The reduced concentration value $N_B$ determined from the slope of the C-V characteristic was $(0.4-6).10^{17} cm^{-3}$ and showed virtually no dependence on temperature.

The width of the space charge region was calculated to be in the range 0.1-0.3 $\mu$m (300 K). The strength of the electric field at zero bias $E_m \simeq 3.10^5$ V/cm (300 K).

## 2.2. I-V characteristics (6H-structures)

Generally, the direct I-V characteristics taken in the region of small currents (V < 2.3 ), consisted of two exponential portions of the from $j = j_0 \exp(qU/nkT)$.

The upper parts of the curves were found to preserve the exponential character for all the temperatures (300-800 K), while the characteristic energy showed linear dependence with temperature, $E = nkT$, where coefficient n is temperature-independent and has the value of 1.19 $\pm$ 0.03. The temperature dependence of the preexponential factor $j_0$ ($j_0 = 10^{-37} A/cm^2$ (300 K)) has revealed thermo-activated behaviour, defined as $j_0 = j_0^* \exp(-E_A/kT)$. Assuming $j_0^*$ to be temperature-independent, the calculated value of $E_A$ is 2.73 $\pm$ 0.03 eV.

Over the temperature range of 370-640 K, the value of the characteristic coefficient n for the lower parts of the I-V curves was close to 2, with the temperature dependence of the preexponent $j_0$ ($j_0 = 10^{-22} A/cm^2$ (300 K)) varying exponentially for the activation energies $E_A$ lying between 1.4 eV and 1.5 eV, that is, corresponding to the value of the energy gap half width in 6H-SiC.

The conclusion that can be drawn is that the experimental results obtained for the upper branches of the I-V curves are consistent with the recombination model involving a multi-level centre in the space charge region, while the lower branches can be interpreted as evidence for a space charge region recombination process via a level located near the middle of the energy gap, in agreement with the Shockley-Noyc-Sah theory.

The reverse current before breakdown (V < 0.8.Vbreakdown) measured approximately < $10^{-9}$ A at room temperature, and were due to leakages off the peripheral regions of the mesa-structures.

## 2.3. pn junction breakdown

The usual type of breakdown for pn junctions grown by COF LPE is that of avalanche microplasma breakdown. At breakdown voltage, the strength of the pn junction electric field had the maximum of about $4.10^6$ V/cm.

Examination of the temperature dependence established for the avalanche breakdown voltage has revealed that the sign of the temperature coefficient of voltage (TCV) for avalanche breakdown as occurs in SiC can be positive, negative, and alternating. The mechanisms that govern this behaviour of the avalanche breakdown TCV in SiC pn junctions are at present unclear.

Thus, it has been shown that electrical characteristics of SiC pn junctions we developed are in accordance with the currently accepted notions of the semiconductor physics.

## 3. SiC-devices

The development of COF LPE made possible fabrication two groups of the devices: (1) optoelectronic devices (Dmitriev et al 1986a, 1989, Vishnevskaya 1988) and (2) high-temperature devices (Dmitriev et al 1985b, 1986b, 1988a, 1988b).

### 3.1. Optoelectronic devices

We have fabricated green LED's ($\lambda_m$ = 530 nm) with luminous intensity $I_L$ = 5 mcd (at current I = 20 mA, $\theta$ = 15°); blue LED's ($\lambda_m$ = 470 nm) with $I_L$= 9 mcd (I = 20 mA, $\theta$ = 15°); violet LED's ($\lambda_m$ = 425 nm) with $I_L$ = 1 mcd (I = 20 mA, $\theta$ = 15°), and LED's emitting at $\lambda_m$ = 398 nm, with $I_L$= 0.5 mcd (at 100 mA pulsed current; 20 mA average current; $\theta$ = 15°), Fig.1. The LED's matrices and three-color (red-green-blue) display have been developed.

### 3.2. Structures for high-temperature electronics

The devices were fabricated without encapsulation. Temperature dependences were measured in air. The surface of mesa-structures had no protective coating. Operating temperatures (up to 800 K) were limited by heat resistance of Al-contacts.

Rectifying diodes for currents 1-5 A and voltages in excess of 300 V were fabricated. In the best structures the voltage drop at current of 5 A was 5.5 V at 300 K and 4.5V at 800 K. Reverse leakage currents at voltage of 300 V was less than $10^{-9}$ A (300 K) and increased to $10^{-6}$ A at a temperature of 800 K.

Voltage limiters for stabilized voltage range 15-75 V and direct current of 1 A were fabricated. The differential resistance was about 1 Ohm (300-800 K).

Field-effect transistors with n-type channel and a pn junction as the gate were fabricated. FET parameters (300 K): drain current ~12 mA, transconductance ~4 mS/mm, drain-to-gate breakdown voltage ~50 V.

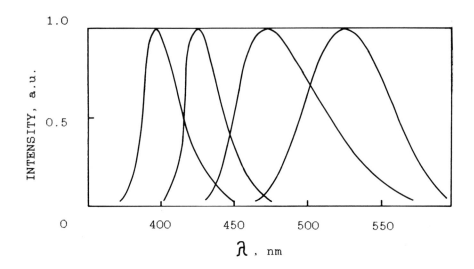

Fig.1. Electroluminescence spectra of SiC-LED's (300 K).

## 4.References

Dmitriev V A, Ivanov P A, Korkin I V, Morozenko Ya V, Popov I V, Sidorova T A, Strel´chuk A M and Chelnokov V.E. 1985a Sov.Tech.Phys.Lett. 11 98

Dmitriev V A, Ivanov P A, Strel´chuk A M, Syrkin A L, Popov I V and Chelnokov V E 1985b Sov.Tech.Phys.Lett. 11 403

Dmitriev V A, Morozenko Ya V, Popov I V, Suvorov A V, Syrkin A L and Chelnokov V E 1986a Sov.Tech.Phys.Lett. 12 221

Dmitriev V A, Ivanov P A, Popov I V, Strel´chuk A M, Syrkin A L and Chelnokov V E 1986b Sov.Tech.Phys.Lett. 12 318

Dmitriev V A, Levinshtein M E, Vainshtein C N and Chelnokov V E 1988a Electron.Lett. 24 1031

Dmitriev V A, Ivanov P A, Il´inskaya N D, Syrkin A L, Tsarenkov B V, Chelnokov V E and Cherenkov A E 1988b Sov. Tech.Phys.Lett. 14 127

Dmitriev V A, Kogan L M, Morozenko Ya V, Tsarenkov B V, Chelnokov V E and Cherenkov A E 1989 Sov.Phys.Semicond. 23 23

Popov I V, Syrkin A L and Chelnokov V E 1986 Sov.Tech.Phys. Lett. 12 99

Shockley W 1949 Bell.Syst.Tech. 28 435

Vishnevskaya B I, Dmitriev V A, Kovalenko I D, Kogan L M, Morozenko Ya V, Rodkin V S, Syrkin A L, Tsarenkov B V and Chelnokov V E 1988 Sov.Phys.Semocond. 22 414

*Paper presented at ESSDERC 90, Nottingham, September 1990*
Session 4C7

# The application of limited reaction processing to the deposition of silicon carbide layers

F H Ruddell, D W McNeill, B M Armstrong and H S Gamble

Institute of Advanced Microelectronics,
School of Electrical Engineering and Computer Science,
The Queen's University of Belfast, Ashby Building,
Stranmillis Road, Belfast, Northern Ireland, BT9 5AH.

Abstract. This paper describes the deposition of microcrystalline silicon carbide in an LRP reactor using silane/propane gas chemistry and discusses the performance of heterojunction bipolar transistors using N-SiC emitters.

1. Introduction

Silicon carbide (SiC) is attracting considerable interest as a wide bandgap emitter material for use in the production of heterojunction bipolar transistors on silicon substrates. There are several advantages associated with this material. The bandgap of B-SiC is 2.2eV and it exhibits high thermal stability, controllable conductivity and high electron saturated drift velocity. However the temperature needed for epitaxial growth of silicon carbide on silicon can be as high as $1300^\circ C$. This is clearly unacceptable for scaled down devices with very narrow base widths.

In a Limited Reaction Processor (LRP) rapid wafer heating using tungsten-halogen lamps is combined with low pressure chemical vapour deposition (LPCVD) technology. The SiC deposition reaction is controlled by temperature and so layer growth is determined by the application of power to the heating lamps. Heating rates of up to $250^\circ C/s$ are used to minimise process thermal budget.

2. Silicon Carbide Deposition

The silicon carbide deposition experiments were carried out in the QUPLAS LRP reactor, developed at Queen's University. The equipment consists of an evacuated quartz reaction chamber in which the single process wafer is supported on quartz pins. The wafer is heated by tungsten-halogen lamps, controlled using a pulse-width modulator, with 9kW of power available. A base pressure of $6 \times 10^{-7}$ mbar is achieved using turbomolecular and rotary pumps. A microwave source is located upstream from the wafer position, allowing an in-situ plasma clean process to remove native oxide from the wafer surface prior to layer deposition without breaking vacuum.

A study was carried out of the temperature, gas flow rates and pressure conditions necessary to achieve silicon carbide layers using silane/propane gas chemistry. Temperatures of 720, 860 and $970^\circ C$ were employed and the deposition rate was characterised by an activation

© 1990 IOP Publishing Ltd

energy of 1.1eV. This indicates that the surface reaction between silicon and carbon is rate limiting.

It was established that a process pressure of 10 mbar and a propane/silane ratio of at least 50 yielded silicon carbide layers at each of the deposition temperatures. Figure 1 compares SIMS analysis of the LRP-deposited material with that of a single-crystal silicon carbide standard, showing that the deposited layer contains 50% carbon and 50% silicon. An XPS study revealed that the silicon and carbon atoms were bonded to form silicon carbide. TEM was employed to observe the grain size and degree of crystallinity obtained in the LRP-deposited layers. An average grain size of around 50 nm, indicating microcrystalline structure, was observed for films deposited at 970°C. Deposition at the highest available temperature (970°C) was favoured as this resulted in the lowest oxygen incorporation in the SiC layers (less than 1%), and minimisation of the SiC-Si transition region due to fast temperature ramp rate.

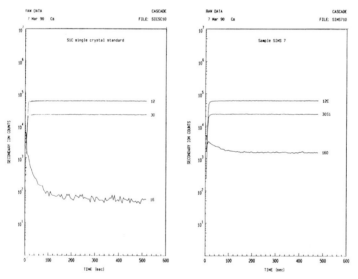

Fig. 1 SIMS analysis comparing silicon carbide standard and LRP-deposited material.

In-situ doped N-type silicon carbide was produced by adding 0.1% phosphine diluted in nitrogen to the deposition ambient. The addition of phosphine to the deposition ambient reduces layer deposition rate due to the high adsorption rate of phosphine on the wafer surface compared to silane. The phosphorus concentration increased with increasing phosphine flow rate up to a maximum of $8 \times 10^{20}$ cm$^{-3}$. The resultant resistivity of the most heavily doped layers, which contained 5% oxygen, was 74 ohm-cm.

3. Silicon Carbide Emitter Bipolar Transistors

The fabrication sequence for silicon carbide heterojunction bipolar transistors was based on that used for polysilicon emitter devices. Figure 2 shows a schematic cross-section of a completed device. The QUPLAS LRP reactor was used to deposit the in-situ doped SiC used for the emitter regions. The process schedule included an in-situ $CF_4$ plasma preclean to remove native oxide from the wafer surface. A 1 minute deposition process was employed at a temperature of $970^\circ C$ to achieve an emitter thickness of 200 nm. A SIMS profile of the device emitter is presented in Figure 3 showing negligible outdiffusion of phosphorus into the transistor base. After SiC deposition a photoresist mask was used to define the emitter, and the SiC was patterned using RIE in a $CF_4/O_2$ ambient.

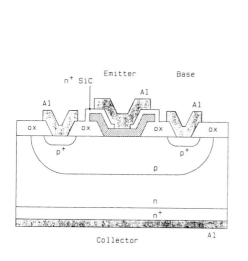

Fig. 2 Schematic cross-section of SiC emitter NPN transistor.

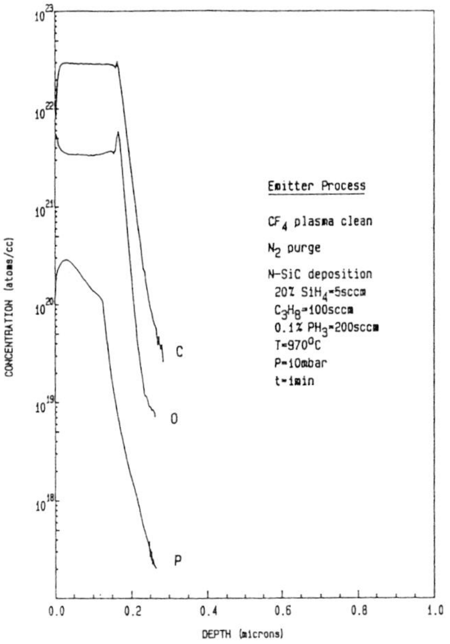

Fig. 3 SIMS profile of SiC emitter.

The current gain of the SiC emitter devices varied between 4 and 14 across a sample wafer. An equivalent device with a silicon emitter would be expected to have a gain of 7. Gummel plots for high and low gain SiC emitter devices are shown in Figure 4, corrected for measured values of emitter resistance. The collector current ideality factor in each case is 1.05. The base current ideality factor of 1.3 indicates a trap density of about $6 \times 10^{11}$ $cm^{-3}$ present in the emitter-base depletion region, perhaps caused by defects in the microcrystalline SiC emitter. It is significant that the gain variation may be attributed solely to change in base current, suggesting heterojunction action.

Figure 3 also shows that the SiC emitter contains up to 10% oxygen. The work of Ghidini and Smith (1984) on the effect of water vapour and oxygen as layer contaminants is shown in Figure 5. This indicates that process gases must contain less than 4 ppm water vapour for oxygen-free silicon layers to be deposited at a temperature of 970°C and a process pressure of 10 mbar. The gases employed in this work were not of the highest grade and therefore this water vapour limit is likely to have been exceeded. It is desirable to minimise oxygen content in the SiC layers in order to draw more meaningful conclusions concerning device operation. Gas purifiers have now been installed and results will be presented on the properties of the resultant layers.

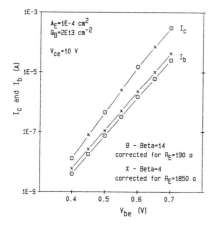

Fig. 4 Gummel plot for SiC emitter transistors.

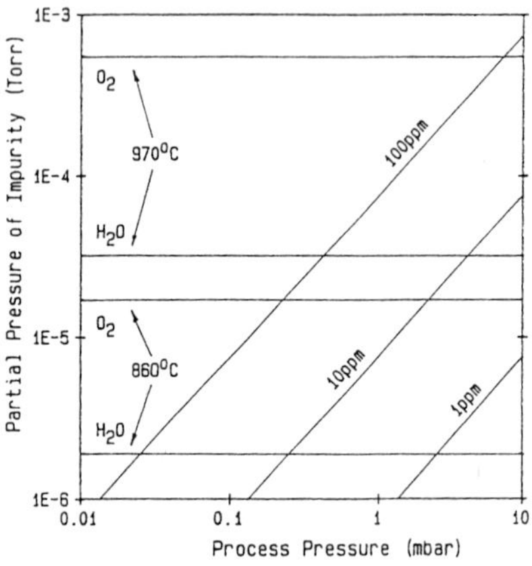

Fig. 5 Partial pressure limit of impurity species.

## 4. Conclusions

Stoichiometric silicon carbide layers have been deposited using Limited Reaction Processing over the temperature range 720 - 970°C. N-SiC layers have been employed as emitters in NPN bipolar transistors, and there is some evidence of heterojunction action. Future work will concentrate on the minimisation of oxygen contamination in the SiC layers.

Reference

Ghidini G and Smith F W 1984 J. Electrochem. Soc. 131 2924

Acknowledgement

The authors wish to acknowledge the financial support of the UK Science and Engineering Research Council.

Paper presented at ESSDERC 90, Nottingham, September 1990
Session 5IP1

# Degradation and wearout of thin dielectric layers during charge injection

Marc M. HEYNS
Interuniversity Microelectronics Centre (IMEC vzw)
Kapeldreef 75, B-3030 Leuven, BELGIUM

ABSTRACT: *The degradation and wearout of thin dielectric layers during charge injection is an important reliability issue. In this paper the oxide field dependence of the trapping and defect generation during injection of electrons or holes is investigated. The generation of slow trapping instabilities during electrical stressing, the charge build-up and degradation during high-field stressing and the trapping characteristics of nitrided oxide layers are also discussed.*

## 1. INTRODUCTION

The continuous scaling down of the minimum device dimensions, without the appropriate scaling of the supply voltage, gives rise to the presence of high fields in small geometry MOS-transistors. This leads to serious reliability problems related with the enhanced probability of charge injection into the gate oxide by hot carriers generated in the transistor channel. Furthermore, the increased oxide fields also enhances the importance of wearout failures. The degradation of the thermal oxide layer ($SiO_2$) used as the gate insulator and of the $Si/SiO_2$ interface during charge injection has, therefore, become an important reliability issue. This paper discusses the field dependence of the trapping and defect generation during electron and hole injection, the slow state generation during electrical stressing, the charge build-up and degradation during high-field stressing and the trapping in (re-oxidized) nitrided oxide layers.

## 2. CHARGE INJECTION AND CHARGE DETECTION TECHNIQUES

Most investigations on the trapping properties of thermal oxide layers [1-3] have used the avalanche injection technique [4] to introduce electrons or holes into the oxide layer. This technique has the advantage that it works on capacitor structures which can be fabricated by a relatively simple process. Its main drawback, however, is that the oxide field ($E_{ox}$) during injection can not be controlled. This important parameter can be controlled (independently from the injected current density) when homogeneous injection in MOS-transistors is used. The injection technique is schematically illustrated in fig.1. Minority carriers are generated in the Si-substrate either by optical means or from an (underlying) diode [5,6]. They are accelerated towards the $Si/SiO_2$ interface by a substrate bias while the source and drain of the transistor are grounded. Under these conditions $E_{ox}$ is determined by the gate voltage. Some carriers gain sufficient energy to overcome the barrier at the $Si/SiO_2$ interface and are injected into the $SiO_2$ layer. The technique works relatively easy for electron injection. The injection of holes, however, is much more difficult due to the shorter inelastic scattering length of holes in the Si-substrate and the higher energy barrier they have to overcome at the $Si/SiO_2$ interface [7].

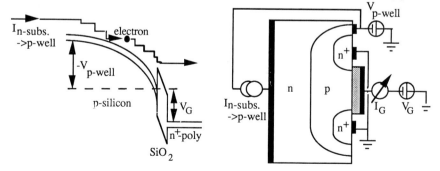

**Fig. 1**: *Schematical representation of the electron injection process in the $Si$-$SiO_2$-poly-$Si$ band diagram (left part) and of the measurement set-up (right part).*

On application of fields above approximately 6 MV/cm substantial currents begin to flow through a thermal oxide layer. This conduction is due to Fowler-Nordheim tunneling of electrons through the triangular barrier at the interface into the oxide conduction band [8]. Tunneling injection has been used for charge trapping studies [9] but the high field necessary to induce injection is a disadvantage. Field-ionization of trapped charge and breakdown of the $SiO_2$ are favoured under these conditions. The technique is, however, very well suited to study degradation phenomena occurring at high fields and prior to breakdown.

## 3. EXPERIMENTAL CONDITIONS

The electron and hole injection experiments described in this paper were performed on poly-Si gated transistors with a gate oxide thickness of respectively 26 and 20 nm. Minority carriers are generated in the bulk of the semiconductor by illuminating with a high-flux low-energy photon source. The interface state density ($D_{it}$) is measured using the charge pumping technique [10], applied with constant pulses and a varying base level. The density of the trapped oxide charge is obtained from the gate voltage shift for a fixed drain current level in deep subthreshold at a fixed drain voltage of -0.1V. As the charge centroid is not known, only an effective density of trapped charge $N_{teff}=C_{ox}\Delta V_G/q$ is given, with $C_{ox}$ the oxide capacitance and q the elementary charge. It was demonstrated [7,11] that the effect of the $D_{it}$ generation on this measurement can be neglected. More experimental details can be found in ref. 7 and 11.

The capacitors used in the high-field stress experiments were fabricated on either p-type or n-type <100> silicon wafers. Oxidation was performed to thicknesses varying from 20 to 40 nm in dry $O_2$ at a typical temperature of 900°C. Either a thin (transparent) aluminium layer or a polycrystalline silicon layer was deposited as the electrode material. Capacitor structures were defined using standard lithography and processing techniques [12].

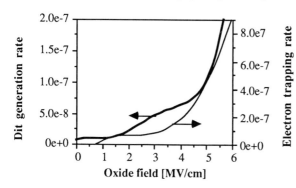

Fig. 2 : *Oxide field dependence of the interface state generation (left scale) and the electron trap generation (right scale) during electron injection.*

## 4. DEFECT GENERATION DURING ELECTRON INJECTION

*4.a. Generation of interface states during electron injection*

The $D_{it}$ generation during electron injection is found to be strongly dependent on $E_{ox}$ [11]. This is shown in fig.2 where the $D_{it}$ generation rate is plotted as a function of $E_{ox}$. In this linear plot two 'threshold' fields, at 1.5 and 4 MV/cm, can be observed, suggesting the existence of two generation mechanisms. The energy distribution of the generated interface states shows a peak in the upper part of the Si-bandgap [13]. A threshold of 1.5 MV/cm was already reported to exist for the $D_{it}$ generation during photoinjection of electrons [14]. This threshold is likely to be correlated with the mechanism by which the electrons loose their energy ("thermalize") when injected in the oxide. Below this threshold the scattering mechanism is dominated by LO-phonons while above this field acoustical phonon scattering (non-polar scattering) becomes important [15].

*4.b. Charge trapping during electron injection*

The electron trapping behaviour was also observed to depend on $E_{ox}$ [11]. For fields up to 3.6 MV/cm the charge trapping has a saturating behaviour (for the total charge densities used in these experiments). For larger fields the charge trapping increases with $E_{ox}$. This is attributed to the generation of electron traps during the charge injection. The generation rate of the charged trap centres was calculated from the linear part of the trapped charge versus injected charge curves with the charge trapping at the lowest measured field subtracted from the measurements as background trapping in pre-existing electron traps. This generation rate is plotted as a function of $E_{ox}$ in fig.2. A 'threshold' around 4 MV/cm is observed. This result is in contrast with the 1.5 MV/cm threshold reported earlier [16]. However, this threshold was inferred from plots of the total trapped charge as a function of $E_{ox}$ for various amounts of injected charge and the *probability* of trap generation was not taken into account. Because the trap generation also occurs at low fields (however at a reduced rate) and the generation has a saturating behaviour, the same total amount of electron traps can be generated at these low fields by injection a sufficiently large number of electrons. Therefore, in our opinion, a meaningful threshold can only be defined from the field dependence of the trap generation *rate*.

### 4.c. Correlation between Dit generation and electron trapping

The field dependence of the $D_{it}$ generation and the charge trapping, shown in fig.2, suggests a common origin for both generation mechanisms at fields larger than 4 MV/cm. A correlation between electron trapping and $D_{it}$ generation was already suggested [17] and could occur by the release of hydrogen from water-related electron traps (Si-H or Si-OH) when an electron is captured at this defect. The hydrogen can diffuse towards the Si/SiO2 interface where it can generate a dangling silicon bond, acting as an interface state. Hydrogen certainly is important in the degradation of thermal oxide layers as illustrated, for example, by the evidence for the role of $H^+$ in the $D_{it}$ generation after irradiation [18], by the hydrogen accumulation at the Si/SiO2 interface during electron injection [19], by the relation between hydrogen diffusion and the $D_{it}$ build-up [20] and by the effect of hydrogen anneals on the electron trap and defect generation [21-23]. However, this does not necessarily mean that hydrogen is the prime cause for the defect generation.

A possible origin for the generation of both the interface states and the electron traps is the injection of holes from the anode [24]. The importance of holes in the $D_{it}$ generation has already been clearly demonstrated [25,26] while the generation of electron traps by the recombination of electrons and trapped holes in $SiO_2$ has also been reported [27]. Within this model the $D_{it}$ generation and electron trap generation are not directly correlated but have the same origin, i.e. the injection of holes from the anode. The hot holes are generated by the energy loss of the electrons when they leave the oxide conduction band and enter the electrode. The average electron energy (above the oxide conduction band) at an oxide field of 4 MV/cm is approximately 2.5 to 3 eV [15]. This energy is in good agreement with the threshold energy for damage generation which was reported to be 2.3 eV [28]. Summed with the conduction band offset between the $SiO_2$ layer and the poly-Si electrode (taken as 3.1 eV), this leads to a threshold energy for the injection of holes between 5.6 and 6.1 eV. This energy is larger than the energy barrier for holes at the poly-Si electrode/SiO2 layer interface but smaller than the suggested 7.5 eV threshold for the generation of surface plasmons [24].

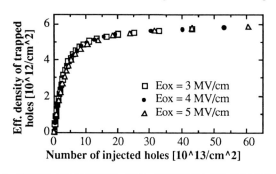

**Fig. 3**: *Effective density of trapped holes as a function of the number of injected holes for various oxide fields.*

## 5. HOLE TRAPPING AND INTERFACE STATE GENERATION

### 5.a Oxide field dependence of hole trapping

The main advantage of using homogeneous hole injection in MOS-transistors is that holes can be injected under conditions of a controlled oxide field and with the total exclusion of the simultaneous injection of electrons. It was demonstrated [7] that the trapping of holes does not depend on the substrate bias (and therefore on the energy of the injected holes) during injection. In fig.3 the hole trapping curves for different average oxide fields are compared. In all cases a very high trapping efficiency is found. In contrast with the results for electron injection there is apparently no strong field dependence for the hole trapping. The small trend towards a decreased trapping at higher fields can be due to either a field dependent capture cross section or to detrapping from shallow hole traps at higher fields [7]. The possibility of electron injection from the cathode at the higher fields, which recombine with the trapped holes, can be excluded. For the oxide fields used in these experiments the probability for Fowler-Nordheim tunneling of electrons is extremely low while also the possibility of 'cathode electron injection', in analogy with anode hole injection [24], can be ruled out [7].

### 5.b Generation of interface states during and after hole injection

The generation of interface states during irradiation, high-field stressing or hot electron injection is still a point of controversy in literature [17,25,26,29-32]. One of the suggested models assumes the $D_{it}$ generation to be a two-step process [25]. The first step is the trapping of holes without any interface state generation. In the second step electrons recombine with the trapped holes causing the creation of interface states. On the other hand, experiments on hot carrier injection in transistor structures observed the maximum $D_{it}$ generation under conditions where electrons and holes are simultaneously injected in the gate oxide layer [33] and no evidence could be found for a two-step mechanism [30,31]. However, in most experiments on transistor structures it is very difficult to investigate the effects of hole injection while completely avoiding the injection of (a small number of) electrons. As the capture efficiency for a trapped hole to capture an electron is very high [34] neutralization of the trapped holes occurs very efficiently and the second step of the two-step process may pass unnoticed. This will lead to the generation of an interface state and a re-structured hole trap which can act as a slow state [35]. The positive charge in these states gives rise to the positive charge observed after the hole injection which can be neutralized without the generation of interface states [30,31].

| $E_{ox}$ [MV/cm] | $N_{teff}$ [1/cm$^2$] | $D_{it}(t=0)$ [cm$^{-2}$.eV$^{-1}$] | $D_{it}(t=7days)$ [cm$^{-2}$.eV$^{-1}$] |
|---|---|---|---|
| 3.0 | 2.46 10$^{12}$ | 1.2 10$^{11}$ | --- |
| 4.0 | 2.51 10$^{12}$ | 1.4 10$^{11}$ | 5.4 10$^{11}$ |
| 4.5 | 2.77 10$^{12}$ | 1.1 10$^{11}$ | --- |
| 5.0 | 2.46 10$^{12}$ | 1.0 10$^{11}$ | 5.3 10$^{11}$ |
| 6.0 | 2.46 10$^{12}$ | 1.1 10$^{11}$ | 5.8 10$^{11}$ |

**Table 1:** *Trapped hole density $N_{teff}$, generated interface states measured directly after the hole injection $D_{it}(t=0)$ and measured after 7 days (storage at room temperature, zero bias applied) $D_{it}(t=7days)$ as function of the oxide field $E_{ox}$ during injection.*

The $D_{it}$ generation after homogeneous hole injection for different oxide fields is shown in table 1. It is observed that the amount of generated interface states is only a small fraction of the number of trapped holes and is independent of $E_{ox}$ during the hole injection. These results clearly demonstrate that the efficiency for $D_{it}$ generation when only holes are injected is very low compared to the hole trapping efficiency. Normalizing the number of generated interface states to the number of injected holes results in a rough estimate of about $5.10^{-3}$ for the $D_{it}$ generation efficiency during hole injection, independent of $E_{ox}$ during injection. As a comparison the $D_{it}$ generation efficiency during electron injection is of the order of $5.10^{-6}$ and strongly depends on $E_{ox}$ during injection [11]. This shows that hole injection, even without a two step process, is still about 1000 times more efficient in generating interface states than electron injection, in good agreement with results obtained with gate-controlled diodes [36].

The $D_{it}$ values measured one week after the hole injection (with the devices stored at room temperature with the gate floating) are also given in table 1 for three of the five samples. The observed $D_{it}$ increase after storage goes together with a decrease in the trapped hole density [7]. A detailled study of the time and field dependence of the two phenomena [37], however, leads to the conclusion that there is not a simple one-to-one correlation as could be expected in a two-step model [25] where trapped holes are transformed into interface states.

## 6. SLOW TRAPPING INSTABILITIES

The generation of slow trapping instabilities (or slow states) has been observed in the form of anomalous positive charge near the Si/SiO$_2$ interface during electron avalanche injection [38-44] and high-field stressing [45,46]. This charge responds to changes in the silicon surface potential (and/or the internal field) with time constants typically ranging from seconds to several hours, depending on temperature and on the history of the sample [40]. A variety of physical models have been invoked to explain the build-up of this positive charge: the diffusion of hydrogen [41] or excitons [47] to the $Si/SiO_2$ interface, hot hole injection from the anode [24], electron-hole pair generation via band-to-band [48,49] or trap-to-band impact ionization [50] and field-stripping of electrons from valence band orbitals [51]. Using a precise quantification technique [40] based on the charging/discharging characteristics of these slow states their generation kinetics were studied during avalanche injection of electrons or holes and during high-field stressing [12,35,40]. The results indicated that there are only a limited and fixed number of sites in the oxide which can be converted into slow states, independent of the stress mode responsible for generating the slow states. It was demonstrated [35] that the positive charge due to slow states is located in initially present hole traps and no new hole traps are generated during electrical stressing. The hole trapping and slow state generation is not a fully reversible process. After the capture and detrapping of a hole the capture cross section of the centre for re-capturing a hole is strongly increased, indicating a change in the local structure of the hole trap. From this a model was proposed [35] according to which slow states originate from hole traps upon sequential trapping of a hole and an electron.

## 7. CHARGE AND DEFECT CREATION DURING HIGH-FIELD STRESSING

### 7.a General results

The degradation and charge build-up in the $SiO_2$ layer during high-field injection provides information on the wearout and breakdown mechanisms of these layers. The total charge-to-breakdown during these tests is found to depend on the electrode material, the injected current density and the stress conditions [52]. The midgap voltage shift ($\Delta V_{MG}$) during high-field stressing often indicates the generation of positive charge. This charge was demonstrated [45] to be located near the $Si/SiO_2$ interface in slow states. Under similar stress conditions these slow states form to a much smaller extent in poly-Si than in Al-gate structures. The exact shape of the $\Delta V_{MG}$-vs-time curve during stressing is the net result of various charge components in the $SiO_2$-layer and at the $Si/SiO_2$ interface and is qualitatively dominated by the charge state of the slow traps during the high-field stress [53]. $\Delta V_{MG}$ can, therefore, not be directly interpreted in terms of the charge build-up in the oxide layer, as is often done. From more detailed measurements it was concluded [54] that no charge is built up in the bulk of the oxide layer during high-field stressing, the reason being that the high applied field (9-11 MV/cm) causes any trapped charge to be detrapped so that no bulk oxide charge remains.

### 7.b The generation of electron traps during high-field stressing

The generation of electron traps during high-field stressing was postulated in order to explain the observations made during constant current stress experiments [55]. These traps were thought to be located either near the injecting interface [56,57] or throughout the oxide layer [55,58]. However, when only information from Fowler-Nordheim and capacitance-vs-voltage characteristics is used an unambiguous separation between bulk and interface charge is not possible. Therefore, a more detailed experimental procedure was used where stressing (and trap generation), trap filling and charge sensing were performed in consecutive steps of constant-current stress, avalanche injection and internal photoemission measurements [59].

Using this procedure it was demonstrated that during negative high-field stressing on Al-gate capacitors electron traps are generated close to the $Si/SiO_2$ interface [54]. The generation rate was found to display a sublinear regime [53], in contrast with other findings assuming a linear generation rate up to the occurrence of breakdown [56-57,60]. A positive high-field stress on Al-gate capacitors generates a large density of positive charge under the Al-electrode [61] while no electron trap generation close to the $Si/SiO_2$ interface was observed after this stress. Experiments conducted on poly-Si gate capacitors showed that during negative high-field stressing electron traps are generated close to the substrate-$Si/SiO_2$ interface [61], confirming the results obtained on Al-gate MOS-capacitors. Under positive high-field stress conditions electron traps are generated close to the poly-Si electrode [61]. From comparing the results on Al-gate and poly-Si gate capacitors it follows that the positive charge present under the Al-gate after positive stress is directly associated with the Al-gate, because it is not found on poly-Si gate capacitors. Furthermore, it can be concluded that during high-field stressing of MOS-structures electron traps are generated in the vicinity of the non-injecting interface. While the generation of electron traps during high-field stressing could be clearly demonstrated, no evidence could be found for the generation of hole traps during high-field stressing [53].

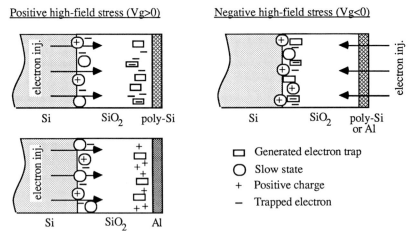

**Fig. 4**: *Schematic illustration of the charge and defect distribution after negative and positive constant current stressing of MOS-capacitors with Al and poly-Si electrodes.*

*7.c Summary model*

The results on the charge build-up and degradation of $SiO_2$ layers during high-field stressing are summarized in fig.4. It is important to notice that all the degradation and charge build-up phenomena encountered during the high-field stressing are occurring at the interfaces and no electrically measurable effects have been seen in the bulk of the oxide in the thickness range investigated here (20-40 nm). This suggests that the quality of the $SiO_2$ layer, as far as wearout and breakdown are concerned, is mainly determined by the quality of these interfaces. Furthermore, all charge build-up and defect generation phenomena were found to display regimes with strongly decreased generation rates before breakdown occurs. It is, therefore, not possible to simply point to one of the charge build-up and defect generation mechanisms as the primary cause for breakdown, for then a continuous increase of such degradation (and eventually run-away) would be expected. Most probably a complex interplay of the different observed phenomena will locally generate a critical condition leading to destructive breakdown or 'local' degradation phenomena, which are not detected by the measurement techniques used in this investigation, are dominating the breakdown behaviour.

## 8. NITRIDED OXIDES

Nitrided oxides and re-oxidized nitrided oxides have been proposed as alternatives to thermal $SiO_2$ layers for very thin gate insulators. The characteristics of these layers are a complicated function of the oxide thickness, the nitridation conditions and the re-oxidation and annealing conditions. A complete discussion of the characteristics of these layers is, therefore, beyond the scope of this paper. In general good breakdown properties, low interface state densities and fixed oxide charge densities have been reported for these layers but usually an increase in the trapping characteristics as compared to thermal $SiO_2$ is observed.

A short rapid thermal nitridation (RTN) step was found to generate a large density of water-related electron traps (with capture cross-sections of ~ $10^{-17}$ and $10^{-18}$ cm$^2$ [62]) in 15 to 30 nm oxide layers [63,64]. More severe nitridation results in the generation of a nitrogen-related $10^{-16}$ cm$^2$ trap [63,64], which is also observed after furnace nitridation [65,66]. Other traps with a variety of capture cross sections (from $10^{-14}$ cm$^2$ to $10^{-17}$ cm$^2$) have been reported [67-70]. Re-oxidation can be used to lower the electron trap density [64,71], most probably because hydrogen can be removed from the films [72]. A strongly increased hole trapping is observed after a short RTN-step [63,64], while prolonged nitridation and re-oxidation decreases the hole trapping again [63,64]. These results are consistent with investigations on furnace nitrided oxide layers [73,74] where it was reported that nitridation at relatively low temperatures and for short times increases the density of hole traps. More severe furnace nitridation conditions were found to result in a hole trap reduction [69,73].

## 9. CONCLUSIONS

It was demonstrated that the electron trap and interface state generation during electron injection increases with increasing oxide field. The trapping of holes is nearly independent on the oxide field and no evidence was found for the generation of hole traps. Holes are much more efficient in generating interface states than electrons but the exact generation mechanism is still unknown. When the oxide field is increased above the Fowler-Nordheim injection threshold other degradation and charge build-up mechanisms start to become important. These results have important consequences in the study of the hot carrier degradation of MOS-transistors. When the results on oxide degradation obtained from investigations on capacitor structures or homogeneous injection on transistor structures are extrapolated towards hot-carrier stress of MOS transistors, it must be realized that the oxide field can strongly differ. Also the validity of accelerated lifetime tests on capacitor structures must be re-evaluated within this framework because the degradation mechanisms during normal operating conditions are different from the mechanisms during accelerated tests, rendering the extrapolation between the two conditions invalid. Nitrided (re-oxidized) oxides are a possible alternative to thermal oxide layers but, in general, they exhibit larger trapping properties.

## REFERENCES

1) D.J. DiMaria in "The Physics of $SiO_2$ and its interfaces", Ed. S.T. Pantelides (Pergamon, New York, 1978), p.160
2) R.F. De Keersmaecker in "Insulating Films On Semiconductors", Eds. J.F. Verweij and D.R. Wolters, (North-Holland, Amsterdam), p. 85 (1983)
3) P. Balk in "Solid State Devices 1983", Ed. E. H. Rhoderick, The Institute of Physics Conf. Ser. No. 69, p. 63 (1984)
4) E.H. Nicollian and C.N. Berglund, J. Appl. Phys. 41, 3052 (1970)
5) J.F. Verwey, J. Appl. Phys. 44, 2681 (1973)
6) T.H. Ning and H.N. Yu, J. Appl. Phys. 45, 5373 (1974)
7) A.v. Schwerin, M.M. Heyns and W. Weber, J. Appl. Phys. 67, 7595 (1990)
8) M. Lenzlinger and E.H. Snow, J. Appl. Phys. 40, 278 (1969)
9) P. Solomon, J. Appl. Phys. 47, 2089 (1976)
10) G. Groeseneken, H.E. Maes, N. Beltran and R.F. De Keersmaecker, IEEE Trans. Electron. Dev. ED-32, 375 (1985)
11) M.M. Heyns, D. Krishna Rao and R.F. DeKeersmaecker, Appl.Surf.Sci. 39, 327 (1989)
12) M.M. Heyns and R.F. De Keersmaecker in "Dielectric layers in semiconductors : novel technologies and devices 1986", Eds. G. Bentini, E. Fogarassy and A. Golanski (Les Editions de Physique, Les Ulis Cedex, France), p. 303 (1986)
13) D. Krishna Rao, M.M. Heyns and R.F. De Keersmaecker, internal IMEC report (1988)
14) V. Zekeryia and T.P. Ma, Appl. Phys. Lett. 43, 95 (1983)
15) M.V. Fischetti, D.J. DiMaria, S.D. Brorson, T.N. Theis and J.R. Kirtley, Phys. Rev. B 31, 8124 (1985)
16) D.J. DiMaria, Appl. Phys. Lett. 51, 655 (1987)
17) L. Do Thanh, M. Aslam and P. Balk, Solid State Electron. 29, 829 (1986)
18) N.S. Saks and D.B. Brown, IEEE Trans. Nucl. Sci. 36, 1848 (1989)
19) R. Gale, F.J. Feigl, C.W. Magee and D.R. Young, J. Appl. Phys. 54, 6938 (1983)
20) D. L. Griscom, J. Appl. Phys. 58, 2524 (1985)
21) Y. Nissan-Cohen and T. Gorczyca, IEEE Electron. Dev. Lett. 9, 287 (1988)
22) Y. Nissan-Cohen, Appl. Surf. Sci. 39, 511 (1989)
23) L. Do Thanh and P. Balk, J. Electrochem. Soc. Vol. 135, No.7, 1797 (1988)
24) M.V. Fischetti, Phys. Rev. B 31, 2099 (1985)
25) S. K. Lai, J. Appl. Phys. 54, 2540 (1983)
26) S.J. Wang, J.M. Sung and S.A. Lyon in "The Physics and Technology of Amorphous $SiO_2$", Ed. R. A. Devine (Plenum, New York, 1988) p. 465 (1988)
27) I.C. Chen, S. Holland and C. Hu, J. Appl. Phys. 61, 4544 (1987)
28) D.J. DiMaria and J.W. Stasiak, J. Appl. Phys. 65, 2342 (1989)
29) S.A. Lyon, Appl. Surf. Sci 39, 552 (1989)
30) P. Heremans, G. Groeseneken and H.E. Maes, paper presented at IEE "Colloquium on hot carrier degradation in short channel MOS", London, Jan. 1987
31) D. Krishna Rao, M.M. Heyns and R.F. De Keersmaecker in "Proc. of ESSDERC 88", Eds. J.-P. Nougier and D. Gasquet, (Les Editions de Physique, France), p. 669 (1988)

32) K. R. Hofmann, C. Werner, W. Weber and G. Dorda, IEEE Trans. Electron Devices, ED-32, 691 (1985)
33) N.S. Saks, P.L. Heremans, L. Van den hove, H.E. Maes, R.F. De Keersmaecker and G.J. Declerck, IEEE Trans. Electron. Dev. ED-33, 1529 (1986)
34) D.J. DiMaria, Z.A. Weinberg and J.M. Aitken, J. Appl. Phys. 48, 898 (1977)
35) M.M. Heyns and R.F. De Keersmaecker in "The Physics and Technology of Amorphous $SiO_2$", Ed. R.A. Devine (Plenum, New York, 1988), p.411 (1988)
36) P. Heremans, R. Bellens, G. Groeseneken and H. Maes, IEEE Trans. Electron. Devices ED-35, 2194 (1988)
37) A.v. Schwerin and M.M. Heyns, to be published
38) D.R. Young, E.A. Irene, D.J. DiMaria, R.F. De Keersmaecker and H.Z. Massoud, J. Appl. Phys. 50, 6366 (1979)
39) S.K. Lai and D.R. Young, J. Appl. Phys. 52, 6321 (1981)
40) M.M. Heyns and R.F. De Keersmaecker, paper presented at the 6th Solid State Device Technol. Symp. (ESSDERC), Toulouse, France, Sept. 1981
41) F.J. Feigl, D.R. Young, D.J. DiMaria, S.K. Lai and J.A. Calise, J. Appl. Phys. 52, 5665 (1981)
42) M.V. Fischetti, R. Gastaldi, F. Maggioni and A. Modelli, J. Appl. Phys. 53, 3129 (1982)
43) M.V. Fischetti, R. Gastaldi, F. Maggioni and A. Modelli, J. Appl. Phys. 53, 3136 (1982)
44) C.T. Sah, J.Y. Sun and J.J. Tzou, J. Appl. Phys. 54, 944 (1983)
45) M.W. Hillen, R.F. De Keersmaecker, M.M. Heyns, S.K. Haywood and I.S. Daraktchiev in "Insulating Films On Semiconductors", Eds. J.F. Verweij and D.R. Wolters (North-Holland, Amsterdam), p. 274 (1983)
46) K.R. Hofmann and G. Dorda in "Insulating Films On Semiconductors", Ed. M. Schulz (Springer, Berlin), p.122 (1981)
47) Z.A. Weinberg and G.W. Rubloff, Appl. Phys. Lett. 32, 184 (1978)
48) T.H. DiStefano and M. Shatzkes, Appl. Phys. Lett. 25, 685 (1974)
49) M. Shatzkes and M. Av-Ron, J. Appl. Phys. 47, 3192 (1976)
50) Y. Nissan-Cohen, J. Shappir and D. Frohman-Bentchkowsky, J. Appl. Phys. 58, 2252 (1985)
51) P. Olivo, B. Ricco and E. Sangiorgi, J. Appl. Phys. 54, 5267 (1983)
52) S. Haywood, M. Heyns and R. De Keersmaecker paper presented at the SISC Conf. 85, Fort Lauderdale, Florida, Dec. 5-7 (1985)
53) M.M. Heyns and R.F. De Keersmaecker, Mat. Res. Symp. Proc. Vol.105, p.205 (1988)
54) M.M. Heyns, R.F. De Keersmaecker and M.W. Hillen, Appl. Phys. Lett. 44, 202 (1984)
55) E. Harari, J. Appl. Phys. 49, 2478 (1978)
56) M.S. Liang and C. Hu, IEEE Int'l Electron Devices Meeting 1981, Technical Digest, p.396 (1981)
57) M.S. Liang, J.C. Choi. P.K. Ko and C. Hu, IEEE Int'l Electron Devices Meeting 1984, Technical Digest p. 152 (1984)
58) C.S. Jenq, T.R. Ranganath, C.H. Huang, H.S. Jones and T.T.L. Chang, IEEE Int'l Electron Devices Meeting 1981, Technical Digest p.388 (1981)
59) D.J. DiMaria, J. Appl. Phys. 47, 1082 (1976)
60) C.-F. Chen and C.-Y. Wu, J. Appl. Phys. 60, 3926 (1986)
61) M.M. Heyns and R.F. De Keersmaecker, J. Appl. Phys. 58, 3936 (1985)
62) A. Hartstein and D.R. Young, J. Appl. Phys. 50, 6321 (1981)
63) E.E. Dooms, M.M. Heyns and R.F. De Keersmaecker, Appl. Surf. Sci. 39, 227 (1989)
64) E.E. Dooms, M.M. Heyns and R.F. De Keersmaecker, internal IMEC-report (1989)
65) M. Severi, M. Impronta and M. Bianconi in "Proc. 17th European Solid State Device Research Conf. ESSDERC 87, p.845 (1987)
66) M. Severi and M. Impronta, Appl. Phys. Lett. 51, 1702 (1987)
67) S. Chang, N.M. Johnson and S.A. Lyon, Appl. Phys. Lett. 44, 316 (1984)
68) F.L. Ferry, P.W. Wyatt, M.L. Naiman, B.P. Mathur, C.T. Kirk and S.D. Senturia, J. Appl. Phys. 57, 2036 (1985)
69) A. Yankova, L. Do Thanh and P. Balk, Solid State Electron. 30, 939 (1987)
70) S.K. Lai, W.D. Dong and A. Hartstein, J. Electrochem. Soc. 129, 2042 (1982)
71) W. Yang, R. Jayaraman and C.G. Sodini, IEEE Trans. Electron. Dev. ED35, 935 (1988)
72) T. Hori and H. Iwasaki, J. Appl. Phys. 65, 629 (1989)
73) M. Severi, M. Impronta, L. Dori and S. Guerri in "Proc. 18th European Solid State Device Research Conf. ESSDERC 88, p.417 (1988)
74) G.J. Dunn, J. Appl. Phys. 65, 4879 (1989)

Paper presented at ESSDERC 90, Nottingham, September 1990
Session 5A1

# Measurement and simulation of degradation effects in high voltage DMOS devices

P. Dickinger, G. Nanz, S. Selberherr
Institute for Microelectronics
Technical University Vienna
Gußhausstraße 27-29, A-1040 Vienna, AUSTRIA

Tel. +43/222/58801-3713

*Abstract* — One of the main constraints for the long-term stability of any kind of integrated circuit technology consists in device degradation. The effect of hot electron induced degradation is a threat to MOSFET reliability when scaling to submicron dimensions. Nevertheless this effect has to be taken into account in high voltage analog MOS circuits as used in telecommunications as well. Various measurements and simulations have been performed in order to improve the behavior of n-channel high voltage DMOS transistors and to analyze the effects responsible for the degradation of these devices.

## Introduction

The power supply in most VLSI circuits is 5 Volt. Scaling down the device feature sizes to submicron dimensions, the electric fields will increase to very high values. This may result in breakdown and degradation effects. Degradation effects play an important role also in high voltage devices as used in telecommunication equipment. Especially in n-channel transistors hot electrons are injected into the oxide and cause device degradation.

We present a comparison of measurements and simulations. The simulated results have been obtained with the two-dimensional device simulator BAMBI which can account for fixed charges in the insulator.

Hot electron and hole generation takes place in high field regions particularly in connection with steep doping gradients [1]. Oxide charges generated by hot carriers [2] influence the current flow in the drift region of the device [3]. The distribution of carriers trapped by oxide defects depends on the previous biasing of the device. Oxide charges are time variant in position and maximum value. Thereby they result in time variant I-V curves changing the device characteristics.

## Degradation of n-channel DMOS transistors

The results of the analysis of submicron MOS transistors are not directly applicable to high voltage n-channel DMOS devices. In Fig. 1 a slightly idealized DMOS structure is shown, the measured real device is too complex for efficient 2D simulation.

For some experiments symmetric device structures must be used [4]. Thus the effects of fixed charges near drain or source (in case of reverse biasing) cannot be observed. For simplicity reasons fast interface states and mobility degradation near the interface between silicon and insulator are not included in our simulations, because the influence of these effects is negligible compared to the oxide charges [1], [5].

In Fig. 2 the measured I-V curve is shown. The measurements from 0V to 40V have

© 1990 IOP Publishing Ltd

been performed before stress, the measurements from 0V to 100V after stress becoming stable after 10 minutes of operating.

The device is designed for a maximum voltage of 120V. This means that the applied voltages should not really provide stress to the structure. Applying a gate voltage of -30V for 12 hours with 90V at drain the electrical behavior changes again as it can be seen in Fig. 3 (lines up to 100V are before applying the high negative gate voltage). Applying again a high negative gate voltage (-40V) the I-V characteristic changes again slightly as it can be seen in Fig. 4 (I-V curves with higher drain current represent the initial condition). In Fig. 5 changes in the characteristics can be observed applying a high positive gate voltage (+30V) with 40V at drain and limiting the drain current to $10\mu A$. The curves up to 100V are the "stable" degraded initial conditions. This means that the carrier distribution in the oxide changes again. Additional electrons are injected into the insulator even for low drain voltages so under normal operating conditions all these effects makes the device unfit for use.

Due to the special geometry the measurement of substrate currents is impossible. In comparison to standard MOS structures [6] no threshold voltage shift can be observed. Thereby it can be concluded that no electrons are injected from the channel region into the oxide. The results of our two-dimensional simulations show high electric fields at the beginning of the drift region. The peak of the field ($3.5 \cdot 10^6$ V/cm) in Fig. 6 indicates that the geometry layout is not appropriate. Degradation occurs because obviously the carriers in this region have enough energy to cross the energy barrier to the oxide. In a modified geometry the values of the field are lowered significantly (Fig. 7). This improvement has been achieved by reducing the overlap of the drain metallization and the drift region. Simulations as proposed by [1] verify the presence of trapped electrons as additional charges in the oxide. Contrarily to MINIMOS we have assumed a two-dimensional Gaussian charge distribution. The maximum of this function lies at the beginning of the drift region. The change of the carrier mobility is not investigated.

In Fig. 8 and Fig. 9 the current densities are shown without and with additional oxide charges, respectively. There is a significant difference in the drift region between these two simulations, the current flow in Fig. 8 is located near the interface. In Fig. 9 the current flow is forced away from the interface by the oxide charges.

In Fig. 10 different simulated I-V curves without charges and with two different peak values of the Gaussian charge distribution in the oxide are shown. These calculated results have essentially qualitative character because only little information is available about the values and the distribution of the oxide charges. For the simulation of hot electron injection no accurate models are available [3], [4].

**Conclusion**

We have measured and simulated the behavior of a high voltage DMOS device to identify the regions where hot electrons are injected into the insulator. The simulated results provide information where to improve the layout of these devices in order to avoid high electric fields near the oxide. Our analysis shows that it is possible to locate critical regions of devices even without simulating the degradation mechanisms itself [2], [5].

## Acknowledgement

This work has been supported by Entwicklungszentrum für Mikroelektronik Ges.m.b.H., Villach, Austria, by the SIEMENS Research Laboratories at Munich, Germany, and by DIGITAL EQUIPMENT CORP. at Hudson, U.S.A.. The authors are indebted to Prof. H. Pötzl for many helpful discussions.

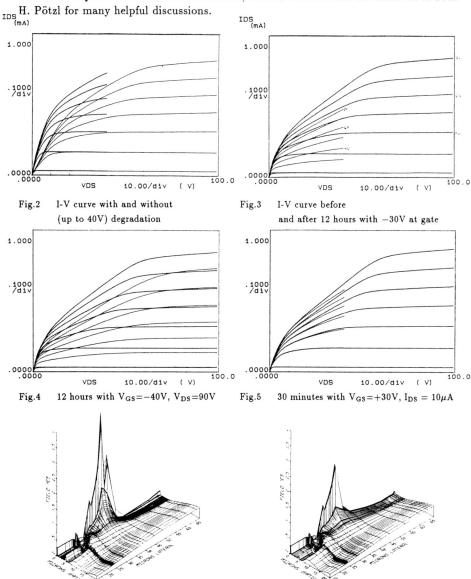

Fig.2  I-V curve with and without (up to 40V) degradation

Fig.3  I-V curve before and after 12 hours with $-30V$ at gate

Fig.4  12 hours with $V_{GS}=-40V$, $V_{DS}=90V$

Fig.5  30 minutes with $V_{GS}=+30V$, $I_{DS}=10\mu A$

Fig.6  Field distribution of a DMOS transistor

Fig.7  Field distribution of a DMOS transistor with improved geometry

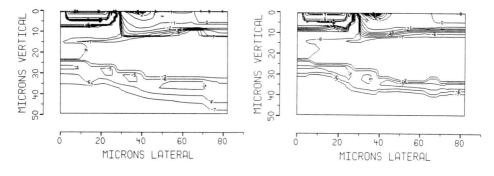

Fig.8 Current density without oxide charges

Fig.9 Current density with oxide charges

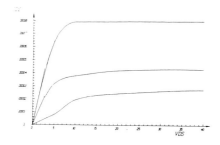

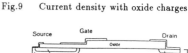

Fig.10 Simulated I-V curve with different oxide charges

Fig.1 simplified geometry of a DMOS transistor

[1] Giebel, T.; Die Alterung bei n-Kanal MOS-Transistoren unter dem Einfluß hochenergetischer Ladungsträger: Der „Hot-Electron" Effekt. Fortsch.-Ber. VDI Reihe 9 Nr. 81, Düsseldorf: VDI Verlag 1988
[2] Tam, S.; Ko, P.K.; Hu, C.; Lucky electron model of channel hot-electron injection in MOSFET's; IEEE Trans. Electron Devices, Vol. ED-31, No. 9, pp. 1116-1125 (1988)
[3] Chan, T.Y.; Chiang, C.L.; Gaw, H.; New insight into hot electron induced degradation of n-MOSFET's; Proc. IEDM Conf. 1988, pp. 196-197
[4] Schwerin, A.; Hänsch, W.; Weber, W.; The relationship between oxide charge and device degradation: a comparative study of n- and p-channel MOSFET's; IEEE Trans. on Electron Devices, Vol. 12, pp. 2493-2500, (1987)
[5] Hofmann, K.R.; Werner, C.; Weber, W.; Dorda, G.; Hot electron and hole-emission effects in short n-channel MOSFET's; IEEE Trans. Electron Devices, Vol. ED-32, No. 3, pp. 691-699 (1985)
[6] Banerjee, S.; Sundaresan, R.; Shichijo, H.; Malhi, S.; Hot-electron degradation of n-channel polysilicon MOSFET's; IEEE Trans. on Electron Devices, Vol. 35, No. 2, pp. 152-157 (1988)
[7] Duvvury, C.; Redwine, D.J.; Stiegler, H.J.; Leakage current degradation in n-MOSFET's due to hot-electron stress; IEEE Electron Device Letters, Vol. 9, No. 11, pp.579-581 (1988)

Paper presented at ESSDERC 90, Nottingham, September 1990
Session 5A2

# Technology and design of SIPOS films used as field plates for high voltage planar devices

D. Jaume*, G. Charitat**, A. Peyre-Lavigne*, P. Rossel**

* Motorola Semiconducteurs S.A. Av Général Einsenhower B.P. 1029 31023 Toulouse Cedex (France).
** Laboratoire d'Automatique et d'Analyse des Systèmes, 7 Avenue du Colonel Roche 31077 Toulouse Cedex (France)

Abstract: In order to improve the voltage handling capability of high voltage and power devices, an efficient technique based on the deposition of a semi-resistive layer acting as a field plate is proposed. The complete design of a high voltage planar transistor (1500V) using a SIPOS layer on $SiO_2$ film is extracted from bidimensional numerical simulations. The evolution of breakdown voltage $BV_{cbo}$ versus critical parameters as oxide thickness, field plate length and field plate-stop channel distance is calculated. A good agreement between theoretical and experimental results is obtained. The breakdown voltage achieved by the devices is near 90% of the ideal planar breakdown voltage without any damage for other electrical parameters.

1. Introduction

In order to improve the voltage handling capability of high voltage and power devices, numerous techniques have been used during the last 25 years, aiming to attain the near-ideal breakdown of a plane junction (Baliga 1987). These include field plates, guard rings, bevels, etch contours, termination extensions, RESURF, semi-resisitive layer... For bipolar power transistors one can utilize guard rings (good efficiency: 80% but space consuming), field plates (low efficiency: 60%), positive or negative etch contour (good efficiency: 80% but complex technology and low yields), junction termination extension (very good efficiency: 95% but space consuming).

The bipolar planar device is depicted in Figure 1. N+ emitter, left-side, is grounded, as well as the P+ base. The collector, at the bottom of the wafer, and the stop-channel, right-side, are biased to the high voltage. The forward blocking capability is provided by the reverse biased one-sided abrupt junction P+N, base/collector. To increase this capability we use the base metallization as a field plate to protect this cylindrical junction and a SIPOS layer to linearize the potential between base and stop channel contacts (Figure 2). This avoids the field crowding near the edge of this field plate (Conti 1972). T\he SIPOS layer is deposited on the top of an oxide layer thus bypassing problems of poor reproducibility and degradation of dynamic behavior encountered when it is deposited directly on silicon (Matsushita 1976). Furthermore, electric discharges between the base and the stop-channel (Selim 1981) and ionic contaminations are circumvented by using a second passivation layer above the SIPOS one. We chose to utilize a second SIPOS film, with a higher resistivity, which favorably replaces a $CVD-SiO_2$ film (Mimura 1985, Jaume 1989).

Regarding breakdown voltage, some of the critical physical and geometrical parameters for this structure are the following: i) epitaxy doping, $C_{epi}$, and thickness, $W_{epi}$, ii) base junction depth, $X_j$, iii) base-stop channel distance, $\Delta L$, iv) base metallization length, F, and v) thickness of the oxide layer under the SIPOS one, $E_{ox}$. For a voltage handling capability of 1500 V the epitaxy characteristics are fixed: $C_{epi}=5.5\ 10^{13}\ cm^{-3}$ and $W_{epi}=150\ \mu m$, which gives a theoretical planar breakdown voltage of 1800 V. The junction depth is given by the standard Motorola technology: $X_j=14\ \mu m$. The evolution of the breakdown voltage of NPN bipolar transistor versus the other parameters are computed using a bidimensional simulator, BIDIM2 (Charitat 1988), aimed for high-voltage and power devices. This software solves the 3 fundamental equations of semiconductors but, for off-state studies, only the Poisson equation is worked out. The breakdown, induced by avalanche phenomenon, is determined using an integral ionization calculus based on Van Overstraeten and De Man (1970) ionization coefficients.

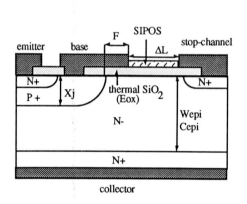

Figure 1: Cross section of a planar transistor using SIPOS layer. Critical parameters.

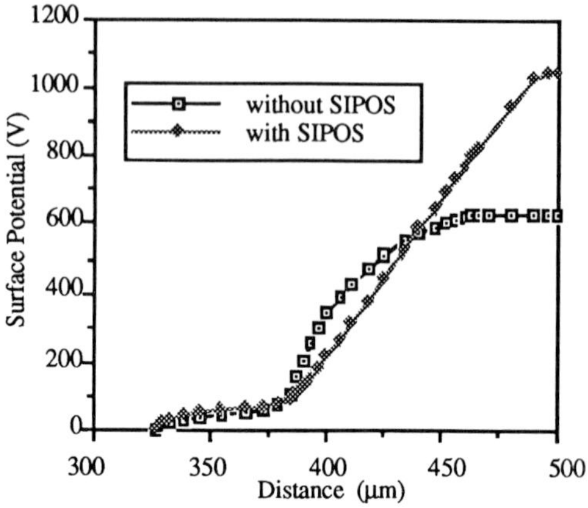

Figure 2: Linearization of surface potential due to SIPOS effect.

## 2. Breakdown voltage versus oxide thickness

To study the evolution of breakdown voltage versus the oxide thickness, $E_{ox}$, we imposed the other parameters on the following values: $\Delta L=210\ \mu m$ and $F=55\ \mu m$. Simulated and measured results are plotted on Figure 3. The numerical simulation gives an important datum about the avalanche breakdown: its location. For $E_{ox}$ less than 1 μm the avalanche phenomenon occurs at the silicon surface just under the end of the field plate (base metallization). Above 1 μm, the breakdown location is switched to the junction curvature: it is important to notice that this value of thickness is much lower than the optimal one deduced from the field plate effect alone, which is near 20 μm for our epitaxy doping if we follow the results of Conti (Conti 1972). This is caused by the SIPOS effect. For thicker oxide, this effect is slowly vanishing and the breakdown potential is decreasing. Theoretical and experimental results agree well: the discrepancy is less than 10% and within the error bar.

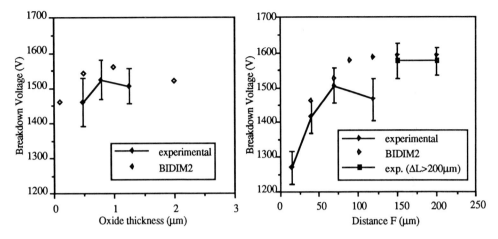

Figure 3: Breakdown voltage versus oxide thickness $E_{ox}$.

Figure 4: Breakdown voltage versus metallization length F.

## 3. Breakdown voltage versus field plate length F

For this study: $E_{ox}$=1.25 µm, $\Delta L$=210 µm for simulation and vary from 265 to 160 µm for devices. This is due to mask designing which is far easier when it is the base-stop-channel distance which is kept constant. Figure 4 gives the theoretical and experimental values of breakdown voltage versus the field plate length F. Regarding the simulation: when this parameter increases, from 15 to 200 µm, the breakdown increases from 1260 to 1600 V exemplifying the growing efficiency of the field plate. The measured values agree very well except on one point: for F=120 µm the distance between the end of the field plate and the stop-channel is here too small, $\Delta L$=160 µm, to avoid a blocking of the space charge on the stop-channel. Consequently the potential linearization (the SIPOS effect) is disturbed and the breakdown voltage lowered. This explanation is confirmed by the two experimental points at F equal 150 and 200 µm, for which the $\Delta L$ distance was increased to 230 µm.

## 4. Breakdown voltage versus base-stop channel distance $\Delta L$

The last investigated parameter is the distance between the end of the field plate, that is the base metallization, and the stop-channel. We have seen in the preceeding section that this distance must be sufficient to allow the spreading of the potential, so that the linearization effect imposed by the SIPOS layer will be operative. Figure 5 shows the calculated and measured breakdown voltages versus $\Delta L$. Once again, good agreement is achieved although it was not possible to explore very large values of $\Delta L$ on our devices, at least for this first set of devices.

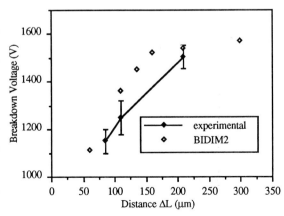

Figure 5: Breakdown voltage versus $\Delta L$.

The theoretical curve gives the minimal value of ΔL to reach the optimal breakdown voltage. This value is near 200 µm, and, above it, the breakdown voltage is constant.

## 5. Conclusion

This paper presents a thorough examination, by mean of bidimensional numerical simulations, of the breakdown phenomena in a planar device using a combination of field plate and semi-resistive layer as junction termination techniques. The complementary of these two methods is an important key to the high efficiency of this design philosophy: the field plate lowers the electric field in the junction curvature and the semi-resistive layer protect the field plate itself from breakdown. The evolution of breakdown voltage versus critical parameters is investigated with BIDIM2, and extensively checked on 1500 V test devices and bipolar transistors fabricated in Motorola Toulouse. Experimental results show that the breakdown voltage achieved by these devices is near 90% of the ideal planar breakdown voltage, without any damage to the other electrical characteristics, such as the $h_{fe}$ gain, and the switching speeds.
Furthermore the chip area gain is about 20% of compared device with standard moat process. We have also developed a new kind of passivation, with two SIPOS layers, to avoid arcing phenomena and ionic contaminations.
This technique can be used for other high voltage devices such as diodes, LDMOS, VDMOS...

References:
B.J. Baliga 1987, "Modern Power Devices", Wiley-Interscience, 1987
G. Charitat, P. Rossel, M. Gharbi, A. Nezar and P. Granadel 1988, "BIDIM2: Software Tool for Power and High Voltage Devices Electrical Simulation", 3rd Symposium on Electronic for Telecommunications, Porto (Portugal), 4-6 May 1988, pp. C8.5-8.7
A. Mimura, M. Oohayashi, S.Murakami, and N. Momma 1985, "High Voltage Planar Structure Using SiO2-SIPOS-SiO2 Film" IEEE Electron Device Letters, EDL-6, 4, April 1985.
F. Conti et M. Conti 1972, "Surface Breakdown in Silicon Planar Diodes Equipped with Field Plate" Solid State Electronics, Vol. 15, pp. 93-105, 1972.
D. Jaume, A. Peyre-Lavigne and G. Charitat 1989, "High Voltage, Semiconductor Device and Fabrication Process" Patent, France Ref N°89.09.897 July 1989, Etats-Unis Ref N°07/437,404, November 1989, Taïwan Ref N°78108782, November 1989.
T. Matsushita, T. Aoki, T. Ohtsu, H. Yamoto, H. Hayashi, M. Okayama and Y. Kawana 1976, "Highly Reliable High-Voltage Transistors by Use of the SIPOS Process" IEEE Trans. on Electron Devices, Vol. ED-23, N°8, August 1976.
F. A. Selim 1981, "High-Voltage, Large-Area Planar Devices" IEEE Electron Device Letters, Vol. EDL-2, N°9, September 1981.
R. Van Overstraeten and H. De Man 1970, "Measurement of the Ionization Rates in diffused Silicon p-n Junctions" Solid-state Electronics, Vol.13, pp. 583-608, 1970.

*Paper presented at ESSDERC 90, Nottingham, September 1990*
*Session 5A3*

# High-speed radiation-hardened ECL circuits on bonded SOI wafers

Katsunobu Ueno, Masanobu Kawano, and Yoshihiro Arimoto[*]

Electronic Device Group, Fujitsu Limited,   [*]Fujitsu Laboratory LTD.
1015 Kamikodanaka, Nakahara-ku, Kawasaki 211, Japan

**Abstract.** This paper describes ECL circuits with completely isolated transistors on bonded silicon-on-insulator (SOI) wafers. We have clarified the superiority of SOI circuits over bulk-silicon circuits in terms of circuit speed and alpha-particle-induced soft errors.

## 1. Introduction

Recent progress in advanced bipolar technology has resulted in high-speed and high-density ECL circuits ( Ueno et al. 1987, Sugiyama et al. 1989, Vora et al. 1984 ).

In high density circuits, especially ECL RAMs over 1K bit, soft errors induced by alpha-particles become a problem ( Sai-Halasz et al. 1983). RAMs over 1K bit are fabricated with a PNP-load cell instead of a Schottky-clamped cell ( Toyoda et al. 1983) because large parasitic capacitances are necessary to decrease the soft errors. Unfortunately, these capacitances also increase an address access time and a write enable time.

On the other hand, high-speed circuits are obtained using a double-polysilicon self-aligned technology. Parasitic capacitances and resistances can be reduced by this technology. In general, radiation-hardened circuits are not compatible with high-speed circuits.

However, we have successfully fabricated high-speed and radiation-hardened circuits, using silicon-on-insulator (SOI) wafers. The SOI structure is one solution to this problem. The alpha-particle-induced charges in the p-substrate cannot reach the collector region, so there is no need to increase parasitic capacitances.

This paper describes the characteristics of three different circuits on bonded SOI wafers. We fabricate a 1K bit ECL RAM with a Schottky-clamped cell and a 16K bit ECL RAM with a PNP-load cell to evaluate alpha-particle induced soft errors and a ring oscillator to measure the gate delay time.

## 2. Wafer Preparation

One oxidized wafer is placed on an SiC-coated graphite strip heater and another is placed on that wafer. A silicon electrode is placed on one wafer to make an electrical contact with the other. Then, wafers are heated up to 800 $^\circ$C and a bonding-pulse of 100 V to 500 V is applied between the electrode and the lower heater for 30 seconds. Bonded wafers are annealed at 1100 $^\circ$C for one hour in nitrogen. After that, the silicon layer on one side of these wafers is thinned down to 3 μm.

Figure 1 shows contour lines of a silicon layer thickness on the buried oxide layer. This silicon layer varied from 2.4 to 2.8 μm thick within the area of 15 mm square. The distribution of that thickness on the

© 1990 IOP Publishing Ltd

SOI wafer is small. This layer changes into the n⁺-buried layer of bipolar transistors. We measured the collector sheet resistance. The deviation of that resistance is less than 2.5 percent in an SOI wafer. The distribution of the silicon layer thickness doesn't affect the transistor characteristics and the circuit performances.

We also measured the tear strength of bonded wafers using a Sebastian Five strength tester. Samples were cut into 1 cm square. A stud coated by epoxy was set on the thinned side. In most samples, the stud was torn off at the surface and the SOI sample was not damaged. Thus, the tear strength of bonded wafers is larger than that of adhesive epoxy and larger than 700 kg/cm².

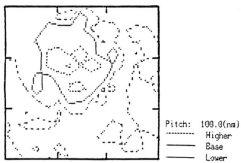

Fig.1. Silicon layer thickness.

The bonded SOI wafers can be treated as conventional wafers. The SOI wafers are fit well with sub-micron processes.

## 3. Characteristics of Fabricated Transistor

We fabricated two ECL RAMs using the IOPⅡ and a ring oscillator using the ESPER. The IOPⅡ is a trench and single polysilicon process (Goto et al. 1982), and the ESPER is a trench and double polysilicon self-aligned process ( Ueno et al. 1987). In both processes, the bottom of the trench reaches the buried oxide layer. Each transistor is completely isolated.

The SEM cross-sectional photograph of a fabricated SOI-ESPER transistor and the characteristics of that are shown in figure 2 and 3, respectively. In this case, the buried oxide layer is 500 nm thick. The n⁺-buried layer and epitaxial layer are formed on that layer. Transistor areas are surrounded by thick oxide layers. The base electrodes are also formed on the thick oxide layer. This is one of the ideal structures for

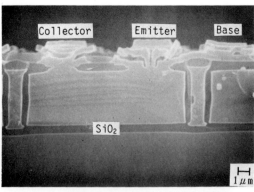

Fig.2. SEM photograph of SOI-ESPER transistor.

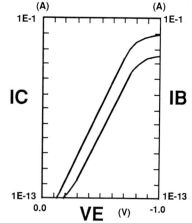

Fig.3. IC/IB vs. VBE for SOI-ESPER transistor.

a bipolar transistor. The characteristics of an SOI-ESPER transistor are the same as those of a bulk-silicon ESPER transistor without a substrate capacitance. The capacitance of an SOI-ESPER transistor is 20 fF, a half of that of a bulk-silicon ESPER transistor. The leakage current is less than 1 pA. The ideality factor is 1.03. The performance of the SOI transistor is much improved.

## 4. Circuit performances

### 4.1 Alpha-particle induced soft-errors

As alpha-particles strike the silicon substrate, they generate electron-hole pairs mainly in the p-substrate. If the generated electrons are collected in the $n^+$-buried layer of an off-side transistor, the collected electrons cause the base voltage of an on-side transistor to drop. So, the on-side transistor changes to an off-side transistor. To immunize the soft-errors, cells have a large capacitance at the collector node. In the PNP-RAM, a diffusion capacitance exists besides a junction capacitance. This capacitance is proportional to the holding current and is larger than the junction capacitance. This is one of the reasons why a PNP load cell is used for RAMs over 16K bit. In SOI structure, generated electrons in the p-substrate are not injected into the $n^+$-buried layer.

We measured the soft-error rate using Am241 with 1 µCi as an alpha-particle source positioned 1 mm above the chip. In the 16K bit SOI-RAM, no soft errors occurred over the irradiation time ( about ten hours ). All parasitic capacitances are reduced. So, the holding current can be reduced to the level of a junction leakage current and a cell size can become as small as possible. We can obtain high-density PNP-RAMs. Figure 4 shows the soft-error rate in the 1K bit RAM with a Schottky-clamped cell. The soft-error for an SOI-RAM occurs four hundreds times later than that for a bulk-silicon RAM. In RAMs with a Schottky-clamped cell, RAM over 1K can be fabricated. We can obtain high-speed and high-density ECL RAMs.

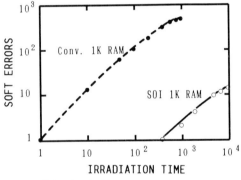

Fig.4. Soft-error rate

### 4.2 Circuit Speed

Table 1 shows the performances of SOI-RAMs. These RAMs were made on a 1 µm-thick oxide layer. The substrate capacitance of SOI-RAMs is a quarter of that of RAMs on conventional wafers. The chip select access time in a 1K bit SOI RAM, which corresponds to the gate delay time, is 13 percent faster than that of a 1K bit RAM on a conventional wafer. Power dissipation measured for SOI RAMs is about 10 percent lower than for RAMs on conventional wafers, with almost the same address access time.

|  | 1K RAM | | 16k RAM | |
| --- | --- | --- | --- | --- |
|  | SOI | Conv. | SOI | Conv. |
| Power dissipation (mW) | 630 | 710 | 175 | 190 |
| Address access time (ns) | 3.1 | 3.2 | 7.8 | 8.0 |
| Write pulse width (ns) | 2.3 | 2.5 | 8.1 | 8.0 |

Table 1. RAM performance

The gate delay time of an ESPER transistor with 0.35 x 10 um$^2$ effective emitter size is also measured, as shown in figure 5. In this case, the SOI-ESPER transistors were made on a 500 nm-thick oxide layer. As determined by SPICE simulation, the gate delay time of an SOI circuit is 6.4 percent faster than that of a bulk-silicon circuit at 1 mA. Our measurements fit well with this result. The gate delay time is 10 percent faster with a 1 μm-thick oxide layer and 12 percent faster with a 2 μm-thick oxide layer than that on a conventional wafer as determined by the simulation. The gate delay time becomes faster with increasing the thickness of the buried oxide layer. However, it is not drastically improved for the buried oxide layers above 1 μm thickness.

Wiring length on an SOI wafer contributes 46 ps/mm to the delay time, 6 percent less than that on a conventional wafer.

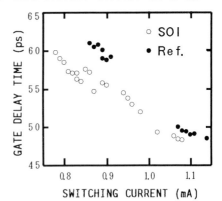

Fig.5. Gate delay time versus switching current.

## 5. Conclusion

We have clarified that the crystalline quality of the bonded SOI wafer is the same as that of a conventional wafer and this SOI wafer can be treated as a conventional wafer. Using these SOI wafers, the radiation-hardened circuits are realized without sacrificing circuit speed by these SOI wafer. No soft errors occurred over the irradiation time in the 16K bit SOI-RAM. The circuit speed on an SOI wafer with a 1 μm-thick buried oxide layer is 10 percent faster than that on a conventional wafer.

The bonded SOI wafer is suitable for sub-micron process and SOI circuits will play an important role in future VLSIs.

## Acknowledgment

The authors would like to thank H.Asuma, Y.Namiki, K.Toyoda, T.Fukuda, K.Monnma, and K.Imaoka for their encouragement and suggests. Special thanks go to T.Nakajima for help with the soft-error measurement.

## References

Goto,H., Takada,T., Abe,R., Kawabe,Y., Oami,K., and Tanaka,M., 1982 IEDM Tech.Dig., 58-61
Sai-Halasz,G.A. and Tang,D.D., 1983 IEDM Tech.Dig., 344-347
Sugiyama,M., Takemura,H., Ogawa,C., Tashiro,T., Morikawa,T., and Nakamae,M., 1989 IEDM Tech. Dig., 221-224
Toyoda,K., Tanaka,M., Isogai,H., Ono,C., Kawabe,Y., and Goto,H., 1983 ISSCC Dig. Tech. Papers, 78-79
Ueno,K., Goto,H., Sugiyama,E., and Tsunoi,H.,1987 IEDM Tech.Dig., 371-374
Vora,M., Chien,F., Burtun,G., Koh,Y.B., Brown,R., Herndon,W., and Heald,R., 1984 IEDM Tech.Dig., 690-693

# Epitaxial regrowth in double-diffused polysilicon emitters

J D Williams, P Ashburn, N E Moisewitsch, D P Gold[*], J Whitehurst[*]
G R Booker[*], G R Wolstenholme.
Dept of Electronics & Computer Science, University of Southampton.
[*]Dept of Materials, University of Oxford.

Abstract

A study is made of epitaxial regrowth of polysilicon in double-diffused polysilicon emitter bipolar transistors. TEM and RBS are used to assess the extent of the epitaxial regrowth, and a simple sheet resistance measurement is used to give a sensitive measure of the amount of regrowth. The temperature of the RTA used for the second diffusion is varied between 950 and 1150°C, with epitaxial regrowth occurring for temperatures at and above 1050°C. Polysilicon emitter bipolar transistors are fabricated to demonstrate the electrical effects of the epitaxial regrowth.

1. Introduction

In order to improve the speed of advanced silicon bipolar transistors, it is necessary to scale the vertical doping profiles. Scaling of the basewidth is of particular importance, but difficult to achieve using ion implantation because of ion channelling. An approach which avoids this problem is the diffusion of the base from a polysilicon layer, which is subsequently used for the emitter diffusion [1]. One potential hazard of this double diffused polysilicon emitter is epitaxial regrowth [2] of the polysilicon either during the first diffusion or the early part of the second diffusion. In this paper, we report on the results of an investigation into the conditions under which epitaxial regrowth of the polysilicon occurs in double-diffused polysilicon emitter transistors.

2. Experimental procedure

Initial experiments were performed on unpatterned p-type (100) wafers, which were dip etched in hydrofluoric acid prior to the deposition of a 400nm layer of polysilicon. Three types of samples were prepared in order to investigate how the base and emitter diffusions influence the structure of the polysilicon. The first were double-diffused (DD) samples, which were implanted with boron ($4.5 \times 10^{14}$cm$^{-2}$ at 40 KeV) and given a first diffusion of 10 mins at 1000°C. In the second type (AD), the boron implant was omitted, but the samples were given the same anneal of 10 mins at 1000°C. The third type were single-diffused (SD) samples in which both the boron implant and the first diffusion were omitted. All wafers were then given an emitter implant, the deposition of a low temperature capping oxide and a rapid thermal emitter diffusion. Both phosphorus and arsenic doped emitters were studied, using an implant dose of $1 \times 10^{16}$ cm$^{-2}$ and energies of 50 and 70 KeV respectively.

© 1990 IOP Publishing Ltd

## 3. Structural results

Fig. 1 shows the sheet resistance results for samples with phosphorus doped emitters. It can be seen that the sheet resistances of the single-diffused (SD) samples are significantly higher at temperatures of 1050 and 1100°C than those of the double-diffused (DD) samples. In addition the sheet resistances of the samples in which the boron implant was omitted (AD) are similar to those of the double diffused. This indicates that it is the first heat treatment, rather than the presence of the boron, which is responsible for the difference in sheet resistance. A further interesting feature of the results in fig. 1 is that a large spread in sheet resistance values (from 16.7 - 40.4$\Omega$/sq) is observed at a temperature of 1050°C for the AD samples.

The explanation for these sheet resistance results is suggested by the Rutherford backscattering data in fig. 2. For an RTA at 1100°C, the spectrum for the SD sample indicates that the emitter is polycrystalline, whereas for the AD sample the emitter is epitaxially aligned with the substrate. Such structures were directly revealed by the TEM observations in figs. 3 and 4, which show 100% regrowth for the AD sample and negligible regrowth for the SD sample. Micro-twin lamellae can be seen in the regrown material for the AD sample.

For an RTA at 1150°C, the sheet resistance data suggests that the SD sample has very nearly regrown. This is confirmed by the RBS results in fig. 2 and the TEM observations in fig. 5. The latter show only a few polysilicon grains near the layer surface, and also twins in the regrown material. From several such micrographs it was estimated that approximately 80% of the polysilicon had regrown.

For an RTA of 1050°C, a wide range of sheet resistance values were obtained for identical AD samples given nominally the same rapid thermal diffusion. The RBS spectra for two samples with sheet resistances of 16.7 (C) and 40.4$\Omega$/sq (D) are shown in fig. 3. The slopes on the spectra indicate epitaxial regrowth close the polysilicon/silicon interface, but by different amounts in the two samples. The TEM observations in figs. 6 and 7 show that approximately 98% of the polysilicon has regrown for the 16.7$\Omega$/sq sample, but only 14% for the 40.4$\Omega$/sq sample. Similar results are obtained for arsenic doped emitters.

## 4. Device results

Polysilicon emitter transistors were fabricated in order to assess the effect of epitaxial alignment on the device characteristics. In order to separate the effect of the regrowth from other effects such as changes in the doping profiles, an ion implanted base was used and an anneal carried out after polysilicon deposition, but before emitter implant. The temperature of this first 10 minute anneal was varied between 900 and 1100°C in order to vary the extent of the break up of the interfacial layer and hence of the epitaxial regrowth during the subsequent emitter diffusion. The emitter was formed from a $1 \times 10^{16} \text{cm}^{-2}$, 50KeV phosphorus implant and an emitter diffusion of 60 mins at 850°C.

The values of base saturation current density are summarised in fig. 8 as a function of the temperature of the first anneal. Results are shown for devices given an HF etch prior to polysilicon deposition and also for comparison devices given an RCA clean. The base saturation current density of the HF devices shows a steady increase as the temperature of the first anneal increase up to 1000°C, and thereafter remains constant. TEM observations show that the region of constant base current corresponds with devices in which complete epitaxial regrowth of the polysilicon has occurred.

Fig. 9 shows the base saturation current for devices given a 1000°C first anneal (device Z) in which the polysilicon has been thinned after the emitter diffusion. The base current is approximately constant for polysilicon thicknesses greater than 500Å, but shows a marked increase for thinner polysilicon layers. This contrasts with the situation for a conventional polysilicon emitter transistor where it has been reported [3] that the base current is insensitive to the polysilicon thickness down to thicknesses as small as 100Å. Thus the epitaxial regrowth of the polysilicon has transformed the device operation from that of a polysilicon emitter, where the gain is controlled by the polysilicon/silicon interface, to that of a more conventional metal-contacted transparent emitter.

## 5. Acknowledgements

The authors would like to thank Dr C Jeynes (University of Surrey) for the RBS data.

## 6. References

1. T Yamaguchi et al, IEEE Trans. Elec. Devices, Vol 35, p1247, 1988.
2. J L Hoyte et al, App Phys. Letts, Vol 50, p751, 1987.
3. T F Meister et al, IEEE BCTM. Tech. Dig., p86, 1989.

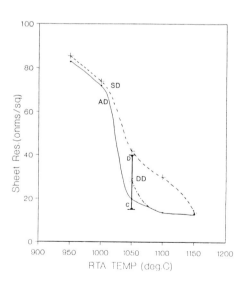

Fig.1 Sheet resistance of Phosphorus implanted polysilicon given 10s RTA diffusion at the temperature indicated for Single Diffused (SD), Double Diffused (DD) & Anneal Diffused (AD) samples.

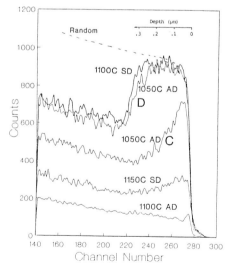

Fig.2 Channelled RBS data for Phosphorus implanted polysilicon given 10s RTA diffusion for SD & AD samples.

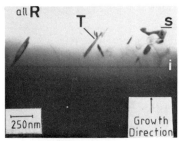

Fig.3 1100°C AD (completely regrown)

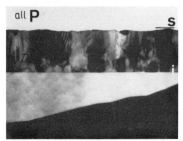

fig.4 1100°C SD (no regrowth)

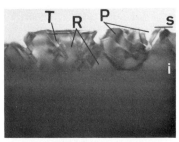

fig.5 1100°C AD (~88% regrown)

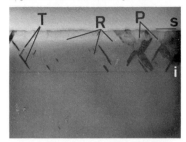

fig.6 1050°C AD 'C' (~98% regrown)

fig.7 1050°C AD 'D' (~14% regrown)

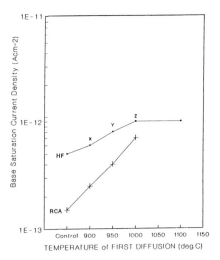

Fig.8 Base saturation current densities for Phoshorus doped polysilicon emitter transistors given the first diffusion at different temperatures, and the second diffusion of 60 mins at 850°C.

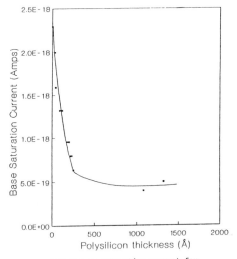

Fig.9 Base saturation current for epitaxially regrown emitters in which the regrown polysilicon has been thinned by different amount.

Figs.3 to 7 show TEM cross-sections of Phosphorus implanted AD samples given 10s RTA at temperatures stated.
[ s-layer surface, R-regrown,
 i-substrate/layer interface,
 P-polysilicon, T-twins.]

# Sidewall effects in submicron bipolar transistors

S Decoutere, L Deferm, C Claeys and G Declerck.

IMEC, Kapeldreef 75, 3030 Leuven, Belgium.

Abstract. Analytical models are presented to calculate the two-dimensional (2D) minority carrier distribution and related currents in the base and emitter. 2D effects on the base and collector current, current gain and charge storage are discussed for different technologies.

## 1. Introduction.

The scaling of the lateral dimensions of bipolar transistors in nowadays and future technologies results in an increasing importance of sidewall effects. The purpose of this paper is to present accurate and fast analytical models taking sidewall effects into account, both for the base and the emitter region. Two-dimensional effects on charge storage and currents will be discussed for poly, epitaxial and metal contacted emitter technologies.

## 2. Model description.

A cross-section of a bipolar transistor is depicted in fig.1. The base region is divided into an intrinsic base (I), a side region (II) and an external base (III). When the impurity concentration in the external base (Nbex) is higher than in the intrinsic base (Nb), the electron injection into the external base is suppressed by the electric field at the low/high transition, such that the effective recombination velocity (Seff) of the low/high transition has minor influence on the charge storage in the side region. In the case of non-negligible emitter outdiffusion (case 1), the side region only slightly affects the charge storage in the intrinsic base region, which implies a one-dimensional (1D) minority carrier distribution in the intrinsic base region. However, in epitaxial emitter technologies, the interaction between the intrinsic and side region requests that both regions are analysed simultaneously (case 2).

*Case 1: non-negligible outdiffusion.*
Using a standard separation technique with boundary conditions described above and for e.g. an npn transistor with an emitter perimeter P, the collector sidewall current $Ic_{side}$ and the minority charge storage in the side region $Q_{side}$ are given by the next equations :

$$Ic_{side} = qn_oPD_n \sum_{i=0}^{\infty} A_i \sinh(sB_i) k_i \frac{\sin(k_iL)}{B_i} \text{ and } Q_{side} = qn_oP \sum_{i=0}^{\infty} A_i \sinh(sB_i) \frac{\sin(k_iL)}{k_iB_i}$$

$$\text{with } k_i = \frac{(2i+1)\pi}{2L}, B_i = (1/L_n^2 + k_i^2)^{1/2} \text{ and } A_i = \frac{\cos(k_ide) - \cos(k_iL)}{k_iwb[\sin(2k_iL)/4 + k_iL/2]\cosh(sB_i)}$$

The electron injection into the external base is : $Ib_{ex} = qn_oP \sum_{i=0}^{\infty} A_i \sin(k_iL)/k_i$ , $L = de + wb$

$n_o$ is the minority carrier concentration in the base at the emitter/base depletion region, Ln (Lnex) and Dn (Dnex) the diffusion length and constant in the internal (external) base region. Since in this case the differential equation for the minority carrier distribution in the side region next to the emitter edge can be approximated as $\partial^2 n'/\partial x^2 + n'/L_{eff}^2 = 0$ with $1/L_{eff}^2 = 1/L_n^2 + 1/(de \cdot wb)$, a single parameter (Leff) can qualitatively describe the 2D effects in the base region. When the emitter outdiffusion and the neutral base width are large, a large amount of minority carriers is stored in the side region next to the emitter, because only a small fraction diffuses towards the collector. In this case, the effective diffusion length approaches its limiting value (Ln). However, when the base width is small, the diffusion current towards the collector is increased, resulting in a decrease of the minority charge storage in the side region. The effective diffusion length becomes smaller than Ln, indicating that minority carriers disappear from the side region because of the enhanced diffusion towards the collector. Additionally, when the emitter outdiffusion is smaller, the collector current is reduced, although the electron current density along the emitter sidewall is increased (smaller Leff).

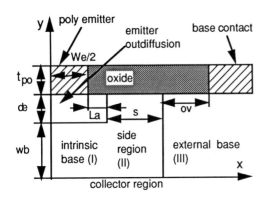
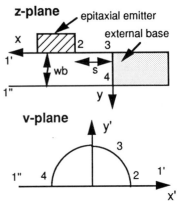

Figure 1 Schematical cross-section of a poly emitter bipolar transistor with non-negligible emitter outdiffusion.

Figure 2 Cross-section of an epitaxial emitter bipolar transistor in the complex z-plane and mapped onto the v-plane.

*Case 2: negligible outdiffusion.*
Using a conformal mapping method, the minority carrier distribution in the intrinsic base region and the side region can be calculated simultaneously. The base region is depicted in the complex z-plane in fig. 2. The z-plane is mapped onto the upper half v-plane minus the unity circle by the consecutive transformations : $w = \cosh(\pi z/wb)$, $u = 2w/(1+t) + (1-t)/(1+t)$, with $t = \cosh(\pi s/wb)$, $v = u + (u^2-1)^{1/2}$.
The minority carrier concentration is the real part of the complex concentration $n(v) = (i/\pi) n_0 \ln(v) + n_0$, and the complex current is $J_n = -q D_n (dn(z)/dz) = -J_{nx} + i J_{ny}$, with $J_{nx}$ and $J_{ny}$ respectively the x and y component of the electron current, and $i = \sqrt{-1}$.

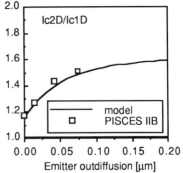

Using the conformal mapping method to extend the range in which the standard separation technique holds, very good agreement is found between our simulations and PISCES IIB simulations, as shown in fig. 3, over a wide range of emitter outdiffusions. In these simulations, the base width was 0.08 µm, the spacing between the emitter edge and the heavily doped external base 0.3 µm, the overlap of the external base on the base contact 0.4 µm, the concentration of the intrinsic base $5.10^{17}/cm^3$ and the concentration of the external base $10^{19}/cm^3$.

Figure 3 Comparison of the 2D collector current (normalized to the 1D value) between the present model and PISCES IIB simulations.

To calculate the minority carrier distribution in the emitter, the method of Hurckx (1987) will be extended to include the minority charge in the polysilicon part of the emitter. The poly/mono interface will be characterized by the interface recombination velocity $\sigma_i = S_i L_p/D_p$. Furthermore, the method will be adapted to include oxide recombination by means of the oxide recombination velocity $\sigma_{ox} = S_{ox}L_p/D_p$. The hole injection in the emitter ($I_{bh}$) and the minority charge in the monocrystalline part ($Q_{mh}$) and the polysilicon part ($Q_{ph}$) of the emitter are given by :

$$I_{bh} = -2qPD_pp_o\{ \sum_{n=0}^{\infty}(A_n+B_n)\frac{\cosh(wde)-1}{w\sinh(wde)} m\sin(mb) + de\ \sinh(\frac{b}{L_p})/L_p\cosh(\frac{b}{L_p}) \}$$

$$-2qD_pp_o\{ \sum_{n=0}^{\infty} \frac{wB_n - wA_n\cosh(wde)}{m\sinh(wde)} (-1)^n \}$$

$$Q_{mh} = 2qPp_o\{ \sum_{n=0}^{\infty}(A_n+B_n)\frac{(\cosh(wde)-1)(-1)^n}{mw\sinh(wde)} + \frac{L_p de\ \sinh(b/L_p)}{\cosh(b/L_p)} \}$$

$$Q_{ph} = q\ P\ C_o\ L_e\ L_{po}\ W_e\ [\cosh(t_{po}/L_{po}) - 1],$$

with : $A_n = \dfrac{2(-1)^n}{bL_p^2 m(L_p^{-2}+m^2)}$, $\sum_{j=0}^{N} M_{nj}B_j = Z_n$, $w = (\dfrac{1}{L_p^2}+m^2)^{1/2}$, $m = (n+\dfrac{1}{2})\dfrac{\pi}{b}$

and $C_o = \dfrac{p_o}{W_e\ \sinh(t_{po}/L_{po})}\{ \sum_{n=0}^{\infty} B_n\dfrac{2\sin(mW_e/2)}{m} + \dfrac{2L_p\ \sinh(W_e/2L_p)}{\cosh(b/L_p)} \}$, $b = W_e + 2L_a$.

$p_o$ is the hole concentration in the emitter at the emitter depletion region, and $S_i = D_{po}\coth(t_{po}/L_{po})/L_{po}$. The coefficients $M_{nj}$ and $Z_n$ are given in the appendix.

## 3. Influence of sidewall effects on scaling.

The non-linear scaling of the collector and base currents is illustrated in figure 4. Without lateral encroachment of the external base into the internal base region, the emitter sidewall results in an increase of the collector current compared to a 1D behaviour. When the emitter outdiffusion is small, the current spreading in the base results in an increase of the collector current. Additionally, the smaller the spacing between the external base and the emitter sidewall, the more the electron current behaves 1D, because the electric field at the low/high transition reflects the electrons injected into the side region, although a part of the injected electrons recombines in the external base region. Serious deviations from 1D behaviour are observed for the current gain and the intrinsic delay $t=(Q_{emitter}+Q_{base})/I_c$ in figure 5.

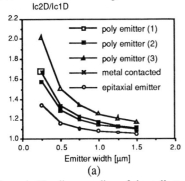
(a)

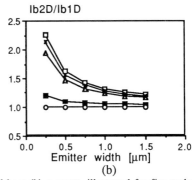
(b)

Figure 4 Non-linear scaling of the collector (a) and base (b) currents, illustrated for five technologies. Emitter outdiffusion and spacing with the external base are listed in table 1. Other parameters are : wb = 0.1 µm, ov = 0.3 µm, concentration intrinsic/external base/emitter $5.10^{17}/10^{19}/10^{20}$ [/cm3], thickness of the poly or epitaxial emitter 0.3 µm and recombination velocity at the poly/mono interface $10^5$ cm/s.

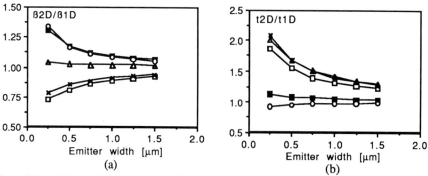

Figure 5 The influence of the non-linear scaling of the collector and base currents on the current gain (a) and the influence of the sidewall minority charge storage on the intrinsic delay t = (Qemitter+Qbase)/Ic (b).

Fair correlation between our calculations and measured values of Wβ is obtained for different emitter technologies (Figure 6). The measured values are taken from the work of Hurckx (1989). Positive (negative) values of Wβ indicate a decreasing (increasing) β for decreasing emitter width. Wβ is the emitter width at which the β deviates by 50% from the 1D value.

TABLE 1 :

|  | poly | metal | epitaxial |
|---|---|---|---|
| emitter out-diffusion [μm] | 0.15(1)(3) 0.05(2) | 0.15 | 0.00 |
| spacing with extern. base[μm] | 0.10(1)(2) 0.30(3) | 0.10 | 0.10 |

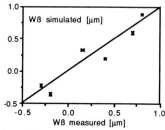

Figure 6 Comparison of our model with measured values of the characteristic emitter width Wβ.

Appendix :

$$M_{nj} = \frac{wde}{\tanh(wde)} + \frac{\sigma_i de}{L_p b}[\frac{1}{2m}\sin(mWe) + \frac{We}{2}] + \frac{\sigma_{ox} de}{L_p b}[\frac{1}{2m}\sin(2mb) - \frac{1}{2m}\sin(mWe) + b - \frac{We}{2}]$$
for n=j

$$M_{nj} = \frac{\sigma_i de}{b L_p}\{\frac{\sin[(m+v)We/2]}{m+v} + \frac{\sin[(m-v)We/2]}{m-v}\}$$

$$+ \frac{\sigma_{ox} de}{L_p b}\{\frac{\sin[(m+v)b]}{m+v} + \frac{\sin[(m-v)b]}{m-v} - \frac{\sin[(m+v)\frac{We}{2}]}{m+v} - \frac{\sin[(m-v)\frac{We}{2}]}{m-v}\}, \text{ for } n \neq j$$

$$Z_n = \frac{wde(-1)^n}{b\sinh(wde)L^2 m(L_p^{-2}+m^2)} - \frac{2\sigma_i de}{L_p b\cosh(\frac{b}{L_p})(L_p^{-2}+m^2)}[m\sin(m\frac{We}{2})\cosh(\frac{We}{2L_p}) +$$

$$\frac{1}{L_p}\cos(m\frac{We}{2})\sinh(\frac{We}{2L_p})] - \frac{2\sigma_{ox} de}{L_p b\cosh(\frac{b}{L_p})(L_p^{-2}+m^2)} \times$$

$$[-\frac{1}{L_p}\cos(m\frac{We}{2})\sinh(\frac{We}{2L_p}) + m(-1)^n\cosh(\frac{b}{L_p}) - m\sin(m\frac{We}{2})\cosh(\frac{We}{2L_p})]$$

References:

Hurckx G A M 1987 IEEE Trans. Electron Dev. **34** pp 1939
Hurckx G A M 1989 Solid-State Electr. **32** pp 397

# BASIC II: a super self-aligned technology for high-performance bipolar applications

A. Pruijmboom, A.C.L. Jansen, H.G.R. Maas, P.H. Kranen, R.A. van Es, R. Dekker and J.W.A. van der Velden

Philips Research Laboratories, P.O. Box 80.000, Eindhoven, The Netherlands

Abstract. A super selfaligned technology is described in this paper, resulting in a sidewall contacted structure. A strongly reduced sidewall base contact dimension of ca. 0.1 µm has been achieved by implementing a novel polysilicon planarisation technology.
$f_T$ and $f_{max}$ of 17 and 15 GHz have been measured for a BASIC-II npn transistor with a 1.5x58 µm$^2$ emitter (at $V_{cb}$= 2 V) which represents a 20-50% increase over conventional BASIC technology.

## 1. Introduction

Recently, BASIC technology (van der Velden et al. 1989) has been shown to give a solution for the base link-up problem, from which most advanced selfaligned bipolar processes suffer (Shiba et al 1989, Li et al 1988, Chuang 1988, Stork and Isaac 1983). Adverse effects, such as low emitter-base breakdown voltages and E-B forward tunneling could be avoided, while still maintaining good analog AC and DC properties, such as high current gain uniformity and a combination of an Early voltage of 80 V with a $f_T$ of 17 GHz (at $V_{cb}$ = 3 V) In this paper, it is shown that further performance improvement is possible, by reducing the sidewall base contact dimension from 0.5 µm to 0.1 µm. This BASIC-II technology leads to a reduced base-collector capacitance and an increased base-collector breakdown voltage.

## 2. Process flow

A novel planarisation technique was merged with standard BASIC processing (van der Velden et al 1989) to achieve a strong reduction of the extrinsic sidewall base contact dimension from 0.5 to 0.1 micron.
The key process steps, which deviate from standard BASIC processing, are shown in Figures 1 and 2. Figure 1 shows a SEM cross-section after formation of the recessed mesa-etched isolation. Locally, layers of p$^{++}$ poly Si are present, which have been obtained by deposition of undoped polysilicon, vertical implantation of boron, annealing and subsequent removal of undoped polysilicon on the vertical sidewalls. Undoped polysilicon can be etched with high selectivity (>50), with respect to p$^{++}$ poly-silicon, in KOH/isopropanol. The remaining p$^{++}$ poly layer serves as an etch mask

© 1990 IOP Publishing Ltd

to remove only part of the nitride layer on the sidewall of the mesa structure, yielding a very small extrinsic base contact. Secondly, it serves as a dopant source for the subsequently deposited polysilicon layer, which will function as a base boost. This second polysilicon layer is planarised in a similar way as described above for the first polysilicon layer. Now, however, boron is introduced into the horizontal surfaces both by implantation and by diffusion from the underlying polysilicon layer.

Again, the same dope selective etchant is used to remove undoped polysilicon from the vertical parts. The result is shown in Figure 2.

Further processing, using similar steps as for the standard BASIC process, including base link-up implant and selfaligned emitter-base formation (van der Velden et al 1989) leads to a BASIC-II npn device.

The final npn BASIC-II device structure is given in Figure 3, showing a 0.6 μm wide emitter formed by RTA diffusion out of $n^{++}$ polysilicon, a 0.8 μm wide base link-up region and a 0.1 μm base contact area on the mesa sidewall.

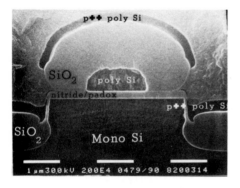

Figure 1.
SEM cross-section after formation of the mesa-etched isolation and selective removal of the first poly-silicon layer from the vertical sidewalls.

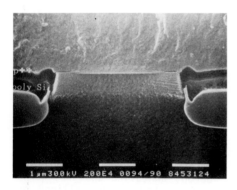

Figure 2.
SEM cross-section after planarisation of the second polysilicon layer.

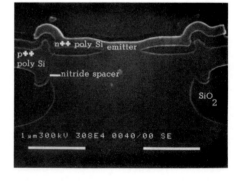

Figure 3.
The final BASIC-II npn device.

## 3. Electrical characteristics

In Table 1, a comparison has been made between BASIC and BASIC-II device characteristics.
Reduction of the base contact area leads to an increase in the collector-base breakdown voltage from 8 to 14 V, due to an increase in the distance between the $p^{++}$ outdiffusion of the base boost and the highly doped buried collector. For the same reason, also the extrinsic collector-base edge capacitance is drastically reduced from 1.1 to 0.7 fF/μm. By using the first $p^{++}$ polysilicon layer as an extra diffusion source for the final polysilicon base boost, a reduction of $p^{++}$ polysilicon sheet resistance has been obtained from 300 to 150 Ohms/square. Both lead to a strong increase in $f_{max}$ from 9 to 15 GHz at $V_{cb}$=2 V.

Other DC characteristics of standard BASIC technology, such as high $BV_{eb0}$ (6 V), could be maintained in BASIC-II technology.

## 4. Conclusion

A novel sidewall contacted super selfaligned bipolar technology has been shown to be successful for further improvement of the AC and DC performance. This has been achieved by strongly reducing the sidewall base contact dimension by applying a simple polysilicon planarisation technique twice. This planarisation technique is based on dope selective etching.

## 5 Acknowledgement

This work has been partly supported by the ESPRIT project 2016 "TIPBASE".

Table 1. Comparison of performance of standard BASIC vs. BASIC-II performance

|  | BASIC | BASIC-II |
|---|---|---|
| Emitter Area (μm$^2$) | 1.5x58 | 1.5x58 |
| $BV_{ce0}$ (V) | 7.5 | 6 |
| $BV_{eb0}$ (V) | 7 | 6 |
| $BV_{cb0}$ (V) | 8 | 14 |
| $C_{eb}$ (fF/μm$^2$) | 3 | 4 |
| $C_{cb}$ (fF/μm$^2$; intr.) | 0.3 | 0.4 |
| $C_{cb}$ (fF/μm; extr.) | 1.1 | 0.7 |
| $f_T$ (max) ($V_{cb}$=2 V; GHz) | 14 | 17 |
| $f_{max}$ ($V_{cb}$=2 V; GHz) | 9 | 15 |
| $h_{FE}$(max) | 100 | 65 |

## 6 References

van der Velden J, Dekker R, van Es R, Jansen S, Koolen M, Kranen P, Maas H, Pruijmboom A, 1989,
IEEE IEDM 89, p. 233

Shiba T, Tamaki Y, Ogiwara I, Kure T, Kobayashi T, Yagi K, Tanebe M, Nakamura T, 1989,
IEEE IEDM 89, p.225

Li G P, Chuang C T, Chen T C, Ning T H, 1988,
IEEE Trans.El.Dev., Vol. 35 (11), p. 1942

Chuang C T, 1988,
IEEE IEDM 88, p. 554

Stork J M C, Isaac R D, 1983,
IEEE Trans.El.Dev., Vol. 30 (11), p. 1527

Paper presented at ESSDERC 90, Nottingham, September 1990
Session 5A7

# Silicon-based pseudo-heterojunction bipolar transistors

Z. A. SHAFI and P. ASHBURN
Dept. of Electronics and Computer Science,
The University, Southampton, SO9 5NH, England.

### Abstract
The structure for a silicon pseudo-heterojunction device incorporating a low-doped emitter region is described. The emitter and base delay and current gain of these devices is calculated and compared to conventional devices. ECL propagation delays are also calculated, and predicted to be 21 ps, compared with 30 ps for the equivalent circuit incorporating conventional silicon devices. Finally, results from fabricated devices, are presented.

### 1. Introduction
The recent interest in silicon-based heterojunction bipolar transistors [1] is fuelled by a desire to trade-off the enhanced gain for an increase in base doping, a decrease in base width, and a decrease in emitter doping. This results in a dramatic decrease in base resistance and increase in $f_T$, without any increase in emitter/base capacitance.

In this paper, an alternative approach is described which is aimed at narrowing the speed advantage of heterojunction transistors over conventional homojunction transistors. The doping profile of this pseudo-heterojunction bipolar transistor is illustrated in fig. 1 and is not dissimilar to that of a conventional heterojunction device [1]. It incorporates a low-doped emitter (LDE), which is thick enough to prevent the E/B depletion region intersecting the $n^+$ region, but thin enough to prevent excessive stored charge. This structure makes it possible to incorporate a heavily doped, heavily band gap narrowed base, without encountering problems with tunnelling leakage at the E/B junction, and without excessive E/B capacitance.

### 2. Predicted performance of pseudo-heterojunction devices
The emitter stored charge delay of the structure has been calculated by considering the emitter as two regions, consisting of a high doped emitter (HDE) region, and a low doped emitter (LDE) region. By defining a recombination velocity at the n/n+ junction, expressions for the HDE delay, LDE delay, and current gain have been derived. In our work, the low doped emitter and thin base regions are considered to be transparent to minority carrier transport. The base delay is calculated from a standard expression, assuming a constant doping profile [2]. The trade-offs involved in emitter delay, base delay and current gain for various device structures are shown in table 1, assuming the latest self-consistent band-gap narrowing, mobility, and lifetime data of del Alamo et al. [3] for n-type silicon and Swirhun et al. [4] for p-type silicon. The table clearly illustrates that an LDE structure, similar to that in fig. 1 should give an acceptable gain, and emitter delay. Figs. 2 & 3 graphically illustrate the trade-offs discussed above. Fig 2 shows the expected fall in gain, and increase in both emitter and base delay with increasing base width, indicating that base widths of less than 0.05 μm are required. Fig 3 again shows the expected decrease in current gain, and increase in both emitter and base delay, but this time with increased base doping. For the pseudo-heterojunction device to be effective, a high base doping, and hence low base resistance is required. The upper limit of base doping for a given base width is controlled by the lowest value of acceptable gain, and upper value of acceptable stored charge delay.

© 1990 IOP Publishing Ltd

### 3. Predicted ECL propagation delay of pseudo-heterojunction circuits

In order to quantify the complicated trade-offs discussed above, we have compared the gate delays of ECL circuits incorporating pseudo-heterojunction and conventional devices (assuming a 1μm technology [5]). We have used the approach of Fang [6] which expresses the gate delay as a weighted summation of R*C time constants which can be calculated from the device geometry and doping profiles [7]. The accuracy of this approach is reported to be within ±5% of direct SPICE simulations. We have placed a lower limit on current gain of 50 for correct circuit operation (although gains as low as 30 have been used in practice [8]) and an upper limit on the base doping for conventional devices of $5 \times 10^{18}$ cm$^{-3}$ to prevent forward-biased E/B tunnelling leakage [9]. Figs.4 & 5 illustrate the predicted propagation delay for conventional and pseudo-heterojunction devices for various base doping concentrations. A minimum gate delay is observed for a given base doping, representing a compromise between high base resistance at narrow base widths and high base delay at wide base widths. The minimum gate delay for gates incorporating conventional bipolar devices is 36 ps at $W_B = 0.05$ μm and $N_B = 5 \times 10^{18}$ cm$^{-3}$. The pseudo-heterojunction circuits exhibit a minimum gate delay of 30 ps at $W_B = 0.02$ μm and $N_B = 5 \times 10^{19}$ cm$^{-3}$. A more dramatic improvement is expected if the transistor collector profile is optimized to suppress the Kirk effect [10], thus increasing the collector current at which the peak $f_T$ occurs. The preceding calculations assume an operating current of $J_C = 2 \times 10^4$ Acm$^{-2}$ for the switching transistors. Fig 5 illustrates how the gate delay varies with the operating current, $J_C$ and predicts a minimum gate delay of 30 ps for circuits incorporating the conventional bipolar devices and 21 ps for the pseudo-heterojunction circuits, The optimum values of base doping from Figs. 4 & 5. have been used in these calculations.

### 4. Practical realization of silicon pseudo-heterojunction devices

In order to demonstrate the practical realization of these LDE structures, transistors have been fabricated, by 'in-situ' doped molecular beam epitaxy, giving constant doping profiles. The epitaxial growth is essential to obtain an LDE region ($7 \times 10^{17}$ cm$^{-3}$, 0.3 μm) above a highly doped base ($2 \times 10^{19}$ cm$^{-3}$, 0.1 μm). To obtain a good contact, and improve the current gain, a top-up implant extending about 0.05μm (SUPREM) into the LDE region was included, making up the HDE region. Fig. 8 shows a gummel plot of the device. Excellent electrical characteristics are obtained, with a peak current gain of 6, compared with the expected gain of 10 for this non-optimized device.

The current gain is not expected to degrade significantly at low temperatures for the LDE devices, since the bandgap narrowing in the base region is larger than in the LDE region. Fig. 9 shows an arrhenius plot for the LDE device and a conventional control polysilicon emitter device. The fall of gain with decreasing temperature is clearly less pronounced in the LDE device, indicating that optimized pseudo-heterojunction devices show great promise for low temperature operation.

### Acknowledgements

The authors would like to thank C. G. Tuppen and C.J. Gibbings of British Telecom Research Labs. for supplying the MBE silicon.

### References

[1] C.A. King, J.L. Hoyt, J.F. Gibbons, IEEE Trans. Electron. Dev. p2093, 1989.
[2] P. Ashburn, 'Design and realization of Bipolar Transistors', J. Wiley, Chichester,1988.
[3] J.A. Del Alamo, S. Swirhun, R.M. Swanson, IEDM Technical Degest p290, 1985.
[4] S.E. Swirhun, Y.H. Kwark, R.M. Swanson, IEDM Technical Digest p24, 1986.
[5] M.C. Wilson, P.C. Hunt, S. Duncan, D.J. Bazley, Electronics Letts., Vol. 24, no. 15, p920, 1988.
[6] W. Fang, IEEE Jnl. Solid State Circuits, Vol. 25 , p572, 1990.
[7] E.F. Chor, A. Brunnschweiler, P. Ashburn, IEEE J. Solid-state Circuits, Vol. SC-23,no. 1, pp251-259 (1988).
[8] T.Sakai, S. Konaka, Y. Yamamoto, M. Suzuki, IEDM Technical Digest, p18, 1985.
[9] J.A. Del Alamo, R.M. Swanson, IEEE Electron. Device Lett., Vol. EDL-7, no. 11, p629, 1986.
[10] C.T. Kirk, IRE Trans. Electron Devices, p164, 1962.

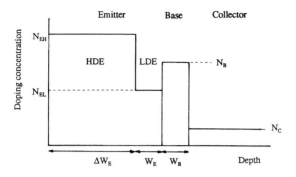

Fig. 1 -    Pseudo- heterojunction structure

| $\Delta W_E$ (μm) | $W_E$ (μm) | $W_B$ (μm) | $N_B$ (cm$^{-3}$) | β | $\tau_E$ (ps) | $\tau_B$ (ps) |
|---|---|---|---|---|---|---|
| 0.3 | - | 0.1 | $10^{18}$ | 178 | 1.2 | 4.8 |
| 0.25 | 0.05 | 0.1 | $10^{18}$ | 137 | 2.3 | 4.8 |
| 0.25 | 0.05 | 0.02 | $10^{19}$ | 206 | 1.3 | 0.3 |
| 0.25 | 0.05 | 0.02 | $5 \times 10^{19}$ | 99 | 2.4 | 0.3 |
| 0.25 | 0.05 | 0.02 | $10^{20}$ | 78 | 3.0 | 0.3 |

Table 1 -    Emitter delay, base delay and current gain for various device structures. $N_{EH}$ and $N_{EL}$ were assumed to be $1 \times 10^{20}$ cm$^{-3}$ and $1 \times 10^{18}$ cm$^{-3}$ respectively.

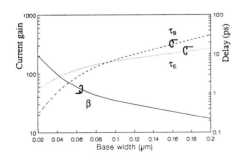

Fig. 2 -    Emitter and base delay / current gain trade-off vs base width. $N_B = 10^{19}$ cm$^{-3}$

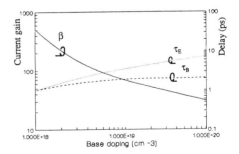

Fig. 3 -    Emitter and base delay / current gain trade-off vs. base doping. $W_B = 0.05$ μm.

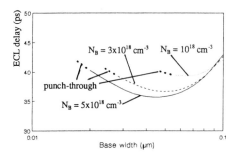

Fig. 4 - Predicted ECL gate delay for circuits incorporating conventional devices vs. base width for various base doping concentrations.

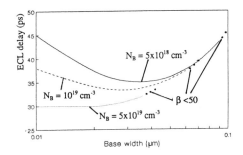

Fig. 5 - Predicted gate delay for ECL circuits incorporating pseudo-heterojunction devices vs. base width for various base doping concentrations.

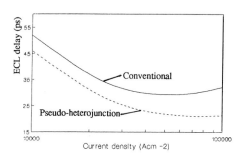

Fig. 6 - ECL gate delay vs. Collector current density at peak $f_T$

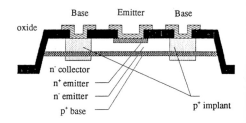

Fig. 7 - Cross-section of fabricated Pseudo-heterojunction device

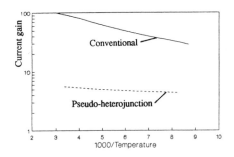

Fig. 9 - Current Gain vs. 1000/Temperature for conventional and pseudo-heterojunction devices.

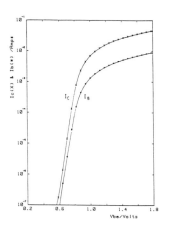

Fig. 8 - Gummel plot of pseudo-heterojunction device.

# A 12V BICMOS technology for mixed analog–digital applications with high performance vertical pnp

C. MALLARDEAU, P. KEEN, A. MONROY, J.C. MARIN, F. DELL'OVA, P.A. BRUNEL, D. CELI, M. ROCHE

SGS-THOMSON MICROELECTRONICS, B.P.217, 38019 GRENOBLE Cedex FRANCE

ABSTRACT
A 1.2μm mixed analog-digital BICMOS technology has been developed and characterized. The objective of this technology is to achieve simultaneously high density CMOS logic (5V) and high performance bipolar transistors for analog applications (12V). High speed NPN and vertical PNP have been implemented for high voltage requirements, together with the need of high frequency performance. Finally the performance of the technology is demonstrated on a RGB decoder circuit, for IDTV applications.

1- INTRODUCTION
BICMOS technology has two major application fields. The first one is for high speed logic circuits like gate arrays or SRAMs; the other one is the one chip integration of complete analog digital systems [1,2,3]. We have developed a mixed analog digital technology in 1.2μm design rules with high density CMOS logic (5V) and high performance bipolar transistors for analog applications (12V). The main challenge in our analog technology is in achieving the proper balance between digital and analog requirements. Our BICMOS technology features NPN bipolar transistors of 8 GHz cut-off frequency and isolated vertical PNPs of 2 GHz cut-off frequency, while guarantying over 12V breakdown voltage and 50V Early voltage for bipolar. This technology is validated on a BICMOS digital to analog videoprocessor, which is a good illustration of a one chip integration of complete analog and digital systems.

2- MAIN FEATURES OF THE PROCESS
A cross sectional view of the 1.2μm BICMOS structure is shown in figure 1. The available components of the technology are NMOS, PMOS, high speed NPN bipolar and high performance vertical PNP.

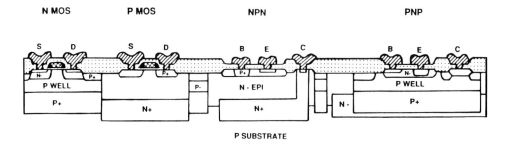

Fig.1- Cross section of the 1.2μm BICMOS technology

The process architecture is described in figure 2. The main features of the 1.2μm CMOS starting point technology are the following: P substrate, twin well structure for high integration, mixed isolation with conventional LOCOS and junction, 1.2μm polygate with silicide (TaSi$_2$), DDD structure for NMOS, with spacer to prevent hot electron degradation, conventional structure for PMOS, BPSG deposition and reflow to smooth down the topography. The double level metal structure consists in Ti/TiN/AlCu for metal 1 to achieve good contact resistance to silicon and to provide a good diffusion barrier between Si and AlCu, a sandwich of PECVD oxide and SOG for planarized mineral interdielectric, and AlCu for metal 2.

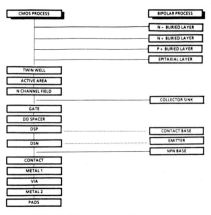

Fig.2- Process architecture

To add the bipolar components to the technology, the following steps have been introduced: N+ buried layer combined with N+ deep collector sink is used for low collector acces resistance, P+ buried layer is added for isolation between components and PNP collector, a 3μm thick N-type epilayer, with higher resistivity than the Nwell, is required for the NPN collector to achieve high performance analog functions. The emitter of the NPN is formed by direct As implantation into monosilicon, also used as drain/source for the NMOS. The base uses an additional mask and is implanted in two steps: a shallow BF$_2$ implantation to reduce the base sheet resistance between emitter and base contact; and a deeper Boron implantation for intrinsic base doping. The emitter-base structure is formed at the end of the process and only sees a final 950°C drive-in , also used as BPSG reflow.

For the vertical PNP, a N- buried layer is used to isolate the transistor from substrate. The base of the PNP is graded; it is implanted and diffused at the same time as the DDD structure of the NMOS. The emitter is implanted at the same time as the drain/source of the PMOS. Therefore no specific masking level is required for the PNP, besides the N- isolation layer.

The intrinsic transistor profiles for both NPN and PNP are plotted in figures 3 and 4 as obtained by spreading resistance measurements. The base width is 140nm for the NPN and 250nm for the PNP.

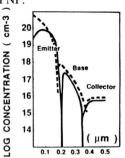

Fig.3- NPN transistor profile
— spreading resistance
- - -simulations

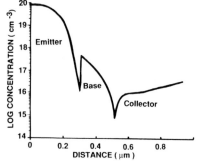

Fig.4- PNP transistor profile obtained by spreading resistance measurements

## 3-DEVICE CHARACTERISTICS

Table1 shows the main device parameters for NMOS, PMOS, NPN and PNP. The analog requirements like high breakdown voltage (>13.2V) and high Early voltage have been achieved for both bipolar transistors.

| C MOS | |
|---|---|
| NMOS ( 50 / 50 ) Vth | : 0.7V |
| PMOS ( 50 / 50 ) Vth | : 1.0V |
| BVDSS | : > 7V |
| td ( fin=fout=1) | : 220 ps |

| NPN | | PNP | |
|---|---|---|---|
| Se | : 3.6 x 3.6µm² | Se | : 3.6 x 3.6µm² |
| hfe | : 120 | hfe | : 90 |
| BVCE0 | : 14V | BVCE0 | : -20V |
| BVCB0 | : 25V | BVCB0 | : -30V |
| BVEB0 | : 5V | BVEB0 | : -5.4V |
| VEA | : 50V | VEA | : -30V |
| Ceb | : 37.5fF | ft | : 2.5 GHz |
| Ccb | : 27fF | | |
| Ccs | : 106fF | | |
| rbb' | : 370Ω | | |
| ft max | : 8 GHz | | |

Table 1- Devices parameters

Moreover the $f_T$ of the NPN was optimized by reducing the base width, and a maximum value of 8GHz was obtained at VCE=5V, as plotted in figure 5. Finally excellent high frequency performances were demonstrated by the vertical PNP, as shown by the plot of cut-off frequency versus collector current in figure 6. At VCE=5V, the maximum cut-off frequency is 2GHz.

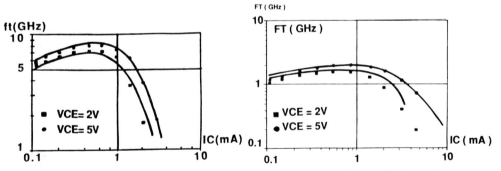

Fig.5- $f_T$ versus Ic for the NPN transistor (Em = 3.6x3.6µm²)

Fig.6- $f_T$ versus Ic for the PNP transistor (Em = 3.6x3.6µm²)

High frequency and speed of NPN bipolar transistors were tested on packaged devices, demonstrating performances of $f_{max}$=890MHz on D flip-flop, and 170ps propagation delay per stage for 21-stage CML ring oscillators. Measurements were in good agreement with the simulations.

## 4- APPLICATION TO A MIXED ANALOG DIGITAL CIRCUIT

The manufacturability of the process has been proven throughout the fabrication of a complex circuit: a RGB decoder, for IDTV applications. The function of the circuit is to implement digital luminance and chrominance to analog RGB components decoding. The chip size is 35mm², and the complexity of the circuit is over 30K transistors. It includes a digital part for signal processing operating at 54MHz, and three digital to analog converters. A microphotograph of the chip is shown in figure 7.

Fig.7- Microphotograph of the RGB decoder chip

The functionnality of the circuit has been demonstrated, and an operating frequency of 54MHz was achieved. Moreover, the digital to analog converters showed excellent analog features. Figure 8 shows the response of the circuit to a digital ramp on the Y input, at a clock frequency of 54MHz.

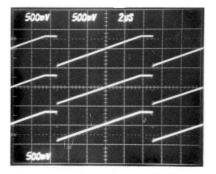

Fig.8- Response to a digital ramp.

## 5-CONCLUSION

A 12V BICMOS process for mixed analog digital applications, with high performance vertical PNP has been developed. The right compromise between digital and analog requirements has been achieved: performant analog features like high Early voltages and high breakdown voltages have been combined together with high speed characteristics for both NPN and PNP bipolar transistors. To validate the performance of the technology, a BICMOS digital to analog video processor has been fabricated successfully. This circuit is a good illustration of a one chip integration of complete analog and digital systems.

## ACKNOWLEDGEMENTS

The authors would like to thank G.TROILLARD and J.MOURIER for their contribution to process set up and device characterisation. Special thanks to all the people of the silicon process facility (ATELIER PROTOTYPE).
This work is supported by the EEC program ESPRIT under contract n°2268

## REFERENCES

1.H.Momose et al , "1µm N -Well CMOS/Bipolar Device Technology "IEEE Trans. ED ,Vol 32 , 1985 p 217
2.P.A.H Hart et al , " BICMOS , ESPRIT 86 : Results and achievement " p 221
3.T.Yamauchi et al , " 20V BICMOS Technology with polysilicon Emitter structure" Electrochem . Soc. Spring Meeting . 1987 , Philadelphia - Abst. n° 286 .

# MBE HEMT-compatible diode lasers

J. Ebner, J.E. Lary, G.W. Eliason, T.K. Plant

Electrical and Computer Engineering Department
Oregon State University

Abstract. Preliminary results are presented on the fabrication of a new HEMT-compatible diode laser structure using MBE growth with plane-selective doping. Light emission from the 2-D electron/hole gas junction has been detected and characterized.

1. Introduction

This paper presents initial results of a unique new device family in which growth of pn junctions between 2-D electron and hole gases promises the possibility of completely HEMT-compatible diode lasers with the HEMT driver an integral part of the device. As these devices are developed, the very low capacitance, high mobility lasers should have extremely wide bandwidth for optical interconnections on chips or for optical communication sources.

2. Device Structure

The device structure combines the benefits of a Graded Index Separate Confinement Heterostructure (GRIN-SCH) laser (layers 1-3 in Figure 1 below) with the high mobility characteristic of a HEMT (layers 3-6). The pn junction, integral to the operation of the laser diode, is formed by the technique of Plane-Selective Doping (Miller 1985). Physically, this is accomplished by etching the semi-insulating {100} GaAs substrate to expose {111}A crystal faces in selected regions prior to the epitaxial growth. Under appropriate MBE growth conditions, the incorporation of Si as a dopant results in n-type material on the {100} plane and p-type material on the {111}A plane as shown in Figure 2. The top layers then result in an n-channel HEMT structure on the {100} planes and an equivalent "p-channel HEMT" on the exposed {111}A surfaces.

This asymmetric structure was chosen initially since it closely resembles a regular HEMT device. The doped AlGaAs layer was made only 500 A thick to assure complete depletion of this layer under unbiased conditions to reduce the parallel leakage in this layer. This obviously causes severe loss of light from the poorly-guided structure and will be remedied in the next iteration when a completely symmetric structure will be made for good optical qualities - perhaps at the expense of some of the electrical properties like parallel conduction. The top heavily-doped GaAs layer is for making good ohmic contacts and is etched off once the contacts are made in order to eliminate this serious parallel leakage path. A cross section of the completed device is shown in Figure 3.

| | | | |
|---|---|---|---|
| 500 Å | GaAs | $5 \text{ E}18 \text{ cm}^{-3}$ Si | #6 |
| 500 Å | $Al_{0.3}Ga_{0.7}As$ | $1 \text{ E}18 \text{ cm}^{-3}$ Si | #5 |
| 50 Å | $Al_{0.3}Ga_{0.7}As$ | undoped | #4 |
| 200 Å | GaAs | undoped | #3 |
| 0.2 μ | graded $Al_x Ga_{1-x}As$ x=0.3–0.5 | undoped | #2 |
| 1.25 μ | $Al_{0.5}Ga_{0.5}As$ | undoped | #1 |
| | GaAs substrate | semi-insulating | |

Fig. 1  The MBE layers grown for the HEMT laser device

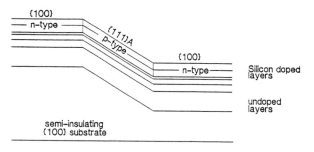

Fig. 2  Plane-selectively doped layers after growth

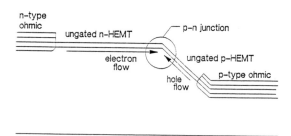

Fig. 3  HEMT-compatible laser after final processing

Such a device was fabricated in our laboratories. The completed substrate was thinned to 150 um and cleaved in to potential laser diode chips 400 um long. The 300K and 77K diode curves are shown in Figure 4. While there is a reduction in the

leaving a small n-region in series with the p-contact. This reverse-biased junction could be the main contribution to the excess resistance. The devices were driven with a 1% duty cycle pulsed current at a 1 kHz repetition rate at 77K. The peak optical output signal is shown in Figure 5 as a function of pulsed drive current level. The corresponding spectral scans are shown in Figure 6.

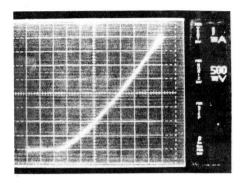
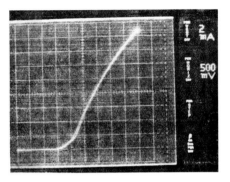

Fig. 4   Diode I-V curves at 300K and 77K for 10 sec etch

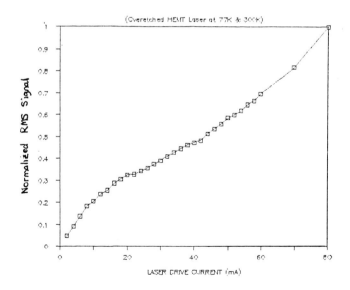

Fig. 5   Optical output signal variation with pulsed drive level

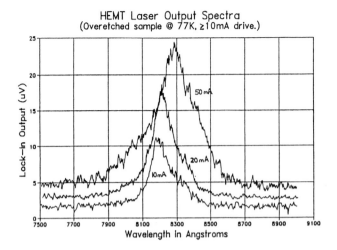

Fig. 6  HEMT-compatible laser output spectrum with drive current

## Conclusions

By showing optical emission at GaAs bandgap wavelength and no emission from the surrounding AlGaAs in the samples with the GaAs cap layer completely removed, we are confident that we are seeing emission from the 2-D electron/hole gas pn diode layer. Unfortunately, the large series resistance and the poor optical confinement near the surface make the achievement of lasing threshold impossible. The bending over of the light output curve at high drive levels and the broadening of the emission spectra are evidence of heating in the active region. This is probably due to the large series resistance.

## Future Work

These promising early results have lead us to new designs which will provide tighter optical confinement and less chance for parasitic parallel conductance paths around the 2D pn junction. Ion-implantation will be used to provide deeper, lower resistance ohmic contacts in the next device. Results of this work will also be reported.

## References

Miller D L 1985 *Appl. Phys. Lett.* **47** 1309

# Optical characterisation of InGaAs/InP and InGaAs/InGaAsP MQW structures for optoelectronic applications

Wolter K, Schwedler R, Reinhardt F, Kersting R, Zhou X Q, Grützmacher D, Kurz H

Institute of Semiconductor Electronics - Basislabor, RWTH Aachen, Sommerfeldstr., D-5100 Aachen, FRG

Abstract. We performed CW and time resolved photoluminescence (PL) measurements on a series of conventional InGaAs/InP MQW structures grown by low pressure metal organic vapor phase epitaxy (LP-MOVPE). Variation of sample temperature allows to distinct between regions with different well widths in CW experiments. The relative sizes of areas with different well widths are determined. In time resolved PL experiments the carrier capture from the barriers into the quantum wells has been investigated. An ambipolar capture time constant of $5 \pm 2$ ps is evaluated. Finally the optical gain of these structures is compared with InGaAs/InP- and InGaAsP/InP-MQW-separate-confinement structures.

## 1. Introduction

There is great interest to develop tunable semiconductor lasers for application in the optical communication systems in the 1.3 and 1.55 µm range, where optical fibers have zero dispersion respectively the attenuation minimum. Multi-Quantum-Well (MQW) laser diodes exhibit principally several advantages compared to conventional double-heterostructure diodes with an active bulk layer as demonstrated already successfully on short wavelength GaAs/AlGaAs lasers (Tsang 1987).

In the case of InGaAs/InP MQW's the emission wavelength of 1.3 µm at room temperature corresponds to a well width $L_z$ of 2.5 nm. For such thin QW's the interfaces start to play an important role. The efficient semiconductor laser action depends critically on the interface quality of these structures. The most revealing physical properties are linewidth and intensity of emission, the carrier capture from the barriers into the quantum wells and the gain efficiency at the desired wavelengths.

## 2. Experimental

The MQW structures investigated in this study were grown by low pressure (LP) metal organic vapor phase epitaxy (MOVPE) with growth interruption sequences (GIS) at both interfaces. Under a total pressure of 20 hPa and high gas velocity of 1.3 m/s a controlled growth down to a fraction of a monolayer could be achieved (Grützmacher et al. 1990). Camassel et al. (1990) found a systematic departure from the nominal well width by, typically, 4 monolayers (MLs). This departure seems to be inherently caused by GIS. It corresponds to the build up of 2 MLs of InAsP at the lower interface and 2 MLs of InGaAs:P at the upper interface. In the following these layers are not taken into account for the assigned nominal well widths.

© 1990 IOP Publishing Ltd

Temperature resolved PL measurements were performed using a combined type of He bath and gas flow cryostat allowing the continuous variation of specimen temperature between 2 and 300 K. The excitation density has been 100 mW/cm$^2$ at 1.96 eV (HeNe). Time resolved measurements were performed at 300 K using luminescence upconversion technique. The samples have been excited with 2 eV pulses of 100 fs pulse duration extracted from a colliding pulse modelocked ring (CPM) laser. Area densities between $10^{11}$ and $10^{13}$ cm$^{-2}$ per well have been achieved. The optical gain has been investigated by exciting a narrow stripe of adjustable length L with 50 ps pulses from a Nd:YLF laser ($\lambda$=1.053 µm). To determine the optical gain coefficient the PL intensity emitted at 90° with respect to the exciting beam has been measured as a function of excitation length.

## 3. Results and Discussion

In fig. 1 the temperature resolved PL spectra of an InGaAs/InP MQW structure with a nominal well width of one monolayer and 30 nm InP barrier are shown. The shape of the luminescence spectrum changes from a broad peak around 1.25 eV at 2 K into two peaks at higher temperatures. We attribute the peaks to transitions within areas of 1 and 2 ML nominal well thickness respectively. The spreading in the transition energy is about 60 meV. The observed intensity variations cannot be caused by thermal activation therefore. The spectra have to be explained by excitonic transport from the initial region to the recombination sites. Previously Grützmacher et al. (1990) have found that the lateral diffusion of QW excitons appears to be enhanced in the temperature range between 80 and 120 K. We conclude from the lowering of 1 ML luminescence peak at 80 K that the mean distance between the 2 ML islands is less than the mean free path of the exciton. On the other hand the observation of a distinct 2 ML peak leads to the conclusion that the size of the 2 ML islands has to be larger than the exciton radius. The drastic quenching of the 1 ML peak indicates an homogeneous distribution of 2 MLs islands on a scale of the exciton mean free path.

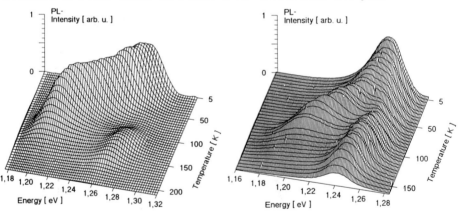

Fig. 1: Temperature resolved luminescence spectra of a InGaAs MQW with a nominal well width of 1 ML.

Fig. 2: Temperature resolved luminescence spectra of a InGaAs MQW with a nominal well width below 1 ML.

Extending these results to a sample with a nominal well width below 1 ML (fig. 2) the 2 MLs fraction is expected to be reduced. The emission peaks appear at the same energies, but with different intensity ratios of the two branches. The 1 ML peak does not disappear completely at 80 K. Thus, in this sample the mean distance between the 2 ML islands has to be significantly larger than the exciton mean free path.

Carrier capture processes have been investigated in conventional MQW samples with 30 nm InP barriers and InGaAs well widths between 0.5 and 4.0 nm by the luminescence upconversion technique. In the case of thin wells and/or high excitation densities bandfilling of the wells plays a dominant role. As shown in fig. 3 the decrease of barrier luminescence depends strongly at the excitation density in this case. To avoid bandfilling phenomena the excitation density has to be kept far below $10^{12}$ cm$^{-2}$ per well. Then the decay time of the InP barrier luminescence tends toward 4 ps independent of the well width in the investigated range (fig. 4). This time is much shorter than the decay in bulk InP (130 ps) reflecting the dominant role of carrier capture within the first picoseconds. Comparing the measured data of the barrier luminescence with a theoretical fit using cooling rates for electrons and holes determined from measurements on bulk InP an ambipolar capture time constant of 5±2 ps can be deduced.

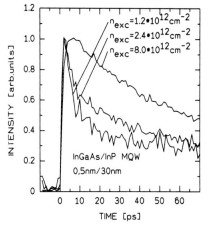

Fig. 3: Temporal evolution of the barrier luminescence at 1.35 eV at different area densities per well.

Fig. 4: Decay times of the barrier luminescence at 1.35 eV versus well width at two different area densities per well.

For MQW laser diodes composed of InGaAs wells and InP barrier layers the lowest values for threshold current densities reported so far have not been lower than those of conventional double-heterostructure lasers employing a quaternary InGaAsP active layer. Rosenzweig et al. (1990) have observed that the threshold current densities of InGaAs/InGaAsP-MQW-separate-confinement (SC) lasers with medium and long cavities were substantially lower than those of conventional double-heterostructure lasers.

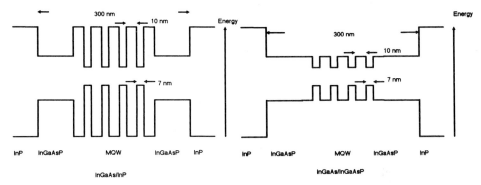

Fig. 5: InGaAs/InGaAsP-separate-confinement-layer structure (schematical)

Fig. 6: InGaAs/InP-separate-confinement-layer structure (schematical)

In this report we compare this InGaAs/InGaAsP-MQW-SC structure (fig. 5) with an InGaAs/InP-MQW-SC structure (fig. 6) by spontaneous emission measurements and gain spectroscopy. The spontaneous emission of this two structures is shown in fig. 7. Only in the case of the InGaAs/InP-MQW-SC structure a strong signal at 0.96 eV is observed in addition to the QW luminescence at 0.8 eV. It corresponds to the bandedge of the waveguiding layers. The capture of carriers from the waveguiding layers into the quantum wells is blocked obviously by the InP barriers.

Furthermore, the gain behaviour of the two structures (fig. 8) confirms the superior laser behaviour of the InGaAs/InGaAsP-MQW-SC structure since it has a higher gain coefficient at 0.8 eV and no gain at 0.96 eV as expected. In contradiction the gain coefficient of the InGaAs/InP-MQW-SC structure remains constant from 0.8 eV up to 1.05 eV due to the carrier blocking.

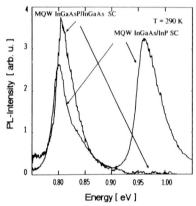 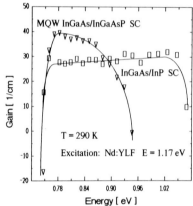

Fig. 7: Spontaneous emission spectra of both MQW-SC structures at room temperature.

Fig. 8: Gain spectra of both MQW-SC structures at room temperature.

## 4. Conclusion

By a comparative study of temperature dependences of CW emission spectra and time resolved luminescence measurements important details on growth induced ML-variations and dynamic processes are obtained which are correlated to the gain performance of device structures.

## Acknowledgement

This work was supported by the Ministry of Research and Technology of the Federal Republic of Germany (project TK 446). The Nd:YLF laser was supported by OTTO-JUNKER-STIFTUNG and the ministry of sciences and research of Nordrhein-Westfalen.

## References

Camassel J, Laurenti J P, Juillaguet S, Reinhardt F, Wolter K, Kurz H, Grützmacher D 1990 *Proc. 5th Int. Conf. on MOVPE* to be published in *J. Cryst. Growth*

Grützmacher D, Hergeth J, Reinhardt F, Wolter K and Balk P 1990 *J. Electron. Mat.* **19** 471

Rosenzweig M, Ebert W, Franke D, Grote N, Sartorius B, Wolfram P 1990 *Proc. 5th Int. Conf. on MOVPE* to be published in *J. Cryst. Growth*

Tsang W T 1987 *Semiconductors and Semimetals* **24** 397

# Strongly directional emission from AlGaAs/GaAs light emitting diodes

A.Köck, and E.Gornik
Walter Schottky Institut, TU München, 8046 Garching, Germany
M.Hauser, and W.Beinstingl
Institut für Experimentalphysik, Universität Innsbruck, 6020 Innsbruck, Austria

Abstract We show for the first time that strongly directional emission of defined polarization can be achieved from conventional AlGaAs/GaAs double heterostructure surface emitting LEDs via coupling to surface plasmons. By microstructuring the surface, we have fabricated LEDs with a beam divergence of less than $4^0$ and a drastically increased quantum efficiency. We prove that surface plasmon excitation and emission mechanism has the potential to overcome basic external quantum efficiency losses and to improve the performance of LEDs.

## 1. Introduction

Infrared LEDs and laser diodes are the most important sources for optical fiber communication. The advantages of LEDs compared to laser diodes include higher temperature operation, smaller temperature dependence of emitted power, simpler device construction and simpler drive circuits. Among the disadvantages are low external quantum efficiency (QE), lower modulation frequency and wide spectral line width. Nondirectional emission is a detrimental property of both laser diodes and LEDs (Sze 1981).

We show for the first time that excitation and light emission of surface plasmons (SP) represent a unique method to improve LED performance with regard to specific technical applications. By the use of an appropriate surface grating we have achieved strongly directional emission and a drastically increased external QE from conventional AlGaAs/GaAs double heterostructure LEDs.

In the past light emission of SP has been investigated in several devices.(Lambe 1976, Jain 1978, Adams 1981, Dawson 1984, Ushioda 1985, Donohue 1986, Watanabe 1988, Suzuki 1989). Sharp and narrowband emission peaks due to the radiative decay of SP were obtained from periodically structured metal–oxide–metal junctions, when appropriately biased (Kirtley 1980a, 1980b, 1983c). Theis et al (1983) achieved light emission of SP from charge injection structures. Recently, we have reported SP enhanced light emission from forward– and reverse biased Ag/n–GaAs Schottky diodes (Köck 1988). If the surface of the diodes is periodically structured, photons generated in the GaAs substrate are first coupled to surface plasmons at the Ag–surface, which then decay into photons. This coupling mechanism has the potential to improve the performance of LEDs. It overcomes basic external quantum efficiency losses, provides strongly directional emission of defined polarization and reduces the linewidth of emitted light.

## 2. Experimental methods and results

The light emission of periodically structured ($LED_1$ and $LED_2$) and flat ($LED_3$) AlGaAs/GaAs double heterostructure surface emitting LEDs is investigated. The gratings were fabricated by electron beam lithography and reactive ion etching. The grating periods of $LED_1$ and $LED_2$ were $\Lambda_1 = 830$ nm and $\Lambda_2 = 1024$ nm, the grating amplitudes were $H_1 = 40$ nm and $H_2 = 60$ nm. An Ag film of 25 nm thickness was evaporated on the front surface to form both an ohmic contact and to provide excitation of SP. All emission measurements were performed at room temperature. As shown in the inset of Fig. 1, the emitted light ($E_0$) was detected in the (x–z) plane, while the angle $\Delta$ was varied.

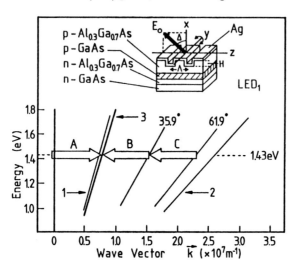

**Fig. 1:** Curves (1) and (2) represent the light lines in air and GaAs, respectively. Curve (3) shows the dispersion relation of SP at the Ag–air interface. The other curves represent light lines of photons, which couple to SP.

The coupling mechanism is explained at maximum emitted wavelength $\lambda = 867$ nm in Fig. 1. Shown are the light lines in air (curve (1)) and in AlGaAs (curve (2)) and the dispersion relation of SP at the Ag–air interface (Johnson 1971) (curve (3)). The other curves represent the light lines of photons, which couple to SP due to the existence of the periodic surface structure.

Two basic loss mechanisms limit the external QE of LEDs: A portion of the photons generated in the active region is reflected at the semiconductor/air interface (Fresnel loss); photons impinging on the interface with an emission angle $\Delta$ larger than the critical angle of internal total reflection ($\Delta c = 16°$) cannot propagate into air (critical angle loss). However, if the surface of the LEDs is periodically structured and coated by a thin metal film, both loss mechanisms can be avoided due to the excitation and emission of SP. Photons with p–polarization (i.e. the magnetic field vector is parallel to the grating grooves) generated in the active layer can interact with SP according to the grating coupling condition

$$k_{SP} = (\omega/c)\, n_{GaAs} \sin \Delta + n\, k_g \qquad (1)$$

with $k_{SP}$ the wave vector of SP, $n_{GaAs}$ the refractive index of GaAs, $k_g = 2\pi/\Lambda$ the grating vector and n an integer. In the case of $LED_1$, photons with $\Delta_1 = 0.4°$

directly pass through the metal film or interact via one grating vector (n = +1, process (A)) with SP. Photons with emission angle $\Delta_2$ slightly larger than $\Delta c$ couple directly to SP (light line for $\Delta_2 = 16.8^0$ intersects the SP dispersion curve (3)). Photons with $\Delta_3 = 35.9^0$ and $\Delta_4 = 61.9^0$ couple to SP for n = −1 (process (B)) and n = −2 (process (C)). Subsequently the excited SP radiate into air according to the grating coupling condition

$$k_{ph} = (\omega/c) \sin \Delta = k_{SP} + n\, k_g \qquad (2)$$

with $k_{ph}$ the wavevector of photons and $\Delta$ the angle of emission. For LED$_1$ coupling occurs only for n = −1 (inverse to process (A)) and results in an emission angle $\Delta = 1.3^0$. For LED$_2$ coupling is possible for n = −1 and n = −2 and emission occurs into two angles $\Delta = 11^0$ and $\Delta = 40.5^0$.

All coupling processes besides process (A) involve photons with emission angles $\Delta$ larger than $\Delta c$. Therefore the SP excitation and emission mechanism drastically reduces losses due to internal total reflection. In addition the Fresnel loss is diminished because in SP resonance the reflection of photons back into the substrate is nearly zero. The result is a drastic increase of the external QE: At an injection current of 20 mA LED$_1$ has an external QE of 1.51%, LED$_2$ of 1.40%, while LED$_3$ has a QE of 1%.

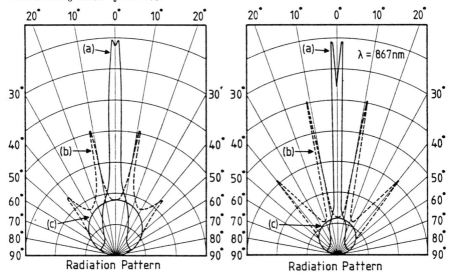

**Fig. 2:** Spatial distribution of the spectral emitted light from structured and flat LEDs. Curves (a) and (b) show the p–polarized radiation pattern of LED$_1$ and LED$_2$. Curve (c) represents the s–polarized radiation pattern of LED$_1$ and LED$_2$.
**Fig. 3:** Spatial distribution at the maximum emitted wavelength of $\lambda = 867$ nm.

The resulting radiation patterns are shown in Fig. 2 for spectral emitted light and in Fig. 3. at the maximum emitted wavelength $\lambda = 867$ nm. Curves (a) and (b) of Fig. 2 show the spatial distribution of p–polarized light emitted from LED$_1$ and LED$_2$ at an injection current of 20 mA. LED$_1$ shows strongly directional emission around $0^0$ with a beam divergence of $4^0$. The small dip at $0^0$ indicates, that the emission peak is composed of two immediately adjacent peaks. A slightly different grating period would result in emission perfectly focused in $0^0$. LED$_2$ shows emission peaks at $11^0$ and $40.5^0$ (curve (b)). Curve (c) represents s–polarized light emitted from both LED$_1$ and LED$_2$. Light emitted from the flat

LED$_3$ is nondirectional and completely unpolarized. Curve (c) also represents the unpolarized radiation pattern of LED$_3$. The radiation patterns at the maximum emitted wavelength of $\lambda = 867$ nm are shown in Fig. 3, respectively.

Excitation and emission of SP decrease the spectral linewidth of the emitted light: The full width at half maximum decreases at maximum emitted intensity from 36 nm for the flat LED$_0$ to 27.5 nm for LED$_1$ and LED$_2$.

## 3. Applications

Based upon the presented results, the performance of conventional LEDs can be designed for certain technical applications such as coupling of light into optical fibers or coupling into waveguides: The grating period determines the angle of directional emission. Therefore the emission angle $\Delta$ can be arbitrarily changed by a proper choice of the grating period. Note that for the present grating configuration, the emission is enhanced only for p–polarized light. The use of a crossed grating provides coupling out of s–polarized light, which will additionally increase the external QE.

Obviously this principle can also be applied to the realization of vertical emitting laser diodes. In the laser situation photons travel parallel to the surface ($\Delta = 90°$). Two requirements have to be met to achieve laser operation via SP. First the distance between the Ag–surface and the active layer has to be small enough to provide sufficient overlap between the electric fields of photons and the SP. Second the grating period has to be adjusted according to equation (2). This technique has a high potential to achieve emission from laser diodes normal to the surface.

The authors are grateful to W. Schlapp and G. Weimann, FTZ der DBP, Darmstadt, Germany, for the growing of the samples. The work was partly sponsored by the "Fond zur Förderung der wissenschaftlichen Forschung" (no. P6129), Austria, the "Stiftung Volkswagenwerk" (no. I61840), Germany, and the Siemens Corporation Munich, "Sonderforschungseinheit TU München", Germany.

A. Adams, and P. K. Hansma, Phys. Rev. B **23**, 3597 (1981)
P. Dawson, D. G. Walmsley, H. A. Quinn, and A. J. L. Ferguson, Phys. Rev. B **30**, 3164 (1984)
J. F. Donohue, and E. Y. Wang, J. Appl. Phys. **59**, 3137 (1986)
R. K. Jain, S. Wagner, and D. H. Olson, Appl. Phys. Lett. **32**, 62 (1978)
P. B. Johnson, and R. W. Christy, Phys. Rev. B **6**, 4374 (1971)
J. R. Kirtley, T. N. Theis, and J. C. Tsang
   1980 Appl. Phys. Lett **37**, 435
   1981 Phys. Rev. B **24**, 5650
   1983 Phys. Rev. B **27**, 4601
A. Köck, W. Beinstingl, K. Berthold, and E. Gornik, Appl. Phys. Lett. **52**, 1164 (1988)
N. Kroo, Zs. Szentirmay, and J. Felszerfalvi, Phys. Lett. A **81**, 399 (1981)
J. Lambe, and S. L. McCarthy, Phys. Rev. Lett. **37**, 923 (1976)
K. Suzuki, J. Watanabe, A. Takeuchi, Y. Uehara, and S. Ushioda, Solid State Comm. **69**, 35 (1989)
S. M. Sze, Physics of Semiconductor Devices, Second edition, John Wiley & Sons, New York (1981), p. 681–739
T. N. Theis, J. R. Kirtley, D. J. DiMaria, and D. W. Dong, Phys. Rev. Lett. **50**, 750 (1983)
S. Ushioda, J. E. Rutledge, and R. M. Pierce, Phys. Rev. Lett. **54**, 224 (1985)
J. Watanabe, A. Takeuchi, Y Uehara, and S. Ushioda, Phys. Rev. B **38**, 12959 (1988)

# Detection of near IR radiation by SiGe material

B. Sopko, J. Pavlu, I. Prochazka, I. Macha

Czech Technical University, Faculty of Nuclear Science and Physical Engineering, Brehova 7, CS-115 19 Prague 1

Abstract. We are reporting on the novel design of the fast photodiode for satellite laser ranging. The new technique, based on Germanium ion implantation into Si substrate, is used to increase the quantum efficiency for infra-red photons. The Single Photon Avalanche Diodes (SPADs) on "pure" silicon are conventionally used for that purposes, but only very low quantum efficiency at the wavelength of $1\,\mu m$ is obtained. Our new advanced technology of the SPADs, allows an increase of the quantum efficiency at the wavelength of $1.07\,\mu m$ by the factor 2-3.

## 1. Introduction

Our group has participated for several years in project Interkosmos by development of high sensitive detectors of light, which are based on the silicon avalanche diodes. Such detectors are used above all in satellite laser ranging as the receiver for returning reflected beam. Its sensitivity enables to recognize signal on single photon level.

The impulse YAG laser at the wavelength of $1.07\,\mu m$ is conventionally used as a source of light. The photomultiplier tubes (PMT) - especially microchannel (MCP PMT) - commonly used as detectors have efficiency of about 6 % at the first harmonic ( $\lambda = 0.53\,\mu m$ ). But their efficiency strongly decays with decreasing of photon energy and at the wavelength about $1\,\mu m$ are inapplicable.

The silicon SPADs have also limited quantum efficiency because of their absorption edge about the wavelength of $1.1\,\mu m$. But the light of $\lambda = 1.07\,\mu m$ can be received with quantum efficiency about 2 % yet.

The main significances of using silicon SPAD in the satellite laser ranging are :
- replacement of expensive photomultiplier by SPAD reduce costs
- only low applied voltage is needed
- TTL or NIM compatible output signal
- the SPAD enable to use the wavelength of $1.07\,\mu m$ at which the earth atmosphere has low absorption
- timing jitter more than twice lower than MCP PMT
- ranging system temporal stability twice higher than MCP PMT
- room temperature operation
- excellent ruggedness (suitable for a spaceborn receiver)

## 2. Structure and operation of SPADs

The structure of typical avalanche diode have to avoid the local maximums of high electric field, because the breakdown must be uniform over the junction area. The avalanche diodes used today have the structure described by Heitz (1964, 1965) and later by Cova (1983). Many attempts were carried out to improve the properties of this diodes. The single or double epitaxial structure were tested by Cova (1988) to decrease carrier diffusion effect for better time resolution (jitter).

Our SPADs have the simple geometry shown in Fig. 1. The avalanche region is surrounded by a n⁻ guard ring. Its breakdown voltage is greater than that of the main n⁺p junction therefore it protect the active region from most of the outer electrons which are generated in the neutral p-type bulk.

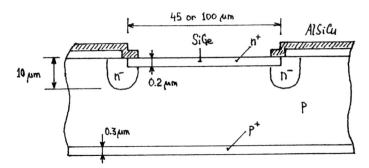

Fig. 1. Schematic cross section of SPAD structure

To obtain the single photon counting operation, the two methods can be used - e.g. Cova (1983):
- the passive quenching
- the active quenching, where the external quenching
  circuit is used to stop the avalanche process

In both techniques the diode is biased above the breakdown voltage $U_B$. Subsequently, the single free carrier in the depletion layer can initiate an avalanche process. During the laser distance measurement, the diode is periodically biased above $U_B$ by the gate signal, which is derived from the laser impulse. The statistics of about 1000 attempts is taken to distinguish between self-triggered breakdown and useful signal caused by returning photon.

## 3. Advanced technology of SPADs

There are three most important parameters to be held in high level by the proper technology of the fabrication :
- the dark count rate $f_D$ (is defined as an inverse
  value of the mean time, for which the diode may be
  biased until the avalanche is self triggered)
- timing jitter (is defined as time indeterminateness
  in output electric pulse caused by returning photon)
- quantum efficiency of photon detection

The last of these parameters depends on used semiconductor material because of its forbidden gap width. The internal light reflection technique described by Muller (1978), or fabrication of germanium avalanche diode (e.g. Fichtner (1976)) can be used for increasing quantum efficiency of infra-red photon detection. To keep low jitter and good technological properties of Si, our latest attempts were focused on decrease gap width of the silicon by ion implantation of Ge atoms.

Our advanced technology of diode, starting from earlier attempts of SPAD - Hamal (1986) -, allows an increase of the diode active area to 100 μm diameter while maintaining acceptable dark count rate at room temperature, and fast time response. In the latest modification of the fabrication process, the step with the ion implantation of Ge have been added.

Fabrication by the planar process (see Fig. 1) :

- substrate 0.3 Ωcm p-type silicon with a thickness of 250 μm
- forming a denuded zone at the surface (48 hours at 1050°C in $O_2$)
- oxide passivation at 1050°C in $O_2$
- **ion implantation of Ge (into front side)**
    dose: $1 \times 10^{16}$ cm$^{-2}$, energy: 150 keV
- ion implantation of P (n⁻ -type guard ring)
    dose: $1 \times 10^{15}$ cm$^{-2}$, energy: 180 keV
- redistribution of P atoms by diffusion at 1050°C
- ion implantation of As (n⁺ -type layer 0.2 μm thick)
    dose: $5 \times 10^{15}$ cm$^{-2}$, energy: 30 keV
- forming the back p⁺ -contact by boron implantation
    dose: $1 \times 10^{16}$ cm$^{-2}$, energy: 180 keV
- rapid isothermal annealing
- metallization with AlSiCu
- mounting by gold bonding to TO-18 headers
- measurement of V-A characteristics and selection

In the Fig. 2. the doping profile of front side is shown. How can be seen, only a part of the depleted region with thickness of about 0.8 μm is into SiGe layer. Under the technological conditions described above, the diodes with a breakdown voltage of about 29 V are obtained.

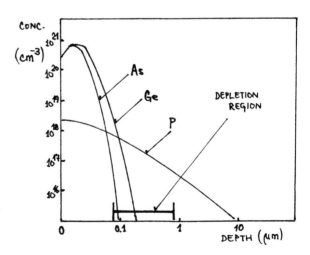

Fig. 2. The doping profile of the front side in advanced SPAD technology. The depletion region at the breakdown voltage is also shown.

We have carried out several experiments with the various modifications of the diode processing sequence described above. The key step to obtaining low jitter and dark count rate as well is the formation of denuded zone, which serves as the gettering layer. Investigation was begun to determine also the optimum annealing time and temperature. The best results was achieved with rapid isothermal annealing at the temperature above 1000°C.

The concentration and the depth of implanted germanium is practically limited by the parameters of the real ion implanters. The thickness of the SiGe layer is about 0.2 μm. The concentration of Ge atoms in this layer is only about 1 % how can be roughly calculated.

## 4. Experimental results

The effect of additional Ge atoms in Si lattice was not measured directly but only through changes in behavior of fabricated SPADs. In the following table the various PMT and MCP PMT are compared with our SPADs (last three items). The last item shows parameters of SiGe diode fabricated according to advanced technological process described above. All data were measured at the room temperature.

| type | material (cathode) | relative quantum eff. 0.53/1.07 µm | jitter [ps] 0.53/1.07 µm | $f_D$ [kHz] |
|---|---|---|---|---|
| RCA 8852 | S20 | 1 / – | 400/ – | 50 |
| RCA 31034A | GaAsInP | 3 / – | 200/ – | 20 |
| Hamamatsu MCP R2287U | S20 | 0.22/ – | 100/ – | <1 |
| SPAD 45 µm | Si | 0.20/0.04 | 30/ 70 | 5 |
| SPAD 100 µm | Si | >0.40/0.05 | 40/ 70 | 200 |
| SPAD 100 µm | SiGe | – /0.15 | – /100 | 400 |

It has to be mentioned that the quantum efficiency is shown in the table only in relative values compared with PMT RCA 8852. The real quantum efficiency of our SPADs has been estimated from comparative measurements and it is about 2 % for Si SPADs and about 6 % for advanced SiGe SPADs at $\lambda = 1.07$ µm. It can be assumed that described new technology of the SPADs, allows an increase of the quantum efficiency at this wavelength by the factor 2-3 while maintaining acceptable dark count rate at room temperature and fast time response.

## 5. Conclusions

An improved single photon avalanche photodiode (SPAD) with high sensitivity at near infra-red radiation has been developed at our workshop and tested at the Satellite Laser Ranging Station with very good result. At actual satellite measurement a ranging jitter of 3 cm when ranging to satellites at the distances up to 7000 km is standard (for the signal strength being within the range of 1 to 10 photoelectrons). But from this point of view the "pure" Si SPADs are about 3 times better. The main advantage of SiGe SPADs is in the decreasing of the number of necessary laser shots.

Our next work will be focused on experiments with SiGe material prepared by metallurgical processes. The greater amount of Ge atoms will make further effective forbidden gap narrowing and the quantum efficiency would be still increased. Simultaneously, the decreasing of the jitter and the dark count rate will be investigated.

References :

Cova S at al 1983 IEEE J. Quantum El. **QE-19** p 630
Cova S at al 1988 Journal de Physique, Colloque C4 p 633
Fichtner W at al 1976 Rev. Sci. Instrum. **47** p 374
Haitz R H 1964 J. Appl. Phys. **35** p 1370
Haitz R H 1965 J. Appl. Phys. **36** p 3123
Hamal K, Prochazka I, Jelinkova H, Sopko B 1986 6.International Workshop on Laser Ranging Instruments, Juan les Pins (F), edited by J.Gaignebet
Muller J 1978 IEEE Trans. on El. Devices **ED-25** p 247

*Paper presented at ESSDERC 90, Nottingham, September 1990*
*Session 5B6*

# High-reliability semiconductor laser amplifiers

S J FISHER, C P SKRIMSHIRE, R N SHAW, J R FARR, P C SPURDENS, H J WICKES, W J DEVLIN

British Telecom Research Laboratories
Ipswich, England

Abstract  This paper presents the first results of a preliminary study into the reliability of Semiconductor-Laser-Amplifiers (SLA's). The results from ageing tests on semiconductor lasers which have one facet coated with a multilayer anti-reflection (AR) coating indicate that the AR coatings present no additional hazard. Preliminary data from devices having both facets AR coated (full amplifier structures) show minimal degradation in the main parameters. No degradation mechanisms, other than those occurring in standard buried heterostructure lasers, have been identified to date.

1. Introduction

Recent laboratory and field demonstrations of optical transmission systems using semiconductor laser amplifiers (SLA's), such as that of Taga *et al* (1989), have underlined the potential of these devices to replace electronic repeaters in medium and long-haul systems, and to improve distribution capacity in optical access networks by compensating for branching losses. SLA's are now being seriously considered for deployment in future submarine transmission systems, where their reliability will be of prime importance. This paper reports, for the first time, initial results of a study into the reliability of SLA's.

2. Device description

High-gain (25dB), polarisation-insensitive planar buried heterostructure laser amplifier structures have been developed at British Telecom Research Labs (BTRL) (Cole *et al* (1989)). The SLA devices described in this work are similar to the BH lasers reported by Nelson *et al* (1985). These devices were shown, by Sim *et al* (1988), to have excellent reliability. The devices used in this study are designed for operation at 1.5 µm and incorporate a second quarternary layer above the active region in order to improve the symmetry of the horizontal and vertical confinement factors. This feature reduces the gain difference $\Delta G_p$ between TE and TM modes ($\Delta G_p \sim 1.6$dB has been achieved in this device). Devices are fabricated, as described by Nelson *et al* (1985), using atmospheric-pressure MOVPE, a process which allows the growth of high-uniformity, large-area wafers. The laser facets were coated with a specially developed three-layer dielectric anti-reflection (AR) coating having typical plane-wave reflectivities in the range $10^{-3} - 10^{-4}$ over bandwidths in excess of 200nm (Devlin, 1990). The coating materials used ($Al_2O_3$, $SiO_2$ and $Si$) remain stable during deposition and are each known to have good long-term stability.

The device chips are 500µm long and are bonded p-side down to a diamond heatsink using AuSn solder. The p-side and n-side metallisations are TiPtAu and TiAu respectively. Amplifiers for lifetests are bonded to a header and mounted in a lifetest sub-carrier, the temperature of which can be accurately controlled.

© 1990 IOP Publishing Ltd

## 3. SLA lifetests and characterisation measurements

In order to resolve degradation effects into those caused by the chip and those caused by the AR coating, devices having one, both or neither of their facets AR coated have been life-tested. Degradation in the AR coating on a one-facet-coated (1-fc) device manifests itself as a change in the tracking ratio[1]. The devices with no coatings provide a reference against which the coated lasers can be compared. On the 1-fc and uncoated devices, threshold current, forward voltage at threshold and tracking ratio are monitored. The devices with both facets coated have extremely high threshold currents which are difficult to determine accurately from the shallow light-current characteristic, and this parameter is therefore not measured. Instead, changes in the spontaneous outputs at selected currents are tracked throughout the lifetest as a measure of degradation. Optical gains under various combinations of selected wavelength, current, input optical polarisation and input optical power are also measured at regular intervals. The experimental set-up used to characterize the amplifiers is shown in figure 1.

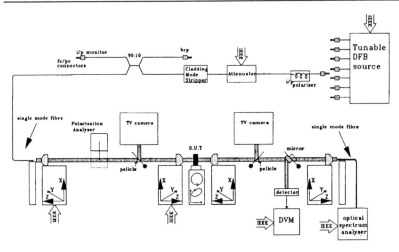

Figure 1: Characterisation equipment for measurement of STWLA devices

In the characterisation system (fig.1) light is launched into the test device using precision AR coated lenses, an arrangement which facilitates automation of the measurements and avoids the risk of damaging device facets with fibre ends.

The lifetests were divided into stages. Stage 1 was an initial medium-stress test of 5,000 hours at 80°C and 100mA device current. Stage 2 was a high stress test of 100°C, 100mA. During both stages, the devices were operated in an atmosphere of flowing dry nitrogen.

## 4. Results

The stage 1 lifetests on single facet coated devices all indicate excellent chip and AR coating stability. Stage 1 tests on uncoated chips (i.e. BH lasers) yielded similar degradation rates, indicating that the addition of AR coatings does not affect the normal chip degradation mechanisms. The forward voltage at threshold exhibited only slight increases during stage 1 testing which shows that the metallisation and die bonding have remained stable. None of the single facet coated devices on test showed any variation in tracking ratio, implying good stability of the AR coating.

---

[1] Tracking ratio is the ratio of the external slope efficiency, dL/dI, of the uncoated facet to that of the coated facet.

The variation in threshold current of a representative sample during a stage 2 test is shown in figure 2, which shows a gradual increase in threshold current associated with normal laser ageing.

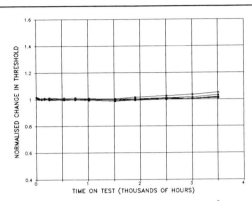

Figure 2: Variation in threshold current of 1-fc chips during stage 2 testing

The variation in tracking ratio of a representative batch during the stage 2 test is shown in figure 3. The tracking ratio is constant to within measurement error. No significant degradation has been seen after 5,000 hours of stage 1 and 3,500 hours of stage 2 tests. This is equivalent to $10^5$ hours of operation at 25°C, assuming a pessimistic activation energy of 0.5eV.

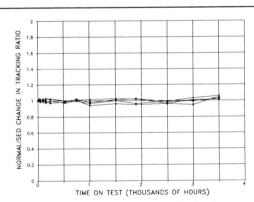

Figure 3: Variation in tracking ratio of 1-fc chips during stage 2 testing

The technique used to measure tracking ratio is prone to a small amount of measurement fluctuation, giving rise to the small changes evident in figure 3. It is likely that these are small random errors in the measurements which are accentuated by the differentiation algorithm used.

A typical small-optical-signal gain versus current plot from one of the 2-fc devices (i.e. a full travelling wave amplifier), for both a peak and a trough in the residual Fabry-Perot (FP) ripple characteristic, is shown in figure 4.

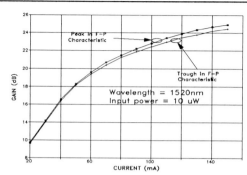

Figure 4: Gain versus current plot from BTRL BH STWLA devices

Preliminary stage 1 testing of 2-facet-coated devices has shown no significant variation in any measured parameter during the first 260 hours. Figure 5 shows the variation of small-optical-signal gain at 100mA bias during the first 260 hours of the stage 1 lifetesting of a sample batch of 2-fc devices.

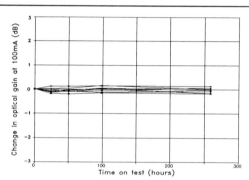

Figure 5: Variation in optical gain at 100mA of 2-fc chips during stage 1 testing

5. Conclusions

The low rate of increase in threshold current seen in lifetests on single-facet-coated devices is similar to that seen in the tests on uncoated devices and indicates that no extra degradation is caused by the addition of AR coatings. The tracking ratio stability of the same devices shows that the coatings appear to be intrinsically reliable. Lifetests to date on single facet coated devices represent reliable operation for service times of more than $10^5$ hours at 25°C, assuming a pessimistic activation energy of 0.5eV. Preliminary data from 2-fc devices during the first few hundred hours of lifetesting at 80°C and 100mA shows no measurable degradation in the amplifying characteristics.

6. References

Taga et al     1989   Trans. IEICE, E72, p1061$ff$
Cole et al     1989   Electron. Lett., 25, p314$ff$
Nelson et al    1985   Electron. Lett., 21, p888$ff$
Devlin et al    1990   Paper submitted to IEEE 2nd conf. On InP
   & related compounds, Denver, Colorado, April 1990
Sim et al      1988   ECOC 14, Vol. 1, p398$ff$

Paper presented at ESSDERC 90, Nottingham, September 1990
Session 5B7

# Band to band absorption coefficients in heavily doped Si and SiGe

A. Nathan, S. C. Jain[*], D. R. Briglio, J. M. McGregor, and D. J. Roulston

Department of Electrical and Computer Engineering
University of Waterloo, Waterloo, Ontario, Canada N2L 3G1

[*]Clarendon Laboratory, Oxford OX1 3PU, England, UK

**Abstract**: The band to band absorption coefficients in heavily doped Si and $Si_{1-x}Ge_x$ have been computed (for Ge fraction, $x \leq 0.3$) using recent values of impurity concentration dependent band gap narrowing. The presence of Ge in highly doped B:Si appears to reduce the band to band absorption coefficients. This can be attributed to the relatively smaller gap shrinkage and the higher Fermi levels in highly doped SiGe.

## 1. Introduction

For realistic simulations of Si or SiGe bipolar photodetectors[1], the band to band absorption coefficients in heavily doped Si or SiGe, are required. In addition, these coefficients can be used to interpret measurement data for extraction of pertinent physical parameters such as minority carrier lifetime and diffusion length in the heavily doped material.

We report in this paper, the values of fundamental band to band absorption coefficients, $\alpha$ of heavily doped Si and SiGe bulk layers. Following the suggestion of Pantelides et al.[2], we have used known values of band gap narrowing due to heavy doping effects. The calculation of band to band absorption coefficients is necessary since it is difficult to extract reliable values of the coefficients from experiments because of the unknown free carrier absorption that is always superposed in the observed data[2]. We have used a simplified form of the theory of optical absorption given by Macfarlane[3] (see also the review of Maclean[4]). This theory has been used by numerous authors to model band to band absorption (see e.g. Haas[5], Pankove and Aigrain[6], and Schmid[7]).

## 2. The Method

Two models for optical absorption in highly doped Si have been given by Haas[5], and Pankove and Aigrain[6]. In a simplified form, Haas' model describing the contribution to absorption by phonon assisted transitions can be expressed as[8]

$$\alpha_{ph} = \hat{A}^2 [(h\upsilon - E_g + E_0)^2 I^+ + (h\upsilon - E_g - E_0)^2 I^-], \tag{1}$$

where

$$I^+ = \int_0^1 \frac{\{x(1-x)\}^{1/2} dx}{1+\exp[E_F - x(h\upsilon - E_g + E_0)]/kT} \tag{2}$$

and $I^-$ is obtained by replacing the optical phonon energy, $E_0$ by $-E_0$. In (1), $\hat{A}$ is independent of doping and its value is taken as 65 cm$^{-1/2}$eV$^{-1}$ (see Ref. 8). The value of $E_0$ is assumed to be 58 meV. Note that eqn. (1) involves the impurity concentration dependent band gap $E_g$, and Fermi energy $E_F$, thus accounting for the effects of high doping. According to the model proposed by Pankove and Aigrain[6], the optical absorption can be expressed as $\alpha = B_2^2 \alpha_{ph}$,

© 1990 IOP Publishing Ltd

where $B_2^2$ is independent of energy but depends on impurity concentration and species. Using realistic models of $E_g(N)$ and $E_F(N)$ [see Ref. 8], the phonon contribution to absorption in Si, is computed and fitted with Schmid's measured $\alpha$ values at sufficiently high energy (1.5 eV to 1.8 eV) where the free carrier absorption can be assumed negligible. The coefficient $B_2^2$ (N) is adjusted for the known value of $Â^2$, so that the calculated value of $\alpha$ at the high energy range is consistent with the observed data of Schmid[7]. The extracted values of $B_2^2$ (N) for Si are subsequently used for the calculation of absorption coefficients in bulk SiGe layers. Here, the dependence of the band gap and hole effective mass (for p-SiGe) on doping concentration and Ge fraction x, is assumed to obey the model recently proposed by Jain and Roulston[9].

## 3. Results and Discussion

The fundamental band to band absorption calculated in this way for Si doped with boron and arsenic are shown in Figs. 1 and 2, respectively. The effect of impurity concentration on the absorption coefficients is more pronounced in the case of Si doped with boron as opposed to arsenic. For concentrations greater than $10^{19}$ cm$^{-3}$, the $\alpha$ values for Si:As appear to remain practically independent of impurity concentration. The intercepts on the $h\nu$ axis are only weakly dependent on doping concentration (the corresponding change in threshold is hardly noticeable on the scale used in Figures). The optical band gap (bandgap plus Fermi level measured from the majority carrier edge) changes by approximately 11 meV when the concentration of As changes from $6\times10^{18}$ to $2.8\times10^{19}$ cm$^{-3}$. The computed band to band absorption values (Figs. 1 and 2) can be subtracted from the spectra observed by Schmid[10], and this yields the corresponding free carrier absorption (see Ref. 8).

Apart from the modification of $E_g(N)$ and $E_F(N)$ and consequently the absorption, due to the presence of impurity, there are also additional transitions taking place in which momentum is conserved by impurity or free carrier scattering. The effect of enhanced transitions is contained in the impurity concentration dependent coefficient $B_2^2$ (N). For the case of boron doped Si at room temperature, this coefficient can be approximated as[8]

$$B_2^2 (N) = 0.8 + 5.0\times10^{-5}(N/N_o) - 3.2\times10^{-6}(N/N_o)^2 + 5.7\times10^{-9}(N/N_o)^3 + 9.0\times10^{-10}(N/N_o)^4 \quad (3)$$

where $N_o$ is $10^{18}$ cm$^{-3}$. This is in contrast to Si:As, where $B_2^2$ (N) is virtually independent of doping concentration[8]. Using the above expression for $B_2^2$ (N), the band to band absorption coefficients for boron doped SiGe bulk alloy are computed for two different Ge compositions: x=0.2 (Fig. 3) and x=0.3 (Fig. 4). For a given composition x, a linear interpolation of the hole effective mass between the Si and Ge values is employed in the calculations of band gap and Fermi level. The Fermi level calculations are based on the conventional density of states formulation. Although there is no significant change in energy gap values going from x=0.2 to x=0.3, there is a noticeable change in the Fermi level: $E_F$ for composition x=0.2 is lower than its corresponding value when x=0.3, leading to slightly higher absorption. Based on our calculations presented here for B:SiGe, we note that both the cases (x=0.2, 0.3) lead to reduced absorption compared to Si:B (Fig. 1). In the case of Si:B, the Fermi levels are lower and furthermore the shrinkage of the gap in p-doped Si is more pronounced than p-SiGe alloys (for x≤0.3, see Ref. 9) and consequently, the band to band absorption is marginally higher. It should be noted, however, following the recent conclusions regarding effective density of states in the valence band (see Green[10] and Ref. 9), that our computed Fermi levels may have been overestimated (e.g. a boron concentration of $1.2\times10^{20}$ leads to $\Delta E_F\sim 50$ meV). If these Fermi level corrections are taken into account, there is a significant increase in band to band absorption particularly at high impurity concentrations and at the low photon energies.

Based on the calculated band to band absorption coefficients for highly doped Si, numerically fitted empirical relations for room temperature have been obtained, viz.,

Si:B  $\alpha = b_0 + b_1 E + b_2 E^2$

$b_0 = 0.4 (N/N_o)^2 - 44.0 (N/N_o) + 5305.3$

$b_1 = -0.6 (N/N_o)^2 + 70.9 (N/N_o) - 9574.0$

$b_2 = 0.3 (N/N_o)^2 - 28.2 (N/N_o) + 4303.0$  (4)

Si:As  $\alpha = a_0 + a_1 E + a_2 E^2$

$a_0 = -18.4 (N/N_o) + 5261.7$

$a_1 = 23.2 (N/N_o) - 9604.1$

$a_2 = -6.4 (N/N_o) + 4365.5$  (5)

These relations are useful for the computer aided design of optical detectors and solar cells. Similar relations have been obtained for free carrier absorption (see Ref. 8). The empirical relations (4) and (5) have been incorporated into the BIPOLE computer program and simulation results illustrating the difference in the photocurrent in a p$^+$-n photodiode using the conventional absorption coefficient (the value being that for intrinsic Si) and the new relations (4) and (5), are shown in Figs. 5 and 6. The doping profile is gaussian with a surface concentration of $3 \times 10^{20}$ cm$^{-3}$. The n-substrate doping is $10^{16}$ cm$^{-3}$. The incident photon flux normal to the surface is $10^{18}$ cm$^{-2}$ and the applied voltage is 5 V. As expected, with the new relations, the photocurrent density in a photodiode ($x_j \sim 0.3$ μm) is larger at higher energies ($\lambda < 1.1$ μm) (see Fig. 5). In Fig. 6, we illustrate the current density in photodiodes of various junction depths. For diodes with very small junction depth ($x_j < 50$ nm), there is hardly a discrepancy (photogeneration in the intrinsic region dominates) but the discrepancy widens and eventually saturates when most of the photogeneration takes place in the highly doped region (when $x_j > 300$ nm).

We are currently extending our calculations to SiGe strained layers. In optical detectors with SiGe bases, the strain further modifies the bandgap and the optical absorption. The confinement energy in the thin layers of superlattice and its effect on the absorption coefficient will be modeled along the lines presented in Ref. 11. Using these computed $\alpha$ values, the Ge fraction x and the thickness of the base of the superlattice layers can be optimised.

## References

[1] H. Temkin, A. Antreasyan, N. A. Olsson, T. P. Pearsall, J. C. Bean, Appl. Phys. Lett. **49**, 809, 1986.
[2] S. Pantelides, A. Selloni, R. Car, Solid-State Electronics **28**, 17, 1985.
[3] G. G. Macfarlane, T. P. Maclean, J. E. Quarrington, V. Roberts, Phys. Rev. **111**, 1245, 1958.
[4] T. P. McLean in: *Progress in Semiconductors*, Eds. A. F. Gibson, F. A. Korger, R. E. Burgess, Heywood:London, **53**, 1960.
[5] C. Hass, Physical Review **125**, 1965, 1962.
[6] J. I. Pankove, P. Aigrain, Physical Review **126**, 956, 1962.
[7] P. E. Schmid, Physical Review B **23**, 5531, 1981.
[8] S. C. Jain, A. Nathan, D. R. Briglio, D. J. Roulston, C. R. Selvakumar, T. Yang, J. Appl. Phys., submitted.
[9] S. C. Jain, D. J. Roulston, Solid-State Electronics, submitted.
[10] M. A. Green, J. Appl. Phys. **67**, 2924, 1990.
[11] F. Cerdeira, A. Pinczuk, J. C. Bean, B. Batlogg, B. A. Wilson, Appl. Phys. Lett. 45, 1138, 1984.

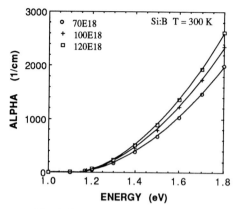

Fig. 1  Band to band absorption for boron doped Si at room temperature.

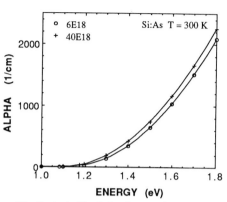

Fig. 2  As in Fig. 1, but for arsenic doped Si

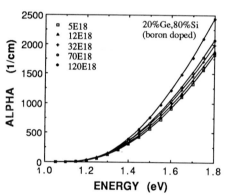

Fig. 3  As in Fig. 1, but for boron doped bulk SiGe alloy (Ge fraction = 0.2)

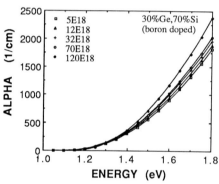

Fig. 4  As in Fig. 3, but Ge fraction = 0.3

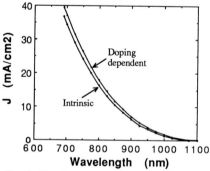

Fig. 5  The photocurrent density as a function of wavelength of incident radiation

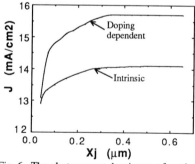

Fig. 6  The photocurrent density as a function of photodiode junction depth

## Design considerations for 0·5 micron ultra thin film submicron SOI transistors by two-dimensional simulation

G.A.Armstrong, W.D.French, J.R.Davis[*]

Department of Electrical Engineering, Queen's University Belfast, N.Ireland
* British Telecom Research Laboratories, Ipswich, United Kingdom

**Abstract**
    Two dimensional device simulation is used to investigate problems associated with the design of 0.5 micron gate length SOI MOSFETs. The choice of SOI film thickness and doping to give acceptable values of threshold voltage, inverse subthreshold slope and breakdown voltage is considered for both n– and p–channel transistors.

**Introduction**
The advantages of thin film or fully depleted silicon on insulator (SOI) CMOS transistors have been described[1]. In addition to low parasitic capacitance, high packing density and simple processing provided by the more traditional thick film (or partially depleted) SOI transistors, thin film devices offer further advantages such as reduced short channel effects, higher drive currents and an absence of the "kink" effect[2]. The ouput characteristics of n–channel thin film SOI transistors have however been shown to be degraded by the effect of bipolar snapback[3]. Holes generated by impact ionisation at the drain serve as a base current for the lateral bipolar transistor and the breakdown characteristic is governed by a lowering of the potential barrier at the source junction. The breakdown mechanism for thin films is similar to the $V_{ceo}$ breakdown in a bipolar transistor[4].

**Two–dimensional simulation**
    A two–dimensional device simulator, based on a modified version of MINIMOS4 [5], has been developed for thin film SOI, and used to model the breakdown characteristics caused by bipolar snapback[6]. The simulation geometry, applicable to SOI transistors, is shown in Fig.1. At source, drain, gate and substrate contacts, fixed electrostatic and Fermi potentials are applied. The source and drain contacts may be extended a specified distance $T_{si}$ below the gate oxide interface to model the action of a silicide contact as an ideal ohmic contact. The inclusion of a substrate contact below the buried insulator allows the effect of substrate bias on the characteristics of SIMOX transistors to be modelled. Full details of the various parameter models, solution algorithms etc. of this simulator are presented in [7]. Impact ionisation is modelled as a localised effect governed by a Chynoweth law, using appropriate coefficients optimised to give good agreement of simulated breakdown voltage with measured devices[9].

**N–channel transistor design**
    To design n–channel transistors with a gate length of 0.5 $\mu$m requires optimisation of film thickness and doping level to achieve a balance between the requirement for low threshold voltage ($< 0.6$V), low inverse subthreshold slope ($< 85$ mV per decade) and high breakdown voltage ($> 4$V). The simulation model employed an n$^+$ polysilicon gate, with a buried oxide thickness, $T_i = 0.4$ $\mu$m. The source and drain were implanted at 90 keV with a dose of $5 \times 10^{15}$ arsenic ions. Simulation of the anneal predicted a sideways diffusion of 0.11 $\mu$m for both source and drain. A fixed positive charge of $10^{11}$ cm$^{-2}$ was used to model the imperfections of the lower interface between the SOI film and the buried oxide, while the gate oxide

interface was assumed ideal. Fig.2 shows the relationship between inverse subthreshold slope, film thickness and gate oxide thickness for a fixed threshold voltage. At a given film thickness, an increase in gate capacitance with decreasing gate oxide thickness is balanced by an increase in depletion capacitance as the film doping is increased to maintain a constant threshold voltage. For a range of film thicknesses down to 50 nm, the use of thinner gate oxide is shown to give a degradation rather than an improvement in subthreshold slope. In addition, Fig.3 shows that for a fixed threshold voltage, a thicker, more lightly doped SOI film is desirable to increase the breakdown voltage. It appears, however, that even with a thicker film, it is not possible to achieve a simulated breakdown voltage exceeding 3.2V for a 0.5 $\mu$m gate length, without using some form of drain engineering. To increase the breakdown voltage, the peak electric field in the SOI film must be reduced. This implies a reduction in the doping gradient at the drain junction either by reduction of the drain doping or else by introduction of a lightly doped drain (LDD) structure. Fig.4 shows the effect of the incorporation of an LDD on the simulated output characteristics of a 0.5 $\mu$m transistor. The oxide thickness was 20 nm, substrate doping $1.1 \times 10^{17}$ cm$^{-3}$, corresponding to a threshold voltage of 0.6V. The characteristics are shown both with and without an LDD implant dose of $10^{13}$ cm$^{-2}$ phosphorous ions at 50 keV using a 0.2 $\mu$m oxide spacer. The hole and electron lifetime was assumed to be 0.1 $\mu$s, independent of doping. Simulated breakdown voltage has been defined as the value of drain voltage for which the drain current is 1$\mu$A per $\mu$m width for $V_g = 0$. Inclusion of the LDD gives rise to a significantly less abrupt breakdown characteristic and an increase in breakdown voltage of more than 1V.

Further additional improvement in breakdown voltage can be achieved by reduction in the drain implant dose from $5 \times 10^{15}$ to $5 \times 10^{14}$ cm$^{-2}$ at 90keV. Fig.5 shows the variation of breakdown voltage as a function of carrier lifetime in the SOI film. A simple empirical model[8] of the lateral bipolar transistor shows that breakdown occurs when the product of bipolar common emitter current gain $\beta$ and peak avalanche generation rate tends to unity. By reducing carrier lifetime, increased recombination occurs in the source (emitter) depletion region, leading to a reduction in emitter efficiency, a lower value of $\beta$. and hence a higher breakdown voltage.

### P—channel transistor design

The major problem in the design of the p—channel transistor is to achieve a low threshold voltage ($> -1V$), and to ensure that the back channel does not turn on in the subthreshold region with negative substrate bias. Fig.6 shows the variation of threshold voltage and subthreshold slope for a 0.5 micron p—channel transistor as a function of film doping, (assumed constant purely to enable simple comparison), for $V_{sub} = -3V$. In a full design however, a depth dependent implant profile, with higher doping at the lower interface, can be incorporated in the simulation.

It is evident that it is not possible to obtain a threshold voltage smaller than $-1V$, while still achieving an acceptable subthreshold slope. The high value of subthreshold slope at low film doping does not occur for a 1 $\mu$m gate length, where the subthreshold slope is less than 70 mV per decade, as long as the doping is less than $1.5 \times 10^{17}$cm$^{-3}$. The degradation in subthreshold slope for ultra short gate length can hence be attributed to back channel leakage, enhanced by the action of the negative substrate bias. This leakage may be suppressed by increasing the film doping, until at $2 \times 10^{17}$ cm$^{-3}$, a low subthreshold slope of less than 70 mV per decade is obtained, corresponding to the condition associated with complete depletion in the film. Such a doping is however not feasible in practice, since the threshold voltage is now larger than $-2.5V$. While the gate oxide thickness does not have a significant effect on threshold voltage, the use of the thinner gate oxide does give improvement in subthreshold slope, but only if the film doping is less than $5 \times 10^{16}$ cm$^{-3}$.

If the gate material is changed from n⁺ to p⁺ polysilicon, with a constant film n–type doping of $1.1\times10^{17}$ cm⁻³, a threshold voltage of $-0.60$V and a subthreshold slope of 85 mV per decade is predicted. By careful optimisation of channel implant profile for the n⁺ polysilicon gate, it is possible to achieve a threshold voltage of $-0.93$ and subthreshold slope of 93 mV/decade. In the latter case, however, the simulated punchthrough voltage is only $-3.5$V.

## Conclusions

Two–dimensional device simulation has shown that it is possible to design 0.5 μm gate p– and n– channel transistors with low threshold voltage and subthreshold slope. Simulations indicate a breakdown voltage of almost 5V is achievable for the n–channel transistor with optimum drain engineering.

## References

1. J.P.Colinge, IEDM Tech. Dig., pp.817–820, 1989.
2. K.Kato, T.Wada and K.Taniguchi, IEEE Trans ED–32, 1985, pp.458–465.
3. A.J.Auberton–Herve, Proc. IEEE SOS/SOI Technology Workshop, Georgia, pp. 55, 1988.
4. M.Haond and J.P.Colinge, Electonics Letters, 1989, 25, pp.1640–1641.
5. W.Hansch and S.Selberherr, IEEE Trans. ED–34, 1987, pp. 1074–1080.
6. G.A.Armstrong, N.J.Thomas and J.R.Davis, Proc. IEEE SOS/SOI Technology Workshop, Nevada, pp.44–45, 1989.
7. S.Selberherr, "Analysis and simulation of semiconductor devices", Springer Verlag, Vienna, 1984.
8. K.K.Young and J.A.Burns, IEEE Trans ED–35, 1988, pp.426–431.
9. G.A.Armstrong, R.S.Ferguson and J.R.Davis, In " Simulation of semiconductor devices and processes", Vol.2, K.Board Ed.,Pineridge Press, pp.446–465, 1986.

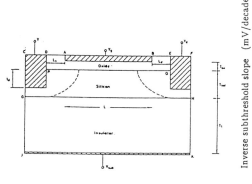

Fig.1 SOI simulation geometry

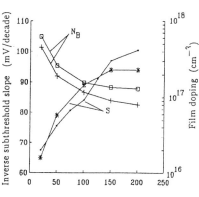

Fig.2 Dependence of inverse subthreshold slope S and film doping $N_b$ on SOI film thickness for a fixed threshold voltage of 0.6V

+ $T_{ox} = 20$ nm    * $T_{ox} = 14$ nm

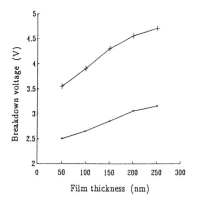

Fig.3 Dependence of breakdown voltage on SOI film thickness for a fixed threshold voltage of 0.6V

$T_{ox} = 20$ nm   + $L = 1.0\ \mu m$   · $L = 0.5\ \mu m$

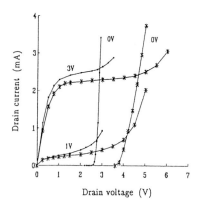

Fig.4 Simulated output characteristics for 0.5 $\mu m$ gate length n–channel SOI transistor

* LDD     $N_B = 1.1 \times 10^{17} cm^{-3}$
· non LDD $T_{ox} = 20$ nm
          $T_{soi} = 100$ nm
          $W = 10\ \mu m$

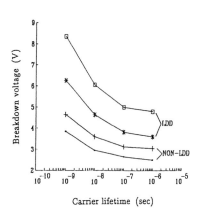

Fig.5 Dependence of simulated breakdown voltage on lifetime for 0.5 $\mu m$ LDD and non–LDD transistors

□
+  Drain implant dose $5 \times 10^{14} cm^{-2}$
*
·  Drain implant dose $5 \times 10^{15} cm^{-2}$

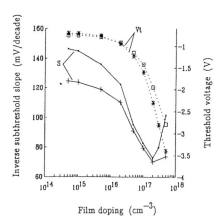

Fig.6 Dependence of inverse subthreshold slope S and threshold voltage $V_t$ on film doping for 0.5 $\mu m$ p–channel transistor

$V_{sub} = -3V$

□
+  $T_{ox} = 14$ nm   $T_{soi} = 100$ nm
·  $T_{ox} = 20$ nm

## The influence of substrate bias fixed charge in the buried insulator on the gain of the parasitic bipolar inherent in silicon-on-insulator MOSFETs

L.J. McDaid, S. Hall, W. Eccleston and J.C. Alderman*

The University of Liverpool, Department of Electrical Engineering and Electronics, Brownlow Hill, P.O. Box 147, Liverpool, L69 3BX.

*Plessey Research (Caswell), Towcester, Northants, NN12 8EQ, U.K.

**Abstract** When Silicon-On-Insulator (SOI) MOSFETS are operated at high drain voltages, the drain-source current becomes strongly influenced by the common emitter current gain of the parasitic, lateral bipolar transistor. For small channel lengths, the gain is limited only by the emitter efficiency and bipolar action can then give rise to high off-state leakage. From measurements performed on lateral SOI diodes it is demonstrated that the recombination rate in the source/body depletion region can be enhanced by the application of a substrate bias sufficient to cause depletion at the body/buried oxide interface. This results in a reduction in the common emitter current gain and from Gummel plots a reduction in gain of 20% was measured for an increase in substrate voltage of 5 volts. Also, fixed positive charge in the buried oxide is expected to produce the same effect for an n-channel transistor, as a depletion region will form automatically at the body/buried oxide interface.

## Introduction

The single most important adverse phenomenon in SOI MOSFETS is the so-called single transistor latch [1] which results in a loss of gate control causing the transistor to remain on. Latch occurs in thick and thin-film transistors and also in p-channel and n-channel devices. However, n-channel transistors are more prone to latch owing to the higher ionisation coefficient for electrons. The effect originates from lateral parasitic bipolar action where the source, body and drain form the emitter, base and collector respectively. With a view to suppressing bipolar action, we have examined previously the characteristics of the body/drain junction under reverse bias and showed that the generation leakage current can be enhanced by the electric field in the drain [2]. Metallic impurities gettered to the junction are likely to be responsible for this parasitic current which leads to high off-state leakage, or in the worst case, so-called latch. For short channel transistors the breakdown voltage, which we define as the lowest drain voltage required to cause the transistor to enter the latch state, is significantly reduced from that associated with the drain/body junction. This is because of the high common emitter current gain of the parasitic bipolar. Therefore if latch is to be suppressed for drain voltages typical of that used in circuit applications ($V_d$=5 volts) then the gain must be reduced and to achieve this aim, the factors which influence this parameter in

© 1990 IOP Publishing Ltd

these transistors must be identified. If SOI is to compete with
conventional MOSFETS then the gain must be suppressed for channel lengths
in the submicron regime. For such transistors, recombination in the body
is negligible and the common emitter current gain depends solely on the
emitter efficiency. It is shown in this work that a significant
reduction in the emitter efficiency is obtained when a depletion region
exists close to the body/buried oxide interface. This condition can be
achieved when a substrate bias is applied of sufficient magnitude such
that depletion occurs at this interface. Note that the bias is not
sufficient to form a back channel. Depletion arising from fixed positive
charge in the buried oxide will exist automatically for an n-channel
transistor but deliberately increasing this charge to effect suppression
of the bipolar action is clearly not desirable. For a p-channel device,
oxide charge accumulates the back interface and the emitter efficiency is
then determined by the bulk properties of the source/body p-n junction.
In the following section, measurements on lateral diodes and transistors
demonstrate that the ideality factor and the common emitter current gain
are both influenced significantly by the substrate voltage. The
correlation between these results verify the dependence of common emitter
current gain on recombination at the body/buried oxide interface.

### Results and discussion

Figure 1 shows a schematic cross-section of a $p^+$-n gated diode
fabricated using Separation by Implantation of Oxygen (SIMOX) technology.
The dose was $1.8 \times 10^{18}$ $O^+$ ions $cm^{-2}$, implanted at 200 keV followed by an
anneal at 1300°C for 6 hours. For all measurements the top gate bias $V_g$,
is zero and the current is measured as a function of the forward voltage
$V_f$, with the substrate voltage $V_s$, as a parameter. Figure 2 shows
typical forward characteristics for $p^+$-n and $n^+$-p diodes with $V_s$ biased
positively to induce negative charge in the body. This results in an
accumulation of electrons at the body/buried oxide interface if the body
is n-type, and depletion of holes if the body is p-type. The ideality
factors are 1.41 and 1.87 for the $n^+$-p and $p^+$-n junctions respectively
and the larger value for the latter junction can be explained as follows.
In the case of the $p^+$-n diode, accumulation of electrons at the
body/buried oxide interface causes a reduction in the width of the space
charge region at that interface. This significantly reduces the
recombination rate and the characteristics of the diode are now
determined by recombination in the bulk of the depletion region. For the
case of a p-type body ($n^+$-p diode), positive substrate bias results in a
widening of the depletion region and electrons injected from the $n^+$
source effectively recombine with holes over a wider region of the
interface causing a reduction in the ideality factor. This behaviour
explains qualitatively the form of the graphs in figure 2. Figure 3
shows the diode current as a function of forward voltage for a $p^+$-n
junction with varying substrate bias. The deterioration in the $\log_e (I_f)$
v $V_f$ plots at low currents with increasingly negative substrate voltage,
is caused by the shift from accumulation to depletion thereby enhancing
hole recombination at the body/buried oxide interface.

To demonstrate the effect of substrate bias on the current gain of
the parasitic bipolar, figure 4 shows the Gummel plot taken for an
n-channel transistor with an aspect ratio of 20:1 and $V_s$=0 and 5 volts.
For $V_s$=0 the gain at low currents is reduced by recombination in the
source/body depletion region. At higher currents the gain peaks at

approximately 380 and then falls rapidly due to the effects of source/drain series resistance and high injection. Increasing Vs to 5 volts leaves the general features of the plot unaltered except that the gain has reduced to approximately 300 owing to enhanced recombination in the source/body depletion region. Clearly an applied substrate bias and/or fixed charge in the buried oxide can increase the recombination rate in the source/body space-charge region and consequently reduce the gain of the parasitic bipolar. The recombination lifetime could be further reduced by external means, without increasing generation rate, by introducing a suitable impurity (for example Pt) into the silicon, so as to produce a deep level lying several kT away from the centre of the energy gap. Such a level would reduce drastically the recombination rate without significantly enhancing generation and would result in a further reduction in the gain of the parasitic bipolar, without compromising off-current leakage.

## Conclusion

It has been shown that substrate bias can reduce the common emitter current gain by allowing recombination to occur in the source/body p-n junction at the interface with the buried oxide. Because n-channel transistors are more prone to latch than p-channel transistors the gain must be reduced for the former device and therefore, for CMOS applications the substrate bias must be positive. In transistors with thick buried oxides this bias has little influence on the multiplication factor and therefore the net effect is to increase the breakdown voltage. Degrading the quality of the back interface is likely to enhance the level of oxide charge and increase the negative surface generation velocity. Therefore, this could further reduce the common emitter current gain without introducing other adverse effects. Finally it should be noted that radiation hard SOI circuits benefit from a negative substrate bias which serves to attract holes generated within the oxide, away from the active devices. The recommendation for a positive substrate bias only applies therefore to high performance SOI CMOS.

## Acknowledgements

The work was funded by the Procurement Executive, Ministry of Defence (RSRE) under contract RP009/341. The authors are grateful to the Directors of the Plessey Company for permission to publish.

## References

[1] C.E. Daniel Chen et al, IEEE Electron Dev. Lett, 9, 636 (1988).

[2] L.J. McDaid, S. Hall, W. Eccleston and J.C. Alderman, Springer Verlag, Proc. of ESSDERC (1989) Berlin, p. 759, 1989.

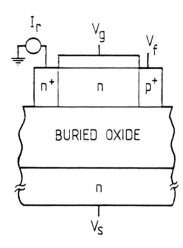

Fig. 1. Schematic cross-section of a $p^+$-n gated diode showing bias arrangement.

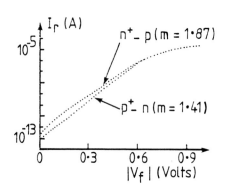

Fig. 2. Forward characteristics of an $n^+$-p and a $p^+$-n gated diode.

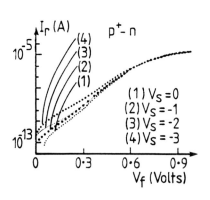

Fig. 3. Forward characteristics of a $p^+$-n gated diode with $V_s$ as a parameter.

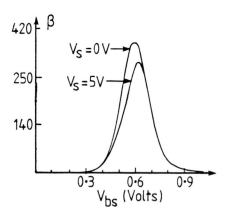

Fig. 4. Common emitter current gain as a function of source/body bias with $V_s$ as a parameter.

Paper presented at ESSDERC 90, Nottingham, September 1990
Session 5C3

# Single-transistor latch induced degradation in thin-film SOI MOSFETs: implications for sub-micron SOI MOSFETs

R. J. T. Bunyan[1], M. J. Uren[1], L. McDaid[2], S. Hall[2], W. Eccleston[2],
N. Thomas[3], J. R. Davis[3]

[1]Royal Signals And Radar Establishment, Great Malvern, Worcs., U.K.

[2]University of Liverpool, Dept. of Electrical Engineering and Electronics, Liverpool, U.K.

[3]British Telecom Research Labs., Martlesham Heath, Ipswich, U.K.

Abstract The presence of a parasitic bipolar between the source and drain in high quality thin-film SOI MOSFETs can result in a single transistor latch condition. In this paper, we will present results that indicate that the latch condition results in hot carrier induced degradation and therefore limits the maximum drain voltage that can be used. A simple model for the length dependence is presented indicating the crucial nature of this problem as lengths are scaled down to the sub-micron regime.

1. Introduction.

One of the outstanding problems observed in thin film SOI MOSFETs is the so called "snap" or "single transistor latch" (Chen *et al* 1988). For large enough drain voltages the subthreshold gate voltage characteristic shows a near infinite subthreshold slope as the gate voltage is increased. As the gate voltage is swept back the characteristic shows some hysteresis before switching off and in the extreme case the device remains on and is said to be in a latched state, see Figure 1. This drain breakdown mechanism results in a reduced voltage for breakdown compared with bulk devices fabricated using a similar process, with the minimum drain breakdown voltage occurring for gate biasses in the subthreshold regime.

In this paper we present results demonstrating that operation of a SOI MOSFET above this "latch" voltage results in degradation of the gate oxide due to the injection of hot holes into the gate oxide adjacent to the drain. A simple model is also presented which predicts the critical nature of this problem as devices are scaled into the sub-micron regime.

2. Device Details.

The devices used in this work were fabricated using standard SOI processes with no low doped drains being employed. The main device parameters are summarised in Table 1. The degradation experiments were carried out on devices made using the FIPOS (Full Isolation by Porous Oxiodation of Silicon) starting wafers. The length dependence data was measured on both FIPOS and SIMOX devices.

© 1990 Crown Copyright

Table 1. Main parameters of devices used in this work.

|       | $t_{gate}$ (nm) | $t_{Si}$ ($\mu$m) | $t_{ins}$ ($\mu$m) |
|-------|-----------------|-------------------|--------------------|
| FIPOS | 26              | 0.1               | 0.7                |
| SIMOX | 19.5            | 0.2               | 0.42               |

## 3. Degradation Of The Gate Oxide During Latch

To assess the lifetime implications of latch we used the charge pumping technique to monitor the change in interface states and fixed positive charge at the gate oxide interface after stressing (Maes et al 1989). A back bias was used to accumulate the back interface to give a lower resistivity substrate contact and to eliminate any contribution to the measured charge pumping current from the back interface.

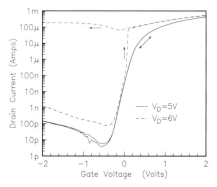

Figure 1. Subthreshold gate voltage characteristic of an $L_{eff}=1.8\mu$m n-channel thin-film FIPOS MOSFET showing normal operation for a drain voltage of 5V and latch for a drain voltage of 6V.

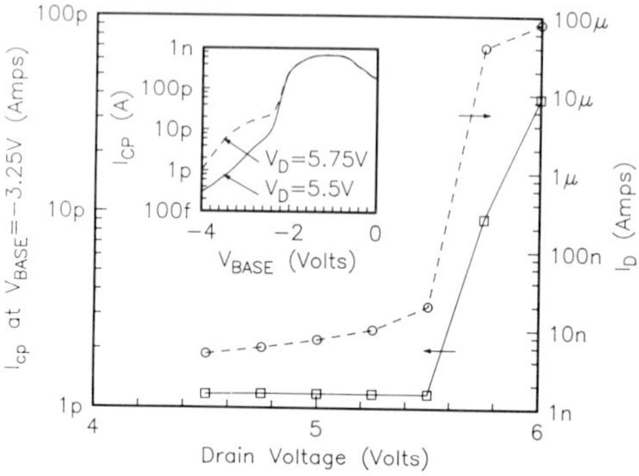

Figure 2. Stress condition $V_G=-0.3$V, $V_{bg}=0$V, time 2000s per point, body left floating. Charge pumping current after stress (□) measured with $\Delta V_G=2.5$V, $V_{base}=-3.25$V, f=400kHz, and drain current during stress (○) are correlated.

Figure 2 shows the effect of sequentially increasing the drain stress voltage on a thin-film FIPOS SOI five terminal MOSFET which is biassed in subthreshold. The electrical dimensions were length 1.8$\mu$m and width 5$\mu$m. We can see that as soon as drain breakdown (latch) occurs the charge pumping current, $I_{cp}$, increases indicating degradation of the gate oxide adjacent to the drain. The inset to Figure 2 shows the charge pumping characteristic before and after latch. We can deduce from the sign of the shift in the rising edge of the characteristic that this degradation results from the trapping of holes in the gate oxide near the drain.

It is clear from these results that the maximum drain voltage that can be used before degradation of the gate oxide occurs is governed by the latch behaviour of the device.

## 4. Length Dependence Model For Latch.

Figure 3 shows the current contributions in the SOI MOSFET when it is biassed such that avalanche occurs at the drain end of the channel. The channel current, $I_{ch}$, flows into the high field region near the drain where avalanche multiplication occurs. The resultant hole current, $I_h$, flows back towards the source where it functions as a base current for the parasitic lateral bipolar inducing an extra electron current, $I_e$, to flow towards the drain. A fraction 1/k of this will pass through the avalanche region along with the channel current so the current that now passes through the high field drain region is $I_{ch}+(1/k)I_e$. Hence we have a positive feed-back process. It can be shown using simple bipolar arguments that avalanche induced breakdown occurs when $\beta(M-1)=k$, (Young and Burns 1988), where $\beta$ is the lateral bipolar gain of the parasitic bipolar transistor and (M-1) is the multiplication factor $I_h/I_{ch}$.

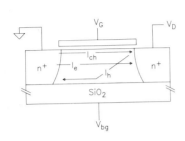
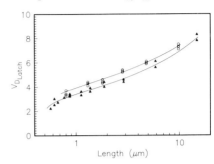

Figure 3. Current components in an SOI MOSFET as avalanche at the drain end of the channel occurs.

Figure 4. Length dependence of drain voltage for latch for SIMOX (▲) and FIPOS (○) devices. Solid lines represent fits using $\beta(M-1)=k$ condition(3).

Given that the multiplication coefficient for electrons is $a\exp(-b/E_{max})$ where $E_{max}$ is the maximum field in the channel, then the multiplication factor at the drain of a MOSFET can be written as (Chan et al 1984),

$$(M-1) = (a/b)(V_D-V_{sat}-V_{sub}).\exp(-b/A(V_D-V_{sat}-V_{sub})) \qquad (1)$$

The substrate voltage term, $V_{sub}$, has been added because the avalanche hole current acts so as to increase the substrate potential, thus forward biasing the source/substrate pn junction and reducing the potential difference between the substrate and drain. The constant $A$ is process dependent and is a function of the oxide and silicon thicknesses present in the device. As latch normally occurs for gate voltages below threshold, there will be no inversion layer beneath the gate to support any voltage drop along the channel, therefore the entire drain voltage is dropped at the drain and the $V_{sat}$ term can be dropped from (1).

If we assume that the bipolar gain, $\beta$, is dominated by the emitter efficiency of the device, i.e. the diffusion length of minority carriers in the base region, $L_B$, is very much greater than 15$\mu$m (the longest device tested) then

$$\beta \approx (n_B D_B L_E/p_E D_E L_B)\coth(L_{base}/L_B)$$
$$\sim \beta_o L_B/L_{base} \qquad (2)$$

where $L_{base}$ is the effective bipolar "width" (Sze 1981).

Combining these two equations gives,

$$L_{as\ drawn} = (a\beta_o L_B/bk).(V_{Dlatch}-V_{sub})\exp(-b/(V_{Dlatch}-V_{sub}))+L \qquad (3)$$

where $V_{Dlatch}$ is the maximum drain voltage before latch occurs, $L$ is the sum of the lateral diffusion under the gate and the widths of the depletion regions at the source and drain. It has been assumed that the drain dependence of this latter quantity is negligible compared to the other terms.

The latch voltage, $V_{Dlatch}$, was measured as a function of gate length for both FIPOS and SIMOX fully depleted devices with the results being presented in Figure 4. This was defined as being the smallest drain voltage for which the subthreshold characteristic showed a near infinite subthreshold slope. A five terminal device was used to measure the substrate potential just before latch and this was found to be typically ~0.4V. The solid lines represent the best fit to the measured data obtained using (3) with the parameters shown in Table 2. The parameters $b$ and $L$ are reasonably close to what we would have expected given the layer thicknesses present in the device.

Table 2. Fitting parameters for length dependence data.

|  | $(a\beta_o L_B/bk)$ | $b$ (V) | $L$ ($\mu$m) |
|---|---|---|---|
| FIPOS | 13.4 | 16 | 0.5 |
| SIMOX | 9.2 | 12 | 0.5 |

We can clearly see that as we enter the sub-micron regime the rapidly increasing bipolar gain has an increasingly important effect on the latch voltage. It is therefore obvious that great care is going to be required in engineering the drain and parasitic bipolar so that this problem is avoided.

5. Acknowledgements.

We would like to express our thanks to J.M. Keen of R.S.R.E., J. Alderman of Plessey (Caswell) and the staff of BT Research Laboratories who processed the devices used in this work.

6. References.

Chan T Y, Ko P K and Hu C 1984 *IEEE Electon Device Letters, EDL-9, No.12,* pp505-507

Chen C-E D etal 1988 *IEEE Electronic Device Letters, EDL-9, No.12, pp636-638*

Maes H E etal 1989 *Applied Surface Science, vol. 39, p523*

Sze S M 1981 *Physics of Semiconductor Devices, 2nd Edition, (New York: Wiley),* pp140-141

Young K K and Burns J A 1988 *IEEE Trans. Electron Devices, Vol. 35, No.4,* pp426-431

Copyright © Controller HMSO, London, 1990.

# Intrinsic gate capacitances of SOI MOSFETs: measurement, modelling, floating substrate effects

D. Flandre [*], F. Van de Wiele [*], P.G.A. Jespers [*] and M. Haond [**].
[*] Laboratoire de Microélectronique, Université Catholique de Louvain, Place du Levant 3, 1348 Louvain-la-Neuve, Belgium.    [**] CNET, BP 98, 38243 Meylan Cedex, France.

Abstract. Measurements of intrinsic gate capacitances of SOI MOSFET's are described and are shown to provide valuable information for characterization purposes as well as to cast some light on dynamic floating substrate effects. An accurate charge-based analytical model valid in linear operation is also presented.

1. Introduction

Until now, little interest has been paid to the dynamic and small-signal characteristics of SOI MOSFET's. However, the characterization and modelling of these features are of great importance in order to understand and simulate properly the transient or high-frequency behaviour of SOI devices. In the present paper, we focus our attention on intrinsic gate capacitances of SOI n-MOSFET's : measurements performed in linear operation are shown to provide valuable information for characterization purposes, a new charge-based model which predicts very accurately the intrinsic capacitance characteristics of SOI MOSFET's operated in linear operation is presented and finally, measurements performed in saturation are described. Differences with conventional bulk characteristics are observed and related to floating substrate effects unique to SOI MOSFET's.

2. Measurements in linear operation

Measurements of front gate-to-source capacitances $Cgs$ in function of front gate potential $Vgs$, performed at low drain-to-source voltage drop $Vds$ for various back contact potentials $Vbs$ (Figure 1), have been described in a previous publication (Flandre et al 1990).

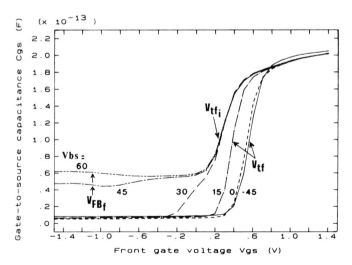

Figure 1 : Measured Cgs capacitance of a 19/18.4 μm n-MOSFET realized in a 0.25 μm-thick SOI film, for various values of Vbs and for Vds equal to 50 mV.

The turn-on $C_{gs}$ capacitance exhibits the classical dependence of SOI MOSFET's front threshold voltage $V_{tf}$ on the back contact potential, while the turn-off $C_{gs}$ capacitance exhibits an astonishing behaviour which may be explained by particular capacitive couplings through the floating body between front/back oxide capacitances, depletion capacitances and source/drain junction capacitances. It has also been shown that the Cgs-Vgs technique may supplement static I-V measurements, since, for example, it allows the direct extraction, on the same 4-terminal device, of the front flatband voltage $V_{FBf}$ and of the front threshold voltage in presence of a back inversion region in the SOI film $V_{tfi}$.

### 3. Charge-based modelling in linear operation

Using a charge-based formulation (Ward 1981), we have extended our physical analytical model valid and accurate for SOI MOSFET's operated in static linear operation (Flandre et al 1989), to quasi-static dynamic regimes. The capacitive behaviour of the floating body has been taken into account as follows, in the case of a P-type body :
- the electron and hole quasi-fermi levels in the SOI film are differentiated, the former being fixed by the source potential $V_s$, the latter by the body potential $V_p$.
- $V_p$ is determined according to the charge conservation within the film, which states that the time-variations of the body hole charge $Q_p$ are balanced by currents through the source and drain junctions :
$$\Delta Q_p / \Delta t + I_{js} + I_{jd} + \Delta Q_{js} / \Delta t + \Delta Q_{jd} / \Delta t = 0.$$

The static, $I_{js}$ and $I_{jd}$, and dynamic, $\Delta Q_{js}/\Delta t$ and $\Delta Q_{jd}/\Delta t$, junction currents are estimated from conventional junction equations, for example,
$$I_{js} = A.J.(e^{\beta(V_p-V_s)/\eta}-1)$$
and $Q_{js} = -A.\sqrt{2.\varepsilon_s.q.N_{af}.(V_{bi}+V_s-V_p)}$,
with $A$ equal to an effective junction area, $J$ to an effective saturation current density, $\eta$ to an effective recombination factor, $V_{bi}$ to the junction built-in potential and $N_{af}$ to the film doping level. The charge conservation equation is solved simultaneously and consistently with the whole system of non-linear equations which describes the SOI MOSFET static behaviour (Flandre et al 1989). In order to compute the value of, for example $C_{gs}$, at a particular point of operation, we apply small increments of source potential and of time, $\Delta V_s$ and $\Delta t$, to the model inputs and analyze the front gate charge variations $\Delta Q_g$,
i.e. $C_{gs} = -(\Delta Q_g / \Delta t) / (\Delta V_s / \Delta t)$.

Using optimized sets of electrical parameters, fairly good agreements have been obtained between experimental and modelling results for thin- (Figures 1 and 2) as well as thick-film (Figure 3.a) SOI MOSFET's. One should note the large value of the subthreshold intrinsic capacitance of the thick-film transistor resulting from the increase of the junction capacitances with the film thickness.

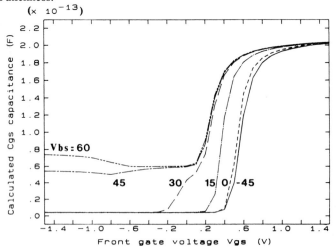

Figure 2 : Calculated Cgs capacitance using our analytical model with the electrical parameters optimized for the device of figure 1.

Figure 3 also shows that false modelling may result from :
- the omission of the junction dynamic currents (Figure 3.b).
- the use of the conventional quasi-static approximation for the computation of the intrinsic capacitances, i.e. Cgs = -$\Delta$Qg / $\Delta$Vs with $\Delta$t $\rightarrow \infty$. This approximation fails in the case of SOI MOSFET's because of the charge conservation equation (Figure 3.c).
- the omission of the drain junction.

These important considerations for the correct simulation of SOI MOS circuits were not established in a previously-published charge-based model (Lim et al 1985).

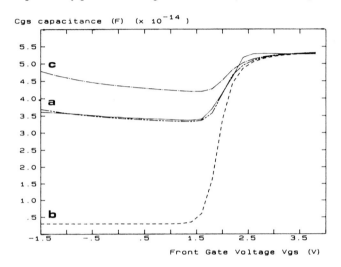

Figure 3 : Cgs capacitance of a 48/6 μm n-MOSFET realized in a 415 nm-thick SOI film with Vbs equal to -50 V and Vds to 50 mV: measured (———), modelled with complete charge equation (— — —), with the omission of the terms $\Delta$Qj/$\Delta$t (- - - - -), with $\Delta$t $\rightarrow \infty$ (— · —).

4. Measurements at high drain-to-source voltage drops

Measurements of front gate-to-source $Cgs$ and -drain $Cgd$ capacitances at very high drain-to-source voltage drops (Figures 4 and 5) also exhibit differences with conventional bulk characteristics :
- the $Cgs$ curve of partially depleted SOI MOSFET's far exceeds, in saturation, the conventional constant value equal to two-thirds of the oxide capacitance $Cox$ (Figure 4, Vbs = -45 V). A first increase (Vds < 3 V) is due to the parallel combination of the conventional intrinsic component associated with the channel inversion charge and of a novel intrinsic component. The latter results from a source-to-gate coupling through the floating body and is approximately equal to the series combination of the source junction capacitance and of the body-to-gate one which is non-zero in saturation. A second increase (Vds > 3 V) can be related to the well-known kink effect, which also causes negative $Cgd$ values at very high drain voltages.
- in subthreshold operation (Figure 5), anomalous characteristics may be induced by impact ionization at the drain, in the same way that abnormally steep subthreshold slopes.
- we have finally demonstrated the reduction of these floating substrate effects when the SOI film is fully depleted (Figure 4, Vbs = 15 V), which yields another advantage of thin-film SOI MOSFET's.

In conclusion, new measurement and modelling results on the small-signal behaviour of SOI MOSFET's have been presented and their importance for characterization and circuit simulation purposes have been discussed.

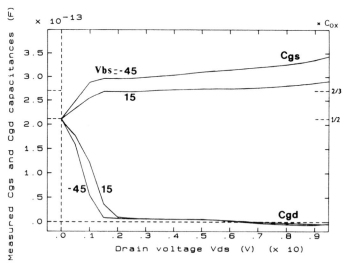

Figure 4 : Measured Cgs and Cgd capacitances of a SOI n-MOSFET similar to the device of figure 1. Vgs is equal to 2 V.

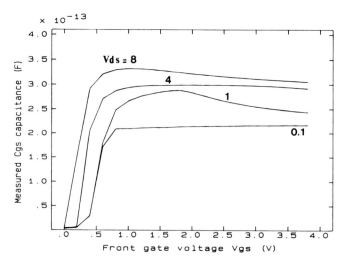

Figure 5 : Measured Cgs capacitance of a SOI n-MOSFET similar to the device of figure 1. Vbs is equal to -45 V.

References :

Flandre D, Van de Wiele F, Jespers P G A and Haond M 1990 "Measurement of intrinsic gate capacitances of SOI MOSFET's" to be published in *IEEE Electron Device Lett.* **9** 7

Flandre D and Van de Wiele F 1989 "A physical model for the characterization of SOI MOSFET's in linear operation" *Proc. 19th ESSDERC* (Berlin: Springer-Verlag) pp 755-8

Lim H K and Fossum J G 1985 "A charge-based large-signal model for the thin-film SOI MOSFET's" *IEEE Trans. Electron Devices* **32** 2 pp 446-57

Ward D E 1981 "Charge-based modeling of capacitance in MOS transistors" *Ph. D. Thesis* (Stanford University: Tech. Rep. G201-11)

Paper presented at ESSDERC 90, Nottingham, September 1990
Session 5C5

# Identification and characterisation of noise sources in SIMOX MOSFETs

T Elewa, B Boukriss, H Haddara*, A Chovet, S Cristoloveanu

Laboratoire de Physique des Composants à Semiconducteurs (UA-CNRS), INPG, ENSERG, BP 257, 38016 Grenoble Cedex, France.
* Department of Electronics, Ain Shams University, Cairo, Egypt.

> **Abstract**. A simple model and an adequate experimental procedure are proposed for the separation of the noise sources related to the front interface, back interface and film volume. This allows the evaluation of the bulk defects and interface traps which are key parameters in the optimization of CMOS-SOI processes.

## 1. Introduction

The advantages of Silicon On Insulator (SOI) materials are now being demonstrated by the fabrication of very fast and complex VLSI circuits. The outstanding features of SOI components are critically related to the crystalline quality of the Si film and its front and back Si-SiO$_2$ interfaces (Elewa et al 1990). The low frequency noise analysis in SOI-MOSFET's stands as an original and valuable method to improve present knowledge on SOI multi-interface materials and devices (Chovet et al 1987). In this paper, we show how to discriminate and identify the various low-frequency noise contributions of bulk and surface sources. *Surface noise* will be discussed in terms of equivalent gate voltage noise $S_{V_g}(f) = S_I/g_m^2$, which is directly related to the densities of interface traps or defects. It is indeed worth noting that the observed noise spectra can strongly depend on technological parameters.

## 2. Model

We consider a depletion-mode MOS transistor fabricated on an SOI film which is partially depleted. This means that even when the two Si-SiO$_2$ interfaces are biased in strong inversion, the corresponding depletion regions are still separated by a neutral region (insert of Fig.1). We can therefore assume a parallel combination of three independent sources of noise: (i) $S_{I_f}$ stands for the noise arising from the carrier interaction with the front interface traps, deep traps in the gate oxide and traps in the top depletion layer, (ii) similarly, $S_{I_b}$ is the noise generated at/near the back interface, and (iii) $S_{I_{vol}}$ is the noise arising from fluctuations in the neutral volume. The total power spectral density $S_I$ of the drain current noise is given by the sum of these three contributions.

Making use of the relation between the power spectral density of drain current noise $S_{I_{f,b}}$ and that of the equivalent gate voltage noise $S_{V_{gf,b}} = S_{I_{f,b}}/g_{mf,b}^2$, we have:

$$S_I = g_{mf}^2 S_{V_{gf}} + g_{mb}^2 S_{V_{gb}} + S_{I_{vol}} \tag{1}$$

where $g_{mf,b}$ are the transconductances associated with the front/back current variations induced by the gate voltages $V_{gf,b}$. In a depletion-mode transistor biased in strong

© 1990 IOP Publishing Ltd

inversion regime, the interface traps are shielded by the inversion layer. They cannot interact with the free carriers in the channel and do not contribute as a source of noise. Therefore, when both interfaces are biased in strong inversion, Eq.(1) reduces to a *minimum* noise level $S_I = S_{I_{volmin}}$, because (i) the carrier fluctuations in the volume represent the sole source of noise and (ii) the active volume thickness is minimized.

Remark that the measurement and calculation of $S_I$ are necessary to separate the different sources and estimate the importance of volume noise. The equivalent gate noise $S_{V_g}$ is also very useful because it is more directly related to the densities of traps. It should be underlined that this model of noise separation is quite general, regardless of the frequency dependence of the noise.

## 3. Thick Film Transistors

SIMOX material was synthesized by molecular oxygen implantation (energy 400 $keV$, dose $1.8 \times 10^{18}$ $cm^{-2}$, current 100 $\mu A$) in n-type Si, followed by annealing at $1405°C$ for 30 minutes. After the processing of depletion-mode edgeless MOS transistors, the film doping was $N_D \simeq 3.5 \times 10^{17}$ $cm^{-3}$ and the thicknesses of the buried oxide, gate oxide and Si film were, respectively, 400 nm, 27 nm and 300 nm. Noise measurements were performed using a low noise amplifier PAR–113 with a high input impedance (100$M\Omega$), a digital spectrum analyzer Takeda and a drain load resistance $R_D$ in the conventional biasing circuit.

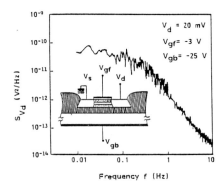

**Fig.1** Drain voltage fluctuations showing a typical Lorentzian noise spectrum.

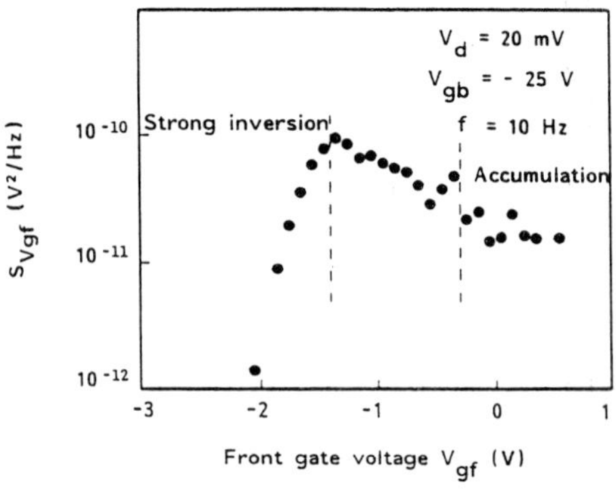

**Fig.2** Spectral density of input equivalent gate voltage fluctuations in the whole bias range (from accumulation to inversion).

A plateau is observed on $S_I(V_{gf,b})$ curves when the volume contribution is dominant, i.e. when the two interfaces are in strong inversion. We deduce that the minimum value is $S_{I_{volmin}} \simeq 10^{-22} A^2/Hz$. As the noise presents a typical generation-recombination spectrum (Fig.1), the density of traps $N_{T_v}$ in the volume can be calculated by adapting the bulk relation to the neutral part of the film:

$$\frac{S_{I_{volmin}}}{I_{volmin}^2} = \frac{N_{T_v}}{N_D^2} \times \frac{f_t(1-f_t)}{ZL(T_f - 2W_{max})} \times \frac{4\tau}{1+\omega^2\tau^2} \qquad (2)$$

where $\omega = 2\pi f$ and $\tau = 1/2\pi f_c$, with $f_c$ the corner frequency of the Lorentzian spectrum, $f_t \simeq 0.5$ is the fractional occupancy of traps around the Fermi level, $W_{max}$ is the maximum depletion depth at both interfaces, $T_f$ is the film thickness and $Z \times L$ is the gate area. A large density $N_{T_v} \simeq 10^{18}$ $cm^{-3}$ is obtained which might be connected with dislocations and/or contamination in the film.

The value of $S_{I_{vol}}$ at any gate voltages can be calculated by taking into account the volume thickness modulation due to the spreading of the two depletion layers. This modulation affects the drain current as well, so that for inversion or depletion regimes, we have: $S_{I_{vol}} = (I/I_{volmin})S_{I_{volmin}}$. It follows that each interface can be characterized independently by measuring $S_I$ as a function of the corresponding gate voltage, while keeping the opposite interface in strong inversion. After subtracting $S_{I_{vol}}$ from $S_I$ we obtain $S_{I_{f,b}}$ and then calculate $S_{V_{gf,b}}$. Figure 2 shows that $S_{V_{gf}}$ increases as $V_{gf}$ is scanned from accumulation to weak inversion. This indicates that the effective density of slow traps increases slightly towards the valence band. In strong inversion, the contribution of the interface noise decreases as a result of the screening effect of minority carriers.

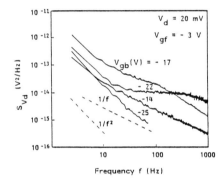

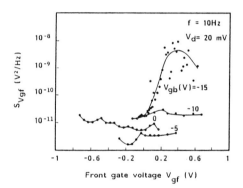

**Fig.3** *Spectral density of drain voltage fluctuations with the back gate bias as a parameter.*

**Fig.4** *Variations of $S_{V_{gf}}(V_{gf})$ with the back gate bias as a parameter.*

Similar conclusions hold also for the back interface, but only in the low-frequency range, because above 10-100 Hz, the noise spectrum is qualitatively changed. The emerging noise is presumably governed by another type of localized traps. A very interesting result is indeed obtained by plotting $S_{V_d}(f)$ with $V_{gb}$ as a parameter and the front interface screened (i.e. biased in strong inversion) (Fig.3). For values of $V_{gb} \simeq -25V$, corresponding to strong inversion, a $1/f^2$ spectrum is recorded. However, as $V_{gb}$ is reduced to $-22V$, a region of constant noise shows up and reveals the presence of generation-recombination noise due to a localized trap. The plateau level as well as the corner frequency decrease with $V_{gb}$ until it disappears totally for $V_{gb} = -15V$. An advantage of this noise analysis is that the measured time constants (1–2 ms) correspond directly to the trap kinetics. The density of defects localized at energy $E_T$ is given by (Gentil et Chaussé 1977):

$$N_{T_s}(E_T) = S_{I_d} \times \frac{C_{ox}^2 ZL}{4q^2 g_{mb}^2 f_t(1-f_t)} \times \frac{1+\omega^2\tau^2}{\tau} \qquad (3)$$

A density of $N_{T_s} \simeq 10^{12} cm^{-2}$ has been calculated, which confirms previous charge pumping and dynamic transconductance data (Elewa et al 1988). Measurements on enhancement-mode transistors have resulted in similar behaviour and data.

## 4. Thin Film Transistor

Fully-depleted transistors, with LOCOS edges, have been fabricated on SIMOX wafers ($1.7 \times 10^{18} cm^{-2}$, $150 keV$, $530°C$), annealed at low temperature ($1250°C$). The Si film was 100 nm thick with a doping of $N_D \simeq 10^{17}$ $cm^{-3}$.

In totally-depleted transistors it is not possible to extract the volume noise by the method described above. Nevertheless, $S_{I_{vol}}$ is directly proportional with the volume current $I_{vol}$, so that for ultra-thin films $S_{I_{vol}}$ is in general negligible with respect to $S_I$. Shown in Fig.4 is the equivalent gate noise $S_{V_{gf}}$ as a function of $V_{gf}$. Since the noise spectrum is $1/f$, we can calculate the effective density of traps using (Stegherr 1984):

$$N_{it} = S_{V_g} \frac{C_{ox}^2 Z L \log(\tau_2/\tau_1)}{q^2 kT} \qquad (4)$$

where $\tau_{1,2}$ are the minimum and maximum response times of traps and other notations are conventional. We found a low density of traps ($N_{it} \simeq 2 \times 10^{10} cm^{-2} eV^{-1}$) when the influence of the back interface and edges was minimized (for $V_{gb} \geq -5V$).

## 5. Conclusions

By adjusting the front and back gate voltages it is easy to isolate bulk and surface sources of noise and characterize the defects responsible for them. The present results confirm the localization of additional single-level traps at the buried interface and suggest a possible influence of dislocations.

**Acknowledgements.** This work is part of the EEC Esprit Basic Research Action #3017. Drs. P. Hemment, J-P. Colinge, J. Davis and G. Celler are thanked for providing the SIMOX devices and Miss M. Gri for technical help.

## References

Chovet A, Boukriss B, Elewa T and Cristoloveanu S 1987 *Noise in Physical Systems (Traneck: World Scientific)* pp *457–60*

Elewa T, Haddara H, Cristoloveanu S and Bruel M 1988 *ESSDERC'88, Proc. in Journal de Physique* **49** pp *137–40*

Elewa T, Balestra F, Cristoloveanu S, Hafez I, Colinge J-P, Auberton-Hervé A-J and Davis J 1990 *IEEE Trans. on Elect. Dev.* **ED-37** pp *1007–19*

Gentil P and Chaussé S 1977 *Solid-St. Electron.* **20** pp *935–40*

Stegherr M 1984 *Solid-St. Electron.* **27** pp *1055–6*

*Paper presented at ESSDERC 90, Nottingham, September 1990*
*Session 5C6*

# Improvement of output impedance in SOI MOSFETs

M -H Gao, J -P Colinge, S -H Wu and C Claeys

IMEC, Kapeldreef 75, B-3030 Leuven, Belgium

Abstract. A common gate dual MOSFET structure is proposed to improve the output impedance of SOI MOS transistors. The device consists of two transistors in series with a common gate but operates as a single device. Kink effect can be confined to the upper transistor while the overall output characteristics of the dual device are dominated by the lower transistor. As a consequence, the kink effect can be effectively suppressed in the overall output characteristics. Furthermore, the output breakdown characteristics and the saturation drain output impedance are also much improved.

## 1. Introduction

The well-known kink effect in SOI MOS transistors (Tihanyi 1975), especially in nMOS transistors, is restricting their applications in analog circuits. In addition, when the gate length gets shorter, the parasitic bipolar effect in SOI nMOSFET's significantly lowers both the output breakdown characteristics and the saturation drain output impedance (Haond 1989). Kink-free operation of SOI MOSFET's can be achieved in devices made in fully depleted "thin" silicon films (Colinge 1988), but such fully depleted "thin-film" SOI MOSFET's do not meet the radiation tolerance requirements imposed to systems such as spacecraft electronics (Mayer 1989).

In this paper, a common gate, Dual-MOSFET structure is proposed to suppress the kink effect and to improve the output characteristics of the SOI MOSFET's. The structure applies to both p- and nMOSFET's. However, only nMOSFET's are dealt with here as the kink effect is usually only significant in the latter. Apparently, the proposed structure looks similar to a cascode MOS configuration, however, the physics behind the device operation is different as will be explained below. Cascode configurations have been reported to achieve high voltage performance in SOS/MOS devices (Ronen 1976) and to improve the output characteristics of Thin Film MOSFET's in analog circuit applications ( Lewis 1988).

The Dual-MOSFET structure consists of two SOI nMOSFET's $T_1$ and $T_2$ in series with a common gate, but operates as a single nMOSFET, as shown in Figure 1-a. The source of the device is the source of $T_1$ while the drain of the device is the drain of $T_2$. The $N^+$ region in between $T_1$ and $T_2$ is kept floating. This structure can confine the kink effect only to the upper (or "slave") transistor $T_2$ and thus can effectively keep the lower (or "master") transistor $T_1$ from pinch-off, impact ionization and the kink effect. If the channel length of $T_1$ is made longer than that of $T_2$, $T_1$ will dominate the overall output characteristics of the device. Therefore the kink effect will be effectively eli-minated from the overall output characteristics. And, due to the $N^+$ region in between the twin transistors, the output breakdown characteristics will be much improved.

Furthermore, this structure can also confine the bipolar effect only to occur in the upper $T_2$. Since the holes in the base region (body) of the $T_2$ will recombine in the middle $N^+$ region, they can not reach the base region of $T_1$. That means the bipolar current of the $T_2$ becomes a component of the total channel current which is dominated by the lower transistor $T_1$. As long as $T_1$ predominates in the overall output characteristics of the device (e. g., when $L_1 > L_2$ and $V_{DS} < 6V$), the bipolar current can be effectively suppressed and the saturation drain output impedance of the SOI nMOS transistors can be significantly increased.

© 1990 IOP Publishing Ltd

The proposed common gate Dual-MOSFET structure can easily be designed as a compact single SOI device with reach-through junctions as shown schematically in Fig.1-b.

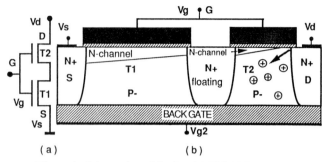

## 2. Experimental

The measurements were made on non-fully depleted nMOSFET's fabricated on SIMOX wafers. The measured transistors in the L-array have constant gate width of 50μm, having a common gate and a common source electrode. The gate lengths range from 2 to 0.4 μm. It is convenient to use the L-array either to measure the single transistor to observe the kink effect, or to measure the combined dual structure to assess the kink suppression effect with different configurations of $T_1$ and $T_2$. All the measurements were made at room temperature.

Figure 1: Schematics of the Dual-MOSFET structure.

## 3. Results and Discussions

The output characteristics of the single SOI nMOS transistors show pronounced kinks. Typical output curves of both single transistors of $T_1$ and $T_2$ are shown in Figure 2 with dashed lines (W/L=50μm/1μm for $T_1$ and W/L=50μm/0.6μm for $T_2$). Also in the same Figure 2 are shown in solid lines the output curves of their combined dual configuration of W/L=50μm/(1+0.6) μm. It is observed that when $T_2$ is added into the drain end of $T_1$ in series with it, the kinks disappear from the overall output characteristics. Different configurations show similar results when $L_1 > L_2$. Generally speaking, the larger the ratio of $L_1 / L_2$ is, the more reduced the kink effect will be.

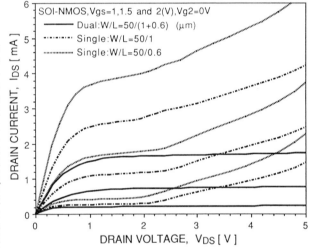

Figure 2: Measured output characteristics for $V_{GS}$=1, 1.5 and 2V of the single n-channel devices (L=1 and 0.6μm) and of their combined Dual-MOSFET struc-ture with L = $L_1 + L_2$ = 1μm + 0.6μm configuration.

The drain current of an nMOS transistor operating in saturation is, in first approximation, nearly constant: $I_{Dsat}=V_{Dsat}/R_L$, because the effective channel resistance $R_L$ can be regarded as a constant once saturation is reached. Kink effect in SOI nMOS transistors is caused by the accumulated holes in the body which are generated through impact ionization from the pinch-off region near the drain. They increase the body potential and decrease the threshold voltage (Kato 1985), thus reduce the $R_L$, and the kink effect is observed in the output characteristics. In the case of the Dual-MOSFET structure, as seen from in Figure 1, the chan-nel is separated into two parts by the floating $N^+$ region, and $R_L$ is thus separated into two parts: $R_L=R_{L1}+R_{L2}$ corresponding to $T_1$ and $T_2$, respectively. As pinch-off only occurs in the upper transistor $T_2$, therefore the $V_T$ of $T_1$ will not change, so that $R_{L1}$ will not change either.

Only $R_{L2}$ of $T_2$ will be reduced due to the kink effect in $T_2$. If $R_{L1} > R_{L2}$, an approximate relationship can be derived between $\Delta I_{kink(dual)}$ and $\Delta I_{kink(single)}$ for transistors having the same total gate length (for example: a single transistor with L=3μm and a dual-transistor with L = $L_1 + L_2$ = 2μm +1μm):

$$\frac{\Delta I_{kink(dual)}}{\Delta I_{kink(single)}} = \frac{R_{L2}}{R_{L1}+R_{L2}} \quad (1)$$

In the case of the dual-transistor configuration, $R_{L2}$ must be smaller than $R_{L1}$, not only because $L_1 > L_2$, but also because $V_{T2} < V_{T1}$ due to the kink effect occurring in $T_2$. Therefore the kink effect in the overall output characteristics will be reduced according this formula. Considering the suppression of the parasitic bipolar effect, the output curves will be even much flatter than that predicted by this formula as will be dealt with below.

Figure 3 shows the breakdown characteristics of both single and dual devices in a larger scale of $V_{DS}$ close to the breakdown value. For a better comparison, another neighbouring single transistor with L=2μm is also measured. The results show that the breakdown voltage of the dual device of L = $L_1 + L_2$ = (1 + 0.6)μm is much higher than that of the corresponding single ones, even much higher than the single one with L = 2μm.

Figure 4 shows the characteristics of the saturation drain output impedance of the SOI nMOS dual-device compared to the single SOI nMOS transistors (one on the same structure and another fully depleted transistor

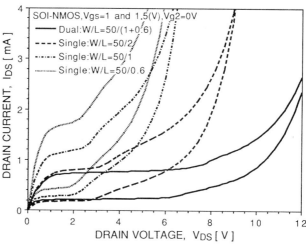

Figure 3: Measured output characteristics for $V_{GS}$=1 and 1.5V of three single devices (L=0.6,1.0 and 2μm) compared to the dual structure of L=$L_1+L_2$ = 1μm + 0.6μm.

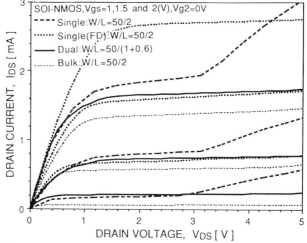

Figure 4: Measured output characteristics showing the saturation drain output impedance of the dual structure compared to the single SOI transistors (thick film as well as fully depleted thin film) and a bulk nMOS transistor from the same run.

from another wafer of the same run) and a bulk nMOS transistor on the test wafer of the same run. It is obvious that the slopes of the Dual-Device in the saturation region are the smallest. This is not only because the Dual-MOSFET structure effectively suppresses the parasitic bipolar behaviour in the lower master transistor $T_1$, but also because the bipolar current appearing in the upper transistor $T_2$ becomes beneficial for improving the output impedance.

In the case of the Dual-MOSFET structure, not only the parasitic bipolar behaviour in the lower transistor $T_1$ can be effectively eliminated, but also can the multiplication in the upper $T_2$ be much reduced. As the bipolar current $I_{C2}$ occurs in the upper $T_2$, the total $I_{Dsat}$ will be the sum of $I_{C2}$ and the channel current of $T_2$ ($I_{D2}$). As long as $I_{Dsat}$ is dominated by $T_1$, we have:

$$I_{Dsat} = I_{D1} = I_{D2} + I_{C2} \quad (2)$$

The occurrence of $I_{C2}$ will surely reduce the channel current of $T_2$ ($I_{D2}$) in a negative feedback since $I_{D1}$ is a constant. As a result the multiplication in $T_2$ will be reduced to some extent. This leads to very flat saturation output characteristics, even flatter than that of the bulk counterpart in the same process.

Also due to the suppression of the bipolar effect, the dual structure can successfully suppress the bipolar "snap-back" (Matsunaga 1979, Armstrong 1989) or the Latch (Chen 1988) in SOI nMOS-FET's which usually happens when the gate is short. Figure 5 shows that the "snapback" and latch occur in a single nMOSFET (L=2μm). The Dual-Device (L= 1+0.6μm), on the other hand, does not exhibit any snap-back or latch behaviour.

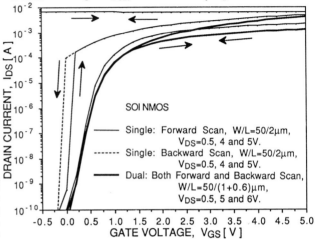

Figure 5: Measured $I_{DS}$ versus $V_{GS}$ curves (with both forward and backward scan of $V_{GS}$) of a single SOI nMOS transistor with W/L=50/2μm compared to a dual device with W/L=50/(1+0.6)μm. Three $V_{DS}$ values are chosen. For the single case, when $V_{DS}$= 4V, snap-back and hysteresis occur, when $V_{DS}$=5V, latch-up occurs. For the dual case, for $V_{DS}$=0.5, 5 and 6V, no "snap-back", hysteresis and latch-up occur. Note that the total gate length of the dual device is even shorter (1.6μm) than that of the single one (2μm).

## 4. Conclusions

The proposed SOI common gate Dual-MOSFET structure can effectively eliminate the kink effect in the output characteristics when the channel length of the lower transistor is longer than that of the upper one ($L_1 > L_2$). The Larger the ratio of $L_1/L_2$, the more reduced the kink effect will be. The overall output breakdown characteristics are much increased. The structure can also effectively suppress the parasitic bipolar effect in SOI MOSFET, therefore the saturation drain output impedance is significantly improved and the bipolar induced "snap-back" or latch-up is also eliminated (e.g., for L=1+0.6μm, a $V_{DS}$ up to 6 volts can be used). The Dual-MOSFET structure can easily be designed as a compact single SOI device with reach-through junctions.

## References

Armstrong G A, Thomas N J and Davis R J 1989 Proc. IEEE SOS/SOI Technol. Conf. p 44
Chen Daniel C -E et. al. 1988 IEEE Electron Device Letters **9** 12 p 636
Colinge J P 1988 IEEE Electron Device Letters EDL-**9** p 97
Haond M and Colinge J P 1989 Electron. Lett. **25** 24 p 1640
Kato K, Wada T, and Taniguchi K 1985 IEEE J. of Solid-Sate Circuits SC-**20** 1 Feb. p 378
Lewis A G et. al. 1988 IEDM Technical Digest p 264
Matsunaga J et. al. 1979 11th Conf. on Solid-State Device Tokyo Technol. Digest p 45
Mayer D C 1989 Proc. IEEE SOS/SOI Technology Conf. p 52
Ronen R S et. al. 1976 IEEE J. of Solid-State Circuits Vol. SC-**11** No. 4 p 431
Tihanyi J and Schloetterer H 1975 IEEE Trans. Electron Dev. ED-**22** p 1017

# Temperature behaviour of CMOS devices built on SIMOX substrates

J Belz, G Burbach, H Vogt, W Zimmermann

Fraunhofer Institute of Microelectronic Circuits and Systems, Finkenstr. 61, D-4100 Duisburg, FRG

Abstract. CMOS devices have been built on SIMOX substrates. The device characteristics are examined in the temperature range from room temperature up to $300°C$. The degradation of the threshold voltage and the carrier mobility are comparable for bulk- and SOI-devices, but at $300°C$ the off-state currents of the SOI-devices are three orders of magnitude less than those of the bulk counterparts. The capability of SOI-devices for high temperature applications will be shown.

1. Introduction

SIMOX-substrates ( S eparation by IM planted OX ygen) /Izumi78/ have become the most promising SOI-substrates suitable for future VLSI. SIMOX enables the realization of homogeneous, very thin silicon-on-insulator (SOI)-layers ($0.05 - 0.25$ μm) which are single crystalline and of low defect density ($< 10^4$ $cm^{-2}$) /Fechner89/. Using these SOI-substrates latch up in CMOS circuits can totally be prevented, the dynamic power consumption can be reduced as well as the sensitivity to ionizing radiation. Speed enhancement and higher packing densities are additional advantages /Belz90,Celler89,Omura90/. This paper deals with the advantages of SIMOX-devices in the field of high temperature applications. In SIMOX devices the pn-junction area is reduced compared to bulk devices so that a reduction of the leakage currents is aspected. But as it will be shown in this paper the scaling of the temperature dependance is different for SIMOX and bulk devices.

2. Experimental

SIMOX substrates have been prepared using an EATON NV 200 high current oxygen implanter located at the Fraunhofer-Institute IMS-Duisburg. The implantations were carried out at $620°C$, running the machine at $200$ $kV$ and $60$ $mA$ accelarated $O^+$-beam. The typical implanted oxygen dose was $1.8 \times 10^{18}$ $cm^{-2}$. The implanted (100)-oriented, p-type substrates have been annealed at $1250°C$ for 8 hrs in $Ar/O_2$ ambient. This kind of sample preparation yields a SOI-structure with sharp silicon to insulator interfaces and a dislocation density in the silicon surface layer which is less than $10^6$ $cm^{-2}$. The layer thicknesses are about 250 $nm$ silicon on top of about 350 $nm$ burried insulator. Oxidation and etchback is used to reduce the silicon film thickness down to 100 $nm$ in case of fully depleted devices. A LOCOS technique used for lateral insulation yields silicon islands suitable for integrated silicon-gate-CMOS-transistors with channel length down to 0.6 μm. Because of the high quality of the silicon film there is no need for an additional epitaxial layer on top of the thin film. The gate oxide thickness is 25 $nm$. The doping level of the channel region of the devices is about $1 - 2 \times 10^{17}$ $cm^{-3}$. The device characteristics were examined in the temperature range from room temperature up to $300°C$ on a hot chuck.

© 1990 IOP Publishing Ltd

## 3. Results

In the model of Fossum et al. /Fossum84/ the front channel threshold voltage is described by

$$U_T^f = U_{FB}^f + (1 + \frac{C_{Si}}{C_{Ox}^f}) \times 2\Phi_F + q\frac{N_{Si}t_{Si}}{2C_{Ox}^f}$$

| | | | |
|---|---|---|---|
| $U_T^f$ | front channel threshold voltage | $U_{FB}^f$ | flatband voltage |
| $C_{Si}$ | capacitance of the depl. film | $\Phi_F$ | inversion potential |
| $N_{Si}$ | carrier density in the film | $t_{Si}$ | silicon film thickness |
| $C_{Ox}^f$ | gate oxide capacitance | | |

To the first order the inversion potential shows the dominant temperature dependance. It can be described as

$$\Phi_F = \frac{kT}{q} \ln(\frac{N_A}{n_i})$$

| | | | |
|---|---|---|---|
| $k$ | Boltzmann constant | $T$ | Temperature in Kelvin |
| $q$ | elementary charge | $N_A$ | doping concentration |
| $n_i$ | intrinsic charge density | | |

and $n_i = n_{i_0} kT^{\frac{3}{2}} e^{-\frac{E_G}{2kT}}$  $E_G$ band gap energy

In figure 1 the prediction of this crude modell is compared to the measurements for a n-channel device. The temperature dependance is well described.

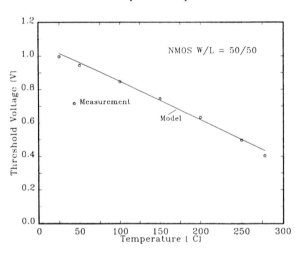

Fig.1: Temperature Dependance of the Threshold Voltage

Figure 2 represents the temperature dependance of the low field mobility of n-channel devices. This behaviour is comparable to bulk devices. The mobility follows the formula:

$$\mu = \mu_0 (\frac{T}{T_0})^{-1.70}$$

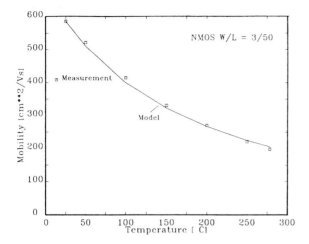

Fig. 2: Temperature Depandance of the low field mobility

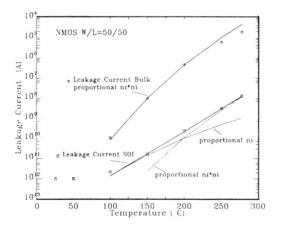

Fig.3: Comparison of the leakage currents of bulk- and of SOI-devices
Solid lines are fits to the data

Figure 3 shows the comparison of the off-state-current (leakage current) of bulk-devices and of SOI-devices. The measurements are compared to a calculation which take regard to the two main sources of pn-junction leakage. The first source is caused by a diffusion component in the neutral region and the second is the generation component in the depletion region. The current density of the diffusion component can be described by

$$J_{RD} = q \left(\frac{D}{\tau_p}\right)^{\frac{1}{2}} \frac{1}{N_A} \times n_i^2$$

$J_{RD}$     diffusion current density     $D$     diffusion constant
$\tau_p$     lifetime

and the generation current density by

$$J_{RG} = qn_i\frac{W}{\tau_e}$$

$J_{RG}$ generation current density     $W$ depletion width
$\tau_e$ generation lifetime

The dominant temperature dependant term is the intrinsic carrier density. If we ignore the temperature dependance of all the other quantities we get the following conclusions. In the case of bulk devices the data are well reproduced by a term proportional to $n_i^2$. But in the case of SOI devices up to $150°C$ the data show a $n_i$-scaling and above $200°C$ the term proportional to $n_i^2$ becomes dominant. This different behaviour results in a leakage current which is 3 decades less for SOI-devices than for the bulk-counterparts.

## 4. Conclusions

We have demonstrated that CMOS devices built on SIMOX substrates have a better high temperature behaviour than the bulk counterparts. At $300°C$ the leackage current of the devices is three decades less in SOI than in bulk technology. So it is possible to operate integrated circuits in the range between $200°C$ and $300°C$ on SOI. Amoung the well known other advantages this high temperature applications can be a strong driving force to introduce SIMOX into the commercial IC-fabrication.

## 5. Acknowledgement

This work has been supported by the Federal Ministry of Research and Technology (BMFT) of the Federal Republic of Germany. The authors alone are reponsible for the content.

## References

Belz J. et al , 1990 , "Characterization of CMOS Devices and Circuits build on SIMOX-Substrates", Proc. 4th Int. Symp.on Silicon-on-Insulator Technology and Devices, Montreal, May 6-11, 1990,p 518

Celler G. K. et al, 1989 , "A 6.2 GHz Digital CMOS Circuit in Thin SIMOX Films", Proc. IEEE SOS/SOI Techn. Conf., Nevada, Oct. 3-5, 1989, p 139

Fechner P. S. et al, 1989 , "Physical Characterization of Low defect SIMOX material",Proc. IEEE SOS/SOI Techn. Conf., Nevada, Oct. 3-5, 1989, p 70

Fossum J. G. and Lim H. K., 1984 , "Current-Voltage Characteristics of Thin Film SOI MOSFETs in Strong Inversion", Trans. Electron Devices, Vol ED 31, pp 401-408, April 1984

Izumi K. et al , 1978 , "MOS devices fabricated on burried $SiO_2$ layers formed by oxygen implantation into silicon", Electron. Lett., Vol 14, p 593,1978

Omura Y. et al., 1990, "A 21.5 ps/stage Ultrathin Film CMOS/SIMOX at 2.5 V using a 0.25 μm Polysilicon Gate", Proc. 4th Int. Symp.on Silicon-on-Insulator Technology and Devices, Montreal, May 6-11, 1990,p 509

# Thin SIMOX SOI material for half-micron CMOS

H. Lifka, P.H. Woerlee

Philips Research Laboratories, 5600 JA Eindhoven, The Netherlands.

**Abstract.** The properties of half-micron CMOS devices fabricated on thin film SIMOX SOI with different material quality will be presented. The gate oxide quality, diode leakage current and breakdown voltage of transistors will be shown. The influence of LDD dope and $TiSi_2$ salicide on the parasitic bipolar transistor breakdown is presented. Temperature measurements on SOI and bulk transistors are presented which show an increased heating effect for thin film SOI transistors.

## 1 Introduction

Thin film SIMOX silicon on insulator (SOI) has interesting features for half-micron CMOS. For instance, an improved device and circuit performance has been reported by Colinge (1987) and Vasudev (1988). Of special importance for thin film SOI is the material quality, which affects the circuit yield and device properties. The material quality has improved enormously during the last few years. The crystal defect levels on SIMOX material decreased from $1 \times 10^9$ to below $1 \times 10^5$ defects/cm$^2$ and contaminations levels decreased below detection limits of SIMS.

In this study, the gate oxide quality, diode leakage and breakdown voltage of devices fabricated on SOI material with different contamination levels will be compared to devices on bulk material. Furthermore the influence of LDD drain structure and $TiSi_2$ salicidation on the breakdown voltage will be presented.

A negative differential output conductance effect can be observed for thin film SIMOX transistors at high gate bias in NMOS device charteristics and to a lesser degree in the device charateristics of the PMOS devices. Mc Daid et al (1989) proposed a self heating effect for SOI devices to explain this effect. A temperature measurement on bulk and SOI NMOS transistors with liquid crystal will be presented.

## 2 Experimental

Bulk floating zone material and two different SIMOX lots were compared. The SIMOX lots had crystal defects below $10^5$ defects/cm$^2$. SIMS measurements showed Fe concentration below $2 \times 10^{16}/cm^3$ and Cu below $2 \times 10^{17}/cm^3$. The first lot had a Carbon contamination of $10^{19}/cm^3$ at a depth of 100 nm which increased to $10^{20}/cm^3$ at the buried Si-$SiO_2$ interface. The second lot had a Carbon contamination below $4 \times 10^{18}/cm^3$. (SIMS detection limit) The top silicon thicknesses of the lots were 0.16 and 0.20 $\mu$m respectively.

The gate oxide test structures were processed in a 0.5 $\mu$m NMOS SOI process. The other test structures were processed in a full 0.5 $\mu$m CMOS SOI process. The 0.5 $\mu$m CMOS SOI process has been described by Woerlee et al (1989).

The self heating effect of NMOS transistors (W/L 1.25 $\mu$m/8.0 $\mu$m) on bulk and 0.16 $\mu$m thick SOI material was measured. Liquid crystal ZLI2900 was spun on a wafer with NMOS transistors. The temperature was controlled with a thermochuck and the turn over point of the crystaline phase to the amorphous phase was determined with a microscope with polarisor. The temperature at the turn over point for $V_g$=4.0 Volt with $V_{sd}$=4.0 Volt was measured.

## 3 Results

### 3.1 Influence of Carbon contamination on device properties

In figure 1 the breakdown field of gated diodes with 12.5 nm thick gate oxide is plotted for different SOI lots and bulk material. The criterion for breakdown is the electric field which

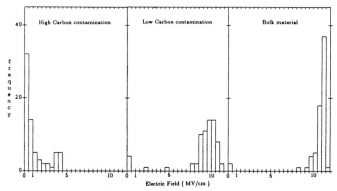

Figure 1: *Distribution of gate oxide breakdown field of SIMOX with high and low Carbon contamination and bulk material.*

is needed to force a current of 100 mA/cm$^2$ through the gate oxide. The gate oxide area was $4.8 \times 10^{-3} cm^2$. As can be seen the material with the high Carbon contamination has a very low breakdown field. The low Carbon contaminated material approaches the bulk values but it is still worse. Jastrzebski et al (1988) found similar correlation of gate oxide quality with Carbon contamination results for thicker (35 nm) gate oxides.

In figure 2 the subthreshold leakage current of transistors fabricated in the high Carbon contaminated material and the low contaminated material are shown. This leakage current showed up in about 30 % of the devices of the contaminated material and not in the improved material.

The diode leakage current was measured on lateral N$^+$-P-P$^+$ diodes with a 1 cm long N$^+$-P perimeter in 0.20 $\mu$m thick SOI. The diode leakage for this structure was below 5 fA/$\mu$m at 5 volt reverse bias (see fig 3 ), which is comparable to edge diode leakage values on bulk wafers.

### 3.2 Breakdown voltage of transistors

A disadvantage of high quality material is the reduced parasitic bipolar breakdown of NMOS devices due to large carrier lifetimes. This breakdown decreases as the SIMOX material approaches bulk material quality as has been observed by woerlee et al (1989). This breakdown voltage is below the maximum operation voltage for 0.5 $\mu$m NMOST. To increase the breakdown voltage LDD NMOS transistors were fabricated with N$^-$ implant doses between $1 \times 10^{12}$ to $4 \times 10^{13}$ Phosphorus at 30 KeV. The drain structures with these N$^-$ implants did not lead to a significant increase of the breakdown voltage. However salicidation of the source-drain junction increased the breakdown voltage by 0.5 Volt. (See fig 4 )

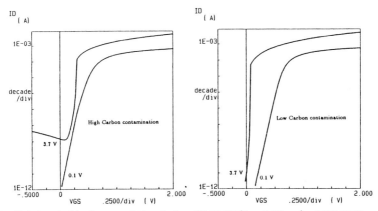

Figure 2: Subthreshold characteristic of ring NMOST($L_g=1.25\mu m$) of SIMOX material with high resp. low Carbon contamination with $V_{DS}$ of 0.1 V and 3.7 V

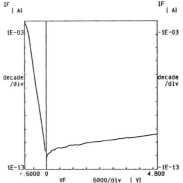

Figure 3: Diode leakage current of $N^+$-$P^-$-$P^+$ diode with 1 cm long $N^+$-P perimeter in low Carbon contaminated material.

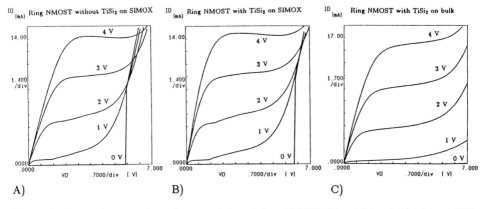

Figure 4: Saturation transistor characteristics of ring NMOST ($L_g=1.25\mu m$) A) without $TiSi_2$, B) with $TiSi_2$ on SIMOX and C) bulk ring NMOST with $TiSi_2$. The gate voltage is indicated in the figures.

## 3.3 Self-heating

The temperature of the SOI transistors could not be measured directly due to the large temperature difference. The temperature was calculated by measuring at several input powers the turn over temperature of the liquid crystal and extrapolating to the input power of the bulk transistor at $V_g, V_{sd}=4.0$ Volt. The results of the temperature measurements are presented in table 1. For the SOI transistor a temperature increase of 59°C was found while for the bulk

| Material | W/L $\mu$m | Input Power mWatt | $\Delta$T °C |
|---|---|---|---|
| Bulk | 8/1.25 | 6.5 | 4 |
| SOI | 8/1.25 | 6.5 | 59 |

Table 1: *Temperature difference of bulk and SOI transistors*

transistor only 4°C was observed.

# 4 Discussion

As can be seen, the improved material with low Carbon contamination improves the leakage current, diode characteristics and gate oxide quality of the devices compared to the old material. However the improved material quality also has adverse effects. Due to the increased carrier lifetime, the parasitic bipolar transistor action becomes increasingly important. With $TiSi_2$ salicide the breakdown voltage can be increased by about 0.5 Volt. This is probably caused by a degraded emitter efficiency of the parasitic bipolar device. For half micron CMOS the bipolar breakdown needs further attention.

As has been shown with the liquid crystal temperature measurement, the rise in temperature of the SOI transistors during operation is significantly higher than for bulk transistors. These measurements are another indication of the significant Joule self heating of the transistors in thin film SOI.

## Acknowledgments

We which to thank P.C. Zalm for the SIMS measurements and P. Damink for the liquid crystal measurements.

## References

Colinge J P, 1987, IEEE Circ. and Dev. Mag., 5 , pp 18
Jastrzebski L, Ipri A C, 1988, IEEE Electron Device Letters, Vol.9, no.3
Mc Daid L J et al, 1989, Proceedings of ESSDERC'89, Berlin, pp 885
Vasudev P K, 1988, Symp. VLSI Tech. Dig., 61
Woerlee P H et al, 1989, Technical digest IEDM , pp 821

# A new stacked capacitor cell for 64 Mbit DRAMs

Cheon Soo Kim, Jin Ho Lee, Kyu Hong Lee, Dae Yong Kim, Jin Hyo Lee and Chung Duk Kim

Memory Div., Electronics & Telecommunication Research Institute
P.O.Box 8, Daedog Science Town, Daejeon, Korea

## Abstract
A new stacked capacitor memory cell named a Cup-shaped Storage Node (CSN) is proposed for high density DRAMs. The storage capacitance of the cell is enhanced by stacking the cup shaped polysilicon layer. So single and double layered stacked cell structures can be fabricated with simple extra process steps. The usefulness of the fabrication process and electrical characteristics of this novel cell structure has been demonstrated.

## 1. Introduction
In megabit-class DRAMs, three dimensional cell structures represented by stacked and trenched capacitor cell are necessary to get sufficient storage capacitance in a limited area. Stacked capacitor cell [1-6] has been extensively studied to achieve high density DRAM, for its advantage in high immunity against soft error, and for its simplicity in the fabrication procedure. The key issue of stacked cell is to obtain enough storage capacitance in a limited cell area to operate with a reasonable signal to noise ratio[5].

This paper describes a new stacked capacitor memory cell having cup-shaped storage node, which attains sufficient cell capacitance and reduces storage node height. The one feature of the CSN cell is that the cup-shaped structure is introduced into storage node, and that both in- and out-sides of storage node are used as cell capacitor. The other is that a 2'nd storage node can be stacked inside the 1'st storage node to enhance capacitance greatly with simple extra process steps. The fabrication procedure and detailed electrical characteristics of the CSN cell are discussed.

## 2. Experimental
The fabrication procedure of the CSN cell are shown in Fig.1. After the transfer transistor formation, polycide bit line is defined before storage capacitor formation. Silicon nitride and thick CVD oxide are deposited according to the required storage node height. Then storage node region which is packed with minimum space is defined by oxide nitride etching (Fig. 1-a). Subsequently polysilicon film which is used for storage node is deposited, and each storage node is formed by etch-back process of the polysilicon (CSN I Type)(Fig. 1-c). Without precise patterning in cell array after plate delineation, large capacitor area and good scalability can be obtained easily. For enhancement of storage capacitance, another polysilicon film is deposited which is isolated by self-aligned oxide sidewall spacer(Fig. 1-b).

So a double cup shaped cell structure can be formed(CSN II Type). Finally the CVD oxide and silicon nitride is removed by wet etching, CSN structure can be obtained as shown in Fig. 2. $SiO_2$-$Si_3N_4$ composite dielectric film and cell plate are deposited on each electrode(Fig. 1-d). So the fabrication procedure for this new cell doesn't need additional mask steps compared with that of conventional stack cell .

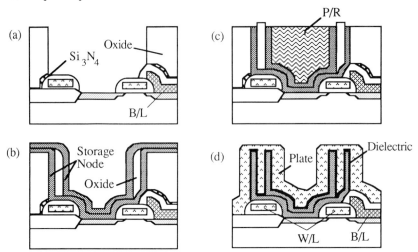

Fig.1 Process sequence of CSN II Type cell. After W/L and B/L formation, (a) Storage node define, (b) Storage node depo./Oxide side wall/Storage node depo., (c) P/R coat/ Etch back, (d) Dielectric depo./Storage plate.

## 3. Results

Two kinds of storage nodes with 1.5 μm height are successfully fabricated. The SEM bird's eye views of CSN cells are shown in Fig.2. Storage node size is 1 μm by 1.5 μm and polysilicon storage node thickness is 220 nm. For 64Mb DRAM application, the capacitor layout size should scale down below 1μ$m^2$. The capacitor area of CSN I structure can be

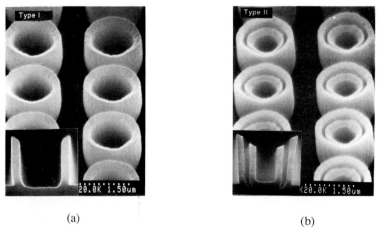

Fig. 2. SEM micrographs of (a)CSN I Type and (b) CSN II Type.

designed below $1 \mu m^2$ with no difficulties. In case of CSN II, the thickness of storage node poly and side wall spacer oxide reduced to 100 nm and 200 nm, respectively.

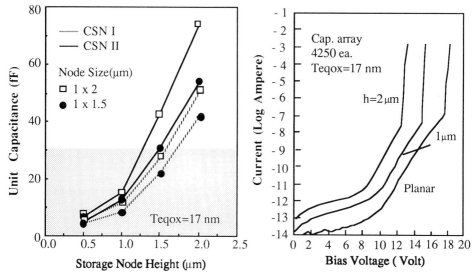

Fig. 3. Measured storage capacitance of CSN cell as a function of the storage node height.

Fig. 4. I-V characteristics of ONO dielectric film with the node height.

The measured capacitance of CSN cells as a function of the storage node height is shown in Fig.3. Above $1 \mu m$ height of storage node, linear increase of capacitance is observed with node height. The capacitance of CSN II Type increase about 50 % compared with that of CSN I Type at same storage node size and height. To obtain a sufficient cell capacitance above 30 fF in 1.5 $\mu m^2$ capacitor area with a 17 nm $SiO_2$ equivalent thickness dielectric film, the storage node height of CSN I and CSN II structure is measured to be 1.7 $\mu m$ and 1.4 $\mu m$, respecti-vely. The I-V characteristics for oxide-nitride-oxide dielectric film with increasing node height is shown in Fig. 4. Leakage current is measured to be 0.23 fA per

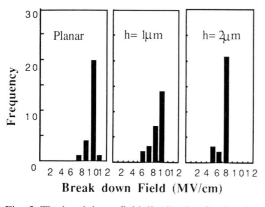

unit capacitor at 5 volts. Fig.5 shows the break down field distribution for the planar capacitor and for the CSNs with 1 $\mu m$ and 2 $\mu m$ storage node height. Breakdown strength of the dielectric film was 10 MV/cm in planar capacitor. Some degradation is observed with the increasing the node height as shown in Fig. 5

Fig. 5. The breakdown field distribution for the planar capacitor and for the CSNs with storage node height.

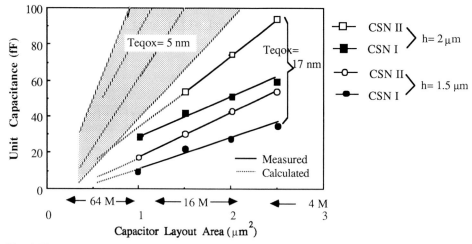

Fig. 6. The measured storage capacitance of various CSN cells with capacitor area.

The measured capacitance of CSN cells as a function of capacitor area is plotted as shown in Fig. 6. Since the CSN cell with a 5 nm $SiO_2$ equivalent thickness dielectric film is expected to have 30 fF capacitance in less 1 $\mu m^2$ size, it is potentially scalable to the 64 Mb generation.

## 4. Conclusions

Experimental results are presented for a new stacked capacitor memory cell structure(CSN) which enhances the storage area greatly for high density DRAMs. Storage capacitance becomes 30 fF in 1.5 $\mu m^2$ capacitor area and 1.7 $\mu m$ node height with a 17 nm $SiO_2$ equivalent thickness dielectric film. If the double shaped structure (CSN II Type) is introduced into the storage node, storage capacitance increases about 50 % compared with that of CSN I Type. These results shows that this new cell structure is suitable for a 64 Mb DRAM or beyond.

## Reference

[1]. W. Wakamiya, Y. Tanaka, H. Kimura, H. Miyatake and S. Satoh, *Symposium on VLSI Technology*, p. 69, 1989.
[2]. T. Ema, S. Kawanago, T. Nishi, S. Yoshida, H. Nishibe, T. Yabu, Y. Kodama, T. Nakano and M. Taguchi, *IEDM Tech. Dig.*, p. 592, 1988.
[3]. H. Watanabe, K. Kurosawa and S. Sawada, *IEDM Tech. Dig.*, p. 600, 1988.
[4]. T. Kisu, S. Kimura, T. Kure, J. yugami, A. Hiraiwa, Y. Kawamoto, M. Aoki and H. Sunami, *Extended Abstract of the 20st Conf. on SSDM*, p.581, 1988.
[5]. N. C. C. Lu, *IEEE Circuits and Device Magazine*, p. 27, 1989.
[6]. D. Kenny, E. Adler, B. Davari, J. DeBrosse, W. Fery, T. Furukawa, P. Geiss, D. Harmon, D. Horak, M. Kerbaugh, C. Koburger, J. Lasky, J. Rembetski, W. Schwittek and E. Sprogis, *Symposium on VLSI Technology*, p. 25, 1988.

Paper presented at ESSDERC 90, Nottingham, September 1990
Session 6A2

# Stacked capacitor cell technology for 16M DRAM using double self-aligned contacts

M. Fukumoto, Y. Naito, K. Matsuyama, H. Ogawa, K. Matsuoka, T. Hori, H. Sakai, I. Nakao, H. Kotani, H. Iwasaki and M. Inoue

Semiconductor Research Center, Matsushita Electric Ind. Co., Ltd.
Moriguchi, Osaka, Japan

Abstract  This paper describes key technology of a small sized stacked capacitor cell for 16MDRAM. The main feature of the technology is unique and highly productive double self-aligned contact process for bit line and for storage node, which is immune against process fluctuations. The cell made by this process showed desirable characteristics for DRAM operation.

## 1. Introduction

Stacked capacitor (STC) cells (Koyanagi 1978) (Ema 1988a) (Kimura 1988b) have grown up to the most promising candidates to realize high density 16MDRAM, because the complicated processes necessary to trench capacitor cells (Sunami 1982) are not needed. Although STC cells are much simpler, process and structure design considerations are still needed to realize the sufficient large storage capacitance in a limited small area. In particular, margins between bit line contact, storage node contact, bit line and word line become very small, which are beyond the pattern alignment accuracy in photolithography. Therefore, it becomes impossible to achieve reliable electrical isolation between them.

To solve this problem, we have introduced two kinds of new self-aligned contact technologies. One is for the bit line contact and the other is for storage capacitor contact. These technologies have the ability to make a STC cell with an area of 4.4μm² or below, and have been confirmed to have good reproducibility and stability.

In this paper, this double self-aligned contact process and the characteristics of the cell with these contacts are described.

## 2. Memory Cell Structure

Fig.1(a) and (b) show the pattern layout and the cross sectional view of the memory cell

© 1990 IOP Publishing Ltd

used to fabricate 16MDRAM with the chip size of 8.1mm×17.5mm, respectively. In the cell, the storage capacitor is placed above the tungsten polycide bit line in order to obtain high storage capacitance (Cs) by the extended capacitor area (Ema 1988a)(Kimura 1988b).

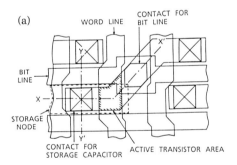

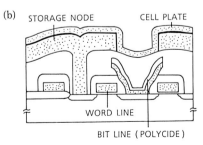

Fig.1 Stacked capacitor cell with double self-aligned contacts.
(a) Layout of the cell   (b) Cross-sectional view of the cell along X-X' line in (a)

Bit line to bit line capacitance coupling is reduced by shielding effects of the storage node poly Si and the cell plate.

The cell area is 4.4µm² (1.6µm × 2.8µm). The size of the contact for the bit line, that for the storage node, and the bit line width are 0.7µm, 0.7µm and 0.5µm, respectively, and word line width is 0.7µm. They are close to allowable minimum size in practical photolithography. Thus, the margins of the bit line contact to the word line and that of the storage capacitor contact to the bit line should be as small as 0.1µm. It is hard for such margins to avoid electrical shorting through the overetching of the contact window or the misalignment in a photolithography. This obstacle is removed by two kinds of newly developed self-aligned contact technologies.

### 3. Double Self-Aligned Contact Technology

Fig.2 shows the process steps of the self-aligned contact for the bit line to substrate. The process is composed of five steps, i.e., (1) deposition of LPCVD HTO, thin $Si_3N_4$ and BPSG films on the poly Si word lines covered with CVD oxide, (2) contact pattern formation, (3) wet etching of BPSG film by buffered HF, (4) anisotropic dry etching of $Si_3N_4$ and HTO films, and (5) BPSG reflow at a high temperature (900°C). SEM cross-sectional view of the completed contact is shown in Fig.3. In the wet etching process (3), the BPSG film at the contact is almost removed and $Si_3N_4$ film stops the wet etching with good controllability and reproducibility to leave only $Si_3N_4$ and HTO films. Because of the thinness of $Si_3N_4$ and HTO films the dry etching time for those films at the step (4) is not so long as to break the cover oxide of the poly Si word line. Thus the electrical shorting of the word line and bit line could be avoided. Moreover, easy bit line patterning can be achieved owing to the surface smoothing by BPSG reflow at the step (d) in Fig.2.

The contact resistance for the bit line to n+ substrate showed low value ( mean value : 110Ω ), although the contact size was reduced to 0.3µm × 0.7µm by the sidewall for the word line LDD formation.

This self-aligned contact has also been successfully applied to peripheral circuits to pack them closely.

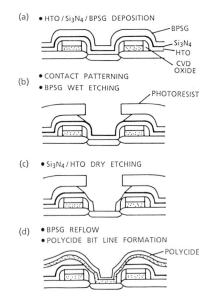

Fig.2 Process sequence of the self-aligned contact for the bit line.

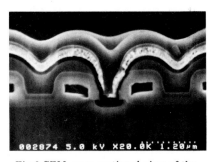

Fig.3 SEM cross-sectional view of the self-aligned contact for the bit line.

Fig.4 shows the process sequence of the self-aligned contact for the storage node poly Si to the substrate and Fig.5 shows SEM cross-sectional view of the contact. The process is featured by the formation of the sidewall insulator by HTO deposition and etch back inside the contact hole. Owing to the sidewall, the electrical short among the word and bit lines and the storage capacitors can be avoided, even when the initial contact hole exposes these lines due to process fluctuations in photolithography and/or contact side etching. In the cell size mentioned above, we confirmed that the electrical isolation was perfect by using 0.2µm thick HTO : the

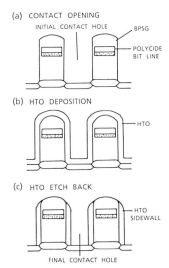

Fig.4 Process sequence of the self-aligned contact for the storage node. Cross-sections are along Y-Y' line in Fig.1(a).

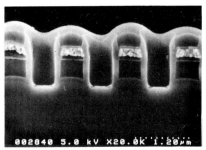

Fig.5 SEM cross-sectional view of the self-aligned contact for the storage node.

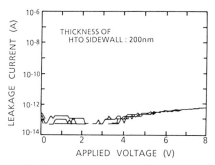

Fig.6 Leakage current through the HTO sidewall of the storage node contact versus applied voltage between bit line and storage node.

conventional one. As evident from Fig.1(a), the active transistor area is not a conventional rectangular shape ( minimum width : 0.7μm ) (Kimura 1988b), because of realizing a minimum LOCOS isolation width, 0.7μm, to eliminate the cell to cell leakage. Furthermore, it is possible that the n+ diffusion depths of source and drain are asymmetric, since one diffusion layer is contacting to the As doped polycide bit line and the other diffusion layer is contacting to phosphorus doped storage node poly Si placed adjacent to the gate. Fig.7 shows the subthreshold characteristics of the cell transistor measured by interchanging source and drain. The figure proves that the shape of the active area and phosphorus diffusion from poly Si to n+ diffusion layer at the storage node contact by following annealing process do not deteriorate the characteristics.

leakage current through the sidewall measured by using several hundreds of 32Kbit test cell array was below the detection limit for the actual operation voltage. One example is demonstrated in Fig.6.

The sidewall decreased the final contact size to 0.3~0.4μm in diameter, but the contact resistance for the storage node poly Si, to substrate n+ diffusion showed considerably low value of ~200Ω.

4. Transistor Characteristics

The transistor used in the cell is not

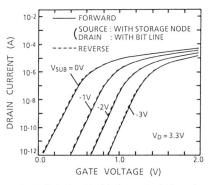

Fig.7 Sub-threshold characteristics of the transistor in the cell.

## 5. Cell Characteristics

For DRAM operation, to obtain sufficiently large output voltage from the memory cell, the ratio of the bit line capacitance ($C_B$) to the storage capacitance ($C_S$) should be as small as possible. The measured $C_S$ of our cell is 35fF (25fF) for 0.5μm (0.3μm) thick storage node poly Si, for $SiO_2$-$Si_3N_4$-$SiO_2$ capacitor insulator of $SiO_2$ equivalent thickness of 7nm. $C_B$ for the bit line connected to 128 cells estimated by 3-dimensional capacitance simulation was 195fF. Hence, quite reasonable value of $C_B/C_S$, 5.6~7.8, for DRAM operation could be achieved. This ratio is smaller compared with a conventional cell in which a storage capacitor is located underneath the bit line. Moreover, it is found that the bit line to bit line capacitance coupling is only 5% of $C_B$ due to the shielding effect of the storage capacitor and the cell plate.

Using the stacked capacitor technology including the double self-aligned contact process, 16MDRAM whose chip size is 8.1mm × 17.5mm has been fabricated. Photomicrograph of the DRAM chip and its characteristics is shown in Fig.8 and Table 1, respectively.

## 6. Conclusion

Stacked capacitor cell technology has been developed, and 16MDRAM with 4.4μm² cell has been produced without any difficulty. The main parts of the technology are two kinds of novel self-aligned contact processes for the bit line and for the storage node. These processes were confirmed to provide reliable contacts free from shorting, even when margins between these contacts and the other parts are as small as 0.1μm. The memory cell and its transistor showed desirable characteristics for DRAM operation. The present technology can realize highly productive 16MDRAM.

## 7. Acknowledgement

The authors would like to thank A.Hiroki and S. Odanaka for 3-dimensional simulation, Y. Ichikawa for the device characterization, and Dr. T. Takemoto for his encouragement.

## 8. References

Koyanagi M, Sunami H, Hashimoto N and Ashikawa M 1978 IEDM Dig. of Tech. Papers pp348-51

Sunami H, Kure T, Hashimoto N, Itoh K, Toyabe T and Asai S 1982 IEDM Dig. of Tech. Papers pp806-8

Ema T, Kawanago S, Nishi T, Yoshida S, Nishibe H, Yabu T, Kodama Y, Nakano T and Taguchi M 1988a IEDM Dig. of Tech. Papers pp592-5

Kimura S, Kawamoto Y, Kure T, Hasegawa N, Etoh J, Aoki M, Takeda E, Sunami H and Itoh K 1988b IEDM Dig. of Tech. Papers pp596-9

Table 1. Typical characteristics of 16M DRAM

| | |
|---|---|
| BIT ORGANIZATION | 16M × 1 / 4M × 4 |
| CHIP SIZE | 8.1 × 17.5mm² |
| CELL SIZE | 1.6 × 2.8μm² |
| POWER SUPPLY | 5.0V  peripheral 5V / cell 3.3V |
| ACCESS TIME | $t_{RAC}$ = 80ns (max) |
| ACTIVE CURRENT | 80mA |
| STANDBY CURRENT | 3mA |
| REFRESH | 4096/64ms |

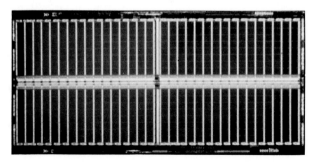

Fig.8 Photomicrograph of 16M DRAM chip

Paper presented at ESSDERC 90, Nottingham, September 1990
Session 6A3

# Buried stacked capacitor cells for 16M and 64M DRAMS

J Dietl, L DoThanh, K H Küsters, L Kusztelan, H M Mülhoff, W Müller, F X Stelz

Siemens AG, Otto-Hahn-Ring 6, 8000 München 83

Abstract. The technology and performance of two trench capacitor cell variants for 16 & 64M DRAM application are assessed. Critical cell leakage mechanisms are determined and means of reducing them further are compared. Trench-transistor separation is shown to be non critical with regard to transistor performance and double channelstop implant technology also allows for further transistor optimisation possibilities.

1. Introduction

The buried Stacked Capacitor cell offers the advantage of a reduced cell topography compared to other variants of the stacked cell. The stacked capacitor is placed in a trench, and the storage node is formed either on Si substrate, STT, Yoshikawa (1989a) or on an oxide isolation layer inside the trench, BSCC, Yoshioka (1987a). The use of these cell concepts for 16 & 64M DRAMs requires process optimization such as through the LOCOS implantation, particularly to improve cell isolation. For the STT cell various limitations due to trench leakage are investigated, and for the BSCC, the adjacent trench sidewall buried contact isolation requirement is discussed. New processes for BSCC (Figure 1) & STT storage node formation are presented which result in significant improvement in performance. Cell Capacitance, ONO dielectric integrity, as well as transistor optimisation possibilities are also assessed. 16M DRAMs have been fabricated based on both variants of stacked capacitor cells. Critical requirements for 64Mbit such as trench isolation and trench to pass transistor separations can be fulfilled by the proposed process.

2. Process Technology

The isolation technique is based on LOCOS. For improved cell performance, a combination of field implants before and after LOCOS formation is employed. In both cell variants, the storage node consists of an As doped polySi layer. In the case of the STT cell, the storage node polysilicon is formed on the Si substrate, which serves to laterally protect the between

Fig.1: 16M-BSCC cell cross section

© 1990 IOP Publishing Ltd

trench LOCOS during trench complex formation processes, and also enables reduced outdiffusion of Trench $n^+$. In situ doped Polysilicon is used to realize a pn junction 45 nm below the trench surface. The BSCC variant (see Figure 1) has an oxide isolation between storage node Poly Si and Si substrate. In this case, the storage node is connected to the transfer gate by a buried trench contact at the sidewall adjoining the active area. For trench contact formation, two processes were employed:

o The trench contact is patterned after trench sidewall oxidation, trench contact depth is determined by lithography (0.2-1um).

o After a shallow trench etch, a nitride spacer is formed at the trench sides adjoining the active area, Figure 2. After a second trench etch, an oxide is grown on the trench wall except where the buried contact is to be formed, following nitride strip. This allows for better control of contact depth.

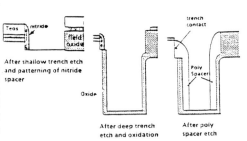

Fig.2: Process Flow for trench contact with double trench etch.

## 3. Cell Characterisation

For the STT concept, a possible bulk punchthrough limitation exists. Figure 3 data, taken from a trench finger comb test structure, however, indicates that at 16M design rules, leakage is surface dominated through its dependance on field implant. In particular, additional through the LOCOS implantation, and use of in-situ doped storage node polysilicon, allows separations along the same bitline down to approx. 0.4um for a surface Pwell concentrations of 1E17 cm-3. The boron surface concentration required at this separation, imposes a limit due to trench to substrate leakage (defect generation).

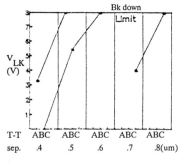

Fig.3: Trench-trench voltage at 1pA/cell leakage.

A. 6E13/4E13 cm-2 Field/Well implant + 8E12 cm-2 additional through LOCOS implant.
B. 6E13/4E13 cm-2 Field/Well implant.
C. 6E13/2E13 cm-2 Field/Well implant.

Trench-trench leakage in the BSCC process, in comparison, is inherently limited to that between adjacent buried contacts. A cell design with quarter pitch bitlines Yoshikawa (1989a) is advantageous, because trench contacts are not directly opposite each other at the trench sidewalls. Figure 4 data, taken from a cell array test structure, shows that a minimum separation between adjacent bitlines of less than 0.4um with normal before LOCOS field Implant doses (i.e. 8E16 cm-3 doping level below field oxide) is achievable for an improved process without significant field oxide thinning. Additionally, no dependance of leakage current on trench contact depth (0.1-0.8um) has been observed.

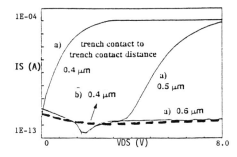

Fig.4: Trench sidewall-Trench sidewall isolation.

A. LOCOS thinned by 100nm during trench contact formation.
B. improved isolation. (minimal oxide etch).

With optimised field implants then, BSCC allows even smaller trench-trench separations, and extends the application of trench type concepts to 64Mbit.

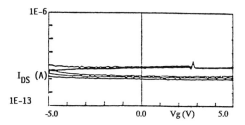

Fig.5: Junction Leakage.
$V_{trench\ n+}$ = 0 to 2V & Vsub = -2.5V

Typical leakage to substrate data, for the BSCC variant, are reproduced in Figure 5. The low, non critical leakage level is reproduced over a number of batches, and with variable trench isolation oxide thickness (40 & 70nm). Hence, the enhanced leakage current due to gated diode effects at the etched trench sidewall as reported in Horiguchi (1987b) have not been observed. The STT variant shows similar low leakage levels, indicating an advantage with respect to the field plate isolation technique, Shen (89b), as the surface gate controlled diode action at the trench edge is suppressed due to the thick LOCOS.

The use of a deposited polysilicon layer as a storage node may adversely affect the ONO dielectric quality, particularly with its use in conjunction with the trench topography. Current-Voltage characteristics for ONO on in-situ As doped PolySi and ONO on $BCl_3$ RIE etched PolySi were, however, similar to those of ONO on single crystal Si. For a 9nm ONO layer on PolySi in trench, a time to breakdown vs. 1/E plot extrapolation yields lifetimes of greater than 10 years at 5MV/cm and 150°C (criterion 10 FIT failure rate).

At this dielectric thickness, an average cell plate capacitance of 55fF for the STT is obtained for a trench diameter of 1um and depth of 4.5um, sufficient for 16M application. A reduction in Cplate of approx. 20 % for the BSCC occurs if a maximum trench isolation oxide thickness of 70nm and a maximum thickness of storage node PolySi of 120nm is used. A thinner ONO will, however, be required for 64 MBit DRAM application. Time zero breakdown data obtained for 6nm planar ONO on PolySi indicate the applicability of this dielectric process to the 64 Mbit generation.

The buried contact process does not impose any restrictions on transistor performance either. Outdiffusion is minimal and the pass transistor may be placed right at the edge of the trench. Figure 6a shows that pass transistor Vt does not decrease even when trench-transistor separations approach zero. Additionally, degradation under hot electron stress, Figure 6b, is not influenced significantly by the buried contact.

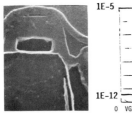
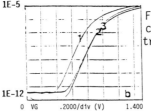

Fig.6a: Cross section of buried contact (d = 0.16um distance to transistor)

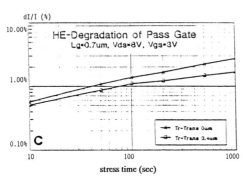

Fig.6b & c: Dependance of Vt & HE degradation on distance between trench and transistor.
1: d = 0.4um
2: d = 0.2um
3: d = 0um

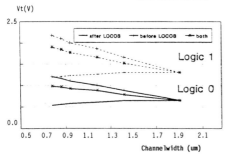

Fig.7: Adjustment of narrow width effect by splitting channelstop implant between "before LOCOS growth" - maximum effect and "through LOCOS" - no or negative effect.

Double channelstop implantation additionally offers considerable process latitude for N-channel transistor optimization. By using implants both before and after LOCOS growth, narrow width effects can be tailored, as shown in Figure 7, with low Vt (0.6V) periphery transistors and high Vt (1V) transfergates. Hence optimization of LOCOS isolation properties is achieved without penalty to transistor performance.

4. References

Horiguchi et al 1987b IEDM Tech. Dig. p.324
Shen     et al 1989b IEDM Tech. Dig. p.27
Yoshikawa et al 1989a Symposium VLSI Technology p.67
Yoshioka  et al 1987a ISSCC Dig. Tech. p.20

*Paper presented at ESSDERC 90, Nottingham, September 1990*
*Session 6A4*

# Coupling of different leakage paths between trench capacitors

W. Bergner and R. Kircher[*]

ZFE SPT 33, SIEMENS AG, Otto-Hahn-Ring 6,
D-8000 Munich 83, Germany, Tel. (+4989) 636-45782.
[*] On leave from Siemens AG, present address:
Research Institute of Electrical Communication,
Tohoku University, Sendai, Japan.

**Abstract** Out from various cell concepts under investigation for the 16 and 64 Megabit DRAM generation the depletion type trench cell is still favoured because of its low process complexity. However, further shrinking of this cell concept, which is been widely used for the 4 Megabit DRAM, may result in leakage problems between neighbouring trench capacitors. The isolation of a modified version of this cell type, the SSP cell, has been investigated by numerical simulation as a function of well implantation dose and the trench to trench separation in the case of three closely spaced trench capacitors.

## 1 Introduction

The SSP trench cell (**S**uper **S**hrink Trench and **P**olysilicon) has the advantage of reduced outdiffusion of the trench As doping compared to a process with TEOS. Fig. 1 shows the schematic cross section of three SSP trenches. The geometry of the trenches has been taken from SEM photographs, which exhibit an elliptical cross section, a steep conical shape near the surface and a smooth cylindrical one in the bulk region. The cell plate is the counterelectrode to the storage node and is connected to a polysilicon layer which acts as a parasitic common gate between the three trenches. The effective thickness of the gate oxide is 450 nm. The doping profiles for the p-well, the field implantation and the As outdiffusion of the storage node have been obtained from SIMS measurements. These data are accurate except for the Boron concentration at the silicon oxide interface.

The first part of the paper shows the influence of the surface concentration on the punch through mechanisms, the second part considers the interaction of trench cells on neighboured bitlines. All simulations have been carried out with the three-dimensional device simulator SITAR [1], which solves the classical semiconductor equations in steady state on a finite difference mesh.

## 2 Volume and Surface Punch Through

In order to take into account the depletion of Boron at the silicon-oxide interface, which is caused by segregation during the field oxidation, the correlation between surface concentration of the p-well and leakage current has been simulated. The structure consists of two trench capacitors with a trench to trench distance $t_{12} = 0.55 \mu m$ corresponding to the design rules of a 16 Mbit DRAM. Fig. 2 shows the punch through voltage (critical current $1pA$) versus the surface concentration, for constant plate voltage $V_p = 1.65V$ and substrate voltage $V_s = -1.5V$. As a result of the simulations, we can observe two different types of punch through: the well

© 1990 IOP Publishing Ltd

known volume punch through in the bulk region, and the surface punch through. This is illustrated in Fig. 3, where the electron distribution in the transition region between surface and volume leakage for the trench voltages $V_{t1} = 0V$, $V_{t2} = 6.5V$ and a surface concentration of $1.0 \cdot 10^{17} cm^{-3}$ is shown.

The volume punch through is caused by the spreading of the space charge regions around the trenches in a depth where the well doping concentration drops below a critical level [2]. To suppress volume punch through, the well profile and the trench etching have to be optimized. Surface punch through may occur due to a gradient of the doping concentration near the silicon surface. In contrary to volume punch through it depends not only on the voltage at the storage node and the substrate bias, but also on the bias of the poly plate. As long as the surface channel is open the entire current flows through the upper channel and therefore there is no punch through in the volume. Although increasing the doping concentration at the surface reduces a surface punch through, it enables the breakdown in the volume channel At the same time. This is expressed in Fig. 3 by the constant punch through voltage for high concentrations.

Because of the limited resolution of SIMS of about 15 nm it is not possible to take the exact surface concentration from the measurements. Therefore we have fitted the p-depletion in such a way that it is possible to reproduce the experimental value of the punch through voltage of $4.3V$ for this device.

## 3  Punch Through between Three Trenches

Further shrinking of the cell makes it neccessary to reduce the bitline spacing. As a result of this the "trench-adjacent bitline leakage" becomes critical. To find an appropriate model we choose three capacitors arranged as shown in Fig. 1. In the simulation the punch through behaviour between the three trenches has been investigated as a function of the distance $t_3$ and the applied voltage. Fig. 4 shows the electron distribution around the trenches under test bias conditions of $V_{t1} = 0V$, $V_{t2} = 4.5V$, $V_{t3} = 0V$, $V_p = 1.65V$ and $V_s = -1.5V$ and the distance $t_{12} = t_{23} = 0.55 \mu m$. Two leakage paths can be detected at the surface with a critical current in the path between trench 2 and 3, and a sub-critical current between trench 1 and 2. In addition a subcritical current flows in the bulk between trench 2 and 3.

A systematical analysis of the leakage characteristics shows two major features: Firstly, the punch through voltage between trench 1 and 2 is independent of the distance $t_{23}$. Secondly, the path between trench 2 and 3 is more sensitive for variations of trench to trench distance. This is caused by the elliptic trench contour and expresses the experience that the punch through is stronger for lower curvature.

## References

[1] Bergner W, Kircher R 1988 *Proc.3th Int. Conf. on Simulation of Semiconductor Devices and Processes* ed G Baccarani (Bologna) pp 165-174.

[2] Bergner W, Kircher R 1990 to be published in *IEEE-CAD*.

DRAM technology    471

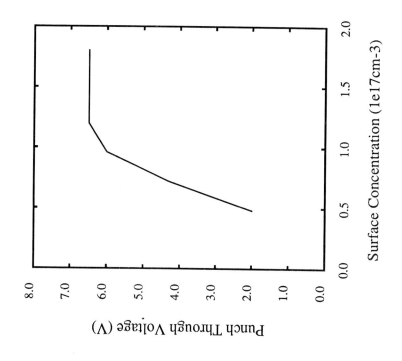

Figure 2: Punch Through Voltage versus surface doping concentration.

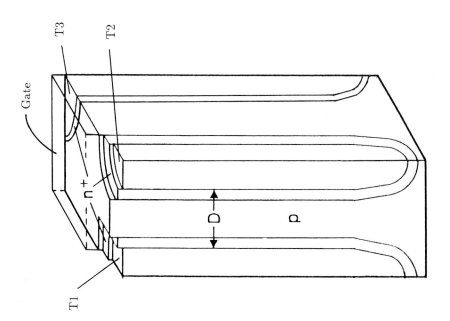

Figure 1: Schematic cross section of the SSP DRAM cell.

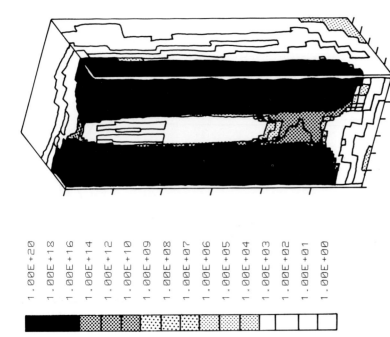

Figure 3: Electron density distribution for $V_{t1} = 0V$, $V_{t2} = 6.5V$, $V_p = 1.65V$ and $V_s = -1.5V$ with $t_{12} = 0.55\mu m$.

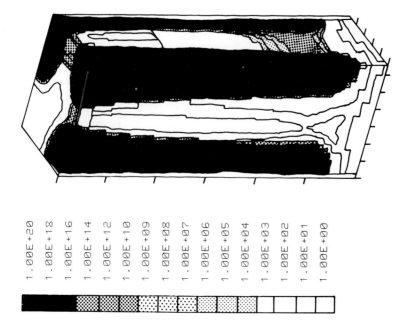

Figure 4: Electron density distribution for $V_{t1} = 0V$, $V_{t2} = 4.5V$, $V_{t3} = 0V$, $V_p = 1.65V$ and $V_s = -1.5V$ with $t_{12} = t_{23} = 0.55\mu m$.

Paper presented at ESSDERC 90, Nottingham, September 1990
Session 6B1

# Temperature dependence of DC characteristics of Si/SiGe heterojunction bipolar transistors

A.S.R.Martin, M.A.Gell, A.A.Reeder, D.J.Godfrey, M.E.Jones, C.J.Gibbings and C.G.Tuppen.

British Telecom Research Laboratories, Martlesham Heath, Ipswich, Suffolk, U.K.

Abstract. Silicon/silicon-germanium heterojunction bipolar transistors incorporating a strained layer of silicon - germanium alloy have been fabricated and their DC characteristics investigated over a wide range of temperatures (80 to 400K). The collector currents were typically found to be ideal over six orders of magnitude at room temperature. Nonideal behaviour observed below 180K is ascribed to surface recombination effects. The mechanisms associated with the nonideal behaviour of the base currents are discussed and the enhanced ideality seen towards 400K assigned to a reduced dominance of recombination mechanisms.

## 1. Introduction

Silicon Heterojunction Bipolar Transistors (HBT's), incorporating a strained layer of $Si_{1-x}Ge_x$ as the narrow band gap base, offer potential advantages over homojunction devices in the drive for high speed, low noise technology. The benefits of a such a silicon process compatible technology has aroused much interest recently. The presence of the $Si/Si_{1-x}Ge_x$ heterojunction results in an increased emitter efficiency which produces an enhanced collector current, when compared to an equivalent homojunction device. An important parameter, which determines the degree of collector current enhancement, is the valence band offset. The resultant increase in gain can be traded for higher base doping, which allows a much lower base resistance and narrower base width to be used. It is these features which provide potential speed and noise advantages compared with silicon homojunction devices. This paper describes work to characterise Si/SiGe HBT's. In particular, the size of the valence band offset and the nonideal mechanisms operating in the devices are discussed.

## 2. Theory

The effects on transport of the presence of a narrow band gap SiGe strained layer as the base region in a silicon bipolar transistor can be modelled, in linear diffusion theory, by introducing an exponential term into the expression for the collector current, $I_c$ (Ashburn 1988)

$$I_c = C_1 \exp([DE_v + DE_{gb} + qV_{be}]/kT). \qquad (1)$$

In this expression, **$DE_v$** is the valence band offset, $DE_{gb}$ is the band gap narrowing in the base, $C_1$ is a pre-exponential factor and all other symbols have their usual meaning (Ashburn 1988). If a plot is made of log $I_c$ against the inverse of temperature (T), for a

© 1990 IOP Publishing Ltd

fixed voltage $V_{be}$, then the slope will be determined by the band gap narrowing and the valence band offset.

## 3. Experimental

Heterojunction bipolar transistors (HBT's) were fabricated from layers grown by molecular beam epitaxy (MBE). Structures with base widths ranging from 25 - 150 nm and germanium content in the alloy ranging from 0 - 30% were studied. The npn devices were produced using a mesa isolation process. A typical structure is shown in figure 1.

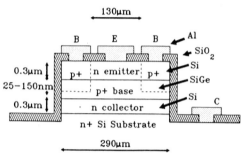

Figure 1: SiGe HBT test structure.

Contact to the boron-doped base region was made either with a local p+ implant or by selectively removing part of the emitter layer. In order to provide control structures, homojunction (silicon base) devices were also produced so that effects related solely to the presence of the SiGe base could be extracted from the results.

The dc behaviour of the terminal currents was studied as a function of $V_{be}$. The variations in the characteristics taken in both forward and reverse configurations were investigated over the temperature range 80-400K with the device structures mounted in a cryostat.

## 4. Results

A typical Gummel plot, obtained from a device with a 15% germanium concentration in the base, is shown in figure 2. This device had a 0.15μm base, doped at $2*10^{19}$ cm$^{-3}$. The emitter and collector dopings were $10^{18}$ cm$^{-3}$ and $10^{16}$ cm$^{-3}$ respectively. Characteristics taken in the conventional and inverted orientations are displayed.

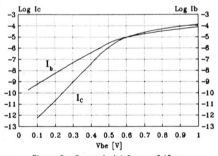

Figure 2a: Gummel plot from a 0.15μm, 15% Ge base HBT. (conventional orientation)

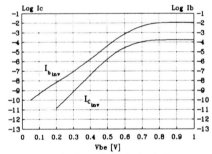

Figure 2b: Gummel plot from a 0.15μm, 15% Ge base HBT (inverted orientation).

Figure 3 compares the collector currents of an HBT and a similar homojunction device.

The variation of the collector current with temperature for a heterojunction device is shown in figure 4. This graph was obtained by extrapolating the linear portion of the collector current characteristic to zero bias and plotting the resulting $I_{co}$ against 1/T on a logarithmic scale. The results of a similar procedure performed on the base currents (measured in forward and inverted orientations) are given in figure 5.

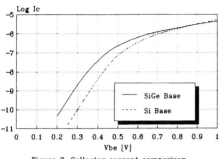
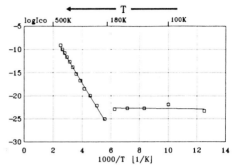

Figure 3: Collector current comparison between a SiGe HBT (15% Ge in base) and the equivalent homojunction device.

Figure 4: Temperature dependence of collector current (Vbe=0) for a 15% Ge base HBT.

The ideality factors of the base/emitter and base/collector junctions, determined from the gradient of the base current characteristic in forward and inverted device orientations respectively, are displayed as a function of inverse temperature in figure 6.

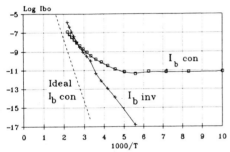
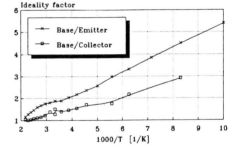

Figure 5: Temperature dependence of base currents for conventional and inverted orientations (extrapolated to Vbe=0).

Figure 6: Ideality factors as a function of temperature for the base/emitter and base/collector junctions.

## 5. Discussion

From Figure 2 the behaviour of the collector current of the HBT's is ideal, typically over six decades at room temperature. Ideality is limited at higher currents by the series resistance associated with the emitter contact, for which no n$^+$ top-up implant was made. It can also be seen from figure 2 that the base current is nonideal, an indication that recombination mechanisms are playing a dominant role. The resulting excess base current may be due to point and extended defects, and/or impurities associated with the low temperature deposition and processing used to produce the devices.

It can be seen from figure 3 that the incorporation of the narrow band gap base results in a collector current enhancement by a factor of about 20. This is in agreement with the predictions of linear diffusion theory.

The variation of collector current with temperature falls into two regimes, as shown in figure 4. Above 180K the slope is constant, as expected from linear diffusion theory. However, below about 180K a markedly nonideal mechanism appears to dominate the transistor action which maybe related to surface recombination effects. The gradient of the graph shown in figure 4 for the range of temperature above 180K yields an estimate for the valence band offset of 94±20 meV, assuming that $DE_{gb}$ is 96±20meV (Ashburn 1988); this is consistent with those obtained from results published by other workers (Patton et al 1988, Gibbons et al 1988 and King et al 1989) and with theoretical predictions based on linear deformation potential theory, combined with results founded on a local density approximation (Van de Walle and Martin 1986). Such measurements, when performed in conjunction with other studies such as SIMS and Raman spectroscopy, could provide valuable information on strain relaxation and outdiffusion of base dopant (Prinz et al 1989).

Examination of figure 5 reveals that the behaviour of the base currents approaches that expected for an ideal, diffusion dominated, mechanism as the temperature increases towards 460K. This suggests that the base current is increasingly controlled by recombination as the temperature is lowered. This is consistent with the observed change in the ideality factors of the base/collector and base/emitter junctions as shown in figure 6. As the temperature is increased to 460K, the ideality factor of both junctions approaches unity. At lower temperatures the ideality factors increase, rising above two. It has been suggested (Ghannam et al 1989) that such behaviour can be ascribed to surface recombination effects.

7. Summary

Si/SiGe HBT's have been fabricated and the DC measurements performed reported. The devices produced ideal collector current characteristics over six orders of magnitude It has been shown that study of the temperature dependence of HBT device characteristics can give a measure of the valence band offset associated with the heterostructure. The measurements made have provided insight into the nonideal mechanisms affecting the performance of the SiGe HBT's. In particular, analysis of the temperature dependence of the base currents has been presented and has provided indications of the reasons for the excess base current observed. Further work is in progress to perform measurements on range of different test structures including devices with an $n^+$ top up implant.

References

Ashburn P 1988, *Design and Realization of Bipolar Transistors,*(Wiley),.
Ghannam M Y and Mertens R P 1989, *IEEE Electron Device Lett*, vol 10, No 6, p242.
Gibbons J F, King C A, Hoyt J L, Noble D B and Gronet C M, 1988, *IEDM Technical Digest*, p566.
King C A, Hoyt J L, Noble D B, Gronet C M, Gibbons J F, Scott M P, Kamins T I, and Laderman S S 1989, *IEEE Electron Device Lett*, vol EDL-10, p52.
Patton G L, Iyer S S, Delage S L, Tiwari S, and Stork J M C 1988, *IEEE Electron Device Lett*, vol EDL-9, p165 .
Prinz E J, Garone P M, Schwartz P V, Xiao X and Sturm J C 1989, *IEDM Technical Digest*, p639.
Van de Walle C G and Martin R M 1986, *Phys Rev B* 34, p5621.

Acknowledgements

The authors wish to thank the members of the thin films analysis and X-ray diffraction groups at BTRL for SIMS and X-ray studies. They are also grateful to P Ashburn for many useful discussions.

# Self aligned Si/SiGe heterojunction bipolar transistors grown by molecular beam epitaxy on diffused $n^+$-buried layer structures

P.Narozny, D.Köhlhoff, H.Kibbel, E.Kasper

Daimler-Benz Research Center Ulm, D-7900 Ulm,
Wilhelm-Runge-Straße 11, FRG, Tel. 49-731-505-2032

<u>Abstract.</u> Si/SiGe heterojunction bipolar transistors (HBT's) were fabricated and characterized. For the first time the layer sequence was deposited by low temperature silicon molecular beam epitaxy (MBE) on locally diffused buried layer structures. To improve the unfavourable ratio between the active transistor area and the inactive base area, self aligned emitter base configuration was introduced for the first time. As a main consequence the power gain cut off frequency was improved. A maximum transit frequency of $f_t$=14GHz and a maximum frequency of oszillation of $f_{max}$=18GHz were measured.

## 1. Introduction

Si/SiGe heterojunction bipolar transistors provide a new design flexibility in the choise of doping levels in the emitter and the base. They have lower base resistance while maintaining high current gain and shorter transit time through the base than those of homojunction devices. A maximum current gain of 5000 has been reported for Si/SiGe HBT's by Schreiber et al (1989). To improve device speed an important issue is the reduction of the base width $W_B$ and an increase in base doping concentration. Further major requirements to achieve superior performances are the minimization of parasitic elements. This paper describes a new HBT structure that has a diffused buried layer collector for reducing the transistor pad capacitances and a self aligned emitter base structure to reduce the extrinsic base resistance. An HBT structure on a buried layer is a prerequisite for later monolithic integration of this device type.

## 2. Growth and device fabrication

The buried layer was realized by $n^+$(P)-double diffusion or $n^+$(As)-diffusion through oxide windows into a high resistivity $p^-$-Si-substrate with a specific resistance of >1000 Ohm*cm. Sheet resistances of <5 Ohm/square ($n^+$(P)) and <12 Ohm/square ($n^+$(As)) were measured for the 1.5um deep diffused buried layers. After removing the oxide mask the HBT layers were deposited by low temperature MBE. The epitaxial parameters are given in Tab.1.

| | | | | |
|---|---|---|---|---|
| Collector | Si | 300nm | $n^- = 4*10^{16} cm^{-3}$ | (Sb) |
| Base | $Si_{0.8}Ge_{0.2}$ | 50nm | $p^+ = 4*10^{19} cm^{-3}$ | (B) |
| Emitter | Si | 200nm | $n = 1*10^{18} cm^{-3}$ | (Sb) |
| Emitter Cap | Si | 100nm | $n^+ = 3*10^{20} cm^{-3}$ | (Sb) |

Tab.1   Epitaxial parameters of the MBE-HBT layers

In early work (Narozny et al 1988), using co-evaporation doping, gallium was used as the primary source of p-type doping. Because the maximum Ga-doping level was limited to the $10^{18}$ cm$^{-3}$ range, a special boron effusion cell was implemented by Kibbel et al (1990). The more convenient dopant boron can now exceed doping levels of $10^{20}$ cm$^{-3}$. The n-collector was doped with antimony, using prebuild-up, secondary implantation (DSI), flash-off technique. The emitter doped with antimony was not flashed off to avoid thermal load above 650°C after base formation. The heavily doped emitter contact was grown by solid phase-MBE (SPE). These techniques (DSI and SPE) are described in text-books on silicon MBE by Kasper et al (1988).

Fig.1 shows a secondary ion mass spectrometry (SIMS) profile of the HBT layer structure with a phosphorus diffused n$^+$-buried collector layer. A strong outdiffusion of phoshorus into the MBE grown n-collector layer can be seen. The out-diffusion is caused by radiation damage produced by the secondary implantation doping technique applied when growing the succeeding MBE n-collector layer.

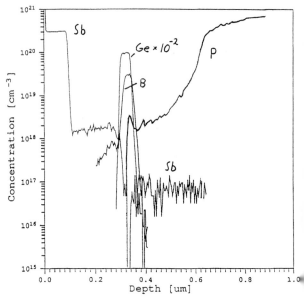

Fig.1: SIMS profile of the HBT structure on P-buried layer

Arsenic used as a dopant material for the diffused buried collector layer is much more localized as can be seen from the corresponding SIMS profile (Fig.2). However, maximum doping concentration is about one magnitude lower compared to the concentration achieved with phosphorus.

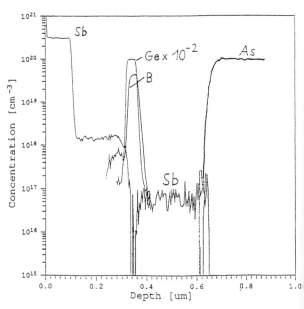

Fig.2: SIMS profile of the HBT structure on As-buried layer

Devices were fabricated by a double mesa approach for quasi planar configuration. Fig.3 shows schematically the basic fabrication steps.

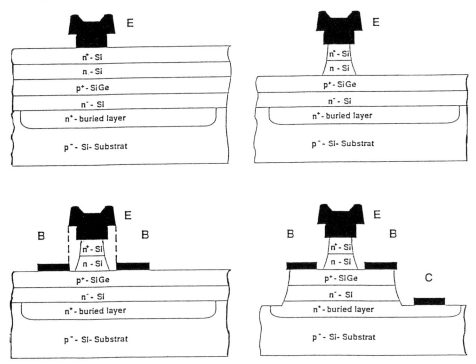

Fig.3: Basic fabrication steps of the Si/SiGe-HBT

A key process in device fabrication is the contact formation to the 50nm thin base. A selective etchant was used for etching off the cap and the emitter layer. An excellent etch stop at the heterointerface was obtained. The T-shaped emitter metal contact was used as an etch mask and provides the separation of the base contact to the emitter contact during evaporation.

Fig.4 shows a SEM photograph of the self aligned emitter base contact. A separation of 0.2um has been achieved resulting in a reduction of the extrinsic base resistance by the factor of 10 compared to a non self aligned structure.

Fig.4: SEM micrograph of the self aligned emitter base contact

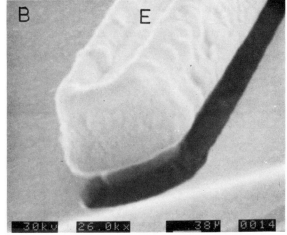

## 3. Device characteristics

The transistors were analysed by DC and RF measurements. I-V characteristics were measured on device structures with different emitter periphery to emitter area ratio. As a result of the high recombination velocity at the etched mesa surface, there is a significant component of base current associated with electron-hole recombination at the edges of the emitter base junction. The influence of this recombination current increases as emitter size shrinks and tends to limit the current gain. The transistor shows a peak current gain of 30 for a small emitter geometry of 1um * 5um. For a structure on $n^+$-substrates a current gain up to 400 was obtained for a large emitter area (Narozny et al 1989).

The self aligned transistor layout was designed for RF-On-Wafer measurement in the common emitter and common base configuration with deembedding structures to eliminate parasitic pad capacitances. The high frequency performance of the devices were investigated by means of S-parameter measurement from 500MHz to 20GHz. From the measured S-parameters a unity current gain frequency ($f_T$) of 14GHz were calculated without deembeding the pad capacitances. The influence of the pad capacitances were measured on special test structures with a vector network analyser. A low value of about 20 fF was found for the pads when using a $p^-$-substrate with a diffused buried layer.

In contrast to the $f_T$ value the unity power gain frequency $f_{max}$ is very strong dependent on the total base resistance. Due to the high base doping concentration (a base sheet resistance of 800 Ohm/square was measured by a TML structure ) and the self aligned structure which minimizes the extrinsic base resistance a maximum frequency of $f_{max}$=18GHz was deduced. That is to our knowledge the highest $f_{max}$ value reported so far for a Si/SiGe HBT.

## 4. Conclusion

Si/SiGe heterojunction bipolar transistors can be fabricated by low temperature MBE on diffused buried layer structures with good DC and RF properties. Due to the buried collector structure the RF performance of the transistor can be measured directly on wafer without the uncertainty of deembedding the pad capacitances. As a main result of the self aligned emitter base structure the $f_{max}$ value exceeds the $f_T$ value. Further work will concentrate on the suppression of the outdiffusion of the phosphorus buried layer and on optimization of the complete transistor structure.

Acknowledgment: This work was supported by the German Ministry for Research and Technology. The authors are responsible for the contents.

## 5. References

Kasper E, Bean J C, 1988 Silicon Molecular Beam Epitaxy,
    CRC Press, Boca Raton, FL
Kibbel H, Kasper E, Narozny P, Schreiber H,
    1989 Thin Solid Films, 184, p163
Narozny P, Hamacher M, Dämbkes H, Kibbel H, Kasper E,
    1988 IEDM Technical Digest, p 562
Narozny P, Dämbkes H, Kibbel H, Kasper E,
    1989 IEEE Trans. on Electr. Dev., 36, p 2363
Schreiber H, Bosch B, 1989 IEDM Technical Digest, p 643

*Paper presented at ESSDERC 90, Nottingham, September 1990*
*Session 6B3*

# Fabrication and characteristics of a MBE-grown InAlAs/InGaAs heterojunction bipolar transistor using an embedded collector

L.M. Su, H. Künzel, R. Gibis, G. Mekonnen, W. Schlaak, N. Grote

Abstract

InAlAs/InGaAs HBTs employing an implanted collector were fabricated and characterized. Current gains of the order of 100 and transit-frequencies of 9-10 GHz were attained. No detrimental effects of implantation on the epitaxial overgrowth were observed.

1. Introduction

Previously, we have proposed a modified heterojunction bipolar transistor structure on InP featuring an embedded collector layer implemented by ion implantation (Su et al 1989). This design offers the following technological and structural advantages over conventional III-V semiconductor-based HBTs using a mesa-type collector:

a) considerable reduction of device height thereby facilitating device processing, especially lithographic procedures

b) elimination of parasitic collector junction capacitance underneath the base contact enhancing the high-frequency performance

c) wider base contacts possible to reduce base contact resistance; less critical formation of base contacts

d) efficient symmetrization of emitter and collector junction areas allowing the achievement of equivalent current gain behaviour with bilateral (i.e. emitter-up or collector-up) transistor operation

In the previous work the feasibility of implanted-collector HBTs was successfully demonstrated on large-area InGaAsP/InP HBT devices grown by LPE. The present paper reports on the fabrication and achieved performance of small-size transistors employing a MBE-grown InAlAs/InGaAs epitaxial structure.

2. Device Structure and Fabrication

In Fig. 1 a schematic cross-section of the implanted-collector InAlAs/InGaAs HBT is shown. The collector layer was formed by selective implantation of Si into a semi-insulating Fe-doped InP layer upon an InP:S substrate. This episubstrate approach (Grote et al. 1989) ensures a well-defined Fe-concentration in the semi-insulating InP layer along with a low density of defects due to the use of a highly S-doped substrate. The InP:Fe layer, the thickness of which was some 7μm, was grown by MOVPE. The Fe-concentration was adjusted to be $1-2 \times 10^{16} \text{cm}^{-3}$. Multiple implantation was

© 1990 IOP Publishing Ltd

employed with the ion energies and the respective doses ranging from 100 keV to 700 keV and $5 \cdot 10^{12}$ cm$^{-3}$ to $5 \cdot 10^{14}$ cm$^{-2}$, respectively. Double-ionized Si was used for the energies > 400 keV. A second implantation step was used to create a highly doped contacting region for the collector. Subsequently, capless annealing of the implanted wafers was carried out in a PH$_3$/H$_2$ ambient at 750 °C for 20 min using a MOVPE reactor.

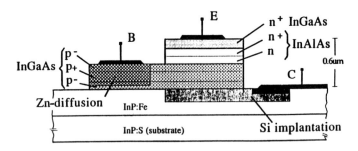

Fig. 1: Cross-sectional view of fabricated InAlAs/InGaAs HBT using an implanted collector

In Fig.2 the resulting electron concentration profile of the implanted collector layer is shown which is composed of a highly doped subcollector region with a doping level of about $2 \times 10^{18}$ cm$^{-3}$ and a lower doped (n = $5 \times 10^{17}$ cm$^{-3}$) region at the surface forming the actual collector of the HBT. The sheet resistance of this n/n$^+$ layer structure exhibiting an effective thickness of more than 1 µm was measured to be approximately 40 Ω/sq.

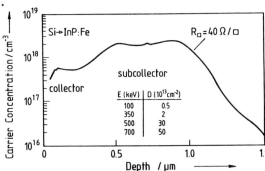

Fig.2: Doping profile of implanted collector layer measured by CV-profiling

Following the implantation process an InAlAs/InGaAs layer structure was grown by conventional MBE to form the emitter and base such that the growth interface coincides with the base/collector junction. Prior to the MBE growth carried out at 500 °C the implanted wafers were slightly etched utilizing a H$_2$SO$_4$:H$_2$O$_2$:H$_2$O(5:1:1) solution. The InGaAs base layer was doped with Be to p$^+$ = $2 \cdot 10^{18}$ cm$^{-3}$, with p$^-$-doped "spacer" regions (p = $2 \cdot 10^{16}$ cm$^{-3}$) adjacent to both heterojunctions. The total thickness of the base was chosen to be 210 nm. The emitter was formed by a Si-doped InAlAs layer (n = $1 \cdot 10^{17}$cm$^{-3}$) which was capped by a highly doped InAlAs/InGaAs (n$^+$ = $2 \cdot 10^{19}$cm$^{-3}$) contact layer structure providing low emitter resistance. The total thickness of the epitaxial layers was 0.6 µm which, however, may be further reduced to result in a fairly planar HBT structure. For comparison, it should be noted that the height of HBTs with a mesa-type collector typically lies well above 1 µm.

The emitter and collector mesas were aligned to the implanted collector region with the help of etch marks in the episubstrate providing an accurracy of 1 μm, or less, and were structured by wet chemical etching. Subsequently, a highly doped shallow contact region was created in the base layer by selective Zn-diffusion to achieve a low resistance ohmic p-contact to the base which, like the n-type emitter and collector contacts, was made of Ti-Au (Kaumanns et al. 1988).

## 3. Device Characteristics

The quality of the growth interface between the implanted region and the epitaxial base layer forming the collector pn-heterojunction is crucial to the electrical performance of the HBT. As a simple way of assessment the quality was judged from the current/voltage forward characteristic of the collector diode. A representative curve is given in Fig. 3. The ideality factors obtained were mostly less than 2, with typical values being around 1.8 which is indicative of a high-quality overgrowth on the implanted layer.

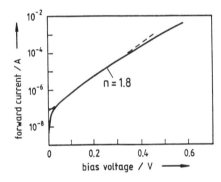

Fig. 3: Forward I/V characteristic of p-InGaAs/n-InP (implanted) collector diode

Fig. 4 shows the common emitter I/V characteristics of a typical fabricated HBT. Current gains of the order of 100 were achieved (Fig.5) in the collector current range between about 1 mA and 50 mA whereas at lower collector currents ($I_c$) the gain was found to decrease following the relationship $I_c \sim V_{ce}^{0.7}$ resulting in still appreciably high values even at currents of as low as 10 μA. A similar behaviour was also found with other InP-based HBT structures recently (Bach et al. 1987, Nottenburg et al. 1987) and may be associated with the low surface and interface recombination velocity found with those materials. The emitter-collector break-down voltage varied between 3-5 V.

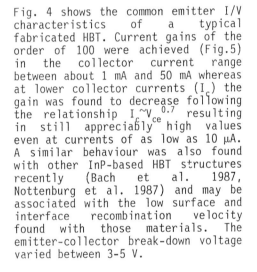

To assess the high-frequency performance S-parameter measurements were carried out "on wafer." As is demonstrated in Fig. 6 a transit-frequency of 9-10 GHz was attained. Results of an equivalent circuit evaluation showed that considerably higher values will be achievable only by further reducing the collector series resistance.

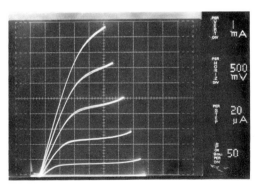

Fig. 4: Common-emitter characteristics of an implanted-collector InAlAs/InGaAs HBT

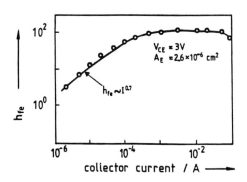

Fig. 5: Current gain vs. collector current

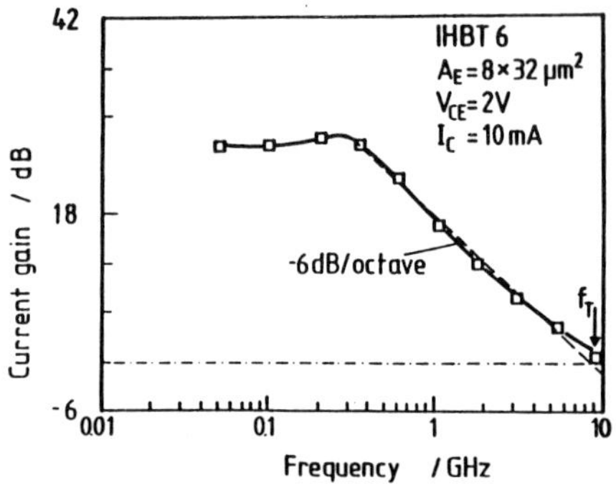

Fig. 6: Small-signal high-frequency characteristic

## 4. Conclusions

A buried collector implemented by implantation was used in an InAlAs/InGaAs HBT providing a relatively planar structure and reduced collector capacitance. Further improvements in the high-frequency performance will be achievable by reducing the sheet resistance of the implanted collector. For this, future work needs to be concentrated on the achievement of low/high doping profiles with maximum doping levels close to the $10^{19}$ cm$^{-3}$ range, and on the related impact on the quality of epitaxial overgrowth.

## 5. Acknowledgements

This work was partially conducted under the ESPRIT programme (project 263).

## 6. References

Bach H G, Grote N and Fiedler F 1987 *Assoc. 17th ESSDERC* ed G Soncini (Elsevier Science Publishers B:V:) pp 883-6
Grote N, Bach H G, Feifel T, Franke D, Harde P, Sartorius B and Wolfram P 1989 *Proc. 19th ESSDERC* ed A. Heiberger, H Ryssel and P Lange (Berlin: Springer Verlag) pp 67-70
Kaumanns R, Grote N, Bach H G and Fidorra F 1988 *Inst. Phys. Conf. Ser.* **90** 501
Nottenburg R N, Panish M B and Temkin H 1987 *Inst. Phys. Conf. Ser.* **83** 483
Su L M, Grote N, Schumacher P and Franke D 1989 *Proc. 19th ESSDERG* ed A Heuberger, H Ryssel and P Lange (Berlin: Springer Verlag) pp 275-8

Paper presented at ESSDERC 90, Nottingham, September 1990
Session 6B4

# Performance of AlGaAs/GaAs heterostructure bipolar transistors grown by MOVPE

B Willén, D Haga, G Landgren

Swedish Institute of Microelectronics
P.O. Box 1084, S-164 21 Kista, Sweden

Abstract A self-aligned fabrication process for AlGaAs/GaAs heterostructure bipolar transistors grown by metal-organic vapor phase epitaxy has been developed. A DC-amplification of 500 has been achieved on large area devices and a maximum frequency of oscillation, $f_{max}$, of more than 30 GHz has been obtained for devices with two stripes of 2.5x10 μm emitter area. We have also characterized the intrinsic and extrinsic time constants and outlined future device optimization expected to improve the performance into the 60 GHz range.

1. Introduction

Heterostructure bipolar transistors (HBT's) are currently of great interest for high frequency devices with a demonstrated performance well over 100 GHz. Some favourable properties for high-speed integrated circuits are the high power handling capacity, large transconductance, low 1/f-noise and well defined threshold voltage. As an example multiplexers and demultiplexers operating over 15 GB/s have been produced, thought to be the highest speed ever reported (Kuriyama 1989). HBT's in the GaAs- or InP-system are also very well suited for opto-electronic integrated circuits (OEIC's), since the material and the vertical structure is similar to a laser diod or a photodetector.

An epitaxial growth technology with exceptional control of doping profiles and heterojunction placement is necessary in order to realize these devices, since the base-emitter junction must coincide with the heterojunction within a few nanometers. Up to now most HBT's have been grown with molecular beam epitaxy (MBE) but promising microwave performance has also been demonstrated for HBT's grown by MOVPE (Bayraktaroglu 1988, Oki 1988, Parton 1989). In principal metal-organic vapor phase epitaxy (MOVPE) offers lower defect density and potentially higher throughput at a lower cost. A disadvantage though is the large diffusion of zinc, commonly used as the p-type dopant. To solve this problem most reports making use of MOVPE have concentrated either on lowering the growth temperature to around 550°C (Enquist 1989), which is comparable to MBE conditions, or on introducing carbon as a substitute for zinc (Bhat 1988, Ito 1989). A potential advantage of keeping the growth temperature as high as 700°C is that the crystal quality of the AlGaAs is greatly improved at this temperature.

The intrinsic speed of HBT's is governed by the epitaxial layers, but subsequent processing of the wafer determines the parasitics. Although these devices can be produced with relaxed optical lithography, since dimensions less then one μm are not required for near optimal performance (Asbeck 1989), the base and the active emitter area patterns must be critically aligned with one another, that is self-aligned, to minimize the extrinsic time delay. It is also very important to achieve low contact resistances and a reduced extrinsic base-collector capacitance.

© 1990 IOP Publishing Ltd

## 2. Device fabrication

The material was grown in a horizontal, atmospheric pressure MOVPE reactor at 700°C with TMA, TMG, AsH$_3$, SiH$_4$ and DEZn as source chemicals. Details of the growth procedure have been published by Nordell 1990. The complete HBT structure is described in Table 1. A base doping as high as $2*10^{19}$ cm$^{-3}$ and a linearly graded aluminum concentration is used in the base to reduce the series and contact resistance and to create a built-in electric field, respectively. A relatively thick spacer layer has been employed to ensure that the zinc doping does not penetrate into the emitter.

A crossection of an HBT is shown in Figure 1. After device isolation by a deep proton implant, a single photoresist pattern is used to define the following steps: Base via holes, which also determines the active emitter width, are formed by wet etching, a shallow proton implant is performed to reduce the base-collector capacitance and Au/Ni/Ti/Pt/Au is evaporated for the p-type contacts and lifted off. This self-aligned process offers a base to emitter distance of about 0.2 µm. In the subsequent lithographic steps wet etching of collector via holes is performed and AuGe/Ni/Ti/Au is evaporated for the n-type contacts and lifted off.

| Layer | Dopant | Doping (cm$^{-3}$) | Thickness (µm) | Al-content |
|---|---|---|---|---|
| Contact | Si | $5*10^{18}$ | 0.10 | |
| Emitter | Si | $5*10^{17}$ | 0.15 | 0.3 |
| Spacer | - | - | 0.10 | 0.03-0.10 |
| Base | Zn | $2*10^{19}$ | 0.05 | 0.00-0-03 |
| Collector | Si | $5*10^{16}$ | 0.50 | |
| Contact | Si | $5*10^{18}$ | 0.50 | |

Table 1. HBT epitaxial structure

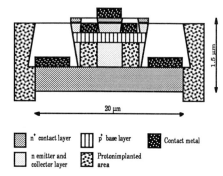

Figure 1. HBT crossection

## 3. Results

The common emitter small signal gain, ß, as a function of collector current is shown in Figure 2 for a device with two fingers of 2.5x10 µm emitter area. From the slope of the curve an ideality factor of 1.5 is derived, indicating a reasonable p/n-junction quality. The dependence of gain on emitter size is illustrated in Figure 3, where $1/ß_{max}$ is plotted as a function of the inverse emitter width. The maximum current gain, $ß_{max}$, is about 500 for large area devices.

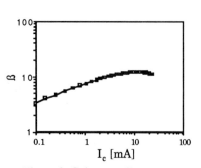

Figure 2. Gain vs. collector current

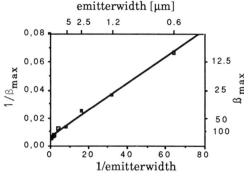

Figure 3. Gain vs. emitterwidth

Heterojunction bipolar technology 487

The microwave response of the transistors has been evaluated on wafer, using a Cascade high frequency wafer probe station and an automatic network analyzer for frequencies up to 18 GHz. The dependence of gain on frequency, extracted from the S-parameters, is shown in Figure 4, for a device with two fingers of 2.5x10 µm emitter area. The maximum frequency of oscillation and current gain cut-off frequency ($f_{max}$ and $f_T$) are extrapolated at 6 dB/oktave to 31 and 21 GHz, respectively.

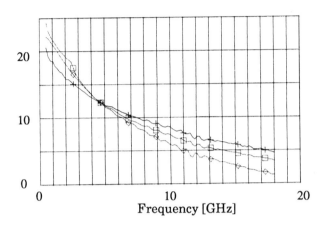

Figure 4. Gain vs. frequency for a device with two stripes of 2.5x10 µm emitter area. $\Diamond = |H_{21}|^2$, [ ] = Maximum Available Gain, + = Unilateral Gain

The cut-off frequensies can be expressed by $f_T = 1/(2\pi\tau_F)$ and $f_{max} = 1/(4\pi\sqrt{\tau_F\tau_{BC}})$, where $\tau_F = \tau_{EC} + \tau_B + \tau_C + \tau_{CC}$ is the total emitter to collector delay time. Here $\tau_{EC}$ is the emitter charging time, $\tau_B$ the base transit time, $\tau_C$ the depleted collector transit time and $\tau_{CC}$ the collector charging time. $\tau_{BC}$ is the base-collector time constant.

An analysis in accordance with the equivalent circuit, shown in Figure 5, has revealed the values shown in Table 2.

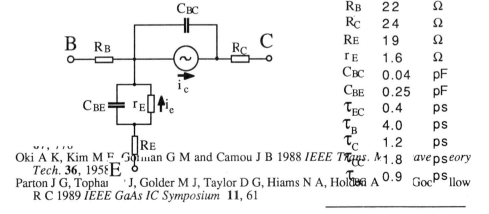

| | | |
|---|---|---|
| $R_B$ | 22 | Ω |
| $R_C$ | 24 | Ω |
| $R_E$ | 19 | Ω |
| $r_E$ | 1.6 | Ω |
| $C_{BC}$ | 0.04 | pF |
| $C_{BE}$ | 0.25 | pF |
| $\tau_{EC}$ | 0.4 | ps |
| $\tau_B$ | 4.0 | ps |
| $\tau_C$ | 1.2 | ps |
| $\tau_{CC}$ | 1.8 | ps |
| $\tau_{BC}$ | 0.9 | ps |

Oki A K, Kim M F, Gorman G M and Camou J B 1988 *IEEE Trans. Micr. Theory Tech.* **36**, 1958

Parton J G, Tophai J, Golder M J, Taylor D G, Hiams N A, Holden A, Goodfellow R C 1989 *IEEE GaAs IC Symposium* **11**, 61

Figure 5. HBT equivalent circuit         Table 2. HBT characteristics

## 4. Discussion

The analysis of the HBT's suggests how to proceed towards higher cutoff frequencies. It is clear that the base transport time, $\tau_B$, is the largest single limitation of the cutoff frequency. This is due to diffusion-induced broadening of the base. However the wide base can be partly compensated for through the use of a compositional grading of AlGaAs. The Al-gradient creates a built-in electric field and thereby improved electron transport through the base. The electron velocity is proportional to the built-in field and a larger gradient would reduce the transit time through the base.

In a device with uniform base the current gain decreases when the emitter area is scaled down. This emitter size effect can be suppressed by the electric field (Nakajima 1985). Due to the strong dependence of gain on emitter size, shown in figure 3, devices with the smallest area have to low gain to be high frequency characterized. The thus required relatively large area contributes to the high $\tau_{CC}$-value.

The built-in field will be improved in the future and together with optimized spacerwidth and better contacts believed to result in $f_{max}$ and $f_T$ values of more than 60 and 40 GHz, respectively.

## 5. Conclusion

We have succesfully produced and characterized AlGaAs/GaAs HBT's grown by MOVPE. A maximum frequency of oscillation, $f_{max}$, of more than 30 GHz has been obtained. We have also outlined future device optimization expected to improve the performance to more than 60 GHz. This frequency indicates sufficient gain for integrated circuits in the 15 to 20 GHz range and since MOVPE offers low defect density and high throughput, HBT's may be a viable alternative to GaAs-FET's and Si-technology for many high frequency applications of current interest.

### Acknowledgment

We would like to thank Nils Nordell and Claes Olsson for performing the epitaxial growth.

### References

Asbeck P M, Chang M-C F, Higgins J A, Sheng N H, Sullivan G J and Wang K-C 1989 *IEEE Electron. Dev. Lett.* **10**, 2032
Bayraktaroglu B, Camilleri N and Lambert S A 1988 *IEEE Trans. Microwave Theory Tech.* **36**, 1869
Bhat R, Hayes J R, Colas E and Esagui R 1988 *IEEE Electron. Dev. Lett.* **9**, 442
Enquist P M, Hutchby J A, Chang M F, Asbeck P M, Sheng N H and Higgins J A 1989 *Electron. Lett.* **25**, 1124
Ito H, Kobayashi T and Ishibashi T 1989 *Electron. Lett.* **19**, 1303
Kuriyama Y, Morizuka K, Akagi J, Asaka M, Tsuda K and Obara M 1989 *IEEE GaAs IC Symposium* **11**, 313
Nakajima O, Nagata K, Ito H and Ishibashi T 1985 *Jpn. J. Appl. Phys.* **24**, 1368
Nordell N, Ojala P, van Berlo W H, Linnarsson M K and Landgren G 1990 *J. Appl. Phys.* **67**, 778
Oki A K, Kim M E, Gorman G M and Camou J B 1988 *IEEE Trans. Microwave Theory Tech.* **36**, 1958
Parton J G, Topham P J, Golder M J, Taylor D G, Hiams N A, Holden A J and Goodfellow R C 1989 *IEEE GaAs IC Symposium* **11**, 61

Paper presented at ESSDERC 90, Nottingham, September 1990
Session 6C1

# A Monte Carlo simulator including generation recombination processes

Lino Reggiani, Tilmann Kuhn, Luca Varani

Dipartimento di Fisica e Centro Interuniversitario di Struttura della Materia, Universita' di Modena, Via Campi 213/A, 41100 Modena, Italy

Daniel Gasquet, Jean Claude Vaissiere, Jean Pierre Nougier

Centre d'Electronique de Montpellier, Université des Sciences et Techniques du Languedoc, 34095 Montpellier Cedex 5, France

> *Abstract.* We present an advanced Monte Carlo code (SIHOLE90) which is applied to p-Si and includes generation recombination processes from shallow impurity levels, impact ionization from neutral impurities and the Poole-Frenkel effect. The very good agreement obtained between calculations and experiments supports the physical reliability of the code which should provide useful information for device modeling.

1. *Introduction*

The fast development in submicron devices has emphasized the importance of high field transport properties of semiconductors (hot-carrier regime). In this respect, the Monte Carlo (MC) simulation is recognized as the most accurate method to provide a microscopic interpretation of the physical processes governing the performance of the final device. This communication presents an advanced MC simulator (code name SIHOLE90) which, in addition to transport in the conducting band, also includes the processes of Generation and Recombination (GR) from shallow impurity levels as well as impact ionization from neutral impurities. In this way we are in the position to control the change in carrier concentration when, because of carrier freeze-out, we are in the presence of a field-assisted ionization. As application we will present a comparison between theoretical results and experiments for the case of p-Si in the temperature range $77 - 160\ K$ and with acceptor concentrations $10^{14} \leq N_A \leq 10^{17}\ cm^{-3}$ for electric fields ranging from Ohmic up to $10^4\ V/cm$. The program runs on a workstation and each simulation requires between $0.5 - 3$ hours of cpu-time on a DECstation 3100 (Digital).

2. *Theoretical model*

We consider a uniform semiconductor sample of cross-sectional area $A$ and length $L$ in which charge transport occurs through a two level system: the valence band and the impurity centers which supply the carriers. Under stationary conditions, the total current $I$ flowing in the sample can be expressed by the two equivalent expressions (Kuhn et al, 1990):

$$I = eAN_A v_d^r = eAN_A u v_d \qquad (1)$$

where $v_d^r$ is the total carrier mean-drift velocity (reduced drift velocity) which accounts for the time spent by the carriers on the impurities, the so called trapping time, $u$ is the fraction of ionized carriers and $v_d$ the mean-drift velocity associated with the carrier motion in the conducting band. We introduce into a standard MC procedure the GR and impact ionization processes as additional scattering mechanisms. The Poole-Frenkel effect is considered for both

© 1990 IOP Publishing Ltd

generation and recombination process. Carrier-carrier interaction is neglected. The parameters entering the simulation are summarized in Table 1. Then, the fraction of ionized carriers $u$ is determined from the ratio between the total time spent in the conducting band and the total time of the simulation. In addition, through the standard stationary algorithm (Jacoboni et al, 1983) we have evaluated both the reduced mean-drift velocity $v_d^r$ and the correlation function of the reduced velocity fluctuations
$< \delta v^r(0) \delta v^r(t) >$ with $\delta v^r(t) = v^r(t) - v_d^r$. From the knowledge of the above quantities we can determine the following transport parameters, which can be directly obtained from experiments: The free carrier concentration $n$, the conductivity $\sigma$, the free carrier mobility $\mu$, the current spectral density at frequency $f$, $S_I(f)$, and the static diffusion coefficient $D$ which are defined as:

$$n = N_A u ; \quad \sigma = \frac{e N_A v_d^r}{E} ; \quad \mu = \frac{v_d^r}{uE} \quad (2)$$

$$S_I(f) = \frac{2e^2 A N_A}{L} \int_{-\infty}^{\infty} exp(i 2\pi f t) < \delta v^r(0) \delta v^r(t) > dt ; \quad D = \frac{L}{4e^2 A N_A} S_I(0) \quad (3)$$

### 3. *Results and conclusions*

We apply the above scheme to the case of lightly doped (Boron) p-type Si, where a direct comparison with experimental results can be carrier out. In the following we consider the case of uncompensated samples, even if the code is also adequate for the compensated case.

Figure 1 shows the values for the fraction of ionized impurities at equilibrium, $u_{eq}$, as function of the acceptor concentration as obtained from MC calculations and compares them with the results given by statistics and experiments (Reggiani et al, 1989).

Figure 2 shows the Ohmic conductivity as a function of the acceptor concentration. The sublinear behavior exhibited by the conductivity is mainly attributed to a decrease in the fraction of ionized impurities at increasing acceptor concentrations. Moreover, this effect is enhanced by the decrease in the mobility associated with the increased efficiency of the ionized impurity scattering. Figure 3 reports the field dependent conductivity for the case of the samples oriented along the $< 100 >$ crystallographic direction. The systematic decreases of the conductivity with the field reflects the prevailence of the decrease in mobility over the increase in free carrier concentration.

TABLE 1 - Parameters used in calculations (Kuhn et al, 1990)

| | |
|---|---|
| effective mass | $m_h = 0.53 \div 1.26\ m_0$ |
| crystal density | $\rho_0 = 2.32\ g\ cm^{-3}$ |
| sound velocity | $s = 6.53 \times 10^5\ cm\ s^{-1}$ |
| optical phonon temperature | $\theta_{op} = 735\ K$ |
| relative static dielectric constant | $\epsilon_0 = 11.7$ |
| acoustic deformation potential | $E_1^0 = 5\ eV$ |
| optical deformation potential | $D_t K = 6 \times 10^8\ eV\ cm^{-1}$ |
| equilibrium volume recombination rate | $\rho_{eq} = 4.2 \times 10^{-6}\ cm^3 s^{-1}$ |
| equilibrium generation rate | $\gamma_{eq} = 2.9 \times 10^9\ s^{-1}$ |
| energy of the acceptor level | $\epsilon_a = 45\ meV$ |
| cross-section for impact ionization | $\sigma_g = 5.02 \times 10^{-14}\ cm^2$ |

These are well known effects associated with a hot-carrier regime (Reggiani, 1985). Indeed, from the MC simulation we have evaluated separately the dependence on the field of the free

carrier concentration and of their mobility. The concentration increases owing to the raise of
the carrier mean energy. Because of that, the low energy population in the carrier distribution
function, which is responsible for capture processes, decreases, thus lowering the average recombination rate at the given field with respect to its equilibrium value. As a consequence, we
have found an enhancement of the free carriers lifetime which in turn leads to a net increase
in their concentration. Concerning the mobility we find a systematic decrease from its Ohmic
value due to an increase of the scattering efficency with increasing fields. At higher doping
levels the effect is smoothed because of the lowering of the Ohmic value which displaces at
higher fields the onset of the hot-electron regime. At the highest fields the mobility becomes
practically independent from the acceptor concentration because of the Coulomb nature of the
ionized impurity scattering which implies a vanishing efficiency of this mechanism at increasing
carrier energies.

Figure 4 reports the field dependent conductivity at increasing temperatures. As a general
feature, at increasing field strengths the conductivity decreases systematically from its Ohmic
value. However, the threshold field for such a decrease increases with temperature. This reflects
the lowering of the Ohmic mobility at increasing temperature which displaces the onset of the
hot-carrier regime to higher fields.

Figure 5 shows the current spectral density $S_I(f)$ as a function of the frequency for an
electric field $E = 2.5 \ kV/cm$. The longitudinal component exhibits two plateau regions associated with the presence of two basic contribution given by GR and Nyquist noise sources while
the transverse component exhibits the Nyquist contribution only. Furthermore, we observe
the different relaxation times associated with the longitudinal and transverse component of
the average velocity of free carriers from the difference in the correspondening high frequency
cut-off values.

Figure 6 reports the longitudinal spectral density at low frequency as a function of the
electric field. Its systematic increase is associated with the presence of a strong GR contribution
which, together with a cross-term (Reggiani et al, 1988), adds to the Nyquist noise-source. We
remark that from Eq. (3) the static diffusion coefficient $D$, associated with Fick's law, can be
directly obtained.

In conclusion, we have presented an advanced Monte Carlo simulator (SIHOLE90) which
satisfactorily interprets transport parameters of p-Si in a wide range of electric fields, temperatures and doping concentration. The agreement found between theory and experiments
ranges from a few per cent up to 50 % in the worst case (second order coefficients like noise
data). In consideration of the lack of arbitrary parameters, this agreement is considered to be
quite satisfactory and, as a consequence, this code should provide useful physical information
to researchers working in the field of submicron electronic engineering.

This work is partially supported by the CEE ESPRIT II BRA 3017 project, the Italian National
Research Council, and the Centro di Calcolo dell'Universita' di Modena.

## References

Jacoboni C and Reggiani L 1983 Rev. Mod. Phys. **55** 645
Kuhn T, Reggiani L, Varani L and Mitin V 1990 Phys. Rev. **B15** to be published
Reggiani L 1985 *Hot Electron Transport in Semiconductors* Topics in Applied Physics,
    Vol. 58 Springer-Verlag (Berlin-Heidelberg)
Reggiani L, Lugli P and Mitin V 1988 Phys. Rev. Lett. **60** 736
Reggiani L, Varani L, Vaissiere J C, Nougier J P and Mitin V 1989 J. Appl. Phys. **66**, 5404

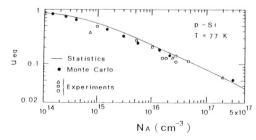

Fig. 1 - Fraction of ionized impurities at equilibrium as a function of the acceptor concentration. The line is obtained from statistics, full circles refer to MC simulation, other symbols to experiments (Reggiani et al, 1989).

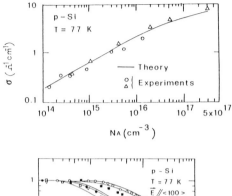

Fig. 2 - Low field (Ohmic) conductivity as a function of the acceptor concentration. Symbols refer to experiments and the line to MC simulation.

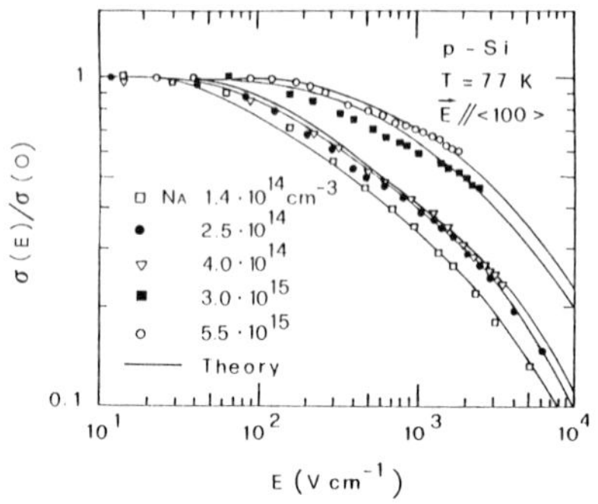

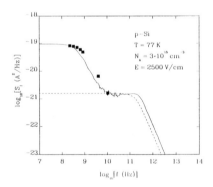

Fig. 3 - Conductivity normalized to its Ohmic value as a function of the electric field for the different acceptor concentrations reported. Symbols refer to experiments, the lines to MC simulations.

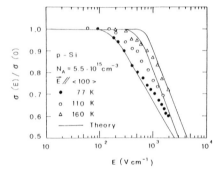

Fig. 4 - Conductivity normalized to its Ohmic value as a function of the electric field for the different temperatures reported. Symbols refer to experiments, the lines to MC simulations

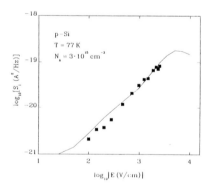

Fig. 5 - Current spectral density as a function of frequency for a sample length $L = 1.5 \times 10^{-2}$ $cm$ and a sample area $A = 3.0 \times 10^{-3} cm^2$. The solid and dot-dashed lines refer to the MC calculations for the longitudinal and transverse spectral-densities. Symbols refer to experiments.

Fig. 6 - Current spectral density at low frequency in the direction of the electric field as a function of field for a sample length $L = 1.5 \times 10^{-2} cm$ and a sample area $A = 3.0 \times 10^{-3} cm^2$. The curve reports MC calculations and symbols refer to experiments.

Paper presented at ESSDERC 90, Nottingham, September 1990
Session 6C2

# Treatment of thermomagnetic effects in semiconductor device modelling

S. Rudin and H. Baltes

Physical Electronics Laboratory, Institute of Quantum Electronics, ETH Zürich,
CH-8093 Zürich, Switzerland

**Abstract.** We present the three-dimensional vector equations of carrier transport by drift, diffusion and thermal diffusion in a small magnetic field. The equations were derived in the framework of Onsager's principles of irreversible thermodynamics.

## 1. Introduction

The galvanomagnetic and thermomagnetic effects must be accounted for in equations used to model e.g. the temperature distribution in a magnetic sensor or the effects of a magnetic field on thermoelectric sensors. In general, these effects are described by tensors of material coefficients defining the relationship between the electrical and thermal current and potential and temperature gradient vectors. We assume isotropic materials, which allows the tensors to be reduced to scalars. In this note, we consider only electrons and use the following coefficients:

The *isothermal heat conductivity* $\kappa_i$ describes the heat current density carried by the electrons caused by a temperature gradient. The *absolute thermoelectric power* $\epsilon$ describes the potential gradient caused by a temperature gradient. The *isothermal electric conductivity* $\sigma_{i0}$ describes the electric current caused by a potential gradient (the index 0 refers to zero magnetic field).

The *isothermal Hall coefficient* $R_i$ characterizes the potential gradient caused by action of the magnetic induction on the electric current. The *Righi-Leduc coefficient* $L$ characterizes the temperature gradient caused by the deflection of carriers flowing 'down a (perpendicular) temperature gradient'. The *isothermal Nernst coefficient* $\eta_i$ characterizes the potential gradient caused by the deflection of carriers flowing 'down a temperature gradient'.

According to Onsager's principles of irreversible thermodynamics, only six coefficients are needed, others can be calculated from these (e.g. the one describing the temperature gradient caused by deflection of carriers flowing 'down a potential gradient').

## 2. Relation to previous work

Figure 1 illustrates our model in comparison with three previous models. Shown are two-dimensional semiconductor samples with the temperature(s) plotted vertically. Please note that all models are actually three-dimensional and some include holes (omitted here for simplicity).

© 1990 IOP Publishing Ltd

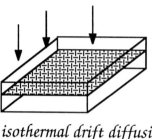
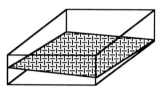

*isothermal drift diffusion*

$T_{el} = T_{la}$ = uniform
Magnetic Field

*Wachutka*

$T_{el} = T_{la}$ = nonuniform

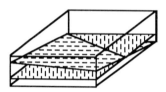
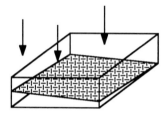

*hydrodynamic model*

$T_{la}$ = uniform
$T_{el}$ = nonuniform

*this paper*

$T_{el} = T_{la}$ = nonuniform
Magnetic Field

*Figure 1*

The hydrodynamic model (see e.g. Rudan and Odeh (1986)) is the only one allowing for hot carrier effects, all others assume electron ($T_{el}$) and lattice ($T_{la}$) temperature to be equal. Previous models accounting for magnetic effects (see e.g. Baltes and Nathan (1989)) did this for uniform temperature throughout the sample; thermomagnetic effects were therefore ignored.

### 3. Form of the equations and approximate coefficients

In the formalism of Onsager's theory, the electrical and thermal currents form a vector dependent on a second vector consisting of the potential and thermal gradients. A 6-by-6 matrix with linear coefficients as elements - factors of the six coefficients given above - describes this relationship. By combining what is known from previous models given by Baltes and Nathan (1989), Callen (1960) and Wachutka (1988), the linear coefficients can be approximated.

### 4. Results

Generalization of the two-dimensional model given by Callen (1960) to three dimensions is not straightforward, but requires the following assumption about the structure of the final equations: The Lorentz force is a cross product; this implies that any equations derived from it will also have such cross products. Assuming this, we are able to complete the linear

## Physical processes in silicon device structures

coefficients and arrange them in the correct order. The result of this synthesis is the following set of equations for n-type semiconductors:

*Electric current*

$$\vec{J}_n = -\sigma_{i0} \vec{\nabla}\phi_n + \sigma_{i0}^2 R_i [\vec{B} \times \vec{\nabla}\phi_n] - \sigma_{i0}^3 R_i^2 [\vec{B} \times (\vec{B} \times \vec{\nabla}\phi_n)]$$
$$+ \sigma_{i0} \epsilon \vec{\nabla}T + (\sigma_{i0} \eta_i - \sigma_{i0}^2 \epsilon R_i)[\vec{B} \times \vec{\nabla}T] + (\sigma_{i0}^3 \epsilon R_i^2 - \sigma_{i0}^2 \eta_i R_i)[\vec{B} \times (\vec{B} \times \vec{\nabla}T)] \quad (1)$$

*Heat current*

$$\vec{Q} = \sigma_{i0} \epsilon T \vec{\nabla}\phi_n + (\sigma_{i0} T \eta_i - \sigma_{i0}^2 \epsilon T R_i)[\vec{B} \times \vec{\nabla}\phi_n]$$
$$+ (\sigma_{i0}^3 \epsilon T R_i^2 - \sigma_{i0}^2 T \eta_i R_i)[\vec{B} \times (\vec{B} \times \vec{\nabla}\phi_n)]$$
$$- (\kappa_{tot} + \sigma_{i0} \epsilon^2 T) \vec{\nabla}T + (\kappa_i \mathcal{L} - 2\sigma_{i0} \epsilon T \eta_i + \sigma_{i0}^2 \epsilon^2 T R_i)[\vec{B} \times \vec{\nabla}T]$$
$$+ (\sigma_{i0}^2 \kappa_i R_i^2 + 2\sigma_{i0}^2 \epsilon T \eta_i R_i - \sigma_{i0}^3 \epsilon^2 T R_i^2 - \sigma_{i0} T \eta_i^2)[\vec{B} \times (\vec{B} \times \vec{\nabla}T)] \quad . \quad (2)$$

where $\kappa_{tot}$ is the sum of the isothermal heat conductivities for the electrons and the lattice.

The *heat conduction equation* becomes (with $c_{tot}$ the sum of the heat capacities for the electrons and the lattice)

$$c_{tot} \frac{\partial T}{\partial t} = \text{div}\left[\kappa_{tot} \vec{\nabla}T - \kappa_i \mathcal{L}(\vec{B} \times \vec{\nabla}T)\right] + H \quad , \quad (3)$$

with the heat generation term,

$$H = H_{\neg B} + H_B \quad , \quad (4)$$

split into one term not influenced by the magnetic field, (5)

$$H_{\neg B} = \frac{1}{\sigma_{i0}}|\vec{J}_{n\neg B}|^2 + q(R - G)\left[T\left(\frac{\partial \phi_n}{\partial T}\right)_n - \phi_n\right] - T\left[\left(\frac{\partial \phi_n}{\partial T}\right)_n - \epsilon\right] \text{div} \vec{J}_n + T(\vec{J}_n \cdot \vec{\nabla}\epsilon) \quad ,$$

and one term existing only in a magnetic field,

$$H_B = \vec{\nabla} \cdot \left[T \eta_i \left\{ \{\vec{B} \times \vec{J}_{n\neg B}\} - \sigma_{i0} R_i [\vec{B} \times (\vec{B} \times \vec{J}_{n\neg B})]\right\}\right] + \\ + \vec{\nabla} \cdot \left[(\sigma_{i0} T \eta_i^2 - \sigma_{i0}^2 \kappa_i R_i^2)[\vec{B} \times (\vec{B} \times \vec{\nabla}T)]\right] \quad , \quad (6)$$

using the undisturbed electric current,

$$\vec{J}_{n \to B} \equiv -\sigma_{i0}\vec{\nabla}\phi_n + \sigma_{i0}\,\epsilon\,\vec{\nabla}T \ . \tag{7}$$

In equation (5) the reader will easily identify Joule heat, generation-recombination heat, Thomson and Peltier heat. The complete model of carrier transport finally consists of the equations (1-7), the continuity equation for electron flow and Poisson's equation.

For further insight into equations (1&2), we apply them (reduced to zero and first order terms) to a few simple cases. In the limit of zero magnetic induction the coefficients for isothermal electric conductivity, isothermal heat conductivity and absolute thermoelectric power come out correctly.

Next we consider magnetic effects with a simple plate geometry: Gradients and currents are in the x or y direction, the magnetic induction in the z direction only. The isothermal Hall effect becomes

$$\frac{-\nabla_y \phi_n}{B_z J_x} = \frac{R_i}{1 + \sigma_{i0}^2 R_i^2 B_z^2}\ , \tag{8}$$

which is very close to $R_i$ for small magnetic fields. The same is true for the isothermal Nernst effect:

$$\frac{\nabla_y \phi_n}{B_z \nabla_x T} = \frac{\eta_i}{1 + \sigma_{i0}^2 R_i^2 B_z^2}\ . \tag{9}$$

The result for the Righi-Leduc effect is somewhat more involved,

$$\frac{\nabla_y T}{B_z \nabla_x T} = L - \frac{\sigma_i^2\,T\,\eta_i^2\,R_i\,B_z^2 + L\,\sigma_{i0}\,T\,\eta_i^2\,B_z^2}{\kappa_i + \kappa_i\,\sigma_{i0}^2\,R_i^2\,B_z^2 + \sigma_{i0}\,T\,\eta_i^2\,B_z^2}\ , \tag{10}$$

but the available data given by Madelung (1964) and Smith (1964) show that the last term on the right hand side of equation (10) can be neglected for small fields.

**References**

Baltes H and Nathan A 1989 *Sensors Vol. 1* ed T Grandke and W H Ko (Weinheim, Germany:VCH Publ.) pp. 45-77

Callen H B 1960 *Thermodynamics* (New York:Wiley) pp. 305-307

Madelung O 1964 *Physics of III-V compounds* (New York:Wiley) pp. 104-220

Smith R A 1964 *Semiconductors* (Cambridge: Cambridge University Press) p. 170

Rudan M and Odeh F 1986 *COMPEL* 5 pp. 149-183

Wachutka G 1988 *Simulation of Semiconductor Devices and Processes Vol. 3* ed G Baccarani and M Rudan (Bologna, Italy: Tecnoprint) pp. 83-95

## On the modelling of mobility in silicon MOS transistors

A. Emrani, G. Ghibaudo and F. Balestra
Laboratoire de Physique des Composants à Semiconducteurs,
ENSERG, 23 rue des martyrs, B.P. 257, 38016 Grenoble, France.

**Abstract** : An original method for the extraction of the depletion charge mobility dependency coefficient $\alpha$ based on the combined exploitation of the body and gate transconductance MOSFET characteristics is presented. This method enables to show that $\alpha$ is not a constant but is a parameter strongly dependent on channel length and on device type (P or N).

### 1. INTRODUCTION

It is generally admitted that the effective mobility in a MOSFET operated in the linear region is a function of an effective transverse electric field $E_{eff}$ as

$$\mu_{eff} = \mu_0/(1 + E_{eff}/E_c) \qquad (1),$$

where $\mu_0$ is a low electric field mobility and $E_c$ a critical field (Sabnis and Clemens 1979, Sun and Plummer 1980, Lin 1985, Krutsick et al 1987). In this relation, the effective electric field is assumed to be both a function of the inversion charge $Q_i$ and of the depletion charge $Q_d$ such as :

$$E_{eff} = (Q_i + \alpha Q_d)/2\epsilon_{si} \qquad (2),$$

with $\epsilon_{si}$ being the silicon permittivity and $\alpha$ a constant which is taken <u>a priori</u> equal to 2 or found equal to 2 and 3 for N and P type devices, respectively (Hairapetian et al 1989).

In this work, an original method for the extraction of $\alpha$ is proposed. This method, which is based on the combined exploitation of the body and gate transconductance characteristics, is applied to obtain the values of $\alpha$ in P and N type MOSFETs with channel lengths from 0.75 $\mu$m to 25 $\mu$m.

### 2. PRINCIPLE OF THE METHOD

As the drain current in ohmic operation reads, $I_d = (W/L)Q_i\mu_{eff}V_d$ ($V_d$ being the drain voltage and W/L the width to channel length ratio), it is easy to prove from (1) and (2) that, at strong inversion where $Q_i \simeq C_{ox}(V_g-V_t)$, one has :

$$I_d = G_m \frac{(V_g - V_t)}{1 + \theta(V_g - V_t^*)} \qquad (3),$$

where $G_m = (W/L)\mu_0 C_{ox} V_d$ is the transconductance parameter, $V_g$ is the gate voltage, $V_t$ is the charge threshold voltage, $\theta$ is the mobility reduction factor ($\theta=C_{ox}/(2\epsilon_{si}E_c)$), $C_{ox}$ is the gate oxide capacitance and $V_t^*=V_t-\alpha Q_d/C_{ox}$. The gate transconductance, $g_m=dI_d/dV_g$, is then obtained as :

$$g_m = G_m \frac{1 + \theta(V_t-V_t^*)}{[1 + \theta(V_g-V_t^*)]^2} \qquad (4).$$

Similarly the body transconductance, $g_b=dI_d/dV_b$ ($V_b$ being the substrate voltage), can be derived from (3) as :

$$g_b = \frac{C_d}{C_{ox}} G_m \frac{1 + \alpha\theta(V_g-V_t) + \theta(V_t-V_t^*)}{[1 + \theta(V_g-V_t^*)]^2} \qquad (5),$$

where $C_d$ is the depletion charge capacitance.

Finally, combining (4) and (5) enables the body to gate transconductance ratio to be expressed as :

$$\frac{g_b}{g_m} = \frac{C_d}{C_{ox}} [1 + \frac{\alpha\theta(V_g-V_t)}{1 + \theta(V_t-V_t^*)}] \qquad (6).$$

As the surface potential is nearly constant at strong inversion, $C_d$ may almost be independent of gate voltage and, as a result, $g_b/g_m$ may vary linearly with gate voltage with a slope proportional to the depletion charge mobility dependency coefficient $\alpha$.

It is worth mentioning that in practical cases, as $(V_t-V_t^*)$ is of the order of 0.1-0.2 V and $\theta \leq 0.1$ /V, the quantity $\theta(V_t-V_t^*)$ in (4) and (6) is much smaller than 1. Therefore, this quantity can be neglected in (4) and (6) so that the function $Y(V_g)=I_d/\sqrt{g_m} \simeq \sqrt{G_m}(V_g-V_t)$ can advantageously be employed to extract the values of $V_t$ and $\mu_0$ (Ghibaudo 1988).

## 3. RESULTS AND DISCUSSION

The above method of $\alpha$ extraction has been tested on conventional P and N type silicon MOSFETs fabricated by SGS-Thomson and having a gate oxide thickness of 28 nm, a channel doping of 1-2x10$^{16}$/cm$^3$, a gate width W=25$\mu$m and a gate length ranging between 0.75 and 25 $\mu$m. The $g_b/g_m(V_g)$ characteristics were deduced after differentiation of a set of $I_d(V_g,V_d,V_b)$ data given by a semiconductor parameter analyzer HP 4145 B.

As a preliminary test, it was checked using the parameter extraction method by Ghibaudo (1988) that $\mu_0$ and $\theta$ are independent of substrate bias $V_b$ and, in turn, of $Q_d$, thereby demonstrating that the effective mobility may only be a function of $Q_d$ through its effective electric field dependence.

Fig. 1 displays a typical plot of the $g_b/g_m(V_g)$ characteristics as obtained on P type MOSFETs with various gate lengths. Similar plots have also been observed for N type devices. As can be seen from Fig. 1, a linear variation of $g_b/g_m$ with gate voltage is observed, in agreement with (6), above threshold. As suggested from (6) and exemplified in Fig. 1, the value of $C_d/C_{ox}$ for each gate length can be deduced from the extrapolation of the linear part of $g_b/g_m(V_g)$ at threshold. The values of $C_d/C_{ox}$ found by this treatment have been plotted as a function of gate length for both device types (see Fig. 2).

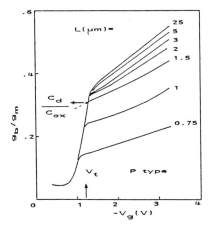

Fig. 1 : Typical variations of the body to gate transconductance ratio $g_b/g_m$ with gate voltage $V_g$ for various gate lengths as obtained on P type devices.

The corresponding values of $\alpha$ have then been derived from the slope of the $g_b/g_m(V_g)$ characteristics above threshold using the $C_d/C_{ox}$ data previously determined and the mobility reduction factors $\theta$ extracted separately for each gate length with Ghibaudo's method (1988).

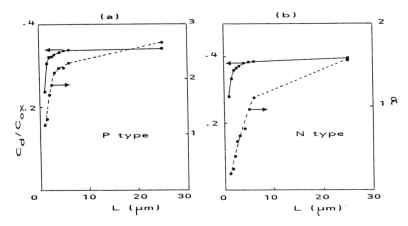

Fig. 2 : Variations of $C_d/C_{ox}$ and $\alpha$ with gate length for (a) P type and (b) N type of device.

This figure clearly shows that the depletion charge mobility dependency coefficient $\alpha$ is not a constant with gate length and is well different from the usually admitted value $\alpha=2$. In contrast, $\alpha$ is found to be very gate length dependent and ranges between 2.6 and 1.1, and, 1.6 and 0.2, for P and N types, respectively. The correlation between the behavior of $\alpha$ with gate length and that of $C_d/C_{ox}$ demonstrates the reduction with gate length due to the charge sharing effect (Hafez et al 1990). In this case, the depletion charge in (2) must be replaced by an effective depletion charge $Q_d^*$ which may account for charge sharing, so that $\alpha$ may read :

$$\alpha = \alpha_0 \, (Q_d^*/Q_d) \qquad (7),$$

where $\alpha_0$ and $Q_d$ refer to the long channel limit values. In this classic charge sharing approach (Sze 1981), $\alpha$ is expected to vary linearly with the inverse of the effective channel length. As shown in Fig. 3, this tendency is in fact relatively well verified for both types of device and especially for the P type. The long channel limits of $\alpha$ deduced from the intercept on the y axis, which are 2.75 and 1.8 for P and N types, respectively, do not any more coincide with the a priori assumed value of $\alpha$ in (2), and are close to values measured on 375$\mu$m long devices using a split current technique (Hairapetian et al 1989).

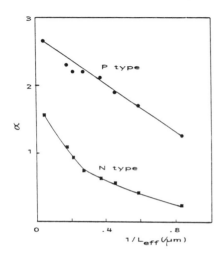

Fig. 3 : Variations of $\alpha$ with the inverse of the effective channel length $1/L_{eff}$ for P and N types of device.

## 4. CONCLUSION

An original method for the extraction of the depletion charge mobility dependency coefficient $\alpha$ based on the combined exploitation of the body and gate MOSFET characteristics has been presented. This method allowed us to show that $\alpha$ is a parameter strongly dependent on channel length and on device type (P or N). It has also been shown that the channel length reduction of $\alpha$ is well correlated to that of $C_d/C_{ox}$ due to the increase of charge sharing when scaling down the devices. Finally, it is worth noting that $\alpha$ cannot be considered as a universal constant but must be regarded as supplementary MOSFET parameter which needs to be extracted for each channel length in order a proper modeling of the effective mobility to be obtained in a MOS transistor .

## REFERENCES
Sabnis A G and Clemens J T 1979 IEDM Technical Digest pp. 18-21.
Sun S C and Plummer J D 1980 IEEE Trans Electron Devices **ED-27** 1497.
Lin M S 1985 IEEE Trans Electron Devices **ED-32** 700.
Krutsick T J, White M, Wong H S, Booth R V 1987 IEEE Trans Electron Devices **ED-34** 1676.
Hairapetian A, Gitlin D, Viswanathan C R 1989 IEEE Trans Electron Devices **ED-36** 1448.
Ghibaudo G 1988 Electronics Letters **24** 543.
Hafez I M, Ghibaudo G, Balestra F 1990 IEEE Electron Device Letters **EDL-11** 120.
Sze S M 1981 Physics of semiconductor devices (New York:Wiley) p. 474.

# The hysteresis behaviour of silicon p-n diodes at liquid helium temperature

B. Dierickx, E. Simoen, L. Deferm and C. Claeys,
IMEC, Kapeldreef 75, B-3030 Leuven, Belgium.

<u>Abstract.</u> *The turn-on behaviour of a Si p-n junction at 4.2 K is dominated by the small injection barrier existing at the "ohmic" contact. The hysteresis typically observed is explained by the build-up of space charge in the lowly doped (frozen out) parts of the junction.*

## 1. Introduction

The behaviour of silicon p-n junctions at 4.2 K has been studied since the early days of semiconductor electronics [Jonscher 1961]. Anomalous forward current/voltage (I/V) characteristics, as in Fig.1, have been reported Jonscher (1961), Maddox (1976), Yang (1984) and Coon (1985), and were explained by different mechanisms.

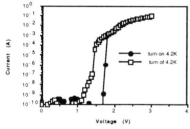

Fig.1. Forward characteristic of a $n^+$pwell diode at 4.2K.

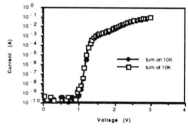

Fig.2. Forward characteristic of a $n^+$pwell diode at 10K.

These anomalies disappear for sufficiently high temperature (T > 10 K), as in Fig.2. Above this temperature, doped contacts to the n- or p-region represent no potential barrier (as current is injected by thermionic emission across the small homojunction barrier Δ), and current starts and stops at a forward voltage that is almost equal to the built-in potential (1.15 V at 4.2 K) [Yang 1984]. The experimentally observed higher turn-on voltage and hysteresis of the Si p-n junction at 4.2 K deserves qualitative modeling. An expression for turn-on and turn-off bias $V_{on}$, resp. $V_{off}$, will be derived, as a function of material and device parameters. Analogous "hysteresis" behaviour is observed in other device structures; it is also essential for understanding of the anomalous CMOS latch-up and thyristor operation observed at liquid helium temperature [Deferm 1990].

## 2. p-n diodes at LHT

The potential and energy band diagram of a n+p⁻p+ diode is represented in fig. 3. At zero bias and thermal equilibrium, we are in a situation highly similar to room temperature, as sketched in (1), with the exception that the lowly doped part is frozen out, and the Fermi energy $E_F$ is close to the band edges.

The bandgap narrowing in the degenerate n+ and p+ regions are somewhat exaggerated. Now when a forward bias is applied to the diode, in first instance no current flows, and the potential is capacitively distributed over the frozen out region (2). By the increasing the electrical field F, the p⁻p+ homojunction barrier starts lowering, until the point where a few free holes tunnel through it, and race to the potential maximum near the diode junction (3). After a finite time, depending on the bias sweep rate, the potential bucket becomes filled, which in turn lowers the junction potential step to the point where holes spill over it, or electrons are injected. Actually, the tunneling of holes is enhanced by the advent of the negative charges of the electrons, or by the creation of ionized acceptor space charge in the p-region.

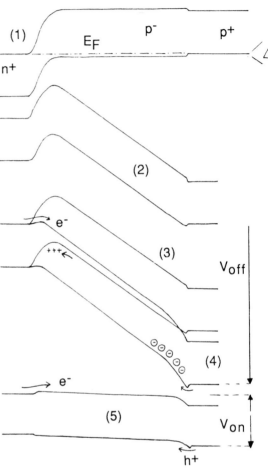

Fig. 3. Schematic representation of bandgap across n+p-p+ diode. Situations (1) till (5) explained in text.

This last process occurs via shallow impact ionization of holes (or electrons) with neutral, frozen-out acceptors. The process is self-catalyzing, so that nearly instantaneously enough free holes are available to fill the potential dip at the diode junction (4). This junction becomes conducting (injection of both holes and electrons) and a high bipolar current density is seen. The electric field in the lowly doped region is virtually constant; the fixed negative acceptor space charge is outnumbered by free electrons and holes.

Reversely, by lowering the diode bias, this bipolar current will not cease, thus explaining the hysteresis of the forward breakdown. At some lower electric field however, the ionized acceptors in the p-region, begin to freeze-out again and trap holes. The current totally ceases when the field somewhere in the p-region drops to zero (5). Due to the possibly remaining fixed space charge in the p-region, and the consequent potential curvature, the turn-off voltage can be somewhat larger than the expected built-in potential.

### 3. Quantitative formulation of $V_{on}$ and $V_{off}$

Starting from the previous explanation, simple expressions can be found for the turn-on voltage. The condition for turn-on is found from tunneling of free holes over the barrier $\Delta$.

In the 10 K - 30 K temperature range, the forward current-voltage (I-V) characteristic of a Si p-n junction (actually a p-i-n junction ), is described by Yang in 1984 based on thermionic emission of carriers across the small potential barrier $\Delta$ existing at a "ohmic" $n^-n^+$ (or $p^-p^+$) contact. The onset of current flow is expected at the built-in potential $V_{bi}$ of about 1.15 V (see also Fig.2) and ideally, no voltage dependence of the forward current is expected, as given by the thermionic equation:

$$J = AT^2 \exp(-\frac{\Delta}{kT}) \qquad (1)$$

with J the current density, A Richardson's and k Boltzmann's constant.

At 4.2 K however, the available thermal energy is not high enough to inject the carriers across the barrier. By applying an electric field the barrier has a more-or-less triangular shape. The quantummechanical tunneling through the triangular barrier is modeled by Kao and Hwang (1981),

$$J = b_1 F^2 \exp(-\frac{b_2 \Delta^{1.5}}{F}) \qquad (2)$$

With F the electric field beyond the barrier, and $b_1$ and $b_2$ material constants. For silicon one finds a very steep threshold field for tunneling $F_{tun}$ in the order of 500000 V/m (corresponding to a typical $\Delta$ of 20 meV). Assuming a homogeneous distribution of the potential drop over the frozen-out p-region, the $V_{on}$-expression becomes

$$V_{on} = V_{bi} + L.F_{tun} - f(\text{space charge}) \qquad (3)$$

$V_{on}$ scales with the distance between the contacts (or with the length L of the frozen-out 'neutral' material). The unknown space charge term accounts for the fact that pre-existing (fixed) space charge can cause barrier lowering at the tunneling contact. The importance of the term is clear from the fact that the hysteresis opening is not exactly reproducible, e.g. it depends on the voltages sweep rate or on the past history of the device. The expression is obviously similar to the turn-on behaviour of a silicon resistor at 4.2 K [Simoen 1990]. For the turn-off voltage one finds:

$$V_{off} = V_{bi} + f(\text{space charge}) \quad (4)$$

Where the undetermined space charge term accounts for the remaining potential drop, due to (fixed...) space charge in the lowly doped region of the diode at the moment where the current is cut off in the injecting junctions. Also this turn-off voltage experimentally is not neatly reproducible. For very long diodes (...100 µ...), electrons and holes recombine, so that in a significant part of the diode the condition for the silicon resistor regime is present. Then, according to [Simoen 1990], the minimal electric field needed for hole conduction is the field threshold for hole-acceptor shallow impact ionization $F_{ii}$, and thus:

$$V_{off} = V_{bi} + L \cdot F_{ii} \quad (5)$$

## 4. Implications for other devices

The anomalous forward behaviour of a Si p-n junction (and of a resistor) is, as clearly demonstrated above, related to the fact that the degenerately doped contact at cryogenic temperatures gives rise to an injection barrier. Once carriers can tunnel through the barrier, shallow-level impact ionization induced material breakdown will occur. These two phenomena will for instance determine the punch-through (breakdown) behaviour of a MOST [Simoen, Vanstraelen] and, as demonstrated in [Deferm 1990] are essential for understanding latch-up at 4.2 K. To a first approximation, the firing of a parasitic four-layer structure (thyristor) can be understood by considering the band-diagrams of Fig.3, mirrored around the lowly doped inner junction.

As shown in [Simoen 1990], the contact barrier is reduced for a scaled down technology, probably because of the higher implantation doses used and the sharper doping profiles obtained.

## References.

D.D. Coon and S.D. Gunapala 1985, J. Appl. Phys. 57, 5525-5528.
D.D. Coon, R.P. Devaty, A.G.U. Perera and R.E. Sherriff 1989, Appl. Phys. Lett. 55, 1738-1740
L. Deferm, E. Simoen and B. Dierickx 1990, presented at the 1st Intern. Low Temperature Electronics Conference, Berkeley (Ca), 23-26 April and to be published in Cryogenics.
A.K. Jonscher 1961, Brit. J. Appl. Phys. 12, 363-371.
K.C.Kao & W.Hwang 1981,"Electrical transport in solids", ch.1, Pergamon Press, Oxford .
R.L. Maddox 1976, IEEE Trans. Electron Devices ED-21, 16-21.
E. Simoen, B. Dierickx, L. Deferm and C. Claeys 1990, presented at the 1st Intern. Low Temperature Electronics Conference, Berkeley (Ca), 23-26 April and to be published in Cryogenics.
E. Simoen, G. Vanstraelen and C. Claeys, paper submitted for publication in IEEE Electron Device Letters.
Y.N. Yang, D.D. Coon and P.F. Shepard 1984, Appl. Phys. Lett. 45, 752-754 .

# Recent trends in InP-based optoelectronic components

O. Hildebrand

SEL-ALCATEL, Research Centre, 7000 Stuttgart 40, FR Germany

## 1. Abstract
This paper gives a comprehensive review on the present status and recent trends in optoelectronic components related to optical fibre based system applications. Focus is laid on transmitting, amplifying and detecting devices including optoelectronic integrated circuits. Systems as well as technology aspects are addressed.

## 2. Introduction: System Trends Related to O/E Components
InP based optoelectronic components have undergone a tremendous number of modifications and diversifications over the last few years. This is due to the continuous need for increasing transmission capacity and the implementation of fibre optics in the subscriber area with much more versatile system architectures than just point-to-point fibre optic transmission.

Two principal ways are used to increase transmission capacity: In the **time domain**, very high speed digital modulation is applied and 16 Gb/s transmission was reported /1/. Speed limitations caused by electronics have been overcome by Optical Time Division Multiplexing (OTDM) techniques and 20 Gb/s transmission over 115 km has been achieved /2/. Trunk length limitations due to chromatic fibre dispersion can be overcome using lasers with reduced wavelength chirping to be discussed below. An interesting approach to minimize the optical spectral width is to use high speed direct frequency modulation of DFB lasers together with optical demodulation and direct detection /3/.

In the optical **frequency domain**, several optical channels are used to increase transmission capacity with direct intensity modulation (Wavelength Division Multiplexing, WDM) /4/, or with coherent modulation/detection schemes (Frequency Division Multiplexing, FDM). Frequency Shift Keying (FSK), Phase Shift Keying (PSK) or even more sophisticated modulation schemes are applied allowing for high sensitivity signal detection and very close channel spacing /5/.

For very long trunk links limited by fibre attenuation coherent detection was regarded as the appropriate choice due to the inherent high detection sensitivity. The recent enormous progress in optical amplifiers, however, has already led to very long links bridged with direct modulation /6/ and further improvements are expected in the near future.

For implementation of fibre optics in the subscriber area the key factor is the overall system cost per subscriber. System architectures under discussion are based on digital transmission with direct modulation even in the multigigabit per second range or with coherent multichannel (CMC) approaches. Optical analog transmission is currently widely discussed /7/, and it could be the lowest cost solution since it uses existing analog equipment at the subscriber (e.g. CATV equipment). Specialized O/E components are needed: Low noise, high linearity and low harmonic distortion are new key requirements for analog lasers.

Independent of transmission scheme a major part of system cost per subscriber is due to the lasers. Therefore, a key towards cost reduction is to share one laser for many subscribers: high optical power is needed, and again the enormous progress in optical amplifiers may offer an attractive solution.

So far, application of optoelectronic components in telecommunication systems and networks is related to transmission only, whereas switching has been the clear domain for electronics. Progress in O/E components with new functions such as optical bistablity, space and wavelength switching has opened prospects towards future telecommunication systems and networks including photonic switching and signal processing /8/.

The systems trends sketched above do not only require further optimization, cost reduction and specialization of "single function" O/E and E/O-converters, but they also require the integration of further functions widely realized to date in hybrid approaches as used in system experiments. In the following, components related status and trends are addressed with focus on transmission applications.

## 3. Laser Devices

Since the invention of semiconductor lasers there has been a large number of different laser structures indicating no clear way towards an "optimum" technology. In the early years Liquid Phase Epitaxy (LPE) was the standard technology not only for the active layer sequence but also for the critical blocking layers regrowth process of buried heterostructure (BH) lasers. Today, there is a clear trend to use large area wafer technologies such as Metal Organic Vapour Phase Epitaxy (MOVPE) /9/ or Gas Source Molecular Beam Epitaxy (GSMBE) /10/, and laser structures are increasingly specialized for system applications.

### 3.1 High Speed Lasers

Very low total chip capacitance is a prerequisite for high speed lasers. Therefore, the conventional lateral current blocking by pn-junctions is no longer appropriate. New approaches are, e.g., undercut mesa structures /11,12/ or the PIQ-BH laser structure using polyimide for isolation /13/. A very promising approach is to replace the pn-junctions by semiinsulating InP material which can be grown either by MOCVD /9/ or by Hydride Vapour Phase Epitaxy (VPE) /14/. The simplicity of the structure makes it potentially low cost; critical undercut etching or exact positioning of blocking pn junctions is not needed. Parasitic capacitances are very low, typically less than 3 pF without additional means such as trench isolation, making this structure very well suited for high speed applications. In addition the lateral electrical isolation makes it ideally suited for OEICs. Several modifications were reported for this basic structure: Semi-Insulating Buried Heterostructure (SIBH) /15/, Etched Mesa Buried Heterostructure (EMBH) /16/, or with additional p-type cap layer, Capped Mesa Buried Heterostructure (CMBH) /16/ or SIPBH /17/.

A 3 dB roll-off frequency of 18 GHz has been reported for SIBH lasers /19/, only exceeded by a world record value of 22 GHz produced with Vapour Phase Regrown (VPR-) BH-lasers with submicron active and burying layer widths /12/. Under direct intensity modulation SIBH lasers have shown suitability up to 10 Gb/s. Further, high output power exceeding 40 mW and very promising stress test data were obtained. /18/.

Apart from residual parasitics induced by the lateral structure the performance of lasers is determined from the active layer design. Recent trends clearly go towards incorporation of Multiple Quantum Well (MQW) structures that allow to improve many relevant laser parameters mainly caused by the higher differential gain of QW material compared to bulk material. Extremely precise control of epitaxial parameters is required for MQW growth and has been demonstrated for MOCVD and for GSMBE technology. Very high optical output power of 95 mW/facet at 1.55 um was reported for all-MOVPE grown Graded Index Separate Confinement Heterostructure (GRINSCH) - MQW lasers /20/. Single mode emission up to 50 mW together with a narrow static linewidth of 1.3 MHz have been obtained with GSMBE grown 1.5 um DFB-MQW lasers, and the differential gain increase by a factor of 3 was observed /21/. Chirp reduction by a factor of 2.5 to 3 was also observed /21,22/, giving important impact for high speed lasers. Extremely low threshold current of 2.5 mA was reported for a MQW-DFB-SIBH laser /23/, allowing for 5 Gb/s direct modulation even with zero bias which simplifies the driving and control electronics.

Several system relevant laser parameters are improved by MQW technology, and especially for high speed direct modulation MQW-DFB lasers with reduced parasitics are considered as the appropriate approach.

### 3.2 Analog Lasers

Lasers for analog data transmission have to meet specific requirements that have been of minor concern for digital systems. For the AM modulation scheme the most stringent requirements are low intensity noise of the laser light, and high linearity of the light-current characteristics in order to keep harmonic and intermodulation distortions low /7/.

Inherently, lasers are nonlinear, and quantum-effects in the electron-to-photon conversion produce noise. In practical structures main origin of nonlinearities is due to DC and AC currents bypassing the active laser stripe, so that careful optimization of laser regrowth technology is a must. In addition, careful optimization of the active layer stripe is needed to minimize intrinsic nonlinearities and noise effects such as mode partitioning, etc.

Although AM transmission experiments have been successfully performed with selected DFB lasers, effort has still to be made towards analog lasers with optimized and reproducible performance.

### 3.3 Wavelength Tunable Lasers

Tunability of emission wavelength is an essential feature required in multichannel WDM and coherent FDM systems. It has been postulated theoretically that wavelength tunable lasers need at least three separately controllable functions in order to obtain wavelength/frequency tuning at constant output power, linewidth and sidemode suppression /24/. Usually this is accomplished by laser structures integrating three sections for gain, frequency and phase control in a linear arrangement. Optimization of the segment structures depends on system requirements: Three electrode DFB lasers with identical segment structures but nonuniform current injection are ideally suited for FSK modulation e.g. for coherent multichannel (CMC) networks. Such devices show excellent FM response sensitivity (0.7 GHz/mA), high bandwidth (> 1 GHz) and continuous tuning range of 2 nm, while maintaining a narrow linewidth of 20 MHz /25/. Recently, the first four channel 565 Mb/s FSK transmission experiment was reported and -44 dBm receiver sensitivity was achieved using these components /25/.

Very narrow linewidth below 1 MHz and a continuous tuning range of several tens of GHz is required e.g. for DPSK heterodyne detection at 565 Mb/s. Although a static linewidth of only 250 kHz can be achieved with MQW-DFB lasers /15/, large tuning efficiency is contradictory to small linewidth for monolithically integrated tunable lasers /26/. The smallest usable linewidth obtained so far within 1.6 nm tuning range was 890 kHz /27/.

A new approach towards integrated tunable lasers with only one single control current is the transversely integrated Tunable Twin-Guide (TTG) - laser. More than 3 nm continuous tuning with linewidth below 60 MHz was recently reported /28/. An additional interesting feature is the low device length down to 200 $\mu$m which increases the number of chips per wafer and thus decreases manufacturing cost.

A novel integrated Y-shaped laser structure was especially designed for very large tuning range /29,30/. This structure overcomes the tuning range limitation inherent for DFB/DBR- approaches, and it also circumvents the problem of fermi level pinning in conventional linear arrangements. For frequency filtering it applies the interferometric principle; it represents (half a) Mach-Zehnder Interferometer with four active sections which can be separately controlled. All segments have the same standard SIBH cross section and no DFB/DBR grating is needed.

Frequency tuning over the enormous range of 22 nm was achieved and any desired wavelength within this range is addressable with additional fine tuning of approximately 1.5 nm. Sidemode suppression of > 20 dB and linewidth < 35 MHz have been achieved with first samples /30/.

## 4. Optical Amplifiers

Enormous progress was recently achieved in fibre optic transmission experiments employing Er-doped optical fibre amplifiers /31/. This new kind of optical amplifier offers several advantages such as high gain, low noise and crosstalk, low insertion loss and polarization insensitivity. High power semiconductor pump lasers are needed with an optimum pump wavelength around 980 nm, or, with lower efficiency, at 1480 nm. The short wavelength requires strained layer InGaAs/GaAs QW pump lasers with high power chip-to-fibre coupling and sufficient reliability which still has to be proven. 1480 nm high power pump lasers can be realized using InP based technology derived from 1550 nm lasers; the use of MQW active layers is appropriate for higher power output. Due to the close vicinity of pump and signal wavelength special means have to be taken to filter out the pump wavelength, and the noise level is higher compared to 980 nm pump wavelength.

Depending on system requirements, semiconductor Travelling Wave Amplifiers (TWAs) might be the appropriate choice: for example, the signal wavelength can be chosen either in the 1300 nm or 1550 nm range, the amplification bandwidth is rather wide, and the technology is compatible to lasers and detectors which also allows to realize monolithic integrated OEICs. Major problems with semiconductor laser amplifiers have been the polarization dependence of gain, and gain ripple caused by residual facet reflectivity. Recent progress has overcome these problems: as an example, polarization-insensitive InGaAsP/InP TWAs realized from GSMBE wafers were reported with an average internal gain as high as 28 dB at only 50 mA pump current. TE/TM gain difference was only 1 dB and gain ripple was negligible./32/.

## 5. Photodetectors

PIN-detectors and Avalanche Photodetectors (APDs) are in production in many companies. Trends are mainly to improve the high speed performance of these devices. For PIN-detectors a bandwidth exceeding 10 GHz can be easily achieved. A standard mesa PIN diode with a bandwidth of 14 GHz mounted in a high speed 50 Ohm impedance matched pigtailed package has been used to detect directly modulated signals up to 20 Gb/s /2/. For even higher speed, small area detectors are used and special optimization has to be employed not to degrade the sensitivity. The highest bandwidth reported so far is 25-26 GHz /33/, and even 60 GHz have been achieved, however, with low quantum efficiency only /34/.

The progress in optical amplifiers has raised discussions about the future needs for APDs. However, optical amplifiers cannot replace APDs for all system requirements. As an example, APDs obviously have very low power consumption and they are much cheaper than semiconductor amplifiers (which need high precision chip-fibre coupling), or fibre amplifiers (which need an expensive pump laser).

Trends in APDs clearly go towards optimization of high speed performance and reliability, on the basis of the well known SAGM structure. LPE technology is replaced by MOVPE /35/, trichloride Vapour Phase Epitaxy (VPE) /36,37/, or Chemical Beam Epitaxy (CBE) /38/. For high speed APDs, optimization of layer structure is very critical in terms of layer thickness (tradeoff between carrier transit time and quantum efficiency), exact positioning of the pn-junction and quality of the InGaAs/InP heterointerface including multiple InGaAsP "grading" layers to avoid hole pile-up effects. A second major task is to realize efficient guard rings with low additional capacitance and dark current.

Gain bandwidth products of APDs have exceeded 70 GHz, with the highest value of 86 GHz which was recently reported for planar VPE grown APDs with a double guard ring formed by ion implantation /36/. Multigigabit operation of APD's was demonstrated up to 10 Gb/s /39,40/.

It is expected that still further optimization of SAGM APDs can lead to gain bandwidth products up to 100 GHz or slightly above. For even higher speed probably new approaches have to be found.

## 6. Optoelectronic Integrated Circuits (OEICs)

Future optoelectronic integrated circuits will not only perform optoelectronic conversion, but also optical and electronic signal processing on the same chip. Expectations of OEIC development are reduced cost and increased reliablility due to high on-chip functionality with minimum number of bond connections and increased performance by low parasitics, and new optical functions.

Two approaches are followed in parallel towards future OEIC's. One is the monolithic integration of lasers or detectors with InP based electronics, and the other is the integration of optical (or photonic) functions.

### 6.1 Photonic Integrated Circuits (PICs)

Photonic integrated circuits, like multisegment/multisection and interferometric lasers were already discussed in this paper. Since lasers need a monitoring photodiode to control the laser output, the integration of laser and detector is obvious. Wafer testable SIBH-laser/monitor OEICs have been realized using the same layer sequence for both components /41,42/. Dry etching technology was applied for the laser mirrors and the tilted monitor facet which avoids optical backreflection.

Remarkable results have been obtained in monolithic integration of WDM transmitters. Recently, a WDM PIC was reported including four wavelength tunable 2 Gbit/s lasers, a passive waveguide combiner and an optical amplifier /43/. Not less remarkable is the realization of first optical tuner PICs for coherent detection, comprising a tunable local oscillator laser, waveguide coupler and two balanced PIN detectors /44,45/. Although function of these devices was demonstrated, performance is still poor compared to hybrid solutions.

Concerning high speed transmitters, monolithic integration of cw operated DFB lasers with electroabsorption modulators was reported /39,46,47,/, and a very low chirp of only 0.1 Å was achieved at 10 Gb/s modulation with 17 mW optical output power /46/. Transmission of 10 Gb/s over 65 km without power penalty was reported using such a laser/modulator IC /39/.

PICs for OTDM applications have only been realized in first steps. As an example, a monolithic mode locked laser producing 4.4 ps pulses at 40 GHz repetition rate was reported /48/, and optical space switches have been realized in different approaches /49-51/.

Future trends in PICs are expected not only towards transmission but also towards switching applications. Optical bistable devices, space and wavelength switches will be key components.

### 6.2 Electronic Integration: Transmitter OEICs

A specific problem related especially to high speed lasers is the matching of low laser chip impedance to the electronic outside world. Hybrid approaches using Si- or GaAs driving ICs within the laser package are limited in performance due to the needed bond connection between IC and laser /52/.

Early reports on transmitter OEICs used LPE based laser structures integrated with up to three HBTs /53,54/ or InP-MESFETs /55/. The first laser/driver/monitor combination was reported in 1986 /55/, using the LPE based DCPBH laser/monitor structure and InP MESFETs.

Recent approaches rely on large area wafer growth technologies such as MOVPE. Laser/FET combinations were realized with active laser layers grown on top of the FET layers. Complex 'gate projection processes' were used to define the FET gates /56/. Self Aligned Constricted Mesa (SACM) laser OEICs with one InP MESFET were operated up to 5 Gb/s /56/. Even higher modulation up to 11 Gb/s was reported for a DFB-SACM-laser integrated with an InGaAs/InAlAs-MODFET /57/.

HBTs might be more appropriate for laser/driver OEICs because of their superior switching and current driving characteristics. An approach towards an SIBH-laser with InP/InGaAs/InP-DHBTs was reported /58/.

Transmitter OEICs combining an InP based laser with three GaAs-MESFETs have been realized by heteroepitaxy using LPE and MBE /59/.

Although multigigabit opertion of transmitter OEICs has been demonstrated, complexity is still very low, and performance or cost advantages compared to hybrid approaches have not been reported: a lot of technology work still has to be done.

### 6.3 Electronic Integration: Receiver OEICs

The performance of an optical receiver essentially is determined by the properties of the photodetector and the first preamplifier stage. Receiver bandwidth and sensitivity are limited by parasitic capacitances. Thus, monolithic integration aims at low parasitic capacitance, and high transconductance input transistors are required with low leakage current.

Much work has been done on InP based receiver OEICs, mostly using PIN detectors and InGaAs JFET's. With a total of eighteen components, a PIN-JFET OEIC was recently reported with a receiver sensitivity of - 28 dBm at 1.2 Gb/s /60/.

Only recently the In(Ga)AlAs compound has reached a high material quality, and it offers the possibility to use high barrier Schottky contact technology for InP based components. High Electron Mobility Transistors (HEMTs) were realized from InAlAs/InGaAs/InP layers, and very high transconductance ($g_m$ > 1000 mS) and bandwidth exceeding 350 GHz were reported /61/. Based on this new technology, PIN-InAlAs/InGaAs HEMT OEICs were realized /62,63/ and the yet highest OEIC receiver sensitivity of - 30.4 dBm at 1.2 Gb/s was achieved /63/.

An MSM-HEMT OEIC with fourteen integrated components was recently reported to operate at 5 Gb/s /64/.

Depending on device parameters, at very high bit rates there might be an advantage in the use of HBTs instead of FETs. InP/InGaAs PIN-HBT OEICs have reached -26 dBm at 1 Gb/s using three HBTs and five NiCr resistors /65/.

Heteroepitaxial approaches with InGaAs-PIN/GaAs-MESFETs have reached -26 dBm at 1.2 Gb/s /59/.

Although much progress has been achieved in the last years, the best reported receiver OEICs are still worse in sensitivity by several dB, as compared to present hybrid or flip-chip receiver circuits, and cost advantages are still to be demonstrated.

### 7. Conclusion

Trends in InP based optoelectronic components are closely related to trends in fibre optic transmission systems. The continuous need for ever increasing transmission capacity and the low cost demand for subscriber systems have stimulated components optimization and specialization. Much progress has been made towards large area wafer technologies such as Metal Organic Vapour Phase Epitaxy (MOVPE) or Gas Source Molecular Beam Epitaxy (GSMBE) replacing the well established Liquid Phase Epitaxy (LPE) which is limited in flexibility, precision and usable wafer area. This technology progress in turn has initiated the realization of new device structures with improved performance, even allowing for new system concepts such as analog transmission.

Much progress has been achieved in technologies for optoelectronic circuits. In general, however, OEICs have not yet reached competitive performance and costs, compared to hybrid approaches as used in systems experiments. For realization of future OEICs with high functionality including optical signal processing a lot of work still has to be done especially concerning compatible integration technologies, reproducibility and yield.

## 8. References

/1/ A. H. Gnauck, R.M. Jopson, J.D. Evankow, C.A. Burrus, S.-J. Wang, N.K. Dutta, H.M. Presby: 1 Tbit/skm Transmission Experiment at 16 Gbit/s using conventional Fibre. Electron. Lett. 25, 1695-1696 (1989)

/2/ B. Wedding, Th. Pfeiffer: 20 Gb/s optical pattern generation, amplification and 115 km fibre propagation using optical time division multiplexing. To be published in Proc. ECOC '90

/3/ R.S. Vodhanel, A.F. Elrefaie: Performance of direct frequency modulation DFB lasers in multigigabit per second ASK, FSK, and DPSK lightwave systems. Techn.Digest OFC '90, paper TUI6

/4/ H. Taga, Y. Yoshida, N. Edagawa, S. Yamamoto, H. Wakabayashi: 459 km, 2.4 Gbit/s 4 Wavelength Multiplexing Optical Fiber Transmission Experiment using 6 Er doped Fiber Amplifiers. Techn. Digest OFC '90, paper PD9

/5/ Y. Yamamoto, T. Kimura: Coherent Optical Fiber Transmission Systems. IEEE J. Quantum Electron. QE-17, 919-934 (1981)

/6/ N. Edagawa, Y. Yoshida, H. Taga, S. Yamamoto, K. Mochizuki, H. Wakabayashi: 904 km, 1.2 Gbit/s Non-regenerative Optical Transmission Experiment using 12 Er-doped Fibre Amplifiers. Proc. ECOC '89, paper PDA-8

/7/ Th.E. Darcie, J. Lipson, Ch.B. Roxlo, C.J. McGrath: Fiber Optic Device Technology for Broadband Analog Video Systems. IEEE LCS Magazine 1 (1), 46-52 (1990)

/8/ H.S. Hinton: Architectural Considerations for Photonic Switching Networks. IEEE J. Selected Areas in Communications 6, 1209-1226 (1988)

/9/ P. Speier, K. Wünstel, F.J. Tegude: MOVPE Studies for the Development of InGaAsP/InP Lasers with Semi-Insulating InP Blocking Layers. Electron. Lett. 23, 1363-1365 (1987)

/10/ B. Fernier, C. Artigue, D. Bonnevie, L. Goldstein, A. Perales, J. Benoit: Low-threshold 1.5 $\mu$m DFB Laser Grown by GSMBE. Electron. Lett. 25, 768-769 (1989)

/11/ Y. Hirayama, H. Furuyama, M. Morinaga, N. Suzuki, T. Nishibe, K. Eguchi, M. Nakamura: High speed 1.5 $\mu$m self-aligned constricted mesa DFB lasers grown by MOCVD with highly doped p-sheet layers. Techn. Digest OFC '89, paper TUC5

/12/ R. Olshansky, W. Powazinik, P. Hill, V. Lanzisera, R.B. Lauer: InGaAsP Buried Heterostructure Laser with 22 GHz Bandwidth and High Modulation Efficiency. Electron. Lett 23, 839-841 (1987)

/13/ K. Uomi, H. Nakano, N. Chinone: Ultrahigh Speed 1.55 $\mu$m $\lambda/4$-Shifted DFB PIQ-BH Lasers with Bandwidth of 17 GHz. Electron. Lett. 25, 668-669 (1989)

/14/ Y. Koizumi, S. Sugou, N. Kuroda, S. Asada, T. Uji: High efficiency and high power Fe-doped InP embedded 1.3 $\mu$m Lasers with reduced double injection leakage current. Proc. ECOC '89, paper TUB 10-2

/15/ H.Yamazaki, T.Sasaki, N.Kida, M.Kitamura, I.Mito: 250 kHz Linewidth Operation in Long Cavity 1.5 $\mu$m Multiple Quantum Well DFB-LDs with Reduced Linewidth Enhancement Factor. Techn. Digest OFC '90, paper PD33

/16/ N.K. Dutta, A.B. Piccirilli: Observation of anomalous far-field intensity distributions in semiconductor lases and their explanation, J.Appl. Phys. 66 (10), 4621-4624 (1989)

/17/ C.E.Zah, C.Caneau, S.G.Menocal, A.S.Gozdz, P.S.D.Lin, F.Favire, A.Yi-Yan, T.P.Lee, A.G.Dentai, C.H.Joyner: Performance of 1.5 $\mu$m $\lambda/4$ shifted DFB-SIPBH Laser Diodes with Electron Beam Defined and Reactive Ion Etched Gratings, Electron. Lett. 25, 650-651 (1989)

/18/ N.K.Dutta, S.J.Wang, A.B.Piccirilli, R.F.Karlicek Jr., R.L.Brown, M.Washington, U.K.Chakrabarti, A. Gnauck: Wide-bandwidth and high-power InGaAsP distributed feed-back lasers, J. Appl. Phys. 66 (10), 4640-4644 (1989)

/19/ R. Weinmann, J. Bouayad-Amine, M. Klenk, U. Koerner, H.P. Mayer, F. Schuler, H. Schweizer, P. Speier, K. Wünstel, E. Zielinski: InGaAsP/InP-laser with semi-insulting current blocking layers for ultra high speed applications. Post deadline paper at "2nd Int. Conf. on InP and Related Compounds", Denver (1990)

/20/ D.M.Cooper, C.P.Seltzer, M.Aylett, D.J.Elton, M.Harlow, H.Wickes, D.L.Murrell: High-power 1.5 $\mu$m All-MOVPE Buried Heterostructure Graded Index Separate Confinement Multiple Quantum Well Lasers. Electron. Lett. 25, 1635-1636 (1989)

/21/ A.Perales, L.Goldstein, A.Accard, B.Fernier, F.Leblond, C.Gourdain, P.Brosson: High performance DFB-MQW-Lasers at 1.5 $\mu$m grown by GSMBE. Proc. ECOC '89, Paper PDB-6

/22/ S. Kakimoto, Y. Nakajima, Y. Sakakibara, H. Watanabe, A. Takemoto, N. Yoshida: Narrow spectrum linewidth and low chirp of 1.5 $\mu$m InGaAs MQW-DFB-PPIBH laser diodes. Proc. ECOC '89, paper TUB10-4

/23/ T. Sasaki, N. Henmi, H. Yamazaki, H. Yamada, M.Yamaguchi, M. Kitamura, I. Mito: A 2.5 mA threshold current operation and 5 Gbit/s zero-bias current modulation of 1.5 $\mu$m MQW-DFB laser diodes. Techn. Digest OFC '90, paper FE 1

/24/ L.A. Coldren, S.W. Corzine:Continuously Tunable Single Frequency Semiconductor Lasers. IEEE J. Quantum Electron. QE-23, 903 (1987)

/25/ N.Flaaronning, J.O.Frorud, M.Solom, G.Vendome, G.DaLoura, J.M.Gabriagues, J.Jacquet, D.Leclerc, J.Benoit: Multichannel FSK Transmission Experiment at 565 Mbit/s using Tunable Three-Electrode DFB Lasers. To be published in Electron Lett.

/26/ M.-C.Amann, R.Schimpe: Excess Linewidth Broadening in Wavelength-Tunable Laser Diodes. Electron. Lett. 26 279-280 (1990)

/27/ Y.Kotaki, T.Fujii, S.Ogita, M.Matsuda, H.Ishikawa: Narrow Linewidth and Wavelength Tunable Multiple Quantum Well $\lambda$/4 Shifted Distributed Feedback Laser. Techn. Digest OFC '90, paper THE 3

/28/ M.C.Amann, C.F.Schanen, S.Illek, H.Lang, W.Thulke: 1.55 $\mu$m Tunable Twin-Guide Laser with Large Continuous Tuning Range and Narrow Spectral Linewidth. Proc. ECOC '89,paper PDB-1

/29/ M.Schilling, H.Schweizer, K.Dütting, W.Idler, E.Kühn, A.Nowitzki, K.Wünstel: Widely Tunable Y-coupled cavity integrated interferometric injection laser. Electron. Lett. 26, 243-244 (1990)

/30/ W. Idler, M. Schilling, H. Schweizer, E. Kühn, G. Laube, K. Wünstel, O. Hildebrand: High speed integrated interferometric injection laser with 22 nm tuning range. To be published in Proc. ECOC '90

/31/ E.Desurvire, C.R.Giles, J.L.Zyskind, J.R.Simpson, P.C.Becker, N. Anders Olsson: Recent advances in erbium doped fiber amplifiers at 1.5 $\mu$m, Techn. Digest OFC '90, paper FA1

/32/ B.Mersali, G.Gelly, A.Accard, J.L.Lafragette, P.Doussiere, M.Lambert, B.Fernier: 1.55 $\mu$m High Gain Polarisation-Insensitive Semiconductor Travelling Wave Amplifier with Low Driving Current. Electron Lett. 26, 124-125 (1990)

/33/ D. Wake, R.H. Walling, I.D. Henning, D.G. Parker: Planar-Junction, Top-Iluminated GaInAs/InP pin Photodiode with Bandwidth of 25 GHz. Electron. Lett. 25 967-969 (1989)

/34/ J.E. Bowers, C.A. Burrus, F. Mitschke: Millimetre-Waveguide-Mounted InGaAs Photodetectors. Electron. Letters 22, 633-635 (1986)

/35/ P.M. Rodgers, M.D.A. MacBean, R.H. Walling, L. Davis, M.J. Robertson: InGaAs/InP Planar Avalanche Photodiodes with Separate Absorption and Grading Regions Grown by Atmospheric Pressure MOVPE. Proc. SIOE '88

/36/ J.N. Hollenhorst: Fabrication and performance of high speed InGaAs APDs. Techn. Digest OFC '90, paper THB6

/37/ Y.Liu, S.R.Forrest, V.S.Ban, U.M.Woodruff, J.Colosi, G.C.Ericson, M.J.Lange, G.H.Olsen: Simple, very low dark current, planar long-wavelength avalanche photodiode. Appl. Phys. Lett. 53, 1311-1313 (1988)

/38/ J.C.Campbell, W.T.Tsang, G.J.Qua, B.C.Johnson, J.E.Bowers: InP/InGaAsP/InGaAs avalanche photodiodes with 70 GHz gain-bandwidth product grown by chemical beam epitaxy. Techn. Digest OFC '88, paper TUC3

/39/ T.Okiyama, I.Yokota, H.Nishimoto, K.Hironishi, T.Horimatsu, T.Touge, H.Soda: A 10 Gb/s, 65 km optical fiber transmission experiment using a monolithic electro-absorption Modulator/DFB laser light source. Proc. ECOC '89, paper MoA1-3

/40/ T.Torikai, K.Makita, S.Fujita, H.Iwasaki, K.Kobayashi: Small area planar InGaAs avalanche photodiode with 7.5 GHz wide bandwidth. Techn. Digest OFC '88, paper TUC5

/41/ K. Wünstel, K. Dütting, J. Bouayad-Amine, W. Idler: InGaAsP/InP-Laser mit monolithisch integrierter Rücklichtdiode. ITG-Fachbericht 112, VDE-Verlag Berlin, 145-149 (1990)

/42/ H. Saito, Y. Noguchi: GaInAsP/InP Laser with monolithically integrated monitoring photodiode fabricated by inclined reactive ion etching. Electron. Lett. 25, 719-720 (1989)

/43/ A.H. Gnauck, U. Koren, T.L. Koch, F.S. Choa, G. Raybon, C.A. Burrus, G. Eisenstein: Four-Channel WDM Transmission Experiment Using a Photonic-Integrated-Circuit Transmitter. Techn. Dig. OFC'90, paper PD261

/44/ H. Takeuchi, K. Kasaya, Y. Kondo, H. Yasaka, K. Oe, Y. Imamura: Monolithic Integrated Coherent Receiver on InP Substrate. IEEE Photon. Technol. Lett.1, 398-400 (1989)

/45/ T.L. Koch, U. Koren, R.P. Gnall, F.S. Choa, F. Hernandez-Gil, C.A. Burrus, M.G. Young, M. Oron, B.I. Miller: GaInAs/GaInAsP Multiple-Quantum-Well Integrated Heterodyne Receiver. Electron. Lett 25, 1621-1623 (1989)

/46/ H. Soda, M. Furutsu, K. Sato, N. Okazaki, S. Yamazaki, I. Yokota, T. Okiyama, H. Nishimoto, H. Ishikawa: High-power semi-insulating BH structure monolithic electro-absorption modulator/DFB laser light source operating at 10 Gbit/s. Proc. IOOC '89, paper 20 PDB-5

/47/ H. Tanaka, M. Suzuki, H. Taga, M. Usami, S. Yamamoto, Y. Matsushima: Five-gigabit/s performance of integrated light sources consisting of $\lambda/4$ shifted DFB laser and E-A modulator with S.I. InP BH structure. Proc. OFC'90, THI3

/48/ R.S. Tucker, U. Koren, G. Raybon, C.A. Burrus, B.I. Miller, T.L. Koch, G. Eisenstein: 40 GHz active mode-locking in a 1.5 $\mu$m monolithic extended-cavity laser, Electron. Lett. 25 (10), 621-622 (1989)

/49/ T. Kikugawa, K.G. Ravikumar, K. Shimomura, A. Izumi, K. Matsubara, Y. Miymoto, S. Arai, Y. Suematsu: Switching Operation in OMVPE Grown GaInAs/InP MQW Intersectional Optical Switch Structures. IEEE Photon. Technol. Lett. 1, 126-128 (1989)

/50/ I. Kotaka, K. Wakita, O. Mitomi, H. Asai, Y. Kawamura: High-Speed InGaAlAs/InAlAs Multiple Quantum Well Optical Modulators with Bandwidths in Excess of 20 GHz at 1.55 µm. IEEE Photon. Technol. Lett. 1, 100-101 (1989)

/51/ J.E. Zucker, K.L. Jones, M.G. Young, B.I. Miller, U. Koren: Compact directional coupler switches using quantum well electrorefraction. Appl. Phys. Lett. 55 2280-2282 (1989)

/52/ M. Nakamura, N. Suzuki, T. Ozeki: The Superiority of Optoelectronic Integration for High-Speed Laser Diode Modulation. IEEE J.Quantum Electr. QE-22, 822-826(1986)

/53/ J. Shibata, I. Nakao, Y. Sasai, S. Kimura, N. Hase H. Serizawa: Monolithic integration of an InGaAsP/InP Laserdiode with heterojunction bipolar transistors. Appl. Phys. Lett. 45, 191-193 (1984)

/54/ K. Kasahara, A. Suzuki, S. Fujita, Y.Inomoto, T.Terakado, M. Shikada: InGaAsP/InP long wavelength transmitter and receiver OEICs for high speed optical transmission systems. Proc. ECOC '86, 119-122

/55/ K. Kasahara, T. Terakado, A. Suzuki, and S. Murata: Monolithically integrated high-speed light source using 1.3 µm wavelength DFB-DCPBH laser. IEEE J. Lightwave Techn. LT-4, 906-912 (1986)

/56/ N. Suzuki, H. Furuyama, Y. Hirayama, M. Morinaga, K. Eguchi, M. Kushibe, M. Funamizu, M. Nakamura:High-Speed Operation of 1.5 µm GaInAsP/InP Optoelectronic Integrated Laser Drivers. Electron. Lett. 24, 467-468 (1988)

/57/ Y.H. Lo, P. Grabbe, J.L. Gimlett, R. Bhat, P.S.D. Lin, J.C. Young, A.S. Gozdz, M.A. Koza, T.P. Lee: A Very High Speed 1.5 µm DFB OEIC Transmitter Grown by OMVPE. Techn. Digest OFC'90, paper PD24-1

/58/ F.J. Tegude, C. Hache, E. Kühn, G. Laube, M. Schilling, F. Schuler, P. Speier: Herstellung und Charakterisierung einer monolithisch integrierten InGaAsP/InGaAs/-InP Laser/ Treiber-Kombination. ITG-Fachbericht 112, VDE-Verlag Berlin, 31-36 (1990)

/59/ Y. Inomoto, S. Fujita, T. Terakado, K. Kasahara, T.Suzuki, K. Asano, T. Torikai, T. Itoh, M. Shikada, A. Suzuki, K. Kobayashi: A 1.2 Gbit/s-52.5 km Optical Fiber Transmission Experiment using OEICs on GaAs-on-InP Heterostructures. Techn. Digest OFC'88, paper PD9-1

/60/ W.S. Lee, D.A.H. Spear, M.J. Agnew, S.W. Bland: A 1.2 Gbit/s fully integrated transimpedance optical receiver OEIC for 1.3 - 1.55 µm transmission systems. Techn. Digest. OFC '90, paper PD 28-1

/61/ P. Ho, P.C. Chao, K.H.G. Duh, A.A. Jabra, J.M.Ballingal, P.M.Smith: Extremely high gain low noise GaInAs/AlInAs HEMTs grown by MBE. IEDM Techn. Dig., 184 (1988)

/62/ H. Nobuhara, H. Hamaguchi, T.Fuji, O.Aoki, M. Makiuchi, O. Wada: Monolithic pin-HEMT receiver for long wavelength optical communications. Electron. Lett. 24, 1246-1248 (1988)

/63/ H. Yano, K. Aga, N. Shiga, G. Sasaki, H. Hayashi: Low noise current optoelectronic integrated receiver with an internal equalizer for Gbit/s long wavelength optical communications. Techn. Digest. OFC '90, paper WB1

/64/ G.-K. Chang, W.P. Hong, J.L. Gimlett, R. Bhat, C.K. Nguyen, G. Sasaki, J.C. Young: High-speed InAlAs/-InGaAs MSM-HEMT transimpedance amplifier for long wavelength receivers. Techn. Digest OFC'90, paper WB2

/65/ S. Chandrasekhar, B.C. Johnson, M. Bonnemason, E. Tokumitsu, A.H. Gnauck, A.G. Dentai, C.H. Joyner, J.S. Perino, G.J. Qua: Monolithically integrated InP/InGaAs pin/HBT transimpedance photoreceiver. Techn. Digest OFC '90, paper PD 27-1 (1990)

# A high-energy ion implanted BICMOS process with compatible EPROM structures

R C M Wijburg, G J Hemink, J Middelhoek and H Wallinga

University of Twente, subdepartment IC-technology and Electronics
P.O. box 217, 7500AE Enschede (NL)

Abstract. A $1.5\,\mu m$ high-energy ion implanted BiCMOS process is proposed. This process offers NMOS, PMOS, vertical npn transistors and VIPMOS-EPROM devices. The process structure is modular in order to achieve CMOS and bipolar transistors in uncompromised forms. Consequently, the N-well and collector regions are defined separately. Conflicting requirements for the collector doping profile have been optimized towards practical electrical device characteristics.

1. Introduction

In general BiCMOS processes employ epitaxy combined with buried layers to ensure low collector and well resistances. A promising alternative is offered by high-energy ion implantation. The high-energy collector or well implantation can be performed in a self-aligned way to the field oxide, thereby making area-consuming stopper implantations and buried layers superfluous. Besides giving process simplifications, ion implantation enables a good reproducibility by its precise control over the implanted dose and energy.

In a first proposal to use high energy ion implantation in BiCMOS processing, Volz and Blossfeld (1988) have reported a technology in which the N-well for the P-channel device is to be shared with the collector for the bipolar device. The N-well and collector, however, do have conflicting implantation requirements. Therefore, the N-well and collector implantations should be separated to reap the full benefit of the BiCMOS technology.

2. Process Structure

The developed BiCMOS process has a modular structure, which is settled in a CMOS, a bipolar and an EPROM part. A beneficial result, achieved by the modular approach, is that the process parts can be optimized individually without severely affecting other parts. The CMOS process, proposed by Stolmeijer (1986), forms the basic part. It covers the fabrication of n-channel and p-channel MOS devices. The bipolar part leads to the realization of a vertical npn transistor with polysilicon emitter. The EPROM part makes the application of VIPMOS-EPROM's (Wijburg et al. (1989)) for integrated circuit design possible. The acronym VIPMOS stands for Vertical Injection Punch-through-based MOS.

High-energy ion implantation is a key-technique for the realization of each of these three parts. A schematic cross-section of the process is shown in Figure 1. The retrograde N-well

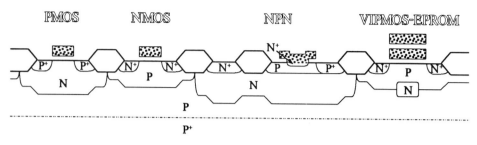

Fig. 1 *A schematic cross-section of the high-energy ion implanted BiCMOS process.*

and P-well are implanted with 1 MeV phosphorus and 350 keV boron ions, respectively. The injector of the VIPMOS-EPROM is made by a local overlap of the P-well and N-well. 1.5 MeV phosphorus ions are implanted to form the collector.

Featuring two polysilicon layers and one metal layer, the total number of masks used is 14. The realization of bipolar transistor involves three extra masks compared to the 10 masks for the basic process. Only one additional mask is needed for the VIPMOS-EPROM, in order to counter-dope trunks from the buried injector up to the surface (Wijburg et al. (1990)). A process flow is illustrated in Figure 2. The retrograde P-well and N-well are implanted after the LOCOS growth. A PT-masked (Preventing Trunk) boron implantation, covering all the injector edges in active area, is carried out to prevent short-circuits between the injector and channel. Then the gate oxide is grown. The first polysilicon layer provides the gates of normal CMOS devices as well as the floating gate of the EPROM. After the gate formation, the base and collector are implanted. Subsequently, an emitter window is opened in the gate-oxide. The polysilicon emitter and control gate for the EPROM are defined with a second polysilicon layer. An implanted collector plug ensures a low-ohmic connection between the collector and its metallic contact. The source and drain of the normal MOS devices employ DDD-structures. Because of the use of the Substrate Hot Electron (SHE) injection technique, the VIPMOS-EPROM can use the same advanced source and drain structures. The source/drain implantations for the PMOS device are used for the base contact, whereas the source/drain implantations for the NMOS device are employed for the collector contact.

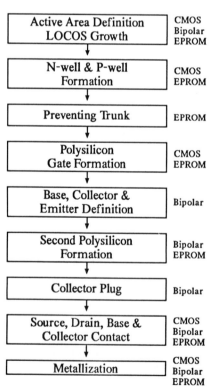

Fig. 2 *The process flow.*

## 3. Collector Optimization

Since even high-energy ion implantation covers a restricted depth, the use of a polysilicon emitter offers profitable space in the vertical direction. Notwithstanding the very shallow emitter-base junction, there are conflicting collector requirements. This is illustrated in Figure 3. It compares

Silicon circuits 517

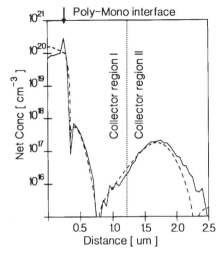

Fig. 3 Typical simulated (dashed) and measured (solid) 1D-doping profiles of the npn transistor.

the 1D-doping profile, as simulated with SUPREM III, and that extracted from SIMS measurements. The collector was realized by means of a 1.5 MeV phosphorus implantation. A dotted line separates the collector in two virtual regions.

On one hand the collector doping concentration in region I should be low because of parameters such as the collector-emitter breakdown voltage $V_{ceo}$, the collector-base junction capacitance $C_{cb}$, and the Early-voltage $V_A$. On the other hand the doping concentration in region II should be high, in order to achieve a low collector resistance $R_c$. Unfortunately, the doping concentration in the collector regions I and II cannot be chosen independently. The above-mentioned problem is aggravated by the demand of a sufficient high doping concentration underneath the LOCOS to avoid possible parasitic MOS-action. These issues are addressed in Figure 4. $V_{ce,imp}$ is defined as the collector-emitter voltage $V_{ce}$, at which the increased collector current, generated through impact ionization in the collector-base space charge region, causes a 25 % decrease of the base current. $V_{th}$ is the threshold voltage of parasitic MOS transistor underneath the 600 nm thick LOCOS.

Aiming at a 5 V supply voltage, it can be concluded from Figure 4 that the 1 MeV implantation hardly suffices. In the case of a 1.5 MeV implantation, the allowable collector dose ranges from about $2 \times 10^{13}$ to $5 \times 10^{13}$ cm$^{-2}$. The lower limit is determined by the threshold voltage of the parasitic MOS transistor; the upper limit by impact ionization. As a result, the minimum achievable collector sheet resistivity is 290 $\Omega/\square$.

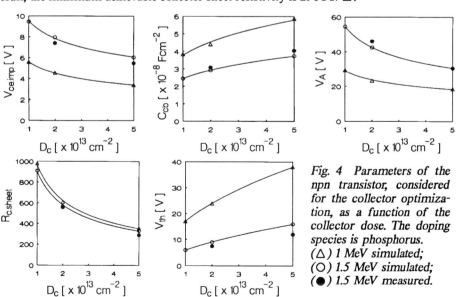

Fig. 4 Parameters of the npn transistor, considered for the collector optimization, as a function of the collector dose. The doping species is phosphorus.
($\triangle$) 1 MeV simulated;
($\bigcirc$) 1.5 MeV simulated;
($\bullet$) 1.5 MeV measured.

## 4. Device Characteristics

Table 1 shows the main parameters of the normal MOS devices and the npn transistor with a maximum collector dose. In Figure 5 the corresponding Gummel plot is displayed. The collector and base current exhibit an ideal behaviour. The current gain is 100 and nearly constant over 6 decades. Cut-off frequencies as high as 6 GHz have been measured.

| (20/20) | | NMOS | PMOS |
|---|---|---|---|
| $V_{th}$ | (V) | 0.70 | -0.72 |
| $\gamma$ | ($V^{1/2}$) | 0.59 | 0.58 |
| S | (mV/dec) | 88 | 82 |
| $BV_{DSS}$ | (V) | >10 | >10 |

| | | NPN |
|---|---|---|
| Beta | | 100 |
| $V_A$ | (V) | 30 |
| $f_{Tmax}$ | (GHz) | 6.0 ($V_{cb}$=4 V) |
| $V_{ceo}$ | (V) | 6.7 |
| $V_{ebo}$ | (V) | 8.6 |

Table 1 Device parameters.

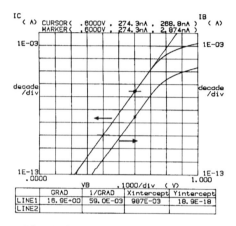

Fig. 5 Gummel plot of the optimized npn transistor ($A_E$ is 5x5 $\mu m^2$).

## 5. Conclusions

A 1.5 $\mu$m high energy ion implanted BiCMOS process has been proposed. This process relies on the modular adding of bipolar and EPROM structures to an existing retrograde twin-well CMOS process. It demonstrates the versatility of high-energy ion implantation. Using a polysilicon emitter and a separate collector implantation, the collector doping profile has been optimized. The optimization problem has been discussed in greater depth for the implantation energy range 1 - 1.5 MeV.

### Acknowledgments

The authors appreciate the assistance of Dr. van der Vlist during the high frequency measurements. They thank G. Boom and A. Kooy for their skilful processing. This work was supported by the Foundation for Fundamental Research on Matter (FOM) and the Netherlands Technology Foundation (STW).

### References

Stolmeijer A 1986 *IEEE Trans. Electron Devices* ED-33 pp 450-457
Volz C and Blossfeld L 1988 *IEEE Trans. Electron Devices* ED-35 pp 1861-1865
Wijburg R C M, Hemink G J, Praamsma L and Middelhoek J 1989 *Proc. ESSDERC* pp 915-918
Wijburg R C M, Hemink G J and Middelhoek J 1990 *IEEE Trans. Electron Devices* ED-37 pp 79-87

# A high performance VLSI structure–SOI/SDB complementary buried channel MOS (CBCMOS) IC

TONG Qin-Yi, XU Xiao-Li and ZHANG Hui-Zhen

Microelectronics Center, Southeast University, Nanjing 210018, China

ABSTRACT. A Complementary Buried Channel MOS(CBCMOS) integrated circuit has been fabricated on a SOI(Silicon On Insulator) substrate prepared by SDB(Silicon wafer Direct Bonding) technology. CBCMOS is a "fully depleted buried channel" MOSFET which is different from conventional "surface buried channel"MOS device in which a pn junction exists at channel surface. Moreover, pn junction is totally eliminated in CBCMOS device structure. Simulation and experimental results have demonstrated that small geometry effects, breakdown voltage, speed etc. have been significantly improved. Kink effect has been totally eliminated. Futhermore, the fabrication process is fully compatible with conventional CMOS/SOI technology.

## 1. Introduction

Silicon On Insulator (SOI) technology has advantages over bulk silicon counterpart. CMOS/SOI demonstartes higher speed, lower power consumption, higher integration density, better relaibility and radiation hardness. However, short and narrow channel effects and hot carrier degradation are not much improved. Moreover, the floating substrate effect(Kink effect) has brought new problems such as current overshoots[1] and reduction of device breakdown voltage[2]. Buried channel structure could be adopted to improve hot carrier effect but suffers from higher leakage current[3]. It is clear that the origin of causing all above mentioned detrimental effects is the presence of pn junctions in the device. Therefore, it is highly desirable to structure the CMOS device having buried channel and at the same time to eliminate any pn junctions in the device. Silicon wafer Direct Bonding (SDB) is a promising technology for high quality SOI preparation[4]. Two oxidized silicon wafers with mirror surfaces contacted face to face at room temperature following a suitable hydrophilizing treatment can be chemically bonded after thermal annealing. SOI film is then produced by thining one side of the bonded pair to a desirable thickness. Not only the SOI quality is superior but also the interfaces on two sides of buried oxide are excellent.

Based on SOI/SDB preparation and characterization[5], a new SOI/SDB complementary buried channel MOS device (CBCMOS) has been developed. The basic CBCMOS/SDB configuration is shown in Fig.1. Similar structrure was reported in [6] but it was based on a special SOI/SIMOX(Separation by IMplanted OXygen) substrate having a electric-field-shielding layer and was for high voltage IC application. The main structure features of CBCMOS/SDB are:
(a) the work function difference

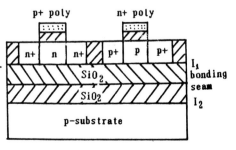

Fig.1 Cross section of CBCMOS

between polysilicon gate and channel region produces a depletion layer. If the thickness of the layer is equal or greater than that of SOI film at zero gate voltage(i.e fully depleted channel region), the normolly-off operation is resulted. The excellent interface properties of SOI/SDB make the enhanced mode device feasible. Unlike the conventional buried channel device, there is no pn junction existing at surface of the channel region in CBCMOS.(b)the input impedence of the device is high, identical to normal MOS structure while the output conductance is similar to that of MESFET configuration. This structure benifits from advantages of both device kinds.

In this paper the CBCMOS device modelling and experimental results are presented and the potential for VLSI application is discussed.

## 2. CBCMOS Simulation

The performance features of CBCMOS are as follows,(a)the absence of pn junction in the device eliminates the physical source of DIBL(Drain Induced Barrier Lowering) effect[7] in OFF state and DFE (Drain Field Effect) effect[8] in ON state of the device. The short and narrow channel effects are thus suppressed effectively.(b) kink effect is diminished since no potential barrier exists at the source of the device.(c) hot carrier effect is weakened and drain breakdown voltage is increased due to the lower field at drain region.(d) parasitic capacitances are smaller owing to no longer exsting pn junction capacitances.(e) there are two conducting mechanisms involved in the device. In n-buried channel MOS device(NBCMOS), increasing gate voltage from zero reduces depletion layer thickness and a conduction buried channel is formed at the SOI/buried oxide interface. At high gate voltage in addition to the bulk conduction channel, the surface accumulation conduction is the dominated conductiong mechanism which enhances device transconductance.

A two-dimensional device analyser has been developed to predict the device performance. Fig. 2 shows simulation results of current density distribution in SOI island (doping=1E14/cm3) of 0.5 μm NBCMOS device at various gate voltages. The gate oxide thickness is 30 nm, buried oxide thickness is 0.5μm and effective oxide charge density is 4.5E10/cm2. These results are applicable to p channel device(PBCMOS). Therefore a logic inverter similar to the common CMOS inverter is realised. In our case, p type lightly doped silicon is adopted as the supporting substrate for SOI film although n type is also possible. The substrate is grounded during device operation. The work function difference between p-substrate and n-SOI island as well as the interface charge at SOI/buried oxide are taken into account. The net effect is either to reduce the effective thickness of SOI island(with small interface charge density) or to form a bottom conducting channel (in the case of large interface charge density). In later case, leakage current is resulted and normally-off operation is prevented.

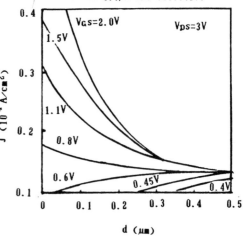

Fig. 2 Current density distribution in NBCMOS SOI island

Analytical expression of threshold voltage of normally-off NBCMOS device is,

$$V_T = \frac{KT}{q}\ln\left[\frac{N_D N_{p+}}{n_i^2}\right] - \frac{Q_{ox1}}{C_{ox}} - \frac{qN_D d}{C_{ox}} - \frac{qN_D d}{2\varepsilon_o\varepsilon_{si}}$$

where the effective SOI thickness d is given by

$$d = \left\{\frac{2\varepsilon_o\varepsilon_{si}\left(\frac{KT}{q}\ln\frac{N_D N_s}{n_i^2} - \frac{Q_{ox2}}{2\varepsilon_o\varepsilon_{si}qN_D}\right)}{qN_D}\right\}^{1/2}$$

where $N_{p+}, N_D$ are doping concentration of poly gate and of SOI island, respectively. $N_s$ is substrate doping, $Q_{ox1}, Q_{ox2}$ are effective charge density at gate oxide/SOI and of SOI/buried oxide interfaces, respectively.

Fig.3 presents a schematical potential distribution of the device at zero gate voltage(Fig.3a) and at threshold voltage(Fig.3b). The drain-source voltage is 3 V and source and substrate are both grounded. The SOI island has doping concentration of $1E14/cm^3$ with thickness of 0.5 μm. The gate oxide thickness is 30 nm and the gate length is 1 μm. As can be seen in Fig.3 that there is no potential well between source and drain at gate voltage greater than threshold voltage to allow holes generated near the drain, if any, to be accumulated. It is therefore expected that no significant changes in potential distribution will occur in SOI island when holes are generated. Simulated device output characteristics indeed has shown that no kink effect is evident at high drain voltage.

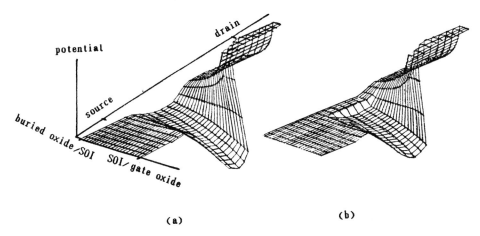

(a)   (b)

Fig.3 Potential distribution of (a) fully depleted mode($V_{GS}$ = 0 V) and (b) turn-on mode($V_{GS} = V_T$, threshold voltage). This is a schematic diagram in arbitarary units.

3. Experimental Results

Devices were fabricated by a modified version of standard CMOS/SOI process in SOI film prepared by SDB technology. Aluminiun was used as a mask during high dose implantation of poly doping. At the stage of S/D implantation nitride-oxide compound layer was prepared on poly gates to prevent from counterdoping. The compound layer was also used to mask poly gates from RIE etching during gate contact holes openning. Totally, 9 lithographgy steps and 5 implats were required in processing.

Fig. 4 shows output characteristics of 1 um NBCMOS device. It is evident that the device is free from kink effect. Preliminary results showed that the transconductance of 1 μm n- and p-channel device was 48 mS/mm and 31 mS/mm, respectively. 0.21 ns delay per stage was measured from 1 μm 21-stage CBCMOS ring oscillator as can be seen in Fig. 5. The comparison of experimental device performance of 1 μm SOI CMOS/SDB(d=0.25 μm) with that of 1 um CBCMOS/SDB(d=0.2 μm) has shown that CBCMOS provides 46 % increasment source-drain breakdown voltage(22 V to 15 V), 380 % reduction of delay per gate(0.21 ns to 0.79 ns), 25 % increasment of transconductance of both n-(48 mS/mm to 38 mS/mm) and p- devive(31 mS/mm to 25 mS/mm). The channel length at which threshold voltage shows obvious shift has moved down from 2 μm in CMOS to 0.8 μm in CBCMOS.

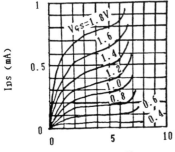

Fig. 4 Output Characteristics of 1 μm NBCMOS (W/L = 20/1)

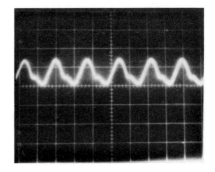

Fig. 5 Output waveform of 1 μm 21-stage CBCMOS ring oscillator
(X, 5ns/div., Y, 2V/div., $V_{DS}$=5V)

4. Conclusion

Numerical simulation and practical fabrication show that CBCMOS can extend the scaling limits of CMOS further down which is attributed to absence of pn junction in the device. Great gain in performance is expected to CBCMOS device with ultra thin SOI film(<100 nm) and operating at low temperature. High speed combined with high reliability of this device makes it very attractive for VLSI application.

5. References

1. K. Kato et al 1986 IEEE Trans. Elec. Dev. ED-33 133
2. K. Kato et al 1985 IEEE J. Sol. Sta. Cir. SC-20 1378
3. L. C. Parillo et al 1892 Tech. Dig. of Intern. Elec. Dev. Meet. 706
4. J. B. Lasky 1986 J. App. Phys. 48 78
5. X. L. Xu et al 1989 Elec. Lett. 25 394
6. S. Nakashima et al 1983 Elect. Lett. 19 568
7. R. Troutman 1977 IEEE Trans. Elec. Dev. ED-24 1266
8. K. N. Ratnakmar etal 1982 IEEE J. Sol. Sta. Cir. SC-17 937

## A well concept for field implant free isolation and width independent n-MOSFET threshold and reliability

Ch. Zeller, C. Mazure, A. Lill and M. Kerber

Siemens AG, Corporate Research and Development, Otto-Hahn-Ring 6, D-8000 München 83, F.R.G.

> Abstract: Boron segregation effects in the p-well during field oxidation can be compensated by splitting the well drive-in into two parts before and after field oxidation. In consequence sufficient field isolation can be achieved without channel stop implants even for moderate well implantation doses, keeping parasitic junction capacitances and body factors low. The smaller doping gradient between field and active areas leads to nearly neutral transistor-width dependence and width independent n-channel degradation.

1. Introduction
Conventional CMOS technologies include a channel stop implantation to assure sufficient isolation. The lateral out-diffusion of field dopants into the active regions leads for narrow transistors to $V_T$ increase, $I_{DS}$ decrease, and due to the higher electric fields at the field oxide edge to enhanced transistor degradation. Field implants can be avoided and process complexity reduced, if well doping concentrations are high enough to guarantee sufficient isolation. Especially for the p-well the implanted boron dose has to make up for the boron loss due to segregation effects during field oxidation. This leads to a high doping level in the active regions and consequently increases body factor and parasitic capacitances. In our paper we demonstrate that by splitting the well drive-in into two parts before and after field oxidation it is possible to compensate segregation effects and to achieve simultaneously field-implant free isolation with moderate well implantation doses and improvement of device performance and reliability.

2. Experimental
A self-aligned twin-well CMOS process with 20nm gate oxide, n$^+$ polysilicon gate and phosphorus LDD n-MOSFETs is used. The phosphorus well is implanted with a nitride mask. The boron well implantation is performed after growth of a thick masking oxide in the nitride windows and removal of the nitride mask. The well drive-in at 1150°C is split symmetrically in two parts before and after field oxidation. A poly-buffered LOCOS is used for isolation (Burmester et al (1988)) with a final field oxide thickness of 600nm. Field oxide threshold voltages are defined by a leakage current criterion of 0.5pA/µm-width at room temperature for a drain voltage $V_{DS}$ of 8V. The n-MOSFET hot electron stability is characterized by stress measurements at the gate voltage $V_G$ corresponding to the maximum substrate current for a given $V_{DS}$ at room temperature.

## 3. Results

Fig.1 shows SIMS profiles of the field isolation region. With the conventional drive-in boron segregation leads to strong dopant depletion. The boron concentration at the silicon/field oxide interface is only 35% of the concentration in the active area. This is an unfavourable constellation in a CMOS process, where high doping levels are required in the field area to achieve satisfactory isolation and low doping concentrations in the transistor region in order to minimize parasitic parameters. In our approach during the second part of the well drive-in boron diffuses back to the surface. The doping concentration underneath the field oxide is increased by 100%. In consequence sufficient isolation can be achieved with a 30% lower p-well implantation dose (fig.2).

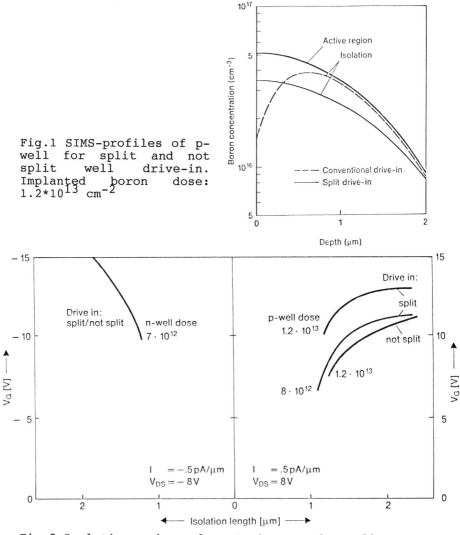

Fig.1 SIMS-profiles of p-well for split and not split well drive-in. Implanted boron dose: $1.2*10^{13}$ cm$^{-2}$

Fig.2 Isolation under poly-gate in n- and p-well

Simultaneously body factor and parasitic junction capacitances are reduced by 15%, improving overall device performance. If the dose is kept constant, on the other hand, the split drive-in allows to gain about 0.2μm in isolation length and to increase the field oxide threshold by 3V, a significant advantage for special applications like EEPROMs. In the phosphorus doped n-well no measurable difference was found between split and not-split drive-in, due to the absence of major segregation effects (fig.2).
Fig.3 shows the relative $V_T$ change for the nMOSFET as a function of transistor width $w_G$ for the cases with channel stops and with split drive-in. In case of field implantation the usual $V_T$ increase is observed due to the lateral outdiffusion of the field doping. For the split drive-in approach diffusion is reversed and the doping gradient at the field oxide edge is reduced. Consequently a nearly neutral transistor width dependence is observed down to $w_G=0.8\mu m$. The identical subthreshold slopes for wide and narrow transistors (fig.4) demonstrate that device leakage is not affected by the field-implant free isolation concept.

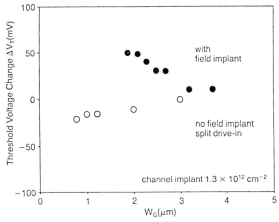

Fig.3 Threshold voltage versus channel width for nMOSFET transistors with channel length $L_G=1.0\mu m$

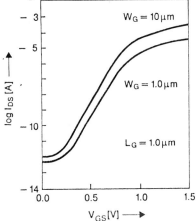

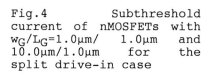

Fig.4 Subthreshold current of nMOSFETs with $w_G/L_G=1.0\mu m/1.0\mu m$ and $10.0\mu m/1.0\mu m$ for the split drive-in case

Fig.5 shows the nMOSFET lifetime as a function of $1/w_G$, normalized to a transistor with $w_G=10\mu m$. In case of field implantation the increasing doping concentration near the field oxide edges leads to enhanced lateral electric fields. In consequence narrow transistor lifetime is deteriorated by more than one order of magnitude. Note that for the split drive-in case the degradation is width independent due to the dopant reduction near the field oxide edges.
Defect analysis has shown that no crystal defects are created by the well drive-in after field oxidation. The quality of $n^+/p$ and $p^+/n$ diodes and the extrinsic defect density of the gate oxide are not affected by the $1150^\circ C$ step after the isolation complex. Charge to breakdown values are even improved by the split drive-in by one order of magnitude (fig.6). These results demonstrate that our drive-in scheme is compatible with CMOS or DRAM processes.

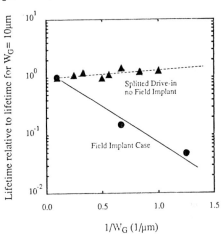

Fig.5 Lifetime relative to the wide transistor lifetime versus reciprocal channel width at $V_{DS}=7V$ and $V_G=V_G(I_{sub}^{peak})$ for n-MOSFETS with $L_G=1.0\mu m$

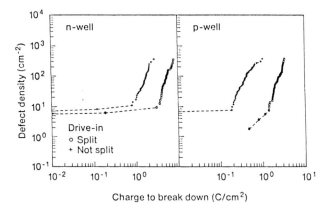

Fig.6 Intrinsic charge to breakdown for split and not-split drive-in

Burmester R, Kerber M and Zeller Ch 1988, ESSDERC Conf. Proc. (Montpellier, France) pp C4 545-548

## Simulation of EPROM writing

### C. Fiegna, F. Venturi, M. Melanotte[+], E. Sangiorgi[†], and B. Riccò

DEIS University of Bologna, Bologna Italy
[+] SGS-THOMSON Microelectronics, Agrate Brianza (Milano) Italy
[†] Department of Physics, University of Udine, Udine Italy

Abstract. This paper presents a simple and efficient model for first order simulation of n-channel EPROMs programming that allows to calculate electron injection into the gate insulator of the cell transistor accounting (at first order) for both the non Maxwellian form of the electron energy distribution and the non local nature of carrier heating. The model has been implemented as a post-processor of a conventional two dimensional device simulator.

### 1. Introduction

Erasable-Programmable Read Only Memories (EPROM) exploit channel hot electrons for cell writing obtained by charging the floating gate (FG) with energetic carriers injected over the energy barrier at the $Si - SiO_2$ interface. This work presents an efficient simulation tool tackling the problem of EPROM writing and including original treatments for the non local process of electron heating and for the energy distribution of hot carriers. Such a program consists of a two dimensional (2-D) drift diffusion (DD) device simulator, featuring charge boundary conditions on the FG, coupled with a post-processor used to compute electron mean energy along the channel and charge injection into the FG.

### 2. The simulator

As mentioned in the introduction, the simulation of EPROM writing presented in this work consists of a 2-D device analysis followed by post-processing of the results. The quasi-stationary procedure used to simulate EPROM cell writing, described by the flow-chart of Fig.1, is divided into time steps of variable duration ($\Delta t_i$). At each step, first a DD calculation provides electric field and carrier concentrations, successively a post-processor computes the electron temperature ($T_e$), the gate current $I_G$ and the charge injected into the FG. The total injected charge is then updated and the procedure is iterated. The post-processor for hot electron injection

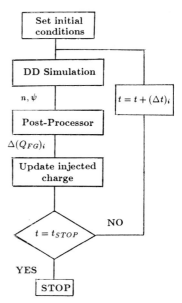

Fig.1 Flow chart of the procedure

into the oxide makes use of a new thermionic emission formula taking into account the non-Maxwellian form of the electron energy distribution (EED) for the high energy region (Cassi and Riccò 1990) namely:

$$J_G = -qnA \int_{E_B}^{+\infty} E^{1.5} e^{(-\frac{\chi E^3}{F^{1.5}})} dE, \quad (1)$$

where: $J_G$ is the injection current density, $E$ is the electron energy, $F$ is the electric field, $E_B$ is the height of the energy barrier for the injection in the oxide, $A$ and $\chi$ are two constants that, as will be shown later, are determined by fitting experimental data of gate current in MOSFET's. Non local effects have been accounted in eq. (1) by an effective field ($F_{EFF}$) calculated, as a function of $T_e$, by means of a relationship holding for the homogeneous case at high fields ($> 6 \cdot 10^4 V/cm$)

$$T_e = T_L + \frac{2q}{3K_B} \tau_w v_{sat} F_{EFF} \simeq T_L + 1.5 \cdot 10^{-2} F_{EFF}, \quad (2)$$

where: $T_L, K_B, \tau_w$ and $v_{sat}$ denote lattice temperature, Boltzmann constant, energy relaxation time and electron saturation velocity, respectively. In essence $F_{EFF}$ represents the field that would correspond to $T_e$ in the homogeneous case, namely that for which eq.1 has been derived. Eq.1 provides a simple, physically based expression for the gate current that can be used once $n$ and $T_e$ are known. When starting from the results of a DD simulator, $T_e$ is calculated as a function of field and carrier concentration following a slightly improved version of the approach of Goldsman and Frey (1988). In particular we calculate $T_e$ using the simplified energy-balance equation

$$\frac{dw}{dx} = -\frac{21}{20}qF - \frac{1}{2}\frac{w}{n}\frac{dn}{dx} - \frac{9}{20}\sqrt{\frac{40}{9}\frac{m^*}{\tau_p \tau_w}(w - w_L)} + \left(qF + \frac{10}{9}\frac{w}{n}\frac{dn}{dx}\right)^2, \quad (3)$$

where: $w$ is the electron mean energy ($\simeq \frac{3}{2}K_B T_e$), $w_L = w(T_L)$ and $\tau_p$ is the momentum relaxation time. $\tau_w$ and $\tau_p$ have been calculated as functions of $w$ by means of MC simulations of homogeneous silicon bars. Eq.3, differently from the similar one derived in Goldsman and Frey (1988), accounts for the effects of concentration gradients. By integrating eq.3 along the current paths (found by backtracing the gradient of the electron quasi-Fermi level) $w$, hence $T_e$, is calculated at any grid point.

3. Simulation of MOSFET's gate current

As already mentioned, our model for $I_G$ features two adjustable parameters ( $A$ and $\chi$) whose values have been determined by fitting gate currents measured on transistors obtained by shorting control- and floating-gate of EPROM cells. The value of $\chi$, that plays a dominant role in determining the main dependence of $I_G$ on $F_{EFF}$ ( hence on $T_e$ ), has been obtained from experimental results plotted in the form $ln(I_G)$ vs $F_{EFFmax}^{-1.5}$, where $F_{EFFmax}$ is the maximum value of $F_{EFF}$ along the

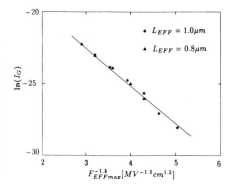

Fig.2  $ln(I_G)$ vs $F_{EFFmax}^{-1.5}$ for several values of $V_{DS}$ and $V_{GS} = V_{DS} + V_{TH}$

channel (i.e. in the point providing the maximum contribution to $I_G$) calculated by applying eqns. 2 and 3 to the results of DD device simulation.
As can be seen in Fig. 2, the function $ln(I_G)$ vs $F_{EFFmax}^{-1.5}$ provides a straight line. By simplifying eq.1 as $I_G \propto exp(-\chi E_B^3/F_{EFFmax}^{1.5})$, the slope of such a curve is simply given by $-\chi E_B^3$, hence $\chi$ can be easily determined. In particular we find $\chi \simeq 1.3 \times 10^8 V^{1.5} cm^{-1.5} eV^{-3}$ in good agreement with the results of Monte-Carlo simulations of silicon resistive slabs (Cassi and Riccó 1990). This substantial agreement and the fact that in this work $\chi$ is found to be independent of channel length, provide significant evidence supporting our model. The results obtained fitting measured transistor gate currents, with a single set of values for the parameters $A$ ($=487 sec^{-1} m \; eV^{-2.5}$) and $\chi$, are reported in Figs.3 and 4 clearly indicating that a good agreement can be achieved (lines and points represent the results of simulations and experiments respectively).

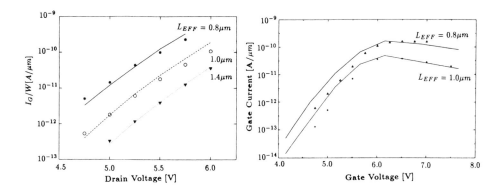

Fig.3 Normalized gate current vs $V_{DS}$ and $V_{GS} = V_{DS} + V_{TH}$.

Fig.4 Normalized gate current vs $V_{GS}$ and $V_{DS} = 5.5V$.

## 4. Simulation of EPROM writing

The model of this work has been compared with experimental writing characteristics of EPROM cells with $L_{eff}$ ranging from 0.8 to 0.5$\mu m$ under typical writing bias conditions: $V_{DS} = 5 \div 6.5V$, $V_{GS} = 10.5 \div 12.5V$, $V_{BULK} = 0V$. All the experimental data has been simulated with the same values of $\chi$ and $A$ determined as explained above. The achieved agreement is quite good for all considered devices and programming voltages as clearly indicated by the typical examples of Figures 5 and 6 showing the dependence of the writing transients on applied bias and channel length, respectively.

With regard to the efficiency of the developed tool, the time needed for post-processing the results of the DD simulation is always a small fraction ($< 5\%$) of the whole simulation time. An important role in determining a good trade-off between CPU time and accuracy is played by the length of the time intervals. In particular, coarse time discretizations inevitably worsen the agreement between experiments and simulations because of the simplifying assumption of constant $I_G$. On the other hand, shorter intervals require more simulation steps. An upper bound to $\Delta t_i$ is imposed by the necessity to limit the electric charge injected during the time interval, hence the variations of the charge boundary conditions, in order to avoid convergence problems

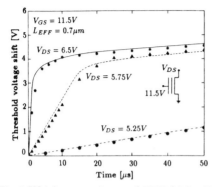

Fig.5 Writing transients of EPROM cells Lines:experiments; points:simulations

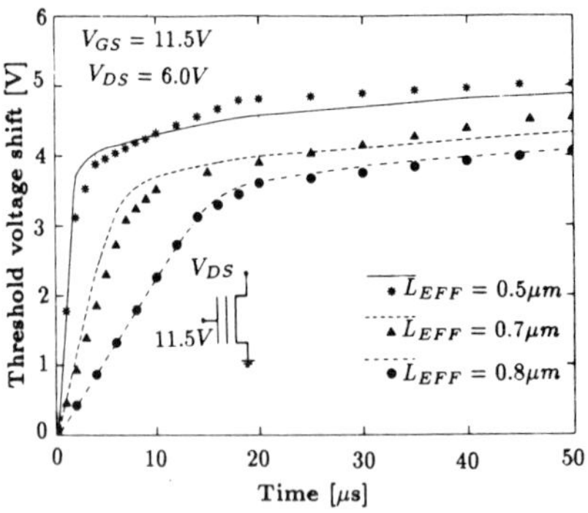

Fig.6 Writing transients of EPROM cells Lines:experiments; points:simulations

for the next DD simulation. An optimal strategy obviously consists in using adaptive time intervals, namely shorter ones in the first part of the simulation where $I_G$ is high and the larger part of threshold voltage shift takes place, and longer ones when the transistor has reached the low-injection region (i.e. $V_{GS} < V_{DS}$).

For the simulations presented in this work, $\Delta t_i$ has been made to vary in the range $0.5 \div 5 \cdot 10^{-6} sec$ and a complete writing characteristic (such a one of the curves reported in Figs.5 and 6) typically requires, on a SUN SPARC-station 330, $\simeq$ 30 CPU minutes.

5. Conclusions

This work has presented a simple and efficient model for the simulation of EPROM cell writing that starts from the electric field and carrier concentration given by a 2-D device simulator and accounts for the non-Maxwellian form of the hot-electron energy distribution as well as for non-local carrier heating. The model, minimally depending on empirical fitting parameters, has been validated by means of a comparison with experimental data obtained with transistors featuring channel lengths in the range $0.8 - 1.4 \mu m$ and EPROM cells with length between 0.5 and $0.8 \mu m$ and wide and realistic range of bias conditions for EPROM writing. Such a comparison has shown that the model provides accurate results and can therefore be used as an efficient tool to support development of new devices with channel length down to $\simeq 0.5 \mu m$.

REFERENCES

Cassi D and Riccò B 1990 to appear on IEEE Trans. ED.
Goldsman N and Frey J 1988, IEEE Trans. ED, Vol.35, N.9, p.1524

# Field isolation and active devices for 16 Mbit DRAMs

H.-M. Mühlhoff, J. Dietl, P. Küpper and R. Lemme
Siemens AG, Semiconductor Division
Otto-Hahn-Ring 6, München, West Germany

Abstract. Results are presented showing that conventional LOCOS Isolation can be used for memory cells with pitches less than 1.5µm. Careful tuning of LOCOS overetch and channel stop implantation as well as optimization of the cell transistor yield satisfactory isolation of 0.7µm LOCOS and half-micron active devices with active area widths of 0.25µm.

Introduction. In 16Mbit DRAMs with cell sizes less than 5µm$^2$, wordline and bitline pitches are approaching 1.5µm /1/. These narrow pitches require submicron active area width and isolation. Using conventional poly-buffered LOCOS is difficult because the isolation thickness decreases sharply with decreasing isolation length and the outdiffusion of the channel stopper implant into the active area raises the threshold voltage of the transfer gate. Fig.1 shows SEM cross-sections of the 16Mbit DRAM cell parallel to the wordline (a) and parallel to the bitline (b). The wordline and bitline pitches are 1.5µm/1.6µm respectively. As is evident from Fig.1b, the sum of active area width and minimum isolation have to equal the bitline pitch.

Sample Preparation. Samples were prepared on 5ohm-cm p-doped substrate. Channel stop implantation was performed either before LOCOS growth at 80keV with active areas masked or after LOCOS growth at 180keV. The implant at 180keV substitutes as deep channel implant in active area NMOS devices. Fig.2 shows the doping profiles under LOCOS measured by SIMS for both doping techniques. LOCOS was grown to an initial thickness of 720nm followed by several backetches and sacrificial oxidations. Final field oxide thickness was varied between 390nm and 490nm. The actual thickness of each FOX-transistor was measured using SEM. The results have been plotted in Fig.3 and Fig.4. Due to lack of oxidizing species in narrow spacings, the oxide thickness is reduced for decreasing FOX length. Final oxide thickness for small structures depends on the oxide thickness of large structures yielding LOCOS isolation of maximum thickness where overetching has been kept to a minimum. Fig.4 shows that the isolation length scales linearly with designed length as long as the final oxide thickness is kept above 440nm. The SEM of a 0.7µm FOX-transistor is shown in Fig.5.

© 1990 IOP Publishing Ltd

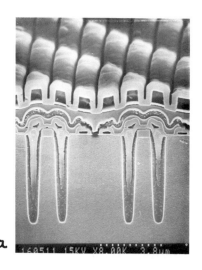 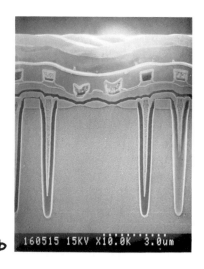

Fig.1 SEM cross-section through 16Mbit DRAM cell, (a) parallel to to bitline, (b) parallel to wordline.

Electrical Results of Isolation: Threshold voltages of polygate FOX transistors for large area FOX thicknesses between 390 and 490nm have been plotted in Fig.6. Punchthrough of gateless LOCOS structures is shown in Fig.7. The results show that channel stop implantation through the LOCOS is as good as or superior to field doping before LOCOS growth. At 0V substrate bias, threshold voltages of 5V can be achieved with a 0.7µm FOX transistor. At 1.5V substrate bias, $V_T$s are raised to above 12V. For punchthrough (Fig.7), isolation is mainly determined by the effective distance between the highly doped regions, which is a critical function of LDD dose/energy, temperature budget and outdiffusion of trench doping. Improvement, however, can be made when the channel implantation is performed through the oxide. Using this doping technique, 0.7µm LOCOS can provide isolation up to 5V.

Performance of Active Transistor: 16nm gate oxide was grown for active NMOSFETs, whose gate lengths range from 0.6 to 3µm. After LDD implantation a 200nm TEOS spacer was formed. Minimum width of active area is 0.15µm. The narrow channel effect is demonstrated in Fig.8. The observed narrow channel effect is due to 2 mechanisms. Outdiffusion of the channelstopper implant into the active area raises $V_T$ until the doping level of active and field area are equal /2/. This occurs at about 1µm width (s.Fig.2). Because diffusion is minimal after LOCOS growth, this effect is absent in devices with channel implantation after LOCOS. Further increase of $V_T$ for W<1µm can be modeled by the lateral spreading of the depletion region /3/

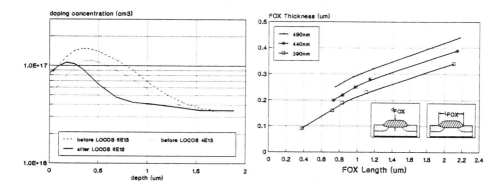

Fig.2 Doping profiles of Channel stopper implanted before LOCOS and after LOCOS at 180keV.

Fig.3 Dependence of FOX thickness on its length.

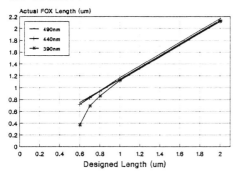

Fig.4 Actual FOX length vs designed length.

Fig.5 SEM of 0.7μm LOCOS.

$$V_T = V_{FB} + 2\phi_b + \{2qe_sN_a\,(2\phi_b+V_{BS})\}^{1/2}/C_{ox}\,(1 + a\,W_{depl}/W)$$

with a=0.35. Satisfactory transistor performance is obtained at channel widths down to 0.25μm. At 0.15μm the nitride mask was lifted during field oxidation eliminating the active area.

In DRAM cell areas, refresh requirements demand minimum size transistors with low leakage currents. $V_T$s of these transistors must be higher than for transistors in peripheral circuits. Fig.9 shows the short channel effect of NMOSFETs with varying active area widths. Transfer gates with W/L=0.45/0.6 having a $V_T$ of 1V and MOSFETs for peripheral circuits with L=0.7μm having a $V_T$ of 0.6V can be used. Mixing both doping techniques for field isolation, both $V_T$s can be adjusted independently with no effect on isolation properties.

Conclusion: It has been shown that 0.7μm LOCOS isolation and half-micron transistors with 0.25μm width can be used in DRAM memory cells. Field doping by ion implantation through

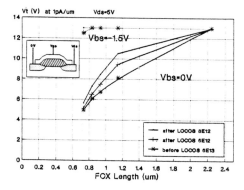

Fig.6 Threshold voltage of FOX transistor (d=490nm at L=10μm).

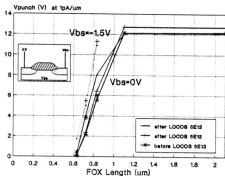

Fig.7 Punchthrough voltage of FOX transistor (d=440 at L=10μm).

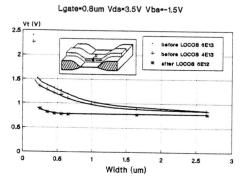

Fig.8 Narrow Width effect of transfer gate.

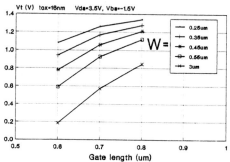

Fig.9 Dependence of short-channel effect on width. Field implant before LOCOS 4E13.

the LOCOS at 180keV improves isolation properties and nearly eliminates the narrow channel effect. These devices permit pitches of 1μm and are suitable to carry 16Mbit concepts into the next generation.

References:

1 K.H.Küsters, L.DoThanh, F.X.Stelz, W.-U.Kellner, H.-M.Mühlhoff and W.Müller, Proceedings of ESSDERC, 907 (1989)

2 H.Iwai, K.Taniguchi, M.Konaka, S.Maeda and Y.Nishi, IEEE Trans. Electron. Dev. ED-29, 625 (1982)

3 G.Merckel, Solid State Electron. 21, 1207 (1980)

# Dimensional characterisation of poly buffer LOCOS in comparison with suppressed LOCOS

N.A.H. Wils, P.A. van der Plas and A.H. Montree

Philips Research Laboratories, P.O. BOX 80000, 5600 JA Eindhoven, The Netherlands

**Abstract**

The main disadvantage when using a LOCOS isolation technique is the formation of the so called bird's beak and its geometric dependence. In this paper we demonstrate that for Poly Buffer LOCOS (PBLOCOS), in comparison with Suppressed LOCOS (SLOCOS), the bird's beak is not only smaller, but is also less dependent on geometry. The comparison between PBLOCOS and SLOCOS will be focussed on four geometric configurations.

## 1 - Introduction

When going to submicron dimensions, it is more and more difficult to find a suitable field isolation technique. LOCOS isolation is still a first choice because of its process simplicity and good performance. Poly Buffer LOCOS (PBLOCOS) is an alternative isolation scheme, which reduces the field oxide encroachment (bird's beak, BB in Figure 2) into active regions. The feasibility of the PBLOCOS isolation has already been presented (Chapman et.al., 1987), however, the geometric dependence of the bird's beak has not been previously reported. We will show that significant geometric dependence, as reported for suppressed LOCOS (SLOCOS) (Plas et.al.,1987 and 1989), is not observed for PBLOCOS. This makes it an attractive isolation method for submicron processes.

## 2 - Experimental

In the experiments the PBLOCOS oxidation stack consisted of 8 nm silicon dioxide, 50 nm poly silicon and 250 nm silicon nitride. For SLOCOS an oxidation stack of 6 nm silicon dioxide, 50 nm oxynitride and 100 nm silicon nitride was used. A field oxide of 500 nm thickness was grown at $1000°C$. The total encroachment of the field oxide under the oxidation mask was characterized by top view SEM and cross section SEM and TEM micrographs.

## 3 - Results

Cross section TEM micrographs after field oxide formation are shown in Figures 1 and 2 for PBLOCOS and SLOCOS respectively. It is clear that for PBLOCOS a smaller bird's beak results. Also the shape of the bird's beak is different, compared with the one observed in a SLOCOS process. In the case of PBLOCOS the bird's beak is for the main part formed above the silicon surface, because the poly silicon layer acts as a source of lateral oxidation.

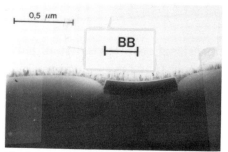

Fig.1: Cross section TEM micrograph of the Poly Buffer LOCOS oxidation scheme after fieldoxide formation

Fig.2: Cross section TEM micrograph of the Suppressed LOCOS isolation after fieldoxide formation

To study the influence of the geometry on the bird's beak, four different situations were distinguished (Plas et.al., 1989):

## 1 - narrow mask width (stack stiffening effect)

A narrow oxidation mask will be more resistent to bending than a wide mask and this can have an effect on the bird's beak dimensions. Figure 3 shows the bird's beak length as a function of the width of the oxidation mask for PBLOCOS and SLOCOS. For the latter process the bird's beak length decreases for widths of the oxidation mask below 1 $\mu$m. For PBLOCOS the bird's beak length is nearly constant.

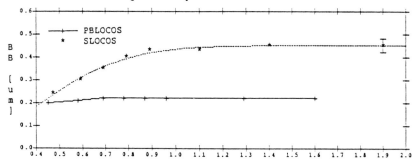

Fig.3: Bird's beak length (BB) as a function of the width of the oxidation mask (W_mask) for Poly Buffer LOCOS and Suppressed LOCOS

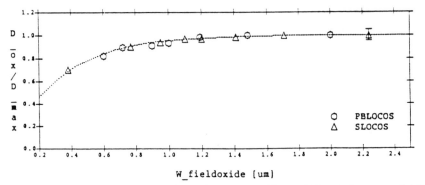

Fig.4: Fieldoxide thinning effect for Poly Buffer LOCOS and Suppressed LOCOS. The curve shows the relative fieldoxide thickness ($D\_ox/D\_max$) versus the opening width of the oxidation mask (W_fieldoxide)

## 2 - narrow mask opening (field oxide thinning effect)

The field oxide thickness in narrow openings will be thinner than in wide openings (Hui et.al.,1987). This effect, known as field oxide thinning, is shown in Figure 4 for PBLOCOS and SLOCOS. In this figure the ratio of the field oxide thickness measured in narrow openings, to the nominal thickness (in wide areas) is plotted. The curves show that the thinning effect is not influenced by the oxidation stack used. This was also demonstrated by Coulman et.al.(1989) for standard LOCOS and SLOCOS.

## 3 - Convex corners (enlarged bird's beak formation)

At the tip of a narrow oxidation mask line (convex corners in the oxidation mask) an enlarged bird's beak is observed. This effect is illustrated in Figure 5 for SLOCOS.

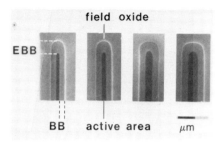

Fig.5: Top view SEM micrograph showing enlarged birds' beaks (EBB) at the tip of narrow active areas. BB refers to the normal bird's beak

In Figure 6 the length of this enlarged bird's beak (after removal of the oxidation stack) is shown versus the width of the oxidation mask for PBLOCOS and SLOCOS. For PBLOCOS the enlarged bird's beak is smaller than for SLOCOS. However the ratio of enlarged bird's beak to nominal bird's beak is the same for PBLOCOS and for SLOCOS, e.g. for an oxidation mask width of 0.5 $\mu$m the enlarged bird's beak is three times the nominal bird's beak for both processes.

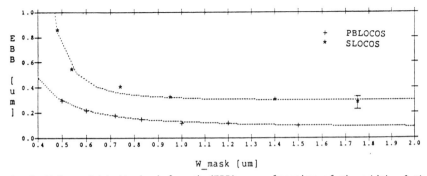

Fig.6: Enlarged bird's beak length (EBB) as a function of the width of the oxidation mask (W_mask) for Poly Buffer LOCOS and Suppressed LOCOS

## 4 - Concave corners (reduced bird's beak formation)

At the tip of narrow field oxide lines (concave corners in the oxidation mask), an opposite effect to the one described in section 3 is observed. The bird's beak that is formed at concave corners is shorter than the nominal beak. Figure 7 shows the length of the bird's beak at the tip of a field oxide bar versus the width of the field oxide. For SLOCOS the bird's beak is reduced at the tip of narrow field oxide bars. This effect is not observed within the level of experimental significance for PBLOCOS, for field oxide widths down to 0.6 $\mu$m.

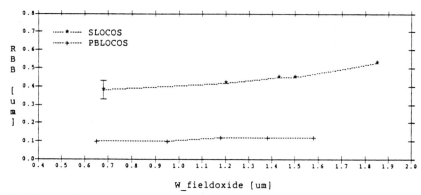

Fig.7: Length of reduced bird's beak (RBB) as a function of the opening width of the oxidation mask (W_fieldoxide) for Poly Buffer LOCOS and Suppressed LOCOS.

## 4 - Conclusions

We have demonstrated that for a PBLOCOS field isolation process a shorter bird's beak is obtained when compared to SLOCOS. Furthermore, we have shown that the geometric dependence of the bird's beak is smaller for the PBLOCOS isolation technique. This facilitates 1) design rule shrinking, required to obtain further reduction of cell sizes and increase of packing densities and 2) improved process control. Both are needed in order to realize ULSI devices.

### Acknowledgements

The authors wish to acknowledge R. Hokke for TEM work.

### Literature

Chapman R.A. et.al., 1987, Proc. IEDM p. 362-365
Coulman B. et.al., 1989, Proc. Symp. on ULSI, Electrochem. Soc., Abstract no 199
Hui J., Voorde van de P. and Moll J., 1987, IEEE trans. ED-34 2255
Plas van der P.A. et.al., 1989, ESSDERC, p. 131-134
Plas van der P.A. et.al., 1987, Digest of Symp. on VLSI techn., III-4,p.19-20

Paper presented at ESSDERC 90, Nottingham, September 1990
Session 7A7

# Modelling and simulation of silicon-on-sapphire MOSFETs for analogue circuit design

R. Howes[1], W. Redman-White[1], K.G. Nichols[1], S.J. Murray[2], P.J. Mole[3].

[1] Department of Electronics and Computer Science, The University, Southampton, U.K.
[2] Marconi Electronic Devices Ltd., Lincoln, U.K.
[3] GEC Hirst Research Centre, Wembley, U.K. (Now at STC Technology, Harlow, U.K.)

## ABSTRACT

A simple physically based CAD model for the SOS MOSFET is described suitable for the design of analogue as well as digital circuits. The model has been implemented in the SPICE2 program. A comparison between simulated and measured amplifier characteristics shows significant improvements in modelling accuracy compared with a bulk MOS model. Simulation times for the new model are presented.

## INTRODUCTION

There has lately been increasing interest in silicon-on-sapphire (SOS) and silicon-on-insulator (SOI) technology, both for VLSI as well as for the more traditional radiation environments. Most applications have been digital, benefiting from the complete isolation of devices by the insulating substrate which ensures latch-up immunity under transient radiation. An additional advantage is the very low parasitic capacitances compared to bulk MOS technologies which imparts good circuit speed. There is, however, increasing pressure to include analogue circuits on the same substrate. Potential advantages of SOS technology for the analogue designer include the possibility of incorporating MOS and bipolar stages on the same substrate, increased flexibility, high quality capacitors and, with careful design, high radiation tolerance (Redman-White *et al* 1990).

However, the design of analogue circuits in SOS presents a number of problems which must be overcome if good performance is to be achieved. These problems arise largely from the kink effect (Tihanyi *et al* 1975). Impact ionisation current generated in the high field region close to the drain when the channel is pinched off, is swept into the undepleted body region where charge accumulates. The body potential therefore rises to the point where the body-source pn-junction becomes forward biased. This leads to a reduction in the device threshold voltage, $V_T$, and an increase in drain current. The effect on the output conductance is profound for analogue circuit performance; at the onset of the kink a sharp increase in conductance of more than an order of magnitude is possible. This causes difficulties in many conventional circuit blocks; current mirror matching deteriorates significantly and the gain of amplifier stages becomes very low and non-linear.

CAD models for SOI are now beginning to appear (Veeraraghavan *et al* 1988). For SOS, modelling has so far been based on bulk device type characteristics ignoring the kink effect. This may be acceptable for digital circuits, but for analogue design accurate modelling of the $g_{DS}$ characteristic is crucial; even the dc operating conditions of an amplifier cannot be predicted otherwise.

In this work we describe a simple physically based CAD model suitable for the design of analogue as well as digital circuits.

© 1990 Crown Copyright

## MODEL FORMULATION

A large signal equivalent circuit for the new SOS device model is illustrated in figure 1.

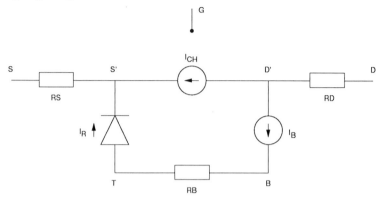

Figure 1. Large signal equivalent circuit for the SOS MOSFET.

$I_{CH}$ is the strong inversion channel current which is derived in conventional form (Tsividis 1987). The drain saturation voltage, $V_{DSAT}$, is chosen to ensure continuity of the drain conductance between linear and saturation regions. Of particular importance is the kink effect. Impact ionisation current is modelled by the current source $I_B$ (El-Mansy et al 1975) and $I_R$ models recombination at the body-source junction. Under steady state conditions:

$$I_B = I_R \tag{1}$$

where

$$I_B = I_{CH} K1 \exp\left(\frac{-K2}{V_{DS} - V_{DSAT}}\right) \tag{2}$$

$$I_R = I_{RO}\left\{\exp\left(\frac{V_{TS}}{\phi_t}\right) - 1\right\} \tag{3}$$

K1 and K2 are treated as model parameters, but are related to the impact ionisation coefficients. $I_{RO}$ is the reverse saturation current of the body-source junction and $\phi_t$ is the thermal voltage. Normal channel shortening is included for accurate $g_{DS}$ values below the kink. In most devices the conductance above the kink point is considerable. This is mainly due to the flow of impact ionisation current through the high resistivity undepleted body region. We therefore include a body resistance, RB, and define the effective body potential, for the modification of $V_T$, to be at node B:

$$V_T = V_{TO} - FB V_{BS} \tag{4}$$

$V_{TO}$ is the zero body bias threshold voltage and FB, the substrate bias factor, is treated as a model parameter. Equation (4) differs from the usual expression for threshold voltage dependence on back bias (Liu et al 1982) but is numerically stable for positive values of $V_{BS}$ and therefore more suitable for use in a circuit simulator. RD and RS are parasitic resistances due to the drain and source diffusions respectively.

This model, combined with reciprocal intrinsic capacitances has been implemented in the SPICE2 circuit simulation program (Nagel 1975) by the inclusion of new modelling subroutines and the modification of existing subroutines.

## MODEL EVALUATION

A comparison of measured and simulated drain current characteristics for a long n-channel device is shown in figure 2. Throughout the saturation region good modelling of the output conductance is obtained.

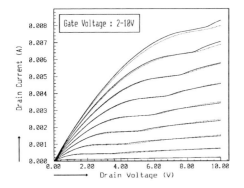

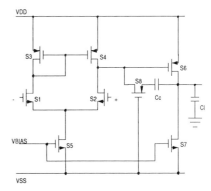

Figure 2. Comparison of measured (dashed line) and simulated (solid line) drain current characteristics for a 100μm/10μm n-channel SOS MOSFET.

Figure 3. Two-stage op-amp circuit used for model evaluation.

The model has been evaluated further by comparing simulation results with measurements of complete amplifier circuits fabricated in a 3μm SOS process. A 2-stage op-amp is illustrated in figure 3. Figure 4(a) shows a measured dc transfer characteristic for this circuit. The influence of the kink can be seen to dramatically reduce the circuit gain over part of the characteristic and to shift the circuit operating point. A corresponding simulated characteristic is shown in figure 4(b).

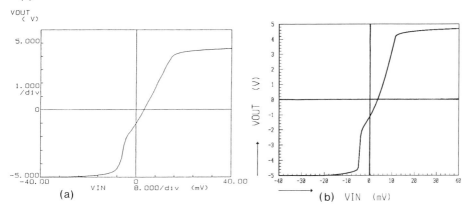

Figure 4. Large signal transfer characteristics for 2-stage op-amp circuit in figure 3. (a) Measured characteristic, (b) simulated characteristic using new SPICE model.

The improvement in modelling accuracy is highlighted in table 1, which compares simulation results of the new model with that of a bulk MOSFET model which does not include the kink effect. The frequency dependence of the output conductance due to charge storage in the substrate (Eaton et al 1975) is reflected in the comparison of simulated and measured values of circuit bandwidth.

Table 1
Comparison of measured and simulated values of circuit gain and bandwidth for a single stage op-amp.

| Circuit Parameter | Measured value | Simulated value (new model) | Simulated value (bulk MOS model) |
|---|---|---|---|
| Open-loop dc gain | 49 dB | 56 dB | 77 dB |
| Unity gain bandwidth | 7.8 MHz | 8.8 MHz | 12 MHz |

Table 2 compares simulation times and convergence performance of the new model with SPICE2 level 1 MOS model. The complexity of the new model is greater than the level 1 but less than the level 3 MOS model. This accounts for the longer time per iteration taken to update the branch current matrix and nodal admittance matrix. Capacitive coupling between the gate and the substrate during the transient analysis is responsible for the slower convergence of the Newton-Raphson algorithm.

Table 2
Comparison of simulation time and convergence performance for a 2-stage op-amp circuit containing 8 transistors.

| | | DC Transfer Curve Analysis | | | Transient Analysis | | |
|---|---|---|---|---|---|---|---|
| Model | Nodes | Output points | Iterations | Time/iteration /transistor | Timepoints | Iterations | Time/iteration /transistor |
| SOS model | 43 | 160 | 395 | 3.29ms | 200 | 524 | 5.70ms |
| MOS level 1 | 29 | 160 | 355 | 2.59ms | 200 | 255 | 5.04ms |

## CONCLUSIONS

SOS and SOI technology offer advantages for both radiation and non-radiation environment applications. However, to achieve the optimum design of analogue circuits in SOS technology requires the use of a circuit simulation model which accounts for the floating substrate effects. Such a model has been presented and shown to offer significant improvements in predicting circuit performance.

## ACKNOWLEDGEMENTS

This work was supported by the procurement executive, Ministry of Defence, and by Marconi Electronic Devices Ltd.

## REFERENCES

Eaton S.S, Lalevic B, (1978) "The effect of a floating substrate on the operation of silicon-on-sapphire transistors", *IEEE Trans. Electron Devices*, **ED-25**, pp 907-912.

El-Mansy Y.A, Caughey D.M, (1975) "Modelling weak avalanche multiplication currents in IGFETS and SOS transistors for CAD", *in IEDM Tech. Digest*, pp 31-34.

Liu S, Nagel L.W, (1982) "Small-signal MOSFET models for analogue circuit design", *IEEE J. Solid-State Circuits*, **SC-17**, pp 983-998.

Nagel L.W, (1975) "SPICE2: A computer program to simulate semiconductor circuits", *Memorandum no. ERL-M520*, Electronics Research Laboratory, University of California, Berkeley.

Redman-White W, Dunn R, Lucas R, Smithers P, (1990) "A Radiation-Hard AGC Stabilized SOS Crystal Oscillator", *IEEE J. Solid-State Circuits*, **SC-25**, pp 282-288.

Tihanyi J, Schlotterer H (1975), "Influence of the floating substrate potential on the characteristics of ESFI SOS transistors", *Solid-State Electronics*, **18**, pp 309-314.

Tsividis Y.P, (1987) Operation and modelling of the MOS transistor, (New York : McGraw-Hill), pp 123-130.

Veeraraghavan S, Fossum J.G, (1988) "A physical short-channel model for the thin-film SOI MOSFET applicable to device and circuit CAD", *IEEE Trans. Electron Devices*, **ED-35**, pp 1866-1875.

# ELSIMA: ELDO short-channel IGFET model for analog applications

T. PEDRON, G. MERCKEL, CNET-CNS, Chemin du vieux Chêne, BP 98 38243 MEYLAN CEDEX FRANCE

**ABSTRACT.**
A new MOSFET model for analog circuits is presented. It allows analog circuit designers to obtain accurate simulation of Operational Amplifiers performances (gain, offset, temperature and frequency behaviour). The continuity of the current and conductance is achieved between the linear and saturation regime. The weak avalanche effect is taken into account.

## 1. INTRODUCTION.
Operational Amplifiers and Current Mirrors are the main basic circuits for analog applications. Classical Switch Capacitors are outside the scope of this paper. For that purpose, a Charge Control Model had been previously implemented in the Electrical Simulator ELDO [1] (besides the physical aspect, the charges are considered as state variables).
Modern MOS Operational Amplifiers in integrated circuits have moderate voltage gain (100 to 200) and a high speed.
From the modelling point of view, a good prediction of the voltage gain requires accurate simulation of the device conductance. In classical NMOS models the inaccuracy is mainly due to bad transitions between the linear and saturation regimes, bad simulation results of the conductance, and to the fact that weak avalanche is not taken into account. In modern technologies (0.7 µm) the last effect appears for drain voltages close to 3.5 V, even for long channel devices.
The bias conditions near the threshold voltage require, first an adaptative method of parameter extraction ( the choice of cost function to be minimized becomes very critical), and secondly an accurate weak inversion model, together with a continuous transition between weak and strong inversion. This aspect will be developed in an other paper to be published.
In an Op. Amps. the offset voltage can be correctly represented if the dependency of the threshold voltage versus the bulk voltage is well modelled. Most of the models neglect this aspect and the threshold voltage is roughly simulated using one or two body coefficients.
The frequency behaviour depends on the correct evaluation of the MOSFET capacitors : source and drain diffusion, overlap, interconnect capacitors.

## 2. A NEW ANALOG MODEL.
The new analog model is based on a previously published logical model [2], with many improvements.
### 2.1 The Threshold Model.
Our previous published threshold voltage models [3] have been simplified and take into account the bulk doping profile and the drain static feedback. A first order expansion has been achieved so that the contribution of the drain voltage is linearised:
$$Vt = Vt(Vds=0) - Dvt(Vbs) \cdot Vds \quad (1)$$
So doing, the analytical expression of the saturation voltage is greatly made easy. It can be noticed that the drain static feedback parameter Dvt is substrate bias dependent. Neglecting this last dependency causes an additional error of 10% on the threshold voltage (Vbs variing from 0 to -5V). Fig. 1 shows a comparison between the model (lines) and experiments (circles) (Lengths ranging from 1.1 µm to 24.6 µm and widths from 4.3 µm to 1.3 µm).

### 2.2 The linear regime.
This regime is modelled by the expression (2). The effect of the vertical (ET) and lateral (EL) electrical fields on mobility are taken into account.

The result obtained for the current is expressed as:

$$Id = \frac{M0 \cdot (Vgs - Vt - (1-d) \cdot Vds/2) \cdot Vds}{\left[ [1 + Tg \cdot (Vgs + Vt - 2 \cdot (Vfb + Dphif) - (1-d) \cdot Vds/2]^b + [Td \cdot Vds]^b \right]^{1/b}} \quad (2)$$

b=1 holds for P-Channel, and 2 for N-Channel.

Hence the access resistance is taken into account, this expression leads to computed M0 and Tg which has been proved to be independent of Vbs and the channel width. It has been shown that a so-called "3/2 model" is not necessary for the range of applied voltages usually used in VLSI circuits. Such a model complicates the algorithm, without making any noticeable improvement. Moreover, anticipating the saturation regime, this formulation of the mobility, besides the fact that it has been proved to be more physical [4], leads to a direct calculation of the saturation voltage. In other words, a Newton method is not necessary to obtain the saturation voltage. (It is often done in sophisticated analog models which assume a continuity of the conductance at the saturation point).

### 2.2 The saturation regime.

In the saturation region a pseudo 2-D solution of the Poisson equation is used (Fig. 2). If the continuity of the potential, electrical field, and it's derivative is assumed at the pinch-off point (P), and if the mobility is constant, then the current, the conductance and the derivative of the conductance are continuous at the saturation point (Vds=Vdsat). So, a self consistently smooth transition exists between linear and saturation regime for the conductance. For electrical field dependent mobility, this characteristic is only true if a full numerical calculation is performed. So the only mean to obtain a smooth transistion for the conductance is to assume, as an hypothesis, the continuity of the current, the conductance and it's first derivative. This result has been obtained without applying any smoothing procedure (as in [5] ), with only one additional parameter in saturation (y0).
According to this, the current is expressed as (N-channel):

$$Ids = Idss \cdot \frac{[1 + (teta \cdot Vdsat)^2]^{1/2}}{[(1 - ld/L)^2 + (teta \cdot Vdsat)^2]^{1/2}} \quad (3)$$

with:

$$ld = y0 \cdot Log \left\{ \frac{Ud + (Ud^2 - U0^2 + (Y0 \cdot Ep)^2)^2}{U0 + y0 \cdot Ep} \right\} \quad (4)$$

y0 is a measured parameter ( oxide thickness and junction depth dependent), Ep=f(Vgs,L,Teta,gdsat), the electrical field at the point P for Vds=Vdsat, is an internal parameter.

$Ud = Vds - V0$, $U0 = Vdsat - V0$, $V0 = Vgs - Vfb - D2v \cdot y0^2$,

D2v is an internal parameter which insure the continuity of the derivative of the conductance at the saturation point.
Teta is a simple function taking into account the gate (ET) and drain (EL) electrical field effect on the mobility.
Idss is the saturation current for Vds=Vdsat (expression (1)), the threshold voltage Vt is Vds dependent (drain static feedback effect), even for Vds>Vdsat. For long channels, Idss is not Vds dependent.

### 2.3 The Weak-Avalanche regime.

The weak avalanche effect on the threshold voltage is taken into account by a CAD version of a more sophisticated physical model [6]. The substrate current is not an extra current, so as to keep the architecture of the models implemented in the ELDO simulator. The parameters related to the substrate current are extracted between Gdmin and Vdmax, the real range of interest (see Fig.3 for Lelec=1.1 µm ).

It is well known that the substrate current effect is maximum for Vgs-Vt=Vds/2. For classical VLSI circuits it leads to an important effect for Vgs close to 2V. For short channels this effect comes mainly, in the classical range of voltage operation, from an internal and positive polarisation of the substrate, generating a negative threshold voltage shift. Obviously the substrate current is added to the Channel current. The effect is generally weak on drain current, whereas it can be noticeable on the gds characteristic for Vds>3.5V (the conductance can be multiplied by a factor of two for Vdmax, leading a reduction of the voltage gain ). For long channels the effect on gds is also present, but, as far as the conductance is weak, it is mainly due to the derivative of Isub versus Vds (see Fig.4 ). So, for modern technologies (thin gate oxide, shallow junctions) the substrate current can no longer be neglected, even for long channels .

## 4. PARAMETER EXTRACTION AND EXPERIMENTAL RESULTS.

It is well known that accurate parameter extraction is as important as the quality of the model itself. The extraction procedure takes into account the error on current as well as on conductance. Moreover these errors are weighted by a function, wei(i), which can be adjusted so that a specific experimental curve is favoured: for example the curve corresponding to a given polarisation point (i.e 0.5V above Vthreshold).

A modified-Gauss method and the Levenberg-Marquard methods are used together with a cyclic relaxation one (following the penalty function which is used and the parameters which are opimized).

Typically 4 parameters are computed in the Ids(Vds) and Gds(Vds) curves. We can choose to optimize the whole set of parameters on the curves or to "decouple" the optimization cycle.

Fig. 5 shows a comparison between the modelled (lines) and measured (circles) results for a channel length of 1.1 µm . The average error is close to 1 to 2 % for the drain current, and 4 to 9 % for the conductance.

Fig. 6 shows a comparison theory-experiment for a P-channel (L=1.05 µm ).

One can notice that, to our knowledge, only very few CAD models can sustain such a comparison between theory and experiment.

[1] ELDO: Simulateur Electrique à relaxation pour VLSI logiques et Analogiques.
    *J.Y Chan Yan Fong, B. Hennion, T. Pédron, P. Senn. Conférence sur la Fiabilité, Bayonne 1986.*
[2] An improved CAD MOS transistor MOdel, extraction procedure and characterization of the parameters.
    *T. Pédron, C. Denat, G. Merckel, ESSDERC, Aachen, 1985.*
[3] The voltage-doping transformation: A new approach to the modeling of MOSFET short-channel effets.
    *T. Skotnicki, G. Merckel, T. Pédron, IEEE electron Device letters, March 1988.*
[4] The hot-electron problem in small semiconductor devices
    *W. Hänsch, M. Miura-Mattausch, J. Appl.Phys. 60 (2), 15 july 1986, p.650.*
[5] Compact modelling of the MOSFET drain Conductance
    *F.M Klaasen, R.M.D Velghe, ESSDERC, Berlin, 1989.*
[6] A re-examination of the physics of multiplication-induced breakdown in MOSFET's.
    *T. Skotnicki, G. Merckel, A. Merrachi, IEDM 1989, p.87.*

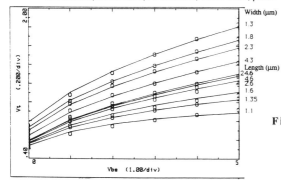

Fig.1 Threshold voltage versus Vbs
several different
Lengths and Widths
experiment (o), theory (-)

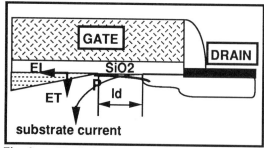

Fig. 2 : Physical phenomenons affecting the conductance in saturation.

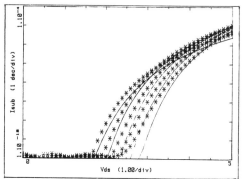

Fig. 3  Isubstrate versus Vds (L=1.1 µm) experiment (*), theory (-)

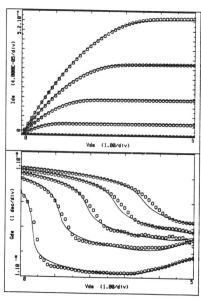

Fig.4  Current and conductance for a long channel (L=24.6 µm) experiment (o), theory (-)

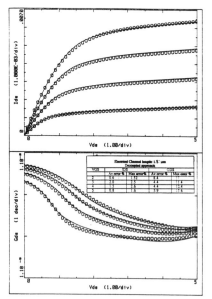

Fig.5  Ids and Gds, Vgs=2,3,4,5 V L=1.1 µm, N-channel experiment (o), theory (-)

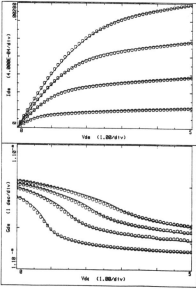

Fig.6  Ids and Gds, Vgs=-2,3,4,5 V L=1.05 µm, P-channel experiment (o), theory (-)

# A fully modular 1 μm CMOS technology incorporating EEPROM, EPROM and interpoly capacitors

Philip J Cacharelis, Michael J Hart, Graham R Wolstenholme,
Roger D Carpenter, Ian F Johnson and Martin H Manley.

National Semiconductor Corporation, Santa Clara, CA, 95052-8090, USA.

## Abstract

This paper will describe a modular technology which uses a novel integration scheme to include double poly EEPROM, single poly EPROM and an interpoly capacitor. The single poly EPROM [1] has been adopted to simplify the integration issues; the three modules (EEPROM, EPROM and A/D) can be combined in any combination without affecting their electrical performance.

## Introduction

Scaling of CMOS technologies has allowed integration of whole systems onto one chip. The differing technology requirements of the various circuit blocks has led to a 'modular' approach to technology development [2]. One example of a highly integrated system is the microcontroller emulator which uses EPROM program memory (to allow simple updates of operating code), and EEPROM data storage (to allow non-volatile storage of trimming variables and error flags). These non-volatile modules have been merged into a single integration scheme so as to retain their independent performance and allow the generation of a standard cell library.

## Process Technology

The modular technology is derived from a baseline 1 μm single polysilicon, double metal CMOS logic process, which has a 200 Å gate oxide and a conventional LDD NMOS transistor. A cross-section of the modular devices and the processing sequence are shown in figure 1 and table 1, respectively.

The EEPROM module consists of all process steps shown in table 1, except those specifically for the EPROM (♦); process steps shared by both modules (*♦) are also retained. The two EPROM mask and implant steps have no impact on EEPROM or interpoly capacitor performance. The module includes a non-self-aligned double polysilicon memory cell [3] and high voltage NMOS enhancement and depletion transistors. The high voltage transistors have a 325 Å gate oxide and DDD junctions. A first layer of polysilicon (2750 Å) is added to the baseline process to serve as the floating gate of the EEPROM cell. An ONO interpoly dielectric provides capacitive coupling between the control gate and the floating gate. The high and low voltage device gates and the control gates of the EEPROM cells are formed with the second (baseline) layer of polysilicon (4000 Å). The low voltage devices are formed as in the baseline process. The interpoly capacitor structure is an integral part of the EEPROM process and does not require additional masks.

A single polysilicon EPROM device was used instead of a conventional stacked-gate double-polysilicon EPROM to ensure simpler integration with the EEPROM process. The EPROM module contains all baseline processes plus two masked implants (♦) and processes shared with the EEPROM (* ♦) [table 1]. The cell uses the second layer of polysilicon as its floating gate and the baseline gate oxide of 200Å. The baseline CMOS Vt implant serves as the programming implant of the cell. An N+ implant forms the diffused control gate, which is coupled to the floating gate via a 235 Å oxide grown during the gate oxidation cycle. An EPROM N⁺ source/drain implant is used to optimize programming speed and disturb margin [4].

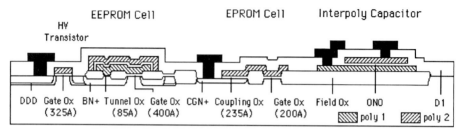

Figure 1. Cross-section of Modular Elements.

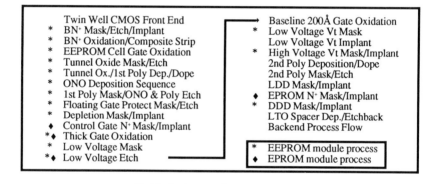

Table 1. Process Sequence

## EEPROM Cell and High Voltage Transistors

The memory cell is a standard floating gate design using buried N⁺ (BN⁺) source/drain regions and a control gate which fully overlaps the floating gate. The cell gate oxide is 400 Å, the tunnel oxide is 85 Å and the ONO oxide equivalent thickness is 260 Å. The cell area is 72 µm², the coupling ratio is 0.77 and the memory transistor W/L is 2.0/1.8 µm. The cell threshold voltage window versus log time for various programming and erase voltages is shown in figure 2. The cell is programmed to a Vt window of 4.4 V with 14.5V, 10ms programming pulses. The cell read characteristics for a Vt of 0 V are plotted in figure 3 for a range of read bias. The typical read current is 110 µA under nominal bias conditions. The data loss for a 160 hour, 250 °C bake is less than 200 mV. The memory cell endurance characteristic is shown in figure 4. The total Vt window decay is 0.6 V for 1.0E5 cycles.

The high voltage enhancement transistor has a gate oxide of 325 Å and DDD (Phosphorus/Arsenic) source/drain junctions. This device is required to provide a high junction breakdown voltage (15.0V @ 1 nA) to allow the use of an on-chip charge pump. A high voltage depletion transistor serves as the word-line select gate in the EEPROM array. This transistor allows the full programming voltage to reach the control gates of the array without a threshold voltage drop.

Silicon circuits 549

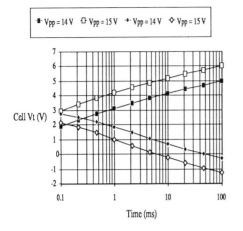

Figure 2. EEPROM Cell Programming Window

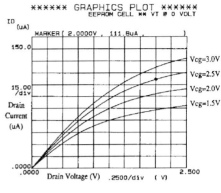

Figure 3. EEPROM Cell Read Characteristic.

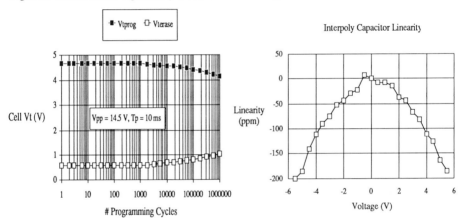

Figure 4. EEPROM Cell Endurance Characteristic

Figure 5. Interpoly C-V Characteristic.

Interpoly Capacitor

The ONO interpoly capacitor provides a high capacitance per unit area, high dielectric field strength and low voltage coefficient for A/D circuits. This capacitor exhibits lower parasitic capacitance than a polysilicon to junction capacitor. The ONO film (260 Å oxide equivalent) exhibits an equivalent dielectric field strength of 8.0 MV.cm$^{-1}$ for a current flux of 10µA.cm$^{-2}$ and has a capacitance per unit area of 1.3 fF.µm-2. A capacitance versus voltage curve for biases of ±5.5 V is shown in figure 5. The capacitor exhibits a worst case voltage coefficient of 35 ppm.V$^{-1}$ over the full Vcc range of ±5.5V.

EPROM Cell

In the single polysilicon EPROM cell, a 235 Å coupling oxide is grown on the control gate N$^+$ region during the 200 Å gate oxidation. The EPROM cell area is 30 µm², it has a W/L of 1.0/0.9 µm and a coupling ratio of 0.7. The programming and read characteristics for the cell are shown in figures 6 and 7, respectively. The cell requires a pulse length of 200 µs to reach

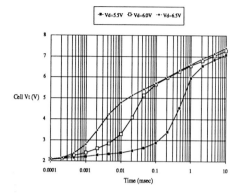
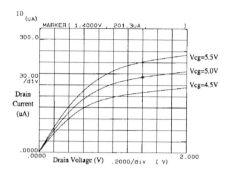

Figure 6. EPROM Cell Programming Characteristic.    Figure 7. EPROM Cell Read Characteristic

a Vt of 6.0 V under typical programming conditions of Vd=6.0V, Vcg=12.0V. This is accomplished without the use of an additional programming implant. The read current is 200 µA under nominal bias conditions and the data loss for a 160 hour, 250 °C bake is less than 400mV.

Conclusion

A novel fully modular 1 µm CMOS process flow; incorporating EEPROM, EPROM, and interpoly capacitors, has been described. The use of a single poly EPROM cell simplifies the integration issues. The two non-volatile modules and the interpoly capacitor retain the same device characteristics and design rules whether they are used individually or in a combined technology. This allows the formation of a standard cell library of these devices and their support circuitry. This technology and its modules is designed for the fabrication of non-volatile 16-bit microcontrollers.

Acknowledgments

The authors wish to thank the Fairchild Research Center's Non-volatile Memory Process Development and CMOS Unit Process Development groups, and the FRC Production Staff.

References

[1] P. Cacharelis et al., 1988 IEEE IEDM Technical Digest, pp. 60-63.

[2] J. Paterson, 1989 IEEE IEDM Technical Digest, pp. 413-416.

[3] W. Johnson et al., 1980 IEEE ISSCC Technical Digest, pp. 152-153.

[4] M. Hart et al., 10th IEEE Non-Volatile Semiconductor Memory Workshop, 1989.

Paper presented at ESSDERC 90, Nottingham, September 1990
Session 7B1

# Optical switches and heterojunction bipolar transistors in InP for monolithic integration

N Shaw, P J Topham and M J Wale

Plessey Research Caswell Ltd
Towcester, Northants, NN12 8EQ, UK

## Abstract

We report optical switching devices and heterojunction bipolar transistors (HBTs) in the InGaAsP/InP material system which are designed to be fully compatible with monolithic integration. One early application envisaged for these integrated structures is an optical switch whose external interfaces employ standard logic levels (e.g. ECL), instead of the higher drive voltages (10-20V) of the electro-optic devices themselves. In the longer term, we envisage optical switching components which are capable of performing self-routing functions based on information contained within the data stream. The heterojunction bipolar transistor is an excellent driver for the optical switch, since very high gain and transition frequency can be achieved.

## 1. Introduction

Optical technologies are playing an increasingly important role in the development of telecommunication networks and the use of fibre optics in transmission has created a demand for other functions (e.g. switching) to be carried out in the optical domain. An efficient form of interfacing is required between the electronic and optical functions [1,2], and so the monolithic integration of optical switching with electronic control functions is an attractive prospect. The InP based material system is an ideal candidate for such integration as it can support the optical source, switch and detector at the fibre optic transmission wavelengths of 1.3 and 1.55 µm, along with the necessary drive transistors. Heterojunction bipolar transistors (HBTs) are particularly suitable for this application because of their high transconductance, high speed and low output conductance. In addition, HBTs do not require the submicron linewidths characteristic of other transistors of comparable speeds (e.g. FETs) and are thus more amenable to fabrication on the non-planar surfaces which are likely to occur in complex circuits. Expertise in using InP based HBTs in complex high speed electronic integrated circuits has already been demonstrated [3,4].

In this paper we specifically investigate the possibility of the monolithic integration of an optical switch with a heterojunction bipolar drive transistor. In this initial experiment the two components were fabricated on separate MOVPE wafers using identical epi-layer structures, thus demonstrating the potential for monolithic integration with a single epitaxial growth.

## 2. Device Design

The layout of the 2x2 switch is shown in figure 1. Two waveguides 2.5 or 3.0 µm in width form a coupler (guide separation 2.5 - 3.5 µm, length 1-5 mm), the waveguides diverging at the input and output to allow for fibre attachment. Electrode connections to the coupler are for reverse delta-beta operation of the switch.

The layer structure used for the switch and HBT is shown in figure 2. The low n-doped InGaAsP layer which provides the guiding core region of the waveguides is set in low n-

© 1990 IOP Publishing Ltd

doped InP cladding layers. A highly doped p-type layer provides contact to the metal electrodes on top of the device. The n-type contact is made to the highly doped n-type layer below the waveguide layers. Lateral confinement of the optical beam is achieved by etching a rib into the upper cladding layer. The switch is operated by reverse biasing the p-n junction and thereby applying an electrical field across the optical beam. The resulting electro-optic effect is used to effect a phase shift between the two modes of the coupler and thus switch the power between the two coupled guides.

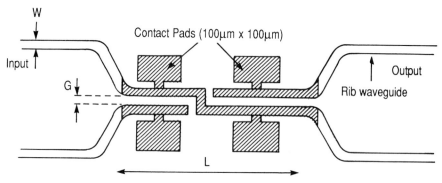

Figure 1. Layout of 2x2 switch

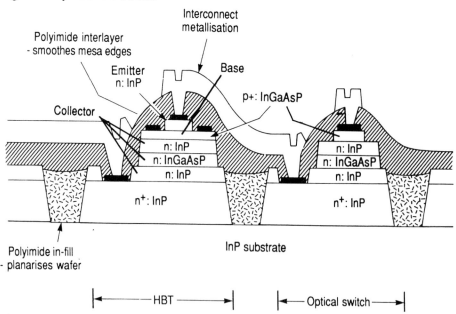

Figure 2. Outline of optical switch/HBT integration scheme

The HBT in fabricated on an InP/InGaAsP epi-layer structure identical to that used for the switches. As can be seen from figure 2 the core and cladding layers of the switch are incorporated into the n-type collector of the transistor. The p-type contact layer of the switch forms the base of the transistor. An InGaAsP alloy of bandgap 1.15 μm was chosen for the base instead of the GaInAs previously used [3,4], since this layer also appears within the switch structure and it is therefore important to avoid absorption of light at the required operation wavelengths of 1.3 or 1.55 μm. The n-doped InP emitter layer which is grown on top of the base layer is readily removed from the switch regions during fabrication. The use of a wider bandgap material (InP) for the emitter than for the base introduces a heterojunction which blocks the reverse injection of carriers from the base, thus allowing the base to be heavily doped without compromising the current gain ($\beta$) characteristics of the transistor. The mesa profile shown in figure 2 enables the necessary electrode connections to be made to the base, emitter and collector layers of the transistor.

The integration scheme employs circuit resistors in nichrome with two levels of interconnect metallisation and is an adaptation of our previous work in GaAs HBTs [5]. Isolation of the devices is achieved by etching through the n-type contact layer of the devices. Two layers of polyimide are used to planarise the surface of the wafer and form an isolation layer, allowing a second layer of metallisation to provide the necessary interconnects to the devices.

### 3. Experimental results

The HBTs were tested in the common base mode, in which the combination of high transconductance and low output conductance yields the effective high voltage gains appropriate for this application. Figure 3 shows the characteristics of the transistors.
The HBTs are capable of providing the 20V necessary to drive the transistors and the common base current gain is close to unity. This particular device has a maximum common emitter current gain of 218 at a collector voltage of 10V and a collector current of 1.3 mA.

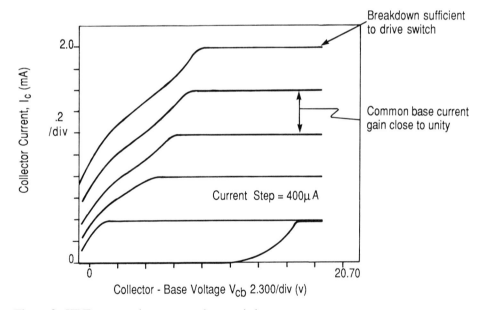

Figure 3. HBT common base output characteristics

Optical assessment of the switches was carried out using a 1.54 µm fibre-pigtailed laser. Figure 4 shows the response of device with L = 3 mm , G = 2.5 µm and W = 2.5 µm. The switching voltage from bar to cross state is 19.8V. This result is comparable to those of switches fabricated in other InP/InGaAsP structures and thus confirms the ability of the integrated structure to support efficient switching.

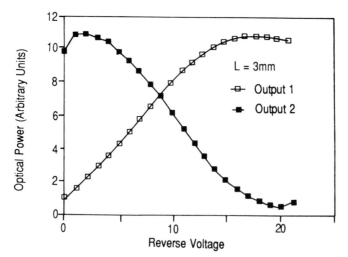

Figure 4. Performance of optical switch using HBT compatible layer structure

## 4. Conclusion

A scheme for the monolithic integration of 2x2 optical switches and HBTs has been presented. The two functions have been demonstrated separately in identical InP/InGaAsP layer structures. The maximum gain of the transistors of over 200 is more than adequate and the switching voltage of 20V for the switch can be derived in a single stage from standard logic levels.

It is anticipated that this advance in InP technology will provide a basis for the full integration of optical switching functions with advanced drive circuits and self-routing functions.

## 5. Acknowledgments

The authors wish to thank A K Wood for the epitaxial growth of the material and I G Griffith and J C Hendy for device fabrication. This work is supported in part by GEC-Plessey Telecommunications Ltd and by the CEC under RACE project 1033 (OSCAR).

## 6. References

[1] Whitehead, N. and Parsons, N., "Network applications of photonic switching", Proc. 15th European Conf. Optical Communications, Gothenburg, Sweden, 1989, vol. 2, pp. 110-113.
[2] Bennion, I., Thylen, l. and Whitehead, N.F., "Electro-optic switching systems", Proc. IEEE Int. Conf. Communications, Atlanta, USA, 1990, pp 1130-1134.
[3] Topham, P.J., Griffith, I., Riffat J., and Goodfellow R.C., "Heterojunction bipolar transistors using GaInAs/InP", IEE Colloquium on heterojunction and quantum well devices, London, 27 Oct 1988.
[4] Topham, P.J, Thompson, J., Griffith, I., Hollis, B.A., Hiams, N.A., Parton, J.G. and Goodfellow, R.C., "Digital integrated circuit using GaInAs/InP HBTs", Electron. Lett., **25** (17), pp. 1116-1117 (1989).
[5] Parton, J.G., Topham, P.J., Golder, M.J., Taylor, D.G., Hiams, N.A., Holden, A.J. and Goodfellow, R.C., "An HBT programmable hexadecimal counter IC clocked at 2.6 GHz", GaAs IC Symposium 89, San Diego, USA, Tech. Digest, pp. 61-63.

*Paper presented at ESSDERC 90, Nottingham, September 1990*
Session 7B2

## Numerical modelling based comparison of the submicrometre III–V compounds MESFET's performance

V.Ryzhii and G.Khrenov

Institute of Physics and Technology USSR Academy of Sciences
Krasikov Str. 25a, Moscow, 117218 USSR

**Abstract.** Performances of submicrometer scaled-down MESFETs based on $Ga_{0.47}In_{0.53}As$, InP and GaAs are investigated by Monte-Carlo two-dimensional numerical simulation method.

Recently the sufficient advances were achieved in the fabrication techniques of semiconductor elements for monolitic microwave circuits and high-speed digital integrated circuits with picosecond delay-time range. As a rule GaAs is used for fabrication of these elements. Unfortunately, rather small energy gap between central and satellite valleys prevents future improvement of the submicrometer gate MESFET's characteristics. For this reason, the expected improvement of the submicrometer-gate MESFET's characteristics could be realized by the using of such semiconductor materials, as $Ga_{1-x}In_xAs$ and InP, with sufficient intervalley gap. In this paper characteristics of the sibmicrometer-gate MESFETs based on GaAs, InP and $Ga_{1-x}In_xAs$ are investigated by two-dimensional numerical simulation method.

In our numerical simulation the electron transport in the MESFET's channel is described with a kinetic model. This model takes into account non parabolic multi-valley electron spectrum in materials and also all essential scattering mechanisms. The presence and recharge of the deep traps in semiinsulating substrate is considered too. The combined particle method (which include Monte Carlo particle method for electron description and stochastic dynamic method for description of the deep traps recharge in the semiinsulating substrate) is used. This combined technique shows high computational efficiency. The Poisson's equation for the potential of the selfconsistent electric field is solved by direct technique with the aim of "capacitance matrix" method.

Scaled-down sets of the MESFET's based on the mentioned above materials with gate lengths from 1 to 0.1 µm are investigated. MESFETs on $Ga_{1-x}In_xAs$ (x=0.53) are assumed to be fabricated on the InP semiinsulating substrate. MESFET with gate length of 1 µm, channel thickness of 0,3 µm and donor concentration in the channel of $2,5 \cdot 10^{16} cm^{-3}$ is chosen as a prototype for scaling.

The distributions of the average electron velocity along the source-to-drain axis for MESFETs with $L_g=0.5$ μm are presented on Fig.1. As shown on Fig.1 the distributions of the average electron velocity for GaAs and InP are similar and they differs only for their maximum value. Larger value of the peak average electron velocity for InP MESFETs is concerned with the late electron transition from central to satellite valleys. At the same time the electron velocity on the initial stage for InP MESFET is smaller than the one for GaAs MESFET. This fact may be explained as follows. Electron scattering rate on polar optical phonons for InP is larger than the one for GaAs. Because of that the dynamic properties of the electron gas in InP is worse in comparison with the one in GaAs, and hence the average electron velocity for InP is smaller.

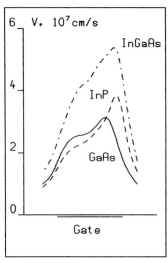

Fig.1

Calculations show that in spite of larger value of maximum electron velocity, the transient time for InP MESFET with gate length 0.5 μm is approximately equal to the one for the GaAs MESFET with the same size.

The variations of the transient time τ as a function of the drain voltage $V_d$ are presented on Fig.2. The analysis of the dependencies shows that for InGaAs MESFETs τ quickly decreases with the increase of $V_d$, and then reaches the saturation value. Saturation is mainly concerned with the strong intervalley scattering, and it begins when the drain voltage amount to value corresponding the intervalley gap energy. For InP and GaAs transistors transition to saturation is more smooth. The latter fact is explained by the greater value of the scattering rate in this materials. Fig.3 shows the dependencies of efficient electron velocity $V_{eff}=L_g/\tau$ as a function of the gate length for considered material. One can see on Fig.3, that for all gate length the electrons in the channel of the InGaAs MESFETs have the largest efficient velocity in comparison with other materials. Comparison of

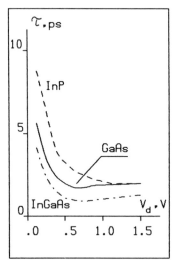

Fig.2

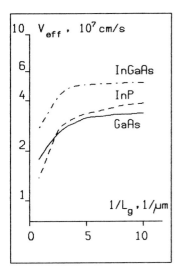

Fig.3

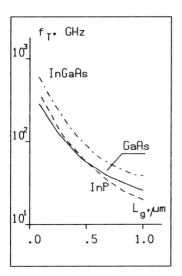

Fig.4

the GaAs and InP MESFETs shows that the InP MESFETs have small advantages when the gate length becomes less than 0.5 μm, and lose their advantages with the larger gate lengths. This phenomenon may be explained as follows. The gain of the average electron velocity due to greater intervalley energy gap for InP can either be masked in long-channel MESFETs by loss of the velocity on the initial stage of the electron acceleration or give essential contribution to value of efficient electron velocity in short-channel MESFET.

Variations of the cut-off frequency $f_T$ as a function of the gate length are shown on the Fig.4. It is found that InGaAs MESFETs are most high-speed among other. High frequency performances of the GaAs and InP MESFETs are founded to be similar. The cut-off frequency of InP MESFETs is only slightly better when the gate length becomes smaller than 0.5 μm. Now it is clear, that for the further increasing of the MESFET's operation speed it is necessary to use semiconductor materials having not only large intervalley energy gap but also fine dynamic properties of the electron gas.

Paper presented at ESSDERC 90, Nottingham, September 1990
Session 7B3

# Small signal analysis of resonant tunnel diodes in the bistable mode

A Zarea, A Sellai and M S Raven, Department of Electrical and Electronic Engineering, University of Nottingham and D P Steenson, J M Chamberlain, M Henini and O H Hughes, Department of Physics, University of Nottingham

Abstract. The small signal impedance of GaAs/Al(Ga)As double-barrier resonant tunnel diodes has been measured and analyzed over a wide range of frequencies and for various d.c. bias values. The RTD equivalent circuit elements were obtained by measuring the impedance in the bistable region. The cut-off frequency and the self-resonant frequency were calculated and compared with direct measurements.

1. Introduction

Since the first proposal of the resonant tunnel diode by Tsu and Esaki (1973), there has been much interest in RTDs because of their potential for high-speed applications in analogue and digital circuits. The electrical properties of RTDs can be calculated using an equivalent circuit model. In this paper we analyse the impedance of RTDs and evaluate the equivalent circuit elements using direct measurements of the impedance in the bistable region of the RTD characteristic.

2. Device Structure

The RTDs used in this study were grown by MBE on $n^+$ Si-doped GaAs substrates. Device NU195 comprised a 5 nm GaAs well, isolated by 5.6 nm AlGaAs barriers, a 2.5 nm spacer and a 50 nm buffer layer Si-doped to $2 \times 10^{16} cm^{-3}$, a bottom 2 um and a top 0.2 um layer of GaAs Si-doped to $2 \times 10^{18} cm^{-3}$. NU195 had a mesa diameter of 50 microns and was mounted in a TO5 package. Device NU298 comprised a 4.3 nm GaAs well, isolated by 1.7nm AlAs barriers, a bottom 420 nm and a top 500 nm layer of GaAs Si-doped to $2 \times 10^{17} cm^{-3}$, and a 3 um bottom buffer layer of GaAs Si-doped to $2 \times 10^{18} cm^{-3}$. NU298 had a mesa diameter of 20 microns and was mounted in a S4 package. Figure 1 shows the I-V characteristic of the two devices.

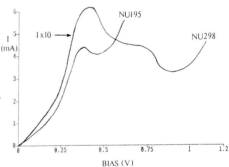

Fig 1 The I-V characteristic of RTDs NU195 (GaAs/AlGaAs) and NU298 (GaAs/AlAs) at room temperature.

© 1990 IOP Publishing Ltd

## 3. Measurement Technique

RTDs and calibration devices were mounted in a co-axial attachment and the impedance of the RTDs obtained using a HP4191 Impedance Analyzer, frequency range 1–1000 MHz and a −20 dBm signal and a HP8510 Network Analyzer for frequencies above 1 GHz. Both machines were calibrated by using an open circuit, a short circuit and a standard 50 ohm load. In order to minimize the package effect the calibration used the same type of package for the open and short circuits as the device under test. The RTDs were biased at different voltages using an external power supply. The equivalent capacitance ($C_{eq}$) and the resistive ($R_e$) and reactive ($X_e$) parts of the RTD impedance were measured over a frequency range of 1 MHz to 1 GHz at various dc bias values throughout the bistable region.

## 4. Results and Analysis

The small signal RTD equivalent circuit is shown in figure 2. This equivalent circuit holds well at frequencies up to 1 GHz Gering et al (1987) and Brown et al (1988) and verified independly as shown in Figure 3. The parasitic impedance associated with the co-axial attachment and device package was automatically subtracted from the measured data following calibration.

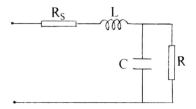

Fig 2 The small signal equivalent circuit of the resonant tunnel diode.

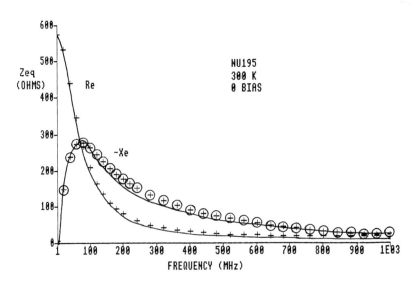

Fig 3. The measured (+) and calculated (−) resistive and reactive parts of the RTD NU195 versus frequency.

Hence R, C, L and $R_s$ in figure 2 are mainly due to the intrinsic properties of the RTD. Plots of the measured $R_e$, $X_e$ and $C_{eq}$ against d.c. bias at 100 MHz are given in figure 4.

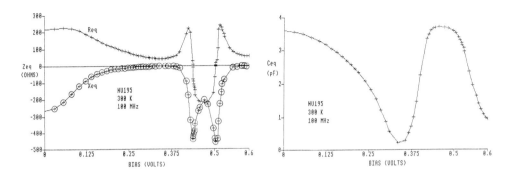

Fig 4 The measured $R_e$ and $X_e$ (a) and $C_{eq}$ (b) of RTD NU195 plotted against dc bias at 100MHz.

The resistive and reactive parts of the impedance of figure 2 are given by

$$R_e = R_s + R/[1 + (\omega C R)^2] \quad \ldots 1$$

$$X_e = \omega (L - C R^2/[1 + (\omega C R)^2]) \quad \ldots 2$$

The equivalent capacitance is

$$C_{eq} = X_e / \omega (X_e^2 + R_e^2) \quad \ldots 3$$

At the peak and the valley of the I-V curve R is infinite and equations 1 and 2 become

$$R_e = R_s \quad \ldots 4$$

$$X_e = \omega L - 1/\omega C \quad \ldots 5$$

Using the values of $R_e$ and $X_e$ at the peak or valley of the I-V characteristic curve $R_s$ was obtained from equation 4. C and L were determined using equation 5 at two different frequencies. R was determined by finding the reciprocal of the negative slope of the I-V curve in the NDR region. For RTD NU195 the values for R, $R_s$, L and C were 4 $\Omega$, -350 $\Omega$, 3.12 nH and 3.58 pF; and for RTD NU298 were 3.6 $\Omega$, - 30 $\Omega$, .75 nH and 1.5 pF respectively. Using these parameters $f_{co}$ and $f_{sr}$ were calculated using the following equations

$$f_{co} = (1/2\pi C) (-1/RR_s - 1/R^2)^{1/2} \quad \ldots 6$$

$$f_{sr} = (1/2\pi C) (C/L - 1/R^2)^{1/2} \quad \ldots 7$$

Table 1 is a comparison between the calculated and the measured $f_{sr}$ and $f_{co}$. The measured values of $f_{sr}$ and $f_{co}$ were determined directly from the measured impedance plotted on Smith charts for $X_e = 0$ and $R_e = 0$ respectively.

Table 1

|  | Calculated GHz | | Measured GHz | |
|---|---|---|---|---|
|  | $f_{sr}$ | $f_{co}$ | $f_{sr}$ | $f_{co}$ |
| NU195 | 1.50 | 1.18 | 1.48 | 1.10 |
| NU298 | 3.16 | 9.58 | 3.01 | 9.25 |

## 5. Discussion and Conclusions

The small signal impedance of GaAs/Al(Ga)As double barrier resonant tunneling heterostructure diodes was measured and the equivalent circuit elements were determined using direct measurements of the impedance in the bistable region of the I-V characteristic curve. $f_{sr}$ and $f_{co}$ were calculated using the determined equivalent circuit elements then compared with values measured directly using Smith charts. Although in the bistable mode $R_s$, L and C were found to be almost independent of bias, the impedance of the RTD is dominated by R which is strongly dependent on the bias. The variation in R causes the equivalent capacitance to increase and the resistive and the reactive parts of the impedance to have sharp peaks at the peak and valley of the I-V curve in the bistable region of the RTD characteristic. The good agreement between the measured and the calculated values for both $f_{sr}$ and $f_{co}$ suggest that the values of the lumped equivalent circuit elements are close to the intrinsic values of the device structure. L is close to the intrinsic value predicted using an empirical equation relating L to the barrier thickness suggested by Gering et al (1987). However this value is larger than the inductance estimated using the electron delay in the quantum well.

Acknowledgement: We thank the SERC for partial support and the Saudi and Algerian Ministries of Higher Education for research studentships. We also thank Dr M Davies and Dr M Heath for device processing.

REFERENCES

Tsu R and Esaki L 1973 Appl. Phys. Lett. 22   562-4
Chang L L, Esaki L and Tsu R 1974 Appl. Phys. Lett. 24 593-5
Luryi S 1985 Appl. Phys. Lett. (USA) 47   490-2
Gering J M. Crim D A, Morgan D G, Colman P D, Kopp W and Morkoc H 1987 J. App. Phys. 61   271-6
Brown E R, Goodhue W D and Sollner T C L G 1988 J. App. Phys. 64   1519-29

## Planar InP/InGaAs avalanche photodiodes fabricated without a guard ring using silicon implantation and two-stage atmospheric pressure MOVPE

M D A MacBean, P M Rodgers, T G Lynch, M D Learmouth and R H Walling

British Telecom Research Laboratories, Martlesham Heath, Suffolk, U.K.

Abstract. We report the design, fabrication and performance of separate absorption, grading and multiplication layer avalanche photodetectors formed without a guard-ring by using localised implantation of silicon to define the field control layer and atmospheric pressure MOVPE to obtain large areas of uniform low doped material. The described process is potentially more controlled than conventional avalanche photodetector fabrication schedules, and yields high gain, low noise devices capable of operating at bit rates up to at least 2Gbit/sec.

1. Introduction

Future high bit rate communications systems require detectors with fast response, low noise and high responsivity in the wavelength range 1.3-1.55µm. InP/InGaAs avalanche photodiodes (APDs) satisfy these requirements (Kasper et al 1987). Most current designs are based on a separate absorption, grading and multiplication (SAGM) structure which places a very tight constraint on the doping and thickness of the doped InP layer used to control the field in the InGaAs absorption region (Forrest et al 1982). Epitaxial growth techniques can have difficulty meeting these tight tolerances. An alternative is to form the field controlling layer from implanted silicon (Donnelly et al 1979, and Webb et al 1987). Implantation allows a much more precise control of the amount of dopant in the field controlling layer and can thus potentially improve yield and wafer uniformity. An additional advantage of this implantation technique is that it allows the localised implantation of the field controlling layer thus effectively defining the gain region and avoiding edge breakdown without the use of a guard ring (Donnelly et al 1979, and Webb et al 1987).

In this paper we describe the design, fabrication and performance of such SAGM APDs using atmospheric pressure MOVPE to obtain large areas of uniform low doped material and silicon implantation to define the field control layer.

2. Design

The Si APD layer structure is shown in Figure 1. Several criteria (Forrest et al 1982) were taken into account in arriving at this design: eg electric field strength at the grading layer was designed to be less than $\approx 1.5 \times 10^5$ V/cm to avoid tunnelling currents in the InGaAs absorption layer but higher than $\approx 8 \times 10^4$ V/cm to avoid hole trapping; junction depth was chosen to be greater than 2.5µm to minimise the problem of enhanced gain at the device perimeter due to junction curvature; and the absorption layer thickness was limited to 3µm to ensure full depletion at the operating voltage and thus eliminate slow diffusion tails from carriers absorbed in low field regions. The InP layer into which the Si is implanted was designed to be thick enough that no silicon would penetrate into the absorption layer, the implantation energies of 100-200keV used in this work resulting in a Si peak a few tenths of a micron into this layer.

© 1990 IOP Publishing Ltd

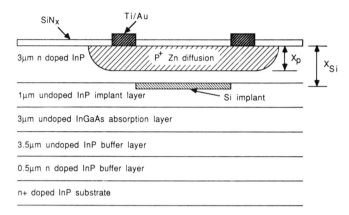

Figure 1. A schematic diagram of the silicon implanted InP/InGaAs APD design.

A detailed one dimensional numerical model of the device was developed using published ionisation rate data, an abrupt p-n junction model, and a silicon impurity profile obtained by SIMS analysis of implanted wafers. The model calculates gain and field profile as a function of applied voltage, silicon implant dose, junction position and epitaxial layer structure. For the layer structure shown, Figure 2 shows that an active silicon dose in the 2.0-2.5x$10^{12}$cm$^{-2}$ range is optimum. A higher dose than this optimum means incomplete depletion of the absorption layer, and a lower dose means high breakdown voltages and the risk of tunnelling in the InGaAs absorption layer. Figure 2 also shows the limits imposed on the junction depth from consideration of the fields within the ternary layer. A junction position of $X_{si}$-$X_p$≈0.8-1.2μm appears to be optimum: a higher value results in incomplete depletion of the absorption layer and a lower value results in tunnelling currents from the field in the absorption layer outside the silicon implanted region.

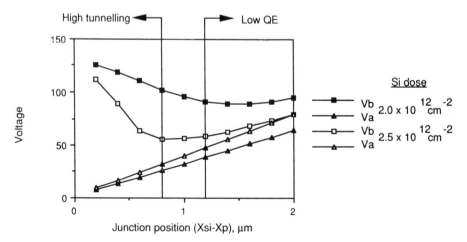

Figure 2. A plot of the voltage at which the depletion region reaches the heterointerface (Vt) and the breakdown voltage (Vb) versus distance between the p-n junction and the peak of the silicon implant.

## 3. Fabrication

Due to the limited implantation depth of Si at the implant energies used, epitaxial growth is carried out in two stages: the first stage growing to the top of the 'implant layer' of Figure 1 and the second stage growing the top InP layer after implantation has taken place. This overgrowth stage also anneals the implant. Atmospheric pressure MOVPE is used in both stages due to its ability to produce large areas of uniform, low doped material, the technology having been developed from that for PIN detectors with details given elsewhere (Nelson et al 1988). Growth takes place on 2" diameter, (100) orientation, $n^+$ InP substrates yielding thickness and doping variations of less than ±3% and ±10% respectively over the whole wafer. Plasma deposited silicon nitride is used throughout as diffusion mask/passivation layer and anti-reflection coating and the abrupt $p^+$ junction is formed by Zn diffusion through a silicon nitride mask.

## 4. Device Characteristics

Device characteristics are broadly as predicted by the model described above. Figure 3 shows a typical plot of dark current and gain in an APD implanted with silicon at a dose of $2.5 \times 10^{12} cm^{-2}$. Maximum multiplication factors as high as 200 have been obtained with typical values being around 40-50. Quantum efficiencies generally exceed 80% while total dark current is usually in the region of 100nA (0.9Vb, 50μm diameter active area) and multiplied dark current in the region of 3-4nA. These results indicate: (1) that the re-growth interface is of good quality, allowing high gain and low dark current; (2) that the silicon implant is highly activated by the regrowth stage.

In order to check the correct operation of the devices we have developed an optical beam induced current (OBIC) system to measure the gain uniformity of devices illuminated at a wavelength of 1.3μm. Figure 4 shows the gain scan of a typical device at various values of gain. It can be seen that no edge enhanced gain is observed at voltages up to breakdown, despite there being no separately formed guard-ring.

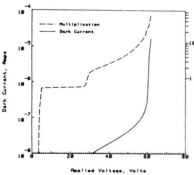

Figure 3. Dark current and gain versus voltage in a typical Si implanted APD.

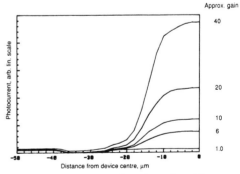

Figure 4. Gain as a function of position across a typical Si implanted device at $\lambda=1.3\mu m$.

The excess noise factor (F) was measured at a frequency of 1MHz and bandwidth of 100kHz. The result for a typical device is plotted in Figure 5 as a function of gain. Also plotted are values of McIntyre's equation (McIntyre 1966) for several values of $k=\alpha/\beta$. The measured value of k is approximately 0.4 with F<5 at M=10. Typically noise currents were less than 1.0pA/√Hz at M=10. This is comparable with the best results from conventional

APDs with separate guard-rings and will allow the APDs to operate at significant gains even when used in very low noise receiver designs (MacBean et al 1990).

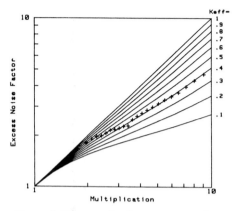

Figure 5. Excess noise factor versus gain for a typical Si implanted device.

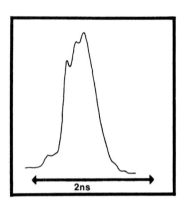

Figure 6. Pulse response of a typical Si implanted APD at 2Gbit/sec with a DC gain of 25 at $\lambda=1.3\mu m$.

The pulse response of these devices has been measured using a 2Gbit/sec pulse generator with a 250ps response time, and a conventional 1.3μm BH semiconductor laser. The resulting APD pulse response at a DC gain of 25 is shown in Figure 6. There is no evidence of a slow time-scale tail to the pulse response due to hole pile-up at the hetero-interface, and the measured response time is typically ≈330ps. These devices are thus capable of high bit rate operation.

6. Conclusions

It has been found that good quality InP/InGaAs APDs can be fabricated using silicon implantation and two stage MOVPE growth with the potential of a very controlled, high yield process. Devices show consistent, high, spatially uniform gains even without a separately formed guard-ring. The devices are capable of operating at bit rates up to at least 2Gbit/sec and show low excess noise.

Acknowledgements

We thank J.C.C. Shaw, R.F. Cobbold, J.S. Gardiner, D. Jacobs, and R. Gooding for there help in the processing of these devices and for measurements carried out.

References

Donnelly J P, Armiento C A, Diadiuk V and Groves S H 1979 *Appl. Phys. Lett.* **35** 74
Forrest S R, Smith R G, Kim O K 1982 *J. Quantum Electronics* **QE-18** 2040
Kasper B L, Campbell J C, Talman J R, Gnauck A H, Bowers J E and Holden W S 1987 *J. Lightwave Technol.* **LT-5** 344
Nelson A W, Spurdens P C et al 1988 *J. Crystal Growth* **93** 792
MacBean M D A 1990 *IEE Colloquium Digest on Optical Detectors* **014** 7/1
McIntyre R J 1966 *IEEE Trans. Electron. Devices* **ED-13** 164
Webb P P, McIntyre R J, Holunga M and Vanderwel T 1987 *SPIE Proc. Components for Fiber Optic Applications II* **839** 148

Paper presented at ESSDERC 90, Nottingham, September 1990
Session 7B5

# Estimation of noise figure for conventional and multilayered avalanche photodiodes using the lucky drift model

J.S. Marsland, R.C. Woods*, C.A. Brownhill* and S. Gould*.

Department of Electrical Engineering and Electronics,
University of Liverpool, P.O. Box 147, Liverpool, L69 3BX.

*Department of Electronic and Electrical Engineering,
University of Sheffield, Mappin Street, Sheffield, S1 3JD.

Abstract  A new technique for estimating the excess noise factor in conventional and multilayered avalanche photodiodes has been developed. It is based upon a computer simulation of carrier motion using lucky drift concepts. In conventional photodiodes the importance of the dead space is demonstrated. In multilayered photodiodes preliminary results show good agreement with other theoretical work.

1. Introduction

There has been much recent interest in the design of avalanche photodiodes (APDs) which have artificially enhanced values of $k=\beta/\alpha$, where $\alpha$ and $\beta$ are the electron and hole ionization coefficients respectively. This has arisen because of the theorem of McIntyre (1966) that the excess noise factor is minimized when $\beta \gg \alpha$ (for electron injection) or when $\beta \ll \alpha$ (for hole injection). However, this result is strictly only applicable to the case of a device where ionization events can occur at any point in the depletion region. This is not the case in a conventional APD where the dead space, or the distance that a carrier travels to achieve the threshold energy, takes up a significant fraction of the depletion region as has been demonstrated by van Vliet et al (1979) and Marsland (1990). Neither is it the case in multilayered APDs where impact ionization is localized at the heterojunctions as shown by Capasso et al (1983) and Teich et al (1986).

2. Conventional Avalanche Photodiodes

In a conventional APD the effect of the dead space on the excess noise factor can be estimated in various ways including the numerical calculation of Marsland (1990). An alternative approach reported in this paper utilizes the lucky drift model of carrier dynamics originally developed by Ridley (1983). In this model the true motion of carriers is represented by a schematic motion determined by two parameters, the ballistic mean free path, $\lambda$, and the mean free path for energy relaxing collisions, $\lambda_E$. A carrier is assumed to start with no energy and then to travel some distance without any collisions and a further distance in lucky drift mode, that is relaxing momentum but not energy. Finally the

© 1990 IOP Publishing Ltd

carrier either gains sufficient energy to impact ionize or loses all its energy in an idealized energy relaxing collision. The kinetic energy at the end of each lucky drift is determined by the distance covered since the last energy relaxing collision and the electric field strength. If this energy is greater than the threshold energy for impact ionization then (taking the hard threshold approximation) the carrier is assumed to ionize. Both the ballistic and lucky drift parts of the motion have a negative exponential distribution of path lengths with $\lambda$ or $\lambda_E$ as the mean value.

In this work lucky drift concepts are used in a computer simulation of the carrier transport so that the random variations in carrier multiplication which constitute the noise generation mechanism in APDs can be estimated. An electron is started from one edge of a high field region and its motion is computed using a pseudo-random number generator to select the path lengths for ballistic and lucky drift motion. The location of each point where the carrier energy exceeds the threshold energy is noted and after the first electron has traversed the high field region an electron and a hole are started from each of these locations. All of these secondary carriers are followed through the depletion region noting if and where any further impact ionizations occur. This procedure is repeated until all carriers resulting from the initial electron have left the depletion region. At this point the total number of impact ionizations, plus one, gives the multiplication, M, for that trial. A large number of trials are conducted and the excess noise factor, F(M), is calculated by the following equation where <M> is the mean multiplication and <$M^2$> is the mean square multiplication.

$$F(M) = <M^2>/<M>^2 \tag{1}$$

The calculation continues until the variation in F(M) between trials is small. The lucky drift model automatically includes the dead space and initial predictions of F(M) for k=0 and k=1 were made to establish the accuracy of this method. The depletion layer width, the electric field strength and the phonon energy were taken to by $1\mu m$, $2\times 10^4$ $kVm^{-1}$ and 27.5 meV for all these calculations. The electron threshold energy, $E_I$, was taken to be either 1 eV or 2 eV and the electron mean free path was varied in order to achieve a range of multiplications. In the calculation for k=0 the hole threshold energy and mean free path were taken to be 100 eV and 100 Å in order to inhibit hole ionization. Figure 1 shows the results for k=0 compared with the work of Marsland (1990) which assumed a dead space of $0.1\mu m$. It can be seen that the results for $E_I$= 1 eV are in much better agreement than those for $E_I$=2 eV. The same is true of the results for k=1. This demonstrates that the dead space is approximately twice the distance a carrier requires to achieve the ionization threshold ballistically, a result previously reported by Kim and Hess (1986). In Figure 2 results are shown for electron and hole threshold energies of 1 eV and mean free paths varied to produce different values of k and M. Also shown is F(M) calculated by the theory of McIntyre (1966) and by the numerical technique of Marsland (1990). The results computed here are in much closer agreement with the work of Marsland (1990) and are always less than those given by McIntyre (1966) which neglect the dead space effect.

*Physics of optical and related devices* 569

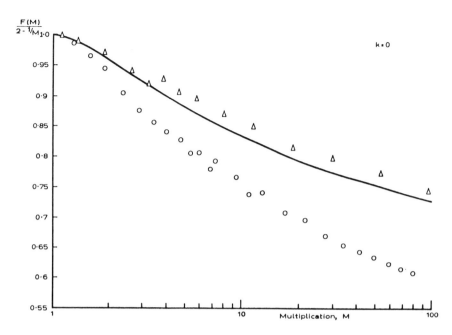

**Figure 1:** Results for k=0 with $E_I$=1 eV (triangles) and 2 eV (circles) compared with Marsland (1990) (solid line).

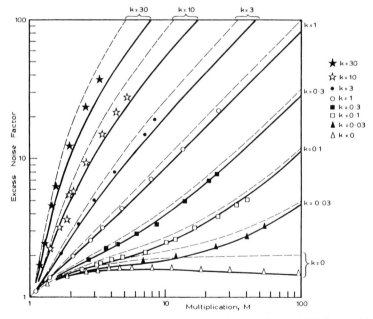

**Figure 2:** Results for $E_I$=1 eV compared with Marsland (1990) (solid lines) and McIntyre (1966) (dashed lines).

## 3. Multilayered Avalanche Photodiode

The lucky drift model has been extended to consider multilayered structures by Marsland and Woods (1987). Similarly the excess noise factor in multilayered APDs has been calculated by a simple extension of the method for conventional APDs. Preliminary results of the lucky drift calculation for a four stage single carrier initiated single carrier multiplication of staircase APD are shown in Figure 3 and are in good agreement with the theoretical predictions of Capasso et al (1983) and Teich et al (1986) showing the flexibility of this novel technique.

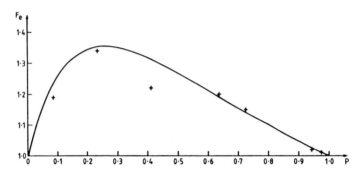

**Figure 3:** Results for a 4 stage staircase APD (crosses) compared to work of Capasso et al (1983) (solid line).

## 4. Conclusions

Agreement between the present calculation and results of Marsland (1990) was found to be good and comparison of the two methods showed that the dead space length was approximately twice the distance a ballistic carrier needs to achieve the ionization threshold energy. This new method has been established as a flexible alternative to other methods of estimating the excess noise factor.

## 5. References

Capasso, F., Tsang, W.T. and Williams, G.F., 1983, IEEE Trans. Electron Dev. ED30 3811.

Kim, K. and Hess, K., 1986, J. Appl. Phys. 60 2626.

McIntyre, R.J., 1966, IEEE Trans. Electron Dev. ED13 164.

Marsland, J.S., 1990, J. Appl. Phys. 67 1929

Marsland, J.S. and Woods, R.C., 1987, IEE Proc. J. 134 313.

Ridley, B.K., 1987, J. Phys. C. 16 3373.

Teich, M.C., Matsuo, K. and Saleh, B.E.A., 1986, IEEE J. Quantum Electron, QE22 1184.

van Vliet, K.M., Friedmann, A. and Rucker, L.M., 1979, IEEE Trans. Electron Dev. ED26 752.

Paper presented at ESSDERC 90, Nottingham, September 1990
Session 7B6

# 10 Gbit/s on-chip photodetection with self-aligned silicon bipolar transistors

J. Popp, H. v. Philipsborn[*]

SIEMENS AG, Corporate Research and Development, Munich, FRG
[*] University of Regensburg, Regensburg, FRG

## ABSTRACT :

The photoresponse of a standard silicon high - speed self-aligned bipolar transistor was analysed. With a sensitivity of 3.2 A/W it is possible to detect modulated laser light up to data rates of 1.25 Gbit/s and by using the transistor in a photo diode mode data rates up to 10 Gbit/s are obtained.

## 1. INTRODUCTION :

Nowadays, silicon technology enables the production of very complex and fast circuits. However, the complicated wiring which is necessary for complex chips restricts the system speed. Thus it is advantageous to trigger electronic devices by direct on-chip photodetection. For example, it is conceivable to detect a clock signal distribution optically by one or several transistors somewhere on the chip without a phase shift due to transit time differencies.

## 2. EXPERIMENTAL :

Vertical npn transistors with a polysilicon emitter fabricated by a self-aligning process [1] were analyzed with respect to their photoresponse. Figure 1 shows the cross section of the transistor and gives the most important dimensions and transistor parameters.

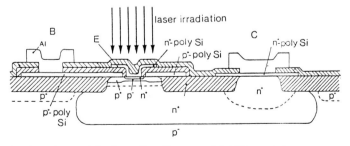

Emitter mask size $A_E = 1.4 * 3$ μm$^2$, Base width $W_B \approx 140$ nm, $R_B \approx 100\,\Omega$, $R_C \approx 30\,\Omega$, $C_{je} \approx 20$ fF, $C_{jC} \approx 40$ fF, $C_{jS} \approx 85$ fF

Fig. 1  Cross section and data of the self-aligned transistor.

© 1990 IOP Publishing Ltd

To improve the optical sensitivity, the emitter contact is located about 10 µm away from the intrinsic transistor so that the emitter window enables light penetration into the active transistor area.

For the optical excitation, a GaAlAs semiconductor laser with a wavelength of $\lambda = 840$ nm is used, of the kind currently under consideration for the realisation of short-range optical links.

By optical transistor operation the base collector diode acts as the internal source for the base current corresponding to the current injection in the electrical transistor mode. This photocurrent is amplified by the transistor with an optical current gain $\beta_{opt} = \beta_{electrical} + 1$. Figure 2 shows the measured DC photocurrent of the transistor and the base collector diode as a function of incident optical power.

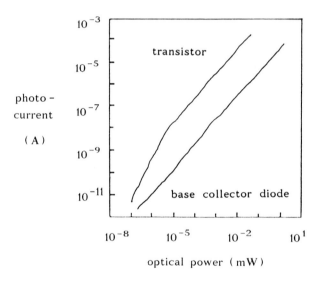

Fig. 2 Photocurrent of the transistor and the base collector diode versus incident optical power.

The photocurrent of the diode increases linearly within the range of $10^{-7}$ mW to 1 mW of incident light power. The sensitivity of 48 mA/W is in agreement with the small vertical dimensions of the device and the light penetration depths $1/\alpha \approx 20 \mu m$. $\alpha$ is the absorption coefficient of silicon at the wavelengths used. The optical gain of the transistor is around 70 - as in the electrical operation mode. Only at the lowest radiation intensity is the optical gain lower due to surface and interface currents. To analyse the time dependency of the photodetection, the transistor signal is considered as a response to a rectangular light pulse.

In Figure 3a, the decay of transistor response to light pulse excitation is plotted logarithmically as $I(t)/I_o$, where $I_o$ is the steady state signal from the DC photon flux.

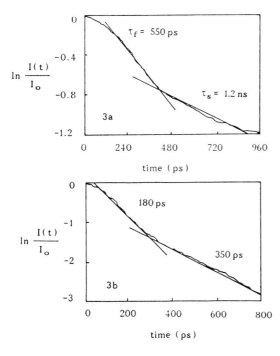

Two different time components are clearly seen. There is a fast component defining the first half of the total signal stroke with a time constant of $\tau_f$ = 550 ps. The photocurrent approaches the final value with an exponential time constant of $\tau_s$ = 1.2 ns from a slow component. In optical transistor operation, the time dependence of the base collector diode has to be taken into account. The result for measuring signal decay is shown in Figure 3b. The fast component, which follows the laser pulse instantaneously has a time constant of 180 ps and the slow component one of 350 ps.

*Fig. 3 : Decay of the photosignal I(t) normalized to the peak signal $I_o$ of the transistor (a) and base collector diode (b).*

## 3. DISCUSSION :

To understand the time behaviour for optical transistor operation, it is useful to take the charge control relation [2] :

$$I_b = \frac{\Delta Q}{\tau} + \frac{d}{dt}\left(Q_{je} + Q_{jc} + \Delta Q_b\right) = C_{je}\frac{dU_{be}}{dt} + C_{jc}\frac{dU_{bc}}{dt} + \tau_b\frac{dI_c}{dt}$$

The term $\frac{\Delta Q}{\tau}$ represents the recombination rate in the base and is negligible for this device, the other terms are junction and diffusion capacitances. On the other hand, the base current is $I_b = I_{b,contact} + I_{b,emitter} + I_{b,collector}$, where $I_{b,collector}$ represents the photocurrent through the base collector diode.

Furthermore $I_{b,contact} = 0$, since the base contact is not connected. The term $I_{b,emitter}$ can be written as $I_{b,emitter} = i_s \exp(U_{be}/nU_T)$ as the standard diode current- voltage relation. This results in :

$$i_s \exp(U_{be}/nU_T) + I_{ph} = C_{je}\frac{dU_{be}}{dt} + C_{jc}\frac{dU_{bc}}{dt} + \tau_b\frac{dI_c}{dt} \quad ;$$

For elemination of $U_{be}$, the Gummel formula [3] for the collector current is used:
$$I_c = ekTA/Q_b \left[\exp(U_{be}/U_T) - \exp(U_{bc}/U_T)\right],$$
where A is the base area and $Q_b$ is the base charge. Finally, a first - order differential equation for the collector current had to be solved with $I_{ph}$ as parameter. Assuming $I_{ph} = I_o \exp(-t/\tau)$ with $\tau$ as the exponential time constants measured in Figure 3b the calculated decay of $I_c$ compares with the measured decay shown in Figure 3a.

The time behaviour of the base collector diode of a similar device is discussed in the literature [4]. The potential for possible applications as receivers for short-range optical data transmission links is demonstrated in Figure 4.

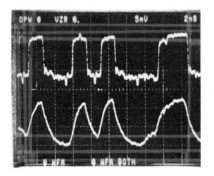 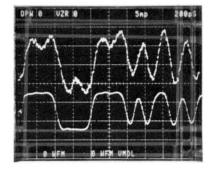

*Fig. 4 Photosignal of the transistor at 1.25 Gbit/s (left) and the base collector diode at 10 Gbit/s (right) compared to the incident light wave forms.*

Fig. 4 compares the photocurrent waveforms of the transistor and the base collector diode with those of the incident optical signals. It proves that data rates up to 1,25 Gbit/s can be detected by phototransistor operation; for the base collector diode, data rates as high as 10 Gbit/s are obtained.

## 4. SUMMARY :

Modern self-aligned silicon bipolar transistors are promising devices for on-chip detection of optical signals with wavelengths smaller than about 900 nm and bit rates up to 10 Gbit/s.

## REFERENCES :

[1] H. Kabza et al.: IEEE ED Letters, vol. 10, No. 8 (1989) pp.344 - 6
[2] Beaufoy, Sparkes: ATE J. (London), 13, No.4, October 1957, pp.310 - 324
[3] H. Gummel : Bell System Technical Journal 1/1970, pp.115 - 120
[4] W. Bock : Essderc 1988, Journal de Physique, C4 - 89, septembre 1988

# A sensitive MNMOS structure for optical storage

M.S.Shivaraman and O.Engström
Department of Solid State Electronics, Chalmers University of Technology
S-412 96 Göteborg, Sweden

Abstract A sensitive device structure for the storage of optical signals is presented. It is based on an indium tin oxide - silicon nitride - aluminum - silicon dioxide - silicon (MNMOS) combination. The light signal is transformed to electric charge through optical excitation of charge carriers in the two metal layers. The charge is transported in the nitride conduction and valence bands to the aluminum. The latter layer is confined between the two insulator layers, thus forming a potential well for the storage of the optically induced charges. Light exposure times down to 0.1 ms are enough to store readable data by a light power of 90 $\mu W/cm^2$ at 2.2 eV photon energy.

## 1. Introduction

In recent development of devices for optical data storage, different combinations of MIS structures have been taken into consideration. MNOS structures were incorporated into traditional circuit solutions for microelectronic image sensors by Yamasaki and Ando (1985). Also, the electro-optical properties of MNOS and MNMOS capacitors for direct optical writing and optical read-out have been suggested as candidates for the same purpose (Engström and Rydén 1985, Shivaraman and Engström 1988). The latter structure was integrated to a focal plane matrix with 65000 picture elements, each consisting of an MNMOS capacitor. This image storage device has the potential advantage of being used in combination with a recently demonstrated technique for the parallel processing of optical information (Engström and Carlsson-Gylemo, 1985).

In the present work we have obtained an MNMOS structure with very high sensitivity to light. The working principle is based on the storage of photon induced charge in the inner metal layer of the two present in the device. The sensitivity, therefore, depends on the combination of material properties of the two metal layers and the silicon nitride layer separating them. Indium-tin oxide (ITO), $In_2O_3$ heavily doped with Sn, was used for the outer metal layer. ITO has a high transparency in the visible region of the spectrum of light and further, at high doping levels it approaches metallic behavior.

## 2. Experimental details

MNMOS structures with geometries as demonstrated in Fig. 1 were fabricated on 1-10 ohmcm, p-type, (100) silicon. A thermal dry oxide was prepared with a thickness of 1000Å on the silicon wafers. Aluminum was evaporated to a thickness of 90Å through a mechanical mask giving metal islands with diameters of 3 mm, constituting the floating gates, M2 (Fig.1). On top of this structure a 1150Å thick layer of silicon nitride was deposited by plasma enhanced CVD-technique. The outer gate metal, M1 of indium-tin-oxide, was prepared by reactive sputtering from an In/Sn target in oxygen atmosphere. Backside ohmic contacts were prepared by evaporating aluminum.

The evaluations were performed by measuring the change in flat band voltage of the MNMOS structure after illuminating the outer metal M1 from a light emitting diode with a peak intensity at a photon energy of 2.2 eV. The light power illuminating the MNMOS device was measured to be 90 µW/cm². The time dependence of the flat band voltage, which is a result of the light induced charging of M2 and gives a measure of the sensitivity, was quantified by measuring the high frequency capacitance as a function of voltage between M1 and the backside contact.

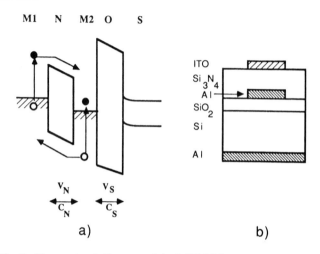

**Fig. 1.** Energy band diagram of the MNMOS structure (a) and geometric configuration (b). Voltages are defined positive when M1 and M2 are positive relative to the backside contact and when M1 is positive relative to M2

### 3. Optical charge generation in MNMOS structures

When the MNMOS device is negatively biased and illuminated on the metal layer M1, electrons may be excited from M1 to an energy level above the conduction band edge of the silicon nitride as shown in Fig. 1. Simultanously holes can be excited in M2 to the valence band of the nitride. Due to the electric field present in the nitride, those carriers which are injected from M1 and M2 are drifted across the nitride layer, which results in the build up of negative charge in M2. This optically induced charge is captured in the potential well governed by the two insulators on each side of M2. Correspondingly, if M1 is positively biased, a positive charge will be created in the potential well. The charge on M2 will determine the flat band voltage of the MNMOS capacitor, which can be measured by conventional CV-technique.

Optical excitations also take place in the silicon region when the device is illuminated on M1 thus creating excess charge carriers close to the oxide-silicon interface. Therefore, the voltage $V$, applied between M1 and the backside contact, is mainly divided between the voltage drops $V_N$ across the nitride layer and $V_S$ across the oxide layer (Fig.1) such that $V = V_N + V_S$. The charge $Q_M$ transferred to M2 during illumination is the difference between the charges collected at the (M2)O interface and the (M2)N interface. It can be expressed as

$$Q_M = (C_N + C_S)V_S - C_N V = C_S V - (C_N + C_S)V_N \qquad (1)$$

where $C_N$ and $C_S$ are the capacitances of the (M1)N(M2) and the (M2)OS systems, respectively. $V_N$ and $V_S$ are the corresponding voltages, and $V$ is the voltage applied to the structure between M1 and the back contact. When $V$ equals the flat band voltage, $V_{FB}$, of the entire structure, $V_S$ also approaches the flat band voltage $V_{FBS}$ of the (M2)OS system. During the illumination interval, $Q_M$ and $V_{FB}$ will change with time while $V_{FBS}$ is a constant, only depending on the fixed charge of the (M2)OS system. For this special case, Eq.(1) reads

$$Q_M(t) = (C_N + C_S) V_{FBS} - C_N V_{FB}(t) \qquad (2)$$

By combining Eq.(1) and (2) we eliminate $Q_M$ and find

$$V_{FB}(t) = \frac{C_N + C_S}{C_N} (V_N(t) + V_{FBS}) - \frac{C_S}{C_N} V \qquad (3)$$

The voltage $V_N$ across the nitride layer depends on the charge $Q_M$ on M2 and approaches zero when more charge is collected. Since the excited charge carriers are drifted across the nitride layer in order to be captured in the potential well at M2, the charging process will stop when $V_N$ equals zero voltage. Thus, a transient in $V_{FB}$ is obtained with a saturation value $V_{FB}(\infty)$ given by Eq.(3) as

$$V_{FB}(\infty) = \frac{C_N + C_S}{C_N} V_{FBS} - \frac{C_S}{C_N} V \qquad (4)$$

Hence, a plot of the saturation value $V_{FB}(\infty)$ as a function of the applied voltage, $V$, would give a straight line with a slope determined by the ratio $C_S/C_N$. Eq.(4) shows that the final value and the swing in the flatband voltage when the structure is illuminated depends linearly on the applied voltage and that the sensitivity can be varied by varying the ratio $C_S/C_N$.

4. Experimental results

Figure 2 shows the measured change in flat band voltage, $V_{FB}$, as a function of time for the MNMOS capacitor illuminated by a power of 90 µW/cm² at a photon energy of 2.2 eV. Data is shown for two different applied voltages, V=40 V and V=-40 V. Steep changes at the beginning of the cycle are followed by smaller slopes at larger time values before saturation. Under these illumination conditions, about 4-5 V changes in $V_{FB}$ are obtained within 1 second. Fig.3 illustrates the short time behavior of the device. The initial slopes in this plot are about 100V/s.

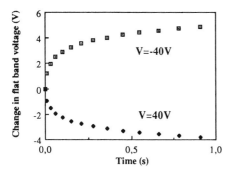
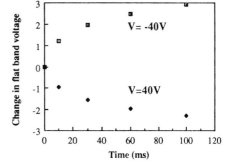

**Fig. 2.** Change in flat band voltage as a function of time for two different bias values

**Fig. 3.** Short time behavior of the flat band voltage for two different bias values

In Fig. 4, the saturation value of the flat band voltage, $V_{FB}(\infty)$, is plotted as a function of applied bias voltage, $V$. The straight line follows the behavior predicted in Eq.(4) giving a slope of -0.53. This is in acceptable agreement with the expected slope of -0.58 calculated using relative dielectric constants of 3.85 for silicon dioxide and 7.6 for silicon nitride, and the insulator thicknesses given above.

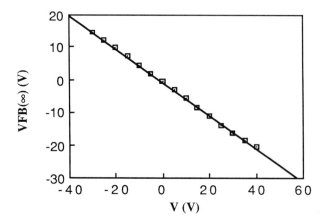

**Fig. 4.** The saturation value of the flat band voltage, $V_{FB}(\infty)$, as a function of the applied bias, $V$

During the first 5 hours of storage a small charge displacement is often observed. This is measured as a change in flat band voltage of about 5% during this time interval followed by a constant value for periods of days. A possible reason for this change is a minor thermal emission of charge carriers from traps in the silicon nitride.

5. Discussion

The optical sensitivity as demonstrated by Fig.3 is more than one order of magnitude higher than earlier obtained values where gold and aluminum were used as materials for layers $M_1$ and M2 (Shivaraman and Engström 1988). Devices where the optical signal is detected by a p-n junction normally give higher sensitivities than the values presented here (Yamasaki and Ando 1985). However, such a concept requires that the optically induced charge is transported through a transfer gate into a conventional floating gate memory element where it is stored. As the present structure both acts as an optical detector and a storage element, a focal plane matrix can be made simpler than for a traditional solution. Also, when a matrix of MNMOS structures of this kind is formed, a device is obtained which can be used for image recognition by the method suggested by Engström and Carlsson-Gylemo (1985). When optical read-out is used (Engström and Rydén 1985), changes in flat band voltage down to about 10 mV can be detected. With the initial slopes of $V_{FB}(t)$ at 100 V/s shown in Fig. 3, this means that an exposure time of 0.1 ms can be used to store optical data when the MNMOS structures are illuminated by a light power of 90 µW at 2.2 eV photon energy.

References

Engström O and Carlsson-Gylemo A 1985 IEEE Trans. Electron. Dev. **ED-32**, 2438
Engström O and Rydén K H 1985 Physica **129B**, 506
Shivaraman M S and Engström O 1988 SPIE (ECO1) **Vol.1027**, 30
Yamasaki H and Ando T 1985 IEEE Trans. Electron.Dev. **ED-32**, 738

# A new charge pumping procedure to measure interface trap energy distributions on MOSFETs

G. Van den bosch, G. Groeseneken, P. Heremans[*] and H. E. Maes

IMEC vzw - Kapeldreef 75 - B3030 Leuven - Belgium

Abstract  A new approach to the application of the well-known charge pumping technique is proposed as a tool for the measurement of the energy distribution of interface traps in small area MOSFET's. In the classical procedure, the band gap is scanned by applying pulses with variable transition times to the gate of the transistor, defining different energy windows from which the charge pumping signal is measured. The new approach is spectroscopic in nature, *i.e.*, only one energy window is defined and forced to move through the band gap by changing the sample temperature. This method has the advantages of addressing a larger part of the band gap as compared to the classical approach, to reduce complication in the processing of the data and to yield information about the hole and electron capture cross sections separately.

## 1. Introduction

When applying rectangular pulses to the gate of a MOSFET, a dc recombination current proportional to the Si-SiO$_2$ interface trap density is measured at the substrate contact. By changing the transition time of the applied pulses, the carrier emission processes occurring in the band gap states can be controlled, so that different energy levels are addressed. In this way an energy distribution over (part of) the band gap can be obtained (Groeseneken *et al.* 1984). The use of a three-level gate wave form, where emission is taking place at a constant, intermediate pulse level was proposed by Przyrembel *et al.* (1987). Direct control over the emission process is obtained by the mid-level width.
However, due to minimum current signal restrictions, the ability of these techniques to determine the energy distribution was limited until now to a relatively small energy range in the band gap. The purpose of this work is to present an alternative procedure to determine interface trap energy distributions on MOSFET's. It involves the use of a hole and an electron emission rate window, which are scanned over the band gap by changing the sample temperature. Our procedure therefore combines the spectroscopic nature of DLTS (Deep Level Transient Spectroscopy) with the high sensitivity of charge pumping.

## 2. Temperature dependence of the charge pumping process

A detailed examination of the process of carrier emission from interface traps non-uniformly distributed over the energy gap has been made, and special attention is given to the temperature dependence. Under appropriate conditions of emission time and temperature the following relations hold (Groeseneken *et al.* 1984):

$$E_{emh}(t_{emh}, T) = E_i + kT \cdot \ln \left[ \sigma_p v_{th} n_i t_{emh} \right] \quad (1)$$

$$E_{eme}(t_{eme}, T) = E_i - kT \cdot \ln \left[ \sigma_n v_{th} n_i t_{eme} \right] \quad (2)$$

---

[*] P. Heremans is a Senior Research Assistant of the Belgian Fund for Scientific Research

where $E_{em}$ is the emission energy level reached after emission time $t_{em}$, at a temperature T, and $E_i$ is the intrinsic energy level. The emission level is thus a logarithmic function of emission time, which lies at the basis of an energy distribution version of the CP technique (Groeseneken et al. 1984, Przyrembel et al. 1987). When keeping the emission time (i.e. the pulse transition time) fixed but changing the temperature, the emission levels vary qualitatively in the same way. In practice however, the accessible energy range turns out to be considerably larger in the latter case.

Experimentally the variation of the emission level with temperature is reflected in the temperature dependence of the charge pumping current Icp. It can be shown that

$$I_{cp} = q f A_g \overline{D_{it}} \cdot [E_{emh}(T) - E_{eme}(T)] \qquad (3)$$

where f is the pulse frequency, $A_g$ the gate area, and $\overline{D_{it}}$ denotes the average interface trap density in the contributing energy interval. Fig. 1 shows the maximum of variable base level curves recorded as a function of temperature, on an nMOSFET of 2 µm gate length and 100 µm gate width, at a frequency of 50 kHz. It can be calculated from eqs. (1) - (3) that, due to the temperature dependence of $E_{emh}-E_{eme}$, the current decreases with increasing T according to the following expression:

$$I_{cp}(T) = - a T - b T \ln T + c \qquad (4)$$

where a, b and c are a function of transistor and gate pulse parameters. The full line in Fig. 1 is fitted to the measurement data using eq. (4) and yields values of $D_{it} = 1.8 \times 10^{10}$ cm$^{-2}$eV$^{-1}$ and $\sigma = 1.3 \times 10^{-15}$ cm$^2$. The CP current can thus be succesfully fitted with constant $D_{it}$ and σ over a temperature range of approximately 300 K.

This result is confirmed by measuring the pumped charge per cycle $Q_{cp} = I_{cp} / f$ vs. frequency in the same temperature range. In this type of measurement, the logarithmic slope of the $Q_{cp}$-f curve is proportional to $D_{it}$, while the frequency axis intercept $f_0$ is proportional to $\sigma = \sqrt{\sigma_n \sigma_p}$ (Groeseneken et al. 1984):

$$\frac{dQ_{cp}}{d \ln f} = 2q kT \overline{D_{it}} A_g \qquad f_0 = \frac{\sigma v_{th} n_i |V_{th}-V_{fb}|}{2 \Delta V_g} \qquad (5,6)$$

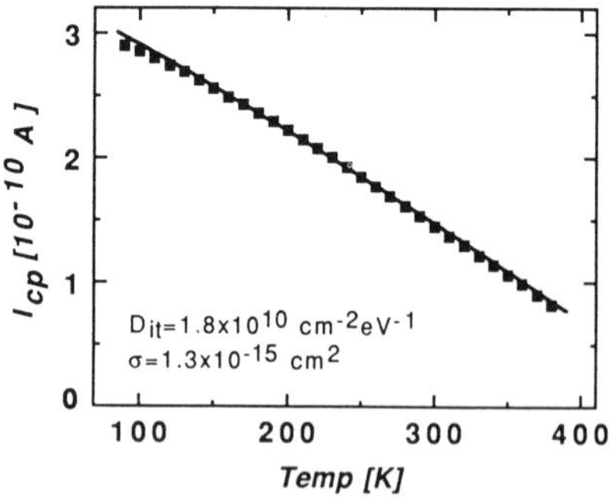

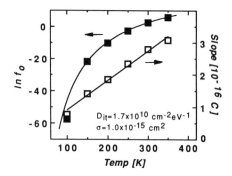

Fig. 1 Maximum of $I_{cp}$ vs. base level recorded as a function of measurement temperature. The full line is eq. (4) fitted to the measurement data with a constant value for $D_{it}$ and σ.

Fig. 2 Slope (□) and frequency axis intercept (■) of $Q_{cp}$ vs. frequency, as a function of temperature. Eq. (5) resp. (6) is fitted to the data (full lines) with a constant value for $D_{it}$ resp. σ.

The temperature dependence of these two parameters is plotted in Fig. 2. It can be concluded that the slope and frequency axis intercept as a function of temperature are adequately fitted with a constant $D_{it} = 1.7 \times 10^{10}$ cm$^{-2}$eV$^{-1}$ and a constant $\sigma = 1 \times 10^{-15}$ cm$^2$.
From these measurements we conclude that the model originally proposed by Groeseneken *et al.* (1984) remains valid in the temperature range 80 K to 400 K. Moreover, for our devices, no evidence for a strong variation of the interface trap density with energy, and of the capture cross section with energy and/or temperature has been found.

## 3. Spectroscopic charge pumping

The spectroscopic approach to the CP technique will now be introduced based on the previous examination. In this approach, the CP current $I_{cp}$ is monitored for two distinct values $t_{1r}$ and $t_{2r}$ ($t_{1f}$ and $t_{2f}$) of the emission time in the rising (falling) edge of the gate pulse, while keeping the falling (rising) edge constant; in this way the relevant emission processes are confined to a portion ('window') of the lower (upper) half of the bandgap, determined by $t_{1r}$ and $t_{2r}$ ($t_{1f}$ and $t_{2f}$). The lower (upper) part of the band gap is then scanned by changing the sample temperature.
The spectroscopic signals are defined as

$$S_r = I_{cp}(t_{1r}) - I_{cp}(t_{2r}) \quad , \quad S_f = I_{cp}(t_{1f}) - I_{cp}(t_{2f}) \qquad (7,8)$$

Calculations show that the following relations hold:

$$S_r = qfA_gD_{it}(E_{or})kT\ln\frac{t_{2r}}{t_{1r}} \quad , \quad S_f = qfA_gD_{it}(E_{of})kT\ln\frac{t_{2f}}{t_{1f}} \qquad (9,10)$$

where $E_{or}$ ($E_{of}$) is the mean energy of the emission window corresponding to the emission time pair $t_{1r}$, $t_{2r}$ ($t_{1f}$, $t_{2f}$). The quantity $kT\ln(t_2/t_1)$ is the width of the trap emission distribution function as originally derived in CC-DLTS theory (Johnson 1982). It can be interpreted as the width of the energy interval contributing to the spectroscopic signal, and thus determines the energy resolution of the technique. $D_{it}(E)$ is linearly proportional to the measured spectroscopic signal.

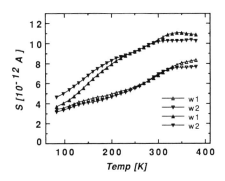
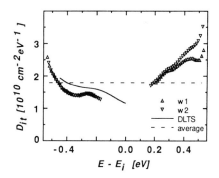

Fig. 3 Spectroscopic signals from two adjacent windows in the lower (empty symbols) and upper part (full symbols) of the band gap as a function of temperature; $t_r$ and $t_f$ resp. 0.158, 0.5, 1.58 μs.

Fig. 4 $D_{it}(E)$ distributions corresponding to the signals of Fig. 3 (symbols). Also shown: $D_{it}(E)$ from DLTS measurement on p-type capacitor (full line) and average $D_{it}$ value from Fig. 1 (dashed line).

Fig. 3 shows a number of spectroscopic signals obtained on the nMOSFET used in the previous experiments. These signals have been obtained using a conventional rectangular wave form, at a frequency of 50 kHz. The temperature dependence of S is determined by the width of the emission window on the one hand, and by the actual $D_{it}(E)$ distribution on the other hand. Measuring S with two different windows results in two curves with similar shape but shifted on the temperature axis: $S_{r1}$ and $S_{r2}$ are the variable rise time signals (emission of holes in the lower part of the band gap), $S_{f1}$ and $S_{f2}$ are the variable fall time signals (emission of electrons in the upper part of the band gap). Fig. 4 shows the $D_{it}(E)$ distribution as calculated from the different signals S of Fig. 3. Information on $\sigma_p$ ($\sigma_n$), needed for the transformation of the temperature scale of Fig. 3 to the energy scale of Fig. 4, can be deduced from the shift between $S_{1r}$ and $S_{2r}$ ($S_{1f}$ and $S_{2f}$). From our measurements we find that $\sigma_p$ and $\sigma_n$ are equal to within an order of magnitude ($\approx 1 \times 10^{-15}$ cm$^2$). Fig. 4 also shows the average $D_{it}$ midgap value as obtained from the fit of Fig. 1, and the result of a DLTS measurement on a large p-type capacitor on the same chip. These additional measurements confirm the results of spectroscopic CP. The use of a three-level wave form yields similar results.

On Fig. 4, it is readily seen that the accessible energy range is large as compared to other techniques such as CV (Capacitance Voltage) and the conventional CP technique. The minimum midgap energy that can be reached is limited by diode reverse leakage current at high temperature and by the lowest allowable measurement frequency (however, an average midgap value for $D_{it}$ can be obtained by the variable base level CP technique). For a given measurement frequency larger emission times can be obtained with the three-level pulse: the emission time is equal to the mid-level width, while for the conventional square pulse the emission time is only the fraction of the transition time spent between threshold and flatband voltage of the transistor (the temperature dependence of $V_{th}$-$V_{fb}$ has been taken into account in the experiments). The measurement sensitivity is in the range $10^9$ cm$^{-2}$eV$^{-1}$, depending on the size of the transistor under test.

Although similar to the DLTS technique for MOS capacitors, our new procedure has considerable advantages over the latter. Firstly, since the measured signal is directly proportional to $D_{it}(E)$, no complex signal processing is needed. Secondly, the measurements are carried out on fully processed small area transistors instead of on large area capacitors, allowing to obtain a direct correlation with IC processing related phenomena. A third advantage is that both parts of the band gap can be reached on the same transistor and within the same measurement.

## 4. Conclusions

It has been experimentally shown that the simple emission model for charge pumping remains valid in the temperature range 80 K - 400 K. No evidence has been found for a strong variation of $D_{it}$ and $\sigma$ with energy and/or temperature within this range. A new spectroscopic CP procedure allows to obtain $D_{it}(E)$ over most of the band gap. It is the equivalent of CC-DLTS for small area MOSFET's, with the same resolution but a much higher sensitivity.

## References

Groeseneken G., Maes H. E., Beltran N. and De Keersmaecker R. F. 1984,
 *IEEE Trans. Electron Devices* **31**, p. 42
Johnson N.M. 1982, *J. Vac. Sci. Technol.* **21**, p. 303
Przyrembel G., Krautschneider W., Soppa W. and Wagemann H. G. 1987,
 *Proc. ESSDERC '87*, p. 687

# Interface states extracted from gated diode and charge pumping measurements

F. Hofmann

**Siemens AG, Central Research and Development**
**Otto-Hahn-Ring 6, D-8000 Munich 83, West Germany**

Abstract: This paper applies the gated diode and the charge pumping method for the extraction of interface states in MOS transistors and gives a comparison of the two techniques. The gated-diode arrangement with floating source will be critically examined. With both methods the interface state densities are extracted at midgap for virgin and for homogeneously stressed MOS transistors. Good agreement for both methods is obtained. The temperature dependence of the surface generation current is also measured.

## I. Introduction

A comparison is presented of the extracted interface states from gated-diode (GD) and charge pumping (CP) measurements. Both methods work with the same teststructure in the same electrical circuit. Also both techniques rely on the Shockley-Read-Hall theory of the recombination-generation process, and should thus yield comparable results.
As a first step the principles of the gated-diode measurements will be explained and the difficulties in the floating source arrangement will be discussed. Furthermore the temperature dependence of the surface generation current will be shown. The results will be compared with those of the charge pumping method.

## II. Principles of gated diode measurements

Since the beginning of MOS technology the gated diode structure was used to measure the recombination current at the Si-SiO2 interface, which was suggested by Grove (1966). The reverse p-n junction current is given by the generation of electron hole pairs at generation centres in the depletion zone. This current is directly related to the number of traps in the depletion zone. The volume of the depletion zone and thus also the generation current depends on the gate voltage. There are three different regimes which contribute to the generation current: the metallurgical junction, the field-induced junction and the surface. When the substrate under the gate is in accumulation, the depletion zones exist only at the p-n junctions at source and drain. The measured current is given by the leakage current at the junctions: $I_{GEN,MJ}$ in Figure 1a. When the flatband condition is exceeded, a depletion

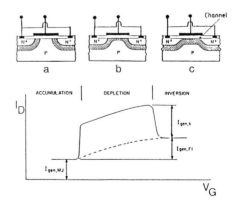

Figure 1: Reverse current of gated diode schematically after Grove (1966)

zone is created under the gate. Now there are additional generation centers at the SiO2 surface, which contribute by their electron hole generation to the GD-current: IGEN,S in Figure 1b. A further increase of the gate voltage builts up an inversion channel. At this condition the surface states are separated from the depletion zone by the channel and these states will no longer contribute to the GD-current. If the gate voltage is increased, the depletion zone varies now only in the volume of the silicon and the reverse current stays constant: IGEN,Fi in Figure 1c. This is a very simple picture for the GD-current in the VG regime without tunnelling effects

## III. Floating source arrangement

Grove (1966) applied his method originally to gated diodes. Some recent publications as Giebel (1989) and Speckbacher (1990) report gated diode measurements on MOSFETs with floating source or drain. They have not taken into account, however, that additional capacitances contribute to the apparent leakage current. Figure 2 shows some measurements with floating source, varying the gate voltage from accumulation to inversion condition. There are different plateaus to be seen, depending on the connection to source: connection of source and drain, source floating with tips lifted off and with floating tips at the source pad. These results in can only be understood if the charging of the floating source and other parasitic capacitances, like the contact pad and the electrical connection to the contact tips, from the drain via the high-resistive channel is taken into account. This current depends also on the time of changing the gate voltage and is absent when source and drain are connected. When the surface is forced from inversion to accumulation, this charging current plateau is also missing. Figure 2 shows the measurements on a single MOS-transistor; here the contribution of these parasitic capacitance is shown clearly. The leakage current in this experimental arrangement was about 4 fA; this is a larger value as $I_{GEN,S}$ so the GD-plateau could hardly be seen. Measurements with floating source depend extremely on the charging of the floating terminal.

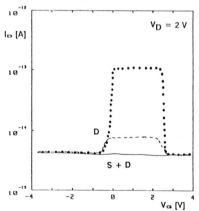

Figure 2: $I_D$ of a single transistor
— S + D are connected
- - D only, S floating, tips are lifted off
o o -"- , -"- , with tips

## IV. Temperature dependence of the generation current

The following results were obtained on n-channel multi transistor (14000 LDD nMOS transistors, L = 0.8 μm, W = 10 μm). Figure 3 shows the temperature dependence of the generation current in the gated diode setup (source and drain are connected), before and after stress. Here the surface generation current can be clearly seen. The plateau of the virgin device starts at flatband condition (-1 V) and exists till the onset of inversion, in this example at $V_G = 2.0$ V (as source and drain are at the same potential: $V_S = V_D = 2.0$ V). If the gate voltages increases, the surface generation is stopped and the $I_{GD}$ remains constant at a lower level. At this voltage condition only $I_{GEN,MJ} + I_{GEN,Fi}$ are detected, because now the current generation takes place only at the volume traps in the depletion zone. The contribution of $I_{GEN,Fi}$ (from traps in the depletion zone below the gate) to the GD-current can be neglected in modern transistors. On the other side, at accumulation condition, $I_{GD}$ increases with decreasing gate voltage. This result could be explained by the expansion of the depletion zone in the LDD regime under the gate; here we have more surface states and thus a higher generation current. The evaluation of the surface generation current after Grove (1966) for single level centers is given by:

$$I_{GEN,S} = q\ U_S\ A_S \quad \text{with} \quad U_S = \frac{\sigma_n \sigma_p\ v_{th}\ N_{st}\ n_i}{\sigma_n + \sigma_p}$$

$A_S$: gate area and $\sigma_n$, $\sigma_p$: capture cross-section, $5 \cdot 10^{-16}$ cm$^{-2}$ each, $v_{th}$: thermal velocity
$N_{st}$: density of surface states

The restriction to single level states is allowed, because only the traps close to midgap are very effective generation centers.

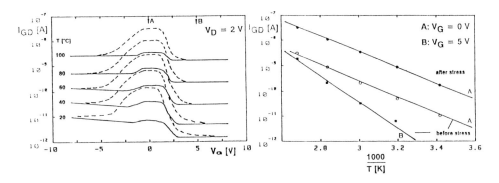

Figure 3: Temperature dependence of $I_{GD}$
—— before stress  -- after stress

Figure 4: Activation energy plot

Figure 4 shows $I_{GD}$ plotted versus 1/T for the virgin device at two different $V_G$ conditions: $V_G = 0V$ (point A, plateau in $I_{GD}$) and $V_G = 5V$ (point B). The slope of the curve gives 0.65 eV as an activation energy for surface generation; this is consistent with the model, where the main contribution for $I_{GEN,S}$ comes from midgap states. The second activation energy for $I_{GEN,MJ}$ and $I_{GEN,Fi}$ is 1.2 eV. Here the electron hole generation has to overcome the whole bandgap.
This multi transistor was stressed homogeneously by a large negative gate voltage (E = 9 MV/cm). The result is a very strong increase of interface states leading to the enlarged current peak in the gated diode measurements. Furthermore the onset of the plateau in the accumulation phase is shifted to more negative gate voltages. This is due to the generation of fixed charge during the stress which changes the flatband condition, which is in agreement with the shift measured in the subthreshhold regime. The activation energy, taken at the GD-plateau, is the same as in the virgin device. The homogeneous stress has mainly built up interface states near midgap, this can be seen clearly in the energetic distribution of the interface states extracted from charge pumping measurements.

V. Charge pumping measurements

The same multi-transistor was also characterized with the charge pumping technique. This is a very sensitive method to get the interface states. A pulse train with constant amplitude and a defined offset voltage level is applied at the gate of the transistor. When the transistor is pulsed into inversion, electrons will flow from the source and drain into the channel, where some of them will be captured by surface states. When the gate is driving the semiconductor surface to accumulation, the mobile charges will flow back to source and drain, but the charges trapped in the interface states will recombine with the majority carriers from the substrate; this flow of charge is called the charge pumping current $I_{cp}$ as discussed by Groeseneken (1984).

$$I_{cp} = 2\ q\ f\ A_S\ \underline{D}_{it}\ kT\ \ln(v_{th}\ n_i\ \sqrt{\sigma_n \sigma_p}\ \sqrt{t_f\ t_r})$$

$t_f$, $t_r$ : fall, rise time of the gate pulse, $\underline{D}_{it}$ : average value of the interface state density

This plateau in the charge pumping current shown in Figure 5 will remain if the gate voltage fulfils the following condition: $U_{FB} > U_{G,base}$ and $U_{TH} < U_{G,high}$. In this method, the current $I_{cp}$ is directly correlated with the interfacial charges and is proportional to the pulse frequency and gate area. The maximum of the $I_{cp}$ current gives the mean value of the energetic distribution of interface traps. The result is shown in Figure 6. The evaluation of the charge pumping method results in a higher $\underline{D}_{it}$ than the gated diode measurements. The $\underline{D}_{it}$ value extracted from charge pumping is averaged over an energy interval of ± 0.3 eV around midgap shown by Hofmann (1989b), whereas the gated diode result is only given at kT around $E_i$. With the charge pumping technique, information can also be obtained about the energetic distribution of the interface state density in a potential range mentioned above, this is shown in detail by Groeseneken (1984) and Hofmann (1989a).

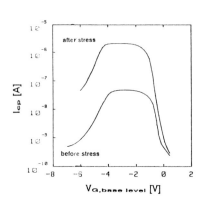
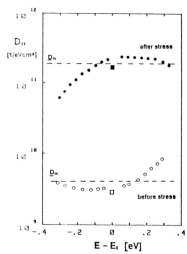

Figure 5: Charge pumping current before and after stress

Figure 6: Distribution of the interface state densties $\underline{D}_{it}$ as a result from maximum of $I_{cp}$
■ □ gated diode results

The extraction of $D_{it}(E)$ is also shown in Figure 6. Now there is a good agreement of CP and GD measurements for the interface states situated at midgap.

The same charge pumping measurements were also done after the stress. The Icp plateau in Figure 5 shows a very strong increase of interface states. The $D_{it}(E)$-curve of the unstressed device has an U-shape whereas the one of the stressed device shows a peak at midgap (Figure 6). This result points out that most interface states were created at midgap. The squares represent the results of the gated diode measurements. It can be seen that the interface trap densities obtained with the different techniques are in very good agreement.

VI. Conclusion

Gated diode measurements on MOS transistors ( without floating source or drain contacts ) are a simple method to detect the interface state density at midgap. The results are in very good agreement with the distribution of the interface states extracted from charge pumping measurements.

References:

Giebel T and Goser K 1989, IEEE EDL 10, p76
Groeseneken G et al 1984, IEEE Trans. Electr. Dev. ED-31, p 42
Grove A S and Fitzgerald D J 1966, Solid State Electronics, p 783
Hofmann F and Krautschneider W 1989a, J. Appl. Phys. vol 65, p 1358
Hofmann F and Hänsch W 1989b, J. Appl. Phys. vol 66, p3092
Speckbacher S et al, IEEE EDL 11, p95

# Extended static CV-procedure to investigate minority carrier generation in MOS capacitors

M. Kerber and U. Schwalke

Siemens AG., Corporate Research and Development,
Otto Hahn Ring 6, 8000 Munich 83, F.R.Germany

> Abstract. An extension of the conventional static CV procedure is presented which explicitely monitors the equilibrium of MOS capacitors throughout the CV sweep. Simultaneously to the static capacitance the minority carrier generation current is deduced by analyzing the time dependent displacement charge.

## 1. Introduction

The minority carrier generation rate as a monitor of substrate quality is of primary interest in modern CMOS process development. Most conveniently such measurements are performed on MOS capacitors due to the simple processing and short learning cycles. Furthermore no junction implants are needed which may introduce additional defects or cause other leakage sources. The established techniques to investigate bulk properties in MOS capacitors such as DLTS (Lang 1974) or Zerbst analysis (Zerbst 1966) are not very simple and easy to use. They require special measurement equipment or complex evaluation procedures.

In this work we extend the conventional quasi-static CV procedure, which is a well established and frequently used diagnostic tool, in order to analyze the bulk minority generation rate. Therefore no extra measurement setup is required to provide this valuable information.

## 2. Measurement Procedure

We used the experimental setup of the 'feedback charge' method (Mego 1986) which uses a staircase instead of the commonly implemented gate voltage ramp (Kuhn 1970) to measure the quasistatic CV curve of MOS capacitors. In contrast to previously published procedures we permit the time delay between individual gate voltage steps to be variabel and adjusted, so that the system can return to equilibrium after each step. This is done by monitoring the displacement charge $\delta Q_G(t)$ continuously until its increment drops below a preset value which was set to

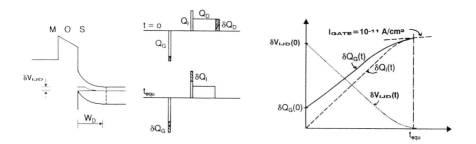

Fig. 1 Band diagram and charge distribution of the MOS capacitor and the induced junction diode (IJD) right after a gate bias step and in equilibrium.

Fig. 2 Schematic time evolution of $\delta Q_G(t)$, $\delta Q_I(t)$ and $\delta V_{IJD}(t)$ from applying a gate bias step $\delta V_G$ to equilibrium defined for a displacement current $I_{GATE} \leq 10^{-11}$ A/cm$^2$

$10^{-11}$ A/cm$^2$ in this study. Then the sweep is continued with the next step. This guarantees that the system is in equilibrium throughout the sweep.

In accumulation the time $t_{equ}$ to settle equlibrium is very short according to dielectric relaxation of majority carriers. In inversion, however, minority carrier generation is the limiting process to built up the inversion charge and therefore considerably rises $t_{equ}$. From the recorded $\delta Q_G(t)$ the generation process of bulk minority carriers can be evaluated.

## 3. Evaluation of the IV characteristics

Immediately after a gate voltage step the MOS capacitor is forced to a non-equilibrium state. If the sweep is performed from accumulation towards inversion, which is preferable by practical reasons, then the small gate voltage step $\delta V_G$ slightly reverse biases the induced junction diode (IJD) between interface and substrate as illustrated by the band diagram of fig. 1. The now flowing reverse current increases the inversion charge and reduces the diode voltage $\delta V_{IJD}(t)$ until equilibrium is reestablished. So, after each gate voltage step a I/V curve is traced. Fig. 2 summarizes the time dependent relation between the quantities $\delta Q_G(t)$, $\delta Q_I(t)$ and $\delta V_{IJD}(t)$. To deduce $V_{IJD}$ and $I_{IJD}$ from the measured quantity $\delta Q_G(t)$ refer to fig. 1. The voltage of the diode is directly related to $\delta Q_G(t)$ by

$$\delta V_{IJD}(t) = \delta V_G - \delta Q_G(t)/C_{ox}.$$

and the current is given by $\quad I_{IJD}(t) = \delta Q_I(t)/\delta t$

Overall charge balance leads to $\quad \delta Q_I(t) = \delta Q_G(t) - \delta Q_D(t)$

and by assuming only small variations of depletion width $w_D$ as a result of $\delta V_{IJD}$

$$\delta Q_D(t) = \delta V_{IJD}(t) \, \varepsilon_{Si}/w_D.$$

$w_D$ can be deduced either from the CV curve or by the condition that in equilibrium $\quad \delta Q_G(t_{equ}) = \delta Q_I(t_{equ}).$

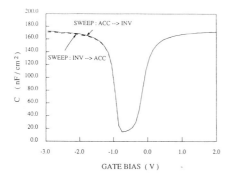
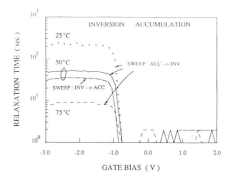

Fig. 3 Static CV-curve of a MOS capacitor ($d_{OX}$ = 20nm, n-substrate: 1.2 $10^{15}$ cm$^{-3}$, area = 1 mm$^2$) with the sweep performed from accumulation to inversion (full line) and vice versa (broken line).

Fig. 4 Relaxation time ($t_{equ}$) after bias steps of 100mV versus gate bias for different substrate temperatures and sweep directions.

## 4. Experimental Results

Some typical experimental results are discussed in the following section. The conventional static CV curve of a MOS capacitor with 20 nm gate oxide on n type substrate is shown in fig 3. Note that the result is independent of the sweep direction as expected, since the system is in equilibrium throughout the sweep. In addition to the static capacitance the corresponding $t_{equ}$ to maintain equilibrium is plotted in fig. 4 which clearly shows the impact of minority generation. As the substrate temperature rises from 25°C to 75°C $t_{equ}$ decreases by more than an order of magnitude which also shows a thermal activated generation process.

Finally the I/V curves are evaluated from the measured $\delta Q_G(t)$. The induced diode characteristics for the individual steps in the inversion range are plotted in fig 5. The resulting reverse currents are essentially independent of inversion bias. This agrees with the nearly voltage independent $t_{equ}$ in inversion as illustrated in fig 4. With increasing temperature the reverse current increases correspondingly. To

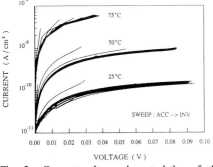
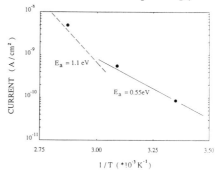
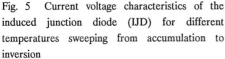

Fig. 5 Current voltage characteristics of the induced junction diode (IJD) for different temperatures sweeping from accumulation to inversion

Fig. 6 Arrhenius plot of the reverse current in fig 5 at 0.03 Volt.

deduce the activation energy the average diode current at 0.03 Volt is plottet versus 1/T in fig. 6. A comparison with the slopes for $E_a = 1.1$ and 0.55 eV indicates that at lower temperatures the reverse current is dominated by minority carrier generation whereas at higher temperature diffusion current becomes dominant.

If the sweep is performed from inversion to accumulation then the above developed arguments hold with the only difference that the IJD is biased in forward direction. The resulting I/V characteristics for 50°C is shown in fig 7. The slope of the forward characteristics indicates a diode quality factor of $n = 1.3$. This value is consistent with the activation energy of the reverse characteristics in fig. 6. The forward diode current is higher than the corresponding reverse current and therefore leads to a shorter relaxation time (fig. 4).

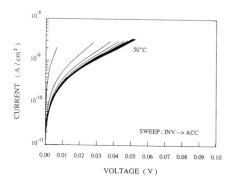

Fig. 7    Current voltage characteristics of the forward biased IJD (sweeping from inversion to accumulation) at 50°C.

## 5. Conclusion

In conclusion we have presented an extension of the static CV procedure. This allows to investigate the substrate quality by monitoring the generation current simultanously to the static CV curve. Reversing the sweep direction offers to fully characterize the induced junction diode already on incomplete processed wafers e.g. without any junction implantations.

## References

M.Kuhn, Solid-State Electr., **13**, 873 (1970)
D.V.Lang, J. Appl. Phys., **45**, 3023 (1974)
T.J.Mego, Rev.Sci.Instrum., **57**, 2798 (1986)
M.Zerbst, Z. Angew. Phys., **22**, 30 (1966)

# Local temperature distribution in Si-MOSFETs studied by micro-raman spectroscopy

R. Ostermeir, K. Brunner, G. Abstreiter
Walter Schottky Institut, Technische Universität München,
D-8046 Garching, FRG

W. Weber
Siemens AG, Corporate Research and Development, Microelectronics,
D-8000 München 83, FRG

Abstract. The local rise of lattice temperatures in enhancement MOSFET's is analyzed by the frequency shift of the silicon optical phonon using Raman spectroscopy with submicron spatial resolution. Operating the devices beyond saturation a source drain asymmetry is observed corresponding to the heat dissipation profile peaked at the pinch off region. A reduction of channel length at standard conditions leads to a local temperature increase due to the enhanced electrical power. An anisotropic behaviour of the temperature distribution is also found in the substrate surrounding the MOSFET, which is related to the geometric shape of the heat source. The experimental results are in good agreement with a simple model calculation based on device parameters.

With increasing density of devices in VLSI circuits thermal effects on device operation and reliability become more and more of a problem. The high power density in a short channel MOSFET causes drain current reduction due to self heating as reported for example by Takacs et al (1987). This may influence the operation of adjacent circuits. The devices studied here are enhancement n-MOSFET's with LDD source drain profiles from a CMOS process. The poly silicon gates are isolated from the channel region by 20 nm gate oxides. The gates have a width of 10 µm and lengths between 2.5µm and 0.6µm. Micro Raman spectroscopy offers an adequate local temperature probe to characterize common devices under operation.
An argon laser beam ( $\lambda$ = 514.5 nm ), which is focused onto the sample by a microscope objective ( N.A. = 0.75 ) to a gaussian spot of 0.5 µm ( FWHM ) offers a spatial resolution which enables us to investigate even submicron devices.
The samples, which are mounted on a piezo driven x-y-stage are positioned with high precision ( $\approx$ 0.1 µm ) by monitoring the photocurrents induced by the laser transmitted through the spacers adjacent to the gate.
At a laser power below 0.5mW heating by the laser spot as well as an additional photoinduced drain current of less than 1% do not affect operation. The Stokes shifted Raman light, which is due to the creation of LO phonons in the silicon ( $\omega_{ph} \approx$ 520 $cm^{-1}$ at T = 300 K ), is analyzed using a triple grating spectrometer in the high dispersion mode ( DILOR, f = 3*500mm ) with a multichannel detector system.

© 1990 IOP Publishing Ltd

The setup is shown in Fig.1.

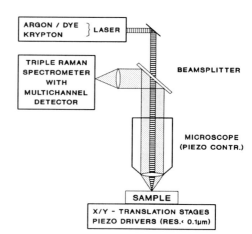

Fig.1. Micro Raman setup ( schematically ).

In agreement with published data ( see for example Balkansi et al ( 1983 )), we found a wavenumber shift of the LO phonon in poly silicon gates of $\Delta\omega/\Delta T = -0.021$ cm$^{-1}$/K for T = 300 - 360 °K by heating the whole samples homogeneously with a thermochuck. The Raman spectra are fitted to a Lorentzian lineshape function. The phonon frequency shift is determined with an accuracy of $\pm$ 0.04 cm$^{-1}$ corresponding to $\Delta T = \pm$ 2K.

Operating the transistors beyond saturation leads to high energy dissipation in the pinch-off region. This results in an asymmetric power density distribution. The asymmetry is reproduced in the observed temperature profile in a 2.5 μm transistor by scanning the laser beam along the channel ( Fig.2 ). A drain voltage $V_d = 6.5$ V and a gate voltage $V_g = 6$V causes a dissipation of 13.3 mW . The temperature distribution peaks at the pinch-off region ( $\Delta T = 12$ K ). At the source side the temperature rise is only about 5 K. Under the same dc operating conditions but with reverse biased source drain voltage the observed temperature profile is just reversed.

The solid line represents a calculation based on a simulation program which was developed recently considering a semi-infinte silicon crystal ( Mautry 1990 ). The heat source within the channel is modelled by a sum of 10μm long line sources which provide a well known analytical solution for the temperature field ( Smith et al 1986 ). The heat generation rate is assumed to increase linearly from source to drain with a parabolic peak in the pinch-off region ( length 200 nm ) imaging the longitudinal electric field. The generated heat is divided into the channel and the pinch-off heat rate $P_{ch} = I_d \cdot V_{dsat}$ and $P_{p.o.} = I_d (V_d - V_{dsat})$, respectively.

The asymmetry caused by pinch-off is also confirmed by measuring temperatures at the source and drain edge of the gate versus applied $V_d$ ( Fig.3 ).

Beyond the onset of drain current saturation at $V_d > V_{dsat} \approx 2.5$ V the temperature at the drain side is significantly higher than at the source side. The temperature difference increases with the applied drain voltage and is in good agreement with our model calculations ( solid line in Fig.3 ).

Even at a 0 μm MOSFET a distinct source drain asymmetry in the normal as well as in the reverse mode is observed outside the gate in the spacer regions. In the direct vicinity of the heat sources we obtain $\Delta T \approx 17$K; $V_g = V_d = 5$V and $I_d = 3.29$ mA again in good agreement with the calculated temperature rise.

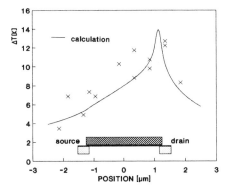
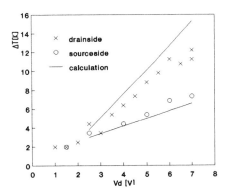

Fig.2. Measured and calculated temperature profile in a 2,5 μm transistor with $V_d$ = 6.5V ; $V_g$ = 6V ; $I_d$ = 2.1 mA.

Fig.3. Drain and source temperatures in a 2.5 μm transistor ($V_g$ = 6V) versus drain voltage $V_d$.

Outside the 0.6 μm MOSFET the temperature distribution shows a rather abrupt decay of temperature at the edge of the inversion layer in gate width direction (Position x = 5 μm). The temperature decreases within a typical range of 3 μm (Fig.4).

Fig.4. Temperature profile of a 0.6 μm FET ($V_g$ = $V_d$ = 6V; $I_d$ = 4.15 mA) scanning from the center of the channel (Position 0 μm) in gate width direction (see insert) compared with the model calculation (solid line).

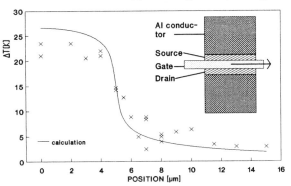

In perpendicular direction the temperature decay is about three times slower. This is shown in Fig.5 together with the scan direction which starts 1μm outside the channel. The observed asymmetry is attributed to the stripe shaped heat source and agrees with the temperature profiles calculated for the specific scan paths. The model predicts confocal elliptic isothermes revealing temperature gradients which are much weaker in channel direction than in gate width direction (Smith et al 1986). At high package densities this implies stronger temperature effects at neighbouring devices in channel direction than perpendicular.

Another thermal problem with decreasing design dimensions is the self-heating of the channel due to the increased drain currents. At standard operating conditions ΔT increases from 5 K up to 17 K for channel lenghts of 2.3 μm to 0.4 μm, shown in Fig.6. The measured ΔT's are somewhat lower than the calculated values at the pinch-off, but show the same general behaviour. According to the power density a 1/L and a $1/L^2$ temperature increase is expected for the pinch-off region and the channel, respectively. Short channel effects, however, cause a deviation from the $I_d \sim 1/L$ law. For a more quantitative analysis the ΔT/P ratio has to be considered. ΔT/P should

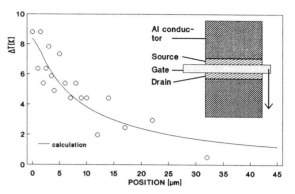

Fig.5. Temperature profile observed under the same conditions as shown in Fig.4 but scanned in the perpendicular direction. ( different length scales in Fig.4 and 5).

be constant for the pinch-off region. The measured values increase from 0.65 K/mW to 1.0 K/mW when the channel length decreases from 2.3 µm to 0.4 µm. We attribute this to the limited spatial resolution, which is given by the laser probe and which leads to an increasing influence of the channel heat source on the measured maximum temperatures in submicron MOSFET's. Thus the observed temperatures show a mixture of the expected behaviour at source and drain ( Fig.6 ).

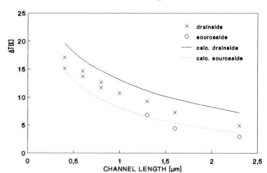

Fig.6. Temperature versus channel length.
$V_d = V_g = 5V$ conditions.

In conclusion, micro Raman spectroscopy is an adequate probe to determine local temperature rises in silicon MOSFET's. The temperatures can be measured within an accuracy of a few degrees Kelvin and a spatial resolution of about ( 0.5µm )³ ( light scattering volume ). Although the temperatures are measured mainly within the poly gate electrode, they agree with the calculated channel temperatures, because of the good thermal coupling to the channel. The peak temperatures, however, are not completely resolved due to the broadening of the temperature profile caused by heat diffusion and by the limits of spatial resolution in the submicron range.

This work was supported financially by the Siemens AG and the Deutsche Forschungsgemeinschaft.

References

Takacs D, Trager J. Schmitt-Landsiedel D 1988 IEEE Proc. on Microelectronic Test Structures Vol. 1, No 1 pp 50-55

Balkanski M, Wallis R F, Haro E 1983 Phys. Rev. B 28 pp 1928-1934

Mautry P 1990 doctoral thesis ( in preparation ) Techn. Univ. Munich

Smith D H, Fraser A, O'Neil J 1986 Measurement and Prediction of Operating Temperatures for GaAs IC's SEMI-THERM Symp. Scottsdale, Arizona

# Defects in highly doped silicon investigated by combined current and capacitance DLTS

Gert I. Andersson and Olof Engström
Department of Solid State Electronics, Chalmers University of Technology,
S-41296 Göteborg, Sweden

Abstract By using a recently developed method based on Deep-Level Transient Spectroscopy (DLTS) we can characterize electrically active defects in highly doped regions of silicon devices. The main advantage of the method is its applicability to nonabrupt $p$-$n$ junctions for investigating impurities, where traditional DLTS fails. The structures investigated were emitter-base $p$-$n$ junctions with phosphorus doped emitter and gallium or boron doped $p$ base with varying concentrations. Using this DLTS method we have studied the development of generation centers gettered to phosphorus emitter regions as they go from isolated centers at lower phosphorus concentrations into defect clusters at higher phosphorus concentrations.

1. Introduction

Crystalline defects and impurities which are likely to be present in the highly doped emitter regions of silicon devices have a considerable influence on the electronic properties of this kind of structure. Their action as recombination-generation centers lowers the injection efficiencies of charge carriers in bipolar devices. Highly doped regions in silicon are known to act as effective getters for metallic impurities. A large portion of the defects introduced during the high temperature steps of processing, are collected by the highly doped emitter regions of the devices. As the gettering action introduced by the emitter diffusions increases the charge carrier lifetimes in the bulk regions of the devices, the emitter region itself becomes densely populated by impurities. This is the reason for the need to characterize the properties of impurities in highly doped regions of devices.

Deep-Level Transient Spectroscopy (DLTS) (Lang 1974) has become the most common method of investigating the basic thermal emission and capture properties of impurity centers with deep lying energy levels in semiconductors. In the conventional version the method requires the use of perfectly abrupt $p^+$-$n$ or $n^+$-$p$ junctions. When impurities are to be investigated in diffused $p$-$n$ junctions from a production facility, this requirement is seldom fulfilled. Recently, we developed a method (Andersson and Engström, 1990) of solving this measurement problem by combining DLTS based on current signals with capacitance signals and information about the doping profile of the $p$-$n$ junction. With this method highly doped emitter-base junctions can be investigated, where earlier only methods utilizing the electric steady state parameters have been used.

In the present work we investigated the properties of impurities gettered to the emitter base region of Gate Turn-Off (GTO) thyristors. It is desirable to have a low resistance in the $p$ doped gate layer of GTO thyristors in order to increase the turn-off capability. But $p$ base dopings greater than $10^{17}$ cm$^{-3}$ have been shown to have negative effects on the emitter injection efficiencies. This correlates with our data indicating that the generation centers tend to cluster at higher phosphorus concentrations and give rise to high charge carrier generation due to tunneling processes. The latter originate from high local electric fields in the vicinity of the centers. The tunneling contribution was found to increase with the phosphorus concentration, suggesting that these phenomena become more severe at higher emitter base doping.

© 1990 IOP Publishing Ltd

## 2. Experimental details

The *p-n* junction samples investigated were taken from the gate-cathode junction at GTO thyristors prepared from 300 ohmcm, (111) oriented FZ silicon wafers. The phosphorus doped cathode had approximately the same doping profile for all samples with a maximum electrically active concentration of dopants measured by spreading-resistance to $5 \cdot 10^{19}$ cm$^{-3}$. The maximum concentration of gallium and boron in the gate region varied from $7 \cdot 10^{16}$ to $2 \cdot 10^{18}$ cm$^{-3}$ among the samples. Both the 10 μm deep phosphorus profile and the 50 μm deep gallium profile were diffused into the wafer at constant surface concentration and therefore predicted to be Gaussian. The small extension of the space-charge region at these doping levels makes it possible to consider the *p* base doping constant and equal to the maximum concentration when the doping profile is calculated from capacitance data using a method presented earlier (Andersson and Engström 1990).

The experiments were performed as stationary measurements of current-voltage and capacitance-voltage characteristics in combination with current DLTS measurements by using a specially designed computerized system (Andersson and Engström 1990). From this technique the generation properties of the centers occurring in the space charge region of the *p-n* junctions are obtained as emission rates and activation energies as well as the position of the emission region. The latter quantity is important in order to establish whether a certain signal originates from electron emission on the *n* side or from hole emission on the *p* side of the *p-n* junction. For the nonabrupt *p-n* junctions investigated here, this information cannot be obtained by using conventional DLTS analysis. Our method is based on computer-simulated capacitance data obtained from our knowledge of the doping profiles and data from the thermal current-emission measurements and their comparison with measured capacitance data.

## 3. Experimental results

A general characteristic of all the investigated samples is a contribution of field dependent components to the electron emission rate, $e_n$, and the hole emission rate, $e_p$, at low temperatures. As shown in Figure 1, one such contribution, $e_{n,p}^{FN}$, may come from the direct tunneling, or Fowler-Nordheim tunneling, from the trap ground state to the semiconductor energy bands. A second contribution, $e_{n,p}^{PAT}$, due to phonons in the tunneling process may also be added (Vincent et al 1979 and Rosencher et al 1984). The Fowler-Nordheim tunneling emission rate, $e_{n,p}^{FN}$ and the phonon assisted tunneling emission rate, $e_{n,p}^{PAT}$ add to the thermal emission rate $e_{n,p}^{t}$ and give the total emission rate $e_{n,p}$ observed during a measurement. This can be written in a general expression as

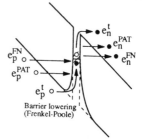

$$e_{n,p} = e_{n,p}^{t} + e_{n,p}^{PAT} + e_{n,p}^{FN}.$$

**Fig. 1.** The energy band diagram and the thermal and the electric-field-assisted transition processes between a deep impurity and band states using the model of Vincent et al (1979) and Rosencher et al (1984). A delta function potential is assumed for the electrons captured in the emission center and a Coulomb potential for the holes.

Figure 2 shows the electron emission rate measured by current DLTS and the corresponding hole emission rate calculated using the steady state reverse current data for sample *B* with a maximum *p* base gallium concentration of $9 \cdot 10^{16}$ cm$^{-3}$. If the measured emission rate is solely thermally activated a straight line is obtained in an Arrhenius plot. At low temperatures the magnitudes of the field dependent components are equal to or larger than the pure thermal emission rate. As the field dependent components have lower thermal activation the slopes of the curves in Figure 2 bend toward lower values at low temperatures.

The broken lines in Figure 2 were calculated using a theory of Vincent (1979) *et al* and Rosencher *et al* (1984). The best fit is obtained by using a delta function potential for the electrons captured in the emission center and a Coulomb potential for the holes. The ratio between the electric field, F, and the square root of the ratio between charge carrier effective mass, $m^*$, and the electron mass, $m_0$, is used as a fitting parameter due difficulties in defining the former quantities. For the electron emission, the delta potential gives an excellent fit when $F/(m^*/m_0)^{1/2}$ is $1\cdot 10^8$ V/m. Three different values of $F/(m^*/m_0)^{1/2}$ are used in the comparison for the hole emission. It is obvious from Figure 2 that even if a Coulomb potential gives a better fit for the hole emission than a delta potential, the agreement with the measured curve is less good than for the electron emission data for all values of the fitting parameter. However, one may conclude that the potential for hole emission is broader than the delta potential for the electron emission. This indicates that the defect center responsible for the charge carrier emission is of acceptor type (Lax 1960) and is in agreement with the result obtained when calculating charge carrier type and emission region (Andersson and Engström 1990).

The tunneling contribution increases as the maximum $p$ base concentration increases. This is clearly demonstrated by the thermal activation plot of the reverse current at different voltages in Figure 3 for samples $A$ and $D$ with maximum $p$ base gallium concentrations $7\cdot 10^{16}$ and $2\cdot 10^{18}$ cm$^{-3}$, respectively. Both samples show a bending in activation energy which is characteristic for all these devices. For sample $A$ an asymptotic slope independent of the voltage can be recognized at higher temperatures corresponding to the activation of the solely thermal emission. At low temperatures emission due to phonon assisted tunneling dominates. The reverse current for sample $B$ shows a pronounced dependence of the voltage at all temperatures. At the highest reverse bias and at low temperatures the reverse current is almost independent of the temperature. This indicates the presence of Fowler-Nordheim tunneling. However, the average electric field calculated from the doping profile is not sufficient to explain the tunneling.

In Figure 4 the reverse current versus average electric field at 150 Kelvin is shown for samples $A$ and $D$ in Figure 3 and sample $C$ with maximum gallium

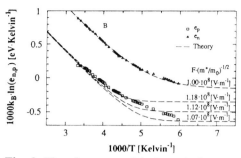

Fig. 2. The electron and hole emission rate calculated from current DLTS data and the reverse current for sample $B$, with maximum $p$ base gallium concentration $9\cdot 10^{16}$ cm$^{-3}$. The broken lines correspond to the theoretically calculated emission rate with different values of $F/(m^*/m_0)^{1/2}$ using the theory of Vincent et al (1979) and Rosencher et al (1984). A delta function potential is used for the electrons captured in the emission center and a Coulomb potential for the holes.

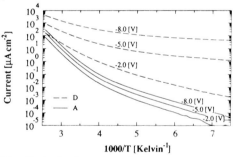

Fig. 3. Arrhenius plot of the reverse current at three different reverse voltages for samples $A$ and $D$ with maximum $p$ base gallium concentrations $7\cdot 10^{16}$ and $2\cdot 10^{18}$ cm$^{-3}$, respectively.

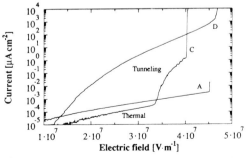

Fig. 4. Current versus electric field at 150 Kelvin for samples $A$, $C$ and $D$ with maximum $p$ base gallium concentrations $7\cdot 10^{16}$, $9\cdot 10^{16}$ and $2\cdot 10^{18}$ cm$^{-3}$, respectively.

concentration $7 \cdot 10^{17}$ cm$^{-3}$. For sample A the current increases moderately with increasing space-charge region up to the field where avalanche breakdown takes place. The reverse current for sample C exhibits two sharp knees, one at the avalanche field and one where local fields give rise to a tunneling component that dominates over the homogeneous generation present. Sample D exhibits a tunneling path for all fields. This demonstrates that tunneling components totally dominate the reverse characteristic at this temperature. In Figure 5 DLTS spectra from all these devices are plotted in the same diagram for comparison. It is seen that the heights of the DLTS spectra, which are proportional to the concentrations of impurity centers, decrease as the maximum gallium concentration increases.

### 4. Discussion and conclusion

In all essential parts the experimental results presented apply also for devices with boron $p$ base doping. Samples with comparable boron and gallium $p$ base doping profiles exhibit almost identical emission properties. This supports the argument that the $p$ base profile itself is of minor importance. By increasing the $p$ base gallium or boron concentration, the intercept with the phosphorus profile takes place higher on the profile. Consequently, a region higher on the phosphorus profile is probed during the measurements. Further, the analysis of the DLTS data shows that the dominating impurity center is positioned on the phosphorus side of the $p$-$n$ junction in all the investigated samples. Phosphorus is know to getter metal impurities very efficiently. It is therefore not especially surprising to find a high concentration of impurity centers in the phosphorus emitter. However, it has only recently been possible to determine whether the DLTS signal from nonabrupt $p$-$n$ junctions is due to electron emission on the $n$ side or hole emission on the $p$ side (Andersson and Engström 1990). Previously the DLTS technique was restricted to one-sided abrupt $p$-$n$ junctions.

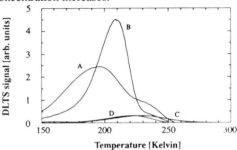

**Fig. 5.** Current DLTS spectra with rate window setting 10 s$^{-1}$ for samples A-D, with maximum $p$ base gallium concentrations $7 \cdot 10^{16}$, $9 \cdot 10^{16}$, $8 \cdot 10^{17}$ and $2 \cdot 10^{18}$ cm$^{-3}$, respectively.

The decrease of emission centers at higher phosphorus concentration (Figure 5) indicates that when we climb higher on the phosphorus profile the impurity centers cluster into larger agglomerates or precipitate. This agglomeration affects the electric field distribution in the space-charge region (Figure 4). Local fields larger than the average field are present and increase the tunneling contribution to the carrier emission rates (Figure 2) and the generation current (Figure 3).

In summary, the emission of charge carriers from the impurity centers is governed by thermal processes and tunneling processes. As the $p$-$n$ junctions climb higher on the phosphorus profile the thermal contribution decreases and the tunneling emission increases. This effect is related to the occurrence of high local electric fields and indicates that the emission centers cluster into more pronounced agglomerates for higher concentrations of phosphorus.

### 5. Acknowledgement

The authors are indebted to Dr. M. Mikes-Lindbäck and Dr. M. Bakowski (Asea-Brown Boveri Drives, Västerås, Sweden) for providing the samples used in this study. This work was financed by the Swedish National Board for Technical Development.

### References

Andersson G. I. and Engström O., J. Appl. Phys., **67**, 3500 (1990).
Lang D. V., J. Appl. Phys., **45**, 3023 (1974).
Lax M., Phys. Rev., **119**, 1502 (1960).
Rosencher E., Mosser V. and Vincent G., Phys. Rev. B, **29**, 1135 (1984).
Vincent G., Chantre A. and Bois D., J. Appl. Phys., **50**, 5484 (1979).

Paper presented at ESSDERC 90, Nottingham, September 1990
Session 7C6

# A four million pixel CCD image sensor

T. H. Lee, B. C. Burkey, and R. P Khosla

Microelectronics Technology Division, Eastman Kodak Company
Rochester, New York, USA 14650-2008

ABSTRACT An ultra-high resolution image sensor was developed for industrial and scientific applications. The imager is a full-frame CCD sensor, consisting of 2048 x 2048 pixels, and its image area measures 18.43 mm x 18.43 mm. The pixel size is 9 microns x 9 microns. The sensor has dual readout registers to increase the data rate. The sensor could be operated in the single or dual readout register mode depending on the user's frame rate requirements and the data capture system. The architecture of this imager is suitable for accumulation mode operation, which results in low dark current of less than 10 pA/cm$^2$ at room temperature. The charge transfer efficiency is 0.99999 for horizontal clock rate up to 20 MHz.

## 1. INTRODUCTION

Continuing advances in VLSI technology have made possible image sensors with high resolution. We have previously reported a CCD imager with 1.4 million pixels (Stevens 1985 and Nichols 1987). Higher resolution imagers will further enhance the electronic still, astronomy, scientific and machine vision applications. To serve the ever increasing resolution needs, we have developed an ultrahigh-resolution CCD image sensor with 4 million pixels.

## 2. DEVICE DESIGN

The image sensor is a full-frame CCD with an N buried channel. We used an advanced two-phase CCD architecture that has two poly silicon levels and a single metal level. Figure 1 shows the configuration of the device. The image area has 2048 x 2048 pixels. Each pixel is 9 microns x 9 microns so that the total image area is 18.43 mm x 18.43 mm.

In a two-phase CCD, each phase consists of a storage region to hold the charge, and a barrier (or transfer) region

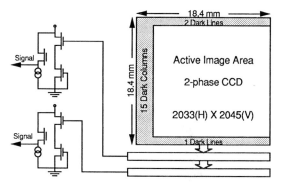

Figure 1. The 4-million pixel CCD image sensor architecture.

© 1990 Eastman Kodak Company

to assure the directional charge flow. The conventional two-phase CCD has one electrode for each of the two regions. The potential difference between the two regions is created by electrical or electrochemical means. Hence, a single pixel of a conventional two-phase CCD has a total of four electrodes, or gates. Therefore, the design rules requirement for the conventional two-phase CCD is very similar to that for the four-phase CCD, as shown in Figures 2a and b. We have developed an advanced two-phase CCD process that uses only one electrode for each phase. As shown in Figure 2c, instead of using a separated gate for the transfer region, the potential difference is formed electrochemically by implanting in a self-aligned manner. This architecture allows higher pixel density without tightening the design rules. This two-phase structure also alleviates the intra-level shorting problems between the CCD electrodes since these gates are common electrically. The architecture of the pixels is shown in Figure 3. This advanced two-phase architecture is particularly suitable for accumulation mode operation suggested by Saks (1980), which will result in very low dark current. There are two horizontal CCD registers, named A and B, to increase the data rate. The image could be operated in the single or dual readout register mode. The mode of operation would depend on the user's frame rate requirements and the data capture system. Each register has 2060 stages and can receive a complete line of signal. After photocharges are integrated in the imaging area, they are transferred, line by line, into the horizontal registers. In the dual line readout mode, an odd line is first transferred to A register, then from A to B register. The subsequent even line is then transferred to A register; now both lines

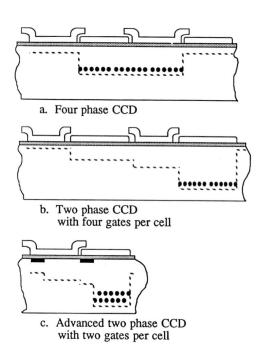

a. Four phase CCD

b. Two phase CCD with four gates per cell

c. Advanced two phase CCD with two gates per cell

Figure 2. A comparison of four-phase, two-phase, and advance two-phase CCD gate structures.

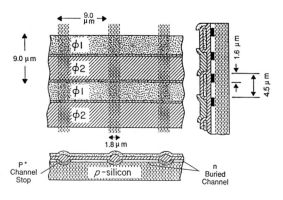

Figure 3. Top and cross sectional schematic views of CCD pixels.

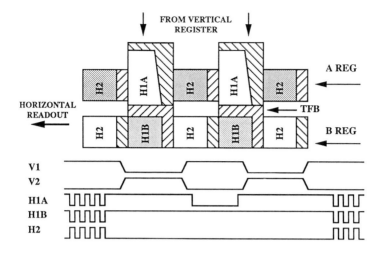

Figure 4. A schematic of horizontal CCD registers and the clock diagram for transferring two rows of signal charges into horizontal registers gates associated with shaded areas are made of second poly silicon.

are ready to be transferred to the floating diffusions for sensing. Figure 4 shows schematically the structure of A-to-B transfer. By taking advantage of the advanced two-phase process, we are able to have a very simple layout for the A-to-B transfer region. We need only five clock lines for both vertical and horizontal transfer, as shown by the clock diagram for dual line transfer in Figure 4.

## 3. EXPERIMENTAL RESULTS

Figure 5 shows the spectral response of the device, which is typical for full-frame or frame-transfer CCDs. The sensor is designed for a horizontal clock rate of 20 MHz. In the dual-line readout mode, it takes 114 milliseconds to read a frame. The dark current of the sensor is less than 0.5 nA/cm$^2$ at room temperature with conventional clocking. This corresponds to about 25 electrons of dark shot noise at 200-msec frame time, which includes both integration and readout time. Although this dark current level is adequate for higher fram rate the device would require cooling for slow scan application if

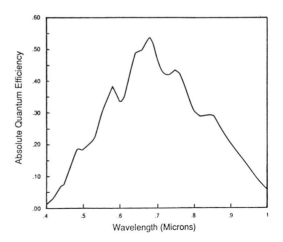

Figure 5. Spectral Response of the 4-million pixel CCD imager.

the conventional clocking method would be used. To avoid the system complication of cooling, the device could be operated in the accumulation mode of clocking. The advanced two-phase CCD process used to fabricate the sensor allows the device to be operated in the accumulation mode without loss in charge capacity. In this mode of operation, the vertical clocks V1 and V2 are held in their negative-voltage state to pin the surface potential at the $Si-SiO_2$ interface to substrate potential, except during charge transferring period. The dark current is reduced to less than 10 $pA/cm^2$ at room temperature. The measured accumulation mode dark current versus reciprocal temperature is shown in Figure 6 for the temperature range of 0 to 120°C. The activation energy of this temperature dependence is calculated to be 1.05 eV, indicating that the bulk diffusion current is dominant.

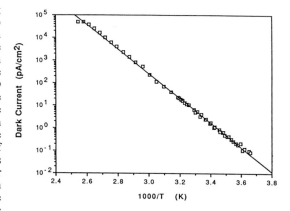

Figure 6. Dark current versus reciprocal temperature in accumulation mode operation.

The output amplifier is a two-stage source follower and has a sensitivity of 10 microvolts per electron. Figure 7 is the measured noise spectral density of the output amplifier. With correlated double sampling, the output amplifier contributes about 10 rms noise electrons at the 20 MHz data rate. The dark current noise is negligible with accumulation mode operation. The charge capacity of CCD is 85,000 electrons, this gives a dynamic range of 8000. The charge transfer efficiency is 0.99999 for horizontal clock rate up to 20 MHz.

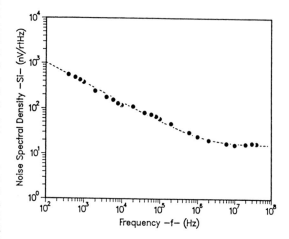

Figure 7. The measured input referred noise spectral density of the output amplifier.

## 4. REFERENCES

Nichols D N, Chang W C, Burkey B C, Stevens E G, Trabka E M, Losee D L, Tredwell T J, Stancampiano C V, Kelly T M, Khosla R P and Lee T H 1987 *Int. Electron Device Meeting Technical Digest* pp. 120-3

Saks N S 1980 *IEEE Electron Device Letters* **EDL-1** pp. 131-3

Stevens E G, Lee T H, Nichols D N, Anagnostopoulos C N, Chang W C, Kelly T M, Khosla R P, Losee D L, and Tredwell T J 1985 *Int. Solid-State Circuits Conf. Digest of Technical Papers* pp. 114-5

*Paper presented at ESSDERC 90, Nottingham, September 1990*
Session 7C7

# Unified model of the enhancement-mode MOS transistor

W J Kordalski

Technical University of Gdańsk, Institute of Electronic Technology,
Majakowskiego 11/12, PL 80-952 Gdańsk, Poland

> Abstract. A new unified fully analytical dc model of the enhancement-
> mode uniformly doped MOS transistor resulting from a wide device-physics-
> -oriented theoretical analysis. The model deals with the triode and
> saturation region of the transistor operation as a whole, and its
> validity is preserved for the long- and short- channel MOS transistors.

1. Introduction

Numerous analytical, as well as an increasing number of semi- or even
fully-empirical models used for describing MOS transistor operation show
impasse in analytical modelling of the insulated gate unipolar transistor.
The impasse arises because all the fundamental implications of the physical
MOS FET model have not been taken into account jointly.

Basing on numerical analysis. Yamaguchi (1979) has demonstrated that in
order to show the principle of MOS transistor operation one should take
into account such effects as: the dwo-dimensional (2-D) nature of the
carrier motion and the dependence of the carrier mobility upon the fields
both perpendicular and parallel to the channel. Although some of these
questions were discussed, they nevertheless were not treated in a
sufficiently comprehensive way to solve the problem of the analytical MOS
transistor modelling. E.g. Murphy (1980) has analysed the effect of carrier
velocity saturation in the channel on the drain curren in unipolar
transistors. Recently, Zhang and Schroder (1987) have taken an importnat
step towards the analytical formulating of the 2-D effects, however, the
model derived by them is not fully analytical.

The purpose of this paper is to present a new unified model of the enhancement-mode MOS transistor with uniformly doped substrate. The model has
resulted from the author's theory of the MOS transistor which is currently
preparing for publication. In this model the MOSFET is treated as a two-
dimensional object(gradual channel approximation is abandoned)in which
the channel has also two-dimensional nature. The unified model of the MOS
transistor is free from such non-realistic terms as: "pinch-off" and
"shortening channel effect".

2. Physical implications

Both in the triode and saturation range of operation the physics of the MOS
transistor is mainly determined by the following phenomena: (1) field effect
 inducing the channel by electric field in the gate insulator, (2) modulation of the channel charge by means of a physical channel-substrate
junction, (3) creation of an "additional" channel charge resulting from

varying the longitudinal electric field in the drain-source region - so called 2-D effect. (4) the variation of the carrier mobility caused by changing the electrophysical conditions between the source and the drain (mainly determined by varying the longitudinal and transversal field in the channel). It is the four above mentioned groups of phenomena related to each other that are the fundamental implications of the unified model of the MOS transistor.

So far, in the analytical models, the regrouping of carriers in the transistor channel has not been taken into consideration. Contrary to those, in the unified MOSFET theory by the author the channel is treated as a two-dimensional object (narrow-width effect is neglected) so the possibility of carrier regrouping in the channel is assumed as a rule. This regrouping results from an interaction of the first three from the four above mentioned phenomena. It is this interaction that leads to <u>Gradual Channel Detachment Effect</u> (GCDE), which consequently changes locally the magnitude of the carrier mobility in any element of the channel because the carrier mobility depends on both their position in the channel and the strength of the electric field components in a given point of the channel. As the drain-source voltage increases, the GCDE intensifies and consequently the dominating surface scattering mechanism tends towards the bulk-type scattering mechanism.

In the MOS transistor modelling the mobility must be treated as a locally determined value because of inhomogeneity of electrophysical conditions (the strength of the electric field components, the impurity concentration, the number and the kind of scattering centres,etc.) in the channel. Moreover, the variations of the transistor biasing voltages cause the variations in electric field distribution as well as carrier density profile in the channel. Therefore, the term "effective mobility" is of importance only for strictly determined boundary conditions and consequently it can not be treated as a constant value over the entire range of the transistor operation.

The dependence of the surface density of the inversion layer charge, $Q_n$, on the gate-source voltage is the important problem in an MOS transistor theory. Generally, this dependance is nonlinear, which showed Brews (1978) and Sze (1981). However, for $V_{GS} \gg V_T$ it becomes linear. This dependence, derived even from approximate analytical solutions, is mathematically very complex. Therefore, it is reasonable to use an approximate expression:

$$Q_n = C_{ox}(V_{GS}-V_T)^2 \left[V_C + (V_{GS}-V_T)\right]^{-1} \tag{1}$$

where: $V_C$ is a characteristic quantity (measured in volts) for this problem, $C_{ox}$ - oxide capacitance per unit area, $V_{GS}$ and $V_T$ are the gate-source and threshold voltages, respectively.

3. <u>Mathematical model</u>

To reach the main equation of the unified model it was necessary: (i) to analyse 2-D effects in the most general case, (ii) to show the occuring of the GCDE, (iii) to model the variations of the carrier mobility in the channel, (iv) to calculate the charge induced by the 2-D effects. The equation has the form:

$$I_{DS} = \frac{E_C W \mu_o C_{ox}}{L\left[E_C + \frac{V_{DS}}{L}\right]} \frac{(V_{GS}-V_T)}{V_C + (V_{GS}-V_T)} \left\{(V_{GS}-V_T)V_{DS} - \frac{hkV_{DS}}{t(1+kV_{DS})}(V_{GS}-V_T)^2 \ln\left[1 + \frac{tV_{DS}}{V_{GS}-V_T}\right]\right\} \tag{2}$$

in which:

$$E_C = \frac{v_{sat}}{\mu_0}, \quad h = 1 - \left[1 + \alpha E_\perp \frac{1}{1+bV_{DS}}\right]^{-1/2}, \quad t = d + m \, V_{GS} - V_T, \quad E_\perp = \frac{C_{ox}(V_{GS}-V_{FB})}{\varepsilon_s}$$

where: W and L are the channel width and length, respectively, $\mu_0$ is the low-field mobility, $v_{sat}$ is the saturation velocity of carriers, $\alpha$ is a coefficient representing the properties of surface mobility, b, d, m and k are coefficients describing the GCDE and variation of the carrier mobility, $\varepsilon_{ox}$ and $\varepsilon_s$ are the dielectric constants of the oxide and semiconductor, respectively, $V_{FB}$ and $V_{DS}$ are the flat-band and drain--source voltages, respectively.

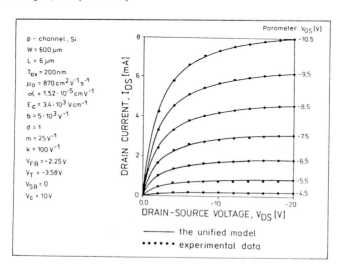

Fig. 1. Experimental and theoretical characteristics of a enhancement MOS transistor

In the extreme case when $V_{DS}$ is large enough equation (2) tends asymptotically to:

$$\lim_{V_{DS} \to \infty} I_{DS} = v_{sat} W C_{ox} (V_{GS}-V_T)^2 \left[V_C + (V_{GS}-V_T)\right]^{-1} \qquad (3)$$

The above expression coincides with a well known feature of the MOS transistor and states that in the limit of carrier velocity being fully saturated at $v_{sat}$, the drain current is independent of the channel length. What is more, equation (3), satisfied in the saturation regime, is quadratic function of $V_{GS}$ for $(V_{GS}-V_T) \ll V_C$ and linear for $(V_{GS}-V_T) \gg V_C$ - compare Yamaguchi (1979).
Agreement between measured and modelled results is excellent - see Figure 1. For the sake of brevity, it has not been discussed all the features of the model in detail  comprehensive discussion of this model will be reported elsewhere in the near future . The main of them are mentioned in the next section.

## 4. Conclusion

In contrast to known models this one treats the MOSFET as a two-dimensional (2-D) object (gradual channel approximation is abandoned) in which the channel has also 2-D nature. The physics of the model bases on the following phenomena: 1) the field effect, 2) modulation of the channel charge by means of the physical channel-substrate junction, 3) creation of an "additional" channel charge resulting from the variations of the longitudinal electric field in the drain-source region: so called 2-D effect, 4) the influence of the gate voltage on the carrier mobility, 5) the carrier velocity saturation effect, and 6) the monlinear dependence of the inversion layer charge on the gate voltage for small $V_{GS}$. The model is from such terms as:"pinch-off", "effective mobility",and "shorting-channel effects".

A novel phenomenon called <u>Gradual Channel Detachment Effect</u> (GCDE)is shown to occur in the MOS transistor. It results from the 2-D nature of the carrier flow between the source and drain. As the drain voltage increases, the GCDE intensifies and the surface scattering mechanism tends towards the bulk-type scattering.

The principal advantages of the unified model are as follws:
(1) The model is true for both the long- and the short-channel transistor.
(2) It is valid over the entire range of the variations of the drain,gate and substrate voltages except the subthreshod region.
(3) Each parameter of the expression for the drain current reflects the corresponding process of the physical MOS transistor model.
(4) The drain current, transconductance and small-signal drain conductance are smooth functions of all the voltages.
(5) The saturation (plateau) of the drain current with the drain voltage is reached only as $V_{DS}$ tends to infinity.
(6) In the saturation regime the drain current is independent of the channel length.
(7) The formula (2) converges to the well-known expression for the long--channel transistor models if only drain-source voltage is small enough and $V_C=0$.
(8) The transfer characteristics go up shift toward lower gate voltages with both decreasing channel length and increasing drain voltage.
(9) The transconductance is a monotone increasing function with respect to $V_{DS}$ on an arbitrarily chosen interval of $V_{DS}$.
(10) The drain conductance $g_d$ increases with gate-source voltage in saturation. It is smooth and monotone decreasing function of $V_{DS}$.

Finally, the new unified physics-oriented fully analytical model enables to gain a deeper insight into the essence of the MOS transistor. Agreement between measured and modelled results is excellent. The model is very useful for circuit-analysis and simulation purposes. The unified model is believed to clear the way for nonuniform doping effects.

## 5. References

Brews J R 1978 *Solid State Electronic 21 pp 345-355*
Kordalski W J *"A theory of the MOS transistor" in preparation*
Murphy B T 1990 *IEEE J. Solid-State Circuits*
   *SC-15 pp 325-328*
Sze S M 1981 *Physics of semiconductor devices*
   */Wiley, New York/ pp 434-438*
Yamaguchi K 1979 *IEEE Trans. Electron Devices*
   *ED-26 pp 1068-1074*

# Directions for optoelectronic circuits

J.R.Hayes, J.Gimlett, W-P.Hong, G-K.Chang and J.B.D.Soole and R. Bhat

BELLCORE
Red Bank, NJ 07701, USA

Abstract

Optoelectronic Integrated Circuits, comprising the integration of optical and electronic components, have the potential to address a wide range of applications involving optical signal distribution, processing and detection with high performance at low cost. In this article we review the present position of OEICs and the trends that indicate future directions

1. Introduction

Optoelectronic Integrated Circuits have the potential to address a wide range of applications involving optical signal processing for telecommunications, computing and military applications. In the early 1980's most research on optical components for telecommunications was aimed at addressing long haul applications where the bit-rate/distance product was the important parameter.

More recently, applications such as the local-loop, computer interconnects and optical computing are being researched; these applications typically require high numbers of optoelectronic components at a low cost. The use of an integrated technology provides a potential means to meet these two critical objectives. In telecommunications it is envisioned that within the decade each subscriber will have access to information rates up to 150Mbits/s, compatible with switched digital HDTV signals for example.[1] However, the bandwidth of optical fibers can support many thousands of such channels and it is a challenge to systems designers to make efficient use of this bandwidth. There are several transmission approaches that have the potential to utilize the available fiber bandwidth and in most instances integrated optoelectronic components can provide advantages. It is to this end, therefore, that OEICs are being investigated. Among the mutiplexing schemes designed to take advantage of fiber bandwidth are time division multiplexing (TDM), wavelength division multiplexing (WDM) and coherent detection. Each of these will require low cost high performance optoelectronic components.

While in the local loop telecommunication systems are expected to utilize fiber and optical components operating at $1.3-1.55\mu m$, in the computing environment the distributed networks may not span such a wide area. It is, therefore unnecessary to operate at the region of lowest fiber attenuation (although dispersion can still be a problem). Processing speeds are increasing to such an extent that it appears attractive to use space or fiber optics to interconnect computer boards. For many of these applications short wavelength ($0.8\mu m$) GaAs devices can be used.[2] However, optoelectronic research is concentrating more and more on the long wavelength region and some WDM networks have been proposed even for computer applications which are based on InP components[3]. If this trend continues, it is expected that ultimately optical communications will take place between computing elements in the long wavelength region.

© 1990 IOP Publishing Ltd

## 2. Discrete devices

It is widely recognized that the III-V semiconductors are the most attractive materials for high speed electronics. This stems in part from the fact that the intrinsic material properties applicable to transport in GaAs and InGaAs are superior to those of silicon. There has been a large amount of research aimed at realizing an optimum device structure in both heterojunction bipolar transistors (HBT) and field effect transistors (FET) to realize these advantages. The essential design principles for achieving high performance devices from III-V semiconductors, were established in GaAs a number of years ago and recently advances in the performance of heterojunction bipolar transistors and field effect transistors have to a large extent been obtained by translating designs into the InP/InGaAs material system[4]: Fig. 1. In general the design of integrated circuits to take full advantage of the speed and low noise performance of the discrete components remains a research challenge. To date, small scale circuits fabricated with these transistor do not demonstrate significant advantages over conventional IC technologies based on Si bipolar devices.

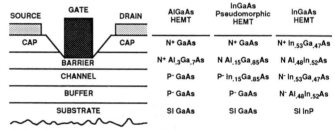

Fig 1. HEMT improvements have been obtained by essentially translating device design from GaAs to InGaAs.

There are a wide range of photodetector devices that can be integrated on an OEIC for receiver applications such as pin detectors, MSM detectors, etc. Avalanche photodiodes (APDs),[5] which are used in applications where the best sensitivities are required require high voltages and are not presently receiving much attention for integration with electronics. For integration purposes it is important to have a detector that is planar with both electrodes on the top surface. This has been achieved using either planarized pin structures[6,7] or metal-semiconductor-metal (MSM) detectors that have interdigitated metal finger electrodes. Although the quantum efficiency of MSM's is not as high as pin detectors, due to metal electrode shadowing, their relatively low capacitance allows relatively large area devices to be used for the same net input capacitance (an important parameter in determining receiver sensitivity). Although the long wavelength MSM technology is not as mature as that of pin photodetectors, recently both high speed and low leakage currents have been obtained from these devices[8,9]: Fig. 2.

## 3. OEICs

Optoelectronic circuits can comprise both electronic, optoelectronic and optical components that are monolithically integrated onto a common substrate. In order to achieve monolithic integration of diverse components that will comprise an OEIC, a material technology must be established that allows the integration of components that have very different material requirements. In addition, for integration purposes it is desirable that the chip surface be near planar. However, the intrinsic device requirements are often significantly different in structure and thickness requiring that a material deposition strategy be developed that yields near planarity. This requires either multiple regrowths, the incorporation of trenches or such techniques as quantum well disordering.[10] Another alternative is the use of a technology that compromises the performance of each structure by using a generic layer set. The heterojunction bipolar transistor may be the most suitable for this approach since the device has been demonstrated as a laser[11], a photodector as well as an electronic amplifier.

Another approach is "hybrid integration," a technique based on epitaxial lift-off[12], that combines the advantages of the hybrid approach with those of monolithic integration. Here, an epitaxial film of either GaAs or InP based material is selectively removed from its growth substrate, is attached to a new substrate through van der Waals forces, and is then used to fabricate devices (Fig. 3).[13] In this way, devices requiring different material structures can be readily incorporated into an OEIC without regard to epitaxial growth constraints such as lattice matching, like in hybrids, yet they can be batch fabricated and closely packed to minimize parasitics, like in monolithic integration. Because the new substrate does not have to be semiconductor, this technique also opens the possibility of integrating semiconductor devices with ones that are not semiconductors.

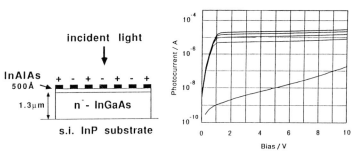

Fig. 2  Schematic diagram and characteristics of an AlInAs/InGaAs MSM photodetector.

A schematic diagram showing the integration of GaAs MSM photodetectors with LiNbO$_3$ waveguides as an example of hybrid integration using epitaxial lift-off, is shown in Fig. 3. In the upper left branch, a GaAs film grown on a GaAs substrate with a sacrificial AlAs layer is removed from its substrate by waxing the surface for mechanical support and selectively etching the AlAs in hydrofluoric acid. Waveguides are formed in the LiNbO$_3$ by proton exchange through a patterned aluminum mask as shown in the upper right. The lifted-off GaAs film is then attached to the LiNbO$_3$ by van der Waals forces and the wax removed. MSM detectors are finally made from the GaAs film using conventional techniques.

The level of electronic integration which will be required for OEICs varies greatly, depending upon the application. For example, simple analog optical-to-electronic conversion, as is achieved by a optical receiver/preamplifier, might require only

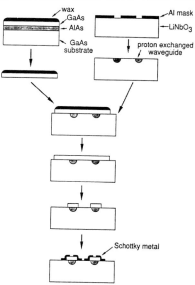

Fig. 3  Illustration of the process for realizing waveguide integrated lift-off devices.

10 to 50 devices. Over one hundred devices might be required for regeneration to an ECL-level digital output, while several hundred devices are necessary for regeneration with retiming. Of course, the ultimate challenge for OEICs is to monolithically integrate onto a single chip all of the circuitry of a fiber optic system which might at present consume an entire optical bench in the laboratory. This will include multiplexing/demultiplexing circuitry for time-division-mulitiplexed systems. However, the signal processing required by the Synchronous Digital Hierarchy for time-division-multiplexed applications requires several thousand gates, and would require a substantial increase in InP/InGaAs device yields and density. This may not be necessary if signals can be conditioned to a sufficient degree on the OEIC using modest levels of integration to allow further post-processing using standard integrated circuit technology. For example, Fig. 4 shows an artist's conception of such a hybrid approach in which the initial high speed processing and first layer of demultiplexing is performed by the OEIC receiver, which is in turn mounted via flip-chip or epitaxial lift-off[13] technology on a silicon integrated circuit which performs the remaining demultiplexing and signal processing. This type of integrated/hybrid technology has the additional benefit of easing some of the packaging bottlenecks associated with very high speed circuits, since the off-chip connections carry only lower speed signals.

## 4. OEIC Receivers

One of the most important applications for OEICs will be in optical receiver/regenerator circuits, including receiver arrays for military applications, neural networks, computer interconnections, and balanced receivers for coherent

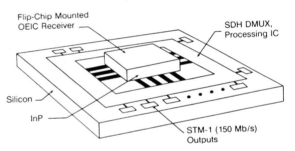

Fig. 4 Flip-chip hybrid approach to optoelectronic integration

communications, as well as for low cost receivers/transceivers for the subscriber loop. In addition to promising greater functionality, reduced size and cost, and higher reliability than hybrid circuits, OEICs offer potentially better performance due to the higher speed and lower noise of InP/InGaAs devices as compared to GaAs and Si devices, and the reduced interconnection parasitics resulting from integration. However, at present the performance of OEIC receivers in terms of receiver sensitivity is somewhat poorer than state-of-the-art hybrid receivers, as shown in Fig. 5. There are many reasons for this performance degradation. For example, many OEIC receivers, particularly early receivers based upon InP JFETs or strained heteroepitaxial GaAs MESFETs on InP, suffer from poor device transconductance. This situation is likely to be remedied in the near future, as newer OEICs based on HBTs or InAlAs/InGaAs HEMTs are achieving transconductances comparable to those of similar-size discrete devices.[14,15] Another disadvantage of present OEICs is that, due to the relative immaturity of the technology, circuits must rely upon conservative lithographies with critical device dimensions typically no smaller than one to two microns. The use of submicron lithography for critical devices will result in higher transconductance and lower input capacitance, and consequently lower receiver noise. Finally, because of technology limitations, most InP based, OEIC receivers consist of only one gain stage and a buffer stage, or even simpler circuit configurations. This results in a non-negligible contribution of noise from the subsequent amplification stages. It is anticipated that the sensitivity of OEIC receivers will approach, or even surpass that of hybrid receivers, as processing technology and circuit designs mature.

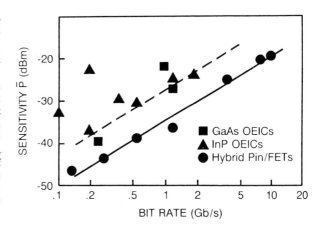

Fig. 5. Representative receiver sensitivities.

Another form of integration currently being actively researched is that of detectors and receivers with optical waveguides. Monolithic integration of a receiver with passive or active waveguide optics into what has recently been termed a Photonic Integrated Circuit (PIC) allows one to perform a last stage of optical processing on the receiver ship itself, doing away with the need for additional bulky optics requiring precise alignment. All the advantages of integration - small size, improved reliability, lower cost, and the elimination of alignment difficulties - result. Many communication architectures require some optical processing just before detection, wavelength division multiplexed (WDM) and coherent communication systems being perhaps the most obvious. For WDM systems, the wavelength filtering can be performed with the use of waveguide based gratings, and for coherent reception all the optics required for either a phase or polarization diversity receiver could be integrated on the receiver chip itself (Figure 6). To date, some significant progress has been made on the latter type with the integration of a laser local oscillator, a waveguide coupler, and two detectors in a balanced configuration on the same substrate.[16,17]

Many problems remain to be solved, however, before all this can be manufacture compatible technology. The problem of device incompatibility is a serious one. As in the case of the integration of a photodetector with receiver electronics discussed

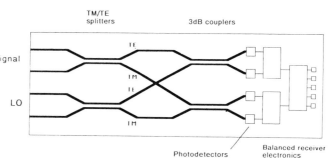

Fig. 6. Schematic of an monolithically integrated polarization diversity receiver, monolithically integrating polarization splitting and L.O. coupling waveguides with balanced receiver circuits on a single substrate.

above, the semiconductor structures required for the waveguide elements - guides, couplers, filters, etc. - are quite different from that used by the detector and receiver electronics, and even differ significantly within themselves; the simultaneous optimization of all component is a formidable task. The processing required by each component is also very different and the overall structure must be designed with fabrication viability as well as performance in mind. Planarity, or near-planarity, of the final structure is important. Semiconductor regrowth is required for some of the guide components and may be used to achieve this, and it is also aided by the use of planar electronic components, such as waveguide couplers using lateral electrodes and the MSM waveguide photodetector.[18]

In conclusion we have attempted to review some of the technologies that are being applied to OEICs. The advent of transmission systems based on either coherent or WDM applications is challenging device researcher to find solutions that will allow a fully integrated approach to realized; as is the case for electronic circuits today.

References

1. E. Nussbaum, J. Selected Areas in Comm. 6, 1036 (1988)
2. J. D. Crow et. al. IEEE., Trans. Elec. Dev., 36, 263 (1989)
3. N. R. Dono, P. E. Green Jr., K. Liu, R. Amaswami and F. Tong, to be published in IEEE, Jour. Selected Area of Communications, August, (1990)
4. J. R. Hayes, J. Vac Science and Tech., (1989)
5. S. R. Forest, in Chapter 14, Optical Fiber CommunicationsII, S. E. Miller and I. P. Kaminow eds (Academic Press, 1988).
6. D. A. H. Spear, P. J. G. Dawe, G. R. Atwell, W. S. Lee and S. W. Bland; Elec. Lett., 25, 156, (1989)
7. S. Miura, H. Kuwatsuka, T. Mikawa and O. Wada; Appl. Phys. Letts., 46, 1522, (1986)
8. J. B. D. Soole, H. Schumacher, H. LeBlanc, R. Bhat, and M. Koza IEEE Photonics Tech. Lett., 1, 250 (1989)
9. .G. K. Chang, W. P. Hong, J. L. Gimlett, R. Bhat nad C. K. Ngyuen, Elec. Lett.,, 25, 1021, (1989)
10. J. Cibert, P. M. Petroff, G. J. Dolan, S. J. Pearton, A. C. Gossard and J. H. English; Appl. Phys. Lett., 49 1275, (1986)
11. J. Shibata, Y. Mori, Y. Sasi, N. Hase, H. Serizawa, and T. Kajiwara; Elec. Letts, 21, 98, (1985)
12. A. Yi-Yan , W. K. Chan, T. J. Gmitter and M. Seto, paper MI1 Integrated Photonics Resaerch Technical Digest, (1990); W. K. Chnan A. Yi-Yan and T. J. Gmitter, to be published in J. Quant. Elec (1990)
13. E. Yablonovitch, T. Gmitter, J. P. Harbison and R. Bhat, Appl. Phys. Letts., 51, 2222-2224, (1987)
14. Fijutsu Oeic
15. G. K. Chang, W. P. Hong, J. L. Gimlett, R. Bhat and C. K. Nguyen, IEEE Photon. Tech. Lett., 2, 197, (1990)
16. T. L. Koch, U. Koren, R. P. Gnall, F. S. Choa, F. Hernandez-Gil, C. A. Burrus, M. G. Young, M. Oron, and B. I. Miller, Electron. Lett., 25, 1621, (1989)
17. H. Takeuchi, K. Kasaya, Y. Kondo, H. Yasaka, K. Oe, and Y. Imamura, IEEE Phot. Tech. Lett., 1, 398, (1989)
18. J. B. D. Soole, H. Schumacher, H. P. LeBlanc, R. Bhat, and M. A. Koza, Appl. Phys. Lett., Vol. 56, 1518, (1989)

# Novel applications of porous silicon

J M KEEN

R.S.R.E. St Andrews Road, Malvern, Worcs. England, WR14 3PS

Porous silicon is best known for its applications in the oxide isolation of silicon integrated circuits. However, porous silicon is a very versatile material and its unique properties are being researched for a surprising range of applications. These include use as a substrate for epitaxial growth, for the production of "smart" dielectrics, for the location of defects in other silicon-on-insulator materials, for the producion of sensors, for conversion to metal and for the emission of visible light. These and others are described.

## 1. Introduction

Porous silicon was discovered by Uhlir (1956) during investigations into the electropolishing of silicon in hydrofluoric acid electrolytes. At anodic voltages lower than those required for true electropolishing, an anodic layer was formed whose origin, composition and structure were the subject of much speculation and many misconceptions for many years. Initially the layer was thought to be redeposited amorphous silicon resulting from the chemical decomposition of $H_2SiF_6$, a byproduct of the electrochemical reaction, and it was not recognised as porous silicon for many years.

## 2. Formation and Properties

Porous silicon is now recognised to be the original single crystalline silicon, strained and riddled with holes (Barla 1984) which have been etched due to localised electrochemical attack rather than the uniform dissolution which occurs during electropolishing. The formation of porous silicon has been shown by L'Ecuyer (1990) to take place via a surface state mediated charge transfer mechanism. The origin of these surface states is linked to the coulombic and dipolar interactions of adsorbed species, most probably fluoride ions, with the silicon lattice. Localised dissolution is initiated at or close to these active adsorption sites and once initiated continues due to geometrical field enhancement effects.

The properties and pore structure of porous silicon can be varied over a wide range to suit the application. The parameters that control these properties are the type and resistivity of the silicon starting material, the electrolyte composition and the precise anodising conditions, particularly the anodising current density (Herino 1987). By controlling these parameters it is possible to reproducibly create structures with pore diameters ranging from less than 2nm to greater than 1 micron and porous densities from near bulk values to less than $0.2$ $gm/cm^3$. The internal surface area can be huge, $500$ $m^2/cm^3$ is not untypical for a

© 1990 Crown Copyright

layer with a 5nm pore diameter. Therefore surface reactions can involve a large proportion of the material. For example, heating such a layer in oxygen for one hour at 300°C results in 25% to 30% of the silicon in the layer being oxidised. This is equivalent to the growth of a monolayer of oxide on the pore walls. This process is often exploited to stabilise the porous structure (Herino 1984). Since the maximum thickness of silicon in the pore walls of typical device material is only a few nanometers, great thicknesses can be completely oxidised at very low temperatures or completely reacted or replaced by other materials (Tsao, 1986). In the case of oxidation, the porous density is chosen so that there is sufficient pore void space to accommodate the volume increase without any change in overall dimensions. Similar considerations are taken into account for other reations in chosing the starting porous structure.

## 3. Composition and Purity

The composition and purity of porous silicon has been the source of controversy. As formed, the internal surfaces of the pores are lined mainly with hydrogen. This hydrogen is strongly bonded and in vacuum or inert ambient, heating to 500°C is required to remove it (Ito, 1988), however, porous silicon slowly oxidises on exposure to air. Characterisation experiments need to take this into account together with the ability of porous silicon to behave in a similar way to a molecular seive due to its pore structure. Metallic contamination is not a problem if porous silicon is produced under clean conditions associated with microelectronic device fabrication contrary to some suggestions. The metallic levels are comparable to the original silicon, less than in SIMOX, and less than those introduced by other commonly used process steps. This is not unexpected since any positive ions, i.e. metals, will deposit on the cathode and not on the anode. Therefore, only the few metals that can chemically plate out onto silicon are a potential problem, and then only if they come into contact with the porous material. Device results confirm these analytical results.

## 4. Applications of Porous Silicon

### 4.1 Full and Lateral Isolation (FIPOS and IPOS)

These are the original and most widely studied applications for porous silicon. The initial target of workers such as Unagami (1977) was the production of a form of trench isolation. This work was probably premature and suffered from the lack of detailed knowledge now available. Sub-micron lateral isolation achievable by this approach was not then necessary and the isolation geometries required were achievable by more conventional approaches.

The major impetus for research into porous silicon came with the demonstration by Imai (1981) that full isolation of silicon regions was possible by a porous approach (FIPOS). This occurred at a time when worries were starting to be felt about the limitations of silicon on sapphire material, especially in the thinner layers required for smaller geometry devices. The initial FIPOS route utilised the ability to selectively anodise and convert into porous silicon, p-type silicon in the presence of n-type. Regions that were to be left were temporarily converted to n by a proton implant and anneal. This approach suffered from a number of inherent drawbacks particularly related to the thickness of the porous silicon which led to warpage problems and to the presence of a large spike, which always contained dislocations, associated with the centres of the islands. Despite these problems, 64k CMOS-SOI SRAM's were produced by Ehara (1985). The problems of this route were overcome by using the ability to selectively anodise $n^+$ silicon in the presence of $n^-$ (Holmstrom, 1983). By creating $n^-/n^+/n^-$ structures the thicknesses of the silicon islands and the buried porous oxides could be independently controlled irrespective of island width, the spike and wafer warpage problems were eliminated and defect free islands formed. Using this approach Thomas (1989) produced fully depleted thin film CMOS-SOI devices and circuits in silicon islands 100nm thick with a performance comparable to SIMOX processed on the same line. The change to this approach required the characterisation of the anodisation of n-type silicon. This knowledge and the $n^-/n^+/n^-$ structures produced can be exploited in novel ways, for example as shown by Tsao (1986), porous silicon can be converted into a metal rather than an oxide to produce buried metal layers. This opens up a whole range of new device possibilities.

4.2 Metallisation of Porous Silicon.

Tungsten hexafluoride can be reduced to tungsten by two different reactions. The first requires silicon:

$$2WF_6 + 3Si = 2W + 3SiF_4$$

but self limits on bulk silicon at a tungsten thickness of 10nm - 30nm (Blewer, 1986). Since the thickness of the silicon in the pore walls is of the order of nanometers the silicon in porous silicon can be consumed entirely and the reaction is unlimited. Pore blockage doesn't occur since the volume of the tungsten deposited is less than that of the silicon consumed enhancing the diffusion of the tungsten hexafluoride into the pores. Once the pores have been lined with tungsten, additional tungsten can be deposited using the hydrogen reduction of $WF_6$. This second reaction is specific to tungsten. The tungsten deposited contains less than 1% silicon and other impurities are also low if suitable precautions are taken. In porous silicon structures lateral penetrations of over 25 microns have been achieved for 0.5 micron thick buried layers with a resistivity of 20 ohm-cm as deposited. Unannealed, the tungsten oxidises in air (Earwaker, 1990).

The potential advantages of this technology is that it enables self aligned tungsten to be inserted into single crystal structures without the need for epitaxial growth at temperatures as low as 200°C. Thus structures can be created that were either impossible previously or required the use of molecular beam epitaxy.

Potential applications are the formation of buried metal underpasses, buried radiation shield grids, metal collectors for bipolar or power devices and novel devices.

Tungsten hexafluoride is the only metal precursor to behave in precisely this way but experiments are being carried out with other metallic precursors in a LPCVD system in order to metallise porous silicon.

Ito (1988) has used a different approach metallisating the porous by the deposition of metals in UHV. The advantage claimed by Ito is the potential reduction of strains due to volume changes before and after the silicidation reaction. Metals studied include Ti, Au and Ag which were deposited onto unheated or heated porous substrates. In this way metal could be deposited onto the porous surface and subsequently diffused to create a silicide utilising the anomalously high diffusion rate in porous silicon. Alternatively the silicide can be formed directly by depositing the metal onto the heated substrate.

This new and exciting field of research is still in its infancy and there could be many important applications.

4.3 Nitridation and oxynitridation of porous silicon and reduced dielectric constant materials.

When porous silicon is oxidised at low temperatures, it can be oxidised over distances of hundreds of microns and "bulk" oxides grown at the porous/bulk silicon interface due to the oxide created being microporous. This allows anomalous diffusion to occur. This property can be exploited to create nitride or oxynitride SOI structures. Using an ammonia atmosphere and temperatures of the order of 1000° C, porous silicon can be directly nitrided or alternatively oxidised porous silicon converted to oxynitride. In a nitrogen atmosphere only the interfaces are converted, presumably by the direct nitridation of the silicon since the nitridation of the oxide is energetically unfavourable. The dielectric layers can be subsequently densified in exactly the same way as oxidised porous silicon.

These dielectrics can be expected to have the same advantages as bulk oxynitrides, i.e. improved hot electron and radiation performance. Non optimised devices with oxynitrided interfaces been fabricated and have a good performance.

To avoid volume changes on oxidation, porous silicon requires sufficient pore volume to adsorb the increased volume of the oxide. If excessive pore volume is deliberately created prior to oxidation, a low density porous oxide is created which should have a reduced dielectric constant. Oxides of this type could be valuable in reducing the RC delays of metal interconnect tracks.

4.4 Epitaxial growth on porous substrates.

Since porous silicon is single crystalline it can, and has been used as a substrate for the growth of defect free epitaxial layers. Porous silicon restructures at elevated temperatures therefore the temperature has to be kept low and the time at temperature kept short. Initially this necessitated the use of MBE (d'Avitaya, 1985) or liquid phase epitaxial techniques (Baumgart, 1984). Laser regrowth (Baumgart, 1982) has also been used. None of these techniques can yet be considered to be roduction processes. Advances in low temperature chemical vapour epitaxial growth of silicon however opens up new commercially viable possibilities. Zero added defect layers have been grown by Oules (1989) in a rapid thermal processing system using silane at 820°C. The initial porosity of the substrate was found to be a critical parameter and this was only achieved in $n^+$ and $p^+$ substrates. This is a limitation to the technique at present. The growth of epitaxial layers onto porous silicon in this way creates new flexibility in the FIPOS approach. In addition to removing circuit layout constraints it has the added advantage of being a cheap route to SOI. By this approach, islands are cut from the epitaxial layer and the underlying porous silicon is then oxidised, or alternatively, metallised.

In addition to silicon, $Ge_xSi_{1-x}$ (Xie, 1990) and $CoSi_2$ (Kao, 1987) have been grown epitaxially on porous substrates by MBE.

In an interesting variant which cannot be epitaxial growth, an oxide layer is producted under the porous silicon by anodic oxidation. This is followed by the restructuring of the single crystal porous silicon into a non porous silicon overlayer (Bomchil, 1989) by a high temperature anneal. It is hoped by this means to retain the monocrystallinity of the silicon overlayer however it is not clear if this has been achieved.

4.5 Other applications

A whole range of other applications have been reported including: porous silicon oxide anti-reflection coatings for solar cells (Prasad,1982), the characterisation of probe points (Gorey, 1979), the fabrication of ink jet nozzles (Li, 1977), as a gettering layer on the backs of wafers and in the production of sensors. Another obvious application is in micromachining. Porous silicon can be designed readily into complex shapes by using suitably doped structures and then dissolved selectively using alkaline etches to leave membranes, beams, diaphrams or trenches.

It is notoriously difficult to reveal defects in the thin SOI layers currently being used for thin film fully depleted CMOS circuits, however an anodising process, similar to that used for the production of porous silicon, has been used by Guilinger (1989) to reveal defects in SIMOX material. This technique will only work for n-type layers but none the less it could be very useful for the rapid evaluation of process variants on SIMOX material quality.

4.6 Optical Emission

The newest development, and one with enormous implications, has been the achievement by Canham (1990) of the emission of visible light at room temperature from a porous silicon structure. The photoluminescence was stimulated by an argon laser and the process appears to be efficient. The emission of light from silicon in this way was previously thought to be impossible! For practical applications electroluminescence is required and that has still to be achieved.

5. Conclusions

Porous silicon has been shown to be a very versatile material and applications outside the conventional silicon on insulator field have hardly begun to be investigated. The material has a unique structure and unusual properties which can be exploited in unexpected ways, one of the most startling of which is the producion of a silicon structure that emits light at room temperature. Other new devices are now a possibility in addition to the improvements that can be made to the mainstream technologies.

6. References

Barla K, Bomchil G, Herino R and Pfister J C  1984 *J. Cryst. Growth* **68** 721
Baumgart H, Frye R C, Phillipp F and Leamy H J 1984 *Mat. Res. Soc. Symp. Proc.* **33** 63
Baumgart H, Frye R C, Trimble L E, Leamy H J and Celler G K 1982 *Laser and Electron-beam interactions with solids* ed B R Appleton and G K Celler
Blewer R S 1986 *Solid State Tech.* **29** 117
Bomchil G and Halimaoui A 1989 *Appl. Surface Sciences* **41/42** 604
Canham L T, Keen, J M and Leong W Y 1990 *U.K. Patent*

d'Avitaya, Barla K, Herino R and Bomchil G 1985 *Proc. Electrochem. Soc.* **85** 323
Earwaker L G, Briggs M J, Farr J P G and Keen J M 1990 *Proc. Microprobe Conf., Melbourne*
Ehra K, Unno H and Muramoto S 1985 *Electrochem. Soc. Extended Abs.* **85** 457
Gorey E F and Poponiak M R 1979 *IBM Tech. Disc. Bul.* **21** 4043
Guilinger T R, Kelly M J, Medernach J W, Tsao S S, Stevenson J O and Jones H D T 1989 *1989 IEEE SOI/SOS Conf. Proc.* p93
Herino R, Perio H, Barla K and Bomchil G 1984 *Mat. Lett.* **2** 519
Herino R, Bomchil G, Barla K and Bertrand C 1987 *J. Electrochem. Soc.* **134** 1994
Holmstrom R P and Chi J Y 1983 *Appl. Phys. Lett.* **42** 386
Imai K 1981 *Solid State Electronics* **24** 159
Ito T, Hiraki A and Satou M 1988 *Appl. Surface Sci.* **33/34** 1127
Ito T, Kato Y and Hiraki A 1988 *Shinku (J. of Vac. Soc. of Japan)* **31** 913
Kao YC, Wang K L, Wu B J, Lin T L, Nieh C Wn Jamieson D and Bai G 1987 *Appl. Phys. Lett.* **51** 1809
L'Ecuyer J D and Farr J P G, 1990 *Proc. 4th Intl. Symp. on Silicon-on-Insulator Tech. and Devices, Montreal*, **Vol. 90-6**, ed D N Schmidt pp 375-383
Li P C, Pliskin W, Poponiak M and Revitz M 1977 *IBM Tech. Disc. Bul.* **20** 573
Oules C, Halimaoui A, Regolini J L, Herino R, Perio A, Bensahel D and Bomchil G 1989 *Mat. Sci. Eng.* **B4** 435
Prasad A, Balakrishnan S, Jain S K and Jain G C 1982 *J. Electrochem. Soc.* **129** 596
Thomas N J, Davis J R, Keen J M, Castledine J G, Brumhead D, Goulding M, Alderman J, Farr J P G, Earwaker L G, L'Ecuyer J, Sturland I M, and Cole J M 1989 *IEEE Electron Device Lett.* **10** 129
Tsao S S, Blewer R S and Tsao J Y 1986 *Appl. Phys. Lett.* **49** 403

Uhlir A 1956 *Bell System Tech. J.* **35** 333
Unagami T and Kato K 1977 *Jap. J. of Appl. Phys.* **16** 1635
Xi Y H, Bean J C 1990 *J. Appl. Phys.* **67** 792


Copyright at HMSO, London, 1990

# Progress in High Frequency Heterojunction Field Effect Transistors

Lester F. Eastman

School of Electrical Engineering and National Nanofabrication
Facility, Cornell University, Phillips Hall,
Ithaca, New York, 14853-5401

Abstract. The concepts, technology and the present experimental limits of performance of heterojunction modulation doped field effect transistors are covered. Both lattice-matched and strained, pseudomorphic quantum well channels on GaAs substrates, as well as lattice-matched quantum well channels on InP substrates are included. Power gain frequency limits to 450 GHz, and noise figures as low as .8-.9 db at 60 GHz are presented for 300K operation of the latter devices.

## 1. Introduction

Microwave and millimeter wave field effect transistors have reached impressive performance levels by the use of heterojunctions. In these structures, a lower band-gap layer holds the electrons, while a higher band-gap layer is doped with donor atoms. As well as lattice-matched combinations on GaAs and InP substrates, thin strained layers have been used to hold the electrons. Very short gates have been realized using electron-beam lithography. This presentation covers the materials used, the fabrication methods, the materials and device assessment methods, and the present performance limits.

## 2. Materials

The initial results used $Al_{.3}Ga_{.7}As/GaAs$, where an undoped .5 - 1.0 μm GaAs buffer layer was grown on the GaAs semi-insulating substrate. The electron sheet density in the GaAs is limited to .8 - .9 x $10^{12}/cm^2$ in such a structure, due to the limited conduction-band potential step of .24V. At low temperatures there is a high electron mobility, ranging from 90,000 $cm^2$/V-s up to 220,000 $cm^2$/V-s at 77K. The latter result is only possible with low electron sheet density and with a substantial (~ 200 Å) undoped region in the

Al,GaAs near the heterojunction. Increasing the fraction of Aluminum does not allow higher electron sheet density because the energy of the donors in the Al,GaAs does not rise more than .16 eV above the GaAs conduction band edge.

In order to increase the donor energy above the conduction band edge, Indium can be added to a thin, pseudomorphic channel layer holding the electrons. With $In_{.15}Ga_{.85}As$ in this channel, electron sheet density values to $1.6 - 1.7 \times 10^{12}/cm^2$ are achievable, and with $In_{.25}Ga_{.75}As$ the sheet density can reach $2.4 - 2.5 \times 10^{12}/cm^2$. In this latter case, the conduction band potential step is .44V. Mobility of electrons is lowered a modest amount at 300K, but is severely lowered at 77K, dropping from 90,000 $cm^2/V$-s for no In, to 18,000 $cm^2/V$-s for 25% Indium.

Using InP substrates, $Al_{.48}In_{.52}As/Ga_{.47}In_{.53}As$ structures have been used to achieve improved results. In this case the conduction band potential step at the heterojunction is .52 eV and the electron sheet density can be at least $3 \times 10^{12}/cm^2$. In addition, electrons have only two thirds as much effective mass, which raises the room temperature mobility by about 50%. The 77K electron mobility is limited by alloy scattering, and to date has not surpassed the 60,000-70,000 $cm^2/V$-s region.

In order to achieve high performance in transistors with short gates, electrons must be confined by a potential barrier from beneath as well as on top. Instead of thick layers of the barrier alloy, a superlattice such as $Al_{.3}Ga_{.7}As/GaAs/Al_{.3}Ga_{.7}As$ etc, is used.

On top of the doped barrier layer, which usually contains Aluminum, there is normally placed a cap layer having no Aluminum. This layer is doped, as needed, to limit the depletion of the desired electron sheet density along the channel. The ohmic contact alloying can more easily penetrate such cap layers having no Aluminum due to the absence of Aluminum oxide.

3. Transistor Structures

The modulation-doped field effect transistors (MODFET's) are fabricated using electron-beam lithography. High performance is achievable with short gates, with ~ .15 µm gate lengths. Such gates are made with a cross section having a mushroom shape, or T-shape, with a short "footprint" but with an enlarged metal conductor region above the semiconductor. Such structures can be fabricated with ≤ 200 Ω/mm resistance along their length. Without the mushroom shape, such gates would have 8-10 times as high a resistance. It would require six gate fingers, each with one sixth the length, to achieve the same performance as two fingers with mushroom shape.

The layout of the MODFET pattern should be designed to have low gate-to-drain feedback capacitance, in order to yield a high value of power gain. This allows the unity-power-gain frequency, $f_{max}$, to be 1.5 - 2.0 times as high as the unity-current-gain frequency $f_T$.

On GaAs substrates, the pseudomorphic $In_yGa_{1-y}As$ channel has interrelated limits on thickness and Indium fraction. Above $y \cong .22$ the surface of the In,GaAs is often rough, lowering yield. At $y = .25$ the maximum thickness of this layer is limited to ~ 100 Å, since dislocations form for thicker layers. As shown in Figure 1, this causes a sharp reduction in mutual transconductance, $g_m$. For quantum well thickness = 0, this is simply a GaAs channel MODFET. The $g_m$ and $f_t$ are both about 50% higher for the optimum $In_{.25}Ga_{.75}As$ quantum well, compared with a GaAs quantum well. Figure 2 shows the current gain versus frequency for a MODFET with an optimum $In_{.25}Ga_{.75}As$ channel and .15 μm x 150 μm mushroom gate. The room-temperature $f_t$ shown is the present record value for pseudomorphic InGaAs/GaAs. (Nguyen, et al, 1989) The room-temperature $f_{max}$ of a MODFET with the same quantum well and .15 μm x 6-0 μm gate was 250 GHz. With lower $f_t$ (100 GHz), $f_{max}$ has been as high as 350 GHz on such devices (Lester et al. 1988).

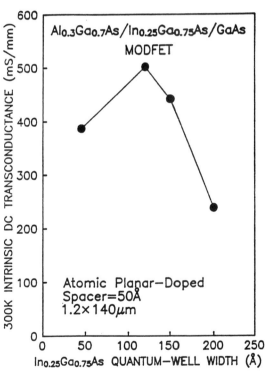

Fig. 1. Mutual transconductance of $In_{.25}Ga_{.75}As$/GaAs MODFET's versus strained-layer thickness.

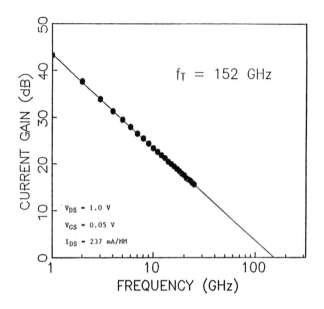

Fig. 2. Current gain performance of $In_{.25}Ga_{.75}As/GaAs$ pseudomorphic 0.15x150μm MODFET versus frequency.

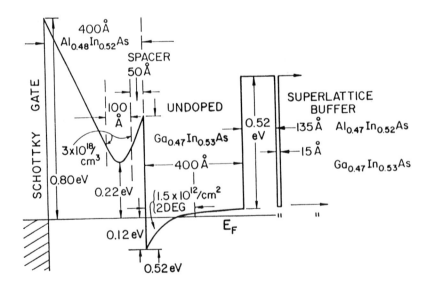

Figure 3. Conduction band potential profile of an $Al_{.48}In_{.52}As/Ga_{.48}In_{.52}As/Ga_{.47}In_{.53}As/InP$ MODFET with superlattice buffer layer.

Devices of this type have yielded 1.6 db noise figure at 60 GHz for .15 μm gates at room temperature.

MODFET's with $Al_{.48}In_{.52}As$ barriers, and $Ga_{.47}In_{.53}As$ channels lattice-matched to InP have yielded the highest frequency performance for a give gate length. Figure 3 shows a scale drawing of the conduction band profile for such a device. The conduction band potential step is .52 V at the interface, allowing electron sheet density up to $\sim 3 \times 10^{12}/cm^2$ as well as ~ 50% higher mobility values than for quantum well channels on GaAs. In addition, electron average transit velocity is at least one third higher. There are still problems related to gate leakage, channel breakdown, electron traps in the AlInAs, and reliability with these devices on InP, but dedicated research to improve these areas has just begun. These devices have yielded .8-.9 db noise figures at 60 GHz for .15 μm gates at room temperature. The best, room temperature values for $f_t$ (Mishra) and $f_{max}$ (Chao and Smith (a)) are 250 and 450 GHz, respectively for these devices. In comparing devices on GaAs and InP substrates, the expected frequency response as a function of gate length is shown in Figure 4. The quantitative values for the maximum electron sheet density, and average electron transit velocity are given in Table 1.

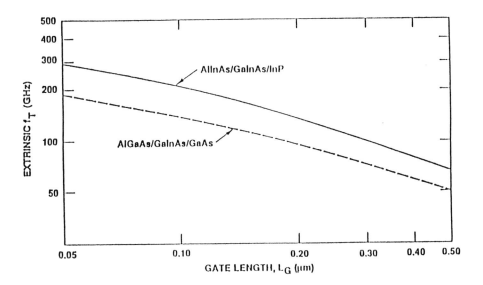

Fig. 4. Expected frequency response of optimum MODFET's on GaAs and InP substrates versus gate length.

TABLE 1

| Channel | Electron Sheet Density | Average Electron Transit Velocity |
|---|---|---|
| GaAs | $.8-.9 \times 10^{12}/cm^2$ | $1.2 \times 10^7$ cm/s |
| $In_{.25}Ga_{.75}As$ | $2.4-2.5 \times 10^{12}/cm^2$ | $1.8 \times 10^7$ cm/s |
| $Ga_{.47}In_{.53}As$ | $3-3.2 \times 10^{12}/cm^2$ | $2.4 \times 10^7$ cm/s |

## 4. Conclusions

It is concluded that MODFET's have the highest frequency response and the lowest noise figures of any transistors. Those MODFET's on InP substrates have the highest $f_t$ and $f_{max}$ and the lowest noise figure, while those MODFET's on GaAs substrates have the highest power (Chao and Smith (b)) and efficiency (Chao and Smith (b)) (.66W/mm at 41% efficiency at 60 GHz and .36 W/mm at 23% efficiency at 94 GHz).

## 5. Acknowledgements

Support from ONR, ARO, AFOSR, Boeing, IBM, Hughes, GE, Motorola and ITT are gratefully acknowledged.

## 6. References

Chao P C and Smith P (a) G.E. Syracuse, *private communication*.
Chao P C and Smith P (b) G.E. Syracuse, *private communication*.
Lester L, Tiberio R, Wolf E, Smith P, Ho P, Chao P and Duh G, 1988, Proc. IEDM, 172-175.
Mishra U, Hughes Research Laboratory, *private communication*.
Nguyen L D, Tasker P J, Radulescu D C and Eastman L F, 1989 *IEEE Trans. Elec. Dev.* **36** 2243-2248.

# Microcontamination—Advanced Manufacturing Process Technologies

Tadahiro Ohmi and Tadashi Shibata

Department of Electronics, Tohoku University Sendai 980, Japan

**Abstract**
The simultaneous establishment of "Ultra Clean Processing Environment," "Ultra Clean Wafer Surface," and "perfect Process Parameter Control," is the key to realize high performance processes for manufacturing deep submicron ULSI's. This has been demonstrated by experimental results of low-temperature silicon epitaxy by low-energy bias sputtering. In order to create ideal interfaces by eliminating native oxide growth, a closed manufacturing system has been proposed as a candidate for future advanced semiconductor manufacturing in which wafers are processed completely isolated from atmosphere.

## 1. Introduction

It is well known that a human brain is an extraordinary complicated and sophisticated system consisting of approximately $1-2 \times 10^{10}$ neurons. Even with the advancement of present day semiconductor technology, however, only $10^3-10^4$ neurons can be implemented on a single chip of silicon. Therefore, the enhancement in the integration density by a factor of more than $10^6$! is still demanded in order to construct a practical artificial intelligence system equivalent to a human brain. Thus we are not at the pinnacle of semiconductor technology development. Rather we are just standing at the starting point for real semiconductor technology development. Empirical semiconductor manufacturing development must be replaced by scientific manufacturing based on surface science technology, i.e., complete control of substrate surfaces.
 The requirements for high performance process technologies for deep-submicron device fabrication are summarized as : **LOW TEMPERATURE; HIGH SELECTIVITY; DAMAGE FREE; STRESS FREE; CONTAMINATION FREE; COMPLETE UNIFORMITY**. In order to achieve these requirements, all possible causes that can induce fluctuations in device fabrication processes must be completely eliminated. The development of Ultra Clean Technology has been conducted for this purpose <1,2>. The Ultra Clean Technology comprehends such a wide range of concept for eliminating not only microcontamination but also all kinds of obstacles for high performance processing, such as temperature variation, micro vibration, process variation and high energy ions in plasma processing. The last is known as energy clean technology in which ion bombardment energy on a wafer surface is precisely controlled to an optimum value for a specific process without causing any damages to the wafer.
 Major contaminations to a wafer surface are : **PARTICULATES; ORGANIC MATERIALS; METALLIC MATERIALS; NATIVE OXIDE; ADSORBED MOLECULES**. Wet

chemical cleaning and drying technologies have been extensively studied to solve the problems associated with the first three contaminants. However, the real problems are the remaining two contaminants, i.e., native oxide and adsorbed impurity molecules. Advanced semiconductor manufacturing line must be designed and installed with equipment in which all these wafer surface contamination must be perfectly eliminated. We will propose a closed manufacturing system for deep-submicron ULSI's instead of the present open manufacturing system.

## 2. Direction for Advanced Semiconductor Manufacturing

The direction for technological development for realizing high performance processes is able to be summarized in the following three concepts : **ULTRA CLEAN PROCESSING ENVIRONMENT; ULTRA CLEAN WAFER SURFACE; PERFECT PROCESS-PARAMETER CONTROL**<3>. The importance of these concepts are demonstrated by experimental data in the following, where low temperature silicon epitaxy by a low-kinetic energy particle process <4,5> is taken as an example. In this process, concurrent Ar ion bombardment of a growing silicon film surface in a very low energy regime is utilized to activate the very surface layer. As a result, a single crystal silicon layer having high crystalline perfection and in situ impurity doping with 100% electrical activation have been successfully grown at temperatures as low as 300°C <4> and 250°C <5>. The process has been realized by using a RF-DC coupled mode bias sputtering system equipped with ultra high vacuum ($10^{-10}$ Torr) and ultra clean gas delivery (1-2ppb moisture level at a point of use) systems shown in Fig.1.

Figure 2 demonstrates the crystallinity changes in epitaxial silicon films depending on the ion bombardment energy. When the energy of individual ion bombardment is precisely controlled to an optimum value of 25eV, perfect epitaxy occurs. However, a change in the energy severely degrades the crystallinity. Bombarding ion energy dependence of the film resistivity is shown in Fig.3. The resistivity is a very sensitive measure of the film crystallinity, and the best quality film was obtained at the minimum of resistivity. Important to note is that apparent change in resistivity, occurs by only 2eV change in the bombarding ion energy, thus showing the importance of PERFECT PROCESS-PARAMETER CONTROL.

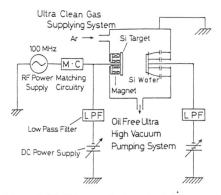

Fig.1. Schematic of RF-DC coupled mode bias sputtering system.

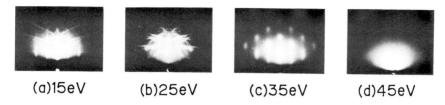

(a)15eV    (b)25eV    (c)35eV    (d)45eV

**Fig.2.** Electron diffraction patterns (top) for epitaxial silicon films formed at 300°C under Ar ion bombardment having four different bombarding energies such as 15eV(a), 25eV(b), 35eV(c) and 45eV(d) (target bias of -120V).

Figure 4 shows the electron diffraction patterns obtained from low temperature grown Si films with(left) or without(right) in situ substrate surface cleaning process. In situ surface cleaning was performed by extremely low energy Ar ion bombardment having an energy of 2eV just before the Si deposition. In this case, substrate surfaces were exposed to the clean room air during a few minutes before loading to the chamber. Although optimum silicon deposition condition was employed in both samples, a perfect crystal was not obtained without the in situ surface cleaning, verifying the importance of ULTRA CLEAN WAFER SURFACE. In Fig.5, the resistivity of the silicon film deposited with or without in situ substrate surface cleaning is shown as a function of air-exposure time. The wafers were wet chemically cleaned with a diluted HF etch at the final step, and exposed to clean air for a certain period of time before setting into the sputtering chamber. Then the silicon growth was carried out using

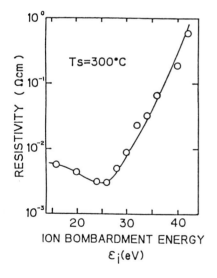

**Fig.3.** Deposited Si film resistivity as a function of the ion bombardment energy.

WITHOUT IN SITU SURFACE CLEANING

WITH IN SITU SURFACE CLEANING

**Fig.4.** Electron diffraction patterns obtained from Si films grown at optimum ion bombardment condition with (left) or without (right) in situ substrate surface cleaning.

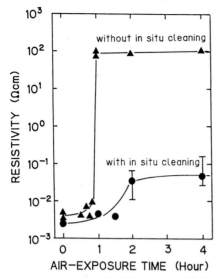

**Fig.5.** Deposited Si film resistivity with or without in situ substrate surface cleaning as a function of air-exposure time.

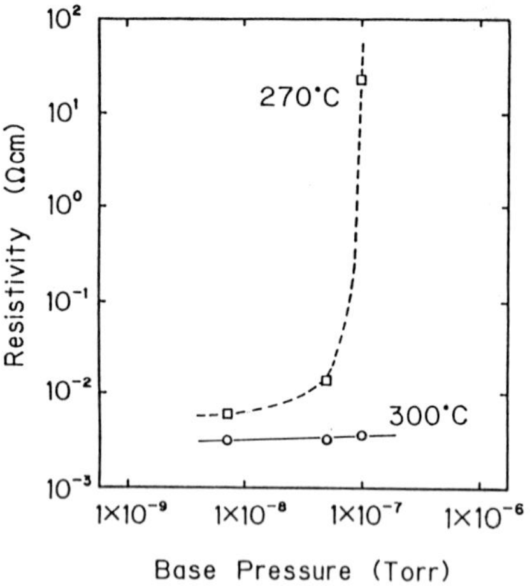

**Fig.6.** Resistivity of a epitaxially grown film as a function of the base pressure before the film growth. The films were grown at 270°C and 300°C under the ion bombardment conditions optimized at 300°C.

the optimum condition for epitaxy at 300°C. Degradation in crystallinity for air-exposure time longer than 1H is evident even with in-situ cleaning. Five orders of magnitude increase in resistivity is observed for samples without in-situ surface cleaning. This has resulted from moisture molecule adsorption and succeeding native oxide formation <6,7>. The fact indicates that the low-energy ion-bombardment surface cleaning disrobes surface adsorbed molecules but cannot remove native oxide. In-situ removal of native oxide should be conducted using the HF gas selective etching process <8,9>. It has been thus confirmed that wafer surfaces change continuously with an elapse of air-exposure time. These results clearly indicate that the present open manufacturing system must be replaced by a closed manufacturing system where wafers are not exposed to the air and are transported in a clean environment such as ultra clean $N_2$ environment.

Figure 6 demonstrates the importance of ULTRA CLEAN PROCESSING ENVIRONMENT, where the film resistivity is shown as a function of the background pressure before film growth for two different substrate temperatures. Increase in the base pressure (increase in the contamination level) result in a severe degradation in the film crystallinity in the case of 270°C. Thus environmental cleanliness is particularly important for lowering a processing temperature.

As discussed above, the impact of the three concepts on the establishment of high performance processes has been experimentally demonstrated. It has been also exemplified by the results of aluminum <10,11> and copper metallization <12> by a low-kinetic energy particle process.

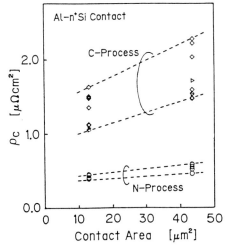

**Fig.7.** Contact resistance of Al/n$^+$-Si contacts without any alloying heat cycles. N-process in the figure means all processes are carried out in a nitrogen ambient from the final stage of wet chemical cleaning up to the transportation and loading to the vacuum chamber. In C-process, wafers are exposed to air before loading to the Al deposition chamber.

The native oxide growth is one of the greatest obstacles in establishing high quality processing. An example of this is shown in Fig. 7, where the contact resistance is shown for two different wafer processes before metallization <13>. The N-process in the figure means all processes were carried out in a nitrogen ambient from the final stage of wet chemical cleaning, i.e., the diluted HF etching, the ultra pure water rinsing and drying up to the transportation and loading to the vacuum chamber. A very low contact resistance of $0.3\mu\Omega$ cm$^2$ has been obtained without any alloying heat cycles. However, in the case of the C-process, where wet chemical cleaning, rinsing, drying and transportation to a sputtering chamber were all carried out in ordinary clean room air, a large increase in the contact resistance as well as large scattering in the data are observed. This is due to the native oxide growth on a wafer surface <6,7>. The growth of native oxide must be completely suppressed to establish high performance processes, and this is only one of numerous examples that show the importance of processes isolated from exposure to air.

Semiconductor devices are constructed as a stacked structure of various thin film, and their interface play essential roles in their operation. The interface integrity as well as the film quality are severely degraded by impurities existing at the interface, especially the native oxide as discussed so far. Successful manufacturing of deep submicron ULSIs can be achieved only by making such interfaces as ideal as possible. In the following, we will propose a closed manufacturing system as a candidate for future advanced semiconductor manufacturing in which wafers are processed completely isolated from atmosphere.

## 3. Closed-Manufacturing Scheme

Figure 8 demonstrates the concept of a manufacturing system we are proposing. Wafers are basically transported in a ultra clean nitrogen filled wafer track from equipment to equipment for processing. Processing equipment that carriers out consecutive processes must be constructed as a multi-chamber system as indicated in the figure. A consecutive process means the sequential thin film deposition processes that constitute principal portions of a device structure such as the silicon/oxide/metal

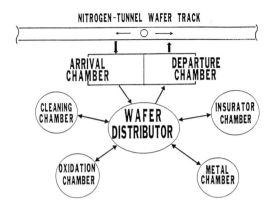

**Fig.8.** Diagram of a closed manufacturing system consisting of a multi-chamber system connected to a nitrogen tunnel wafer track.

structure for MOSFETs, the silicon/metal-1/metal-2 for contacts, the silicon/thin oxide/ferroelectric film/gate metal for nonvolatile memory cells and so forth. All such sequential processes should be conducted under ultra high vacuum using the multi-chamber system. This is the best way to realize ideal interfaces for advanced device structures.

A total LSI manufacturing process is conducted basically repeating a cycle of thin film growth and lithography/ etching processes. From a thin film growth process to lithography/etching processes, wafers are transported in clean nitrogen filled tunnel. We propose the nitrogen gas floatation wafer transportation system in the $N_2$ tunnel, where a wafer is lifted by a nitrogen gas blow and moves along the wafer track free of frictions. Steering the wafer movement is also controlled by $N_2$ gas blow. In the closed manufacturing system, each piece of process equipment interfaces to a common $N_2$-track and wafers are transported from equipment to equipment via the $N_2$ tunnel. Thus the whole LSI manufacturing processes are conducted completely isolated from air from the very beginning of a bare Si wafer up to the final passivation film coating. Although wafers are isolated from atmosphere, such system must be installed in a super clean room environment in order to keep equipment clean, especially during the equipment maintenance work.

## 4. Advanced Plasma Processing Equipment

We proposed advanced plasma processing equipment which is capable of implementing various functions such as low energy bias sputtering (low kinetic energy particle process), plasma CVD, and RIE using an identical hardware configuration <14>. Such equipment is ideal to use in the closed manufacturing system, especially for constructing the multi-chamber system. Double frequency-excitation plasma equipment built based on this concept is schematically shown in Fig.9. RF powers of two different frequencies of $f_1$ and $f_2$ are utilized for plasma excitation and wafer biasing, respectively. The ion impact energy which is roughly equal to the RF amplitude appearing on a powered electrode is proportional to $\sqrt{p/f}$ where p and f are RF power and frequency, respectively. Such a relationship is shown in Fig.10. The system works as RIE or plasma CVD when $f_1$(150-250MHz)>$f_2$(50-100MHz),while it works as a sputtering system when $f_1$(10-

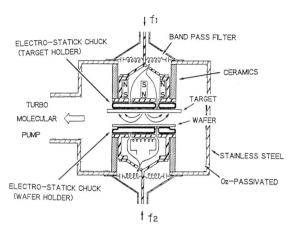

**Fig.9.** Schematic of a double-frequency-excitation plasma process system built based on the concept of advanced plasma process equipment<14>.

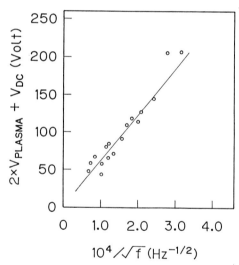

**Fig.10.** Frequency dependence of peak to peak RF amplitude ($2V_p+V_{DC}$) appearing on the powered electrode. Here $V_p$ is the plasma potential determined by probe measurement.

20MHz)<$f_2$(100-150MHz). In either case, the RF power and frequency for wafer susceptor is controlled to give an optimum ion impact energy for each process. One of the key issues of such equipment is to eliminate wafer contaminations due to the sputtering of chamber materials. A new technique has been developed to reduce the plasma potential to a small value so as not to cause any sputtering of materials consisting the process chamber <15>.

## 5. Summary

Fully automated wafer manufacturing line must exhibit complete reproducibility. Therefore all parameters related to the performance of each equipment must be perfectly controlled at designed values. Thus all fluctuations in parameters must be eliminated. We have proposed a closed manufacturing system to eliminate the fluctuations in device characteristics. A future super clean room system employing such a closed manufacturing system along with all associated facilities will be an extraordinarily complex one. So it must be an easy maintenance system, or a maintenance free system, and thus human originated fluctuations must be eliminated. As for the impurity level fluctuations in gasses and chemicals, for instance, high purity levels must be guaranteed at the manufactures and they should be directly delivered to a semiconductor plant where gases and chemicals are used at each point of use with the as-received quality. In this way, the responsibilities to establish fluctuation-free manufacturing must be shared among related expert companies such as gas companies, chemical companies and semiconductor manufacturers. Such an extensive cooperation among industries to create an infra-structure for advanced semiconductor manufacturing is going to be essentially important.

References
<1> T.Ohmi, N.Mikoshiba, and K.Tsubouchi, ULSI Science and Technology/1987ed S.Broydo and C.M.Osburn(The Electrochemical Soc.Inc., Pennington) pp.761-785 (1987).
<2> T.Ohmi, Microcontamination, Vol.6, No.10,49 (1988).
<3> T.Ohmi,IEDM Tech.Dig.,49 (1989).
<4> T.Ohmi, T.Ichikawa, H.Iwabuchi, and T.Shibata, J.Appl.Phys.,66, 4756 (1989).
<5> T.Ohmi, K.Hashimoto, M.Morita, and T.Shibata, IEDM Tech.Dig., 53 (1959).
<6> M.Morita, T.Ohmi, E.Hasegawa, M.Kawasaki, and K.Suma, Dig. Tech.Papers,VLSI Technology Symposium, 75 (1989).
<7> M.Morita, T.Ohmi, E.Hasegawa, M.Kawasaki, and K.Suma,Appl. phys.Lett.,55,562 (1989).
<8> N.Miki, H.Kikuyama, M.Maeno, J.Murota, and T.Ohmi, IEDM Tech. Dig.,730 (1988).
<9> N.Miki, H.Kikuyama, I.Kawanabe, M.Miyashita, and T.Ohmi, IEEE Trans.Electron Devices, 37, 107 (1990).
<10> H.Kuwabara, S.Saito, T.Shibata, and T.Ohmi, Dig.Tech.Papers, VLSI Technology Symp.,71 (1989).
<11> T.Ohmi, H.Kuwabara, S.Saito, and T.Shibata, J.Electrochem. Soc.,137,1008 (1990).
<12> T.Ohmi, T.Saito, T.Shibata, and T.Nitta, Appl. Phys. Lett., 2236 (1988).
<13> M.Miyawaki, S.Yoshitake, T.Saito, S.Saito, and T.Ohmi, Dig. Papers, 1990 MicroProcess Conf., Chiba, Japan (1990).
<14> T.Ohmi, Automated Integrated Circuits Manufacturing, pv 90-3 (The Electrochemical Soc., Pennington, NJ, 1990) pp.3-18.
<15> H.Goto, M.Sasaki, T.Ohmi, T.Shibata, A.Yamagami, N.Okamura , and O.Kamiya, to be presented at 1990 Int. Conf. on Solid State Devices and Materials, Sendai, Japan,August, 1990.

# Author Index

Abstreiter G, *591*
Alderman J C, *429*
Allen R W, *25*
Altrip J L, *221*
Altschul V, *157*
Amin A A M, *177*
Amm D T, *273*
Andersson G I, *595*
Arai N, *169*
Arimoto Y, *377*
Armstrong B M, *357*
Armstrong G A, *425*
Ashburn P, *333, 341, 381, 393*
Ashworth J, *241*
Augustus P D, *97*
Azoulay R, *237*

Bach H G, *33*
Badoz P A, *41*
Bagnoli P E, *113*
Balestra F, *253, 497*
Balk P, *319*
Baltes H, *493*
Beinstingl W, *409*
Belache A, *233*
Belz J, *449*
Bengtsson S, *1*
Benson T M, *5*
Bergner W, *469*
Bergonzoni C, *245*
Berlec F, *17*
Bertenburg R, *105*
Beyer A, *229*
Bez R, *165*
Bhat R, *607*
Biblemont S, *237*
Booker G R, *381*
Borel G, *257*
Boukriss B, *441*
Boys D, *69*
Bricard L, *133*
Briglio D R, *421*
Brijs G, *77*
Brockerhoff W, *105*
Brownhill C A, *567*

Brox M, *291, 295*
Brugger H, *17*
Brunel P A, *397*
Bruni M D, *13*
Brunner K, *591*
Budil M, *201, 205*
Bunyan R J T, *433*
Burbach G, *449*
Burkey B C, *599*

Cacharelis P J, *547*
Callegari A, *113*
Cantarelli D, *165*
Cappelletti P, *165*
Caquot E, *21*
Carpenter R D, *547*
Celi D, *397*
Chamberlain J M, *559*
Chang G-K, *607*
Chantre A, *345*
Charitat G, *373*
Chelnokov V E, *353*
Cherenkov A E, *353*
Chen X, *29*
Cheon Soo Kim, *457*
Chovet A, *441*
Chung Duk Kim, *457*
Circelli N, *81*
Claeys C, *385, 445, 501*
Clei A, *237*
Clements S J, *121*
Colinge J-P, *445*
Cristoloveanu S, *257, 441*

Dae Yong Kim, *457*
Dalla Libera G, *245*
Dangla J, *21*
Davies D E, *5*
Davies R A, *117*
Davis J R, *425, 433*
Declerck G, *385*
Decoutere S, *385*
Deferm L, *385, 501*
Degors N, *345*
De Graaff H C, *49*

Dekker R, *389*
Dell'ova F, *397*
Devlin W J, *417*
Dickinger P, *369*
Dierickx B, *501*
Dietl J, *465, 531*
Dilhac J-M, *65*
Dmitriev V A, *353*
DoThanh L, *465*
Draida N, *237*
Drouot S, *209*
Dubon-Chevallier C, *133*
Duchenois A M, *133*
Ducroquet F, *125*

Eastman L F, *619*
Ebner J, *401*
Eccleston W, *265, 429, 433*
Ehwald K E, *173*
Elewa T, *441*
Elias P J H, *145*
Eliason G W, *401*
Emrani A, *497*
Engström O, *1, 575, 595*
Evans A G R, *221*

Fallon M, *85*
Fantini F, *113*
Farr J R, *417*
Fiegna C, *527*
Filoche M, *21*
Finkman E, *157*
Fisher S J., 417
Flandre D, *437*
French W D, *425*
Fukumoto, M, *461*

Gajewski H, *173*
Galvier J, *13*
Gamble H S, *357*
Gammie W R, *85*
Ganibal C, *65*
Gao M-H, *445*
Gara S, *217*
Gasquet D, *489*
Gell M A, *473*
Gérodolle A, *61, 209, 277*
Ghibaudo G, *253, 497*
Ghione G, *225*
Gibbings C J, *473*
Gibis R, *481*

Gill A, *303*
Gimlett J, *607*
Girard P, *197*
Giroult-Matlakowski G, *345*
Godfrey D J, *473*
Gold D P, *333, 381*
Gong L, *93*
Gornik E, *409*
Gould S, *567*
Goulding M R, *97*
Gourrier S, *233*
Grelsson Ö, *101*
Griswold E, *273*
Groeseneken G, *261, 579*
Grote N, *33, 481*
Gruhle A, *41*
Grützmacher D, *405*
Guegan G, *311*
Gueissaz F, *109*
Guerrero E, *201*
Guillot G, *125*

Habaš P, *161*
Haddara H, *441*
Haga D, *485*
Hage J, *201*
Hall S, *429, 433*
Hamel J S, *333*
Hänsch W, *299*
Haond M, *13, 437*
Hart M J, *547*
Hauser M, *409*
Hayes J R, *607*
Heime K, *105, 229*
Heinemann B, *173*
Heinrich M, *205*
Héliot F, *133*
Hemink G J, *515*
Henini M, *559*
Hensel H J, *33*
Heremans P, *261, 579*
Heyns M M, *361*
Hildebrand O, *505*
Hill C, *53, 69, 97*
Ho H P, *5*
Hobler G, *217*
Hoefflinger B, *9*
Hofmann F, *583*
Holden A J, *25*
Holwill R J, *85*

Hong W-P, *607*
Hönlein W, *307*
Hopfmann Ch, *213*
Hori T, *461*
Houdré R, *109*
Howes R, *539*
Huber D, *201*
Hughes O H, *559*
Hunt T D, *117*
Hurkx G A M, *49*

Ilegems M, *109*
Inoue M, *461*
Iwai H, *73, 149, 287*
Iwasaki H, *461*

Jacobs H, *153*
Jain S C, *421*
Jansen A C L, *389*
Jaume D, *373*
Jeppson K O, *137*
Jespers P G A, *437*
Jeynes C, *221*
Jin Ho Lee, *457*
Jin Hyo Lee, *457*
Johnson D J, *273*
Johnson I F, *547*
Joly C, *237*
Jones M E, *473*
Jones S K, *61, 69*
Joseph M, *229*
Juffermans C A H, *249*

Kaneko Y, *169*
Karlsson P R, *137*
Kasper E, *477*
Kassim N M, *5*
Kawano M, *377*
Keen J, *613*
Keen P, *397*
Kerber M, *299, 523, 587*
Kersting R, *405*
Khosla R P, *599*
Khrenov G, *555*
Kibbel H, *477*
Kightley P, *97*
Kircher R, *469*
Kirsch H, *315*
Klaassen F M, *141, 145, 181*
Kleefstra M, *193*
Kloosterman W J, *49*

Knuvers M P G, *49*
Köck A, *409*
Köhlhoff D, *477*
Konczyzkowska A, *21*
Kordalski W J, *603*
Kotani H, *461*
Kranen P H, *389*
Kraus J, *105*
Kuhn T, *489*
Künemund T, *291*
Künzel H, *481*
Küpper P, *531*
Kurz H, *405*
Küsters K H, *465*
Kusztelan L, *465*
Kyu Hong Lee, *457*

Landgren G, *485*
Lane A A, *327*
Lary J E, *401*
Launay P, *133*
Learmouth M D, *563*
Lee T H, *599*
Lemme R, *531*
Le Néel O, *13*
Lerme M, *311*
Letourneau P, *37*
Lifka H, *249, 453*
Lill A, *299, 523*
Lindorfer P, *241*
Logan J R, *221*
Long A P, *25*
Lorenz J, *93, 217*
Lösch R, *105*
Lubzens D, *157*
Lüth H, *45*
Lynch T G, *563*

Maas H G R, *389*
MacBean M D A, *563*
Macha I, *413*
Maes H E, *261, 579*
Maex K, *77*
Magnusson U, *101*
Mallardeau C, *397*
Manley M H, *547*
Mantl S, *45*
Marin J C, *397*
Marsland J S, *567*
Martin A S R, *473*
Martin S, *277*

Marty A, *345*
Mathewson A, *129*
Matsuoka K, *461*
Matsuyama K, *461*
Matzke W-E, *173*
Maurelli A, *165*
Mazure C, *523*
McDaid L J, *429, 433*
McGregor J M, *421*
McNeill D W, *357*
Mekonnen G, *33, 481*
Melanotte M, *527*
Merckel G, *269, 543*
Merrachi A, *269*
Meschede H, *105, 229*
Metcalfe J G, *25*
Middelhoek J, *515*
Miura-Mattausch M, *153*
Moisewitsch N E, *381*
Mole P J, *539*
Momose H S, *73, 149, 287*
Monroy A, *397*
Montree A H, *535*
Mori S, *169*
Morimoto T, *73, 149, 287*
Morozenko Ya V, *353*
Morrow D, *85*
Mouis M, *37*
Mühlhoff H M, *465, 531*
Müller W, *465*
Murray S J, *539*
Myers F A, *327*

Naito Y, *461*
Nakao I, *461*
Nannini A, *245*
Nanz G, *369*
Narozny P, *477*
Nathan A, *421*
Neppl F, *299*
Nguyen-Duc C, *253*
NiDheasuna C, *129*
Nichols K G, *539*
Nickel H, *105*
Nicklin R, *25*
Nolhier N, *65*
Norström H, *77*
Nouailhat A, *125, 345*
Nouet P, *197*
Nougier J P, *489*

O'Neill A G, *69*
Ogawa H, *461*
Ohmi T, *625*
Ohshima Y, *169*
Onga S, *149*

Orlowski M, *315*
Ostermeir R, *591*
Ouisse T, *257*
Ouwerling G J L, *193*

Paccagnella A, *113*
Papadopoulo A C, *133*
Paraskevopoulos A, *33*
Paulzen G M, *249*
Pavlu J, *413*
Pedron T, *543*
Pelletier J, *209*
Peyre-Lavigne A, *373*
Philipsborn H, *571*
Pistoulet B, *197*
Plant T K, *401*
Polignano M L, *81*
Poncet A, *277*
Pongratz P, *201*
Popp J, *571*
Post I, *341*
Pötzl H W, *201, 205, 217*
Prochazka I, *413*
Prost W, *105*
Pruijmboom A, *389*

Ravazzi L, *165*
Raven M S, *559*
Redman-White W, *341, 539*
Reeder A A, *473*
Reggiani L, *489*
Reimbold G, *257, 311*
Reinhardt F, *405*
Reisinger H, *307*
Renaud J C, *125*
Riccò B, *527*
Richter R, *173*
Roche M, *397*
Rodgers P M, *563*
Roos G, *9*
Rossel P, *373*
Roth B, *229*
Roulston D J, *333, 421*
Ruddell F H, *357*
Rudin S, *493*
Ryssel H, *93*
Ryzhii V, *555*

Sakagami E, *169*
Sakai H, *461*
Sangiorgi E, *527*
Sayer M, *273*
Scheffer F, *229*
Schlaak W, *481*
Schlapp W, *105*
Schlicht B, *337*
Schmitt-Landsiedel D, *291*

Schork R, *217*
Schrems M, *201*
Schroeter-Janßen H, *33*
Schüppen A, *45*
Schwalke U, *299, 587*
Schwedler R, *405*
Selberherr S, *161, 369*
Sellai A, *559*
Selvakumar C R, *333*
Shafi Z A, *393*
Shapiro I Yu, *89*
Shaw N, *551*
Shaw R N, *417*
Shibata T, *625*
Shivaraman M S, *575*
Siabi-Shahrivar N, *341*
Simoen E, *501*
Skotnicki T, *269*
Skrimshire C P, *417*
Smedes T, *141*
Smith U, *77*
Sobolev N A, *89*
Söderbärg, *101*
Sokolov V I, *89*
Soole J B D, *607*
Sopko B, *413*
Spalding T, *129*
Spitzer A, *307*
Spurdens P C, *417*
Steenson D P, *559*
Stelz F X, *465*
Stevenson J T M, *85*
Stingeder G, *217*
Strel'chuk A M, *353*
Strobel L, *337*
Su L M, *481*
Subrahmanyan R, *315*
Swanston D M, *273*

Takagi S, *149, 287*
Taylor S, *265*
Thomas D J, *121*
Thomas N, *433*
Thompson J, *117*
Tong Q-Y, *519*
Topham P J, *551*
Tuppen C G, *473*

Ueno K, *377*
Ulacia F J I, *213*
Uren M J, *433*
Urquhart J, *117*

Vaissiere J C, *489*
Vandervorst W, *77*

Van den bosch G, *579*
van Es R A, *389*
Van Houdt J, *261*
van der Plas P A, *535*
van de Roer G Th, *145*
van der Velden J W A, *389*
Van de Wiele F, *437*
Vanhellemont J, *77*
Varani L, *489*
Vasilyeva E D, *89*
Venturi F, *527*
Vescan L, *45*
Vincent G, *37*
Vogt H, *449*

Wale M J, *551*
Walling R H, *563*
Wallinga H, *515*
Wallis R H, *117*
Walton A J, *85*
Wang Q, *291*
Watkinson P, *265*
Weber W, *291, 295, 591*
Werner Ch, *213*
Whitehurst J, *381*
Wickes H J, *417*
Wijburg R C M, *515*
Willén B, *485*
Williams J D, *381*
Wils N A H, *535*
Wilson M C, *349*
Winkler W, *173*
Witters J S, *261*
Woerlee P H, *249, 453*
Wölk C, *17*
Wolstenholme G R, *381, 547*
Wolter K, *405*
Woltjer R, *249*
Woods R C, *567*
Wu S-H, *445*
Wu Y, *29*

Xu X-L, *519*

Yamabe K, *73, 149, 287*
Yoshikawa K, *169*
Young N D, *265, 303*

Zarea A, *559*
Zeller Ch, *523*
Zhang H-Z, *519*
Zhang J F, *265*
Zhou X Q, *405*
Zimmermann W, *449*
Zingg R, *9*